Foundations of Engineering Mechanics

Series Editors: V.I. Babitsky, J. Wittenburg

V.A. Svetlitsky

Dynamics of Rods

Translated by E. Evseev and K. Ramodanova

With 189 Figures

Springer

Series Editors:

V.I. Babitsky
Department of Mechanical Engineering
Loughborough University
Loughborough LE11 3TU, Leicestershire
United Kingdom

J. Wittenburg
Institut für Technische Mechanik
Universität Karlsruhe (TH)
Kaiserstraße 12
76128 Karlsruhe
Germany

Author:

Prof. Dr. Valery A. Svetlitsky
Bauman Moscow State Technical University
2-nd Baumanskaya 5
107005 Moscow
Russia

Translators:

Evgeny Evseev & Kseniya Ramodanova
Ap. 34, 50 Skhodnenskaya Street
125363 Moscow
Russia
e-mail: *evg_seev@public.mtu.ru*

ISSN 1612-1384

ISBN 3-540-24284-8 **Springer Berlin Heidelberg New York**

Library of Congress Control Number: 2005921897

Springer is a part of Springer Science+Business Media

springeronline.com

Typesetting: Data conversion by the author.
Final processing by PTP-Berlin Protago-TEX-Production GmbH, Germany
Cover-Design: deblik, Berlin
Printed on acid-free paper 62/3141/Yu - 5 4 3 2 1 0

Preface

The volume is devoted to the dynamics of rods, which is a branch of mechanics of deformable bodies. The main goal of the book is to present systematically theoretical fundamentals of the mechanics of rods as well as numerical methods used for practical purposes.

Linear and nonlinear equations governing a rod's oscillations are presented. Methods of determining eigenvalues and eigenfunctions in conservative and non-conservative problems along with numerical methods dealing with forced, parametric, and random oscillations of rods are given. Some issues of interaction of rods with air (liquid) flows and the dynamics of space-curved rods containing flows of liquid are considered.

The book consists of nine chapters and appendices and may be conventionally divided into two parts. That is, Chapters 1 to 6 contain, in the main, theoretical material, whereas Chapters 7 to 9 illustrate the application of the theoretical results to problems of practical interest. Problems for self-study are found in Chapters 3, 5, and 7. The solutions to most of the problems are given in Appendix B.

The monograph is addressed to undergraduate and postgraduate students and teaching staff of technical universities. It may also be useful for scientists and mechanical engineers working in a wide range of industries.

I wish to express my deep appreciation to my colleagues, Dr. S.A. Voronov and C.B. Danilenko, for their help in preparing the manuscript.

Moscow, November 2004 *Valery Svetlitsky*

Table of Contents

Introduction

In practice, rods may experience not only static but also dynamic loads (deterministic and random). Dynamic loads cause oscillations of the rod about its equilibrium configuration. Thus, it is impossible to investigate the dynamical behavior of a rod without determining first its static stress–strain state.

For this reason, the first part (Svetlitsky (2000a)) presented theoretical background and methods of analyses of rod-like structural elements exerted by static forces and moments. The problem of "static" interaction between a rod and an ambient air or liquid flow was considered.

Problems of interaction of rods with flows are both of great practical and theoretical interest because the vibrations of rods may result in damage accumulation. Besides, for rods of non-circular cross section, such vibrations lead to parasitic oscillations (auto-vibrations, flutter), which, in their turn, often cause failure.

In the first part (Svetlitsky (2000a)), problems involving determination of critical forces (those under which a static loss of stability occurs) were considered. However, not every problem on stability analysis of rods can be solved with methods of statics. If some of the forces applied to the rod are follower ones, then the critical loads can not be generally obtained from the static equations of equilibrium. The reason for this is that the dynamical loss of stability may occur (when the forces attain their critical values, the rod starts to oscillate about its equilibrium configuration). Such problems can be explored using methods of dynamics, that is, by considering the equations of small oscillations of the rod about the critical equilibrium configuration.

Some engineering components (structural members of various machines and gauges) are shown in Figs. 0.1–0.12. When designing such components, which can be treated as rods, it is important to take into consideration the dynamical loads on the components.

A spiral spring (Fig. 0.1) serves as an elastic member of most gauges. When designing an elastic member, it is crucial to know how its spectrum depends on the frequencies of the inertial loads (for example, on the magnitude of the acceleration $|\mathbf{a}|$). As the whole structure is always subject to some disturbances, the acceleration vector $\mathbf{a}$ may "leave" the plane containing the spiral's axis and thereby give rise to spatial vibrations of the spring.

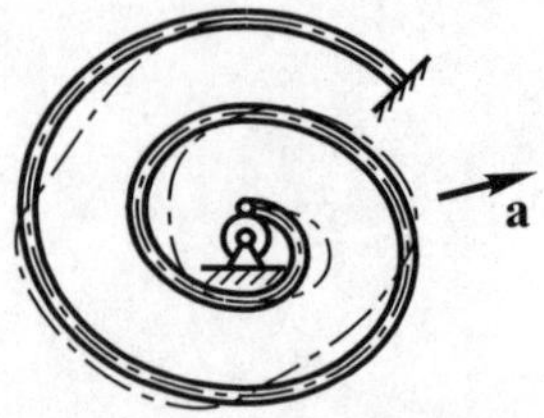

Fig. 0.1.

The spectrum of the spatial (three-dimensional) vibrations may differ dramatically from that of in-plane vibrations. This difference may lead to quite unexpected effects affecting the gauge's accuracy. Thus, effective and reliable numerical methods for determining frequencies of elastic components are absolutely necessary in structural design.

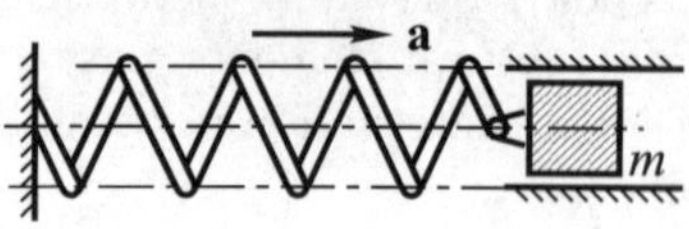

Fig. 0.2.

An accelerometer, a device which measures the acceleration of a moving object, is schematically depicted in Fig. 0.2. To determine the dynamical characteristics (say, the spectrum) more accurately, it is necessary to estimate the influence of the inertial properties of the spring on the accelerometer's reading. To do this, the vibrating spring should be considered as a system with distributed parameters and a concentrated mass m. In addition to the deterministic component, the acceleration **a** may have a random component. In this connection, the problem arises of how seriously this random component affects the accuracy of the accelerometer's reading.

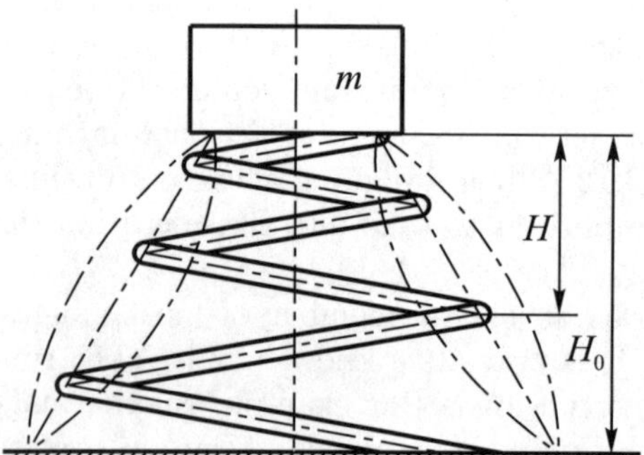

Fig. 0.3.

Various elastic structural components, such as rods, plates and shells, are used in vibroprotection systems. Along with cylindrical springs, conical springs also find their application (Fig. 0.3). The advantage of conical springs over cylindrical ones is that the value of travel of a conical spring H is almost equal to its height H_0. The main difficulty in designing vibroprotection systems with conical springs is as follows: to account for the inertial properties of the spring it is necessary to solve the nonlinear equations governing the *spring plus concentrated mass* system.

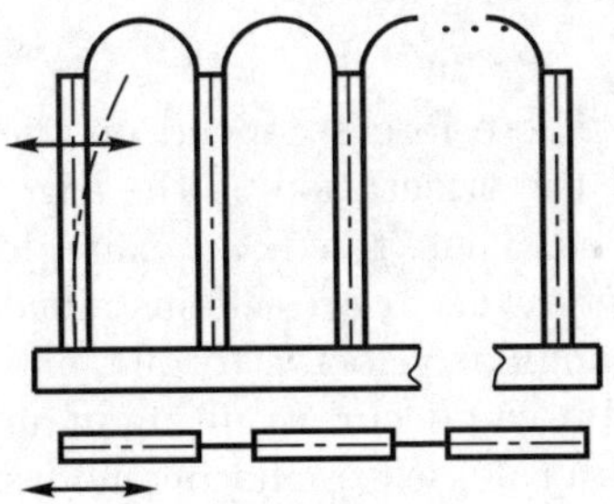

Fig. 0.4.

In various technical devices, electromechanical rod filters (Fig. 0.4) are used to produce low frequency oscillations (low in comparison with electrical oscillations). It is well known that low frequency devices based on radio component parts that use tuned-circuits are not reliable. For this reason, mechanical systems with elastic elements are used as generators of low frequency oscillations.

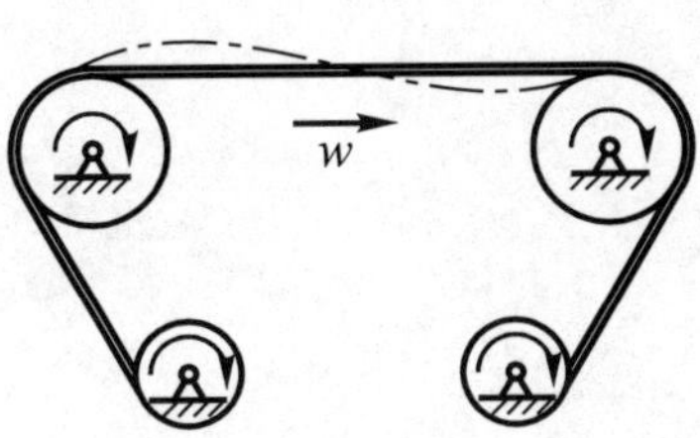

Fig. 0.5.

A tape-drive mechanism is schematically shown in Fig. 0.5. Such drives are used in various systems dealing with storing and retrieving information. Due to random disturbances, lateral oscillations of the tape may occur thereby affecting the quality of the read/write process. Thus, to be able to estimate possible distortions of the recorded information, one needs to know the dependence of the tension and speed of the tape on the frequencies of lateral oscillations.

Drilling (Fig. 0.6) is often accompanied with strong lateral vibrations of the drilling bit. Lateral vibrations lead to the occurrence of a bending

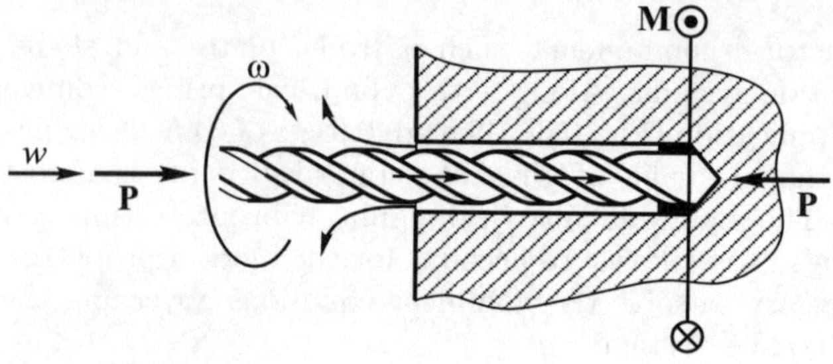

Fig. 0.6.

moment which depends on the cutting forces. These forces depend on the angle of rotation of the bit's cutting edge, while the orientation on the edge, in its turn, is governed by the bending vibrations. Thus, this is an example of a feedback system. Therefore, the oscillations that occur are self-sustained oscillations. For some value of the parameters (angular velocity, torque, and axial force), unstable bending vibrations of the bit can occur, resulting in an intolerable deviation of the drilling hole. Thus, as a rule, self-oscillating modes of drilling must be excluding in course of design of regimes of drilling. At the same time, during deep-hole drilling, bending vibrations can turn out to be useful as they assist in crushing and removing chips with a flow of cooling liquid. In this case, the drilling mode has to be chosen so that self-sustained oscillations appear.

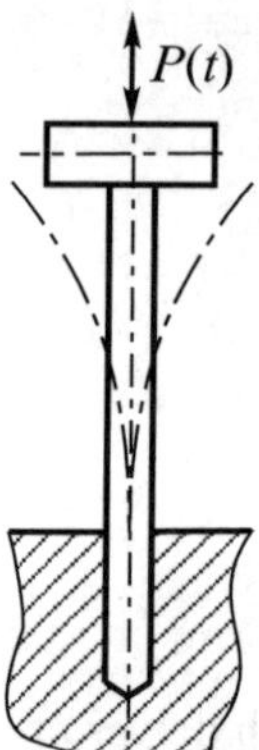

Fig. 0.7.

A pile being driven into the ground by a periodic force $P(t)$ is shown in Fig. 0.7. If the values of amplitude and frequency of the force $P(t)$ are arbitrarily chosen, then absolutely intolerable lateral vibrations of the pile may take place. Thus, the amplitude and frequency of the force $P(t)$ must be carefully selected so that the vibrations do not occur. The matter is that the equations governing lateral vibrations of the pile contain a coefficient that changes periodically with time (such vibrations are known as parametric vibrations). As a result, for some value of the parameters the vibrations are unstable, meaning that when the pile is even slightly deflected out from its straight configuration, the amplitude of vibrations can start growing infinitely. Therefore, the operating parameters of the pile hammer are to be so chosen as to avoid parametric vibrations of the pile.

Problems in which structural components of the rod structural elements are interacting dynamically with liquid and air flows are both of practical and theoretical interest. An oscillating rod immersed in a liquid flow is subject to (among other forces) a so-called "dynamically changing" component of the aerodynamic force. This component depends on the velocity of the points of the rod axis and the dynamical components of the angles of rotation of the rod-fixed frame of reference. The main difficulty in the analysis of such systems is that analytical expressions for the dynamical components of the aerodynamic forces are needed for mathematical simulations. These expressions cannot be obtained with traditional experimental techniques as the aerodynamic forces and moments depend on the orientation of the rod's axis with respect to the flow (evidently, as the rod is oscillating, the orientation is constantly changing). In greater detail, the dynamical components of aerodynamic forces and methods for obtaining estimate analytical formulae for these components are discussed in several chapters of this book. Some examples illustrating rods interacting with flows are shown in Figs. 0.8–0.11.

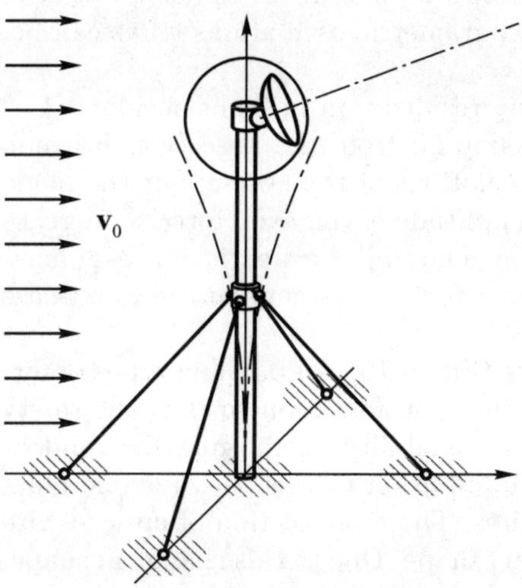

Fig. 0.8.

A tower with an aerial fixed on it is shown in Fig. 0.8. The tower is exerted by an air flow. This flow can be either a gust of wind acting over a finite time interval, or a short-time impulsive loading or an instantaneous shock loading (shock wave). For each type of loading, it is required to determine the maximum deflection of the aerial beam out from the desired direction. (If the deflection exceeds the allowable, the connection may be lost.)

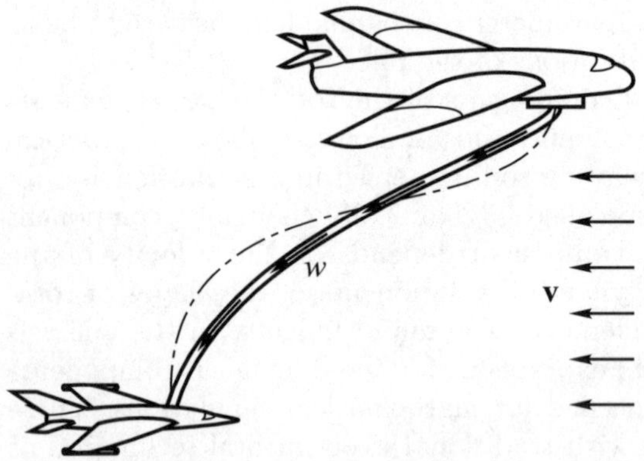

Fig. 0.9.

The hoses used in a refueling in-flight process (Fig. 0.9) can be approximated with rods of small bending and torsional stiffness. The distributed drag force on a hose depends on the aircraft's velocity and thus can take extremely large values. As a result, the axial force applied to the hoses fixed ends may become larger than the design value and thereby cause failure. As a rule, the speed of motion of fuel in a hose and pressure of the fuel flow have periodic components which give rise to parametric oscillations (that can be, as well, unstable).

Unstable parametric oscillations may result in an in-flight accident. Under certain conditions, the vortices peeling off from the hose (von Kármán vortexes) may occur. This results in oscillations of the hose about the plane of drawing. The oscillations cause the amplitude of the axial force to increase and reduce the safety factor of the system. Thus, all the mentioned cases must be taken into account in design of such a refueling system and the worst-case loading must be considered.

In design of power transmission lines (Fig. 0.10), all possible types of airflow loads on the wires must be taken into consideration so that the safety factor (for deterministic loads) and the reliability coefficient (for random loads) remain within the rated range. In addition to this, there is one more source of loading, the ice-covering of wires. The cross section of an iced wire has an arbitrary (in general, non-circular) shape. Due to this change in shape, two additional random loads (lift and torsional moment) occur. It seems to be almost impossible to obtain the probabilistic characteristics of such loads. However, even in this case, a mechanical engineer should be able to estimate possible consequences of random loading. (Theoretical foundations and methods of numerical investigation of mechanical systems with random forces

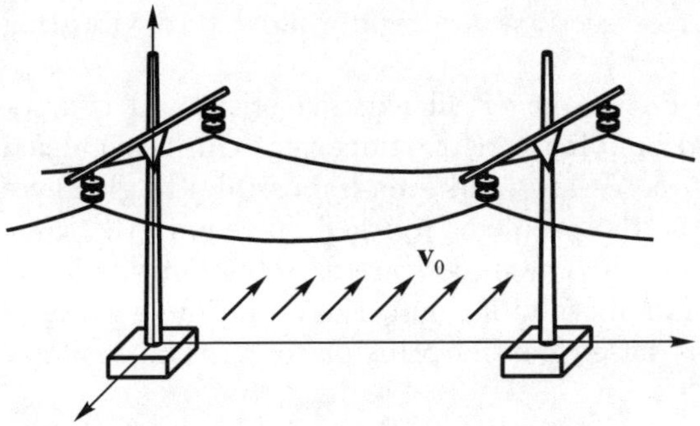

Fig. 0.10.

whose probabilistic characteristics are only partially known are presented in Svetlitsky (2003).

Dealing with a wire coated with ice, it may be reasonable to assume that the wire's cross section is approximately an ellipse. In contrast with wires of circular cross section, the key feature here is that under certain conditions, the aerodynamic lift forces and moments on the wire lead to dynamically unstable oscillations (flutter). As a rule, such oscillations inevitably result in failure.

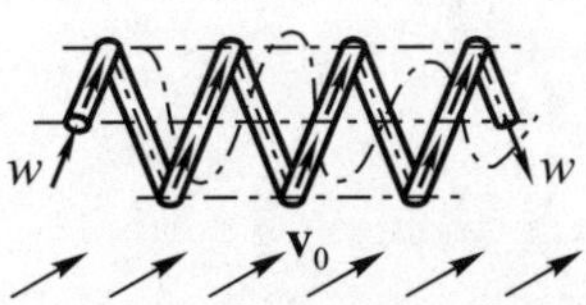

Fig. 0.11.

Pipelines (hollow rods) have numerous applications: in-flight refueling (Fig. 0.9) and cooling of heat-transfer agents in atomic power industry (Fig. 0.11); they are also used in propulsion systems of aircraft and space vehicles (Fig. 0.12).

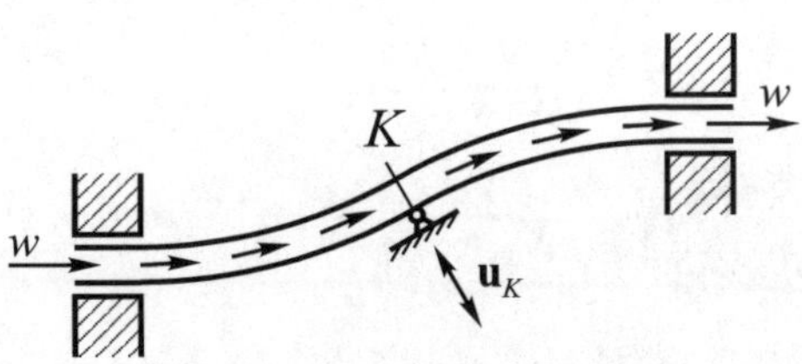

Fig. 0.12.

The normal operation of a pipeline can be disrupted by parametric oscillations (Figs. 0.9, 0.11) and kinematic excitation (Fig. 0.12). The latter occurs when some of the pipeline's cross sections are rigidly fixed to a vibrating foundation.

The examples cited above do not cover all existing structural components of machines, devices, and building constructions that can be modeled as rods. However, this is hardly necessary (and even impossible) to give here an exhaustive list of examples as the problems touched on require the same mathematical tools for their solution. These tools are presented in this book, including the general approximate method for numerical solution of systems of partial differential equations based on the principle of virtual displacements.

The monograph presents the results and algorithms of numerical solution of a number of applied problems from various fields of the mechanics of rods.

1. Kinematics

To derive the equations of motion of a curvilinear rod, it is essential to know the relationships between the generalized displacements and their time derivatives.

1.1 Basic Relationships

1.1.1 The Time Derivatives of the Basis Vectors $\{\mathbf{e}_i\}$

Consider a coordinate frame attached to the curvilinear axial line of a rod. The coordinate axes at time instants t_0 and t_1 are shown in Fig. 1.1. The axial coordinate of any point of the rod axis does not vary with time, in particular, for the axial coordinate of the origin, we have $\dot{s} = 0$. The formulas giving the relationship between the basis vectors at different times t are presented in the Appendix A of Svetlitsky (2000a).

The attached axes may change their location due to the following reasons.

(1) The location of the axes is changed in time as a result of the motion of the rod (the axial coordinate is constant (see Fig. 1.1)).

(2) The change in the position of the axes at a fixed time t_0.

That is, the basis vectors $\{\mathbf{e}_i\}$ are functions of two independent variables t and s.

In the first case, the change in the location of the axes is due to the change in the variable t while s is constant. In the second case, the change in the location of the axes is caused by the change in s while t is constant. The position of the rod axis varies continuously as the rod moves. To describe the motion of the rod and the variation of the form of its axial line, one must know the derivatives of the basis vectors $\{\mathbf{e}_i\}$ with respect to t and s. The derivative of the vector function $\mathbf{e}_j$ is a new vector, which can be represented in the basis $\{\mathbf{e}_i\}$ as follows:

$$\frac{\partial \mathbf{e}_i}{\partial t} = \sum_{j=1}^{3} \omega_{ij}\mathbf{e}_j = \omega_{ij}\mathbf{e}_j \,, \qquad i,\, j = 1,\, 2,\, 3\,;$$

here, ω_{ij} are the elements of the matrix $[\omega_{ij}]$, which is similar to the matrix $[æ_{ij}]$ (A.75) (see Appendix A.2 in Svetlitsky (2000a)). The summation over

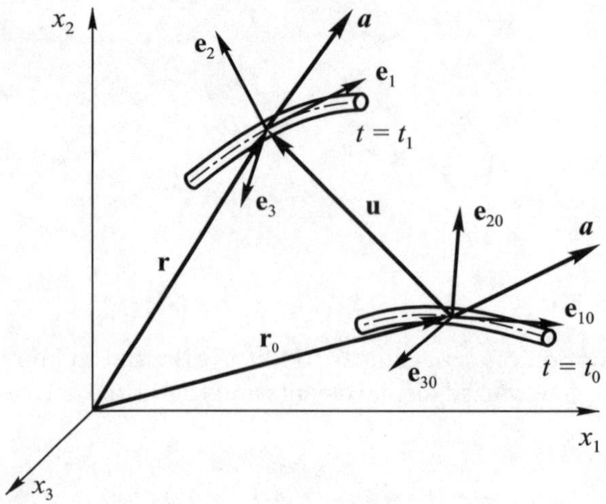

Fig. 1.1.

repeated indices is assumed. The matrix $[\omega_{ij}]$ is skew-symmetric and thus uniquely defined by the three parameters:

$$\mathrm{A}_\omega = [\omega_{ij}] = \begin{bmatrix} 0 & -\omega_3 & \omega_2 \\ \omega_3 & 0 & -\omega_1 \\ -\omega_2 & \omega_1 & 0 \end{bmatrix} .$$

As a result, we obtain

$$\begin{aligned} \frac{\partial \mathbf{e}_1}{\partial t} &= \omega_3 \mathbf{e}_2 - \omega_2 \mathbf{e}_3 \, , \\ \frac{\partial \mathbf{e}_2}{\partial t} &= -\omega_3 \mathbf{e}_1 + \omega_1 \mathbf{e}_3 \, , \\ \frac{\partial \mathbf{e}_3}{\partial t} &= \omega_2 \mathbf{e}_1 - \omega_1 \mathbf{e}_2 \, . \end{aligned} \tag{1.1}$$

We have

$$\mathrm{A}_{\boldsymbol{\omega}} \mathbf{e}_i = \boldsymbol{\omega} \times \mathbf{e}_i \, , \qquad i = 1,\, 2,\, 3 \, ,$$

where the vector $\boldsymbol{\omega} = \omega_1 \mathbf{e}_1 + \omega_2 \mathbf{e}_2 + \omega_3 \mathbf{e}_3$ is called the *vector of angular velocity*. This vector describes the rotation of the attached coordinate frame. Then, the time derivatives of the basis vectors can be written as follows:

$$\frac{\partial \mathbf{e}_1}{\partial t} = \boldsymbol{\omega} \times \mathbf{e}_1 \, , \qquad \frac{\partial \mathbf{e}_2}{\partial t} = \boldsymbol{\omega} \times \mathbf{e}_2 \, , \qquad \frac{\partial \mathbf{e}_3}{\partial t} = \boldsymbol{\omega} \times \mathbf{e}_3 \, . \tag{1.2}$$

Using the Levi-Civita symbols, we get

$$\frac{\partial \mathbf{e}_i}{\partial t} = \varepsilon_{kji} \omega_j \mathbf{e}_k = -\varepsilon_{ijk} \omega_j \mathbf{e}_k \, . \tag{1.3}$$

1.1.2 Absolute and Local Time Derivatives of Vectors

Consider a vector $\mathbf{a}(t)$ referred to the attached (movable) coordinate frame $\{\mathbf{e}_i\}$ (see Fig. 1.1). We have

$$\mathbf{a}(t) = a_1(t)\,\mathbf{e}_1 + a_2(t)\,\mathbf{e}_2 + a_3(t)\,\mathbf{e}_3 \,.$$

The components of the vector $\mathbf{a}$ and the basis vectors $\{\mathbf{e}_i\}$ of the movable coordinate frame are time-dependent variables, hence, in view of (1.1), the absolute time derivative of the vector $\mathbf{a}(s,\, t)$ is

$$\frac{\partial \mathbf{a}}{\partial t} = \frac{\partial a_1}{\partial t}\,\mathbf{e}_1 + \frac{\partial a_2}{\partial t}\,\mathbf{e}_2 + \frac{\partial a_3}{\partial t}\,\mathbf{e}_3 + a_1(\boldsymbol{\omega} \times \mathbf{e}_1) + a_2(\boldsymbol{\omega} \times \mathbf{e}_2) + a_3(\boldsymbol{\omega} \times \mathbf{e}_3) = \frac{\tilde{\partial} \mathbf{a}}{\partial t} + \boldsymbol{\omega} \times \mathbf{a}\,, \tag{1.4}$$

where $\dfrac{\partial \mathbf{a}}{\partial t}$ denotes the absolute derivative and $\dfrac{\tilde{\partial} \mathbf{a}}{\partial t}$ denotes the local partial derivative of the vector $\mathbf{a}$.

The local derivative describes the change in time of the components of $\mathbf{a}$ in the attached basis; the vector $\boldsymbol{\omega} \times \mathbf{a}$ is responsible for the change in time of the vector $\mathbf{a}$ due to the rotation of the coordinate axes.

Let us obtain the formulas relating the components ω_j of the angular velocity $\boldsymbol{\omega}$ to the angles $\vartheta = \vartheta_1$, $\psi = \vartheta_2$, $\varphi = \vartheta_3$. Let

$$\mathbf{e}_i = l_{i\rho}\mathbf{e}_{\rho 0}\,,$$

where $\mathbf{e}_{\rho 0}$ are the unit base vectors at time $t = t_0$. Differentiating the last expression with respect to t, we get

$$\frac{\partial l_{i\rho}}{\partial t}\,\mathbf{e}_{\rho 0} = \varepsilon_{kji}\omega_j\mathbf{e}_k \,. \tag{1.5}$$

Using

$$\mathbf{e}_{\rho 0} = l_{k\rho}\mathbf{e}_k \tag{1.6}$$

and substituting (1.6) into (1.5), we obtain

$$\varepsilon_{kji}\omega_j = \frac{\partial l_{i\rho}}{\partial t}\,l_{k\rho} \,. \tag{1.7}$$

For illustrative purposes, let us derive the expression for ω_1 performing explicitly all the necessary summations. Let $j = 1$, $k = 3$, $i = 2$, and $\varepsilon_{312} = 1$, then, summing over ρ in the right-hand side of (1.7), we get

$$\omega_1 = \frac{\partial l_{21}}{\partial t}\,l_{31} + \frac{\partial l_{22}}{\partial t}\,l_{32} + \frac{\partial l_{23}}{\partial t}\,l_{33} \,. \tag{1.8}$$

Another expression for ω_1 can be obtained if we put $k = 2$, $i = 3$, and $\varepsilon_{213} = -1$:

$$-\omega_1 = \frac{\partial l_{31}}{\partial t} l_{21} + \frac{\partial l_{32}}{\partial t} l_{22} + \frac{\partial l_{33}}{\partial t} l_{23} . \tag{1.9}$$

When representing l_{ij} in terms of the angles, one can use (1.9) to verify the correctness of (1.8).

Finally, we can represent the components of the angular velocity as follows (for brevity, only one possible form of the expression for ω_i is written out):

$$\begin{aligned}
\omega_1 &= \frac{\partial l_{21}}{\partial t} l_{31} + \frac{\partial l_{22}}{\partial t} l_{32} + \frac{\partial l_{23}}{\partial t} l_{33} ; \\
\omega_2 &= \frac{\partial l_{31}}{\partial t} l_{11} + \frac{\partial l_{32}}{\partial t} l_{12} + \frac{\partial l_{33}}{\partial t} l_{13} ; \\
\omega_3 &= \frac{\partial l_{11}}{\partial t} l_{21} + \frac{\partial l_{12}}{\partial t} l_{22} + \frac{\partial l_{13}}{\partial t} l_{23} .
\end{aligned}$$

Using the expressions (A.44) for l_{ij} (see Svetlitsky (2000a)), we get

$$\begin{aligned}
\omega_1 &= \frac{\partial \vartheta_1}{\partial t} \cos\vartheta_2 \cos\vartheta_3 - \frac{\partial \vartheta_3}{\partial t} \sin\vartheta_2 , \\
\omega_2 &= \frac{\partial \vartheta_2}{\partial t} - \frac{\partial \vartheta_1}{\partial t} \sin\vartheta_3 , \\
\omega_3 &= \frac{\partial \vartheta_3}{\partial t} \cos\vartheta_2 + \frac{\partial \vartheta_1}{\partial t} \sin\vartheta_2 \cos\vartheta_3 .
\end{aligned} \tag{1.10}$$

These equations can be written in vector–matrix from

$$\boldsymbol{\omega} = \mathrm{L}_1 \frac{\partial \boldsymbol{\vartheta}}{\partial t} , \tag{1.11}$$

where

$$\mathrm{L}_1 = \begin{bmatrix} \cos\vartheta_2 \cos\vartheta_3 & 0 & -\sin\vartheta_2 \\ -\sin\vartheta_3 & 1 & 0 \\ \sin\vartheta_2 \cos\vartheta_3 & 0 & \cos\vartheta_2 \end{bmatrix} , \qquad \boldsymbol{\vartheta} = \begin{bmatrix} \vartheta_1 \\ \vartheta_2 \\ \vartheta_3 \end{bmatrix} .$$

For small values of the angles ϑ_j, the angular velocity $\boldsymbol{\omega}$ reads

$$\boldsymbol{\omega} = \frac{\partial \boldsymbol{\vartheta}}{\partial t} . \tag{1.12}$$

The absolute and local derivatives of the vectors $\boldsymbol{\omega}$ and $\boldsymbol{æ}$ with respect to t and s, respectively, coincide. This fact follows from

$$\begin{aligned}
\frac{\partial \boldsymbol{\omega}}{\partial t} &= \frac{\tilde{\partial} \boldsymbol{\omega}}{\partial t} + \boldsymbol{\omega} \times \boldsymbol{\omega} = \frac{\tilde{\partial} \boldsymbol{\omega}}{\partial t} , \\
\frac{\partial \boldsymbol{æ}}{\partial s} &= \frac{\tilde{\partial} \boldsymbol{æ}}{\partial s} + \boldsymbol{æ} \times \boldsymbol{æ} = \frac{\tilde{\partial} \boldsymbol{æ}}{\partial s} .
\end{aligned} \tag{1.13}$$

1.1.3 Relationship Between the Vectors $\boldsymbol{\omega}$ and æ

Consider an arbitrary nonzero vector $\mathbf{a}(s,t)$ of constant magnitude. Assume that this vector persistently keeps its direction with respect to the coordinate axes of the rod-fixed frame. Since

$$\frac{\tilde{\partial}\mathbf{a}}{\partial s} = \frac{\tilde{\partial}\mathbf{a}}{\partial t} = 0\,,$$

we see that the absolute derivatives of the vector with respect to t an s are as follows:

$$\frac{\partial \mathbf{a}}{\partial t} = \boldsymbol{\omega} \times \mathbf{a}\,,$$
$$\frac{\partial \mathbf{a}}{\partial s} = \text{æ} \times \mathbf{a}\,. \tag{1.14}$$

Differentiating the first equation with respect to s and the second one with respect to t and equating the mixed partial derivatives, we arrive at

$$\frac{\tilde{\partial}\boldsymbol{\omega}}{\partial s} \times \mathbf{a} + \text{æ} \times (\boldsymbol{\omega} \times \mathbf{a}) = \frac{\tilde{\partial}\text{æ}}{\partial t} \times \mathbf{a} + \boldsymbol{\omega} \times (\text{æ} \times \mathbf{a})\,.$$

Using the Jacobi identity

$$\boldsymbol{\omega} \times (\text{æ} \times \mathbf{a}) + \text{æ} \times (\boldsymbol{\omega} \times \mathbf{a}) + \mathbf{a} \times (\text{æ} \times \boldsymbol{\omega}) = 0\,,$$

we get

$$\frac{\tilde{\partial}\boldsymbol{\omega}}{\partial s} \times \mathbf{a} - \frac{\tilde{\partial}\text{æ}}{\partial t} \times \mathbf{a} - (\boldsymbol{\omega} \times \text{æ}) \times \mathbf{a} = 0 \tag{1.15}$$

or

$$\left[\frac{\tilde{\partial}\boldsymbol{\omega}}{\partial s} - \frac{\tilde{\partial}\text{æ}}{\partial t} - (\boldsymbol{\omega} \times \text{æ})\right] \times \mathbf{a} = 0\,. \tag{1.16}$$

Recall that **a** is an arbitrary vector of nonzero magnitude, hence, the identity (1.16) holds provided

$$\frac{\tilde{\partial}\boldsymbol{\omega}}{\partial s} - \frac{\tilde{\partial}\text{æ}}{\partial t} - \boldsymbol{\omega} \times \text{æ} = 0\,. \tag{1.17}$$

This equation establishes the sought-for relationship between the vectors $\boldsymbol{\omega}$ and æ.

Equation (1.17) can be represented in terms of absolute partial derivatives. We have

$$\frac{\tilde{\partial}\boldsymbol{\omega}}{\partial s} = \frac{\partial \boldsymbol{\omega}}{\partial s} - \text{æ} \times \boldsymbol{\omega}\,,$$
$$\frac{\tilde{\partial}\text{æ}}{\partial t} = \frac{\partial \text{æ}}{\partial t} - \boldsymbol{\omega} \times \text{æ}\,. \tag{1.18}$$

Substituting (1.18) into (1.17), we get

$$\frac{\partial \boldsymbol{\omega}}{\partial s} - \frac{\partial \boldsymbol{æ}}{\partial t} + \boldsymbol{\omega} \times \boldsymbol{æ} = 0\,. \tag{1.19}$$

In terms of projections in the attached coordinate system, the last equation reads

$$\begin{aligned}
&\frac{\partial \omega_1}{\partial s} - \frac{\partial æ_1}{\partial t} + \omega_2 æ_3 - \omega_3 æ_2 = 0\,,\\
&\frac{\partial \omega_2}{\partial s} - \frac{\partial æ_2}{\partial t} + \omega_3 æ_1 - \omega_1 æ_3 = 0\,,\\
&\frac{\partial \omega_3}{\partial s} - \frac{\partial æ_3}{\partial t} + \omega_1 æ_2 - \omega_2 æ_1 = 0\,.
\end{aligned} \tag{1.20}$$

For a plane curve (the vector $\mathbf{e}_3$ is perpendicular to the plane in which the curve lies), we have

$$\frac{\partial \omega_3}{\partial s} = \frac{\partial æ_3}{\partial t}\,. \tag{1.21}$$

Since in the case of a plane curve we have

$$æ_3 = \frac{\partial \vartheta_3}{\partial s}\,, \qquad \omega_3 = \frac{\partial \vartheta_3}{\partial t}\,,$$

it follows that equation (1.21) becomes an identity.

If $æ_i$ and ω_i in (1.20) are represented in terms of the angles ϑ_1, ϑ_2, and ϑ_3, then these equations become identities. The analysis of some problems concerned with oscillation of rods does not necessarily imply the determination of the angles ϑ_1, ϑ_2, and ϑ_3. In this case, ω_i and $æ_i$ are considered as unknowns related by (1.20).

1.1.4 Velocity and Acceleration of Axial Points of a Rod

Consider some basic issues of the kinematics of material points in the context of the dynamics of rods. When a rod is set in motion, each point of its axis has a certain velocity $\mathbf{v}$. This velocity is related to the time derivative of the position vector of the point $\mathbf{r}$ as follows (see Fig. 1.2):

$$\mathbf{v} = \frac{\mathrm{d}\mathbf{r}(s,\,t)}{\mathrm{d}t}\,, \qquad \mathbf{r} = \sum_{i=1}^{3} a_i \mathbf{e}_i\,. \tag{1.22}$$

The vector $\mathbf{v}$ is a tangent to the trajectory of motion of the axis of the rod element. Note that (1.22) differs from the commonly used definition of the velocity of a material point, that is, the vectors $\mathbf{r}$ and $\mathbf{v}$ in (1.22) are functions of two independent variables s and t. For example, if the coordinate s of the point O_1 is constant during the motion, then, using (1.12), we get

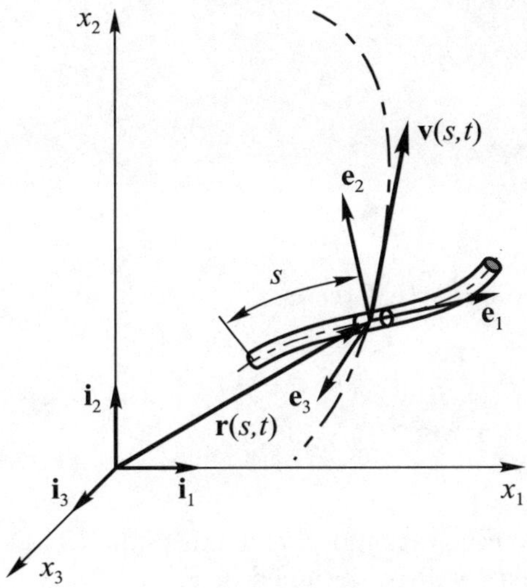

Fig. 1.2.

$$\mathbf{v} = \frac{\partial \mathbf{r}(s,\, t)}{\partial t}\,,$$

where $\dfrac{\partial}{\partial t}$ is the material partial derivative.

In the Cartesian basis, we have

$$\mathbf{v} = \sum_{j=1}^{3} \dot{x}_j \mathbf{i}_j\,.$$

In the attached coordinate system $\{\mathbf{e}_j\}$, we get

$$\mathbf{v} = \frac{\tilde{\partial} \mathbf{r}}{\partial t} + \boldsymbol{\omega} \times \mathbf{r}\,, \tag{1.23}$$

where $\boldsymbol{\omega}(s,\, t)$ is the vector of angular velocity of the trihedron $\{\mathbf{e}_j\}$.

Let $\mathbf{r} = \mathbf{r}_0 + \mathbf{u}$, where $\mathbf{u}$ is the displacement vector of the rod axis points. Equation (1.23) can be represented as follows:

$$\frac{\tilde{\partial} \mathbf{u}}{\partial t} + \boldsymbol{\omega} \times \mathbf{u} = \mathbf{v}\,. \tag{1.24}$$

The acceleration of a rod axis point reads

$$\frac{\mathrm{d}^2 \mathbf{r}}{\mathrm{d}t^2} = \frac{\partial^2 \mathbf{r}}{\partial t^2} = \dot{\mathbf{v}}(s,\, t)\,.$$

In the Cartesian and attached coordinate systems, the acceleration is given by

$$\frac{\partial^2 \mathbf{r}}{\partial t^2} = \sum_{j=1}^{3} \ddot{x}_j \mathbf{i}_j \tag{1.25}$$

and

$$\frac{\partial^2 \mathbf{r}}{\partial t^2} = \frac{\tilde{\partial} \mathbf{v}}{\partial t} + \boldsymbol{\omega} \times \mathbf{v}\,, \tag{1.26}$$

respectively.

Expanded notation of (1.26) in the basis $\{\mathbf{e}_i\}$ reads

$$\frac{\partial^2 \mathbf{r}}{\partial t^2} = \left(\frac{\partial v_1}{\partial t} + \omega_2 v_3 - \omega_3 v_2\right) \mathbf{e}_1 + \left(\frac{\partial v_2}{\partial t} + \omega_3 v_1 - \omega_1 v_3\right) \mathbf{e}_2 + \left(\frac{\partial v_3}{\partial t} + \omega_1 v_2 - \omega_2 v_1\right) \mathbf{e}_3\,. \tag{1.27}$$

The formulas for the velocity and acceleration of a rod axis point for the case when the coordinate s depends on t will be obtained in Sect. 1.2.

1.1.5 Eulerian and Lagrangian Variables in Mechanics of Rods

Lagrangian Variables A moving inextensible rod is shown in Fig. 1.3. The rod is clamped at the end $s = 0$. In this case, the axial coordinate s of any element of the rod of length ds remains unaltered during the motion. Suppose that the initial location of the rod axis points is known and given by $x_{i0}(s, t_0)$; then if the coordinates of the axis points $x_i(s, t)$ at an arbitrary time t are also known, the spatial location of the rod is well defined. Thus, the coordinates of the rod axis points at time t are functions of the initial coordinates of the points, that is, $x_i = x_i(x_{j0}, t)$, $i = 1, 2, 3$. The coordinates x_{i0} of a rod axis point are termed *Lagrangian variables*.

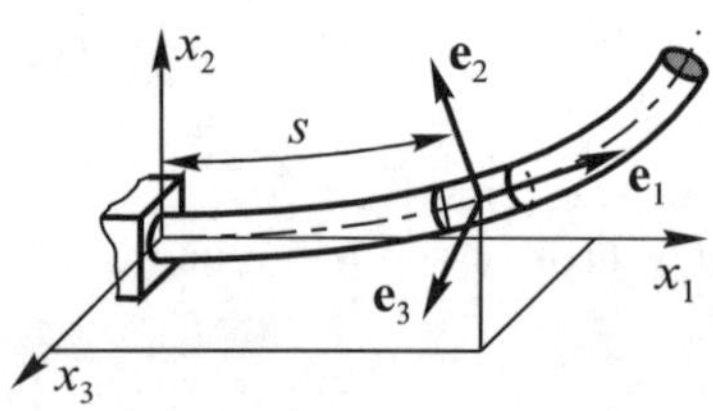

Fig. 1.3.

Within the Lagrangian description (the material description), we observe the motion of a particular element of the rod. The location of the rod axis is defined parametrically as follows: $x_i = x_i(s, t)$; at that, s does not depend on time. To derive the equations of motion, one must know the total derivatives of the coordinates of the axis points, that is,

$$\frac{\mathrm{d}x_i}{\mathrm{d}t} = \frac{\mathrm{d}x_i(x_{j0}, t)}{\mathrm{d}t}$$

or

$$\frac{\mathrm{d}x_i}{\mathrm{d}t} = \frac{\mathrm{d}x_i(s, t)}{\mathrm{d}t} .$$

Since x_{j0} and s are time independent, we see that

$$\frac{\mathrm{d}x_i}{\mathrm{d}t} = \frac{\partial x_i}{\partial t} .$$

Assuming s to be time independent, we used the Lagrangian variables in the derivation of (1.24), (1.25).

Eulerian Variables Consider an inextensible rod that moves in a conduit *1* (Fig. 1.4). Let $\mathbf{w}$ be the velocity of a rod element relative to the conduit. Denote the velocity of the conduit by $\mathbf{v}$. The absolute velocity of the rod element is $\mathbf{V} = \mathbf{v} + \mathbf{w}$.

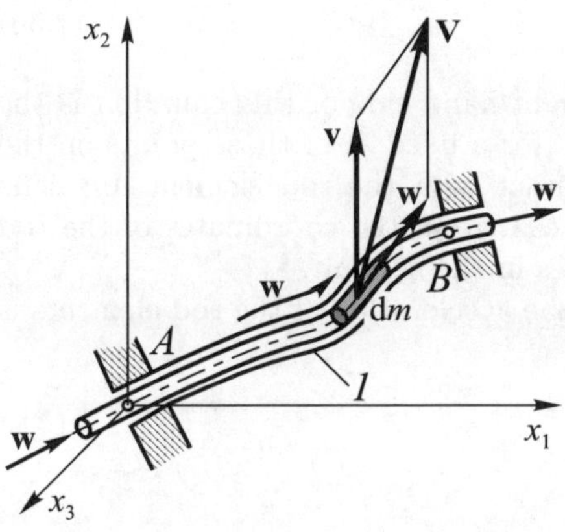

Fig. 1.4.

Since the rod is inextensible, the velocity $\mathbf{w}$ is a function of t alone. The Lagrangian variables are inconvenient when we study the motion of a finite segment AB of the rod because in this case we study the behavior of the segment as a whole but not the behavior of its small element $\mathrm{d}m$.

To clarify the Eulerian description of motion (the spatial description), let us consider an example. Let a rod be placed into an absolutely flexible massless conduit. To describe the behavior of the segment AB, it is sufficient to know the internal stress field in the rod and how the configuration of the conduit varies with time. The decomposition of the motion into the relative motion (described by the velocity $\mathbf{w}$) and the transportation motion (the velocity $\mathbf{v}$) is efficient in the study of dynamics of liquid-filled rods (e.g., pipelines). Within the framework of such problems, the liquid and the pipeline are examined as a whole. For an incompressible fluid of fixed flow rate, the relative velocity $\mathbf{w}$ is insensitive to the motion of the rod.

The rod moves relative to the conduit *1* (Fig. 1.4), hence, the coordinate s of its element $\mathrm{d}m$ varies with time, that is, the total derivative of the vector $\mathbf{r}$ reads

$$\frac{\mathrm{d}\mathbf{r}(s(t),\, t)}{\mathrm{d}t} = \mathbf{V} = \frac{\partial \mathbf{r}}{\partial t} + \frac{\partial \mathbf{r}}{\partial s}\, w\,, \qquad w = |\mathbf{w}| = \frac{\mathrm{d}s}{\mathrm{d}t}\,. \tag{1.28}$$

In view of the formulas

$$\frac{\partial \mathbf{r}}{\partial t} = \mathbf{v}, \qquad \frac{\partial \mathbf{r}}{\partial s} = \mathbf{e}_1\,,$$

we get

$$\frac{\mathrm{d}\mathbf{r}}{\mathrm{d}t} = \mathbf{V} = \mathbf{v} + \mathbf{w}\,. \tag{1.29}$$

In terms of projections in the Cartesian basis, we have

$$v_{x_i} = \frac{\mathrm{d}x_i}{\mathrm{d}t} = \frac{\partial x_i}{\partial t} + \frac{\partial x_i}{\partial s}\, w\,. \tag{1.30}$$

The first term of the sum in the right-hand side of this equation is the time rate of change in the coordinates (for a fixed s) of those points on the conduit axis that currently are in contact with the rod element $\mathrm{d}m$. The second term of the sum describes the change in the coordinates of the rod elements (for a fixed t) as the rod moves in the conduit.

In terms of the Eulerian variables, the acceleration of the rod elements is as follows:

$$\frac{\mathrm{d}\mathbf{V}}{\mathrm{d}t} = \frac{\partial \mathbf{V}}{\partial t} + \frac{\partial \mathbf{V}}{\partial s}\, w \tag{1.31}$$

or

$$\frac{\mathrm{d}^2\mathbf{r}}{\mathrm{d}t^2} = \frac{\partial^2 \mathbf{r}}{\partial t^2} + 2w\,\frac{\partial^2 \mathbf{r}}{\partial t\,\partial s} + w^2\,\frac{\partial^2 \mathbf{r}}{\partial s^2} + \frac{\partial \mathbf{r}}{\partial s}\,\frac{\mathrm{d}w}{\mathrm{d}t}\,. \tag{1.32}$$

In terms of projections in the Cartesian basis, we have

$$\frac{\mathrm{d}^2 x_i(s(t),\, t)}{\mathrm{d}t^2} = \frac{\partial^2 x_i}{\partial t^2} + 2w\,\frac{\partial^2 x_i}{\partial t\,\partial s} + w^2\,\frac{\partial^2 x_i}{\partial s^2} + \frac{\partial x_i}{\partial s}\,\frac{\mathrm{d}w}{\mathrm{d}t}\,. \tag{1.33}$$

If the motion of the rod in the conduit is steady, that is, $|\mathbf{w}| = \mathrm{const}$ and w is a function of s alone, then the acceleration has the following components:

$$\frac{\mathrm{d}^2 x_i}{\mathrm{d}t^2} = \frac{\partial^2 x_i}{\partial t^2} + 2w\,\frac{\partial^2 x_i}{\partial s\,\partial t} + w^2\,\frac{\partial^2 x_i}{\partial s^2}\,. \tag{1.34}$$

Let $\mathbf{e}_i$ be the unit base vectors of the rod-fixed frame. The vectors $\mathbf{e}_i$ are functions of t and $s(t)$, hence, the total derivative of $\mathbf{e}_i$ reads

$$\frac{\mathrm{d}\mathbf{e}_i}{\mathrm{d}t} = \frac{\partial \mathbf{e}_i}{\partial t} + \frac{\partial \mathbf{e}_i}{\partial s} w = \boldsymbol{\omega} \times \mathbf{e}_i + (\text{æ} \times \mathbf{e}_i)\, w = \varepsilon_{kji}(\omega_j + w\text{æ}_j)\, \mathbf{e}_k \tag{1.35}$$

or

$$\begin{aligned}
\frac{\mathrm{d}\mathbf{e}_1}{\mathrm{d}t} &= (\omega_3 + \text{æ}_3 w)\, \mathbf{e}_2 - (\omega_2 + \text{æ}_2 w)\, \mathbf{e}_3 \,, \\
\frac{\mathrm{d}\mathbf{e}_2}{\mathrm{d}t} &= (\omega_1 + \text{æ}_1 w)\, \mathbf{e}_3 - (\omega_3 + \text{æ}_3 w)\, \mathbf{e}_1 \,, \\
\frac{\mathrm{d}\mathbf{e}_3}{\mathrm{d}t} &= (\omega_2 + \text{æ}_2 w)\, \mathbf{e}_1 - (\omega_1 + \text{æ}_1 w)\, \mathbf{e}_2 \,.
\end{aligned} \tag{1.36}$$

As it has been mentioned above, the partial derivatives $\partial \mathbf{e}_i / \partial t$ are the local rate of change of the vectors $\mathbf{e}_i$ with respect to time when s is constant. Thus, it is clear that $\boldsymbol{\omega}$ is the local (that is, angular) velocity of the conduit but not of the rod.

1.2 Kinematic Equations

1.2.1 Kinematic Equations for Velocities

Let us derive an equation that relates $\mathbf{v}$ to $\boldsymbol{\omega}$. First, consider the case when the coordinates s of the rod elements do not vary with time, hence, $\mathbf{w} \equiv 0$ (Fig. 1.3). Consider an element $\mathrm{d}s$ of this rod and find the velocity of the cross section B relative to the cross section A (Fig. 1.5). We have

$$\delta \mathbf{v}_{AB} = (\mathbf{v} + \mathrm{d}\mathbf{v}) - \mathbf{v} = \mathrm{d}\mathbf{v} \,.$$

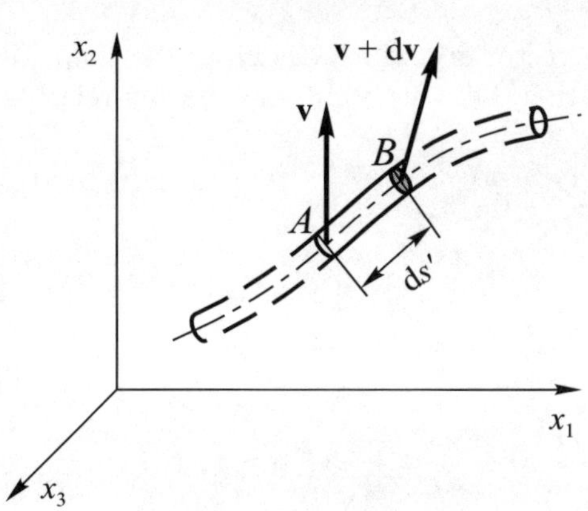

Fig. 1.5.

Introducing the vector of angular velocity, we can represent the relative velocity as

$$\delta \mathbf{v}_{AB} = \boldsymbol{\omega} \times \mathbf{e}_1 \, \mathrm{d}s \,.$$

Equating the right-hand sides of the last two equations, we get

$$\frac{\partial \mathbf{v}}{\partial s} = \boldsymbol{\omega} \times \mathbf{e}_1 \, .$$

Using local derivative notation, we obtain

$$\frac{\tilde{\partial} \mathbf{v}}{\partial s} + \text{æ} \times \mathbf{v} = \boldsymbol{\omega} \times \mathbf{e}_1 \, . \tag{1.37}$$

In terms of projections in the attached coordinate system, the last equation reads

$$\begin{aligned}
&\frac{\tilde{\partial} v_1}{\partial s} + v_3 æ_2 - v_2 æ_3 = 0 \, , \\
&\frac{\tilde{\partial} v_2}{\partial s} + v_1 æ_3 - v_3 æ_1 = \omega_3 \, , \\
&\frac{\tilde{\partial} v_3}{\partial s} + v_2 æ_1 - v_1 æ_2 = -\omega_2 \, .
\end{aligned} \tag{1.38}$$

If the displacements and velocities of the rod axis points (u_j and v_j, respectively) are small, then (1.37) yields

$$\begin{aligned}
&\frac{\tilde{\partial} v_1}{\partial s} + v_3 æ_{20} - v_2 æ_{30} = 0 \, , \\
&\frac{\tilde{\partial} v_2}{\partial s} + v_1 æ_{30} - v_3 æ_{10} = \omega_3 \, , \\
&\frac{\tilde{\partial} v_3}{\partial s} + v_2 æ_{10} - v_1 æ_{20} = -\omega_2 \, .
\end{aligned}$$

Here, $æ_{j0}$ are the components of the vector æ_0, which describes the shape of the rod axis in the natural configuration. Let us obtain an equation relating $\mathbf{v}$ to $\mathbf{u}$. In view of

$$\mathbf{v} = \frac{\partial \mathbf{u}}{\partial t} \, ,$$

using local derivatives, we get

$$\mathbf{v} = \frac{\tilde{\partial} \mathbf{u}}{\partial t} + \boldsymbol{\omega} \times \mathbf{u} \, . \tag{1.39}$$

Therefore,

$$\frac{\partial \mathbf{v}}{\partial s} = \frac{\tilde{\partial}}{\partial s} \left(\frac{\tilde{\partial} \mathbf{u}}{\partial t} + \boldsymbol{\omega} \times \mathbf{u} \right) + \text{æ} \times \left(\frac{\tilde{\partial} \mathbf{u}}{\partial t} + \boldsymbol{\omega} \times \mathbf{u} \right)$$

or

$$\frac{\partial \mathbf{v}}{\partial s} = \frac{\tilde{\partial}^2 \mathbf{u}}{\partial s \, \partial t} + \frac{\tilde{\partial} \boldsymbol{\omega}}{\partial s} \times \mathbf{u} + \boldsymbol{\omega} \times \frac{\tilde{\partial} \mathbf{u}}{\partial s} + \text{æ} \times \frac{\tilde{\partial} \mathbf{u}}{\partial t} + \text{æ} \times (\boldsymbol{\omega} \times \mathbf{u}) \, ,$$

In view of (1.15), we get

$$\frac{\partial \mathbf{v}}{\partial s} = \frac{\tilde{\partial}^2 \mathbf{u}}{\partial s\, \partial t} + \frac{\tilde{\partial} \text{æ}}{\partial t} \times \mathbf{u} + \boldsymbol{\omega} \times \frac{\tilde{\partial} \mathbf{u}}{\partial s} + \text{æ} \times \frac{\tilde{\partial} \mathbf{u}}{\partial t}. \tag{1.40}$$

Suppose that the rod is in longitudinal motion, hence, the coordinate s depends on t. The velocity of the cross section B relative to the cross section A is

$$\delta \mathbf{V}_{AB} = \mathbf{V} + \mathrm{d}\mathbf{V} - \mathbf{V} = \mathbf{v} + \mathrm{d}\mathbf{v} + \mathbf{w} + \mathrm{d}\mathbf{w} - \mathbf{v} - \mathbf{w} = \mathrm{d}\mathbf{v} + \mathrm{d}\mathbf{w}\,.$$

Consider an element of the rod that coincides with an element of the conduit. Denote the angular velocity of the rod element by $\boldsymbol{\omega}_{\mathrm{a}}$. We have

$$\delta \mathbf{V}_{AB} = \boldsymbol{\omega}_{\mathrm{a}} \times \mathbf{e}_1 \,\mathrm{d}s\,.$$

By analogy with the representation for the absolute velocity of a translational motion, we can write

$$\boldsymbol{\omega}_{\mathrm{a}} = \boldsymbol{\omega} + \boldsymbol{\omega}_{\mathrm{r}}\,,$$

where $\boldsymbol{\omega}_{\mathrm{a}}$ is the absolute angular velocity of the rod element, $\boldsymbol{\omega}$ is the angular velocity of the conduit element, and $\boldsymbol{\omega}_{\mathrm{r}}$ is the angular velocity of the rod element. The velocity $\boldsymbol{\omega}_{\mathrm{r}}$ is resulted from the motion of the rod relative to the conduit (Fig. 1.4).

It follows from (1.39) and (1.40) that

$$\frac{\partial \mathbf{v}}{\partial s} + \frac{\partial \mathbf{w}}{\partial s} = \boldsymbol{\omega} \times \mathbf{e}_1 + \boldsymbol{\omega}_{\mathrm{r}} \times \mathbf{e}_1\,. \tag{1.41}$$

Consequently, using (1.41), we get

$$\frac{\partial \mathbf{w}}{\partial s} = \boldsymbol{\omega}_{\mathrm{r}} \times \mathbf{e}_1\,. \tag{1.42}$$

Applying local derivative notation, we obtain

$$\frac{\tilde{\partial} \mathbf{w}}{\partial s} + \text{æ} \times \mathbf{w} = \boldsymbol{\omega}_{\mathrm{r}} \times \mathbf{e}_1\,. \tag{1.43}$$

Since for an inextensible rod $\mathbf{w}$ does not depend on s, we see that the last relation can be rearranged as follows:

$$\text{æ} \times \mathbf{w} = \boldsymbol{\omega}_{\mathrm{r}} \times \mathbf{e}_1\,. \tag{1.44}$$

From this equation we readily obtain the components of the vector $\boldsymbol{\omega}_{\mathrm{r}}$:

$$\omega_{\mathrm{r}\,1} = 0\,, \qquad \omega_{\mathrm{r}\,2} = \text{æ}_3 w\,, \qquad \omega_{\mathrm{r}\,3} = -\text{æ}_2 w\,. \tag{1.45}$$

If the attached axes coincide with the natural ones, then $\text{æ}_2 = 0$ and the vector $\boldsymbol{\omega}_{\mathrm{r}}$ may have only one nonzero component, $\omega_{\mathrm{r}\,2} = \text{æ}_3 w$.

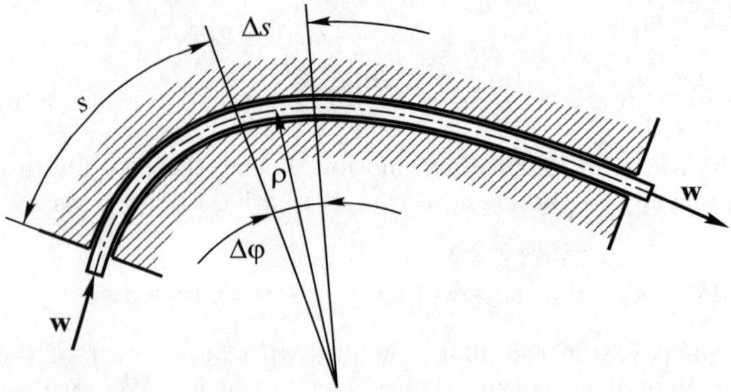

Fig. 1.6.

Once the relationships for $\omega_{\mathrm{r}\,i}$ are obtained, we have an opportunity to determine the rotation of the rod element that is produced by the motion of the rod along the curvilinear conduit.

Consider an example. Suppose that a rod moves in an immovable conduit. The axial line of the conduit is a plane curve (Fig. 1.6). From the figure it is clearly seen that

$$\Delta s = w\,\Delta t = \rho\,\Delta\varphi\,, \qquad æ_3 = \frac{1}{\rho}\,,$$

thus, as $\Delta t \to 0$, we have

$$æ_3 w = \frac{\mathrm{d}\varphi}{\mathrm{d}t} = \omega_{\mathrm{r}\,3}\,.$$

If the rod moves along the conduit, then the kinematic equation for the velocities (1.41) is equivalent to two equations

$$\frac{\partial \mathbf{v}}{\partial s} = \boldsymbol{\omega} \times \mathbf{e}_1\,, \qquad æ \times \mathbf{w} = \boldsymbol{\omega}_{\mathrm{r}} \times \mathbf{e}_1\,.$$

As noted above, if $w = |\mathbf{w}|$ is known, then the components of the vector $\boldsymbol{\omega}_{\mathrm{r}}$ (see (1.45)) can be obtained from the last equation.

1.2.2 Kinematic Equations for Accelerations

In terms of the Eulerian variables, the absolute time derivative of the transportation velocity of a rod element reads (Fig. 1.4)

$$\frac{\mathrm{d}\mathbf{v}}{\mathrm{d}t} = \frac{\partial \mathbf{v}}{\partial t} + \frac{\partial \mathbf{v}}{\partial s}\,w\,,$$

where

$$\mathbf{v} = \sum_{i=1}^{3} v_i \mathbf{e}_i = v_i \mathbf{e}_i \,.$$

Here, $\mathbf{v}$ is the velocity of the conduit axis point that coincides at this very moment with the centroid of the rod element. Thus, for $\boldsymbol{\omega} \neq 0$, $\mathbf{v}$ is the transportation velocity. If the rod does not move along the conduit ($\mathbf{w} = 0$), then $\mathbf{v}$ is the absolute velocity of the rod. By use of (1.4) and (A.129) from Svetlitsky (2000a), the material and local derivatives are related to each other as follows:

$$\frac{\partial \mathbf{v}}{\partial t} = \frac{\tilde{\partial} \mathbf{v}}{\partial t} + \boldsymbol{\omega} \times \mathbf{v} \,, \tag{1.46}$$

$$\frac{\partial \mathbf{v}}{\partial s} = \frac{\tilde{\partial} \mathbf{v}}{\partial s} + \text{æ} \times \mathbf{v} \,. \tag{1.47}$$

Setting $w = 0$ in (1.46), we get the coordinates of the acceleration of the rod element in the attached coordinate system

$$\begin{aligned}
\left(\frac{\mathrm{d}\mathbf{v}}{\mathrm{d}t}\right)_1 &= \frac{\tilde{\partial} v_1}{\partial t} + v_3\omega_2 - v_2\omega_3 \,, \\
\left(\frac{\mathrm{d}\mathbf{v}}{\mathrm{d}t}\right)_2 &= \frac{\tilde{\partial} v_2}{\partial t} + v_1\omega_3 - v_3\omega_1 \,, \\
\left(\frac{\mathrm{d}\mathbf{v}}{\mathrm{d}t}\right)_3 &= \frac{\tilde{\partial} v_3}{\partial t} + v_2\omega_1 - v_1\omega_2 \,.
\end{aligned} \tag{1.48}$$

For $\mathbf{w} \neq 0$, the absolute acceleration of the rod axis points reads

$$\begin{aligned}
\frac{\mathrm{d}(\mathbf{v} + \mathbf{w})}{\mathrm{d}t} &= \frac{\mathrm{d}\mathbf{V}}{\mathrm{d}t} = \frac{\tilde{\partial} \mathbf{v}}{\partial t} + \boldsymbol{\omega} \times \mathbf{v} + \left[\frac{\tilde{\partial} \mathbf{v}}{\partial s} + (\text{æ} \times \mathbf{v})\right] w \\
&+ \frac{\tilde{\partial} \mathbf{w}}{\partial t} + \boldsymbol{\omega} \times \mathbf{w} + \frac{\tilde{\partial} \mathbf{w}}{\partial s} w + (\text{æ} \times \mathbf{w})\, w \,.
\end{aligned} \tag{1.49}$$

Using (1.47), we can eliminate

$$\frac{\tilde{\partial} \mathbf{v}}{\partial s} + \text{æ} \times \mathbf{v}$$

from the last equation; then, we get

$$\frac{\mathrm{d}\mathbf{V}}{\mathrm{d}t} = \frac{\tilde{\partial} \mathbf{v}}{\partial t} + \boldsymbol{\omega} \times \mathbf{v} + 2\,\boldsymbol{\omega} \times \mathbf{w} + \frac{\tilde{\partial} \mathbf{w}}{\partial s} + \frac{\tilde{\partial} \mathbf{w}}{\partial s} w + (\text{æ} \times \mathbf{w})\, w \,. \tag{1.50}$$

For an inextensible rod, we should put

$$\mathbf{w}(s,t) = w(t)\,\mathbf{e}_1, \qquad \frac{\tilde{\partial} \mathbf{w}}{\partial s} = 0, \qquad \frac{\tilde{\partial} \mathbf{w}}{\partial t} \neq 0$$

in (1.50); if the magnitude of the velocity w of the rod is constant, we have

$$\frac{\tilde{\partial}\mathbf{w}}{\partial t} = \frac{\tilde{\partial}\mathbf{w}}{\partial s} = 0\,.$$

In the case of an inextensible rod, in view of (1.50), the projections of the absolute acceleration on the coordinate axes of the conduit-fixed frame are as follows:

$$\begin{aligned}
\left(\frac{\mathrm{d}\mathbf{V}}{\mathrm{d}t}\right)_1 &= \frac{\tilde{\partial} v_1}{\partial s} + v_3\omega_2 - v_2\omega_3 + \frac{\tilde{\partial} w}{\partial t}\,, \\
\left(\frac{\mathrm{d}\mathbf{V}}{\mathrm{d}t}\right)_2 &= \frac{\tilde{\partial} v_2}{\partial s} + v_1\omega_3 - v_3\omega_1 + 2\omega_3 w + w^2 æ_3\,, \\
\left(\frac{\mathrm{d}\mathbf{V}}{\mathrm{d}t}\right)_3 &= \frac{\tilde{\partial} v_3}{\partial s} + v_2\omega_1 - v_1\omega_2 - 2\omega_2 w - w^2 æ_2\,.
\end{aligned} \tag{1.51}$$

If the velocity w does not vary with time, then we should put

$$\frac{\tilde{\partial} w}{\partial t} = 0$$

in (1.51).

Let us derive an equation that relates the acceleration vector of the axis points to the displacement vector. Since

$$\mathbf{v} = \frac{\partial \mathbf{u}}{\partial t} = \frac{\tilde{\partial}\mathbf{u}}{\partial t} + \boldsymbol{\omega}\times\mathbf{u}\,,$$

we have

$$\frac{\mathrm{d}\mathbf{v}}{\mathrm{d}t} = \frac{\tilde{\partial}}{\partial t}\left(\frac{\tilde{\partial}\mathbf{u}}{\partial t} + \boldsymbol{\omega}\times\mathbf{u}\right) + \boldsymbol{\omega}\times\left(\frac{\tilde{\partial}\mathbf{u}}{\partial t} + \boldsymbol{\omega}\times\mathbf{u}\right)$$

or

$$\frac{\mathrm{d}\mathbf{v}}{\mathrm{d}t} = \frac{\tilde{\partial}^2\mathbf{u}}{\partial t^2} + \frac{\tilde{\partial}\boldsymbol{\omega}}{\partial t}\times\mathbf{u} + 2\boldsymbol{\omega}\times\frac{\tilde{\partial}\mathbf{u}}{\partial t} + \boldsymbol{\omega}\times(\boldsymbol{\omega}\times\mathbf{u})\,.$$

Using the formula (A.31) for triple vector product, we finally arrive at

$$\frac{\mathrm{d}\mathbf{v}}{\mathrm{d}t} = \frac{\tilde{\partial}^2\mathbf{u}}{\partial t^2} + 2\boldsymbol{\omega}\times\frac{\tilde{\partial}\mathbf{u}}{\partial t} + \frac{\tilde{\partial}\boldsymbol{\omega}}{\partial t}\times\mathbf{u} + (\boldsymbol{\omega}\cdot\mathbf{u})\times\boldsymbol{\omega} - |\boldsymbol{\omega}|^2\mathbf{u}\,. \tag{1.52}$$

The kinematic equations obtained above are essential in the study of dynamics of rods. The fact is that the derivation and rearrangement of most equations of motion can hardly be performed without resort to these equations.

2. General Equations of Motion

This chapter is concerned with the detailed derivation of the general equations of motion. These equations find their use in the subsequent chapters. During the derivation of the equations, the vector–matrix form of the relationships is used. This makes transformation of the equations more compact and transparent.

First, we derive the general nonlinear equations of motion of a rod. Then, we consider various special types of motion, including a steady-state motion.

2.1 Nonlinear Equations of Motion

2.1.1 Equations of Motion in the Attached Coordinate System

Consider an element of the rod (Fig. 2.1a). The motion of the element can be represented as a composition of two motions, that is, the element moves translationally with a linear velocity $\mathbf{v}$ and rotates with an angular velocity $\boldsymbol{\omega}$. The loads acting on the rod element can be either time-dependent or not. We shall consider both concentrated and distributed forces and moments.

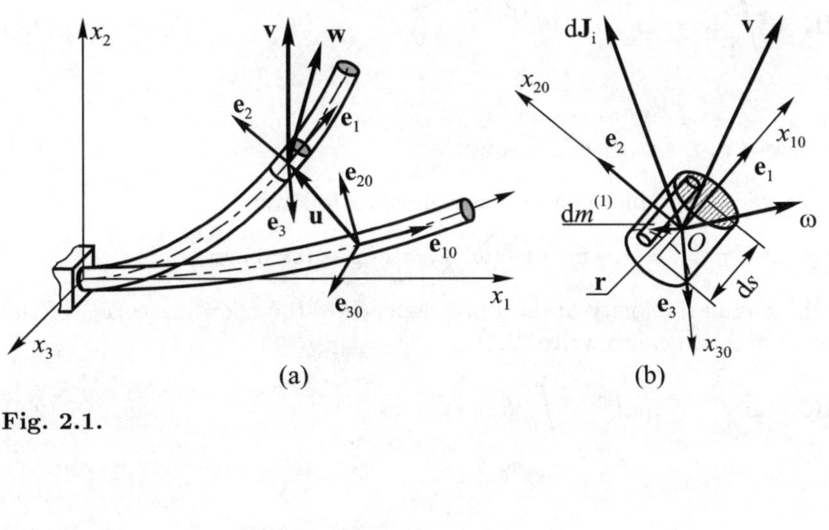

Fig. 2.1.

To clarify the meaning of the notion *time-dependent load*, let us consider an example. A follower load (a load of constant magnitude) is time-independent with respect to the attached basis because the components of such a load in this basis are constant. At the same time, the Cartesian components of a follower load may vary with time.

Studying the motion of a rod, we assume that the vectors of internal loads $\mathbf{Q}$ and $\mathbf{M}$ as well as the vectors æ, $\mathbf{u}$, $\mathbf{v}$, and $\boldsymbol{\omega}$ and the components ϑ_j are functions of both s and t.

Using Lagrangian variables, we can represent the inertial force acting on an element of the rod (Fig. 2.1*b*) as follows:

$$\mathrm{d}\mathbf{J}_{\mathrm{i}} = -\mathrm{d}m\,\frac{\mathrm{d}\mathbf{v}}{\mathrm{d}t}\,, \qquad \mathrm{d}m = m_0\,\mathrm{d}s\,; \tag{2.1}$$

here, m_0 is the mass per unit length of the rod. The velocity vector reads

$$\mathbf{v} = \frac{\mathrm{d}\mathbf{u}}{\mathrm{d}t}\,, \tag{2.2}$$

where $\mathbf{u}$ is the displacement vector of the axial points of the rod (Fig. 2.1*a*). Let us consider the rotary inertia of the element. The principal moment of inertial forces acting on it is

$$\mathrm{d}\boldsymbol{\mu}_{\mathrm{i}} = -\frac{\mathrm{d}}{\mathrm{d}t}\,(\mathrm{J}\boldsymbol{\omega})\,\mathrm{d}s\,, \tag{2.3}$$

where J is a diagonal matrix whose elements are the principal moments of inertia of a rod element of unit length. Consider this matrix in detail. The point O in Fig. 2.1*b* is the center of gravity of the rod element $\mathrm{d}m$. Assuming that the point O is immovable, we can find the principal moment of momentum of the rod element

$$\mathrm{d}\mathbf{K} = \int_{\mathrm{d}V} [\mathbf{r}\times(\boldsymbol{\omega}\times\mathbf{r})]\,\mathrm{d}m^{(1)}\,, \tag{2.4}$$

where

$$\mathbf{v} = \boldsymbol{\omega}\times\mathbf{r}\,, \quad \mathbf{r} = x_{20}\mathbf{e}_2 + x_{30}\mathbf{e}_3\,.$$

Triple vector product can be rearranged as follows:

$$\mathbf{r}\times(\boldsymbol{\omega}\times\mathbf{r}) = \boldsymbol{\omega}\cdot(\mathbf{r}\cdot\mathbf{r}) - \mathbf{r}\cdot(\boldsymbol{\omega}\cdot\mathbf{r}) = r^2\boldsymbol{\omega} - (\boldsymbol{\omega}\cdot\mathbf{r})\cdot\mathbf{r}\,.$$

Since the angular velocity $\boldsymbol{\omega}$ does not depend on the coordinates x_{j0} of the element $\mathrm{d}m^{(1)}$, we can rewrite (2.4) as

$$\mathrm{d}\mathbf{K} = \boldsymbol{\omega}\int_{\mathrm{d}V} r^2\,\mathrm{d}m^{(1)} - \int_{\mathrm{d}V} (\boldsymbol{\omega}\cdot\mathbf{r})\cdot\mathbf{r}\,\mathrm{d}m^{(1)}\,, \tag{2.5}$$

where

$$\mathrm{d}m^{(1)} = \rho\, \mathrm{d}x_{10}\, \mathrm{d}x_{20}\, \mathrm{d}x_{30} = \rho\, \mathrm{d}F\, \mathrm{d}s, \qquad \mathrm{d}s = \mathrm{d}x_{10}\,.$$

In terms of projections in the attached basis, equation (2.5) reads

$$\begin{aligned}
\mathrm{d}K_1 &= \mathrm{d}J_{11}\,\omega_1 - \mathrm{d}J_{12}\,\omega_2 - \mathrm{d}J_{13}\,\omega_3\,,\\
\mathrm{d}K_2 &= -\mathrm{d}J_{21}\,\omega_1 + \mathrm{d}J_{22}\,\omega_2 - \mathrm{d}J_{23}\,\omega_3\,,\\
\mathrm{d}K_3 &= -\mathrm{d}J_{31}\,\omega_1 - \mathrm{d}J_{32}\,\omega_2 + \mathrm{d}J_{33}\,\omega_3\,,
\end{aligned} \tag{2.6}$$

where

$$\begin{aligned}
\mathrm{d}J_{11} &= \rho \int_F (x_{20}^2 + x_{30}^2)\, \mathrm{d}F\, \mathrm{d}s = J_{11}\, \mathrm{d}s\,,\\
\mathrm{d}J_{22} &= \rho \int_F (x_{10}^2 + x_{30}^2)\, \mathrm{d}F\, \mathrm{d}s = J_{22}\, \mathrm{d}s\,,\\
\mathrm{d}J_{33} &= \rho \int_F (x_{10}^2 + x_{20}^2)\, \mathrm{d}F\, \mathrm{d}s = J_{33}\, \mathrm{d}s\,,\\
\mathrm{d}J_{12} &= \rho \int_F x_{10}x_{20}\, \mathrm{d}F\, \mathrm{d}s = J_{12}\, \mathrm{d}s\,,\\
\mathrm{d}J_{13} &= \rho \int_F x_{10}x_{30}\, \mathrm{d}F\, \mathrm{d}s = J_{13}\, \mathrm{d}s\,,\\
\mathrm{d}J_{23} &= \rho \int_F x_{20}x_{30}\, \mathrm{d}F\, \mathrm{d}s = J_{23}\, \mathrm{d}s\,;
\end{aligned}$$

here, ρ is the density of the rod.

Since $x_{10} = 0$, the centrifugal moments of inertia J_{12} and J_{13} are zero and the axial moments of inertia J_{22} and J_{33} are

$$J_{22} = \int_F x_{30}^2\, \mathrm{d}F\,, \qquad J_{33} = \int_F x_{20}^2\, \mathrm{d}F\,, \qquad J_{23} = \rho \int_F x_{20}x_{30}\, \mathrm{d}F\,.$$

Using matrix notation, we can represent (2.6) as

$$\mathrm{d}\mathbf{K} = \mathrm{J}\boldsymbol{\omega}\, \mathrm{d}s\,, \tag{2.7}$$

where

$$\mathrm{J} = \begin{bmatrix} J_{11} & 0 & 0 \\ 0 & J_{22} & -J_{23} \\ 0 & -J_{32} & J_{33} \end{bmatrix}$$

is the tensor of inertia of a rod element of unit length. If the axes x_{i0} coincide with the principal axes of the cross section, then the moments of inertia J_{23} and J_{32} are zero and the matrix J is as follows:

$$\mathrm{J} = \begin{bmatrix} J_{11} & 0 & 0 \\ 0 & J_{22} & 0 \\ 0 & 0 & J_{33} \end{bmatrix}. \tag{2.8}$$

By use of the geometric properties of the cross section, we can represent the moments of inertia J_{ii} as

$$J_{11} = J_1 \rho\,, \qquad J_{22} = J_2 \rho\,, \qquad J_{33} = J_3 \rho\,,$$

where

$$J_1 = \int_F (x_{20}^2 + x_{30}^2)\,\mathrm{d}F\,, \qquad J_2 = \int_F x_{30}^2\,\mathrm{d}F\,, \qquad J_3 = \int_F x_{20}^2\,\mathrm{d}F\,.$$

For a rod of circular cross section, J_1 coincides with the polar moment of inertia of the cross section. The moments J_2 and J_3 are related to the stiffnesses of the rod A_{ii} as follows:

$$A_{11} = GJ_\rho\,, \qquad A_{22} = EJ_2\,, \qquad A_{33} = EJ_3\,.$$

For rods of noncircular cross section, we can represent the torsional stiffness in the form $A_{11} = GJ_{\mathrm{t}}$, where J_{t} is a geometric characteristic of the cross section such that $J_{\mathrm{t}} \neq J_1$.

For a rod of rectangular cross section, we have

$$J_1 = \frac{b^2 + h^2}{12}\,bh\,, \qquad J_{\mathrm{t}} = b^3 h \beta\left(\frac{h}{b}\right),$$

where $\beta(h/b)$ is a dimensionless-valued function of the quotient of the rectangle sides (Feodosiev (1999)). For example, if $h = b$, then we get $J_1 = 0.166h^4$ and $J_{\mathrm{t}} = 0.141h^4$ (thus, $\beta = 0.141$).

The elements of the matrix J are either constant or, in the attached coordinate system, functions of s. For example, the Cartesian coordinates of the center of gravity of the element $\mathrm{d}m^{(1)}$ vary with time as the rod moves; hence, J_{ij} also vary with time. For an inextensible rod, the axial coordinate s of each rod element is constant during motion (provided the rod is not in longitudinal motion). Using the Lagrangian variables, we can rewrite (2.1), (2.2), and (2.3) in the form

$$\mathrm{d}\mathbf{J}_{\mathrm{i}} = -\mathrm{d}m\,\frac{\partial \mathbf{v}}{\partial t}\,, \qquad \mathbf{v} = \frac{\partial \mathbf{u}}{\partial t}\,, \qquad \mathrm{d}\boldsymbol{\mu}_{\mathrm{i}} = -\frac{\partial}{\partial t}\,(\mathrm{J}\boldsymbol{\omega})\,\mathrm{d}s\,.$$

In view of the d'Alembert principle, the vector equation of translational motion of a rod element is as follows:

$$m_0\,\frac{\partial \mathbf{v}}{\partial t} = \frac{\partial \mathbf{Q}}{\partial s} + \mathbf{q} + \sum_{i=1}^{n} \mathbf{P}^{(i)}\,\delta\,(s - s_i)\,. \tag{2.9}$$

Taking into account the rotary inertia, we obtain the equation describing the rotation of a rod element

$$\frac{\partial}{\partial t}\,(\mathrm{J}\boldsymbol{\omega}) = \frac{\partial \mathbf{M}}{\partial s} + \mathbf{e}_1 \times \mathbf{Q} + \boldsymbol{\mu} + \sum_{\nu=1}^{\rho} \mathbf{T}^{(\nu)}\,\delta\,(s - s_\nu)\,. \tag{2.10}$$

The elements of the matrix J are time-independent if calculated in the rod-fixed coordinate system. In terms of local derivatives, the relationships (2.9) and (2.10) can be rearranged as follows (a tilde in the notation of local derivatives is dropped out):

$$m_0\left(\frac{\partial \mathbf{v}}{\partial t}+\boldsymbol{\omega}\times\mathbf{v}\right)=\frac{\partial \mathbf{Q}}{\partial s}+æ\times\mathbf{Q}+\mathbf{P}\,, \tag{2.11}$$

$$\mathrm{J}\,\frac{\partial \boldsymbol{\omega}}{\partial t}+\boldsymbol{\omega}\times\mathrm{J}\boldsymbol{\omega}=\frac{\partial \mathbf{M}}{\partial s}+æ\times\mathbf{M}+\mathbf{e}_1\times\mathbf{Q}+\mathbf{T}\,; \tag{2.12}$$

here,

$$\mathbf{P}=\mathbf{q}+\sum_{i=1}^{n}\mathbf{P}^{(i)}\,\delta\,(s-s_i)\,,\qquad \mathbf{T}=\boldsymbol{\mu}+\sum_{\nu=1}^{\rho}\mathbf{T}^{(\nu)}\,\delta\,(s-s_\nu)\,.$$

It is a common practice in applications that the rotary inertia is neglected, that is, $\mathrm{J}\boldsymbol{\omega}\approx 0$. In this case, equation (2.12) is identical to its static counterpart (see (1.58) of Svetlitsky (2000a))

$$\frac{\partial \mathbf{M}}{\partial s}+æ\times\mathbf{M}+\mathbf{e}_1\times\mathbf{Q}+\mathbf{T}=0\,.$$

2.1.2 Kinematic Relationships and Equations of Motion in Dimensionless Form

Before proceeding to the reduction of the equations to a dimensionless form, let us recall the basic relationships obtained in the previous sections.

In Sect. 1.1, the kinematic equation (1.17) relating the vectors æ and $\boldsymbol{\omega}$ was obtained (a tilde in the notation of local derivatives is dropped out)

$$\frac{\partial \boldsymbol{\omega}}{\partial s}-\frac{\partial æ}{\partial t}=\boldsymbol{\omega}\times æ\,.$$

The relationship between the vector **M** and the increment of the vector æ as well as the equation for the displacement vector **u** are the same as those in the statics of rods, that is,

$$\mathbf{M}=\mathrm{A}(æ-æ_0^{(1)})\,, \tag{2.13}$$

$$\frac{\partial \mathbf{u}}{\partial s}+æ\times\mathbf{u}=\mathbf{e}_1-\mathbf{e}_{10}\,. \tag{2.14}$$

Equation (1.37) reads

$$\frac{\partial \mathbf{v}}{\partial s}+æ\times\mathbf{v}-\boldsymbol{\omega}\times\mathbf{e}_1=0\,; \tag{2.15}$$

it also finds its use in the dynamics of rods.

There are two other equations, namely, the relationship between the components of the vector æ and the angles $\boldsymbol{\vartheta}_j$ (see (1.17))

$$æ = \mathrm{L}_1 \frac{\partial \boldsymbol{\vartheta}}{\partial s} + \mathrm{L} æ_0^{(1)} \tag{2.16}$$

or

$$\mathrm{L}_1 \frac{\partial \boldsymbol{\vartheta}}{\partial s} + \mathrm{L}_2 æ_0^{(1)} - \mathrm{A}^{-1}\mathbf{M} = 0\,, \qquad \mathrm{L}_2 = \mathrm{L} - \mathrm{E}$$

and the equation involving the components of the vector $\boldsymbol{\omega}$ and the angles $\boldsymbol{\vartheta}_j$ (see (1.11))

$$\boldsymbol{\omega} = \mathrm{L}_1 \frac{\partial \boldsymbol{\vartheta}}{\partial t}\,. \tag{2.17}$$

Let us reduce the systems of equations (2.9)–(2.13) and (2.15)–(2.17) to dimensionless form. To do this, we introduce dimensionless variables as follows:

$$\eta = \frac{s}{l}, \qquad \tau = p_0 t\,, \qquad \tilde{\boldsymbol{\omega}} = \frac{\boldsymbol{\omega}}{p_0}, \qquad \tilde{\mathbf{v}} = \frac{\mathbf{v}}{l p_0},$$

$$\tilde{\mathbf{M}} = \frac{\mathbf{M} l}{A_{33}(0)}, \qquad \tilde{\boldsymbol{\mu}} = \frac{\boldsymbol{\mu} l^2}{A_{33}(0)}, \qquad \tilde{\mathbf{Q}} = \frac{\mathbf{Q} l^2}{A_{33}(0)}, \qquad \tilde{\mathbf{q}} = \frac{\mathbf{q} l^3}{A_{33}(0)},$$

$$p_0 = \sqrt{\frac{A_{33}(0)}{m_0(0)\, l^4}}, \qquad \tilde{J}_{ii}(\eta) = \frac{J_{ii}(\eta)}{F_0 l^2}, \qquad \tilde{A}_{ii}(\eta) = \frac{A_{ii}(\eta)}{A_{33}(0)};$$

here, $m_0(0)$ is the mass per unit length at the origin and $A_{33}(0)$ is the rod stiffness at the origin. A tilde is used to denote dimensionless variables.
We can represent the mass per unit length of an arbitrary cross section as

$$m_0(\eta) = m_0(0)\, n_1(\eta) = \rho F_0 n_1(\eta)\,,$$

where $n_1(\eta)$ is a dimensionless function and F_0 is the cross-sectional area at the origin. In dimensionless form, the fundamental equations used in the dynamics of rods (2.9)–(2.13) and (2.15)–(2.17) are as follows (a tilde in the notation of dimensionless quantities is dropped out):

$$n_1 \frac{\partial \mathbf{v}}{\partial \tau} = \frac{\partial \mathbf{Q}}{\partial \eta} + \mathbf{P}\,, \tag{2.18}$$

$$\frac{\partial}{\partial \tau}(\mathrm{J}\boldsymbol{\omega}) = \frac{\partial \mathbf{M}}{\partial \eta} + \mathbf{e}_1 \times \mathbf{Q} + \mathbf{T}\,. \tag{2.19}$$

After some rearrangements, in the attached basis, we have

$$n_1(\eta)\left(\frac{\partial \mathbf{v}}{\partial \tau}+\boldsymbol{\omega}\times\mathbf{v}\right)-\frac{\partial \mathbf{Q}}{\partial \eta}-æ\times\mathbf{Q}-\mathbf{P}=0\,, \tag{2.20}$$

$$\mathrm{J}(\eta)\,\frac{\partial \boldsymbol{\omega}}{\partial \tau}+\boldsymbol{\omega}\times\mathrm{J}(\eta)\,\boldsymbol{\omega}-\frac{\partial \mathbf{M}}{\partial \eta}-æ\times\mathbf{M}-\mathbf{e}_1\times\mathbf{Q}-\mathbf{T}=0\,, \tag{2.21}$$

$$\mathbf{M}=\mathrm{A}(æ-æ_0^{(1)})\,, \tag{2.22}$$

$$\mathrm{L}_1\,\frac{\partial \boldsymbol{\vartheta}}{\partial \eta}+\mathrm{L}æ_0^{(1)}-æ=0\,, \tag{2.23}$$

$$\frac{\partial \mathbf{u}}{\partial \tau}+\boldsymbol{\omega}\times\mathbf{u}-\mathbf{v}=0\,, \tag{2.24}$$

$$\mathrm{L}_1\,\frac{\partial \boldsymbol{\vartheta}}{\partial \tau}-\boldsymbol{\omega}=0\,, \tag{2.25}$$

$$\frac{\partial \mathbf{u}}{\partial \eta}+æ\times\mathbf{u}-(1-l_{11})\mathbf{e}_1+l_{21}\mathbf{e}_2+l_{31}\mathbf{e}_e=0\,, \tag{2.26}$$

where

$$\mathbf{P}=\mathbf{q}+\sum_{i=1}^{n}\mathbf{P}^{(i)}\,\delta\,(\eta-\eta_i)\,,\qquad \mathbf{T}=\boldsymbol{\mu}+\sum_{\nu=1}^{\rho}\mathbf{T}^{(\nu)}\,\delta\,(\eta-\eta_\nu)\,,$$

$$\mathrm{L}_1=\begin{bmatrix}\cos\vartheta_2\cos\vartheta_3 & 0 & -\sin\vartheta_2\\ -\sin\vartheta_3 & 1 & 0\\ \sin\vartheta_2\cos\vartheta_3 & 0 & \cos\vartheta_2\end{bmatrix},$$

$$\mathrm{L}=\begin{bmatrix}\cos\vartheta_2\cos\vartheta_3 & \begin{matrix}\cos\vartheta_2\sin\vartheta_3\cos\vartheta_1\\ +\sin\vartheta_2\sin\vartheta_1\end{matrix} & \begin{matrix}\cos\vartheta_2\sin\vartheta_3\sin\vartheta_1\\ -\sin\vartheta_2\cos\vartheta_1\end{matrix}\\ -\sin\vartheta_3 & \cos\vartheta_1\cos\vartheta_3 & \cos\vartheta_3\sin\vartheta_1\\ \sin\vartheta_2\cos\vartheta_3 & \begin{matrix}\sin\vartheta_2\sin\vartheta_3\cos\vartheta_1\\ -\cos\vartheta_2\sin\vartheta_1\end{matrix} & \begin{matrix}\sin\vartheta_2\sin\vartheta_3\sin\vartheta_1\\ +\cos\vartheta_2\cos\vartheta_1\end{matrix}\end{bmatrix}.$$

Recall that the angles ϑ_j are the angles of rotation of the basis $\{\mathbf{e}_j\}$ relative to the basis $\{\mathbf{e}_{j0}\}$ (see Appendix A.1 of Svetlitsky (2000a)). The elements of the matrices L_1 and L are functions of these angles. To account for forces of viscous damping, one should substitute the terms $\mathrm{B}^{(1)}\mathbf{v}$ and $\mathrm{B}^{(2)}\boldsymbol{\omega}$ into the left-hand sides of (2.18)–(2.21). Here, $\mathrm{B}^{(1)}$ and $\mathrm{B}^{(2)}$ are matrices whose elements are the coefficients of distributed forces and moments of viscous damping.

Consider the formula (2.22) for the moment $\mathbf{M}$ in detail. In the current notation, the components of the vector $æ_0^{(1)}$ are curvatures that are characteristics of the natural configuration of the rod. If the vector $æ_0^{(1)}$ is a characteristic of a static state of the rod, then we have

$$\mathbf{M}=\mathrm{A}(æ-æ_0^{(1)})+\mathbf{M}_0\,,$$

where $\mathbf{M}_0$ is the static moment given by

$$\mathbf{M}_0 = \mathrm{A}(æ_0 - æ_{00}^{(1)})\,;$$

here, the components of the vector $æ_{00}^{(1)}$ are the curvatures of the rod axis in the natural configuration.

Thus, the study of a problem of dynamics of rods implies the determination of the seven unknown vectors $\mathbf{v}$, $\mathbf{Q}$, $\mathbf{M}$, $\mathbf{u}$, æ, $\boldsymbol{\omega}$, and $\boldsymbol{\vartheta}$. To do this, we have the system of seven equations (2.20)–(2.26).

The relationship (1.19) (or (1.20), which is the same as (1.19) but written in components),

$$\frac{\partial \boldsymbol{\omega}}{\partial \eta} - \frac{\partial æ}{\partial \tau} - \boldsymbol{\omega} \times æ = 0\,, \tag{2.27}$$

may also be useful.

Equations of Motion in the Attached Coordinate System Tensor form of the system of equations (2.20)–(2.26) is as follows:

$$\begin{aligned}
&n_1(\eta)\left(\frac{\partial v_k}{\partial \tau} + \varepsilon_{kij}\omega_i v_j\right) - \frac{\partial Q_k}{\partial \eta} - \varepsilon_{kij}æ_i Q_j - P_k = 0\,,\\
&J^0_{k\gamma}\frac{\partial \omega_\gamma}{\partial \tau} + \varepsilon_{kij}\omega_i J^0_{j\gamma}\omega_\gamma - \frac{\partial M_k}{\partial \eta} - \varepsilon_{kij}æ_i M_j + \varepsilon_{ki1}Q_i - T_k = 0\,,\\
&M_k = A_{kk}(æ_k - æ_{k0})\,,\\
&l_{(1)kj}\frac{\partial \vartheta_j}{\partial \eta} - æ_k + l_{kj}æ_{0j} = 0\,,\\
&\frac{\partial u_k}{\partial \tau} + \varepsilon_{kij}\omega_i u_j - v_k = 0\,,\\
&l_{(1)kj}\frac{\partial \vartheta_j}{\partial \tau} - \omega_k = 0\,,\\
&\frac{\partial u_k}{\partial \eta} + \varepsilon_{kij}æ_i u_j - (\delta_{1k} - l_{1k}) = 0\,.
\end{aligned} \tag{2.28}$$

Here, δ_{1k} is the Kronecker delta.

In components, we have

$$\begin{aligned}
&n_1(\eta)\left(\frac{\partial v_1}{\partial \tau} + v_3\omega_2 - v_2\omega_3\right) - \frac{\partial Q_1}{\partial \eta} - Q_3æ_2 + Q_2æ_3 - P_1 = 0\,,\\
&n_1(\eta)\left(\frac{\partial v_2}{\partial \tau} + v_1\omega_3 - v_3\omega_1\right) - \frac{\partial Q_2}{\partial \eta} - Q_1æ_3 + Q_3æ_1 - P_2 = 0\,,\\
&n_1(\eta)\left(\frac{\partial v_3}{\partial \tau} + v_2\omega_1 - v_1\omega_2\right) - \frac{\partial Q_3}{\partial \eta} - Q_2æ_1 + Q_1æ_2 - P_3 = 0\,,
\end{aligned} \tag{2.29}$$

$$
\begin{aligned}
&J_{11}\,\frac{\partial \omega_1}{\partial \tau}+(J_{33}-J_{22})\,\omega_2\omega_3-\frac{\partial M_1}{\partial \eta}+æ_3 M_2-æ_2 M_3-T_1=0\,,\\
&J_{22}\,\frac{\partial \omega_2}{\partial \tau}+(J_{11}-J_{33})\,\omega_1\omega_3-\frac{\partial M_2}{\partial \eta}+æ_1 M_3-æ_3 M_1+Q_3-T_2=0\,,\\
&J_{33}\,\frac{\partial \omega_3}{\partial \tau}+(J_{22}-J_{11})\,\omega_1\omega_2-\frac{\partial M_3}{\partial \eta}+æ_2 M_1-æ_1 M_2-Q_2-T_3=0\,, \qquad (2.30)
\end{aligned}
$$

$$
\begin{aligned}
M_1&=A_{11}(æ_1-æ_{10})\,,\\
M_2&=A_{22}(æ_2-æ_{20})\,,\\
M_3&=A_{33}(æ_3-æ_{30})\,, \qquad (2.31)
\end{aligned}
$$

$$
\begin{aligned}
&l_{(1)11}\,\frac{\partial \vartheta_1}{\partial \eta}+l_{(1)13}\,\frac{\partial \vartheta_3}{\partial \eta}-æ_1+l_{11}æ_{01}+l_{12}æ_{02}+l_{13}æ_{03}=0\,,\\
&l_{(1)21}\,\frac{\partial \vartheta_1}{\partial \eta}+\frac{\partial \vartheta_2}{\partial \eta}-æ_2+l_{21}æ_{01}+l_{22}æ_{20}+l_{23}æ_{03}=0\,,\\
&l_{(1)31}\,\frac{\partial \vartheta_1}{\partial \eta}+l_{(1)33}\,\frac{\partial \vartheta_3}{\partial \eta}-æ_3+l_{31}æ_{01}+l_{32}æ_{02}+l_{33}æ_{03}=0\,, \qquad (2.32)
\end{aligned}
$$

$$
\begin{aligned}
&\frac{\partial u_1}{\partial \tau}+u_3\omega_2-u_2\omega_3-v_1=0\,,\\
&\frac{\partial u_2}{\partial \tau}+u_1\omega_3-u_3\omega_1-v_2=0\,,\\
&\frac{\partial u_3}{\partial \tau}+u_2\omega_1-u_1\omega_2-v_3=0\,, \qquad (2.33)
\end{aligned}
$$

$$
\begin{aligned}
&l_{(1)11}\,\frac{\partial \vartheta_1}{\partial \tau}+l_{(1)13}\,\frac{\partial \vartheta_3}{\partial \tau}-\omega_1=0\,,\\
&l_{(1)21}\,\frac{\partial \vartheta_1}{\partial \tau}+\frac{\partial \vartheta_2}{\partial \tau}-\omega_2=0\,,\\
&l_{(1)31}\,\frac{\partial \vartheta_1}{\partial \tau}+l_{(1)33}\,\frac{\partial \vartheta_3}{\partial \tau}-\omega_3=0\,, \qquad (2.34)
\end{aligned}
$$

$$
\begin{aligned}
&\frac{\partial u_1}{\partial \eta}+æ_2 u_3-æ_3 u_2-1+l_{11}=0\,,\\
&\frac{\partial u_2}{\partial \eta}+æ_3 u_1-æ_1 u_3+l_{21}=0\,,\\
&\frac{\partial u_3}{\partial \eta}+æ_1 u_2-æ_2 u_1-l_{31}=0\,. \qquad (2.35)
\end{aligned}
$$

2.1.3 Equations of Plane Motion

Consider a particular case when the axial line of a rod in the natural configuration is a plane curve. Let one of the principal axes of each cross section of the rod be perpendicular to this plane. If the rod is deflected such that the axial line of the rod lies in this plane, then, if left to its own device, the rod will move in this plane. For a planar motion, some components of the vectors in (2.29)–(2.34) are zero, namely,

$$
\begin{aligned}
&u_3 = v_3 = \omega_1 = \omega_2 = 0\,, \qquad \vartheta_1 = \vartheta_2 = 0\,,\\
&æ_1 = æ_2 = æ_{10} = æ_{20} = 0\,,\\
&Q_3 = M_1 = M_3 = 0\,, \qquad q_3 = \mu_1 = \mu_2 = 0\,.
\end{aligned}
$$

Equations (2.29)–(2.34) yield a system of eight equations in the unknowns v_1, v_2, Q_1, Q_2, M_3, ω_3, $æ_3$, and ϑ_3:

$$
\begin{aligned}
&n_1(\eta)\left(\frac{\partial v_1}{\partial \tau} - v_2\omega_3\right) - \frac{\partial Q_1}{\partial \eta} + Q_2 æ_3 - P_1 = 0\,,\\
&n_1(\eta)\left(\frac{\partial v_2}{\partial \tau} + v_1\omega_3\right) - \frac{\partial Q_2}{\partial \eta} - Q_1 æ_3 - P_2 = 0\,,\\
&J_{33}\frac{\partial \omega_3}{\partial \tau} - \frac{\partial M_3}{\partial \eta} - Q_2 - T_3 = 0\,,\\
&M_3 = A_{33}(æ_3 - æ_{30})\,,\\
&\frac{\partial \vartheta_3}{\partial \eta} - æ_3 = 0\,,\\
&\frac{\partial u_1}{\partial \tau} - \omega_3 u_2 - v_1 = 0\,,\\
&\frac{\partial u_2}{\partial \tau} + \omega_3 u_1 - v_2 = 0\,, \qquad \omega_3 = \frac{\partial \vartheta_3}{\partial \tau}\,,\\
&\frac{\partial u_1}{\partial \eta} - æ_3 u_2 - 1 + \cos\vartheta_3 = 0\,,\\
&\frac{\partial u_2}{\partial \eta} + æ_3 u_1 + \sin\vartheta_3 = 0\,.
\end{aligned}
\tag{2.36}
$$

2.1.4 Equations of Motion of a Rod in Longitudinal Motion

In the general case, the velocity of a rod element is a sum of the transportation velocity $\mathbf{v}$ and the velocity of the longitudinal motion $\mathbf{w}$ (see Fig. 1.4). The velocity vector of each point of the rod between the cross sections A and B may be represented as the vector sum of the transportation velocity $\mathbf{v}$ and the relative velocity $\mathbf{w}$. By applying the d'Alembert principle to an element of the rod, we obtain the equation of motion for a rod of constant cross section

$$\frac{\partial \mathbf{v}}{\partial \tau} + \frac{\partial \mathbf{w}}{\partial \tau} = \frac{\partial \mathbf{Q}}{\partial \eta} + \mathbf{P}\,.$$

In terms of Eulerian variables, we get

$$\frac{\partial \mathbf{v}}{\partial \tau} + \frac{\partial \mathbf{v}}{\partial \eta} w + \frac{\partial \mathbf{w}}{\partial \tau} + \frac{\partial \mathbf{w}}{\partial \eta} w = \frac{\partial \mathbf{Q}}{\partial \eta} + \mathbf{P}\,. \tag{2.37}$$

The dimensionless velocity w is defined as

$$w = \frac{w_\mathrm{d}}{l p_0}\,,$$

where w_d is the dimensional velocity. Introducing the vector

$$\mathbf{Q}^{(1)} = \mathbf{Q} - w^2 \mathbf{e}_1$$

such that

$$\frac{\partial \mathbf{w}}{\partial \eta} w = \frac{\partial}{\partial \eta} \left(w^2 \mathbf{e}_1 \right),$$

we can represent (2.37) as follows:

$$\left(\frac{\partial \mathbf{v}}{\partial \tau} + \frac{\partial \mathbf{v}}{\partial \eta} w + \frac{\partial \mathbf{w}}{\partial \tau} \right) = \frac{\partial \mathbf{Q}^{(1)}}{\partial \eta} + \mathbf{P}\,. \tag{2.38}$$

With regard for the rotary inertia, from (2.11) it follows that

$$\frac{\partial (\mathrm{J}\boldsymbol{\omega}_\mathrm{a})}{\partial \tau} + \frac{\partial (\mathrm{J}\boldsymbol{\omega}_\mathrm{a})}{\partial \eta} w = \frac{\partial \mathbf{M}}{\partial \eta} + \mathbf{e}_1 \times \mathbf{Q}^{(1)} + \mathbf{T}\,. \tag{2.39}$$

The angular velocity $\boldsymbol{\omega}_\mathrm{a}$ in (2.39) is a sum of the angular velocity of the conduit element (that coincides with the rod element at this very moment) and the angular velocity of the rod element itself due to the motion of the rod relative to the conduit. When the conduit is immovable, the angular velocity of the rod element is $\boldsymbol{\omega}_\mathrm{r}$. Thus, according to Sect. 1.2, we have $\boldsymbol{\omega}_\mathrm{a} = \boldsymbol{\omega} + \boldsymbol{\omega}_\mathrm{r}$.

Recall the relationship between $\boldsymbol{\omega}_\mathrm{r}$ and the velocity of relative motion $|\mathbf{w}|$. For an immovable conduit (see Fig. 1.4), we have

$$\frac{\partial \mathbf{e}_i}{\partial \tau} = \frac{\partial \mathbf{e}_i}{\partial \eta} w = w \text{æ} \times \mathbf{e}_i\,. \tag{2.40}$$

Since

$$\frac{\partial \mathbf{e}_i}{\partial \tau} = \boldsymbol{\omega}_\mathrm{r} \times \mathbf{e}_i\,, \tag{2.41}$$

the relationships (2.40) and (2.41) yield

$$\boldsymbol{\omega}_\mathrm{r} = w \text{æ}\,. \tag{2.42}$$

Eliminating

$$\frac{\partial \mathbf{v}}{\partial \eta} = \boldsymbol{\omega} \times \mathbf{e}_1$$

from (2.38) and using the notation of local derivatives, we get

$$\frac{\partial \mathbf{v}}{\partial \tau} + \boldsymbol{\omega} \times \mathbf{v} + 2\boldsymbol{\omega} \times \mathbf{w} + \frac{\partial \mathbf{w}}{\partial \tau} = \frac{\partial \mathbf{Q}^{(1)}}{\partial \eta} + æ \times \mathbf{Q}^{(1)} + \mathbf{P}\,, \tag{2.43}$$

$$\frac{\partial (\mathrm{J}\boldsymbol{\omega}_{\mathrm{a}})}{\partial \tau} + \boldsymbol{\omega} \times \mathrm{J}\boldsymbol{\omega}_{\mathrm{a}} + \frac{\partial (\mathrm{J}\boldsymbol{\omega}_{\mathrm{a}})}{\partial \eta} w + æ \times \mathrm{J}\boldsymbol{\omega}_{\mathrm{a}} w$$
$$= \frac{\partial \mathbf{M}}{\partial \eta} + æ \times \mathbf{M} + \mathbf{e}_1 \times \mathbf{Q}^{(1)} + \mathbf{T}\,. \tag{2.44}$$

The other equations (2.22)–(2.26) remain unchanged.

In tensor form, equations (2.43) and (2.44) read

$$\frac{\partial v_k}{\partial \tau} + \varepsilon_{kij}\omega_i v_j + 2\varepsilon_{ki1} w\omega_i + \varepsilon_{k23}\frac{\partial w}{\partial \tau} - \frac{\partial Q_k^{(1)}}{\partial \eta} - \varepsilon_{kij} æ_i Q_j^{(1)} = P_k\,,$$
$$J_{ki}\frac{\partial \omega_{\mathrm{a}\,i}}{\partial \tau} + \varepsilon_{kij}\omega_i J_{\gamma\nu}\omega_{\mathrm{a}\,\nu} + J_{ki} w \frac{\partial \omega_{\mathrm{a}\,i}}{\partial \eta} + w\varepsilon_{kij} æ_i J_{j\nu}\omega_{\mathrm{a}\,\nu}$$
$$- \frac{\partial M_k}{\partial \eta} - \varepsilon_{kij} æ_i M_j - \varepsilon_{ki1} Q_i^{(1)} = T_k\,. \tag{2.45}$$

In components, we have

$$\frac{\partial v_1}{\partial \tau} + v_3\omega_2 - v_2\omega_3 + \frac{\partial w}{\partial \tau} - \frac{\partial Q_1^{(1)}}{\partial \eta} - Q_3 æ_2 + Q_2 æ_3 = P_1\,,$$
$$\frac{\partial v_2}{\partial \tau} + v_1\omega_3 - v_3\omega_1 + 2\omega_3 w - \frac{\partial Q_2}{\partial \eta} - Q_1^{(1)} æ_3 + Q_3 æ_1 = P_2\,,$$
$$\frac{\partial v_3}{\partial \tau} + v_2\omega_1 - v_1\omega_2 - 2\omega_2 w - \frac{\partial Q_3}{\partial \eta} - Q_2 æ_1 + Q_1^{(1)} æ_2 = P_3\,,$$

$$J_{11}\frac{\partial \omega_1}{\partial \tau} + J_{11} w \frac{\partial æ_1}{\partial \tau} + J_{33}\omega_2(\omega_3 + æ_3 w) - J_{22}\omega_3(\omega_2 + æ_2 w)$$
$$+ \frac{\partial}{\partial \eta}(J_{11}\omega_1 + J_{11}æ_1 w)\, w + J_{33}æ_2(\omega_3 + æ_3 w)\, w - J_{22}æ_3(\omega_2 + æ_2 w)\, w$$
$$- \frac{\partial M_1}{\partial \eta} - æ_2 M_3 + æ_3 M_2 = T_1\,,$$
$$J_{22}\frac{\partial \omega_2}{\partial \tau} + J_{22} w \frac{\partial æ_2}{\partial \tau} + J_{11}\omega_3(\omega_1 + æ_1 w) - J_{33}\omega_1(\omega_3 + æ_3 w)$$
$$+ \frac{\partial}{\partial \eta}(J_{22}\omega_2 + J_{22}æ_2 w)\, w + J_{11}æ_3(\omega_1 + æ_1 w)\, w - J_{33}æ_1(\omega_3 + æ_3 w)\, w$$
$$- \frac{\partial M_2}{\partial \eta} - æ_3 M_1 + æ_1 M_3 + Q_3 = T_2\,,$$

$$J_{33}\frac{\partial \omega_3}{\partial \tau} + J_{33}w\frac{\partial æ_3}{\partial \tau} + J_{22}\omega_1(\omega_2 + æ_2 w) - J_{11}\omega_2(\omega_1 + æ_1 w)$$
$$+ \frac{\partial}{\partial \eta}(J_{33}\omega_3 + J_{33}æ_3 w)\, w + J_{22}æ_1(\omega_2 + æ_2 w)\, w$$
$$- J_{11}æ_2(\omega_1 + æ_1 w)\, w - \frac{\partial M_3}{\partial \eta} - æ_1 M_2 + æ_2 M_1 - Q_2 = T_3\,, \qquad (2.46)$$

where

$$Q_1^{(1)} = Q_1 - w^2\,, \qquad Q_2^{(1)} = Q_2\,, \qquad Q_3^{(1)} = Q_3\,.$$

Equations of Motion for a Rod Rotating Relative to Its Axis
A space-curved rod rotating relative to its axis with angular velocity $\boldsymbol{\omega}_0$ is shown in Fig. 2.2.

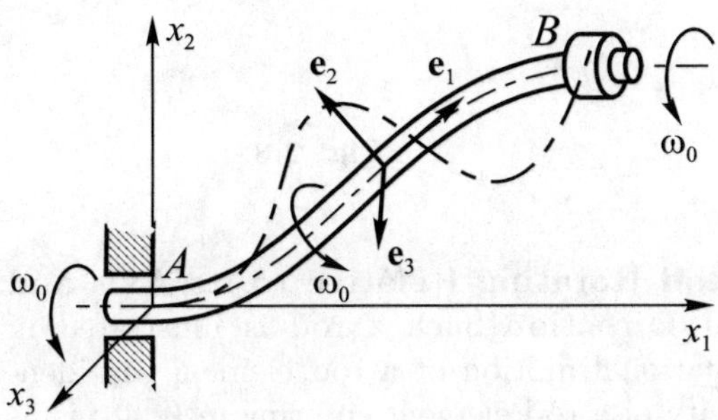

Fig. 2.2.

In applications, such rods are employed to transfer the rotation to some objects whose location in space may change continuously (that is, the coordinates x_i of the point B may vary with time). Imagine that the rod moves in an absolutely flexible, massless conduit such that the rod and the conduit are of exactly the same shape. Then, the absolute angular velocity of a rod element is

$$\boldsymbol{\omega}_a = \boldsymbol{\omega} + \boldsymbol{\omega}_0(\tau)\,, \qquad \boldsymbol{\omega}_0(\tau) = \omega_0(\tau)\,\mathbf{e}_1\,,$$

where $\boldsymbol{\omega}$ is the angular velocity of the natural reference frame associated with the imaginary conduit. The equations of translational motion of a rod element are the same as (2.20) for $\boldsymbol{\omega}_0 = 0$. Substituting the expression for $\boldsymbol{\omega}_0(\tau)$ into (2.21), we get

$$\frac{\partial \mathrm{J}(\boldsymbol{\omega} + \boldsymbol{\omega}_0)}{\partial \tau} + \boldsymbol{\omega} \times \mathrm{J}(\boldsymbol{\omega} + \boldsymbol{\omega}_0) - \frac{\partial \mathbf{M}}{\partial \eta} - æ \times \mathbf{M} - \mathbf{e}_1 \times \mathbf{Q} = \mathbf{T}\,.$$

For a rod of constant cross section, in terms of projections in the coordinate frame associated with the conduit, we have

$$J_{11}\frac{\partial \omega_1}{\partial \tau} + J_{11}\frac{\partial \omega_0}{\partial \tau} + J_{33}\omega_3\omega_2 - J_{22}\omega_2\omega_3 - \frac{\partial M_1}{\partial \eta} - æ_2 M_3 + æ_3 M_2 = T_1\,,$$

$$J_{22}\frac{\partial \omega_2}{\partial \tau} + J_{11}\omega_3(\omega_1 + \omega_0) - J_{33}\omega_1\omega_3 - \frac{\partial M_2}{\partial \eta} - æ_3 M_1 + æ_1 M_3 + Q_3 = T_2\,,$$

$$J_{33}\frac{\partial \omega_3}{\partial \tau} + J_{22}\omega_1\omega_2 - J_{11}\omega_2(\omega_1 + \omega_0) - \frac{\partial M_3}{\partial \eta} - æ_1 M_2 + æ_2 M_1 - Q_2 = T_3\,.$$

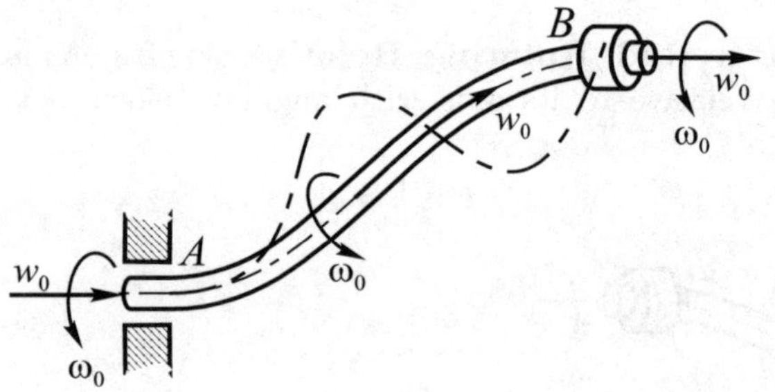

Fig. 2.3.

Equations of Motion for a Rod Rotating Relative to its Axis and Moving in the Longitudinal Direction Such a rod is illustrated in Fig. 2.3. The equations of translational motion of a rod element are identical to (2.43). The angular velocity of a rod element entering in (2.48) is as follows:

$$\boldsymbol{\omega}_a = \boldsymbol{\omega} + \boldsymbol{\omega}_0 + w\boldsymbol{æ}\,, \qquad \boldsymbol{\omega}_0(\tau) = \omega_0(\tau)\,\mathbf{e}_1\,.$$

Combining this formula with (2.44), we arrive at

$$\mathrm{J}\,\frac{\partial(\boldsymbol{\omega} + \boldsymbol{\omega}_0 + w\boldsymbol{æ})}{\partial \tau} + \boldsymbol{\omega} \times \mathrm{J}(\boldsymbol{\omega} + \boldsymbol{\omega}_0 + w\boldsymbol{æ}) + \frac{\partial}{\partial \eta}\left[\mathrm{J}(\boldsymbol{\omega} + \boldsymbol{\omega}_0 + w\boldsymbol{æ})\right] w$$
$$+\, \boldsymbol{æ} \times \mathrm{J}(\boldsymbol{\omega} + \boldsymbol{\omega}_0 + w\boldsymbol{æ})\, w - \frac{\partial \mathbf{M}}{\partial \eta} - \boldsymbol{æ} \times \mathbf{M} - \mathbf{e}_1 \times \mathbf{Q} = \mathbf{T}\,. \qquad (2.47)$$

2.1.5 Equations of Motion in the Cartesian Basis

The analysis of some applied problems of dynamics of rods can be considerably simplified by referring the equations of motion to an immovable coordinate system. In this case, there is no need in local derivatives because the unit vectors of the immovable basis $\{\mathbf{i}_j\}$ do not depend on τ and η. When the rotary inertia of a rod element can be neglected, it is especially profitable to refer the equations to the Cartesian coordinate system. If the rotary inertia is to be taken into consideration, the equations of motion in the Cartesian basis are complicated.

Vector Equations Vectors referred to the Cartesian coordinate system x_j will be supplied with the lower index x as follows: $\mathbf{Q}_x$, $\mathbf{M}_x$, $\mathbf{v}_x$, $\mathbf{u}_x$, $\mathbf{q}_x$, $\mathbf{P}_x^{(1)}$. In components, we have

$$\mathbf{Q}_x = \sum_{j=1}^{3} Q_{x_j} \mathbf{i}_j \,, \qquad \mathbf{M}_x = \sum_{j=1}^{3} M_{x_j} \mathbf{i}_j \,.$$

Unlike the components Q_j and M_j in the attached coordinate system, the physical meaning of the components Q_{x_j} and M_{x_j} of the vectors $\mathbf{Q}_x$ and $\mathbf{M}_x$ in the Cartesian coordinate system is not clearly defined in the general case. Nevertheless, when the equations of motion are solved, the components of the vectors in any coordinate system can be readily found by use of the transformation matrix. For example, to obtain the components of the vectors $\mathbf{Q}$ and $\mathbf{M}$ in the attached basis, we should use the matrix of transformation from the basis $\{\mathbf{i}_j\}$ to the basis $\{\mathbf{e}_j\}$; thus, we get

$$\mathbf{Q}_e = \mathrm{L}^{(1)} \mathbf{Q}_x \,, \qquad \mathbf{M}_e = \mathrm{L}^{(1)} \mathbf{M}_x \,,$$

where

$$\mathbf{Q}_e = \sum_{j=1}^{3} Q_j \mathbf{e}_j \,, \qquad \mathbf{M}_e = \sum_{j=1}^{3} M_j \mathbf{e}_j \,.$$

In view of (A.57) (see Svetlitsky (2000a)), the matrix $\mathrm{L}^{(1)}$ can be represented as a product $\mathrm{L}^{(1)} = \mathrm{L}\mathrm{L}^0$, where L is the matrix of transformation from the basis $\{\mathbf{e}_{j0}\}$ to the basis $\{\mathbf{e}_j\}$ and L^0 is the matrix of transformation from the basis $\{\mathbf{i}_j\}$ to the basis $\{\mathbf{e}_{j0}\}$. The matrix L^0 is assumed to be known. The elements of the matrix L are functions of the three unknown angles ϑ_j.

Equations (2.18) and (2.19) remain unchanged under coordinate transformations because during the derivation of the equations we did not associate the vectors $\mathbf{Q}$, $\mathbf{M}$, and $\boldsymbol{\vartheta}$ with any particular coordinate system. Neglecting the rotary inertia and using (2.18) and (2.19), we get

$$n_1 \frac{\partial \mathbf{v}_x}{\partial \tau} = \frac{\partial \mathbf{Q}_x}{\partial \eta} + \mathbf{P}_x \,, \tag{2.48}$$

$$\frac{\partial M_x}{\partial \eta} + \left(\mathrm{L}^{(1)^{\mathrm{T}}} \mathbf{i}_1 \right) \times \mathbf{Q}_x + \mathbf{T}_x = 0 \,, \tag{2.49}$$

where

$$\mathbf{P}_x = \mathbf{q}_x + \sum_{i=1}^{n} \mathbf{P}_x^{(i)} \, \delta\left(\eta - \eta_i\right) , \qquad \mathbf{T}_x = \boldsymbol{\mu}_x + \sum_{\nu=1}^{\rho} \mathbf{T}_x^{(\nu)} \, \delta\left(\eta - \eta_\nu\right) . \tag{2.50}$$

In the Cartesian basis, we have

$$\mathbf{v}_x = \frac{\partial \mathbf{u}_x}{\partial \tau};$$

hence, (2.48) can be represented as follows:

$$n_1 \frac{\partial^2 \mathbf{u}_x}{\partial \tau^2} = \frac{\partial \mathbf{Q}_x}{\partial \eta} + \mathbf{P}_x . \tag{2.51}$$

Consider (2.22), which is to be represented in the basis $\{\mathbf{i}_j\}$. We cannot simply put $\mathbf{M} = \mathbf{M}_x$ and $\text{æ} = \text{æ}_x$ in (2.22) because these equations are referred to the attached coordinate system. For this reason, we should refer the vector $\mathbf{M}$ to the Cartesian basis. We have $\mathbf{M}_e = \mathrm{L}^{(1)}\mathbf{M}_x$. In view of this relationship, equation (2.22) takes the form

$$\mathbf{M}_x = \mathrm{L}^{(1)^{\mathrm{T}}} \mathrm{A}(\text{æ}_e - \text{æ}_{0e}^{(1)}) . \tag{2.52}$$

Consider vibrations of a rod about its equilibrium configuration. In this case, the vector $\mathbf{M}_x$ reads

$$\mathbf{M}_x = \mathrm{L}^{(1)^{\mathrm{T}}} \mathrm{A}(\text{æ} - \text{æ}_0^{(1)}) + \mathbf{M}_{x0} . \tag{2.52a}$$

The relationship (2.52) allows us to represent the components M_{x_j} in terms of projections of the vectors æ_e and $\text{æ}_{0e}^{(1)}$ in the attached basis. Such a representation of equations, when some of the vectors are written in terms of their projections in the attached basis while the others are referred to the Cartesian basis, is best suited for application of numerical methods. Moreover, the Cartesian components of the vector æ are of no practical interest. If necessary, these components can be found as follows: first, we obtain the projections æ_{ej} in the attached frame and then, by use of the matrix of transformation, we get

$$\text{æ}_x = \mathrm{L}^{(1)^{\mathrm{T}}} \text{æ}_e .$$

Equation (2.23) is associated with the attached coordinate system. Using the expression for the vector $\mathbf{M}_x$, we get

$$\mathrm{L}_1 \frac{\partial \boldsymbol{\vartheta}}{\partial \eta} + \mathrm{L}_2 \text{æ}_0^{(1)} - \mathrm{A}^{-1}\mathrm{L}^{(1)}\mathbf{M}_x = 0 , \tag{2.53}$$

where $\mathrm{L}_2 = \mathrm{L} - \mathrm{E}$.

The components of the vector $\boldsymbol{\vartheta}$ are the angles of rotation of the basis $\{\mathbf{e}_j\}$ relative to the basis $\{\mathbf{e}_{j0}\}$. The elements of the matrices L_1, L_2, and $\mathrm{L}^{(1)}$ are functions of these angles.

Consider now dimensionless form of the kinematic equations (1.37) and (1.11). We have

$$\frac{\partial \mathbf{v}}{\partial \eta} - (\boldsymbol{\omega} \times \mathbf{e}_1) = 0\,, \tag{2.54}$$

$$\mathrm{L}_1 \frac{\partial \boldsymbol{\vartheta}}{\partial \tau} - \boldsymbol{\omega}_e = 0\,. \tag{2.55}$$

Setting $\mathbf{v} = \mathbf{v}_x$, $\boldsymbol{\omega} = \boldsymbol{\omega}_x$, and $\mathbf{e}_1 = \mathrm{L}^{(1)^{\mathrm{T}}} \mathbf{i}_1$ in (2.54), we arrive at

$$\frac{\partial \mathbf{v}_x}{\partial \eta} + \left(\mathrm{L}^{(1)^{\mathrm{T}}} \mathbf{i}_1 \times \boldsymbol{\omega}_x \right) = 0\,. \tag{2.56}$$

Since $\boldsymbol{\omega}_e = \mathrm{L}^{(1)} \boldsymbol{\omega}_x$, equation (2.55) takes the form

$$\mathrm{L}_1 \frac{\partial \boldsymbol{\vartheta}}{\partial \tau} - \mathrm{L}^{(1)} \boldsymbol{\omega}_x = 0\,. \tag{2.57}$$

The displacement vector of the axial points of the rod satisfies the following equation:

$$\frac{\partial \mathbf{u}}{\partial \eta} = \mathbf{e}_1 - \mathbf{e}_{10}\,.$$

Let

$$\mathbf{u} = \mathbf{u}_x = \sum u_{x_j} \mathbf{i}_j\,, \qquad \mathbf{e}_1 = \mathrm{L}^{(1)^{\mathrm{T}}} \mathbf{i}_1\,, \qquad \mathbf{e}_{10} = \mathrm{L}^{(0)^{\mathrm{T}}} \mathbf{i}_1\,.$$

Then, we have

$$\frac{\partial \mathbf{u}_x}{\partial \eta} - \left(\mathrm{L}^{(1)^{\mathrm{T}}} - \mathrm{L}^{(0)^{\mathrm{T}}} \right) \mathbf{i}_1 = 0\,, \tag{2.58}$$

where $\mathrm{L}^{(0)}$ is the matrix of transformation from the basis $\{\mathbf{i}_j\}$ to the basis $\{\mathbf{e}_{j0}\}$. The basis $\{\mathbf{e}_{j0}\}$ is a characteristic of the static configuration of the rod.

Scalar Form of Equations of Motion In components, equations (2.48) and (2.49) read

$$n_1 \frac{\partial v_{x_j}}{\partial \tau} = \frac{\partial Q_{x_j}}{\partial \eta} + P_{x_j}\,, \qquad j = 1,\, 2,\, 3\,,$$

$$\frac{\partial M_{x_j}}{\partial \eta} + \varepsilon_{jmi} l_{1m}^{(1)} Q_{x_i} + T_{x_j} = 0\,,$$

where

$$P_{x_j} = q_{x_j} + \sum_{i=1}^{n} P_{x_j}^{(i)}\, \delta\left(\eta - \eta_i\right), \qquad T_{x_j} = \mu_{x_j} + \sum_{\nu=1}^{\rho} T_{x_j}^{(\nu)}\, \delta\left(\eta - \eta_\nu\right).$$

In components, equations (2.52), (2.53), (2.56), (2.57), and (2.58) take the form

$$M_{x_j} = l_{ji}^{(2)}(æ_i - æ_{i0}) , \qquad j = 1, 2, 3 ,$$

$$l_{(1)ji}\frac{\partial \vartheta_i}{\partial \eta} - l_{ji}^{(3)}M_{x_i} - l_{(2)ji}æ_{0i} = 0 , \qquad \frac{\partial v_{x_j}}{\partial \eta} + \varepsilon_{jmi}l_{1m}^{(1)}\omega_{x_i} = 0 ,$$

$$l_{(1)ji}\frac{\partial \vartheta_i}{\partial \tau} - l_{ji}^{(1)}\omega_{x_i} = 0 , \qquad \frac{\partial u_{x_j}}{\partial \eta} - (l_{1j}^{(1)} - l_{1j}^{(0)}) = 0 ,$$

where $l_{(1)ji}$, $l_{ji}^{(2)}$, $l_{ji}^{(3)}$, $l_{(2)ji}$, and $l_{ji}^{(1)}$ are the elements of the matrices L_1, $\mathrm{L}^{(1)^{\mathrm{T}}}\mathrm{A}$, $\mathrm{A}^{-1}\mathrm{L}^{(1)}$, L^2, and $\mathrm{L}^{(1)}$, respectively.

In this section, the fundamental equations of motion for the most general case of space-curved rods have been developed. These equations were referred to both the Cartesian (immovable) and attached coordinate systems. Both vector and scalar form of the equations have been presented.

2.2 Equations of Motion of a Rod Relative to Equilibrium Configuration

2.2.1 Vector Equations

In the previous section, the motion of a rod relative to its natural (unloaded) configuration was considered. To study the motion relative to an equilibrium configuration (that is not necessarily natural), we should take into consideration the internal static stress field of the rod defined by the vectors $\mathbf{Q}_0$ and $\mathbf{M}_0$. Considering the static stress state of the rod, we can represent the vectors $\mathbf{Q}$ and $\mathbf{M}$ (see the equations of motion in the previous section) as follows:

$$\mathbf{Q} = \mathbf{Q}_0 + \mathbf{Q}_{\mathrm{d}} , \qquad \mathbf{M} = \mathbf{M}_0 + \mathbf{M}_{\mathrm{d}} ; \tag{2.59}$$

here, $\mathbf{Q}_0$, $\mathbf{M}_0$, and $\mathbf{Q}_{\mathrm{d}}$, $\mathbf{M}_{\mathrm{d}}$ are the static and dynamic components of the internal loads, respectively. Similarly, we can represent external loads as follows:

$$\mathbf{P} = \mathbf{P}_0 + \mathbf{P}_{\mathrm{d}} , \qquad \mathbf{T} = \mathbf{T}_0 + \mathbf{T}_{\mathrm{d}} . \tag{2.60}$$

Substituting (2.59) and (2.60) into (2.19), we get

$$n_1\frac{\partial \mathbf{v}}{\partial \tau} = \frac{\partial \mathbf{Q}_0}{\partial \eta} + \frac{\partial \mathbf{Q}_{\mathrm{d}}}{\partial \eta} + \mathbf{P}_0 + \mathbf{P}_{\mathrm{d}} , \tag{2.61}$$

$$\frac{\partial}{\partial \tau}(\mathrm{J}\boldsymbol{\omega}) = \frac{\partial \mathbf{M}_0}{\partial \eta} + \frac{\partial \mathbf{M}_{\mathrm{d}}}{\partial \eta} + \mathbf{e}_1 \times \mathbf{Q}_0 + \mathbf{e}_1 \times \mathbf{Q}_{\mathrm{d}} + \mathbf{T}_0 + \mathbf{T}_{\mathrm{d}} . \tag{2.62}$$

Eliminating the equilibrium equations from (2.61) and (2.62), we obtain

$$n_1 \frac{\partial \mathbf{v}}{\partial \tau} = \frac{\partial \mathbf{Q}_{\mathrm{d}}}{\partial \eta} + \mathbf{P}_{\mathrm{d}} \,, \tag{2.63}$$

$$\frac{\partial}{\partial \tau} (\mathrm{J}\boldsymbol{\omega}) = \frac{\partial \mathbf{M}_{\mathrm{d}}}{\partial \eta} + \mathbf{e}_1 \times \mathbf{Q}_{\mathrm{d}} + \mathbf{T}_{\mathrm{d}} \,. \tag{2.64}$$

From these equations it is seen that the equations of motion written in invariant form do not include the initial stresses.

2.2.2 Equations of Motion in the Attached Coordinate System

Substituting (2.59) and (2.60) into (2.20) and (2.21) and eliminating the equilibrium equations, we get

$$n_1 \left(\frac{\partial \mathbf{v}}{\partial \tau} + \boldsymbol{\omega} \times \mathbf{v} \right) - \frac{\partial \mathbf{Q}_{\mathrm{d}}}{\partial \eta} - æ \times \mathbf{Q}_{\mathrm{d}} - (æ - æ_0^{(1)}) \times \mathbf{Q}_0^{(1)} = \mathbf{P}_{\mathrm{d}} \,,$$

$$\mathrm{J} \frac{\partial \boldsymbol{\omega}}{\partial \tau} + \boldsymbol{\omega} \times \mathrm{J}\boldsymbol{\omega} - \frac{\partial \mathbf{M}_{\mathrm{d}}}{\partial \eta} - æ \times \mathbf{M}_{\mathrm{d}} - \mathbf{e}_1 \times \mathbf{Q}_{\mathrm{d}} - (æ - æ_0^{(1)}) \times \mathbf{M}_0^{(1)} = \mathbf{T}_{\mathrm{d}} \,, \tag{2.65}$$

where

$$\mathbf{Q}_0^{(1)} = \sum_{i=1}^{3} \mathbf{Q}_{i0} \mathbf{e}_i \,, \qquad \mathbf{M}_0^{(1)} = \sum_{i=1}^{3} \mathbf{M}_{i0} \mathbf{e}_i \,.$$

In terms of projections in the attached basis, we have

$$n_1 \frac{\partial v_1}{\partial \tau} + v_3\omega_2 - v_2\omega_3 - \frac{\partial Q_{\mathrm{d}1}}{\partial \eta} - Q_{\mathrm{d}3}æ_2 + Q_{\mathrm{d}2}æ_3 - (æ_2 - æ_{20})\, Q_{30}$$
$$+ (æ_3 - æ_{30})\, Q_{20} = P_{\mathrm{d}1} \,,$$
$$n_1 \frac{\partial v_2}{\partial \tau} + v_1\omega_3 - v_3\omega_1 - \frac{\partial Q_{\mathrm{d}2}}{\partial \eta} - Q_{\mathrm{d}1}æ_3 + Q_{\mathrm{d}3}æ_1 - (æ_3 - æ_{30})\, Q_{10}$$
$$+ (æ_1 - æ_{10})\, Q_{30} = P_{\mathrm{d}2} \,,$$
$$n_1 \frac{\partial v_3}{\partial \tau} + v_2\omega_1 - v_1\omega_2 - \frac{\partial Q_{\mathrm{d}3}}{\partial \eta} - Q_{\mathrm{d}2}æ_1 + Q_{\mathrm{d}1}æ_2 - (æ_1 - æ_{10})\, Q_{20}$$
$$+ (æ_2 - æ_{20})\, Q_{10} = P_{\mathrm{d}3} \,,$$
$$J_{11} \frac{\partial \omega_1}{\partial \tau} + (J_{33} - J_{22})\, \omega_2\omega_3 - \frac{\partial M_{\mathrm{d}1}}{\partial \eta} - æ_2 M_{\mathrm{d}3} + æ_3 M_{\mathrm{d}2} - (æ_2 - æ_{20})\, M_{30}$$
$$+ (æ_3 - æ_{30})\, M_{20} = T_{\mathrm{d}1} \,,$$
$$J_{22} \frac{\partial \omega_2}{\partial \tau} + (J_{11} - J_{33})\, \omega_1\omega_3 - \frac{\partial M_{\mathrm{d}2}}{\partial \eta} - æ_3 M_{\mathrm{d}1} + æ_1 M_{\mathrm{d}3} - (æ_3 - æ_{30})\, M_{10}$$
$$+ (æ_1 - æ_{10})\, M_{30} + Q_{\mathrm{d}3} = T_{\mathrm{d}2} \,,$$
$$J_{33} \frac{\partial \omega_3}{\partial \tau} + (J_{22} - J_{11})\, \omega_1\omega_2 - \frac{\partial M_{\mathrm{d}3}}{\partial \eta} - æ_1 M_{\mathrm{d}2} + æ_2 M_{\mathrm{d}1} - (æ_1 - æ_{10})\, M_{20}$$
$$+ (æ_2 - æ_{20})\, M_{10} - Q_{\mathrm{d}2} = T_{\mathrm{d}3} \,.$$
$$M_{kk} = A_{kk}(æ_k - æ_{k0}) + M_{k0} \,, \qquad k = 1,\, 2,\, 3 \,.$$

The other equations, (2.32)–(2.35), remain the same.

2.3 Equations of Motion of a Rod with Concentrated Masses

2.3.1 Determination of Inertial Forces and Moments

A space-curved rod with concentrated masses m_1 and m_2 is shown in Fig. 2.4. The rotary inertia of the mass m_1 is zero and that of the mass m_2 is nonzero.

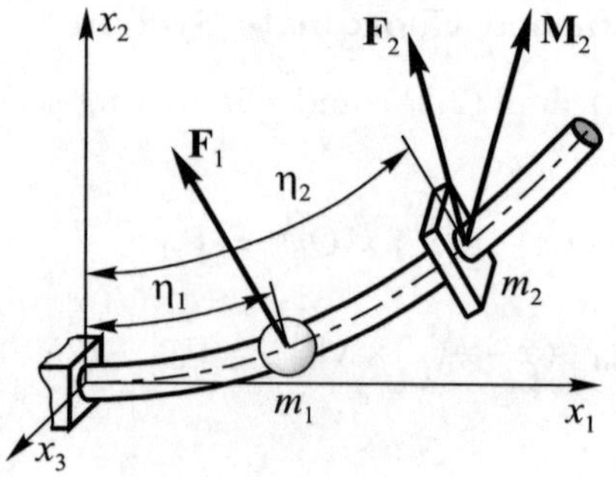

Fig. 2.4.

When the rod is in motion, the concentrated masses are subjected to the inertial forces $\mathbf{F}_i$, $i = 1, 2$, and the moment of inertia $\mathbf{M}_2$. These forces and moment can be added to the equations of motion by use of the δ-function. In dimensionless form, the inertial forces $\mathbf{F}_i$ and moment $\mathbf{M}_2$ are as follows:

$$\mathbf{F}_1 = -\tilde{m}_1 \frac{\partial \mathbf{v}}{\partial \eta} \delta(\eta - \eta_1),$$

$$\mathbf{F}_2 = -\tilde{m}_2 \frac{\partial \mathbf{v}}{\partial \tau} \delta(\eta - \eta_2),$$

$$\mathbf{M}_2 = -\frac{\partial}{\partial \tau} (\mathrm{J}^{(2)} \boldsymbol{\omega})\, \delta(\eta - \eta_2). \tag{2.66}$$

In the attached coordinate system, we have (as before, a tilde in the notation of local derivatives is dropped out)

$$\mathbf{F}_1 = -\tilde{m}_1 \left(\frac{\partial \mathbf{v}}{\partial \eta} + \boldsymbol{\omega} \times \mathbf{v} \right) \delta(\eta - \eta_1), \tag{2.67}$$

$$\mathbf{F}_2 = -\tilde{m}_2 \left(\frac{\partial \mathbf{v}}{\partial \tau} + \boldsymbol{\omega} \times \mathbf{v} \right) \delta(\eta - \eta_2), \tag{2.68}$$

$$\mathbf{M}_2 = -\left(\mathrm{J}^{(2)} \frac{\partial \boldsymbol{\omega}}{\partial \tau} + \boldsymbol{\omega} \times \mathrm{J}_3^{(2)} \boldsymbol{\omega} \right) \delta(\eta - \eta_2). \tag{2.69}$$

Here, the elements of the matrix $\mathrm{J}^{(2)}$ are the dimensionless moments of inertia; $\tilde{m}_i$, $i = 1, 2$, are the dimensionless masses given by

$$\tilde{m}_i = \frac{m_i}{m_0 l},$$

where m_0 is the mass per unit length and l is the length of the rod. If the principal axes of the cross section of the rod coincide with the principal axes of the mass m_2, then the matrix $\mathrm{J}^{(2)}$ is diagonal and identical in form to (2.8). If these axes do not coincide, then, since the equations of motion are referred to the principal axes of a cross section, the matrix $\mathrm{J}^{(2)}$ is as follows:

$$\mathrm{J}^{(2)} = \begin{bmatrix} J_{11}^{(2)} & J_{12}^{(2)} & J_{13}^{(2)} \\ J_{21}^{(2)} & J_{22}^{(2)} & J_{23}^{(2)} \\ J_{31}^{(2)} & J_{32}^{(2)} & J_{33}^{(2)} \end{bmatrix} . \tag{2.70}$$

The dimensionless elements of this matrix are connected with their dimensional counterparts by

$$\tilde{J}_{ij}^{(2)} = \frac{J_{ij}^{(2)}}{m_0 l^3} ,$$

where $\tilde{J}_{ij}^{(2)}$ and $J_{ij}^{(2)}$ are dimensionless and dimensional quantities, respectively.

The expressions for $\mathbf{F}_2$ and $\mathbf{M}_2$ given above are valid provided the size of the mass m_2 measured along the rod axis is small as compared with the total length of the rod. (In the subsequent sections, we will consider examples when the sizes of the concentrated masses cannot be neglected.) Assuming $\eta_2^{(1)} \approx \eta_2 \approx \eta_2^{(2)}$,, where η_2 is the coordinate of the center of gravity of the rod, we can put

$$J_{12}^{(2)} = J_{13}^{(2)} = J_{21}^{(2)} = J_{31}^{(2)} = 0 .$$

2.3.2 Vector Equations of Motion of a Rod with Concentrated Masses

The equations of interest are readily obtained from (2.20) and (2.21) (a tilde in the notation of dimensionless quantities is dropped out):

$$\begin{gathered}[n_1(\eta) + m_1\,\delta(\eta - \eta_1) + m_2\,\delta(\eta - \eta_2)] \left(\frac{\partial \mathbf{v}}{\partial \tau} + \boldsymbol{\omega} \times \mathbf{v} \right) \\ - \frac{\partial \mathbf{Q}}{\partial \eta} - æ \times \mathbf{Q} - \mathbf{P} = 0 , \\ [\mathrm{J}(\eta) + \mathrm{J}^{(2)}\,\delta(\eta - \eta_2)] \frac{\partial \boldsymbol{\omega}}{\partial \tau} + \boldsymbol{\omega} \times [\mathrm{J}(\eta) + \mathrm{J}^{(2)}\,\delta(\eta - \eta_2)]\,\boldsymbol{\omega} \\ - \frac{\partial \mathbf{M}}{\partial \eta} - æ \times \mathbf{M} - \mathbf{e}_1 \times \mathbf{Q} - \mathbf{T} = 0 . \end{gathered} \tag{2.71}$$

The other equations of the system (2.20)–(2.25) remain the same.

2.4 Equations of Stationary Motion of a Rod

2.4.1 Vector Equations

Consider a particular case of the equations obtained in Sect. 2.1. Let us assume that the shape of the rod axis does not vary with time; hence, $\mathbf{v} = \boldsymbol{\omega} = 0$ and $\mathbf{w} \neq 0$. Such a motion is called *stationary*. The most important peculiarity of stationary motion is that, in spite of the presence of longitudinal velocity $\mathbf{w}_0$, the rod as a whole(the part of a rod between the cross sections A and B in Fig. 2.5) seems at rest to an observer associated with an immovable reference frame.

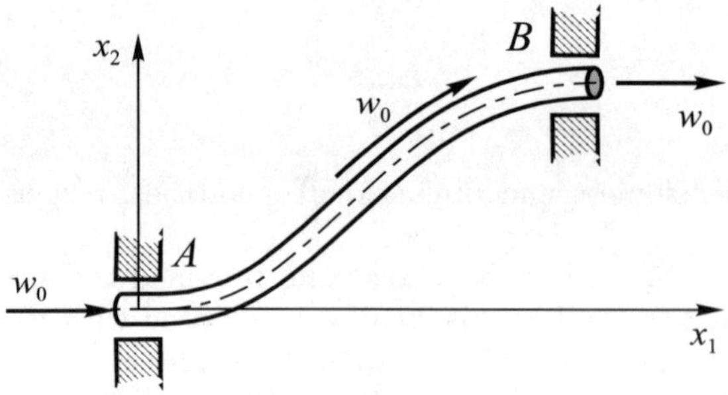

Fig. 2.5.

That is why a stationary motion is sometimes called an *apparent rest.* A stationary motion relative to a movable coordinate system (for example, rotating as shown in Fig. 2.6) may also take place.

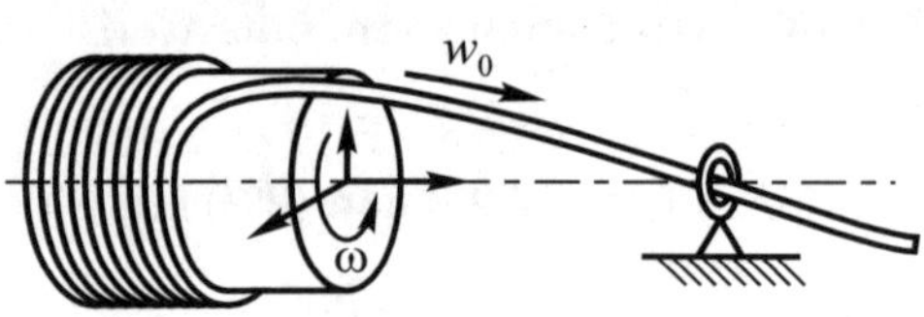

Fig. 2.6.

Mechanisms and devices with flexible rods find industrial application. The classical examples are a belt transmission (Fig. 2.7) and a tape-drive transmission (Fig. 2.8).

Flexible rods in stationary motion are used in most manufacturing processes: winding and reeling in electrical engineering, rolling (Fig. 2.9), textile industry (Fig. 2.6) and some other industries.

Belt-type radiators remove heat from various power units; say, reactors (Fig. 2.10). When the radiator belt is in contact with the reactor, the belt heats up. Then, at free motion, the belt radiates heat to the surrounding medium (in the case of vacuum, the belt radiates heat into the space).

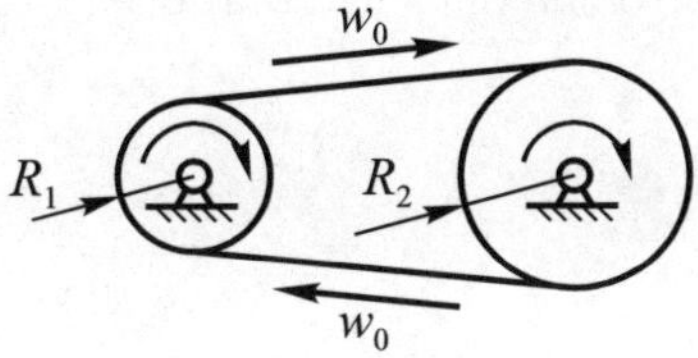

Fig. 2.7.

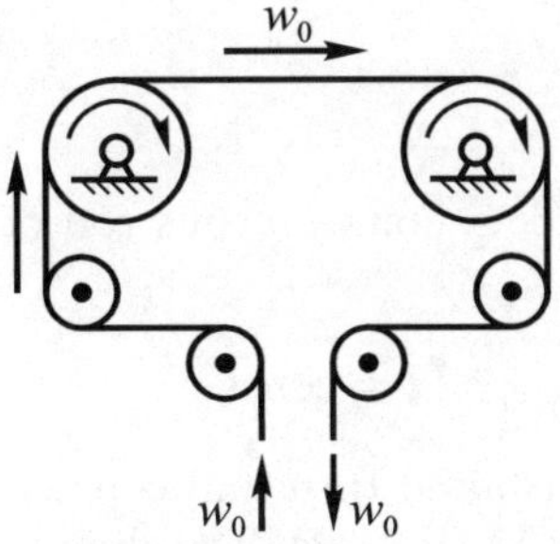

Fig. 2.8.

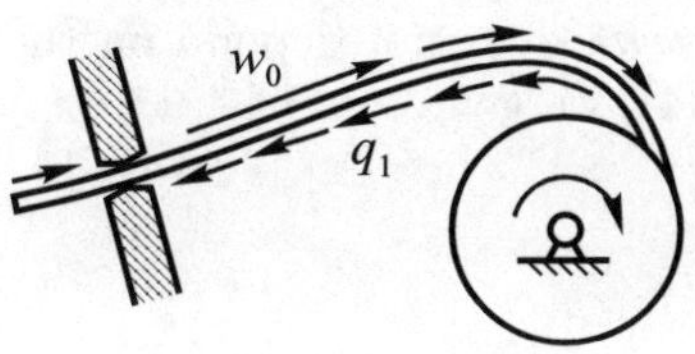

Fig. 2.9.

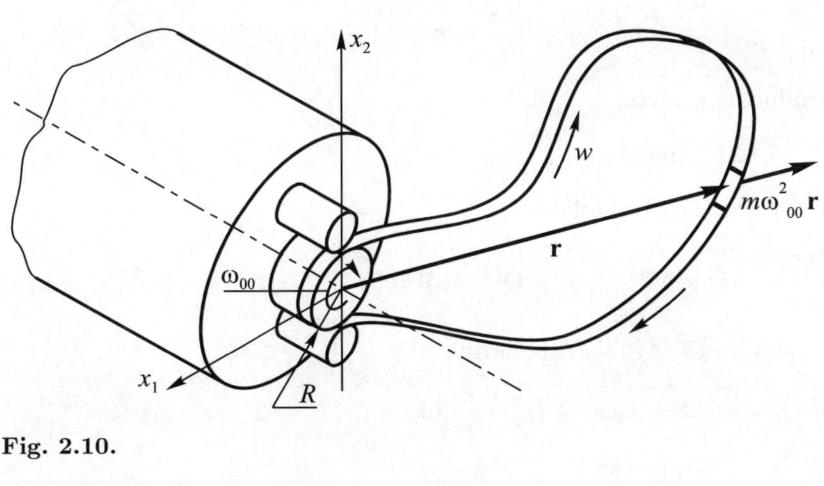

Fig. 2.10.

Problems concerned with statics and dynamics of tubes filled with moving liquid (pipelines) (Fig. 2.11) are closely related to the ones listed above.

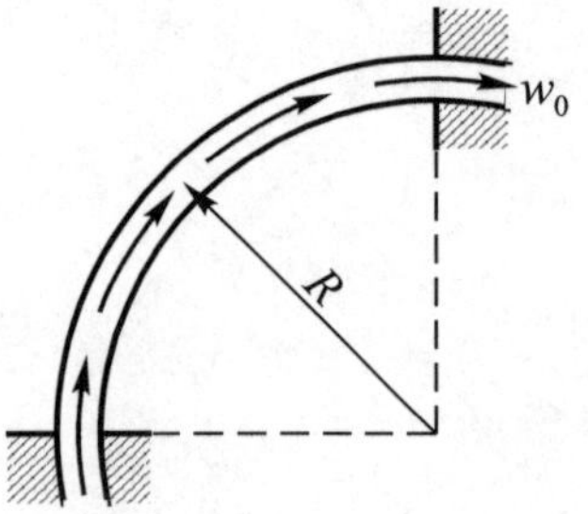

Fig. 2.11.

A stationary motion of a rod is possible only for a homogeneous rod of constant cross section, that is,

$$m_0 = \text{const}\,, \qquad F = \text{const}\,, \qquad A_{ii} = \text{const}\,, \qquad J_{ii} = \text{const}\;.$$

In addition, the loads applied to the rod, both distributed $\mathbf{q}$, $\boldsymbol{\mu}$ and concentrated $\mathbf{P}^{(i)}$, $\mathbf{T}^{(\nu)}$, must be time-independent. For $A_{ii} = \text{const}$, the dimensionless stiffness A_{33} equals unity. In the case of stationary motion, the time derivatives in (2.43) and (2.44) as well as the vectors $\mathbf{v}$ and $\boldsymbol{\omega}$ vanish. For an inextensible rod, the magnitude of the velocity $\mathbf{w}_0$ does not depend on the coordinate η. As a result, using (2.43) and (2.44), we get

$$\frac{\mathrm{d}\mathbf{Q}^{(1)}}{\mathrm{d}\eta} + \text{æ} \times \mathbf{Q}^{(1)} + \mathbf{P} = 0\,,$$

$$\mathbf{Q}^{(1)} = (Q_{10} - w_0^2)\,\mathbf{e}_1 + Q_2\mathbf{e}_2 + Q_3\mathbf{e}_3\,, \tag{2.72}$$

$$\mathrm{J}\,\frac{\mathrm{d}\text{æ}}{\mathrm{d}\eta}\,w_0^2 + (\text{æ} + \mathrm{J}\text{æ})\,w_0^2 - \frac{\mathrm{d}\mathbf{M}}{\mathrm{d}\eta} - \text{æ} \times \mathbf{M} - \mathbf{e}_1 \times \mathbf{Q}^{(1)} - \mathbf{T} = 0\,. \tag{2.73}$$

Introducing a vector

$$\mathbf{M}^{(1)} = \mathbf{M} - \mathrm{J}w_0^2\text{æ}\,, \tag{2.74}$$

we can rewrite (2.73) as follows:

$$\frac{\mathrm{d}\mathbf{M}^{(1)}}{\mathrm{d}\eta} + \text{æ} \times \mathbf{M}^{(1)} + \mathbf{e}_1 \times \mathbf{Q}^{(1)} + \mathbf{T} = 0\,. \tag{2.75}$$

Then, equation (2.74) takes the form

$$\mathbf{M}^{(1)} = \mathrm{A}^{(1)}\text{æ} - \mathrm{A}\text{æ}_0^{(1)}\,,$$

where

$$\mathrm{A}^{(1)} = \begin{bmatrix} A_{11} - J_{11} w_0^2 & 0 & 0 \\ 0 & A_{22} - J_{22} w_0^2 & 0 \\ 0 & 0 & 1 - J_{33} w_0^2 \end{bmatrix} . \tag{2.76}$$

Equations (2.72) and (2.75) are identical to the equilibrium equations for a rod. If æ, $\mathbf{Q}^{(1)}$, and $\mathbf{M}^{(1)}$ are known, then the force $\mathbf{Q}$ and the moment $\mathbf{M}$ at a cross section of the tube in which the rod moves are as follows:

$$\mathbf{Q} = \mathbf{Q}^{(1)} + w_0^2 \mathbf{e}_1 , \qquad \mathbf{M} = \mathbf{M}^{(1)} + \mathrm{J} w_0^2 æ . \tag{2.77}$$

2.4.2 Critical Velocities of Stationary Motion of a Rod

Consider the matrix $\mathrm{A}^{(1)}$ (2.76); recall that the elements of this matrix are the dynamic dimensionless stiffnesses of the rod. From (2.76) it is seen that if the rotary inertia of the rod is taken into consideration, then, for some values of the velocity w_{0*}, the corresponding dynamic stiffnesses vanish. Such values are termed *critical values.* As an example, let us determine these critical velocities for a rod of circular cross section. We get

$$A_{11}^{(1)} = A_{11} - J_{11} w_0^2 = 0 ; \tag{2.78}$$

using dimensional quantities, we can write

$$A_{11} = \frac{(A_{11})_\mathrm{d}}{A_{33}} = \frac{J_1 G}{J_3 E} , \qquad J_{11} = \frac{J_1}{F_0 l^2} , \qquad w_0 = \frac{w_\mathrm{d}}{l p_0} ,$$

$$p_0 = \sqrt{\frac{J_3 E}{m_0 l^4}} , \qquad m_0 = F_0 \rho .$$

For a rod of circular cross section, we have $J_1 = 2\,J_2 = 2\,J_3$. Then, from (2.78) we can obtain the first (dimensional) critical velocity

$$w_{\mathrm{d}*}^{(1)} = \sqrt{\frac{G}{\rho}} . \tag{2.79}$$

The velocity $w_{\mathrm{d}*}^{(1)}$ is the velocity of propagation of shear waves (torsional waves) along the cylindrical rod. Since for rods of circular cross section $A_{22}^{(1)} = A_{33}^{(1)}$, it is sufficient to consider only $A_{33}^{(1)}$. We have

$$A_{33}^{(1)} = 1 - \frac{J_3 w_0^2}{F_0 l^2} = 0 . \tag{2.80}$$

Representing w_0 as a function of the dimensional velocity w_d, we obtain the second (dimensional) critical velocity of the rod

$$w_{\mathrm{d}**}^{(2)} = \sqrt{\frac{E}{\rho}} . \tag{2.81}$$

The velocity $w_{d**}^{(2)}$ is the velocity of propagation of longitudinal waves in the rod. For a rod of rectangular cross section, the critical velocities are as follows:

$$w_{d*}^{(1)} = \sqrt{\beta\left(\frac{h}{b}\right)\frac{12b^2}{(b^2+h^2)}}\sqrt{\frac{G}{\rho}}, \qquad w_{d**}^{(2)} = \sqrt{\frac{E}{\rho}}. \tag{2.82}$$

2.4.3 Particular Cases of Equations of Stationary Motion

Consider the problem of determination of internal loads that arise in a rod moving in an immovable conduit (Fig. 2.12*a*). The geometry of the conduit axis (defined by the components of the vector æ) is known. We assume that originally, before being placed into the conduit, the rod was straight.

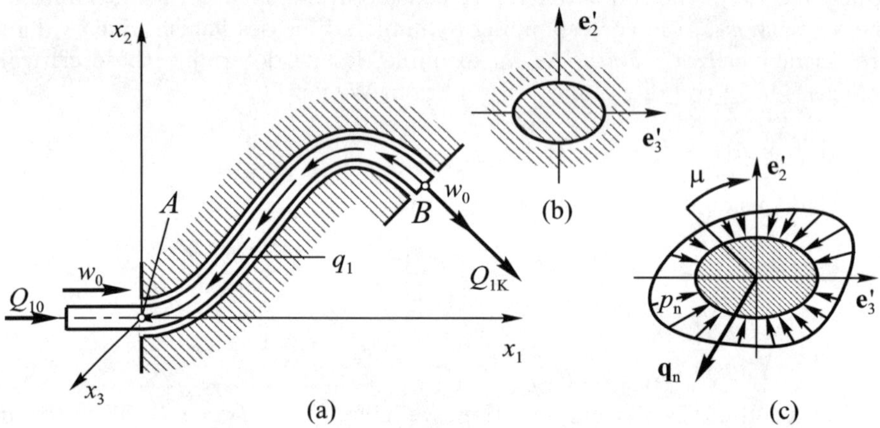

Fig. 2.12.

Let us assume that the distributed force of friction between a rod and the internal surface of a conduit is directed tangentially to the rod axis, that is,

$$\mathbf{q}_1 = -q_{10}\mathbf{e}_1\,, \qquad q_{10} = \text{const}\,.$$

In addition, the walls of the conduit act on the rod with a distributed force normal to the rod axis. Thus, the total distributed force of interaction reads

$$\mathbf{q} = \mathbf{q}_n - q_{10}\mathbf{e}_1\,.$$

For rods of circular cross section, the distributed moment $\boldsymbol{\mu}$ can be considered as negligibly small. For rods of noncircular cross section, say, of elliptic one (Fig. 2.12*b*), the moment $\boldsymbol{\mu}$ depends on $\mathbf{q}_n$ and $\mathbf{q}_n$ depends on the pressure p_n. We put $\boldsymbol{\mu} = \mu_{10}\mathbf{e}_1$ (see Fig. 2.12*c*).

In terms of projections in the attached basis, using (2.72) and (2.75), we get

$$\frac{\mathrm{d}Q_1^{(1)}}{\mathrm{d}\eta} - æ_{30}Q_2 - q_{10} = 0\,,$$
$$\frac{\mathrm{d}Q_2}{\mathrm{d}\eta} + æ_{30}Q_1^{(1)} - æ_{10}Q_3 + q_2 = 0\,,$$
$$\frac{\mathrm{d}Q_3}{\mathrm{d}\eta} + æ_{10}Q_2 + q_3 = 0\,, \tag{2.83}$$

$$\frac{\mathrm{d}M_1^{(1)}}{\mathrm{d}\eta} - æ_{30}M_2^{(1)} - \mu_{10} = 0\,,$$
$$\frac{\mathrm{d}M_2^{(1)}}{\mathrm{d}\eta} + æ_{30}M_1^{(1)} - æ_{10}M_3^{(1)} - Q_3 = 0\,,$$
$$\frac{\mathrm{d}M_3}{\mathrm{d}\eta} + æ_{10}M_2^{(1)} + Q_2 = 0\,. \tag{2.84}$$

Since the rod was originally straight, we have $\boldsymbol{æ}_0^{(1)} \equiv 0$ and therefore $\mathbf{M}^{(1)} = \mathrm{A}^{(1)}\boldsymbol{æ}$.

In the general case, the natural axes of a cross section do not coincide with the principal axes $\{\mathbf{e}_i\}$; hence, we have

$$\boldsymbol{æ} = \left(æ_{10} + \frac{\mathrm{d}\vartheta_{10}}{\mathrm{d}\eta}\right)\mathbf{e}_1^{(1)} + æ_{30}\sin\boldsymbol{\vartheta}_{10}\,\mathbf{e}_2^{(1)} + æ_{30}\cos\boldsymbol{\vartheta}_{10}\,\mathbf{e}_3^{(1)}\,,$$

where ϑ_{10} is the angle between the principal and natural axes of a cross section of the rod.

Suppose that at each cross section of the conduit the principal axes coincide with the natural axes; then, we have

$$\vartheta_{10} = 0\,, \qquad \boldsymbol{æ} = \boldsymbol{æ}_0\,, \qquad \mathbf{M}^{(1)} = \mathrm{A}^{(1)}\boldsymbol{æ}_0\,.$$

Let us examine this special case in some detail. For $\vartheta_{10} = 0$ and $M_0^{(1)} = 0$, we can solve for μ_{10}, Q_2, and Q_3 from (2.84) as follows:

$$\mu_{10} = A_{11}^{(1)}æ'_{10}\,,$$
$$Q_2 = -æ'_{30}A_{33}^{(1)}\,,$$
$$Q_3 = æ_{10}æ_{30}\left(A_{11}^{(1)} - A_{33}^{(1)}\right)\,. \tag{2.85}$$

Recall that a prime is used to denote differentiation with respect to η

The first equation of the system (2.83) can be solved for $Q_1^{(1)}$. Since w_0 does not depend on η, we have

$$\frac{\mathrm{d}Q_1^{(1)}}{\mathrm{d}\eta} = \frac{\mathrm{d}Q_1}{\mathrm{d}\eta}$$

and therefore

$$Q_1(\eta) = q_{10}\eta + \frac{æ_{30}^2 A_{33}^{(1)}}{2} + c_1 \,.$$

If $\eta = 0$, then

$$Q_1 = -Q_{10} \,, \qquad c_1 = Q_{10} - \frac{æ_{30}^2(0)}{2} A_{33}^{(1)} \,.$$

The result is

$$Q_1(\eta) = -Q_{10} + q_{10}\eta + \frac{A_{33}^{(1)}}{2}\left[æ_{30}^2(\eta) - æ_{30}^2(0)\right] . \tag{2.86}$$

For a rod to be in a stationary motion, a force equal in magnitude and opposite in direction to the force

$$Q_1(1) = -Q_{10} + q_{10} + \frac{A_{33}^{(1)}}{2}\left[æ_{30}^2(1) - æ_{31}^2(0)\right]$$

must be applied at $\eta = 1$. If the shape of the conduit axis is so chosen that $æ_{30}(1) = æ_{30}(0)$, then, in view of (2.86), we obtain the dimensional relationship

$$Q_{1\mathrm{d}}(1) + (Q_{10})_\mathrm{d} = (q_{10})_\mathrm{d} l \,,$$

where l is the conduit length.

For some shapes of the conduit axis, a stationary motion is feasible when $Q_1(1) = 0$. From (2.85) and (2.86) it is seen that if, at $\eta = 0$ and $\eta = 1$, the quantities $æ_{10}$, $æ_{30}$, and $æ_{30}'$ are nonzero, then some external concentrated forces are applied to these cross sections.

Consider the process of insertion of a rod of elliptic cross section into a conduit. Let us assume that the principal axes of a cross section of the conduit coincide with the natural axes of this cross section. Then, we have $\vartheta_{10} = 0$. This means that during the insertion the rod is being twisted under the action of the moments μ_{10} such that the principal axes of a cross section of the rod constantly coincide with the natural axes.

Now consider a rod of circular cross section in the course of insertion. If originally the principal axes of a cross section of the rod coincide with the natural axes of the inlet cross section of the conduit, then, as the rod is being advanced into the conduit, these principal axes of the rod may rotate relative to the natural axes of the cross section of the conduit (that is, $\vartheta_{10} \neq 0$). This follows from the fact that, for rods of circular cross section, the moment μ_{10} is zero.

In view of (2.83), the components of the contact load $\mathbf{q}_n$ are

$$q_2 = æ_{30}'' A_{33}^{(1)} - \left[-Q_{10} + q_{10}\eta - w_0^2 + \frac{1}{2} A_{33}^{(1)}\left(æ_{30}^2(\eta) - æ_{30}^2(0)\right)\right] æ_{30}$$

$$+ æ_{10}^2 æ_{30} (A_{11}^{(1)} - A_{33}^{(1)}) \,,$$

$$q_3 = æ_{10} æ_{30}' A_{33}^{(1)} - (æ_{10} æ_{30})' \left(A_{11}^{(1)} - A_{33}^{(1)}\right) .$$

Plane Stationary Motion of a Rod At the beginning of this section we considered a belt-like radiator (Fig. 2.10). Let the angular velocity of the cylinders that press the belt against the barrel be denoted by $\boldsymbol{\omega}_{00}$. Derive the equations of stationary motion of the belt referred to a coordinate system x_1Ox_2 that rotates with the velocity $\boldsymbol{\omega}_{00}$ (Fig. 2.10). The longitudinal velocity of the belt relative to attached system of coordinate is $|\mathbf{w}| = \omega_{00}R$. The belt is subjected to a distributed load $m\omega_{00}^2\mathbf{r}$. From the equilibrium equation, neglecting the weight of the rod, we get

$$\frac{\mathrm{d}\mathbf{Q}^{(1)}}{\mathrm{d}s} + \mathbf{q} = 0\,.$$

Recall that (2.72) contains a local derivative in the notation of which a tilde is dropped out; hence, we have

$$\frac{\mathrm{d}\mathbf{Q}^{(1)}}{\mathrm{d}\eta} = \frac{\tilde{\mathrm{d}}\mathbf{Q}^{(1)}}{\mathrm{d}\eta} + \text{æ} \times \mathbf{Q}^{(1)}\,.$$

In the frame x_1Ox_2, the components of the vectors $\mathbf{Q}$ and $\mathbf{q}$ are

$$\mathbf{Q} = \left(Q_1^{(1)}\frac{\mathrm{d}x_1}{\mathrm{d}s} - Q_2\frac{\mathrm{d}x_2}{\mathrm{d}s}\right)\mathbf{i}_1 + \left(Q_1^{(1)}\frac{\mathrm{d}x_2}{\mathrm{d}s} + Q_2\frac{\mathrm{d}x_2}{\mathrm{d}s}\right)\mathbf{i}_2\,,$$

$$\mathbf{q} = m_0\omega_{00}^2 x_1\mathbf{i}_1 + m_0\omega_{00}^2 x_2\mathbf{i}_2\,,$$

where

$$Q_1^{(1)} = Q_1 - m_0 w^2\,.$$

In dimensionless form, we have

$$\tilde{\mathbf{Q}} = \left(\tilde{Q}_1^{(1)}\frac{\mathrm{d}\tilde{x}_1}{\mathrm{d}\eta} - \tilde{Q}_2\frac{\mathrm{d}\tilde{x}_2}{\mathrm{d}\eta}\right)\mathbf{i}_1 + \left(\tilde{Q}_1^{(1)}\frac{\mathrm{d}\tilde{x}_2}{\mathrm{d}\eta} + \tilde{Q}_2^{(1)}\frac{\tilde{\mathrm{d}}x_1}{\mathrm{d}\eta}\right)\mathbf{i}_2\,,$$

$$\tilde{\mathbf{q}} = (\tilde{\omega}_{00})^2\tilde{x}_1\mathbf{i}_1 + (\tilde{\omega}_{00})^2\tilde{x}_2\mathbf{i}_2\,,$$

where

$$n_{10} = \frac{R}{l}\,, \qquad \tilde{x}_1 = \frac{x_1}{l}\,, \qquad \tilde{x}_2 = \frac{x_2}{l}\,, \qquad \tilde{Q}_1^{(1)} = \tilde{Q}_1 - n_{10}^2(\tilde{\omega}_{00})^2\,.$$

On rearrangement, we obtain the equilibrium equations of the belt (a tilde in the notation of dimensionless quantities is dropped out)

$$\begin{aligned}
&\frac{\mathrm{d}}{\mathrm{d}\eta}\left[Q_1^{(1)}\frac{\mathrm{d}x_1}{\mathrm{d}\eta}-Q_2\frac{\mathrm{d}x_2}{\mathrm{d}\eta}\right]+\omega_{00}^2x_1=0\,,\\
&\frac{\mathrm{d}}{\mathrm{d}\eta}\left[Q_1^{(1)}\frac{\mathrm{d}x_2}{\mathrm{d}\eta}+Q_2\frac{\mathrm{d}x_1}{\mathrm{d}\eta}\right]+\omega_{00}^2x_2=0\,,\\
&\frac{\mathrm{d}M_3}{\mathrm{d}\eta}=-Q_2\,,\\
&M_3=A_{33}\left(\frac{\mathrm{d}^2x_1}{\mathrm{d}\eta^2}\frac{\mathrm{d}x_2}{\mathrm{d}\eta}-\frac{\mathrm{d}^2x_2}{\mathrm{d}\eta^2}\frac{\mathrm{d}x_1}{\mathrm{d}\eta}\right),\\
&\left(\frac{\mathrm{d}x_1}{\mathrm{d}\eta}\right)^2+\left(\frac{\mathrm{d}x_2}{\mathrm{d}\eta}\right)^2=1\,.
\end{aligned}\tag{2.87}$$

To solve these equations numerically, it would be convenient to represent them as

$$\begin{aligned}
&\frac{\mathrm{d}Q_{x1}}{\mathrm{d}\eta}+\omega_{00}^2x_1=0\,,\\
&\frac{\mathrm{d}Q_{x2}}{\mathrm{d}\eta}+\omega_{00}^2x_2=0\,,\\
&\frac{\mathrm{d}M_3}{\mathrm{d}\eta}+Q_{x_2}\cos\vartheta_3-Q_{x_1}\sin\vartheta_3=0\,,\\
&\frac{\mathrm{d}\vartheta_3}{\mathrm{d}\eta}-\frac{M_3}{A_{33}}=0\,,\\
&\frac{\mathrm{d}x_1}{\mathrm{d}\eta}-\cos\vartheta_3=0\,,\\
&\frac{\mathrm{d}x_2}{\mathrm{d}\eta}-\sin\vartheta_3=0\,,
\end{aligned}\tag{2.88}$$

where

$$Q_{x_1}=Q_1^{(1)}x_1'-Q_2x_2'\,,\qquad Q_{x_2}=Q_1^{(1)}x_2'+Q_2x_1'\,.$$

2.4.4 Stationary Motion of a Rod in a Liquid or Air

In the previous subsection, we have considered a stationary motion of a rod with no regard for resisting forces. Real liquids or air possess viscosity; hence, a rod moving in such a liquid or air is subjected to an aerodynamic forces $\mathbf{q}_1$. If the motion of a rod is stationary, that is, the special configuration of its axis remains unchanged during the motion, then these forces are tangent to the rod axis (Fig. 2.13).

We have

$$\mathbf{q}_1=|\mathbf{q}_1|\,\mathbf{e}_1=\sum_{j=1}^{3}q_{10}x_j'\mathbf{i}_j\,,$$

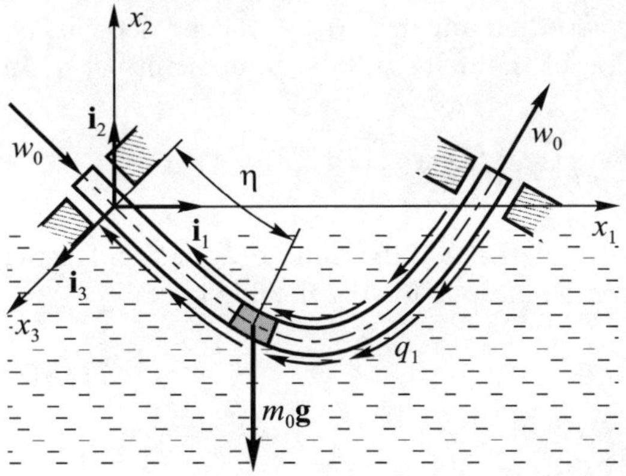

Fig. 2.13.

where $q_{10} = |\mathbf{q}_1|$. Taking into consideration the gravity force and the force of hydrodynamic resistance, we can represent the vector equation of stationary motion of the rod as

$$\frac{\mathrm{d}\mathbf{Q}^{(1)}}{\mathrm{d}\eta} + \mathbf{q}_1 - 1 \cdot \mathbf{i}_2 = 0\,;$$

in components, we have

$$\frac{\mathrm{d}Q^{(1)}_{x_j}}{\mathrm{d}\eta} + q_{1x_j} - \delta_{2j} \cdot 1 = 0\,,$$

where $q_{1x_j} = q_{10}x'_j$. The other equations of the system (2.88) remain the same.

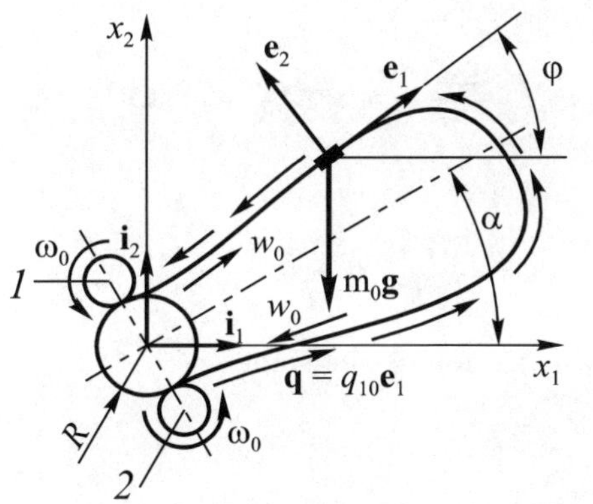

Fig. 2.14.

As an example, consider a moving belt-like radiator under terrestrial conditions (Fig. 2.14). The rotation of the belt relative to the barrel is undesirable. To fix the location of the belt in space, one should fix the rollers *1* and

2 and rotate the barrel with angular velocity ω. Thus, the belt is subjected to the gravity force $m_0\mathbf{g}$ and the aerodynamic drag $\mathbf{q}_1$. The drag depends on the velocity of longitudinal motion of the belt w_0. The components of $\mathbf{q}_1$ in the frame x_1Ox_2 are

$$\mathbf{q}_1 = q_{10}x_1'\mathbf{i}_1 + q_{10}x_2'\mathbf{i}_2\,, \qquad q_{10} = |\mathbf{q}_1|\,\frac{A_{33}}{l^3}\,.$$

Considering the gravity force and the aerodynamic drag $\mathbf{q}_1$ and using (2.88), we arrive at the equilibrium equations for the flexible belt:

$$\begin{aligned}
&\frac{\mathrm{d}Q_{x_1}}{\mathrm{d}\eta} - q_{10}\,\frac{\mathrm{d}x_1}{\mathrm{d}\eta} = 0\,,\\
&\frac{\mathrm{d}Q_{x_2}}{\mathrm{d}\eta} - q_{10}\,\frac{\mathrm{d}x_2}{\mathrm{d}\eta} - 1 = 0\,,\\
&\frac{\mathrm{d}M_3}{\mathrm{d}\eta} + Q_{x_2}\cos\vartheta_3 - Q_{x_1}\sin\vartheta_3 = 0\,,\\
&\frac{\mathrm{d}\vartheta_3}{\mathrm{d}\eta} - \frac{M_{33}}{A_{33}} = 0\,,\\
&\frac{\mathrm{d}x_1}{\mathrm{d}\eta} - \cos\vartheta_3 = 0\,,\\
&\frac{\mathrm{d}x_2}{\mathrm{d}\eta} - \sin\vartheta_3 = 0\,.
\end{aligned}$$

2.4.5 Equations of Stationary Motion of a Rod Rotating Relative to its Axis

In this case (see Fig. 2.3), the point B is immovable. By setting $J_{ii} = \text{const}$ and $|\mathbf{w}| = w_0 = \text{const}$ in (2.44), we get

$$\mathrm{J}\,\frac{\mathrm{d}\text{æ}}{\mathrm{d}\eta}\,w_0^2 + (\text{æ}\times\mathrm{J}\text{æ})\,w_0^2 + \text{æ}\times\mathrm{J}\boldsymbol{\omega}_0 w_0 - \frac{\mathrm{d}\mathbf{M}}{\mathrm{d}\eta} - \text{æ}\times\mathbf{M} - \mathbf{e}_1\times\mathbf{Q}^{(1)} = \boldsymbol{\mu}$$

or

$$\frac{\mathrm{d}\mathbf{M}^{(1)}}{\mathrm{d}\eta} + \text{æ}\times\mathbf{M}^{(1)} - \text{æ}\times\mathrm{J}\boldsymbol{\omega}_0 w_0 + \mathbf{e}_1\times\mathbf{Q}^{(1)} + \boldsymbol{\mu} = 0\,,$$

where $\mathbf{M}^{(1)} = \mathbf{M} - Jw_0^2\text{æ}$. Equation (2.72) remains the same.

3. Small Vibrations of Space-Curved Rods

When a rod is subjected to external disturbances of determinate or random nature, it may fall into vibrations either about a position of equilibrium or with respect to a stationary motion. Generally, such vibrations are to be avoided as they may lead to an accident or disrupt the normal operation of the device.

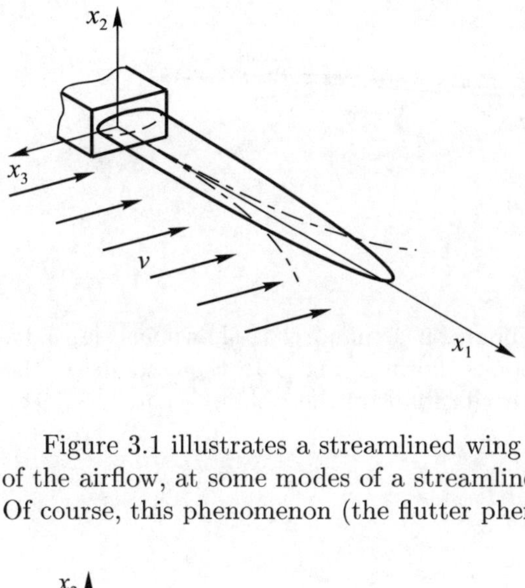

Fig. 3.1.

Figure 3.1 illustrates a streamlined wing of an aircraft. Under the action of the airflow, at some modes of a streamline, the wing may start vibrating. Of course, this phenomenon (the flutter phenomenon) is undesirable.

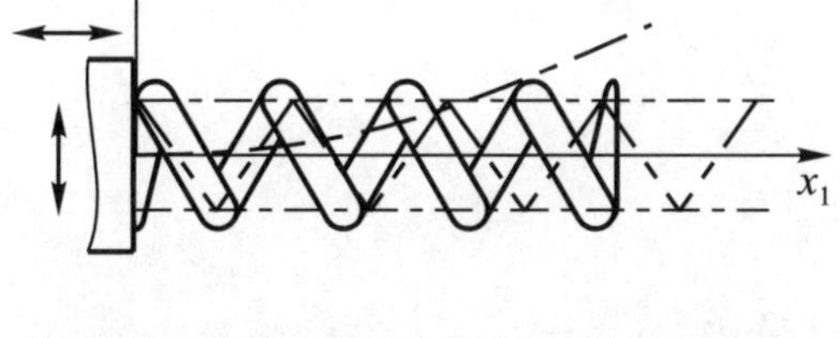

Fig. 3.2.

A cylindrical spring is shown in Fig. 3.2. The spring is rigidly attached to a vibrating foundation. The foundation is in an oscillatory motion in two mutually orthogonal planes; as a result, the coils of the spring oscillate in the direction of the spring axis (the axis x_1) as well as in the transversal direction (the axis x_2).

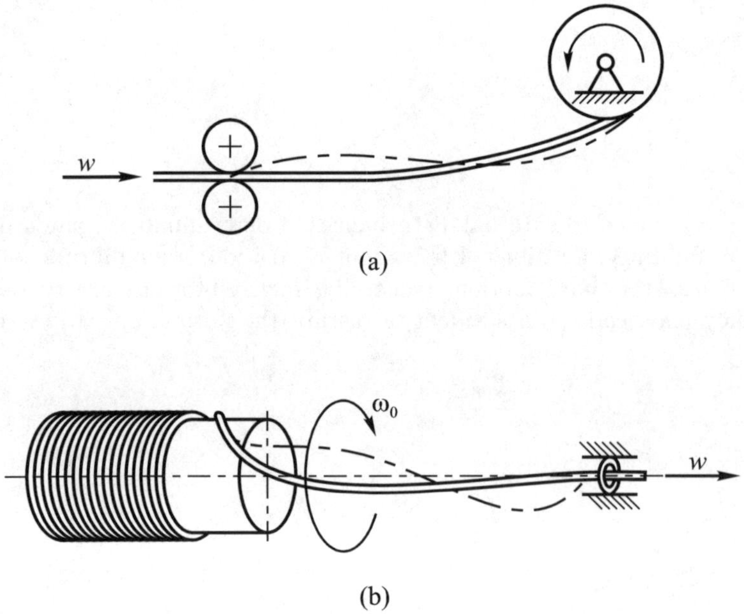

Fig. 3.3.

Winding (reeling on) of a ribbon on a mandrel is shown in Fig. 3.3*a*. Unwinding (reeling off) of a ribbon is shown in Fig. 3.3*b*. Some modes of the motion may be accompanied with vibrations of the ribbon.

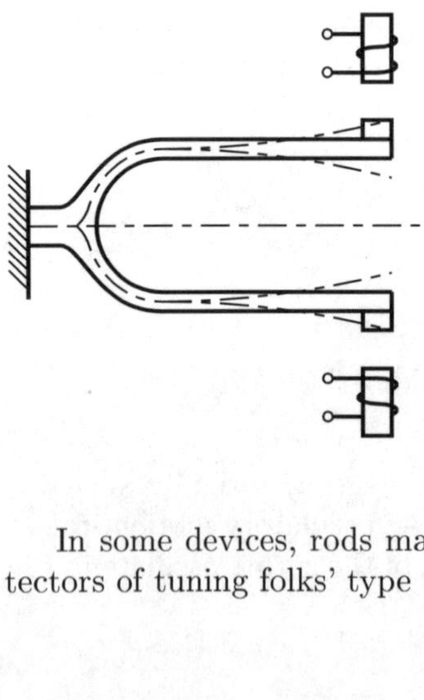

Fig. 3.4.

In some devices, rods may be used as sensing elements of frequency detectors of tuning folks' type (Figs. 3.4 and 3.5).

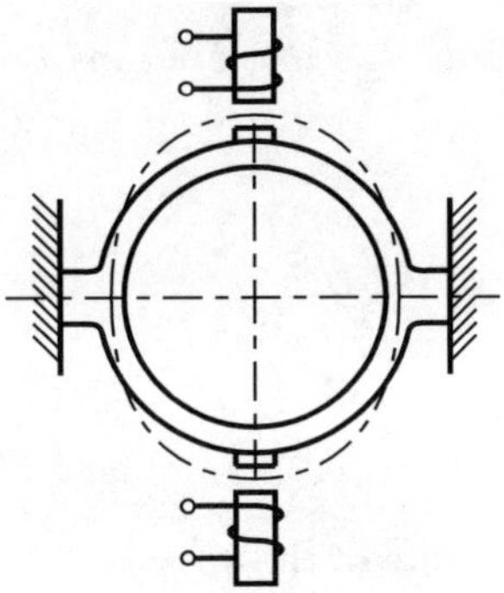

Fig. 3.5.

The operating conditions of such devices correspond to the regime of forced resonance vibrations. In this case, rod vibrations may be of advantage.

Small vibrations of rods fall into the following categories: free vibrations, forced vibrations, parametric vibrations, and self-excited vibrations. For example, a process of pipage is shown in Fig. 3.6. Dangerous parametric vibrations may occur because the pressure and velocity of the flow in the pipeline vary in time.

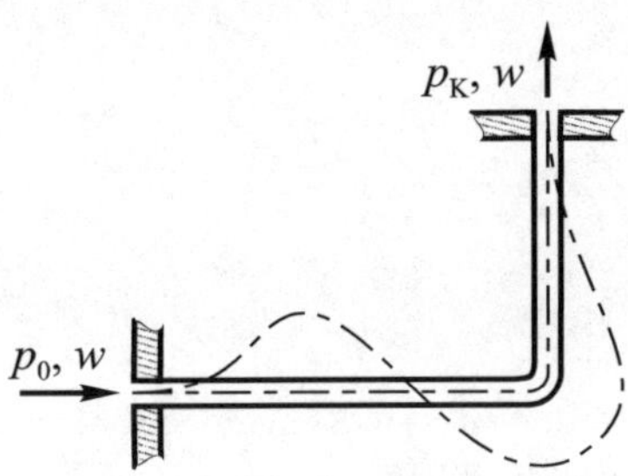

Fig. 3.6.

To examine vibrations of rods, it is essential to derive the appropriate equations of motion. For this purpose, the first few sections of this chapter are concerned with the derivation of the general equations of small vibrations of space-curved rods. Application of these equations to specific problems is considered in the rest of the chapter.

3.1 Equations of Small Vibrations

3.1.1 Auxiliary Relationships

Consider small vibrations of a rod about its equilibrium configuration. Below we derive the equations governing these vibrations. We should emphasize that *the additional internal stresses and displacements due to vibrations are assumed to be small.*

Let

$$\begin{aligned}
&\mathbf{Q} = \mathbf{Q}_0 + \Delta\mathbf{Q}\,, \qquad \mathbf{M} = \mathbf{M}_0 + \Delta\mathbf{M}\,,\\
&\text{æ} = \text{æ}_0 + \Delta\text{æ}\,, \qquad \boldsymbol{\vartheta} = \Delta\boldsymbol{\vartheta}\,,\\
&\mathbf{q} = \mathbf{q}_0 + \Delta\mathbf{q}\,, \qquad \boldsymbol{\mu} = \boldsymbol{\mu}_0 + \Delta\boldsymbol{\mu}\,,\\
&\mathbf{v} = \Delta\mathbf{v}\,, \qquad \mathbf{u} = \Delta\mathbf{u}\,, \qquad \boldsymbol{\omega} = \Delta\boldsymbol{\omega}\,,
\end{aligned} \tag{3.1}$$

where $\mathbf{Q}_0$, $\mathbf{M}_0$, æ_0, $\mathbf{q}_0$, and $\boldsymbol{\mu}_0$ are the static components of the vectors. The components of the vectors $\Delta\mathbf{Q}$, $\Delta\mathbf{M}$, $\Delta\text{æ}$, $\Delta\mathbf{q}$, $\Delta\boldsymbol{\mu}$, $\Delta\boldsymbol{\nu}$, $\Delta\mathbf{u}$, $\Delta\boldsymbol{\omega}$, and $\Delta\boldsymbol{\vartheta}$ are of the first order of smallness. The products of these components will be neglected in the course of mathematical manipulations.

Considering only small vibrations about an equilibrium configuration, we have

$$\mathbf{M} = \mathrm{A}\cdot\Delta\text{æ} + \mathbf{M}_0\,, \qquad \Delta\text{æ} = \text{æ} - \text{æ}_0^{(1)}\,.$$

The total moment reads

$$\mathbf{M} = \mathbf{M}_0 + \Delta\mathbf{M}\,;$$

hence, we get

$$\Delta\mathbf{M} = \mathrm{A}\cdot\Delta\text{æ}\,.$$

Since

$$\mathbf{v} = \frac{\mathrm{d}\mathbf{u}}{\mathrm{d}t}\,,$$

it follows that, in terms of local derivatives, we can write

$$\mathbf{v} = \frac{\tilde{\partial}\mathbf{u}}{\partial\tau} + \boldsymbol{\omega}\times\mathbf{u}\,.$$

For small vibrations, we put $\boldsymbol{\omega}\times\mathbf{u}\approx 0$, therefore,

$$\mathbf{v} = \frac{\tilde{\partial}\mathbf{u}}{\partial\tau}\,, \qquad \frac{\partial\mathbf{v}}{\partial\tau} = \frac{\tilde{\partial}^2\mathbf{u}}{\partial\tau^2}\,. \tag{3.2}$$

Consider the relationships (2.23) and (2.25). For small angles ϑ_i, the matrix L_1 is to be taken as the identity matrix. Hence, we have

$$\boldsymbol{\omega} = \frac{\partial\boldsymbol{\vartheta}}{\partial\tau}\,.$$

Recall that $\boldsymbol{\vartheta}_i$ are the angles of rotation of the axes currently attached to the rod axis relative to their configuration corresponding to the static solution.

3.1.2 Vector Equations of Small Vibrations of a Rod in the Attached Coordinate System

Let us derive the equations of small vibrations of a rod on the basis of the system of equations (2.20)–(2.26). Using the representation (3.1) in these equations and retaining terms linear in the small quantities, we get the following vector equations referred to the attached coordinate system:

$$n_1(\eta)\,\frac{\partial^2 \mathbf{u}}{\partial \tau^2} - \frac{\partial \Delta \mathbf{Q}}{\partial \eta} - \Delta æ \times \mathbf{Q}_0 - æ_0 \times \Delta \mathbf{Q} = \Delta \mathbf{P}\,, \tag{3.3}$$

$$\mathrm{J}\,\frac{\partial^2 \boldsymbol{\vartheta}}{\partial \tau^2} - \frac{\partial \Delta \mathbf{M}}{\partial \eta} - \Delta æ \times \mathbf{M}_0 - æ_0 \times \Delta \mathbf{M} - \mathbf{e}_1 \times \Delta \mathbf{Q} = \Delta \mathbf{T}\,, \tag{3.4}$$

$$\Delta \mathbf{M} = \mathrm{A} \cdot \Delta æ\,, \tag{3.5}$$

$$\frac{\partial \mathbf{u}}{\partial \eta} + æ_0 \times \mathbf{u} - \vartheta_3 \mathbf{e}_2 + \vartheta_2 \mathbf{e}_3 = 0\,, \tag{3.6}$$

$$\frac{\partial \boldsymbol{\vartheta}}{\partial \eta} + æ_0 \times \boldsymbol{\vartheta} - \Delta æ = 0\,, \tag{3.7}$$

$$\frac{\partial \boldsymbol{\vartheta}}{\partial \tau} - \boldsymbol{\omega} = 0\,; \tag{3.8}$$

here,

$$\Delta \mathbf{P} = \Delta \mathbf{q} + \sum_{i=1}^{n} \Delta \mathbf{P}^{(i)}\, \delta\,(\eta - \eta_i)\,,$$

$$\Delta \mathbf{T} = \Delta \boldsymbol{\mu} + \sum_{\nu=1}^{\rho} \Delta \mathbf{T}^{(\nu)}\, \delta\,(\eta - \eta_\nu)\,.$$

Another useful equation is the one establishing the relationship between the vectors $\Delta æ$ and $\boldsymbol{\omega}$:

$$\frac{\partial \boldsymbol{\omega}}{\partial \eta} - \frac{\partial \Delta æ}{\partial \tau} - æ_0 \times \boldsymbol{\omega} = 0\,.$$

This equation can be derived on the basis of (1.19).

The vectors $\mathbf{Q}_0$, $\mathbf{M}_0$, and $æ_0$ in (3.3)–(3.7) are assumed to be known. Therefore, to solve these equations numerically, it is convenient to represent them in vector–matrix form.

In accordance with Sect. 1.4 of Svetlitsky (2000a), let us represent the vector products $\Delta æ \times \mathbf{Q}_0$ and $\Delta æ \times \mathbf{M}_0$ in the form

$$\Delta æ \times \mathbf{Q}_0 = \mathrm{A}_Q\, \Delta æ\,, \qquad \Delta æ \times \mathbf{M}_0 = \mathrm{A}_M\, \Delta æ\,, \tag{3.9}$$

where

$$\mathrm{A}_Q = \begin{bmatrix} 0 & Q_{30} & -Q_{20} \\ -Q_{30} & 0 & Q_{10} \\ Q_{20} & -Q_{10} & 0 \end{bmatrix},$$

$$\mathrm{A}_M = \begin{bmatrix} 0 & M_{30} & -M_{20} \\ -M_{30} & 0 & M_{10} \\ M_{20} & -M_{10} & 0 \end{bmatrix}. \tag{3.10}$$

The other vector products in (3.3)–(3.8) can be obtained in a similar way. On rearrangement, we have

$$n_1(\eta)\,\frac{\partial^2 \mathbf{u}}{\partial \tau^2} - \frac{\partial \Delta \mathbf{Q}}{\partial \eta} - \mathrm{A}_Q\,\Delta æ - \mathrm{A}_{æ}\,\Delta \mathbf{Q} = \Delta \mathbf{P}\,, \tag{3.11}$$

$$\mathrm{J}\,\frac{\partial^2 \boldsymbol{\vartheta}}{\partial \tau^2} - \frac{\partial \Delta \mathbf{M}}{\partial \eta} - \mathrm{A}_M\,\Delta æ - \mathrm{A}_{æ}\,\Delta \mathbf{M} - \mathrm{A}_1\,\Delta \mathbf{Q} = \Delta \mathbf{T}\,, \tag{3.12}$$

$$\frac{\partial \boldsymbol{\vartheta}}{\partial \eta} + \mathrm{A}_{æ}\boldsymbol{\vartheta} - \Delta æ = 0\,, \tag{3.13}$$

$$\frac{\partial \mathbf{u}}{\partial \eta} + \mathrm{A}_{æ}\mathbf{u} + \mathrm{A}_1\boldsymbol{\vartheta} = 0\,, \tag{3.14}$$

$$\Delta \mathbf{M} = \mathrm{A} \cdot \Delta æ\,, \tag{3.15}$$

where

$$\mathrm{A}_{æ} = \begin{bmatrix} 0 & -æ_{30} & æ_{20} \\ æ_{30} & 0 & -æ_{10} \\ -æ_{20} & æ_{10} & 0 \end{bmatrix}, \qquad \mathrm{A}_1 = \begin{bmatrix} 0 & 0 & 0 \\ 0 & 0 & -1 \\ 0 & 1 & 0 \end{bmatrix}. \tag{3.16}$$

Eliminating $\Delta æ$ from (3.11)–(3.15), we can represent these equations as one vector equation, namely,

$$\mathrm{A}^{(1)}\,\frac{\partial^2 \mathbf{Z}}{\partial \tau^2} + \frac{\partial \mathbf{Z}}{\partial \eta} + \mathrm{A}^{(2)}\mathbf{Z} = \Delta \boldsymbol{\Phi}\,, \tag{3.17}$$

where

$$\mathbf{Z} = \begin{bmatrix} \Delta \mathbf{Q} \\ \Delta \mathbf{M} \\ \boldsymbol{\vartheta} \\ \mathbf{u} \end{bmatrix}, \qquad \Delta \boldsymbol{\Phi} = \begin{bmatrix} -\Delta \mathbf{P} \\ -\Delta \mathbf{T} \\ \mathbf{0} \\ \mathbf{0} \end{bmatrix},$$

$$\mathrm{A}^{(1)} = \begin{bmatrix} 0 & 0 & 0 & -n_1\mathrm{E} \\ 0 & 0 & -\mathrm{J} & 0 \\ 0 & 0 & 0 & 0 \\ 0 & 0 & 0 & 0 \end{bmatrix},$$

$$\mathrm{A}^{(2)} = \begin{bmatrix} \mathrm{A}_{æ} & \mathrm{A}_Q\mathrm{A}^{-1} & 0 & 0 \\ \mathrm{A}_1 & (\mathrm{A}_M\mathrm{A}^{-1} + \mathrm{A}_{æ}) & 0 & 0 \\ 0 & -\mathrm{A}^{-1} & \mathrm{A}_{æ} & 0 \\ 0 & 0 & \mathrm{A}_1 & \mathrm{A}_{æ} \end{bmatrix}.$$

Equation (3.17) is of frequent use in the analysis of applied problems.

Setting

$$\frac{\partial}{\partial \tau} = 0 \,,$$

in (3.17), we arrive at the linear equilibrium equation for a rod

$$\frac{\partial \mathbf{Z}}{\partial \eta} + \mathrm{A}^{(2)} \mathbf{Z} = \Delta \boldsymbol{\Phi}_{\mathrm{s}} \,, \tag{3.18}$$

where $\Delta \boldsymbol{\Phi}_{\mathrm{s}}$ is an increment of the static load.

Another form of equations of small vibrations of a space-curved rod can be obtained by eliminating the vectors $\Delta \mathbf{M}$ and Δæ from the system of equations (3.11)–(3.15). On rearrangement, we have

$$\begin{aligned}
&n_1(\eta) \frac{\partial^2 \mathbf{u}}{\partial \tau^2} - \frac{\partial \Delta \mathbf{Q}}{\partial \eta} - \mathrm{A}_Q \frac{\partial \boldsymbol{\vartheta}}{\partial \eta} + \mathrm{A}_Q \mathrm{A}_{æ} \boldsymbol{\vartheta} - \mathrm{A}_{æ} \Delta \mathbf{Q} = \Delta \mathbf{P} \,, \\
&\mathrm{J} \frac{\partial^2 \boldsymbol{\vartheta}}{\partial \tau^2} - \frac{\partial}{\partial \eta} \left[\mathrm{A} \left(\frac{\partial \boldsymbol{\vartheta}}{\partial \eta} + \mathrm{A}_{æ} \boldsymbol{\vartheta} \right) \right] \\
&\qquad - \mathrm{A}_M \left(\frac{\partial \boldsymbol{\vartheta}}{\partial \eta} + \mathrm{A}_{æ} \boldsymbol{\vartheta} \right) + \mathrm{A}_{æ} \mathrm{A}^{-1} \left(\frac{\partial \boldsymbol{\vartheta}}{\partial \eta} + \mathrm{A}_{æ} \boldsymbol{\vartheta} \right) - \mathrm{A}_1 \Delta \mathbf{Q} = \Delta \mathbf{T} \,, \\
&\frac{\partial \mathbf{u}}{\partial \eta} + \mathrm{A}_{æ} \mathbf{u} + \mathrm{A}_1 \boldsymbol{\vartheta} = 0 \,, \\
&\frac{\partial \boldsymbol{\vartheta}}{\partial \tau} - \boldsymbol{\omega} = 0 \,.
\end{aligned} \tag{3.19}$$

Consider in detail the expressions for increments of external forces $\Delta \mathbf{P}$ and $\Delta \mathbf{T}$. (In the case of static loads, such expressions were obtained in Sect. 1.4 of Svetlitsky (2000a).) We have

$$\begin{aligned}
\Delta \mathbf{P} &= \Delta \mathbf{q} + \sum_{i=1}^{n} \Delta \mathbf{P}^{(i)} \, \delta (\eta - \eta_i) \,, \\
\Delta \mathbf{T} &= \Delta \boldsymbol{\mu} + \sum_{\nu=1}^{\rho} \Delta \mathbf{T}^{(\nu)} \, \delta (\eta - \eta_\nu) \,.
\end{aligned} \tag{3.20}$$

In the most general case, we can write

$$\begin{aligned}
\Delta \mathbf{q} &= \mathrm{C}^{(1)} \boldsymbol{\vartheta} + \mathrm{C}^{(2)} \mathbf{u} + \mathrm{C}^{(3)} \mathbf{u}' + \mathrm{C}^{(11)} \dot{\boldsymbol{\vartheta}} + \mathrm{C}^{(21)} \dot{\mathbf{u}} \,, \\
\Delta \mathbf{P}^{(i)} &= \mathrm{B}_i^{(1)} \boldsymbol{\vartheta} + \mathrm{B}_i^{(2)} \mathbf{u} + \mathrm{B}_i^{(3)} \mathbf{u}' + \mathrm{B}_i^{(11)} \dot{\boldsymbol{\vartheta}} + \mathrm{B}_i^{(21)} \dot{\mathbf{u}} \,, \\
\Delta \boldsymbol{\mu} &= \mathrm{C}^{(4)} \boldsymbol{\vartheta} + \mathrm{C}^{(5)} \mathbf{u} + \mathrm{C}^{(6)} \mathbf{u}' + \mathrm{C}^{(41)} \dot{\boldsymbol{\vartheta}} + \mathrm{C}^{(51)} \dot{\mathbf{u}} \,, \\
\Delta \mathbf{T}^{(\nu)} &= \mathrm{B}_\nu^{(4)} \boldsymbol{\vartheta} + \mathrm{B}_\nu^{(5)} \mathbf{u} + \mathrm{B}_\nu^{(6)} \mathbf{u}' + \mathrm{B}_\nu^{(41)} \dot{\boldsymbol{\vartheta}} + \mathrm{B}_\nu^{(51)} \dot{\mathbf{u}} \,.
\end{aligned} \tag{3.21}$$

In these relationships, the elements of the matrices depend on the behavior of the load applied to the rod. In Sect. 1.2 of Svetlitsky (2000a), the matrices

$B^{(1)}$ and $B^{(2)}$ were obtained for some particular cases of external loads; as an example, the matrixes $B^{(1)}$ and $B^{(2)}$ were obtained for an increment of the force $\mathbf{P}$ directed toward a fixed point in space.

It is worse noting that when the rod is in motion, the vectors $\boldsymbol{\vartheta}$ and $\mathbf{u}$ in (3.21) depend on τ, otherwise (when considering a static problem) these vectors are functions of η alone.

3.1.3 Vector Equations of Small Vibrations of a Rod in the Cartesian Coordinates

Let

$$\begin{aligned}&\mathbf{Q}_x = \mathbf{Q}_{x0} + \Delta\mathbf{Q}_x\,, \qquad \mathbf{M}_x = \mathbf{M}_{x0} + \Delta\mathbf{M}_x\,,\\ &\boldsymbol{\vartheta} = \Delta\boldsymbol{\vartheta}\,, \qquad \boldsymbol{\omega}_x = \Delta\boldsymbol{\omega}_x\,, \qquad \mathbf{v}_x = \Delta\mathbf{v}_x\,,\\ &\mathbf{P}_x = \mathbf{P}_{x0} + \Delta\mathbf{P}_x\,, \qquad \mathbf{T}_x = \mathbf{T}_{x0} + \Delta\mathbf{T}_x\,,\end{aligned} \tag{3.22}$$

where the vectors $\mathbf{Q}_{x0}$, $\mathbf{M}_{x0}$, $\mathbf{P}_{x0}$, and $\mathbf{T}_{x0}$ are characteristics of the static equilibrium of the rod.

The vectors $\Delta\mathbf{Q}_x$, $\Delta\mathbf{M}_x$, $\Delta\mathbf{v}_x$, $\Delta\boldsymbol{\omega}_x$, $\Delta\boldsymbol{\vartheta}$, $\Delta\mathbf{P}_x$, and $\Delta\mathbf{T}_x$ describe small vibrations of the rod about this equilibrium configuration.

Let us derive the sought-for equations of small vibrations on the basis of (2.49), (2.51)–(2.53), and (2.57).

First, consider the matrix $L^{(1)}$) entering in (2.49). We have

$$L^{(1)} = LL^0\,,$$

where L^0 is the matrix of transformation from the basis $\{\mathbf{i}_j\}$ to the basis $\{\mathbf{e}_{j0}\}$. (Recall that equation (2.49) includes the transpose $L^{(1)^T}$.) The axes of the basis $\{\mathbf{e}_{j0}\}$ are associated with the axis of the rod in the static equilibrium configuration.

The elements of the matrix L^0 are assumed to be known. The matrix L is the matrix of transformation from the basis $\{\mathbf{e}_{j_0}\}$ to the basis $\{\mathbf{e}_j\}$. The axes of the latter basis are attached to the axial line of the moving rod.

When the attached axes rotate by small angles, the matrix L reads (see (A.47) of Svetlitsky (2000a))

$$L = \begin{bmatrix} 1 & \vartheta_3 & -\vartheta_2 \\ -\vartheta_3 & 1 & \vartheta_1 \\ \vartheta_2 & -\vartheta_1 & 1 \end{bmatrix}. \tag{3.23}$$

Taking the transpose of the equality

$$L^{(1)} = LL^0\,,$$

we get

$$L^{(1)^T} = L^{(0)^T} L^T\,.$$

Carrying out the last multiplication, we arrive at

$$\mathrm{L}^{(1)^{\mathrm{T}}} = \begin{bmatrix} l_{11}^0 + l_{21}^0\vartheta_3 - l_{31}^0\vartheta_2 & -l_{11}^0\vartheta_3 + l_{21}^0 + l_{31}^0\vartheta_1 & l_{11}^0\vartheta_2 - l_{21}^0\vartheta_1 + l_{31}^0 \\ l_{12}^0 + l_{22}^0\vartheta_3 - l_{32}^0\vartheta_2 & -l_{12}^0\vartheta_3 + l_{22}^0 + l_{32}^0\vartheta_1 & l_{12}^0\vartheta_2 - l_{22}^0\vartheta_1 + l_{32}^0 \\ l_{13}^0 + l_{23}^0\vartheta_3 + l_{33}^0\vartheta_2 & -l_{13}^0\vartheta_3 + l_{23}^0 + l_{33}^0\vartheta_1 & l_{13}^0\vartheta_2 - l_{23}^0\vartheta_1 + l_{33}^0 \end{bmatrix}$$

or

$$\mathrm{L}^{(1)^{\mathrm{T}}} = \mathrm{L}^{(0)^{\mathrm{T}}} + \Delta\mathrm{L}^{(1)^{\mathrm{T}}}, \tag{3.24}$$

where

$$(\Delta\mathrm{L}^{(1)})^{\mathrm{T}} = \begin{bmatrix} -l_{31}^0\vartheta_2 + l_{21}^0\vartheta_3 & l_{31}^0\vartheta_1 - l_{11}^0\vartheta_3 & -l_{21}^0\vartheta_1 + l_{11}^0\vartheta_2 \\ -l_{32}^0\vartheta_2 + l_{22}^0\vartheta_3 & l_{32}^0\vartheta_1 - l_{12}^0\vartheta_3 & -l_{22}^0\vartheta_1 + l_{12}^0\vartheta_2 \\ -l_{33}^0\vartheta_2 + l_{23}^0\vartheta_3 & l_{33}^0\vartheta_1 - l_{13}^0\vartheta_3 & -l_{23}^0\vartheta_1 + l_{13}^0\vartheta_2 \end{bmatrix}.$$

By use of (3.24), the vector product

$$\mathbf{a} = (\mathrm{L}^{(1)^{\mathrm{T}}}\mathbf{i}_1) \times (\mathbf{Q}_{x_0} + \Delta\mathbf{Q}_x)$$

in (2.49) is rearranged as follows:

$$\begin{aligned} \mathbf{a} = {} & (\mathrm{L}^{(0)^{\mathrm{T}}}\mathbf{i}_1) \times \mathbf{Q}_{x0} + (\Delta\mathrm{L}^{(1)^{\mathrm{T}}}\mathbf{i}_1) \times \mathbf{Q}_{x0} \\ & + (\mathrm{L}^{(0)^{\mathrm{T}}}\mathbf{i}_1) \times \Delta\mathbf{Q}_x + (\Delta\mathrm{L}^{(1)^{\mathrm{T}}}\mathbf{i}_1) \times \Delta\mathbf{Q}_x\,. \end{aligned} \tag{3.25}$$

From this relationship, we obtain the expression for the increment of the vector $\mathbf{a}$ (in more exact terms, the linear part of the increment) in the form

$$\Delta\mathbf{a} = (\Delta\mathrm{L}\mathbf{i}_1) \times \mathbf{Q}_{x_0} + (\mathrm{L}^{(0)^{\mathrm{T}}}\mathbf{i}_1) \times \Delta\mathbf{Q}_x$$

or

$$\Delta\mathbf{a} = \mathrm{A}_{QL}\boldsymbol{\vartheta} + \mathrm{A}_L\,\Delta\mathbf{Q}_x\,, \tag{3.26}$$

where

$$\mathrm{A}_{QL} = \begin{bmatrix} 0 & l_{33}^0 Q_{x_20} - l_{32}^0 Q_{x_30} & l_{22}^0 Q_{x_30} - l_{23}^0 Q_{x_20} \\ 0 & l_{31}^0 Q_{x_30} - l_{33}^0 Q_{x_10} & l_{23}^0 Q_{x_10} - l_{21}^0 Q_{x_30} \\ 0 & l_{32}^0 Q_{x_10} - l_{31}^0 Q_{x_20} & l_{21}^0 Q_{x_20} - l_{22}^0 Q_{x_10} \end{bmatrix},$$

$$\mathrm{A}_L = \begin{bmatrix} 0 & -l_{13}^0 & -l_{12}^0 \\ l_{13}^0 & 0 & -l_{11}^0 \\ -l_{12}^0 & l_{11}^0 & 0 \end{bmatrix}.$$

Let the initial stress state of the rod be described by Q_{j_0} and M_{j_0}. Then, using (3.22) and (3.26) in the relationships (2.48), (2.49), (2.52), (2.53), and (2.58), we obtain the equations of small vibrations

$$
\begin{aligned}
& n_1 \frac{\partial^2 \mathbf{u}_x}{\partial \tau^2} - \frac{\partial \Delta \mathbf{Q}_x}{\partial \eta} = \Delta \mathbf{P}_x \,, \\
& \mathrm{J} \frac{\partial^2 \boldsymbol{\vartheta}}{\partial \tau^2} - \frac{\partial \Delta \mathbf{M}_x}{\partial \eta} - \mathrm{A}_L \, \Delta \mathbf{Q}_x - \mathrm{A}_M \mathrm{A}^{-1} \mathrm{L}^0 \, \Delta \mathbf{M}_x - \mathrm{A}_{LQ} \boldsymbol{\vartheta} = \Delta \mathbf{T}_x \,, \\
& \frac{\partial \boldsymbol{\vartheta}}{\partial \eta} + \mathrm{A}_{æ} \boldsymbol{\vartheta} - \mathrm{A}^{-1} \mathrm{L}^0 \, \Delta \mathbf{M}_x = 0 \,, \\
& \frac{\partial \mathbf{u}_x}{\partial \eta} + \mathrm{A}_L^{(1)} \boldsymbol{\vartheta} = 0 \,, \\
& \Delta \mathbf{M}_x = \mathrm{L}^{(0)^{\mathrm{T}}} \mathrm{A} \cdot \Delta æ \,,
\end{aligned}
\tag{3.27}
$$

where

$$
\mathrm{A}_L^{(1)} = \begin{bmatrix} 0 & l_{31}^0 & -l_{21}^0 \\ 0 & l_{32}^0 & -l_{22}^0 \\ 0 & l_{33}^0 & -l_{23}^0 \end{bmatrix} .
$$

These equations are the equations of small vibrations of the rod about its static equilibrium configuration; they are referred to the Cartesian coordinates.

It is worth noting that the vector equations governing small vibrations of space-curved rods find extensive application in the analysis of a great number of applied problems related to various branches of engineering.

3.1.4 Equations in Terms of Projections in the Attached Basis

Using (3.8) and (3.11)–(3.15), we can obtain the equations in tensor form:

$$
\begin{aligned}
& n_1(\eta) \frac{\partial^2 u_k}{\partial \tau^2} - \frac{\partial \Delta Q_k}{\partial \eta} - \varepsilon_{kij} \, \Delta æ_i Q_{j0} - \varepsilon_{kij} æ_{i0} \, \Delta Q_j = \Delta P_k \,, \\
& J_{k\gamma} \frac{\partial^2 \vartheta_k}{\partial \tau^2} - \frac{\partial \Delta M_k}{\partial \eta} - \varepsilon_{kij} æ_{i0} \, \Delta M_i - \varepsilon_{kij} \, \Delta æ_i M_{j0} + \varepsilon_{ki1} \, \Delta Q_i = \Delta T_k \,, \\
& \frac{\partial \vartheta_k}{\partial \eta} + \varepsilon_{kij} æ_{i0} \vartheta_j - \Delta æ_k = 0 \,, \\
& \frac{\partial u_k}{\partial \eta} + \varepsilon_{kij} æ_{i0} u_j - \varepsilon_{ki1} \vartheta_i = 0 \,, \\
& \Delta M_k = A_{kk} \, \Delta æ_k \,,
\end{aligned}
\tag{3.28}
$$

The expanded equations read

$$
\begin{aligned}
& n_1 \frac{\partial^2 u_1}{\partial \tau^2} - \frac{\partial \Delta Q_1}{\partial \eta} + Q_{20} \, \Delta æ_3 - Q_{30} \, \Delta æ_2 + æ_{30} \, \Delta Q_2 - æ_{20} \, \Delta Q_3 = \Delta P_1 \,, \\
& n_1 \frac{\partial^2 u_2}{\partial \tau^2} - \frac{\partial \Delta Q_2}{\partial \eta} + Q_{30} \, \Delta æ_1 - Q_{10} \, \Delta æ_3 + æ_{10} \, \Delta Q_3 - æ_{30} \, \Delta Q_1 = \Delta P_2 \,,
\end{aligned}
$$

$$n_1 \frac{\partial^2 u_3}{\partial \tau^2} - \frac{\partial \Delta Q_3}{\partial \eta} + Q_{10}\, \Delta æ_2 - Q_{20}\, \Delta æ_1 + æ_{20}\, \Delta Q_1 - æ_{10}\, \Delta Q_2 = \Delta P_3 \,,$$

$$J_{11} \frac{\partial^2 \vartheta_1}{\partial \tau^2} - \frac{\partial \Delta M_1}{\partial \eta} - æ_{20}\, \Delta M_3 + æ_{30}\, \Delta M_2 - M_{30}\, \Delta æ_2 + M_{20}\, \Delta æ_3 = \Delta T_1 \,,$$

$$J_{22} \frac{\partial^2 \vartheta_2}{\partial \tau^2} - \frac{\partial \Delta M_2}{\partial \eta}$$
$$- æ_{30}\, \Delta M_1 + æ_{10}\, \Delta M_3 - M_{10}\, \Delta æ_3 + M_{30}\, \Delta æ_1 + \Delta Q_3 = \Delta T_2 \,,$$

$$J_{33} \frac{\partial^2 \vartheta_3}{\partial \tau^2} - \frac{\partial \Delta M_3}{\partial \eta}$$
$$- æ_{10}\, \Delta M_2 + æ_{20}\, \Delta M_1 - M_{20}\, \Delta æ_1 + M_{10}\, \Delta æ_2 - \Delta Q_2 = \Delta T_3 \,,$$

$$\frac{\partial \vartheta_1}{\partial \eta} + æ_{20}\vartheta_3 - æ_{30}\vartheta_1 - \Delta æ_1 = 0 \,,$$
$$\frac{\partial \vartheta_2}{\partial \eta} + æ_{30}\vartheta_1 - æ_{10}\vartheta_3 - \Delta æ_2 = 0 \,,$$
$$\frac{\partial \vartheta_3}{\partial \eta} + æ_{10}\vartheta_2 - æ_{20}\vartheta_1 - \Delta æ_3 = 0 \,,$$

$$\frac{\partial u_1}{\partial \eta} + æ_{20} u_3 - æ_{30} u_2 = 0 \,,$$
$$\frac{\partial u_2}{\partial \eta} + æ_{30} u_1 - æ_{10} u_3 - \vartheta_3 = 0 \,,$$
$$\frac{\partial u_3}{\partial \eta} + æ_{10} u_2 - æ_{20} u_1 + \vartheta_2 = 0 \,,$$

$$\Delta M_1 = A_{11}\, \Delta æ_1 \,,$$
$$\Delta M_2 = A_{22}\, \Delta æ_2 \,,$$
$$\Delta M_3 = A_{33}\, \Delta æ_3 \,. \qquad (3.29)$$

By setting

$$\Delta P_j = \Delta T_j = 0 \,, \qquad j = 1,\, 2,\, 3 \,,$$

in (3.29), we can obtain the equations governing free vibrations of a rod about its equilibrium configuration.

3.1.5 Equations in Terms of Projections in the Cartesian Basis

In this case, we get

$$n_1 \frac{\partial^2 u_{x_1}}{\partial \tau^2} - \frac{\partial \Delta Q_{x_1}}{\partial \eta} = \Delta P_{x_1},$$
$$n_1 \frac{\partial^2 u_{x_2}}{\partial \tau^2} - \frac{\partial \Delta Q_{x_2}}{\partial \eta} = \Delta P_{x_2},$$
$$n_1 \frac{\partial^2 u_{x_3}}{\partial \tau^2} - \frac{\partial \Delta Q_{x_3}}{\partial \eta} = \Delta P_{x_3},$$

$$\frac{\partial \Delta M_{x_1}}{\partial \eta} - l_{13}^0\, \Delta Q_{x_2} + l_{12}^0\, \Delta Q_{x_3} + (l_{33}^0 Q_{x_20} - l_{32}^0 Q_{x_30})\, \vartheta_2 + (l_{22}^0 Q_{x_30} - l_{23}^0 Q_{x_20})\, \vartheta_3 = -\Delta T_{x_1},$$
$$\frac{\partial \Delta M_{x_2}}{\partial \eta} + l_{13}^0\, \Delta Q_{x_1} - l_{11}^0\, \Delta Q_{x_3} + (l_{31}^0 Q_{x_30} - l_{33}^0 Q_{x_10})\, \vartheta_2 + (l_{23}^0 Q_{x_10} - l_{12}^0 Q_{x_30})\, \vartheta_3 = -\Delta T_{x_2},$$
$$\frac{\partial \Delta M_{x_3}}{\partial \eta} - l_{12}^0\, \Delta Q_{x_1} + l_{11}^0\, \Delta Q_{x_2} + (l_{32}^0 Q_{x_10} - l_{31}^0 Q_{x_20})\, \vartheta_2 + (l_{21}^0 Q_{x_20} - l_{22}^0 Q_{x_10})\, \vartheta_3 = -\Delta T_{x_3},$$

$$\frac{\partial \vartheta_1}{\partial \eta} - æ_{30}\vartheta_2 + æ_{20}\vartheta_3 - \sum_{j=1}^{3} l_{1j}^0 A_{11}^{-1}\, \Delta M_{x_j} = 0,$$
$$\frac{\partial \vartheta_2}{\partial \eta} + æ_{30}\vartheta_1 - æ_{10}\vartheta_3 - \sum_{j=1}^{3} l_{2j}^0 A_{22}^{-1}\, \Delta M_{x_j} = 0,$$
$$\frac{\partial \vartheta_3}{\partial \eta} - æ_{20}\vartheta_1 + æ_{10}\vartheta_2 - \sum_{j=1}^{3} l_{3j}^0 A_{33}^{-1}\, \Delta M_{x_j} = 0,$$

$$\frac{\partial u_{x_1}}{\partial \eta} + l_{31}^0 \vartheta_2 - l_{21}^0 \vartheta_3 = 0,$$
$$\frac{\partial u_{x_2}}{\partial \eta} + l_{32}^0 \vartheta_2 - l_{22}^0 \vartheta_3 = 0,$$
$$\frac{\partial u_{x_3}}{\partial \eta} + l_{33}^0 \vartheta_2 - l_{23}^0 \vartheta_3 = 0,$$

$$\Delta M_{x_j} = \sum_{j=1}^{3} l_{ij}^0 A_{ii} \, \Delta æ_i \,, \qquad j = 1, 2, 3 \,. \tag{3.30}$$

By setting

$$\Delta P_{x_j} = \Delta T_{x_j} = 0 \,, \qquad j = 1, 2, 3 \,,$$

in (3.30), we can obtain the equations governing free vibrations of a rod about its equilibrium configuration.

3.2 Equations of Small Vibrations of a Rod About its Natural Configuration

Setting $\mathbf{Q}_0 = \mathbf{M}_0 = 0$ in (3.11)–(3.15) and (3.27), we can obtain the equations in the attached and Cartesian bases, respectively.

3.2.1 Vector Equations in the Attached Basis

In the attached basis, we have

$$\begin{aligned}
&n_1 \frac{\partial^2 \mathbf{u}}{\partial \tau^2} - \frac{\partial \Delta \mathbf{Q}}{\partial \eta} - \mathrm{A}_{æ} \, \Delta \mathbf{Q} = \Delta \mathbf{P} \,, \\
&\mathrm{J} \frac{\partial^2 \boldsymbol{\vartheta}}{\partial \tau^2} - \frac{\partial \Delta \mathbf{M}}{\partial \eta} - \mathrm{A}_{æ} \, \Delta \mathbf{M} - \mathrm{A}_1 \, \Delta \mathbf{Q} = \Delta \mathbf{T} \,, \\
&\frac{\partial \boldsymbol{\vartheta}}{\partial \eta} + \mathrm{A}_{æ} \boldsymbol{\vartheta} - \Delta æ = 0 \,, \\
&\frac{\partial \mathbf{u}}{\partial \eta} + \mathrm{A}_{æ} \mathbf{u} + \mathrm{A}_1 \boldsymbol{\vartheta} = 0 \,, \\
&\Delta \mathbf{M} = \mathrm{A} \cdot \Delta æ \,.
\end{aligned} \tag{3.31}$$

3.2.2 Equations in Terms of Projections in the Attached Basis

In that case, instead of the first six equations of the system of equations (3.29), we get

$$\begin{aligned}
&n_1 \frac{\partial^2 u_1}{\partial \tau^2} - \frac{\partial \Delta Q_1}{\partial \eta} + æ_{30} \, \Delta Q_2 - æ_{20} \, \Delta Q_3 = \Delta P_1 \,, \\
&n_1 \frac{\partial^2 u_2}{\partial \tau^2} - \frac{\partial \Delta Q_2}{\partial \eta} + æ_{10} \, \Delta Q_3 - æ_{30} \, \Delta Q_1 = \Delta P_2 \,, \\
&n_1 \frac{\partial^2 u_3}{\partial \tau^2} - \frac{\partial \Delta Q_3}{\partial \eta} + æ_{20} \, \Delta Q_1 - æ_{10} \, \Delta Q_2 = \Delta P_3 \,,
\end{aligned}$$

$$
\begin{aligned}
&J_{11}\,\frac{\partial^2\vartheta_1}{\partial\tau^2}-\frac{\partial\Delta M_1}{\partial\eta}-æ_{20}\,\Delta M_3+æ_{30}\,\Delta M_2=\Delta T_1\,,\\
&J_{22}\,\frac{\partial^2\vartheta_2}{\partial\tau^2}-\frac{\partial\Delta M_2}{\partial\eta}-æ_{30}\,\Delta M_1+æ_{10}\,\Delta M_3+\Delta Q_3=\Delta T_2\,,\\
&J_{33}\,\frac{\partial^2\vartheta_3}{\partial\tau^2}-\frac{\partial\Delta M_3}{\partial\eta}-æ_{10}\,\Delta M_2+æ_{20}\,\Delta M_1-\Delta Q_2=\Delta T_3\,.
\end{aligned}
\tag{3.32}
$$

The other equations of (3.29) are the same.

3.2.3 Vector Equations in the Cartesian Basis

In the Cartesian basis, we have

$$
\begin{aligned}
&\frac{\partial^2\mathbf{u}_x}{\partial\tau^2}-\frac{\partial\Delta\mathbf{Q}_x}{\partial\eta}=\Delta\mathbf{P}_x\,,\\
&\frac{\partial\Delta\mathbf{M}_x}{\partial\eta}+\mathrm{A}_L\,\Delta\mathbf{Q}_x=-\Delta\mathbf{T}\,,\\
&\frac{\partial\boldsymbol{\vartheta}}{\partial\eta}+\mathrm{A}_æ\boldsymbol{\vartheta}-\mathrm{A}^{-1}\mathrm{L}^0\,\Delta\mathbf{M}_x=0\,,\\
&\frac{\partial\mathbf{u}_x}{\partial\eta}+\mathrm{A}_L^{(1)}\boldsymbol{\vartheta}=0\,,\\
&\Delta\mathbf{M}_x=\mathrm{L}^{(0)^{\mathrm{T}}}\mathrm{A}\cdot\Delta æ\,.
\end{aligned}
\tag{3.33}
$$

3.2.4 Equations in Terms of Projections in the Cartesian Basis

All the equations (3.30) remain the same except the equations for ΔM_{x_j}. In the latter equations, one should put $Q_{x_j0}=0$. As a result, we have

$$
\begin{aligned}
&\frac{\partial\Delta M_{x_1}}{\partial\eta}-l_{13}^0\,\Delta Q_{x_2}+l_{12}^0\,\Delta Q_{x_3}=-\Delta T_{x_1}\,,\\
&\frac{\partial\Delta M_{x_2}}{\partial\eta}+l_{13}^0\,\Delta Q_{x_1}-l_{11}^0\,\Delta Q_{x_3}=-\Delta T_{x_2}\,,\\
&\frac{\partial\Delta M_{x_3}}{\partial\eta}-l_{12}^0\,\Delta Q_{x_1}+l_{11}^0\,\Delta Q_{x_2}=-\Delta T_{x_3}\,.
\end{aligned}
\tag{3.34}
$$

By setting

$$
\Delta P_j=\Delta T_j=\Delta P_{x_j}=\Delta T_{x_j}=0\,,\qquad j=1,\,2,\,3\,,
$$

in (3.31), (3.32), and (3.33), we can obtain the equations governing free vibrations of a rod about its equilibrium configuration.

3.2.5 Small Vibrations of a Rod Whose Axial Line is a Plane Curve

A spiral spring is shown in Fig. 3.7. The spring axis is a plane curve both in the natural ($\mathbf{T} = 0$) and loaded ($\mathbf{T} \neq 0$) configurations. When the spring is deflected from its equilibrium configuration, it starts oscillating. If the spring is disturbed in its plane, then, during the vibrations, the spring remains in this plane. Otherwise, when the spring is initially deflected out of the plane, small spatial vibrations of the spring occur.

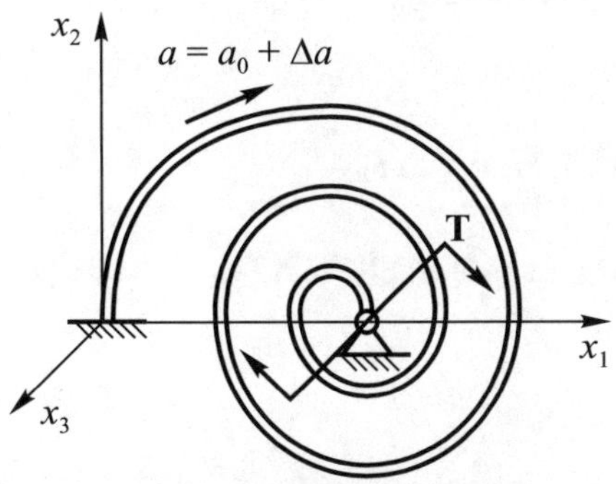

Fig. 3.7.

Consider a spring that is mounted on a moving object (say, the sensing element of a timing device). Suppose the object moves with acceleration and this acceleration have a random component. Due to this random component of the acceleration of the object, $\Delta\mathbf{a}(t)$, forced vibrations of the spring may occur. Generally, these vibrations are spatial. The constant component of acceleration $\mathbf{a}_0$ loads the rod; hence, we have $Q_{10} \neq 0$, $Q_{20} \neq 0$, and $M_{30} \neq 0$.

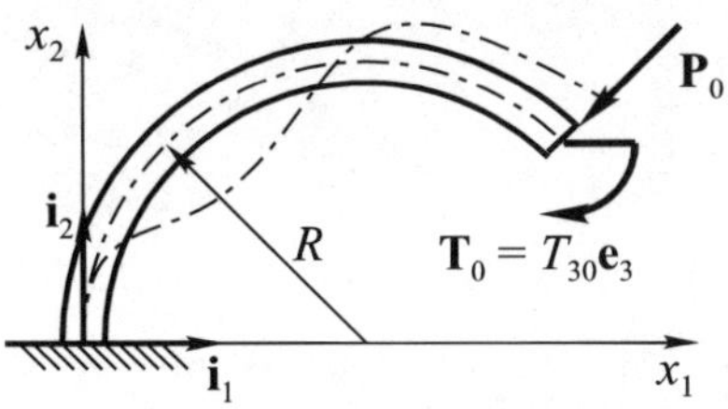

Fig. 3.8.

A rod subjected to static loads $\mathbf{P}_0$ and $\mathbf{T}_0$ is shown in Fig. 3.8. In this case, we also have $Q_{10} \neq 0$, $Q_{20} \neq 0$, and $M_{30} \neq 0$.

Considering a static problem, when the axis of the rod in a loaded configuration is a plane curve, we can state that

$$Q_{10} \neq 0\,, \qquad Q_{20} \neq 0\,, \qquad Q_{30} = 0\,,$$

$$M_{10} = M_{20} = 0\,, \qquad M_{30} \neq 0\,, \qquad æ_{10} = æ_{20} = 0\,, \qquad æ_{30} \neq 0\,.$$

By use of (3.29), we obtain the equations of forced vibrations in the attached axes:

$$n_1 \frac{\partial^2 u_1}{\partial \tau^2} - \frac{\partial \Delta Q_1}{\partial \eta} + Q_{20}\,\Delta æ_3 + æ_{30}\,\Delta Q_2 = \Delta P_1\,,$$
$$n_1 \frac{\partial^2 u_2}{\partial \tau^2} - \frac{\partial \Delta Q_2}{\partial \eta} - Q_{10}\,\Delta æ_3 - æ_{30}\,\Delta Q_1 = \Delta P_2\,,$$
$$n_1 \frac{\partial^2 u_3}{\partial \tau^2} - \frac{\partial \Delta Q_3}{\partial \eta} + Q_{10}\,\Delta æ_2 - Q_{20}\,\Delta æ_1 = \Delta P_3\,,$$

$$J_{11} \frac{\partial^2 \vartheta_1}{\partial \tau^2} - \frac{\partial \Delta M_1}{\partial \eta} + æ_{30}\,\Delta M_2 - M_{30}\,\Delta æ_2 = \Delta T_1\,,$$
$$J_{22} \frac{\partial^2 \vartheta_2}{\partial \tau^2} - \frac{\partial \Delta M_2}{\partial \eta} - æ_{30}\,\Delta M_1 + M_{30}\,\Delta æ_1 + \Delta Q_3 = \Delta T_2\,,$$
$$J_{33} \frac{\partial^2 \vartheta_3}{\partial \tau^2} - \frac{\partial \Delta M_3}{\partial \eta} - \Delta Q_2 = \Delta T_3\,,$$

$$\frac{\partial \vartheta_1}{\partial \eta} - æ_{30}\vartheta_2 - \Delta æ_1 = 0\,,$$
$$\frac{\partial \vartheta_2}{\partial \eta} + æ_{30}\vartheta_1 - \Delta æ_2 = 0\,,$$
$$\frac{\partial \vartheta_3}{\partial \eta} - \Delta æ_3 = 0\,,$$

$$\frac{\partial u_1}{\partial \eta} - æ_{30} u_2 = 0\,,$$
$$\frac{\partial u_2}{\partial \eta} + æ_{30} u_1 - \vartheta_3 = 0\,,$$
$$\frac{\partial u_3}{\partial \eta} + \vartheta_2 = 0\,,$$

$$\Delta M_i = A_{ii}\,\Delta æ_i \qquad i = 1, 2, 3\,. \tag{3.35}$$

Setting

$$\Delta P_i = \Delta T_i = 0\,, \qquad i = 1, 2, 3\,,$$

we can obtain the equations of free vibrations of a rod whose axial line in loaded configuration is a plane curve.

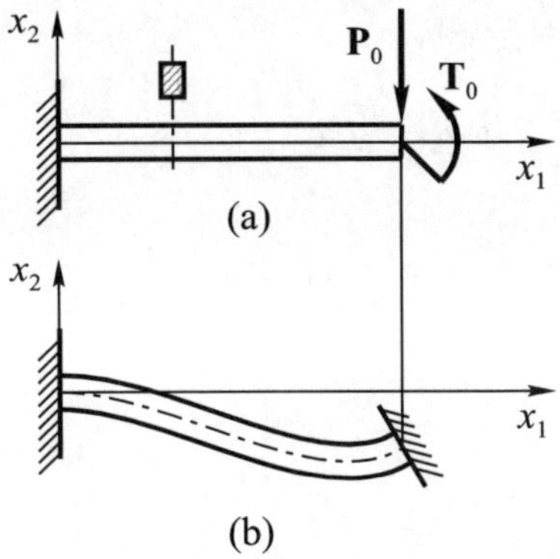

Fig. 3.9.

A rod loaded by a force $\mathbf{P}_0 = -P_2\mathbf{e}_{20}$ and a moment $\mathbf{T} = T_3\mathbf{e}_{30}$ is shown in Fig. 3.9*a*. As soon as the rod has been loaded, the cross section of application of the force $\mathbf{P}_0$ and moment $\mathbf{T}_0$ is fixed (Fig. 3.9*b*). Thus, we have

$$Q_{10} \neq 0\,, \qquad Q_{20} \neq 0\,, \qquad M_{30} \neq 0\,, \qquad \Delta\mathbf{P} = \Delta\mathbf{T} = 0\,.$$

In the case of free vibrations, the system of equations (3.35) splits into two uncoupled systems.

(1) The equations of small vibrations in the plane of the rod axis are

$$\begin{aligned}
&n_1 \frac{\partial^2 u_1}{\partial \tau^2} - \frac{\partial \Delta Q_1}{\partial \eta} + Q_{20}\,\Delta æ_3 + æ_{30}\,\Delta Q_2 = 0\,,\\
&n_1 \frac{\partial^2 u_2}{\partial \tau^2} - \frac{\partial \Delta Q_2}{\partial \eta} - Q_{10}\,\Delta æ_3 - æ_{30}\,\Delta Q_1 = 0\,,\\
&J_{33} \frac{\partial^2 \vartheta_3}{\partial \tau^2} - \frac{\partial \Delta M_3}{\partial \eta} - \Delta Q_2 = 0\,,\\
&\frac{\partial \vartheta_3}{\partial \eta} - \Delta æ_3 = 0\,,\\
&\frac{\partial u_1}{\partial \eta} - æ_{30} u_2 = 0\,,\\
&\frac{\partial u_2}{\partial \eta} + æ_{30} u_1 - \vartheta_3 = 0\,,\\
&\Delta M_3 = A_{33}\,\Delta æ_3\,.
\end{aligned} \tag{3.36}$$

(2) The equations of small out-of-plane vibrations are

$$n_1 \frac{\partial^2 u_3}{\partial \tau^2} - \frac{\partial \Delta Q_3}{\partial \eta} + Q_{10}\,\Delta æ_2 - Q_{20}\,\Delta æ_1 = 0\,,$$

$$
\begin{aligned}
&J_{11}\frac{\partial^2\vartheta_1}{\partial\tau^2}-\frac{\partial\Delta M_1}{\partial\eta}+æ_{30}\,\Delta M_2-M_{30}\,\Delta æ_2=0\,,\\
&J_{22}\frac{\partial^2\vartheta_2}{\partial\tau^2}-\frac{\partial\Delta M_2}{\partial\eta}-æ_{30}\,\Delta M_1+M_{30}\,\Delta æ_1+\Delta Q_3=0\,,\\
&\frac{\partial\vartheta_1}{\partial\eta}-æ_{30}\vartheta_2-\Delta æ_1=0\,,\\
&\frac{\partial\vartheta_2}{\partial\eta}+æ_{30}\vartheta_1-\Delta æ_2=0\,,\\
&\frac{\partial u_3}{\partial\eta}+\vartheta_2=0\,,\\
&\Delta M_1=A_{11}\,\Delta æ_1\,,\\
&\Delta M_2=A_{22}\,\Delta æ_2\,.
\end{aligned}
\tag{3.37}
$$

Consider free vibrations of an unloaded rod (Fig. 3.9*b*). In this case, we have $Q_{10}=Q_{20}=M_{30}=0$. By use of (3.36) and (3.37), we get the following systems of equations.

(1) The equations of small vibrations in the plane of the rod axis are

$$
\begin{aligned}
&n_1\frac{\partial^2 u_1}{\partial\tau^2}-\frac{\partial\Delta Q_1}{\partial\eta}+æ_{30}\,\Delta Q_2=0\,,\\
&n_1\frac{\partial^2 u_2}{\partial\tau^2}-\frac{\partial\Delta Q_2}{\partial\eta}-æ_{30}\,\Delta Q_1=0\,,\\
&J_{33}\frac{\partial^2\vartheta_3}{\partial\tau^2}-\frac{\partial\Delta M_3}{\partial\eta}-\Delta Q_2=0\,,\\
&\frac{\partial\vartheta_3}{\partial\eta}-\Delta æ_3=0\,,\\
&\frac{\partial u_1}{\partial\eta}-æ_{30}u_2=0\,,\\
&\frac{\partial u_2}{\partial\eta}+æ_{30}u_1-\vartheta_3=0\,,\\
&\Delta M_3=A_{33}\,\Delta æ_3\,.
\end{aligned}
\tag{3.38}
$$

(2) The equations of small out-of-plane vibrations are

$$
\begin{aligned}
&n_1\frac{\partial^2 u_3}{\partial\tau^2}-\frac{\partial\Delta Q_3}{\partial\eta}=0\,,\\
&J_{11}\frac{\partial^2\vartheta_1}{\partial\tau^2}-\frac{\partial\Delta M_1}{\partial\eta}-æ_{30}\,\Delta M_2=0\,,\\
&J_{22}\frac{\partial^2\vartheta_2}{\partial\tau^2}-\frac{\partial\Delta M_2}{\partial\eta}-æ_{30}\,\Delta M_1+\Delta Q_3=0\,,
\end{aligned}
$$

$$\frac{\partial \vartheta_1}{\partial \eta} - æ_{30}\vartheta_2 - \Delta æ_1 = 0\,,$$
$$\frac{\partial \vartheta_2}{\partial \eta} + æ_{30}\vartheta_1 - \Delta æ_2 = 0\,,$$
$$\frac{\partial u_3}{\partial \eta} + \vartheta_2 = 0\,,$$
$$\Delta M_1 = A_{11}\,\Delta æ_1\,,$$
$$\Delta M_2 = A_{22}\,\Delta æ_2\,. \tag{3.39}$$

In a particular case of a circular rod, we have $æ_{30} = \text{const}$. Then, the systems of equations (3.36) and (3.37) may be reduced to one equation. For example, let us obtain, using (3.36), the equations of vibrations of a rod in the plane of its axis. Eliminating u_1, ϑ_3, $\Delta æ_3$, ΔQ_1, and ΔQ_2 from these equations, we arrive at the following equation in u_2:

$$\frac{\partial^2 u_2}{\partial \tau^2} - \frac{1}{æ_{30}^2}\frac{\partial^4 u_2}{\partial \eta^2 \partial \tau^2} + J_{33}\frac{\partial}{\partial \tau}\left(\frac{1}{æ_{30}^2}\frac{\partial^4 u_2}{\partial \eta^4} + 2\frac{\partial^2 u_2}{\partial \eta^2} + æ_{30}^2 u_2\right)$$
$$- \left(\frac{1}{æ_{30}^2}\frac{\partial^6 u_2}{\partial \eta^6} + 2\frac{\partial^4 u_2}{\partial \eta^4} + æ_{30}^2\frac{\partial^2 u_2}{\partial \tau^2}\right) - \frac{\partial}{\partial \tau}\left[Q_{10}\left(\frac{\partial^2 u_2}{\partial \tau^2} + æ_{30}^2 u_2\right)\right]$$
$$- æ_{30}\left[Q_{20}\left(\frac{\partial^2 u_2}{\partial \tau^2} + æ_{30}^2 u_2\right)\right] = 0\,. \tag{3.40}$$

The special forms of the general equations of motion considered above are of frequent use in the analysis of applied problems.

3.3 Equations of Small Vibrations of a Rotating Rod

3.3.1 Vector Equations in the Attached Basis

In Sect. 2.1, the equations of motion were obtained for a rod that was either in a translational motion with the velocity $\mathbf{w}$ (Fig. 1.4) or in a rotational motion with the angular velocity $\boldsymbol{\omega}_0$ (Fig. 2.2). After that, the equations were derived for a rod that combined both translational and rotational motion (Fig. 2.3). Let us now derive the equations of small vibrations for the latter case, that is, when $\mathbf{w} \neq 0$ and $\boldsymbol{\omega}_0 \neq 0$.

Recall that the velocities $\mathbf{w}$ and $\boldsymbol{\omega}_0$ enter only in the equations of translational motion of an element of rod (2.43) and in the equations of rotational motion of an element of rod (2.47). Using (2.43) and (2.47) and setting $n_1 = 1$, we arrive at the following vector equations of small vibrations of a rod:

$$\frac{\partial^2 \mathbf{u}}{\partial \tau^2} + 2\frac{\partial \boldsymbol{\vartheta}}{\partial \tau} \times \mathbf{w} + \frac{\partial \mathbf{w}}{\partial \tau} - \frac{\partial \Delta \mathbf{Q}_1^{(1)}}{\partial \eta} - æ_0^{(1)} \times \Delta \mathbf{Q}^{(1)} - \Delta æ \times \mathbf{Q}_0^{(1)} = \Delta \mathbf{P}\,,$$

$$
\begin{aligned}
&\mathrm{J}\,\frac{\partial^2 \boldsymbol{\vartheta}}{\partial \tau^2} + \mathrm{J}\,\frac{\partial \boldsymbol{\omega}_{10}}{\partial \tau}\,\mathbf{e}_1 + \mathrm{J}\boldsymbol{\omega}_{10}\left(\frac{\partial \boldsymbol{\vartheta}}{\partial \tau} \times \mathbf{e}_1\right) + \mathrm{J}w\,\frac{\partial \Delta \text{æ}}{\partial \tau} \\
&\quad + \frac{\partial \boldsymbol{\vartheta}}{\partial \tau} \times \mathrm{J}\,(\boldsymbol{\omega}_{10}\mathbf{e}_1 + w\text{æ}_0^{(1)})\,w + \mathrm{J}\,\frac{\partial^2 \boldsymbol{\vartheta}}{\partial \eta\,\partial \tau}\,w + \mathrm{J}\boldsymbol{\omega}_{10}w\,(\Delta\text{æ} \times \mathbf{e}_1) + \mathrm{J}w^2\,\frac{\partial \Delta \text{æ}}{\partial \eta} \\
&\quad + \text{æ}_0^{(1)} \times \mathrm{J}\left(\frac{\partial \boldsymbol{\vartheta}}{\partial \tau} + w\,\Delta\text{æ}\right) w + \Delta\text{æ} \times \mathrm{J}\,(\boldsymbol{\omega}_{10}\mathbf{e}_1 + w\text{æ}_0^{(1)})\,w \\
&\quad - \frac{\partial \Delta \mathbf{M}}{\partial \eta} - \text{æ}_0^{(1)} \times \Delta\mathbf{M} - \Delta\text{æ} \times \mathbf{M}_0 - \mathbf{e}_1 \times \Delta\mathbf{Q} = \Delta\mathbf{T}\,; \qquad (3.41)
\end{aligned}
$$

here,

$$
\boldsymbol{\omega} = \frac{\partial \boldsymbol{\vartheta}}{\partial \tau}\,.
$$

3.3.2 Equations in Terms of Projections in the Attached Basis

Confine our consideration to the following cases:

(1) $\mathbf{w} \neq 0$ and $\boldsymbol{\omega}_0 = 0$;
(2) $\mathbf{w} = 0$ and $\boldsymbol{\omega}_0 \neq 0$.

(1) In the first case, using (3.41), we get

$$
\begin{aligned}
&\frac{\partial^2 u_1}{\partial \tau^2} + \frac{\partial w}{\partial \tau} - \frac{\partial \Delta Q_1^{(1)}}{\partial \eta} + Q_{20}\,\Delta\text{æ}_3 - Q_{30}\,\Delta\text{æ}_2 + \text{æ}_{30}\,\Delta Q_2 - \text{æ}_{20}\,\Delta Q_3 = \Delta P_1\,, \\
&\frac{\partial^2 u_2}{\partial \tau^2} + 2\,\frac{\partial \vartheta_3}{\partial \tau}\,w - \frac{\partial \Delta Q_2}{\partial \eta} \\
&\quad - Q_{10}^{(1)}\,\Delta\text{æ}_3 + Q_{30}\,\Delta\text{æ}_1 + \text{æ}_{10}\,\Delta Q_3 - \text{æ}_{30}\,\Delta Q_1 = \Delta P_2\,, \\
&\frac{\partial^2 u_3}{\partial \tau^2} - 2\,\frac{\partial \vartheta_2}{\partial \tau}\,w - \frac{\partial \Delta Q_3}{\partial \eta} \\
&\quad - Q_{20}\,\Delta\text{æ}_1 + Q_{10}^{(1)}\,\Delta\text{æ}_2 + \text{æ}_{20}\,\Delta Q_1 - \text{æ}_{10}\,\Delta Q_2 = \Delta P_3\,, \qquad (3.42)
\end{aligned}
$$

$$
\begin{aligned}
&J_{11}\,\frac{\partial^2 \vartheta_1}{\partial \tau^2} + J_{11}w\,\frac{\partial \Delta\text{æ}_1}{\partial \tau} + J_{33}w\,\frac{\partial \vartheta_2}{\partial \tau}\,\text{æ}_{30} - J_{22}w\,\frac{\partial \vartheta_3}{\partial \tau}\,\text{æ}_{20} \\
&\quad + J_{11}\,\frac{\partial^2 \vartheta_1}{\partial \tau\,\partial \eta}\,w + J_{11}w^2\,\frac{\partial \Delta\text{æ}_1}{\partial \eta} + J_{33}\,\Delta\text{æ}_2\text{æ}_{30}w^2 + J_{33}\text{æ}_{20}w\,\frac{\partial \vartheta_3}{\partial \eta} \\
&\quad + J_{33}\text{æ}_{20}w^2\,\Delta\text{æ}_3 - J_{22}\text{æ}_{30}\left(\frac{\partial \vartheta_2}{\partial \tau} + \Delta\text{æ}_2 w\right) w - J_{22}\,\Delta\text{æ}_3\text{æ}_{20}w^2 \\
&\quad - \frac{\partial \Delta M_1}{\partial \eta} - \text{æ}_{20}\,\Delta M_3 + \text{æ}_{30}\,\Delta M_2 - M_{30}\,\Delta\text{æ}_2 + M_{20}\,\Delta\text{æ}_3 = \Delta T_1\,, \\
&J_{22}\,\frac{\partial^2 \vartheta_2}{\partial \tau^2} + J_{22}w\,\frac{\partial \Delta\text{æ}_2}{\partial \tau} + J_{11}\,\frac{\partial \vartheta_3}{\partial \tau}\,\text{æ}_{10}w - J_{33}\,\frac{\partial \vartheta_1}{\partial \tau}\,\text{æ}_{30}w + J_{22}\,\frac{\partial^2 \vartheta_2}{\partial \eta\,\partial \tau}\,w
\end{aligned}
$$

$$+ J_{22} \frac{\partial \Delta æ_2}{\partial \eta} w^2 + J_{11} \Delta æ_3 æ_{10} w^2 + J_{11} æ_{30} \left(\frac{\partial \vartheta_1}{\partial \tau} + \Delta æ_1 w \right) w$$

$$- J_{33} æ_{10} \left(\frac{\partial \vartheta_3}{\partial \tau} + \Delta æ_3 w \right) w - J_{33} \Delta æ_1 æ_{30} w^2$$

$$- \frac{\partial \Delta M_2}{\partial \eta} - æ_{30} \Delta M_1 + æ_{10} \Delta M_3 - M_{10} \Delta æ_3 + M_{30} \Delta æ_1 + \Delta Q_3 = \Delta T_2 ,$$

$$J_{33} \frac{\partial^2 \vartheta_3}{\partial \tau^2} + J_{33} w \frac{\partial \Delta æ_3}{\partial \tau} + J_{22} \frac{\partial \vartheta_1}{\partial \tau} æ_{20} w - J_{11} \frac{\partial \vartheta_2}{\partial \tau} æ_{10} w + J_{33} w \frac{\partial^2 \vartheta_3}{\partial \tau \, \partial \eta}$$

$$+ J_{33} w^2 \frac{\partial \Delta æ_3}{\partial \eta} + J_{22} \Delta æ_1 æ_{20} w^2 + J_{22} æ_{10} w \frac{\partial \vartheta_2}{\partial \tau} - J_{11} \Delta æ_2 æ_{10} w^2$$

$$- J_{11} æ_{20} w \frac{\partial \vartheta_1}{\partial \tau} + J_{22} æ_{10} \Delta æ_2 w^2 - J_{11} æ_{20} \Delta æ_1 w^2 - \frac{\partial \Delta M_3}{\partial \eta}$$

$$- æ_{10} \Delta M_2 + æ_{20} \Delta M_1 - M_{20} \Delta æ_1 + M_{10} \Delta æ_2 - \Delta Q_2 = \Delta T_3 . \quad (3.43)$$

The other equations (2.31)–(2.35) remain the same.

Suppose that the magnitude of the velocity of translational motion of an inextensible rod is given in advance; furthermore, this velocity is constant during the motion. Hence, $\Delta Q_1^{(1)} = \Delta Q_1$ and $\Delta(w^2) = 0$. If the rotary inertia of the rod is disregarded, that is, $J_{ii} = 0$, then the velocity of translational motion enters only into the first three equations of the set (3.42).

(2) In the second case, we have $\mathbf{w} = 0$ and $\boldsymbol{\omega}_0 \neq 0$. Consequently, the equations of translational motion of the rod are identical to the first three equations of the system of equations (3.29). Using (3.41), we get the following equations of rotation of a rod element in terms of projections in the attached basis:

$$J_{11} \frac{\partial^2 \vartheta_1}{\partial \tau^2} + J_{11} \frac{\partial \omega_{10}}{\partial \tau} - \frac{\partial \Delta M_1}{\partial \eta} - æ_{20} \Delta M_3 + æ_{30} \Delta M_2$$

$$- M_{30} \Delta æ_2 + M_{20} \Delta æ_3 = \Delta T_1 ,$$

$$J_{22} \frac{\partial^2 \vartheta_2}{\partial \tau^2} + J_{11} \omega_{10} \frac{\partial \vartheta_3}{\partial \tau} - \frac{\partial \Delta M_2}{\partial \eta} - æ_{30} \Delta M_1 + æ_{10} \Delta M_3$$

$$- M_{10} \Delta æ_3 + M_{30} \Delta æ_1 + \Delta Q_3 = \Delta T_2 ,$$

$$J_{33} \frac{\partial^2 \vartheta_3}{\partial \tau^2} - J_{11} \omega_{10} \frac{\partial \vartheta_2}{\partial \tau} - \frac{\partial \Delta M_3}{\partial \eta} - æ_{10} \Delta M_2$$

$$+ æ_{20} \Delta M_1 - M_{20} \Delta æ_1 + M_{10} \Delta æ_2 - \Delta Q_2 = \Delta T_3 . \quad (3.44)$$

The other equations of the system of equations (3.29) remain the same.

3.4 Small Vibrations About a Stationary Motion

3.4.1 Vector Equations in the Attached Basis

Consider a rod that is in a stationary motion. We have $|\mathbf{w}| = |\mathbf{w}_0| = \text{const}$ and $\boldsymbol{\omega}_0 = 0$. The corresponding nonlinear equations were obtained in Sect. 2.4. Let us now obtain the equations of small vibrations of the rod about the steady-state motion.

It follows from (3.41) that

$$\frac{\partial^2 \mathbf{u}}{\partial \tau^2} + 2\,\frac{\partial \boldsymbol{\vartheta}}{\partial \tau} \times \mathbf{w}_0 - \frac{\partial \Delta \mathbf{Q}}{\partial \eta} - \text{æ}_0^{(1)} \times \Delta \mathbf{Q} - \Delta \text{æ} \times \mathbf{Q}_0^{(1)} = \Delta \mathbf{P}\,,$$

$$\begin{aligned}
&\mathrm{J}\,\frac{\partial^2 \boldsymbol{\vartheta}}{\partial \tau^2} + \mathrm{J}w_0\,\frac{\partial \Delta \text{æ}}{\partial \tau} + \frac{\partial \boldsymbol{\vartheta}}{\partial \eta} \times \mathrm{J}w_0 \text{æ}_0^{(1)} + \mathrm{J}\,\frac{\partial^2 \boldsymbol{\vartheta}}{\partial \eta\,\partial \tau}\,w_0 + \mathrm{J}w_0^2\,\frac{\partial \Delta \text{æ}}{\partial \eta} \\
&\quad + \text{æ}_0^{(1)} \times \mathrm{J}\left(\frac{\partial \boldsymbol{\vartheta}}{\partial \tau} + w_0\,\Delta \text{æ}\right) w_0 + \Delta \text{æ} \times \mathrm{J}w_0 \text{æ} \times \mathrm{J}w_0 \text{æ}_0^{(1)}\,w_0 \\
&\quad - \frac{\partial \Delta \mathbf{M}}{\partial \eta} - \text{æ}_0^{(1)} \times \Delta \mathbf{M} - \Delta \text{æ} \times \mathbf{M}_0 - \mathbf{e}_1 \times \Delta \mathbf{Q} = \Delta \mathbf{T}\,. \qquad (3.45)
\end{aligned}$$

Introducing the matrices that depend on the parameters of the stationary motion, we can rearrange these equations as follows:

$$\frac{\partial^2 \mathbf{u}}{\partial \tau^2} - 2w_0 A_1\,\frac{\partial \boldsymbol{\vartheta}}{\partial \tau} - \frac{\partial \Delta \mathbf{Q}}{\partial \eta} - \mathrm{A}_Q^{(1)}\,\Delta \text{æ} - \mathrm{A}_{\text{æ}}\,\Delta \mathbf{Q} = \Delta \mathbf{P}\,,$$

$$\begin{aligned}
&\mathrm{J}\,\frac{\partial^2 \boldsymbol{\vartheta}}{\partial \tau^2} + \mathrm{J}w_0\,\frac{\partial^2 \boldsymbol{\vartheta}}{\partial \eta\,\partial \tau} + \mathrm{A}_{\text{æ}} w_0\,\frac{\partial \boldsymbol{\vartheta}}{\partial \tau} + \mathrm{J}w_0\,\frac{\partial \Delta \text{æ}}{\partial \tau} + \mathrm{J}w_0^2\,\frac{\partial \Delta \text{æ}}{\partial \eta} + \mathrm{A}_{\text{æJ}} w_0^2\,\Delta \text{æ} \\
&\quad - \frac{\partial \Delta \mathbf{M}}{\partial \eta} - \mathrm{A}_M\,\Delta \text{æ} - \mathrm{A}_{\text{æ}}\,\Delta \mathbf{M} - \mathrm{A}_1\,\Delta \mathbf{Q} = \Delta \mathbf{T}\,; \qquad (3.46)
\end{aligned}$$

here,

$$\mathrm{A}_{\text{æ}J} = \begin{bmatrix} 0 & (J_{33} - J_{22})\,\text{æ}_{30} & (J_{33} - J_{22})\,\text{æ}_{20} \\ (J_{11} - J_{33})\,\text{æ}_{30} & 0 & (J_{11} - J_{33})\,\text{æ}_{10} \\ (J_{22} - J_{11})\,\text{æ}_{20} & (J_{22} - J_{11})\,\text{æ}_{10} & 0 \end{bmatrix},$$

$$\mathrm{A}_Q^{(1)} = \begin{bmatrix} 0 & Q_{30} & -Q_{20} \\ -Q_{30} & 0 & (Q_{10} - w_0^2) \\ Q_{20} & -(Q_{10} - w_0^2) & 0 \end{bmatrix},$$

$$\mathrm{A}_1 = \begin{bmatrix} 0 & 0 & 0 \\ 0 & 0 & -1 \\ 0 & 1 & 0 \end{bmatrix}.$$

In addition to equations (3.46), we have equations identical to (3.13)–(3.15), that is,

$$\frac{\partial \boldsymbol{\vartheta}}{\partial \eta} + \mathrm{A}_{æ}\boldsymbol{\vartheta} - \Delta\boldsymbol{æ} = 0\,,$$
$$\frac{\partial \mathbf{u}}{\partial \eta} + \mathrm{A}_{æ}\mathbf{u} + \mathrm{A}_1\boldsymbol{\vartheta} = 0\,, \qquad \Delta\boldsymbol{æ} - \mathrm{A}^{-1}\,\Delta\mathbf{M} = 0\,. \tag{3.47}$$

Eliminating the vector $\Delta\boldsymbol{æ}$, we can rewrite the system of equations (3.46) and (3.47) as one vector equation

$$\mathrm{A}^{(1)}\frac{\partial^2 \mathbf{Z}}{\partial \tau^2} + \mathrm{A}^{(3)}\frac{\partial \mathbf{Z}}{\partial \tau} + \mathrm{A}^{(4)}\frac{\partial^2 \mathbf{Z}}{\partial \tau\,\partial \eta} + \mathrm{E}^{(1)}\frac{\partial \mathbf{Z}}{\partial \eta} + \mathrm{A}^{(2)}\mathbf{Z} = \Delta\boldsymbol{\Phi}\,, \tag{3.48}$$

where

$$\mathbf{Z} = \begin{bmatrix} \Delta\mathbf{Q} \\ \Delta\mathbf{M} \\ \boldsymbol{\vartheta} \\ \mathbf{u} \end{bmatrix}, \qquad \mathrm{A}^{(1)} = \begin{bmatrix} 0 & 0 & 0 & -\mathrm{E} \\ 0 & 0 & -\mathrm{J} & 0 \\ 0 & 0 & 0 & 0 \\ 0 & 0 & 0 & 0 \end{bmatrix},$$

$$\mathrm{A}^{(3)} = \begin{bmatrix} 0 & 0 & 2w_0\mathrm{A}_1 & 0 \\ 0 & -\mathrm{J}\mathrm{A}^{-1}w_0 & -\mathrm{A}_{æ}\mathrm{J}w_0 & 0 \\ 0 & 0 & 0 & 0 \\ 0 & 0 & 0 & 0 \end{bmatrix},$$

$$\mathrm{A}^{(4)} = \begin{bmatrix} 0 & 0 & 0 & 0 \\ 0 & 0 & -\mathrm{J}w_0 & 0 \\ 0 & 0 & 0 & 0 \\ 0 & 0 & 0 & 0 \end{bmatrix}, \qquad \Delta\boldsymbol{\Phi} = \begin{bmatrix} -\Delta\mathbf{P} \\ -\Delta\mathbf{T} \\ \mathbf{0} \\ \mathbf{0} \end{bmatrix}.$$

$$\mathrm{E}^{(1)} = \begin{bmatrix} \mathrm{E} & 0 & 0 & 0 \\ 0 & -\mathrm{J}w_0^2\mathrm{A}^{-1} + \mathrm{E} & 0 & 0 \\ 0 & 0 & \mathrm{E} & 0 \\ 0 & 0 & 0 & \mathrm{E} \end{bmatrix},$$

$$\mathrm{A}^{(2)} = \begin{bmatrix} \mathrm{A}_{æ} & \mathrm{A}_Q^{(1)}\mathrm{A}^{-1} & 0 & 0 \\ \mathrm{A}_1 & \mathrm{A}_{æ} + \mathrm{A}_M\mathrm{A}^{-1} - \mathrm{A}_{æ\mathrm{J}}\mathrm{A}^{-1}w_0^2 & 0 & 0 \\ 0 & -\mathrm{A}^{-1} & \mathrm{A}_{æ} & 0 \\ 0 & 0 & \mathrm{A}_1 & \mathrm{A}_{æ} \end{bmatrix}.$$

3.4.2 Equations in Terms of Projections in the Attached Basis

Considering only the system of equations (3.46), we get

$$\frac{\partial^2 u_1}{\partial \tau^2} - \frac{\partial \Delta Q_1}{\partial \eta} + Q_{20}\,\Delta æ_3 - Q_{30}\,\Delta æ_2 + æ_{30}\,\Delta Q_2 - æ_{20}\,\Delta Q_3 = \Delta P_1\,,$$
$$\frac{\partial^2 u_2}{\partial \tau^2} + 2w_0\,\frac{\partial \vartheta_3}{\partial \tau} - \frac{\partial \Delta Q_2}{\partial \eta} + Q_{30}\,\Delta æ_1 - Q_{10}^{(1)}\,\Delta æ_3$$
$$+ æ_{10}\,\Delta Q_3 - æ_{30}\,\Delta Q_1 = \Delta P_2\,,$$

$$\frac{\partial^2 u_3}{\partial \tau^2} - 2w_0 \frac{\partial \vartheta_2}{\partial \tau} - \frac{\partial \Delta Q_3}{\partial \eta} + Q_{10}^{(1)} \Delta æ_2 - Q_{20} \Delta æ_1$$

$$+ æ_{20} \Delta Q_1 - æ_{10} \Delta Q_2 = \Delta P_3 , \tag{3.49}$$

$$\Delta J_1 - \frac{\partial \Delta M_1}{\partial \eta} - æ_{20} \Delta M_3 + æ_{30} \Delta M_2 - M_{30} \Delta æ_2 + M_{20} \Delta æ_3 = \Delta T_1 ,$$

$$\Delta J_2 - \frac{\partial \Delta M_2}{\partial \eta} - æ_{30} \Delta M_1 + æ_{10} \Delta M_3 - M_{10} \Delta æ_3 + M_{30} \Delta æ_1 + \Delta Q_3 = \Delta T_2 ,$$

$$\Delta J_3 - \frac{\partial \Delta M_3}{\partial \eta} - æ_{10} \Delta M_2 + æ_{20} \Delta M_1 - M_{20} \Delta æ_1 + M_{10} \Delta æ_2 - \Delta Q_2 = \Delta T_3 , \tag{3.50}$$

where

$$Q_{10}^{(1)} = Q_{10} - w_0^2 ,$$

$$\Delta J_1 = J_{11} \frac{\partial^2 \vartheta_1}{\partial \tau^2} + J_{11} w_0 \frac{\partial^2 \vartheta_1}{\partial \eta \, \partial \tau} + (J_{33} - J_{22})\, æ_{30} w_0 \frac{\partial \vartheta_2}{\partial \tau}$$

$$+ (J_{33} - J_{22})\, æ_{20} w_0 \frac{\partial \vartheta_3}{\partial \tau} + J_{11} w_0 \frac{\partial \Delta æ_1}{\partial \tau}$$

$$+ J_{11} w_0^2 \frac{\partial \Delta æ_1}{\partial \eta} + (J_{33} - J_{22})\, æ_{30} w_0^2 \Delta æ_2 + (J_{33} - J_{22})\, æ_{20} w_0^2 \Delta æ_3 ,$$

$$\Delta J_2 = J_{22} \frac{\partial^2 \vartheta_2}{\partial \tau^2} + J_{22} w_0 \frac{\partial^2 \vartheta_2}{\partial \eta \, \partial \tau} + (J_{11} - J_{33})\, æ_{30} w_0 \frac{\partial \vartheta_1}{\partial \tau}$$

$$+ (J_{11} - J_{33})\, æ_{10} w_0 \frac{\partial \vartheta_3}{\partial \tau} + J_{22} w_0 \frac{\partial \Delta æ_2}{\partial \tau}$$

$$+ J_{22} w_0^2 \frac{\partial \Delta æ_2}{\partial \eta} + (J_{11} - J_{33})\, æ_{30} w_0^2 \Delta æ_1 + (J_{11} - J_{33})\, æ_{10} w_0^2 \Delta æ_3 ,$$

$$\Delta J_3 = J_{33} \frac{\partial^2 \vartheta_3}{\partial \tau^2} + J_{33} w_0 \frac{\partial^2 \vartheta_3}{\partial \eta \, \partial \tau} + (J_{22} - J_{11})\, æ_{20} w_0 \frac{\partial \vartheta_1}{\partial \tau}$$

$$+ (J_{22} - J_{11})\, æ_{10} w_0 \frac{\partial \vartheta_2}{\partial \tau} + J_{33} w_0 \frac{\partial \Delta æ_3}{\partial \tau} + J_{33} w_0^2 \frac{\partial \Delta æ_3}{\partial \eta}$$

$$+ (J_{22} - J_{11})\, æ_{20} w_0^2 \Delta æ_1 + (J_{22} - J_{11})\, æ_{10} w_0^2 \Delta æ_2 .$$

3.4.3 Equations of Small in-Plane Vibrations of a Stationary Moving Rod

These equations can be readily obtained by setting

$$æ_{10} = æ_{20} = 0 , \qquad M_{10} = M_{20} = Q_{30} = 0$$

in (3.49) and (3.50). As a result, we arrive at

$$\begin{aligned}
&\frac{\partial^2 u_1}{\partial \tau^2} - \frac{\partial \Delta Q_1}{\partial \eta} + Q_{20}\,\Delta æ_3 + æ_{30}\,\Delta Q_2 = \Delta P_1\,,\\
&\frac{\partial^2 u_2}{\partial \tau^2} + 2w_0\,\frac{\partial \vartheta_1}{\partial \tau} - \frac{\partial \Delta Q_2}{\partial \eta} - Q_{10}^{(1)}\,\Delta æ_3 - æ_{30}\,\Delta Q_1 = \Delta P_2\,,\\
&\frac{\partial^2 u_3}{\partial \tau^2} - 2w_0\,\frac{\partial \vartheta_2}{\partial \tau} - \frac{\partial \Delta Q_3}{\partial \eta} + Q_{10}^{(1)}\,\Delta æ_2 - Q_{20}\,\Delta æ_1 = \Delta P_3\,,
\end{aligned} \tag{3.51}$$

$$\begin{aligned}
&\Delta J_1 - \frac{\partial \Delta M_1}{\partial \eta} + æ_{30}\,\Delta M_2 - M_{30}\,\Delta æ_2 = \Delta T_1\,,\\
&\Delta J_2 - \frac{\partial \Delta M_2}{\partial \eta} - æ_{30}\,\Delta M_1 + M_{30}\,\Delta æ_1 + \Delta Q_3 = \Delta T_2\,,\\
&\Delta J_3 - \frac{\partial \Delta M_3}{\partial \eta} - \Delta Q_2 = \Delta T_3\,,
\end{aligned} \tag{3.52}$$

where

$$\begin{aligned}
\Delta J_1 &= J_{11}\,\frac{\partial^2 \vartheta_1}{\partial \tau^2} + J_{11} w_0\,\frac{\partial^2 \vartheta_1}{\partial \eta\,\partial \tau} + (J_{33} - J_{22})\,æ_{30} w_0\,\frac{\partial \vartheta_2}{\partial \tau} + J_{11} w_0\,\frac{\partial \Delta æ_1}{\partial \tau}\\
&\quad + J_{11} w_0^2\,\frac{\partial \Delta æ_1}{\partial \eta} + (J_{33} - J_{22})\,æ_{30} w_0^2\,\Delta æ_2\,,\\
\Delta J_2 &= J_{22}\,\frac{\partial^2 \vartheta_2}{\partial \tau^2} + J_{22} w_0\,\frac{\partial^2 \vartheta_2}{\partial \eta\,\partial \tau} - (J_{33} - J_{22})\,æ_{30} w_0\,\frac{\partial \vartheta_1}{\partial \tau} + J_{22} w_0\,\frac{\partial \Delta æ_2}{\partial \tau}\\
&\quad + J_{22} w_0^2\,\frac{\partial \Delta æ_2}{\partial \eta} + (J_{11} - J_{33})\,æ_{30} w_0^2\,\Delta æ_1\,,\\
\Delta J_3 &= J_{33}\,\frac{\partial^2 \vartheta_3}{\partial \tau^2} + J_{33} w_0\,\frac{\partial^2 \vartheta_3}{\partial \eta\,\partial \tau} + J_{33} w_0\,\frac{\partial \Delta æ_3}{\partial \tau} + J_{33} w_0^2\,\frac{\partial \Delta æ_3}{\partial \eta}\,.
\end{aligned} \tag{3.53}$$

If during the rod vibrates in a certain plane, then one should put

$$\Delta æ_1 = \Delta æ_2 = 0\,, \qquad \Delta Q_3 = \Delta M_1 = \Delta M_2 = 0\,, \qquad \vartheta_1 = \vartheta_2 = 0$$

in (3.51)–(3.53). The result is

$$\begin{aligned}
&\frac{\partial^2 u_1}{\partial \tau^2} - \frac{\partial \Delta Q_1}{\partial \eta} + Q_{20}\,\Delta æ_3 + æ_{30}\,\Delta Q_2 = \Delta P_1\,,\\
&\frac{\partial^2 u_2}{\partial \tau^2} + 2w_0\,\frac{\partial \vartheta_3}{\partial \tau} - \frac{\partial \Delta Q_2}{\partial \eta} - Q_{10}^{(1)}\,\Delta æ_3 - æ_{30}\,\Delta Q_1 = \Delta P_2\,,\\
&J_{33}\,\frac{\partial^2 \vartheta_3}{\partial \tau^2} + J_{33} w_0\,\frac{\partial^2 \vartheta_3}{\partial \eta\,\partial \tau} + J_{33} w_0\,\frac{\partial \Delta æ_3}{\partial \tau}\\
&\quad + J_{33} w_0^2\,\frac{\partial \Delta æ_3}{\partial \eta} - \frac{\partial \Delta M_3}{\partial \eta} - \Delta Q_2 = \Delta T_3\,,\\
&\frac{\partial u_1}{\partial \eta} - æ_{30} u_2 = 0\,,
\end{aligned}$$

$$\frac{\partial u_2}{\partial \eta} + æ_{30} u_1 = 0\,,$$

$$\Delta æ_3 = \Delta M_3\,, \qquad A_{33} = 1\,,$$

$$\frac{\partial \vartheta_3}{\partial \eta} = \Delta æ_3\,. \tag{3.54}$$

3.4.4 Small Vibrations of a Rod About its Stationary Rotation

Let us derive the equations governing small vibrations of a rod that rotates with constant angular velocity ω_0 relative to its axial line. Recall that the angular velocity ω_0 enter only in the equation of rotation (2.12) Using (2.12), we get the following vector equation for small vibrations of the rotating rod (stationary rotation at ω_0 = const):

$$\mathrm{J}^0 \frac{\partial \boldsymbol{\omega}}{\partial \tau} + \boldsymbol{\omega} \times \mathrm{J}^0 \boldsymbol{\omega}_0 - \frac{\partial \Delta \mathbf{M}}{\partial \eta} - æ_0 \times \Delta \mathbf{M} - \Delta æ \times \mathbf{M}_0 - \mathbf{e}_{10} \times \mathbf{Q} = \Delta \mathbf{T}\,,$$

$$\boldsymbol{\omega}_0 = \omega_{10} \mathbf{e}_1\,. \tag{3.55}$$

The other equations of the system of equations (3.11), (3.13)–(3.15) remain the same.

Setting

$$\boldsymbol{\omega} = \frac{\partial \boldsymbol{\vartheta}}{\partial \tau}\,,$$

we can represent (3.55) in terms of projections in the attached basis as follows:

$$J_{11}^0 \frac{\partial^2 \vartheta_1}{\partial \tau^2} - \frac{\partial \Delta M_1}{\partial \eta} - æ_{20}\, \Delta M_3 + æ_{30}\, \Delta M_2 - M_{30}\, \Delta æ_2 + M_{20}\, \Delta æ_3 = \Delta T_1\,,$$

$$J_{22}^0 \frac{\partial^2 \vartheta_2}{\partial \tau^2} + J_{11}^0 \omega_{10} \frac{\partial \vartheta_3}{\partial \tau} - \frac{\partial \Delta M_2}{\partial \eta} - æ_{30}\, \Delta M_1 + æ_{10}\, \Delta M_3$$
$$- M_{10}\, \Delta æ_3 + M_{30}\, \Delta æ_1 + \Delta Q_3 = \Delta T_2\,,$$

$$J_{33}^0 \frac{\partial^2 \vartheta_3}{\partial \tau^2} - J_{11}^0 \omega_{10} \frac{\partial \vartheta_2}{\partial \tau} - \frac{\partial \Delta M_3}{\partial \eta} - æ_{10}\, \Delta M_2 + æ_{20}\, \Delta M_1$$
$$- M_{20}\, \Delta æ_1 + M_{10}\, \Delta æ_2 - \Delta Q_2 = \Delta T_3\,. \tag{3.56}$$

These equations, together with equations (3.11) and (3.13)–(3.15), are useful in the analysis of free and forced vibrations of fast-rotating flexible shafts.

3.5 Problems

3.1. A circular rod of constant cross section is mounted on a rotating disk (Fig. 3.10). The angular velocity of the disk ω_0 is constant. Derive the equations of small vibrations of the rod in the plane of drawing.

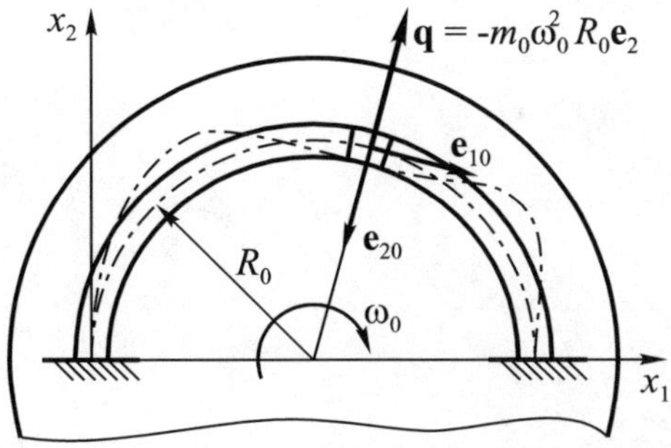

Fig. 3.10.

3.2. A ring of constant circular cross section is shown in Fig. 3.11. The ring is subjected to a follower static load $\mathbf{q}$. Taking into consideration the rotary inertia of the ring, derive the equations of small vibrations relative to the plane of drawing.

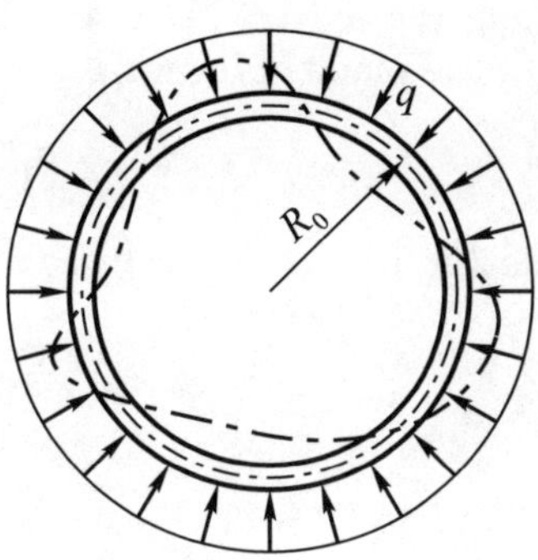

Fig. 3.11.

3.3. Consider an unloaded ring rotating with constant angular velocity ω_0. Derive the equations of small vibrations in the plane of the ring.

3.4. Derive the equations of small vibrations of a rod with concentrated masses m_1 and m_2 (Fig. 3.12). The mass m_1 is a point mass while the mass m_2 possess rotary inertia. The components of the inertia tensor J_0 of the mass m_2 with respect to the principal axes are given. The principal axes of the mass m_2 coincide with the principal axes of the cross section of the rod at the coordinate $\eta = \eta_2$. Assume that the size a of the mass m_2 is negligibly small as compared to the length of the rod l, that is, $a \ll 1$.

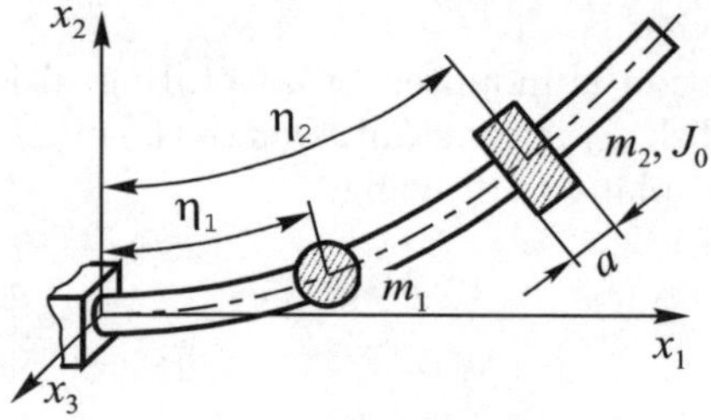

Fig. 3.12.

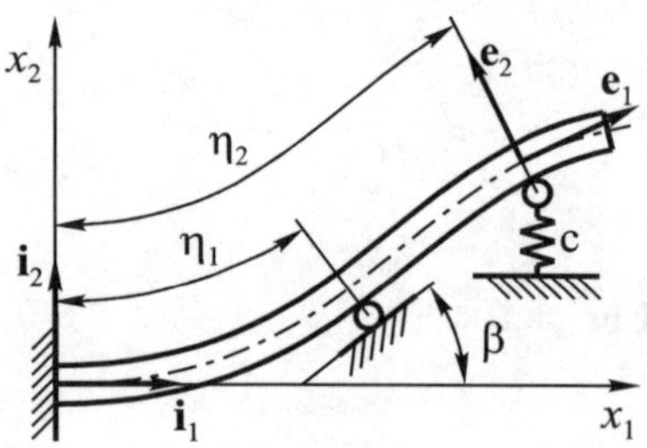

Fig. 3.13.

3.5. A rod of varying cross section with two intermediate supports is shown in Fig. 3.13. The axial coordinate of the hinge support is $\eta = \eta_1$ and that of the elastic support is $\eta = \eta_2$. When the rod oscillates, a reaction force in the elastic support occurs. This force is directed along the axis x_2. Derive the equations governing small vibrations of the rod in the plane of drawing.

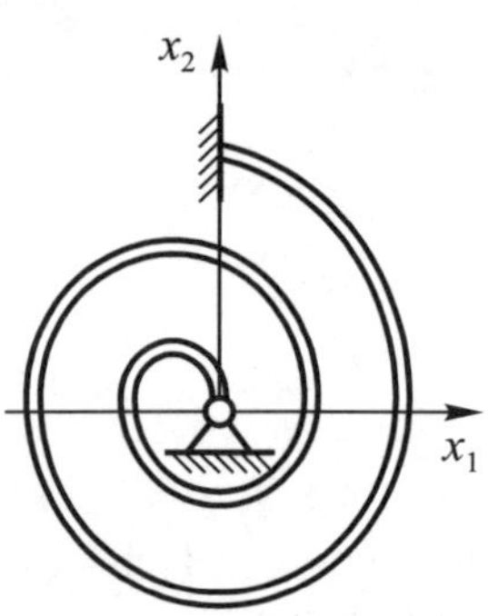

Fig. 3.14.

3.6. Spiral springs find their use in timing devices (Fig. 3.14). Consider such a spiral of constant cross section whose axial line is the Archimedean spiral. Derive the equations governing small vibrations of the spiral in the plane of drawing.

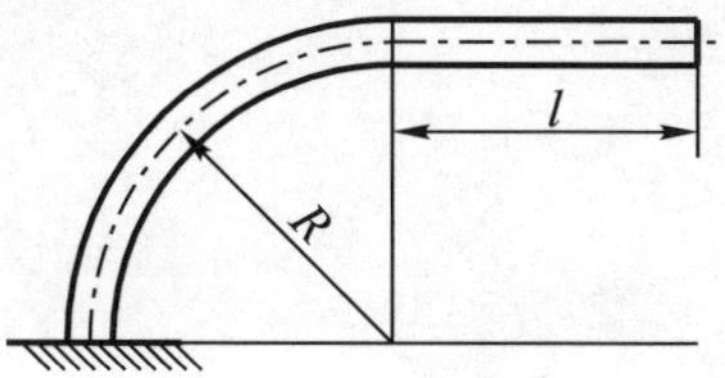

Fig. 3.15.

3.7. The rod of constant cross section comprised of two segments, the straight one and the circular one, is shown in Fig. 3.15. Derive the equations of small vibrations in the plane of drawing.

4. Determination of Eigenvalues and Eigenfunctions

Determination of frequencies and modes of rods' vibrations is a fundamental problem of dynamics of rods. Knowing the frequencies of vibrations of a structure and the frequencies of the external forces acting on it, one can predict the emergence of resonance and therefore avoid it by relevant variation of the design parameters. The design of elastic structural elements, such as frequency detectors (see Figs. 0.7–0.9), involves the determination of the sizes of a rod made of a material of given mechanical characteristics. For example, the sizes of the rod are to be so chosen as to make the first frequency be equal to the value that is assigned in advance. The accuracy of readings of a timing device with an elastic rod as the central part of its construction considerably depends on the accuracy of determination of the spectrum of frequencies of the rod.

Knowing modes of vibrations of a rod, we can apply approximate methods to the analysis of complicated problems (both linear and nonlinear) of dynamics of rods. Nowadays, there exist a lot of methods of determination of the frequencies of vibrations of structures with distributed parameters. These methods fall into two categories: exact methods and approximate methods. In their turn, exact methods fall into two categories.

(1) Exact analytical methods for solving linear equations (for example, the Fourier method that admits the separation of variables) both with constant and variable coefficients; solutions obtained in the last case may be represented in terms of special functions.

(2) Exact numerical methods, for example, the method of initial parameters and the sweep method.

Approximate methods (for example, methods based on the principle of virtual displacements) are preferable when only the frequency of vibrations is to be determined.

4.1 Determination of Eigenvalues (Eigenfrequences) in Conservative Problems

4.1.1 Basic Equations

To study free vibrations of a rod, we should put $\Delta\mathbf{P} = \Delta\mathbf{T} = 0$ in (3.11)–(3.15). Then, eliminating Δæ by use of (3.15), we obtain the following system of homogeneous vector equations:

$$n_1(\eta)\,\frac{\partial^2 \mathbf{u}}{\partial \tau^2} - \frac{\partial \Delta\mathbf{Q}}{\partial \eta} - \mathrm{A}_Q\mathrm{A}^{-1}\,\Delta\mathbf{M} - \mathrm{A}_{æ}\,\Delta\mathbf{Q} = 0\,, \tag{4.1}$$

$$\mathrm{J}\,\frac{\partial^2 \boldsymbol{\vartheta}}{\partial \tau^2} - \frac{\partial \Delta\mathbf{M}}{\partial \eta} - \mathrm{A}_M\mathrm{A}^{-1}\,\Delta\mathbf{M} - \mathrm{A}_{æ}\,\Delta\mathbf{M} - \mathrm{A}_1\,\Delta\mathbf{Q} = 0\,, \tag{4.2}$$

$$\frac{\partial \boldsymbol{\vartheta}}{\partial \eta} + \mathrm{A}_{æ}\boldsymbol{\vartheta} - \mathrm{A}^{-1}\,\Delta\mathbf{M} = 0\,, \tag{4.3}$$

$$\frac{\partial \mathbf{u}}{\partial \eta} + \mathrm{A}_{æ}\mathbf{u} + \mathrm{A}_1\boldsymbol{\vartheta} = 0\,. \tag{4.4}$$

These equations of small vibrations provide a handy way of determining the frequencies. The reason is that the presence of the moment $\Delta\mathbf{M}$ among unknown variables is the necessary condition for satisfying the boundary conditions. The coefficients in these equations are defined through the characteristics of the static stress–strain configuration, that is, these equations govern small vibrations of a space-curved rod about its static equilibrium configuration.

The equations of small vibrations of a space-curved rod about its natural configuration can be obtained by direct specialization of the system of equations (4.1)–(4.4). In this case, the elements of the matrices A_Q and A_M are equal to zero. Therefore, we get

$$n_1(\eta)\,\frac{\partial^2 \mathbf{u}}{\partial \tau^2} - \frac{\partial \Delta\mathbf{Q}}{\partial \eta} - \mathrm{A}_{æ}\,\Delta\mathbf{Q} = 0\,, \tag{4.5}$$

$$\mathrm{J}^0\,\frac{\partial^2 \boldsymbol{\vartheta}}{\partial \tau^2} - \frac{\partial \Delta\mathbf{M}}{\partial \eta} - \mathrm{A}_{æ}\,\Delta\mathbf{M} - \mathrm{A}_1\,\Delta\mathbf{Q} = 0\,, \tag{4.6}$$

$$\frac{\partial \boldsymbol{\vartheta}}{\partial \eta} + \mathrm{A}_{æ}\boldsymbol{\vartheta} - \mathrm{A}^{-1}\,\Delta\mathbf{M} = 0\,, \tag{4.7}$$

$$\frac{\partial \mathbf{u}}{\partial \eta} + \mathrm{A}_{æ}\mathbf{u} + \mathrm{A}_1\boldsymbol{\vartheta} = 0\,. \tag{4.8}$$

We seek the solution of these equations in the form

$$\Delta\mathbf{Q} = \Delta\mathbf{Q}_0(\eta)\,\mathrm{e}^{\mathrm{i}\lambda\tau}\,, \qquad \Delta\mathbf{M} = \Delta\mathbf{M}_0(\eta)\,\mathrm{e}^{\mathrm{i}\lambda\tau}\,,$$
$$\boldsymbol{\vartheta} = \boldsymbol{\vartheta}_0(\eta)\,\mathrm{e}^{\mathrm{i}\lambda\tau}\,, \qquad \mathbf{u} = \mathbf{u}_0(\eta)\,\mathrm{e}^{\mathrm{i}\lambda\tau}\,. \tag{4.9}$$

Recall that damping forces are not taken into consideration in the system of equations (4.1)–(4.4); thus, this system describes small vibrations of a conservative system. Consequently, the eigenvalues (eigenfrequencies) of the boundary value problem λ_j are real numbers. On rearrangement, we arrive at the following system of vector equations in unknowns $\mathbf{u}_0$, $\boldsymbol{\vartheta}_0$, $\Delta\mathbf{Q}_0$, and $\Delta\mathbf{M}_0$:

$$\frac{\mathrm{d}\,\Delta\mathbf{Q}_0}{\mathrm{d}\eta} + \mathrm{A}_Q\mathrm{A}^{-1}\,\Delta\mathbf{M} + \mathrm{A}_{æ}\,\Delta\mathbf{Q}_0 + \lambda^2 n_1 \mathbf{u}_0 = 0\,, \tag{4.10}$$

$$\frac{\mathrm{d}\,\Delta\mathbf{M}_0}{\mathrm{d}\eta} + (\mathrm{A}_M\mathrm{A}^{-1} + \mathrm{A}_{æ})\,\Delta\mathbf{M}_0 + \mathrm{A}_1\,\Delta\mathbf{Q}_0 + \mathrm{J}\lambda^2\boldsymbol{\vartheta}_0 = 0\,, \tag{4.11}$$

$$\frac{\mathrm{d}\boldsymbol{\vartheta}_0}{\mathrm{d}\eta} + \mathrm{A}_{æ}\boldsymbol{\vartheta}_0 - \mathrm{A}^{-1}\,\Delta\mathbf{M}_0 = 0\,, \tag{4.12}$$

$$\frac{\mathrm{d}\mathbf{u}_0}{\mathrm{d}\eta} + \mathrm{A}_{æ}\mathbf{u}_0 + \mathrm{A}_1\boldsymbol{\vartheta}_0 = 0\,. \tag{4.13}$$

These equations can be reduced to one vector equation

$$\frac{\mathrm{d}\mathbf{Z}_0}{\mathrm{d}\eta} + \mathrm{B}(\eta,\,\lambda)\,\mathbf{Z}_0 = 0\,, \tag{4.14}$$

where

$$\mathbf{Z}_0 = \begin{bmatrix} \Delta\mathbf{Q}_0 \\ \Delta\mathbf{M}_0 \\ \boldsymbol{\vartheta}_0 \\ \mathbf{u}_0 \end{bmatrix},$$

$$\mathrm{B}(\eta,\,\lambda) = \begin{bmatrix} \mathrm{A}_{æ} & \mathrm{A}_Q\mathrm{A}^{-1} & 0 & \lambda^2 n_1\mathrm{E} \\ \mathrm{A}_1 & \mathrm{A}_M\mathrm{A}^{-1} + \mathrm{A}_{æ} & \mathrm{J}\lambda^2 & 0 \\ 0 & -\mathrm{A}^{-1} & \mathrm{A}_{æ} & 0 \\ 0 & 0 & \mathrm{A}_1 & \mathrm{A}_{æ} \end{bmatrix}.$$

In expanded form, the matrix $B(\eta,\,\lambda)$ reads

$$
\mathrm{B}(\eta,\,\lambda) = \left[\begin{array}{cccccccccccc}
0 & -\ae_{30} & \ae_{20} & 0 & \dfrac{Q_{30}}{A_{22}} & -\dfrac{Q_{20}}{A_{33}} & 0 & 0 & 0 & n_1\lambda^2 & 0 & 0 \\
\ae_{30} & 0 & -\ae_{10} & -\dfrac{Q_{30}}{A_{11}} & 0 & \dfrac{Q_{10}}{A_{33}} & 0 & 0 & 0 & 0 & n_1\lambda^2 & 0 \\
-\ae_{20} & \ae_{10} & 0 & \dfrac{Q_{20}}{A_{11}} & -\dfrac{Q_{10}}{A_{22}} & 0 & 0 & 0 & 0 & 0 & 0 & n_1\lambda^2 \\
0 & 0 & 0 & 0 & \dfrac{M_{30}}{A_{22}} - \ae_{30} & -\dfrac{M_{20}}{A_{33}} + \ae_{20} & J_{11}\lambda^2 & 0 & 0 & 0 & 0 & 0 \\
0 & 0 & -1 & -\dfrac{M_{30}}{A_{11}} + \ae_{30} & 0 & \dfrac{M_{10}}{A_{33}} - \ae_{10} & 0 & J_{22}\lambda^2 & 0 & 0 & 0 & 0 \\
0 & 1 & 0 & \dfrac{M_{20}}{A_{11}} - \ae_{20} & -\dfrac{M_{10}}{A_{22}} + \ae_{10} & 0 & 0 & 0 & J_{33}\lambda^2 & 0 & 0 & 0 \\
0 & 0 & 0 & -\dfrac{1}{A_{11}} & 0 & 0 & 0 & -\ae_{30} & \ae_{20} & 0 & 0 & 0 \\
0 & 0 & 0 & 0 & -\dfrac{1}{A_{22}} & 0 & \ae_{30} & 0 & -\ae_{10} & 0 & 0 & 0 \\
0 & 0 & 0 & 0 & 0 & -\dfrac{1}{A_{33}} & -\ae_{20} & \ae_{10} & 0 & 0 & 0 & 0 \\
0 & 0 & 0 & 0 & 0 & 0 & 0 & 0 & 0 & 0 & -\ae_{30} & \ae_{20} \\
0 & 0 & 0 & 0 & 0 & 0 & 0 & 0 & -1 & \ae_{30} & 0 & -\ae_{10} \\
0 & 0 & 0 & 0 & 0 & 0 & 0 & 1 & 0 & -\ae_{20} & \ae_{10} & 0
\end{array}\right].
$$

The vector equation (4.14) is equivalent to the system of 12 equations of the first order with variable coefficients. The elements of the matrix B depend on the static stress–strain state of the rod, more precisely, on the elements of the vectors $\mathbf{Q}_0$, $\mathbf{M}_0$, and $\boldsymbol{\ae}_0$. In the general case (in the case of a rod of variable cross section), the matrices J and A are functions of η. Particularly, in the case of free vibrations of an unloaded rod, the elements of the matrices A_Q and A_M equal zero, and the matrix B reads

$$\mathrm{B}(\eta,\,\lambda) = \begin{bmatrix} \mathrm{A}_{æ} & 0 & 0 & \lambda^2 n_1 \mathrm{E} \\ \mathrm{A}_1 & \mathrm{A}_{æ} & \lambda^2 \mathrm{J} & 0 \\ 0 & -\mathrm{A}^{-1} & \mathrm{A}_{æ} & 0 \\ 0 & 0 & \mathrm{A}_1 & \mathrm{A}_{æ} \end{bmatrix}.$$

The solution of the vector equation (4.14) must satisfy 12 boundary conditions: six at $\eta = 0$ and six at $\eta = 1$. In general, the vector equation (4.14) cannot be solved explicitly. Therefore, numerical methods are to be used for the determination of the frequencies.

4.1.2 Method of Initial Parameters

For a given λ, we seek the solution of (4.14) in the form

$$\mathbf{Z}_0 = \mathrm{K}(\eta)\,\mathbf{C}\,, \qquad \mathrm{K}(0) = \mathrm{E}\,, \tag{4.15}$$

where $\mathrm{K}(\eta)$ is the fundamental matrix of the homogeneous equation and E is the identity matrix. To obtain the fundamental matrix $\mathrm{K}(\eta)$, the vector equation (4.14) is to be solved twelve times for the following initial conditions:

$$\mathbf{Z}_{0(1)} = \begin{bmatrix} 1 \\ 0 \\ 0 \\ \vdots \\ 0 \\ 0 \end{bmatrix}, \qquad \mathbf{Z}_{0(2)} = \begin{bmatrix} 0 \\ 1 \\ 0 \\ \vdots \\ 0 \\ 0 \end{bmatrix}, \quad \ldots, \quad \mathbf{Z}_{0(12)} = \begin{bmatrix} 0 \\ 0 \\ 0 \\ \vdots \\ 0 \\ 1 \end{bmatrix}. \tag{4.16}$$

To calculate the components of the matrix $\mathrm{K}(\eta)$ more accurately, let us use the method of iterations (see Sect. 2.1 of Svetlitsky (2000a)). The matrix $\mathrm{K}(\eta)$ satisfies the equation

$$\frac{\mathrm{dK}}{\mathrm{d}\eta} + \mathrm{BK} = 0\,. \tag{4.17}$$

It follows from (4.16) that

$$\mathrm{K}^{(2)}(\eta) = -\int_0^{\eta} \mathrm{BK}^{(1)}(\zeta)\,\mathrm{d}\zeta + \mathrm{E}\,, \tag{4.18}$$

where $\mathrm{K}^{(1)}(\eta)$ is the fundamental matrix obtained by solving the vector equation (4.14) with the initial conditions (4.16). This matrix can be taken as the first approximation for $\mathrm{K}(\eta)$. (The second approximation of $\mathrm{K}(\eta)$ is denoted by $\mathrm{K}^{(2)}(\eta)$.)

Let us consider homogeneous boundary conditions. For $\eta = 0$, six components of the vector $\mathbf{Z}_0$ are equal to zero, say, $\boldsymbol{\vartheta}_0(0) = \mathbf{u}_0(0) = 0$. Consequently, the six components of the vector $\mathbf{C}$ are zero: $c_7 = c_8 = \ldots = c_{12} = 0$. The other six components of $\mathbf{C}$ can be found from the boundary conditions at

$\eta = 1$. In the case of homogeneous boundary conditions, we have the following six homogeneous equations:

$$\sum k_{ij}(1)\, c_j = 0\,. \tag{4.19}$$

The indices i and j take six different values depending on the specific boundary conditions. For example, if the right-hand end of the rod, $\eta = 1$, is free (that is, $\Delta\mathbf{Q}(1) = \Delta\mathbf{M}(1) = 0$), then, in accordance with the order of elements in the vector $\mathbf{Z}_0$, the indices i and j take the values $i, j = 1, \ldots, 6$, and the system (4.19) reads

$$\begin{aligned} &k_{1,1}c_1 + k_{1,2}c_2 + \ldots + k_{1,6}c_6 = 0\,,\\ &\ldots\\ &k_{6,1}c_1 + k_{6,2}c_2 + \ldots + k_{6,6}c_6 = 0\,. \end{aligned} \tag{4.20}$$

To ensure that this system of equations have a nontrivial solution, its determinant must equal zero. Let this determinant be denoted by $|\mathrm{D}(1)|$; hence, we have the equation $|\mathrm{D}(1)| = 0$. Solving (4.14) for some values of λ, we can find numerically λ_* such that the determinant $|\mathrm{D}(1)|$ is assumed to be zero up to the accuracy required. The values λ_i such that $|\mathrm{D}_i(1)| = 0$ are the frequencies of the rod.

For a space-curved rod, various types of boundary condition exist. For example, one end of a rod may be simply supported in one coordinate plane and it is free to move in two other planes of the coordinate system. It is worth emphasizing that the elements of $\mathrm{D}(1)$ are formed by a combination of the elements of $\mathrm{K}(1, \lambda_k)$; this combination may be of various types. However, to determine the matrix D, it is sufficient to know six columns of the matrix $\mathrm{K}(\eta, \lambda_k)$. As a result, the method saves the time of computation.

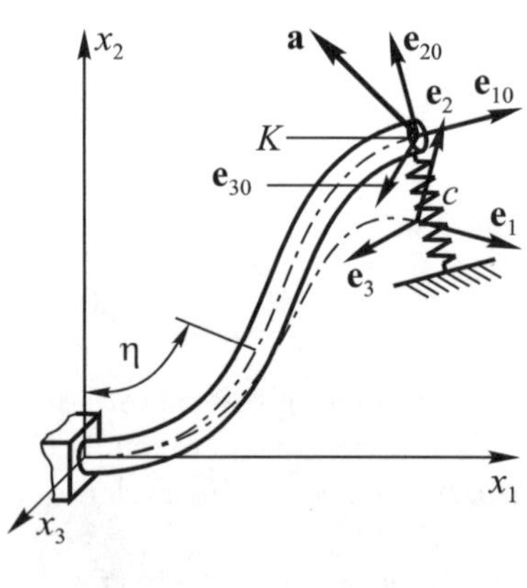

Fig. 4.1.

Consider some examples illustrating the determination of D for particular boundary conditions. A rod elastically supported at its the right-hand end is shown in Fig. 4.1. The direction of the reaction force at the support is defined by the unit vector **a**. Assume that the reaction force **R** does not change its direction under small displacements of the right-hand end of the rod. We have the following boundary conditions:

(1) at $\eta = 0\,, \quad \mathbf{u}_0(0) = \boldsymbol{\vartheta}_0(0) = 0\,;$

(2) at $\eta = 1\,, \quad \Delta\mathbf{M}_0(1) = 0 \quad \text{and} \quad \Delta\mathbf{Q}_0(1) = \mathbf{R}\,.$

It follows from the boundary conditions at $\eta = 0$ that the last six components of the vector **C** equal zero. Therefore, to obtain the frequencies from the boundary conditions at $\eta = 1$, it is sufficient to know six columns of the matrix $\mathrm{K}(\eta,\, \lambda)$.

Consider the boundary conditions at $\eta = 1$ in detail. The vector **R** is defined by

$$\mathbf{R} = -c\,(\mathbf{u}_0(1) \cdot \mathbf{a})\,\mathbf{a}\,, \qquad c = \frac{l^3 c_p}{A_{33}}\,,$$

where c_p is the support stiffness and c is the dimensionless value of c_p.

For an immovable rod, we have (Fig. 4.1)

$$\mathbf{a} = \sum_{i=1}^{3} a_j^0 \mathbf{e}_{0j}\,.$$

During vibrations of the rod, the vectors $\mathbf{e}_{0i}$ can be written in terms of their projections in the basis $\{\mathbf{e}_i\}$ as follows:

$$\mathbf{e}_{0j} = \sum_{j=1}^{3} l_{ji}\mathbf{e}_j\,;$$

here, l_{ji} are the elements of the matrix L; for small angles of rotation of the attached axes ϑ_i, the matrix L is the same as (A.46) (Svetlitsky (2000a)). In the basis $\{\mathbf{e}_i\}$, the vector **a** reads

$$\mathbf{a} = \sum_{i=1}^{3}\left(\sum_{j=1}^{3} a_j^0 l_{ji}\right)\mathbf{e}_i = \sum_{i=1}^{3} a_i\mathbf{e}_i\,, \qquad a_i = \sum a_j^0 l_{ji}\,.$$

Considering the smallness of the components of the vector $\mathbf{u}_0(1)$ and the expressions of l_{ij} for small ϑ_j, on rearrangement, we get

$$\mathbf{R} = -\sum_{j=1}^{3} A_j\mathbf{e}_j\,,$$

where

$$\mathrm{A}_1 = c\left[(a_1^0)^2 u_{01} + a_2^0 a_1^0 u_{02} + a_3^0 a_1^0 u_{03}\right],$$
$$\mathrm{A}_2 = c\left[a_1^0 a_2^0 u_{01} + (a_2^0)^2 u_{02} + a_3^0 a_2^0 u_{03}\right],$$
$$\mathrm{A}_3 = c\left[a_1^0 a_3^0 u_{01} + a_2^0 a_3^0 u_{02} + (a_3^0)^2 u_{03}\right].$$

As a result, we arrive at the boundary conditions

$$\text{at} \quad \eta = 1, \quad \Delta Q_{0j} = -A_j, \quad j = 1, 2, 3,$$

or, in vector–matrix form,

$$\Delta \mathbf{Q}_0(1) = -\tilde{\mathrm{A}}\mathbf{u}_0(1), \tag{4.21}$$

where

$$\tilde{\mathrm{A}} = \begin{bmatrix} c(a_1^0)^2 & c a_2^0 a_1^0 & c a_3^0 a_1^0 \\ c a_1^0 a_2^0 & c(a_2^0)^2 & c a_3^0 a_2^0 \\ c a_1^0 a_3^0 & c a_2^0 a_3^0 & c(a_3^0)^2 \end{bmatrix}.$$

Let us represent the solution (4.15) in the form

$$\mathbf{Z}_0 = \begin{bmatrix} \Delta\mathbf{Q}_0 \\ \Delta\mathbf{M}_0 \\ \boldsymbol{\vartheta}_0 \\ \mathbf{u}_0 \end{bmatrix} = \begin{bmatrix} \mathrm{K}_{11} & \mathrm{K}_{12} & \mathrm{K}_{13} & \mathrm{K}_{14} \\ \mathrm{K}_{21} & \mathrm{K}_{22} & \mathrm{K}_{23} & \mathrm{K}_{24} \\ \mathrm{K}_{31} & \mathrm{K}_{32} & \mathrm{K}_{33} & \mathrm{K}_{34} \\ \mathrm{K}_{41} & \mathrm{K}_{42} & \mathrm{K}_{43} & \mathrm{K}_{44} \end{bmatrix} \cdot \begin{bmatrix} \mathbf{C}_Q \\ \mathbf{C}_M \\ \mathbf{C}_\vartheta \\ \mathbf{C}_u \end{bmatrix}.$$

It follows from the boundary conditions at $\eta = 0$ that $\mathbf{C}_\vartheta = \mathbf{C}_u = 0$; hence, $\Delta\mathbf{Q}_0(1)$ and $\mathbf{u}_0(1)$ in (4.21) are

$$\Delta\mathbf{Q}_0(1) = \mathrm{K}_{11}\mathbf{C}_Q + \mathrm{K}_{12}\mathbf{C}_M, \qquad \mathbf{u}_0(1) = \mathrm{K}_{41}\mathbf{C}_Q + \mathrm{K}_{42}\mathbf{C}_M. \tag{4.22}$$

Finally, using (4.21), we get

$$(\mathrm{K}_{11} + \tilde{\mathrm{A}}\mathrm{K}_{41})\,\mathbf{C}_Q + (\mathrm{K}_{12} + \tilde{\mathrm{A}}\mathrm{K}_{42})\,\mathbf{C}_M = 0. \tag{4.23}$$

From the condition $\Delta\mathbf{M}(1) = 0$ it follows that

$$\mathrm{K}_{21}\mathbf{C}_Q + \mathrm{K}_{22}\mathbf{C}_M = 0. \tag{4.24}$$

The frequencies λ_j can be obtained by equating the determinant of the coefficients of equations (4.23) and (4.24) to zero. Thus, we get

$$D = \det\begin{bmatrix} \mathrm{K}_{11} + \tilde{\mathrm{A}}\mathrm{K}_{41} & \mathrm{K}_{12} + \tilde{\mathrm{A}}\mathrm{K}_{42} \\ \mathrm{K}_{21} & \mathrm{K}_{22} \end{bmatrix} = 0. \tag{4.25}$$

As another example, consider a rod shown in Fig. 4.2. The rod is subjected to a time-independent follower load $\mathbf{P}_0$ and a moment $\mathbf{M}_0$. The axial line of the rod in an equilibrium configuration is a spatial curve (for example, the rod may be straight before loading). There is a concentrated mass m at one of the

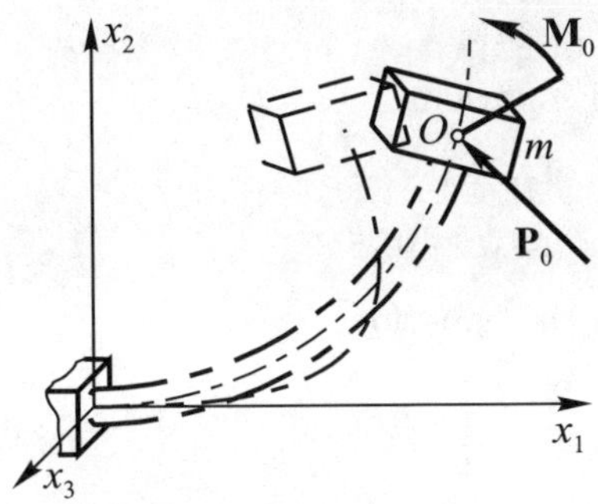

Fig. 4.2.

rod ends. Assume that the point O (the center of gravity of the concentrated mass m) coincides, with the center of the end cross section.

The projections of follower forces and moments do not enter into the equations of motion referred to the attached basis. Consequently, in the case of follower forces, the equations of small vibrations of the rod in this basis are homogeneous. In the case under consideration, the boundary conditions are

$$\begin{aligned}&(1)\quad \text{at}\quad \eta = 0\,,\quad \mathbf{u}_0(0) = 0 \quad\text{and}\quad \boldsymbol{\vartheta}_0(0) = 0\,;\\ &(2)\quad \text{at}\quad \eta = 1\,,\quad \Delta\mathbf{M}(1) + \mathbf{M}_{\mathrm{i}} = 0 \quad\text{and}\quad \Delta\mathbf{Q}(1) + \mathbf{J}_{\mathrm{i}} = 0\,,\end{aligned}$$

where $\mathbf{M}_{\mathrm{i}}$ is the moment of forces of inertia and $\mathbf{J}_{\mathrm{i}}$ is the inertial force applied to the mass m. We have

$$\mathbf{M}_{\mathrm{i}} = -\mathrm{J}_m \left.\frac{\partial^2 \boldsymbol{\vartheta}}{\partial \tau^2}\right|_{\eta=1}, \qquad \mathbf{J}_{\mathrm{i}} = -\tilde{m} \left.\frac{\partial^2 \mathbf{u}}{\partial \tau^2}\right|_{\eta=1}, \qquad \tilde{m} = \frac{m}{m_0 l}\,;$$

here, $\tilde{m}$ is the dimensionless mass, m_0 is the mass per unit length of the rod, and J_m is a matrix whose elements are the dimensionless moments of inertia of the concentrated mass m with respect to the coordinate system with the origin at O. If the axes of the coordinate system, the central axes, are directed along the principal axes, then the matrix J_m is a diagonal one.

Since we seek the solution of (4.10)–(4.13) in the form

$$\begin{aligned}&\Delta\mathbf{Q} = \Delta\mathbf{Q}_0(\eta)\mathrm{e}^{\mathrm{i}\lambda\tau}\,, \qquad \Delta\mathbf{M} = \Delta\mathbf{M}_0(\eta)\mathrm{e}^{\mathrm{i}\lambda\tau}\,,\\ &\mathbf{u} = \mathbf{u}_0(\eta)\mathrm{e}^{\mathrm{i}\lambda\tau}\,, \qquad \boldsymbol{\vartheta} = \boldsymbol{\vartheta}_0(\eta)\mathrm{e}^{\mathrm{i}\lambda\tau}\,,\end{aligned}$$

the boundary conditions at $\eta = 1$ are as follows (from here on, a tilde in the notation of the dimensionless mass is omitted):

$$\Delta\mathbf{M}_0(1) = \mathrm{J}_m\lambda^2\boldsymbol{\vartheta}_0(1)\,, \qquad \Delta\mathbf{Q}_0(1) = -m\lambda^2\mathbf{u}_0(1)\,. \tag{4.26}$$

Since $\mathbf{C}_\vartheta = \mathbf{C}_u = 0$, the vectors appearing in (4.26) are

$$\begin{aligned}&\Delta\mathbf{Q}_0 = \mathrm{K}_{11}\mathbf{C}_Q + \mathrm{K}_{12}\mathbf{C}_M\,, \qquad \Delta\mathbf{M}_0 = \mathrm{K}_{21}\mathbf{C}_Q + \mathrm{K}_{22}\mathbf{C}_M\,,\\ &\boldsymbol{\vartheta}_0 = \mathrm{K}_{31}\mathbf{C}_Q + \mathrm{K}_{32}\mathbf{C}_M\,, \qquad \mathbf{u}_0 = \mathrm{K}_{41}\mathbf{C}_Q + \mathrm{K}_{42}\mathbf{C}_M\,.\end{aligned}$$

These relationships are identical in form to (4.22).

Then, from (4.26), we obtain two vector equations

$$
\begin{aligned}
&(\mathrm{K}_{11} + m\lambda^2 \mathrm{K}_{41})\,\mathbf{C}_Q + (\mathrm{K}_{12} + m\lambda^2 \mathrm{K}_{42})\,\mathbf{C}_M = 0\,,\\
&(\mathrm{K}_{21} + \mathrm{J}_m\lambda^2 \mathrm{K}_{31})\,\mathbf{C}_Q + (\mathrm{K}_{22} + \mathrm{J}_m\lambda^2 \mathrm{K}_{32})\,\mathbf{C}_M = 0\,.
\end{aligned}
\tag{4.27}
$$

Finally, we can find the frequencies λ_j from the equation

$$
D = \det \begin{bmatrix} \mathrm{K}_{11} + m\lambda^2 \mathrm{K}_{41} & \mathrm{K}_{12} + m\lambda^2 \mathrm{K}_{42} \\ \mathrm{K}_{21} + \mathrm{J}_m\lambda^2 \mathrm{K}_{31} & \mathrm{K}_{22} + \mathrm{J}_m\lambda^2 \mathrm{K}_{32} \end{bmatrix} = 0\,. \tag{4.28}
$$

Solving these equation, we can obtain the frequencies of vibrations for a particular case when the concentrated mass m is a point mass. In this case, the rotary inertia of this mass should be neglect and thus we have $\mathrm{J}_m = 0$.

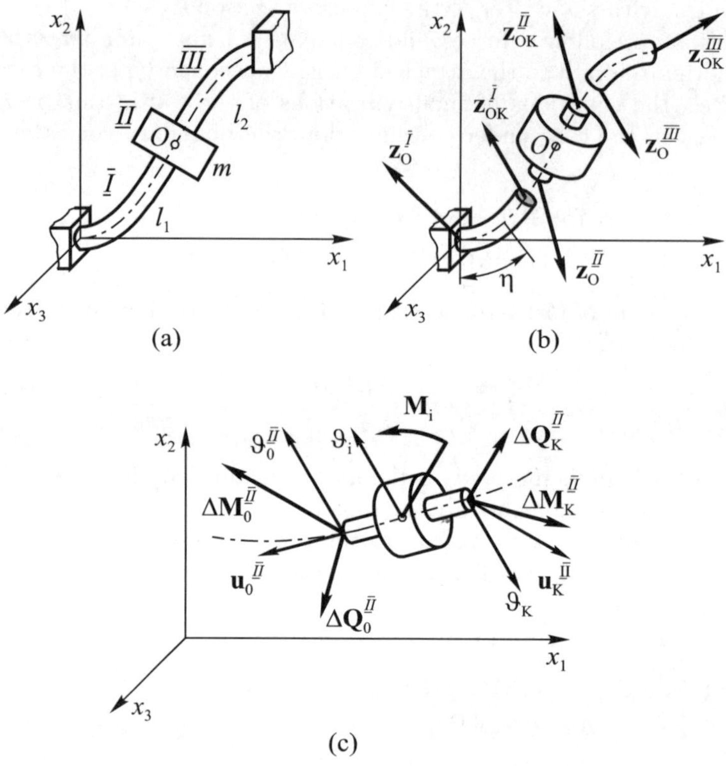

Fig. 4.3.

Assume now that a concentrated mass m is attached to an arbitrary cross section of the rod (Fig. 4.3*a*). Let the rod be divided into three segments, *I*, *II*, and *III* (Fig. 4.3*b*). Let us obtain the transition matrix by which the vectors $\mathbf{Z}_{0\,\mathrm{f}}^{(i)}$ and $\mathbf{Z}_0^{(i)}$, $i = I, II, III$, are related to each other. For the first segment (see (4.14)), we have

$$\mathbf{Z}_0^{\mathrm{I}}(\eta) = \mathrm{K}^{\mathrm{I}}(\eta)\,\mathbf{C}^{\mathrm{I}}\,. \tag{4.29}$$

When solving equation (4.14) numerically, it would be convenient to supply each segment with an individual origin from which the axial coordinates of segment points are measured. The length of the first segment, l_1, can be used in the reduction of the equations to dimensionless form. Therefore, the coordinate η ranges from 0 to 1 for the first segment and from 0 to $l_2/l_1 = \eta_{\mathrm{f}}$ for the third segment. Using (4.29), we obtain the transition matrix for the first segment in the form

$$\mathbf{Z}_{0\,\mathrm{f}}^{\mathrm{I}}(1) = \mathrm{K}^{\mathrm{I}}(1)\,\mathbf{Z}_0^{\mathrm{I}}(0)\,, \qquad \mathrm{A}^{\mathrm{I}} = \mathrm{K}^{\mathrm{I}}(1)\,. \tag{4.30}$$

Let us obtain the matrix that relates the vector $Z_{0\,\mathrm{f}}^{\mathrm{II}}$ to the vector $Z_{0\,\mathrm{f}}^{\mathrm{I}}$. The mass m and all the forces applied to it are shown in Fig. 4.3*c*. The distances of the point O to each of the cross sections of the rod that adjoin the mass m are small and thus are neglected. We have

$$\mathbf{u}_{\mathrm{f}}^{\mathrm{II}} = \mathbf{u}_0^{\mathrm{II}} = \mathbf{u}_{\mathrm{f}}^{\mathrm{I}}\,, \qquad \boldsymbol{\vartheta}_{\mathrm{f}}^{\mathrm{II}} = \boldsymbol{\vartheta}_0^{\mathrm{II}} = \boldsymbol{\vartheta}_{\mathrm{f}}^{\mathrm{I}}\,. \tag{4.31}$$

Using the d'Alembert principle, we get

$$\begin{aligned} &\Delta\mathbf{Q}_{\mathrm{f}}^{\mathrm{II}} + \Delta\mathbf{Q}_0^{\mathrm{II}} + \mathbf{J}_{\mathrm{i}} = 0\,, \qquad \Delta\mathbf{Q}_0^{\mathrm{II}} = -\Delta\mathbf{Q}_{\mathrm{f}}^{\mathrm{I}}\,,\\ &\Delta\mathbf{M}_{\mathrm{f}}^{\mathrm{II}} + \Delta\mathbf{M}_0^{\mathrm{II}} + \mathbf{M}_{\mathrm{i}} = 0\,, \qquad \Delta\mathbf{M}_0^{\mathrm{II}} = -\Delta\mathbf{M}_{\mathrm{f}}^{\mathrm{I}}\,, \end{aligned} \tag{4.32}$$

where

$$\mathbf{J}_{\mathrm{i}} = -m\left.\frac{\partial^2\mathbf{u}^{\mathrm{I}}}{\partial\tau^2}\right|_{\eta=1}, \qquad \mathbf{M}_{\mathrm{i}} = -\mathrm{J}_m\left.\frac{\partial^2\boldsymbol{\vartheta}^{\mathrm{I}}}{\partial\tau^2}\right|_{\eta=1}.$$

Introducing the amplitude values

$$\mathbf{J}_{\mathrm{i}\,0} = m\lambda^2\mathbf{u}_{0\mathrm{f}}^{\mathrm{I}}\,, \qquad \mathbf{M}_{\mathrm{i}\,0} = \mathrm{J}_m\lambda^2\boldsymbol{\vartheta}_{0\,\mathrm{f}}^{\mathrm{I}}\,,$$

we can rewrite (4.31) and (4.32) as follows:

$$\begin{aligned} &\Delta\mathbf{Q}_{0\,\mathrm{f}}^{\mathrm{II}} = \Delta\mathbf{Q}_{0\,\mathrm{f}}^{\mathrm{I}} - m\lambda^2\mathbf{u}_{0\,\mathrm{f}}^{\mathrm{I}}\,, \qquad \Delta\mathbf{M}_{0\,\mathrm{f}}^{\mathrm{II}} = \Delta\mathbf{M}_{0\,\mathrm{f}}^{\mathrm{I}} - \mathrm{J}_m\lambda^2\boldsymbol{\vartheta}_{0\,\mathrm{f}}^{\mathrm{I}}\,,\\ &\boldsymbol{\vartheta}_{0\,\mathrm{f}}^{\mathrm{II}} = \boldsymbol{\vartheta}_{0\,\mathrm{f}}^{\mathrm{I}}\,, \qquad \mathbf{u}_{0\,\mathrm{f}}^{\mathrm{II}} = \mathbf{u}_{0\,\mathrm{f}}^{\mathrm{I}}\,. \end{aligned} \tag{4.33}$$

Finally, we get

$$\mathbf{Z}_{0\,\mathrm{f}}^{\mathrm{II}} = \begin{bmatrix} \mathrm{E} & 0 & 0 & -m\lambda^2\mathrm{E}\\ 0 & \mathrm{E} & -\lambda^2\mathrm{J}_m & 0\\ 0 & 0 & \mathrm{E} & 0\\ 0 & 0 & 0 & \mathrm{E} \end{bmatrix} \mathbf{Z}_{0\,\mathrm{f}}^{\mathrm{I}}\,.$$

Thus, the sought-for transition matrix is

$$\mathrm{A}^{\mathrm{II}} = \begin{bmatrix} \mathrm{E} & 0 & 0 & -m\lambda^2\mathrm{E}\\ 0 & \mathrm{E} & -\lambda^2\mathrm{J}_m & 0\\ 0 & 0 & \mathrm{E} & 0\\ 0 & 0 & 0 & \mathrm{E} \end{bmatrix}. \tag{4.34}$$

The transition matrix for the third segment is identical in form to that for the first segment, that is,

$$\mathrm{A}^{\mathrm{III}} = \mathrm{K}^{\mathrm{III}}(\eta_{\mathrm{f}})\,, \qquad \mathbf{Z}_{0\,\mathrm{f}}^{\mathrm{III}} = \mathrm{K}^{\mathrm{III}}\mathbf{Z}_0^{\mathrm{III}}\,, \tag{4.35}$$

where the vector $\mathbf{Z}_0^{\mathrm{III}}$ is equal to the vector $\mathbf{Z}_{0\,\mathrm{f}}^{\mathrm{II}}$.

Eliminating in sequence the intermediate vectors $\mathbf{Z}_{0\,\mathrm{f}}^{\mathrm{I}}$ and $\mathbf{Z}_{0\,\mathrm{f}}^{\mathrm{II}}$, we get the following relationship between the vectors $\mathbf{Z}_{0\,\mathrm{f}}^{\mathrm{III}}$ and $\mathbf{Z}_0^{\mathrm{I}}$:

$$\mathbf{Z}_{0\,\mathrm{f}}^{\mathrm{III}} = \mathrm{A}^{\mathrm{III}}\mathrm{A}^{\mathrm{II}}\mathrm{A}^{\mathrm{I}}\mathbf{Z}_0^{\mathrm{I}} = \mathrm{A}^{\mathrm{I-III}}\mathbf{Z}_0^{\mathrm{I}}\,, \qquad \mathrm{A}^{\mathrm{I-III}} = \mathrm{A}^{\mathrm{III}}\mathrm{A}^{\mathrm{II}}\mathrm{A}^{\mathrm{I}}\,. \tag{4.36}$$

Combining (4.36) and the appropriate boundary conditions, we obtain the system of homogeneous equations identical to the system (4.20) or (4.28). This system allows us to determine the rod frequencies when the concentrated mass m is taken into account. The boundary conditions for the rod shown in Fig. 4.3 are as follows:

(1) at $\eta = 0\,, \quad \mathbf{u}_0 = 0$ and $\boldsymbol{\vartheta}_0 = 0\,,$
i.e., $\mathbf{Z}_0^{\mathrm{I}} = (\Delta\mathbf{Q}_0(0)\,, \Delta\mathbf{M}_0(0)\,, 0\,, 0)^{\mathrm{T}}\,;$

(2) at $\eta = \eta_{\mathrm{f}}$ (the end of the third segment),
$\mathbf{Z}_{0\,\mathrm{f}}^{\mathrm{III}} = (\Delta\mathbf{Q}_0(\eta_{\mathrm{f}})\,, \Delta\mathbf{M}_0(\eta_{\mathrm{f}})\,, 0\,, 0)^{\mathrm{T}}\,.$

Considering the boundary conditions at the end of the third segment, we get

$$a_{7,1}c_1 + a_{7,2}c_2 + a_{7,3}c_3 + \ldots + a_{7,6}c_6 = 0\,,$$
$$\ldots$$
$$a_{12,1}c_1 + a_{12,2}c_2 + a_{12,3}c_3 + \ldots + a_{12,6}c_6 = 0\,, \tag{4.37}$$

where a_{ij} are the elements of the matrix $\mathrm{A}^{\mathrm{I-III}}$. The frequencies λ_j can be obtained by equating the determinant of the system (4.27) to zero.

In conclusion, consider numerical determination of the frequencies for the helical rod of constant circular cross section shown in Fig. 4.4*a*. There is a concentrated mass m of spherical shape at one of the rod ends. For a rod whose axis is a helical line, the components of the vector æ_0 are

$$\text{æ}_{10} = \frac{\sin 2\alpha}{2R_1^0}\,, \qquad \text{æ}_{20} = 0\,, \qquad \text{æ}_{30} = \frac{\cos^2\alpha}{R_1^0}\,, \qquad R_1^0 = \frac{D}{2l}\,.$$

The initial stress–strain state of the rod caused by the gravity force applied both to the concentrated mass and the rod is disregarded, consequently, we can put $\mathrm{A}_Q = \mathrm{A}_M = 0$. The elements of the matrix B in (4.14) are thus constant. Unlike the previous example (Fig. 4.3), the distance from the center of the concentrated mass O to the end cross section of the rod cannot be neglected. Hence, the boundary conditions need correcting. Figure 4.4*b* illustrates in detail the mass m and the forces applied to it. The condition of balance of forces and moments gives

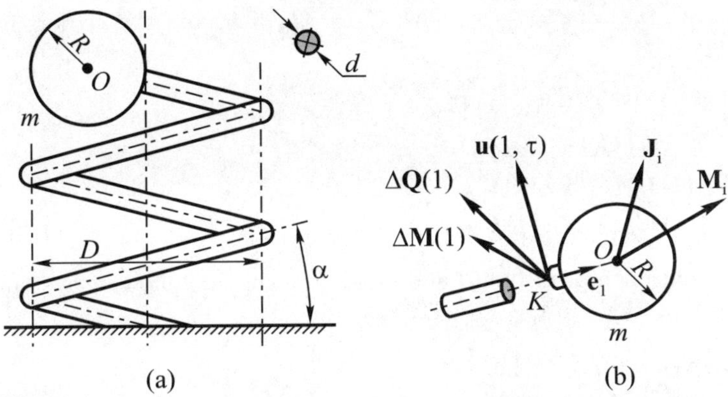

Fig. 4.4.

$$\Delta\mathbf{Q}\,(1) + \mathbf{J}_{\mathrm{i}} = 0\,, \qquad \mathbf{J}_{\mathrm{i}} = -m\,\left.\frac{\partial^2 \mathbf{u}_{01}}{\partial\tau^2}\right|_{\eta=1}, \tag{4.38}$$

$$\Delta\mathbf{M}\,(1) + \mathbf{M}_{\mathrm{i}} + R^0\mathbf{e}_1 \times \Delta\mathbf{Q}\,(1) = 0\,, \qquad \mathbf{M}_{\mathrm{i}} = -\mathrm{J}_m\,\left.\frac{\partial^2 \boldsymbol{\vartheta}}{\partial\tau^2}\right|_{\eta=1}. \tag{4.39}$$

It should be noted that the displacement vector of the point O does not equal to that of the cross section of the rod K (Fig. 4.4*b*), that is,

$$\mathbf{u}_{01} = \mathbf{u}\,(1,\,\tau) + \Delta\mathbf{u}\,(\tau)\,.$$

For small vibrations, we get

$$\Delta\mathbf{u} = \boldsymbol{\vartheta}\,(1,\,\tau) \times R^0\mathbf{e}_1$$

or

$$\Delta\mathbf{u} = -R^0\mathrm{A}_1\boldsymbol{\vartheta}\,(1,\,\tau)\,.$$

The boundary-condition equation (4.38) takes the form

$$\Delta\mathbf{Q}\,(1) = m\,\left.\frac{\partial^2 \mathbf{u}}{\partial\tau^2}\right|_{\eta=1} - R^0 m\mathrm{A}_1\,\left.\frac{\partial^2 \boldsymbol{\vartheta}}{\partial\tau^2}\right|_{\eta=1}.$$

Using the amplitude values in (4.38) and (4.39), we get

$$\Delta\mathbf{Q}_0(1) = -m\lambda^2\mathbf{u}_0(1) + mR^0\mathrm{A}_1\lambda^2\boldsymbol{\vartheta}_0(1)\,, \tag{4.40}$$

$$\Delta\mathbf{M}_0(1) = -\mathrm{J}_m\lambda^2\boldsymbol{\vartheta}_0(1) - R^0\mathrm{A}_1\,\Delta\mathbf{Q}_0(1)\,. \tag{4.41}$$

The solution of (4.14) can be written as

$$\mathbf{Z}_0 = \begin{bmatrix} \Delta\mathbf{Q}_0 \\ \Delta\mathbf{M}_0 \\ \boldsymbol{\vartheta}_0 \\ \mathbf{u}_0 \end{bmatrix} = \begin{bmatrix} \mathrm{K}_{11} & \mathrm{K}_{12} & \mathrm{K}_{13} & \mathrm{K}_{14} \\ \mathrm{K}_{21} & \mathrm{K}_{22} & \mathrm{K}_{23} & \mathrm{K}_{24} \\ \mathrm{K}_{31} & \mathrm{K}_{32} & \mathrm{K}_{33} & \mathrm{K}_{34} \\ \mathrm{K}_{41} & \mathrm{K}_{42} & \mathrm{K}_{43} & \mathrm{K}_{44} \end{bmatrix} \cdot \begin{bmatrix} \mathbf{C}_Q \\ \mathbf{C}_M \\ \mathbf{C}_\vartheta \\ \mathbf{C}_u \end{bmatrix}. \tag{4.42}$$

Since, at $\eta = 0$, $\boldsymbol{\vartheta}_0 = \mathbf{u}_0 = 0$, we have $\mathbf{C}_\vartheta = \mathbf{C}_u = 0$, and the vectors in (4.40) and (4.41) read

$$
\begin{aligned}
\Delta\mathbf{Q}_0(1) &= \mathrm{K}_{11}(1)\,\mathbf{C}_Q + \mathrm{K}_{12}(1)\,\mathbf{C}_M\,,\\
\Delta\mathbf{M}_0(1) &= \mathrm{K}_{21}(1)\,\mathbf{C}_Q + \mathrm{K}_{22}(1)\,\mathbf{C}_M\,,\\
\boldsymbol{\vartheta}_0(1) &= \mathrm{K}_{31}(1)\,\mathbf{C}_Q + \mathrm{K}_{32}(1)\,\mathbf{C}_M\,,\\
\mathbf{u}_0(1) &= \mathrm{K}_{41}(1)\,\mathbf{C}_Q + \mathrm{K}_{42}(1)\,\mathbf{C}_M\,.
\end{aligned}
\tag{4.43}
$$

Using these relationships in (4.40) and (4.41), we get the following system of homogeneous equations in $\mathbf{C}_Q$ and $\mathbf{C}_M$:

$$
\begin{aligned}
&(\mathrm{K}_{11} + m\lambda^2\mathrm{K}_{41} + mR^0\lambda^2\mathrm{A}_1\mathrm{K}_{31})\,\mathbf{C}_Q\\
&\qquad + (\mathrm{K}_{12} + m\lambda^2\mathrm{K}_{42} + mR^0\lambda^2\mathrm{A}_1\mathrm{K}_{32})\,\mathbf{C}_M = 0\,,\\
&(\mathrm{K}_{21} + \mathrm{J}_m\lambda^2\mathrm{K}_{31} + R^0\mathrm{A}_1\mathrm{K}_{11})\,\mathbf{C}_Q\\
&\qquad + (\mathrm{K}_{22} + \mathrm{J}_m\lambda^2\mathrm{K}_{32} + R^0\mathrm{A}_1\mathrm{K}_{12})\,\mathbf{C}_M = 0\,.
\end{aligned}
\tag{4.44}
$$

The eigenvalues of the boundary value problem can be found from the equation

$$
D = \det\,[d_{ij}] = 0\,, \qquad i,\, j = 1,\, 2\,,
\tag{4.45}
$$

where

$$
\begin{aligned}
d_{11} &= \mathrm{K}_{11} + m\lambda^2\mathrm{K}_{41} + mR^0\lambda^2\mathrm{A}_1\mathrm{K}_{31}\,,\\
d_{12} &= \mathrm{K}_{12} + m\lambda^2\mathrm{K}_{42} + mR^0\lambda^2\mathrm{A}_1\mathrm{K}_{32}\,,\\
d_{21} &= \mathrm{K}_{21} + \mathrm{J}_m\lambda^2\mathrm{K}_{31} + R^0\mathrm{A}_1\mathrm{K}_{11}\,,\\
d_{22} &= \mathrm{K}_{22} + \mathrm{J}_m\lambda^2\mathrm{K}_{32} + R^0\mathrm{A}_1\mathrm{K}_{12}\,.
\end{aligned}
$$

To discuss a numerical example, let us first consider in detail the matrix J_m in (4.39) and the matrices A and J. The matrices A and J are elements of the matrix B. The dimensionless elements of the matrices J and A are

$$
\begin{aligned}
&\tilde{J}_{ii} = \frac{J_{ii}}{F_0 l^2}\,, \qquad \tilde{A}_{ii} = \frac{A_{ii}}{A_{33}(0)}\,, \qquad J_{11} = \frac{\pi d^4}{32}\,, \qquad J_{22} = J_{33} = \frac{\pi d^4}{64}\,,\\
&\tilde{A}_{11} = \frac{GJ}{EJ_{33}} = 2\,\frac{G}{E}\,, \qquad \tilde{A}_{22} = \tilde{A}_{33} = 1\,,\\
&\tilde{J}_{11} = \frac{4\pi d^4}{32\pi d^2 l^2} = \frac{d^2}{8l^2}\,, \qquad \tilde{J}_{22} = \tilde{J}_{33} = \frac{d^2}{16l^2}\,,
\end{aligned}
$$

where F_0 is the cross-sectional area of the rod.

The dimensional elements of J_m are

$$
(J_m)_{ii} = \frac{2}{5}\,mR^2\,, \qquad m = \frac{4}{5}\,\pi R^3\gamma
$$

and dimensionless elements of J_m are

$$(\tilde{J}_m)_{ii} = \frac{(\mathrm{J}_m)_{ii}}{\gamma F_0 l^3} = \frac{32}{15}\,(R^0)^3 \left(\frac{R}{d}\right)^2 , \qquad R^0 = \frac{R}{l} .$$

The dimensionless mass of the solid sphere reads

$$\tilde{m} = \frac{m}{m_0 l} = \frac{4\pi R^3 \gamma 4}{3\pi d^2 l \gamma} = \frac{16}{3} R^0 \left(\frac{R}{d}\right)^2 .$$

As an example, let us determine the frequencies of a spring. The values of the dimensional parameters are taken as follows:

$$l = 50\,\mathrm{cm}\,, \qquad R_1 = 10\,\mathrm{cm}\,, \qquad d = 1\,\mathrm{cm}\,,$$
$$R = 5\,\mathrm{cm}\,, \qquad D = 2R_1\,.$$

Then, the values of the dimensionless parameters are:

$$R_1^0 = \frac{D}{2l} = 0.2\,, \qquad \frac{R}{d} = 5.0, \qquad R^0 = \frac{R}{l} = 0.1\,.$$

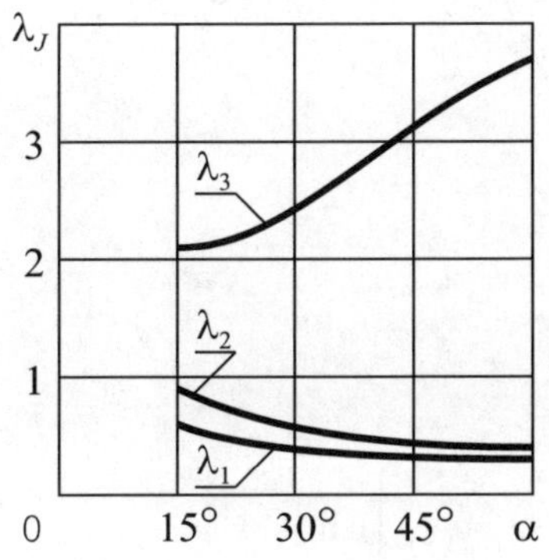

Fig. 4.5.

The first three frequencies are plotted versus the angle of helix at $\alpha = 15°$, $\alpha = 30°$, and $\alpha = 45°$ in Fig. 4.5.

4.1.3 Special Cases

In-Plane Vibrations of a Rod Consider small vibrations of a rod whose axis is a plane curve. We assume that the rod belongs to this plane during the vibrations.

The equations of small vibrations for this case were obtained in Sect. 3.2 (see (3.68)). Recall that these equations were written in terms of projections in the attached basis. Eliminating $\Delta æ_3$ from (3.68) and setting

$$u_j = u_{j0}(\eta)\,\mathrm{e}^{\mathrm{i}\lambda\tau}\,, \qquad \Delta Q_j = \Delta Q_j(\eta)\,\mathrm{e}^{\mathrm{i}\lambda\tau}\,, \qquad j = 1, 2\,,$$
$$\Delta M_3 = \Delta M_{30}(\eta)\,\mathrm{e}^{\mathrm{i}\lambda\tau}\,, \qquad \vartheta_3 = \vartheta_{30}(\eta)\,\mathrm{e}^{\mathrm{i}\lambda\tau}\,, \tag{4.46}$$

we get

$$\begin{aligned}
&\frac{\mathrm{d}\,\Delta Q_1}{\mathrm{d}\eta} - æ_{30}\,\Delta Q_2 - \frac{Q_{20}}{A_{33}}\,\Delta M_3 + n_1\lambda^2 u_{10} = 0\,,\\
&\frac{\mathrm{d}\,\Delta Q_2}{\mathrm{d}\eta} + æ_{30}\,\Delta Q_1 + \frac{Q_{10}}{A_{33}}\,\Delta M_3 + n_1\lambda^2 u_{20} = 0\,,\\
&\frac{\mathrm{d}\,\Delta M_3}{\mathrm{d}\eta} + \Delta Q_2 + J_{33}\lambda^2\vartheta_{30} = 0\,,\\
&\frac{\mathrm{d}\vartheta_{30}}{\mathrm{d}\eta} - \frac{1}{A_{33}}\,\Delta M_3 = 0\,,\\
&\frac{\mathrm{d}u_{10}}{\mathrm{d}\eta} - æ_{30}u_{20} = 0\,,\\
&\frac{\mathrm{d}u_{20}}{\mathrm{d}\eta} + æ_{30}u_{10} - \vartheta_{30} = 0\,.
\end{aligned} \tag{4.47}$$

In vector–matrix form, we have

$$\mathbf{Z}_0' + \mathrm{B}\,\mathbf{Z}_0 = 0\,, \tag{4.48}$$

where

$$\mathbf{Z}_0 = \begin{bmatrix} \Delta Q_{10} \\ \Delta Q_{20} \\ \Delta M_{30} \\ \vartheta_{30} \\ u_{10} \\ u_{20} \end{bmatrix},$$

$$\mathrm{B} = \begin{bmatrix} 0 & -æ_{30} & -\dfrac{Q_{20}}{A_{33}} & 0 & n_1\lambda^2 & 0 \\ æ_{30} & 0 & \dfrac{Q_{10}}{A_{33}} & 0 & 0 & n_1\lambda^2 \\ 0 & 1 & 0 & J_{33}\lambda^2 & 0 & 0 \\ 0 & 0 & -\dfrac{1}{A_{33}} & 0 & 0 & 0 \\ 0 & 0 & 0 & 0 & 0 & -æ_{30} \\ 0 & 0 & 0 & -1 & æ_{30} & 0 \end{bmatrix}.$$

Recall that a prime is used to denote differentiation with respect to η.

Equation (4.48) includes Q_{10}, Q_{20}, and $æ_{30}$, and thus accounts for the static stress–strain state of the rod.

The solution of (4.48) is $\mathbf{Z}_0 = \mathrm{K}(\eta)\mathbf{C}$; it must satisfy six boundary-condition equations (here, we have three boundary-condition equations at each end). Consider some examples of consistent boundary conditions.

Consider an unloaded rod such that

$$Q_{10} = Q_{20} = M_{30} = 0\,.$$

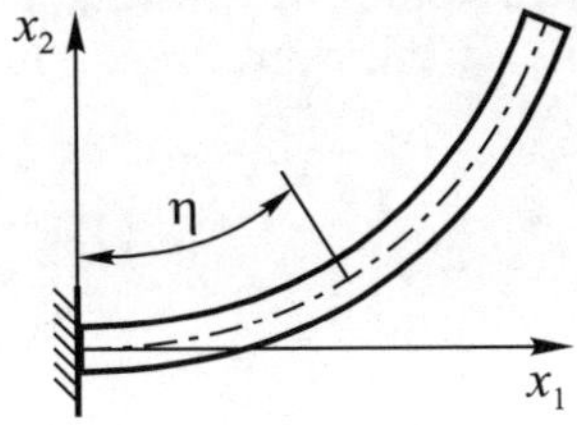

Fig. 4.6.

For the rod clamped at one end (Fig. 4.6), we have

(1) at $\eta = 0\,,\quad u_{10} = u_{20} = 0$ and $\vartheta_{30} = 0\,;$

(2) at $\eta = 1\,,\quad \Delta M_{30} = 0$ and $\Delta Q_{10} = \Delta Q_{20} = 0\,.$

Since it is assumed that the matrix $\mathrm{K}(\eta)$ for $\eta = 0$ is the identity matrix, three components of the vector $\mathbf{C}$ equal zero, namely, $c_4 = c_5 = c_6 = 0$. The boundary conditions at $\eta = 1$ give the relationships

$$
\begin{aligned}
&k_{11}c_1 + k_{12}c_2 + k_{13}c_3 = 0\,,\\
&k_{21}c_1 + k_{22}c_2 + k_{23}c_3 = 0\,,\\
&k_{31}c_1 + k_{32}c_2 + k_{33}c_3 = 0\,.
\end{aligned}
\tag{4.49}
$$

Equating the determinant of this system to zero, we get

$$
D = \det \begin{bmatrix} k_{11} & k_{12} & k_{13} \\ k_{21} & k_{22} & k_{23} \\ k_{31} & k_{32} & k_{33} \end{bmatrix} = 0\,.
$$

This equation can be solved for the frequencies λ_j.

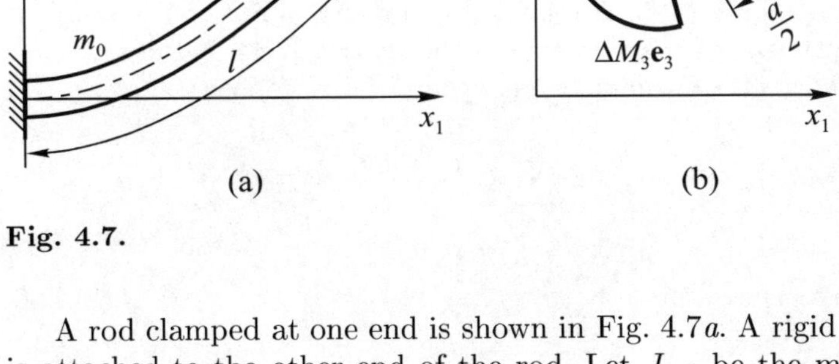

Fig. 4.7.

A rod clamped at one end is shown in Fig. 4.7*a*. A rigid body of mass m is attached to the other end of the rod. Let J_{m0} be the moment of inertia of the body about the central axis perpendicular to the plane of drawing at the point O. It follows from the boundary conditions at the right-hand end of the rod in Fig. 4.7b that

$$\Delta\mathbf{Q}\,(1) + \mathbf{J}_{\mathrm{i}} = 0\,, \tag{4.50}$$

$$\Delta\mathbf{M}_3(1)\,\mathbf{e}_3 + \mathbf{M}_{\mathrm{i}} + \frac{a}{2l}\,\mathbf{e}_1 \times \Delta\mathbf{Q}\,(1) = 0\,, \tag{4.51}$$

where

$$\mathbf{J}_{\mathrm{i}} = -m\,\frac{\partial^2\mathbf{u}_{01}}{\partial\tau^2}\,, \qquad \mathbf{M}_{\mathrm{i}} = -J_{m0}\,\frac{\partial^2\boldsymbol{\vartheta}_3}{\partial\tau^2}\,\mathbf{e}_3\,.$$

Here, m and J_{m0} are dimensionless quantities. The displacement vector of the point O enters in the expression for $\mathbf{J}_{\mathrm{i}}$; it reads

$$\mathbf{u}_{01} = \mathbf{u}(1) + \Delta\mathbf{u}\,,$$

where

$$\Delta\mathbf{u} = \vartheta_3\mathbf{e}_3 \times \frac{a}{2l}\,\mathbf{e}_1 = \frac{a}{2l}\,\vartheta_3\mathbf{e}_2\,.$$

Setting

$$\Delta\mathbf{Q} = \Delta\mathbf{Q}_0\,\mathrm{e}^{\mathrm{i}\lambda\tau}\,, \qquad \Delta M_3 = \Delta M_{30}\,\mathrm{e}^{\mathrm{i}\lambda\tau}\,,$$
$$\mathbf{u} = \mathbf{u}_0\,\mathrm{e}^{\mathrm{i}\lambda\tau}\,, \qquad \boldsymbol{\vartheta}_3 = \boldsymbol{\vartheta}_{30}\,\mathrm{e}^{\mathrm{i}\lambda\tau}\,,$$

and using (4.50) and (4.51), we arrive at

$$\Delta\mathbf{Q}_0(1) + m\lambda^2\mathbf{u}_0(1) + \frac{a}{2l}\,m\lambda^2\vartheta_{30}(1)\,\mathbf{e}_2 = 0\,, \tag{4.52}$$

$$\Delta M_{30}(1) + J_{m0}\lambda^2\vartheta_{30}(1) + \frac{a}{2l}\,\Delta Q_{20}(1) = 0\,. \tag{4.53}$$

Equation (4.52) yields two scalar relationships

$$\Delta Q_{10} + m\lambda^2 u_{10} = 0\,, \tag{4.54}$$

$$\Delta Q_{20} + m\lambda^2 u_{20} + \frac{a}{2l}\,m\lambda^2\vartheta_{30} = 0\,. \tag{4.55}$$

The quantities u_{10}, u_{20}, ϑ_{30}, ΔM_{30}, ΔQ_{10}, and ΔQ_{20} in (4.53)–(4.55) can be represented in terms of the elements of the matrix $\mathrm{K}(1)$ and of the other components of the vector $\mathbf{C}$, namely, c_1, c_2, and c_3, as follows:

$$\Delta Q_{10} = \sum_{j=1}^{3} k_{1j}c_j\,, \qquad \Delta Q_{20} = \sum_{j=1}^{3} k_{2j}c_j\,, \qquad \Delta M_{30} = \sum_{j=1}^{3} k_{3j}c_j\,,$$

$$\vartheta_{30} = \sum_{j=1}^{3} k_{4j}c_j\,, \qquad u_{10} = \sum_{j=1}^{3} k_{5j}c_j\,, \qquad u_{20} = \sum_{j=1}^{3} k_{6j}c_j\,. \tag{4.56}$$

Substituting (4.56) into (4.53)–(4.55), we get three homogeneous equations in c_1, c_2, and c_3:

$$
\begin{aligned}
&b_{11}c_1 + b_{12}c_2 + b_{13}c_3 = 0\,,\\
&b_{21}c_1 + b_{22}c_2 + b_{23}c_3 = 0\,,\\
&b_{31}c_1 + b_{32}c_2 + b_{33}c_3 = 0\,,
\end{aligned}
$$

Finally, we can find the frequencies λ_j from the condition

$$
D = \det \begin{bmatrix} b_{11} & b_{12} & b_{13} \\ b_{21} & b_{22} & b_{23} \\ b_{31} & b_{32} & b_{33} \end{bmatrix} = 0\,.
$$

Out-of-Plane Vibrations of a Rod Let us consider the system of equations (3.69). Setting

$$
\begin{aligned}
&\Delta Q_3 = \Delta Q_{30}\,\mathrm{e}^{\mathrm{i}\lambda\tau}\,, && \Delta M_1 = \Delta M_{10}\,\mathrm{e}^{\mathrm{i}\lambda\tau}\,, && \Delta M_2 = \Delta M_{20}\,\mathrm{e}^{\mathrm{i}\lambda\tau}\,,\\
&\boldsymbol{\vartheta}_1 = \boldsymbol{\vartheta}_{10}\,\mathrm{e}^{\mathrm{i}\lambda\tau}\,, && \boldsymbol{\vartheta}_2 = \boldsymbol{\vartheta}_{20}\,\mathrm{e}^{\mathrm{i}\lambda\tau}\,, && u_3 = u_{30}\,\mathrm{e}^{\mathrm{i}\lambda\tau}\,,
\end{aligned}
$$

and eliminating $\Delta æ_1$ and $\Delta æ_2$, we arrive at

$$
\begin{aligned}
&\frac{\mathrm{d}\,\Delta Q_{30}}{\mathrm{d}\eta} + \frac{Q_{20}}{A_{11}}\,\Delta M_{10} - \frac{Q_{10}}{A_{22}}\,\Delta M_{20} + n_1\lambda^2 u_{30} = 0\,,\\
&\frac{\mathrm{d}\,\Delta M_{10}}{\mathrm{d}\eta} + \left(\frac{M_{30}}{A_{22}} - æ_{30}\right)\Delta M_{20} + J_{11}\lambda^2\vartheta_{10} = 0\,,\\
&\frac{\mathrm{d}\,\Delta M_{20}}{\mathrm{d}\eta} - \Delta Q_3 + \left(æ_{30} - \frac{M_{30}}{A_{11}}\right)\Delta M_{10} + J_{22}\lambda^2\vartheta_{20} = 0\,,\\
&\frac{\mathrm{d}\vartheta_{10}}{\mathrm{d}\eta} - \frac{1}{A_{11}}\,\Delta M_{10} - æ_{30}\vartheta_{20} = 0\,,\\
&\frac{\mathrm{d}\vartheta_{20}}{\mathrm{d}\eta} - \frac{1}{A_{22}}\,\Delta M_{20} + æ_{30}\vartheta_{10} = 0\,,\\
&\frac{\mathrm{d}u_{30}}{\mathrm{d}\eta} + \vartheta_{20} = 0\,,
\end{aligned}
\tag{4.57}
$$

In vector–matrix form, we have

$$
\mathbf{Z}_0' + \mathrm{B}\,\mathbf{Z}_0 = 0\,,
$$

where

$$
\mathbf{Z}_0 = \begin{bmatrix} \Delta Q_{30} \\ \Delta M_{10} \\ \Delta M_{20} \\ \vartheta_{10} \\ \vartheta_{20} \\ u_{30} \end{bmatrix},
$$

$$
\mathrm{B} = \begin{bmatrix}
0 & \dfrac{Q_{20}}{A_{11}} & -\dfrac{Q_{10}}{A_{22}} & 0 & 0 & n_1\lambda^2 \\
0 & 0 & \dfrac{M_{30}}{A_{22}} - æ_{10} & J_{11}\lambda^2 & 0 & 0 \\
-1 & æ_{10} - \dfrac{M_{30}}{A_{11}} & 0 & 0 & J_{22}\lambda^2 & 0 \\
0 & -\dfrac{1}{A_{11}} & 0 & 0 & -æ_{30} & 0 \\
0 & 0 & -\dfrac{1}{A_{22}} & æ_{30} & 0 & 0 \\
0 & 0 & 0 & 0 & 1 & 0
\end{bmatrix}.
$$

4.1.4 Methods Based on the Theory of Distributions

If we deal with a rod with concentrated masses, intermediate supports and composed of segments of different rigidities, then the method of initial parameters discussed above implies multiplication of a considerable amount of transition matrices. This may cause some difficulties of computational nature.

Let us consider a method that does not imply multiplication of matrices when determining numerically frequencies of rods with lateral supports and concentrated masses. This method was applied to some problems of statics of rods (Sect. 2.2 of Svetlitsky (2000a)).

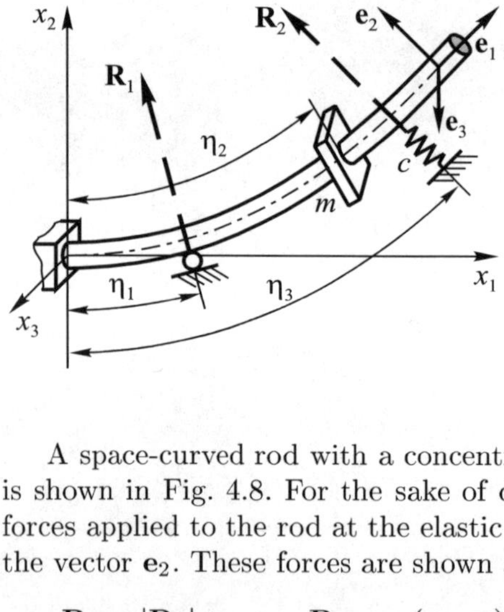

Fig. 4.8.

A space-curved rod with a concentrated mass and intermediate supports is shown in Fig. 4.8. For the sake of definiteness, assume that the reaction forces applied to the rod at the elastic and hinge supports are collinear with the vector $\mathbf{e}_2$. These forces are shown as a dashed line in Fig. 4.8. We have

$$\mathbf{R}_1 = |\mathbf{R}_1|\,\mathbf{e}_2\,, \qquad \mathbf{R}_2 = -(\mathbf{u}\cdot\mathbf{e}_2)\,c\mathbf{e}_2 = -u_2 c\,\mathbf{e}_2\,.$$

In the general case, the reaction forces can be of arbitrary but specified direction defined by the unit vectors $\mathbf{e}_{R_1}$ and $\mathbf{e}_{R_2}$. The coordinates of these vectors in the attached basis are assumed to be known. Let

$$\mathbf{e}_{R_j} = \sum_{j=1}^{3} a_j \mathbf{e}_j \,.$$

Therefore, the reaction forces with respect to the attached basis read

$$\mathbf{R}_1 = |\mathbf{R}_1|\,\mathbf{e}_{R_1} = \sum_{j=1}^{3} |\mathbf{R}_1|\,a_j \mathbf{e}_j = \sum_{i=1}^{3} R_i \mathbf{e}_i \,, \tag{4.58}$$

$$\mathbf{R}_2 = -c\,(\mathbf{u}(\eta_2)\cdot\mathbf{e}_{R_2})\mathbf{e}_{R_2} = -\tilde{\mathrm{A}}\mathbf{u}(\eta_2)\,, \tag{4.59}$$

where

$$\tilde{\mathrm{A}} = \begin{bmatrix} ca_1^2 & ca_1a_2 & ca_1a_3 \\ ca_2a_1 & ca_2^2 & ca_2a_3 \\ ca_3a_1 & ca_3a_2 & ca_3^2 \end{bmatrix} .$$

Eliminating $\Delta æ$ from (3.11)–(3.14) and taking into consideration the concentrated forces $\mathbf{R}_1$ and $\mathbf{R}_2$ and the inertia of the concentrated mass m, we get

$$n_1 \frac{\partial^2 \mathbf{u}}{\partial \tau^2} - \frac{\partial \Delta\mathbf{Q}}{\partial \eta} - \mathrm{A}_{æ}\,\Delta\mathbf{Q} - \mathrm{A}_Q\mathrm{A}^{-1}\,\Delta\mathbf{M} = \mathbf{R}_1\,\delta(\eta-\eta_1) - m\,\frac{\partial^2 \mathbf{u}}{\partial \tau^2}\,\delta(\eta-\eta_2) + \mathbf{R}_2\,\delta(\eta-\eta_3)\,, \tag{4.60}$$

$$\mathrm{J}\,\frac{\partial^2 \boldsymbol{\vartheta}}{\partial \tau^2} - \frac{\partial \Delta\mathbf{M}}{\partial \eta} - \mathrm{A}_1\,\Delta\mathbf{Q} - (\mathrm{A}_M\mathrm{A}^{-1} + \mathrm{A}_{æ})\,\Delta\mathbf{M} = -\mathrm{J}_m\,\frac{\partial^2 \boldsymbol{\vartheta}}{\partial \tau^2}\,\delta(\eta-\eta_2)\,, \tag{4.61}$$

$$\frac{\partial \boldsymbol{\vartheta}}{\partial \eta} + \mathrm{A}_{æ}\boldsymbol{\vartheta} - \mathrm{A}^{-1}\,\Delta\mathbf{M} = 0\,, \tag{4.62}$$

$$\frac{\partial \mathbf{u}}{\partial \eta} + \mathrm{A}_{æ}\mathbf{u} + \mathrm{A}_1\boldsymbol{\vartheta} = 0\,. \tag{4.63}$$

Here, $\delta(\eta)$ is the Dirac delta-function.

Setting

$$\mathbf{u} = \mathbf{u}_0(\eta)\,\mathrm{e}^{\mathrm{i}\lambda\tau}\,, \qquad \Delta\mathbf{Q} = \Delta\mathbf{Q}_0(\eta)\,\mathrm{e}^{\mathrm{i}\lambda\tau}\,, \qquad \Delta\mathbf{M} = \Delta\mathbf{M}_0(\eta)\,\mathrm{e}^{\mathrm{i}\lambda\tau}\,,$$
$$\boldsymbol{\vartheta} = \boldsymbol{\vartheta}_0(\eta)\,\mathrm{e}^{\mathrm{i}\lambda\tau}\,, \qquad \mathbf{R}_1 = \mathbf{R}_{10}(\eta)\,\mathrm{e}^{\mathrm{i}\lambda\tau}\,, \qquad \mathbf{R}_2 = \mathbf{R}_{20}(\eta)\,\mathrm{e}^{\mathrm{i}\lambda\tau}\,,$$

and assuming $\mathbf{R}_{20} = -cu_{20}\mathbf{e}_2$, we obtain the following vector equations:

$$\mathbf{Z}_0' + \mathrm{B}\,\mathbf{Z}_0 = \Delta\boldsymbol{\Phi}_0\,; \tag{4.64}$$

here,

$$\mathbf{Z}_0 = \begin{bmatrix} \Delta\mathbf{Q}_0 \\ \Delta\mathbf{M}_0 \\ \boldsymbol{\vartheta}_0 \\ \mathbf{u}_0 \end{bmatrix}, \qquad \mathrm{B} = \begin{bmatrix} \mathrm{A}_{æ} & \mathrm{A}_Q\mathrm{A}^{-1} & 0 & 0 \\ \mathrm{A}_1 & \mathrm{A}_M\mathrm{A}^{-1} + \mathrm{A}_{æ} & 0 & 0 \\ 0 & -\mathrm{A}^{-1} & \mathrm{A}_{æ} & 0 \\ 0 & 0 & \mathrm{A}_1 & \mathrm{A}_{æ} \end{bmatrix},$$

$$\Delta\boldsymbol{\Phi}_0 = \begin{bmatrix} -\mathbf{R}_1\,\delta(\eta-\eta_1) - m\lambda^2\mathbf{u}_0\,\delta(\eta-\eta_2) + u_{20}c\mathbf{e}_2\,\delta(\eta-\eta_3) \\ -\mathrm{J}_m\lambda^2\boldsymbol{\vartheta}_0\,\delta(\eta-\eta_2) \\ \mathbf{0} \\ \mathbf{0} \end{bmatrix}.$$

The solution of (4.64) is

$$\mathbf{Z}_0 = \mathrm{K}(\eta)\,\mathbf{C} + \int_0^\eta \mathrm{G}(\eta,\,\zeta)\,\boldsymbol{\Phi}_1\,\delta(\zeta-\eta_1)\,\mathrm{d}\zeta + \int_0^\eta \mathrm{G}(\eta,\,\zeta)\,\boldsymbol{\Phi}_2\,\delta(\zeta-\eta_2)\,\mathrm{d}\zeta + \int_0^\eta \mathrm{G}(\eta,\,\zeta)\,\boldsymbol{\Phi}_3\,\delta(\zeta-\eta_3)\,\mathrm{d}\zeta\,, \tag{4.65}$$

where

$$\mathrm{G}(\eta,\,\zeta) = \mathrm{K}(\eta)\,\mathrm{K}^{-1}(\zeta)\,,$$

$$\boldsymbol{\Phi}_1 = \begin{bmatrix} -\mathbf{R}_{10} \\ \mathbf{0} \\ \mathbf{0} \\ \mathbf{0} \end{bmatrix}, \qquad \boldsymbol{\Phi}_2 = \begin{bmatrix} -m\lambda^2\mathbf{u}_0(\eta_2) \\ -\mathrm{J}_m\lambda^2\boldsymbol{\vartheta}_0(\eta_2) \\ \mathbf{0} \\ \mathbf{0} \end{bmatrix},$$

$$\boldsymbol{\Phi}_3 = \begin{bmatrix} u_{20}(\eta_3)\,c\mathbf{e}_2 \\ \mathbf{0} \\ \mathbf{0} \\ \mathbf{0} \end{bmatrix}.$$

Performing the integration in the right-hand side of (4.65), we arrive at

$$\mathbf{Z}_0 = \mathrm{K}(\eta)\,\mathbf{C} + \mathrm{G}^{(1)}(\eta,\,\eta_1)\,\boldsymbol{\Phi}_1(\eta_1)\,H(\eta-\eta_1) + \mathrm{G}^{(2)}(\eta,\,\eta_2)\,\boldsymbol{\Phi}_2(\eta_2)\,H(\eta-\eta_2) + \mathrm{G}^{(3)}(\eta,\,\eta_3)\,\boldsymbol{\Phi}_3(\eta_3)\,H(\eta-\eta_3)\,. \tag{4.66}$$

Recall that $H(\eta)$ is the Heaviside function.

Let us represent the matrices $\mathrm{K}(\eta)$ and $\mathrm{G}^{(i)}(\eta,\,\eta_1)$ and the vector $\mathbf{C}$ as follows:

$$\mathrm{K}(\eta) = \begin{bmatrix} \mathrm{K}_{11} & \mathrm{K}_{12} & \mathrm{K}_{13} & \mathrm{K}_{14} \\ \mathrm{K}_{21} & \mathrm{K}_{22} & \mathrm{K}_{23} & \mathrm{K}_{24} \\ \mathrm{K}_{31} & \mathrm{K}_{32} & \mathrm{K}_{33} & \mathrm{K}_{34} \\ \mathrm{K}_{41} & \mathrm{K}_{42} & \mathrm{K}_{43} & \mathrm{K}_{44} \end{bmatrix}, \qquad \mathbf{C} = \begin{bmatrix} \mathbf{C}_Q \\ \mathbf{C}_M \\ \mathbf{C}_\vartheta \\ \mathbf{C}_u \end{bmatrix},$$

$$\mathrm{G}^{(i)} = \begin{bmatrix} \mathrm{G}_{11}^{(i)} & \mathrm{G}_{12}^{(i)} & \mathrm{G}_{13}^{(i)} & \mathrm{G}_{14}^{(i)} \\ \mathrm{G}_{21}^{(i)} & \mathrm{G}_{22}^{(i)} & \mathrm{G}_{23}^{(i)} & \mathrm{G}_{24}^{(i)} \\ \mathrm{G}_{31}^{(i)} & \mathrm{G}_{32}^{(i)} & \mathrm{G}_{33}^{(i)} & \mathrm{G}_{34}^{(i)} \\ \mathrm{G}_{41}^{(i)} & \mathrm{G}_{42}^{(i)} & \mathrm{G}_{43}^{(i)} & \mathrm{G}_{44}^{(i)} \end{bmatrix}; \tag{4.67}$$

here, K_{ij} and $\mathrm{G}_{mn}^{(i)}$ are 3×3 matrices and the vectors $\mathbf{C}_Q$, $\mathbf{C}_M$, $\mathbf{C}_\vartheta$, and $\mathbf{C}_u$ have three components.

For the rod shown in Fig. 4.8, we have the following boundary conditions.

(1) At $\eta = 0$,

$$\boldsymbol{\vartheta}_0(0) = 0 \quad \text{and} \quad \mathbf{u}_0(0) = 0\,; \quad \text{hence,} \quad \mathbf{C}_\vartheta = \mathbf{C}_u = 0\,.$$

(2) At $\eta = \eta_1$,

$$(\mathbf{u}_0 \cdot \mathbf{e}_2) = 0 \quad \text{or}$$

$$(\mathbf{u}_0 \cdot \mathbf{e}_2) = \sum_{i=1}^{3} k_{11,\,i} C_{Qi} + \sum_{i=1}^{3} k_{11,\,(3+i)} C_{Mi} = 0\,. \tag{4.68}$$

(3) At $\eta = \eta_2$,

$$\boldsymbol{\vartheta}_0(\eta_2) = \mathrm{K}_{31}(\eta_2)\,\mathbf{C}_Q + \mathrm{K}_{32}(\eta_2)\,\mathbf{C}_M - \mathrm{G}_{31}^{(1)}(\eta_2,\,\eta_1)\,\mathbf{R}_{10}\,, \tag{4.69}$$

$$\mathbf{u}_0(\eta_2) = \mathrm{K}_{41}(\eta_2)\,\mathbf{C}_Q + \mathrm{K}_{42}(\eta_2)\,\mathbf{C}_M - \mathrm{G}_{41}^{(1)}(\eta_2,\,\eta_1)\,\mathbf{R}_{10}\,. \tag{4.70}$$

(4) At $\eta = \eta_3$,

$$u_{20}(\eta_3) = \Big[\mathrm{K}_{11}(\eta_3)\,\mathbf{C}_Q + \mathrm{K}_{42}(\eta_3)\,\mathbf{C}_M - \mathrm{G}_{41}^{(1)}(\eta_3,\,\eta_1)\,\mathbf{R}_{10} - \mathrm{G}_{41}^{(2)}(\eta_3,\,\eta_2)\,m\lambda^2\mathbf{u}_0(\eta_2) + \mathrm{G}_{42}^{(2)}(\eta_3,\,\eta_2)\,\mathrm{J}_m\lambda^2\boldsymbol{\vartheta}_0(\eta_2)\Big]\,\mathbf{e}_2\,. \tag{4.71}$$

(5) At $\eta = 1$,

$$\Delta\mathbf{Q}_0(1) = \Delta\mathbf{M}_0(1) = 0 \quad \text{or}$$

$$\mathrm{K}_{11}(1)\,\mathbf{C}_Q + \mathrm{K}_{12}(1)\,\mathbf{C}_M - \mathrm{G}_{11}^{(1)}(1,\,\eta_1)\,\mathbf{R}_{10} - \mathrm{G}_{11}^{(2)}(1,\,\eta_2)\,m\lambda^2\mathbf{u}_0(\eta_2) - \mathrm{G}_{12}^{(2)}(1,\,\eta_2)\,\mathrm{J}_m\lambda^2\boldsymbol{\vartheta}_0(\eta_2) + \mathrm{G}_{11}^{(3)}(1,\,\eta_3)\,cu_{20}(\eta_3)\,\mathbf{e}_2 = 0\,, \tag{4.72}$$

$$\mathrm{K}_{21}(1)\,\mathbf{C}_Q + \mathrm{K}_{22}(1)\,\mathbf{C}_M - \mathrm{G}_{21}^{(1)}(1,\,\eta_1)\,\mathbf{R}_{10} - \mathrm{G}_{21}^{(2)}(1,\,\eta_2)\,m\lambda^2\mathbf{u}_0(\eta_2) - \mathrm{G}_{22}^{(2)}(1,\,\eta_2)\,\mathrm{J}_m\lambda^2\boldsymbol{\vartheta}_0(\eta_2) + \mathrm{G}_{21}^{(3)}(1,\,\eta_3)\,cu_{20}(\eta_3)\,\mathbf{e}_2 = 0\,. \tag{4.73}$$

Note that the vector $\mathbf{R}_{10} = R_{10}\mathbf{e}_2$ has only one nonzero component.

In scalar form, the system of equations (4.68)–(4.71) in 14 unknowns C_{Qj}, C_{Mj}, $u_{0j}(\eta_2)$, $v_{0j}(\eta_2)$, $j = 1, 2, 3$, R_{10} and $u_{20}(\eta_3)$, consists of 14 linear algebraic equations.

Equations (4.68)–(4.73) can be rearranged as follows:

$$\mathrm{P}(\lambda)\,\mathbf{C}_1 = 0\,; \tag{4.74}$$

here, $\mathrm{P}(\lambda)$ is a matrix whose elements are entries of the matrices entering in the system (4.68)–(4.73). The components of the vector $\mathbf{C}_1$ are those of the vectors $\mathbf{C}_Q$, $\mathbf{C}_M$, $\mathbf{u}_0(\eta_2)$, and $\boldsymbol{\vartheta}_0(\eta_2)$ and the parameter R_{10} and $u_{20}(\eta_3)$. The frequencies λ_j can be found from the relationship

$$D = \det[\mathrm{P}] = 0\,. \tag{4.75}$$

Consider the procedure of determination of frequencies. When we use the method of initial parameters, we are to find the product of five transition matrices of the dimensions 12×12.

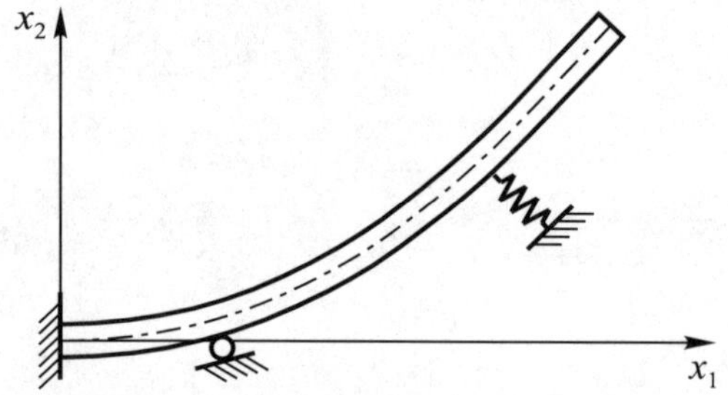

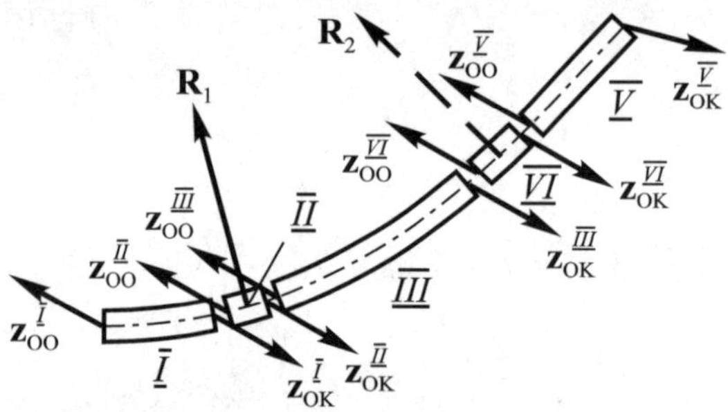

Fig. 4.9.

Consider, as an example, the transition matrices for the easy case of $m = 0$. Let us divide mentally the rod shown in Fig. 4.9*a* into five segments as shown in Fig. 4.9*b*. Let $\mathbf{Z}_{00}^{(i)}$ and $\mathbf{Z}_{0\,\mathrm{f}}^{(i)}$ be the state vectors at the beginning and the end of the ith segment, respectively. The matrix $\mathrm{A}^{(i)}$ such that

$$\mathbf{Z}_{0\,\mathrm{f}}^{(i)} = \mathrm{A}^{(i)}\mathbf{Z}_{00}^{(i)}$$

is called the transition matrix for the ith segment. For the five segments, we have

$$\begin{aligned} &\mathbf{Z}_{0\,\mathrm{f}}^{\mathrm{I}} = \mathrm{A}^{(1)}\mathbf{Z}_{00}^{\mathrm{I}}\,, \qquad \mathbf{Z}_{0\,\mathrm{f}}^{\mathrm{II}} = \mathrm{A}^{(2)}\mathbf{Z}_{00}^{\mathrm{II}}\,, \qquad \mathbf{Z}_{0\,\mathrm{f}}^{\mathrm{III}} = \mathrm{A}^{(3)}\mathbf{Z}_{00}^{\mathrm{III}}\,, \\ &\mathbf{Z}_{0\,\mathrm{f}}^{\mathrm{IV}} = \mathrm{A}^{(4)}\mathbf{Z}_{00}^{\mathrm{IV}}\,, \qquad \mathbf{Z}_{0\,\mathrm{f}}^{\mathrm{V}} = \mathrm{A}^{(5)}\mathbf{Z}_{00}^{\mathrm{V}}\,. \end{aligned} \tag{4.76}$$

Since

$$\mathbf{Z}_{00}^{\mathrm{II}} = \mathbf{Z}_{0\,\mathrm{f}}^{\mathrm{I}}\,, \qquad \mathbf{Z}_{00}^{\mathrm{III}} = \mathbf{Z}_{0\,\mathrm{f}}^{\mathrm{II}}\,, \qquad \mathbf{Z}_{00}^{\mathrm{IV}} = \mathbf{Z}_{0\,\mathrm{f}}^{\mathrm{III}}\,, \qquad \mathbf{Z}_{00}^{\mathrm{V}} = \mathbf{Z}_{0\,\mathrm{f}}^{\mathrm{IV}}\,,$$

the relationship (4.76) can be rewritten in the form

$$\mathbf{Z}^{\mathrm{V}}_{0\,\mathrm{f}} = \mathrm{A}^{(5)}\mathrm{A}^{(4)}\mathrm{A}^{(3)}\mathrm{A}^{(2)}\mathrm{A}^{(1)}\mathbf{Z}^{\mathrm{I}}_{00} = \mathrm{A}\,\mathbf{Z}^{\mathrm{I}}_{00}\,, \tag{4.77}$$

where

$$\mathbf{Z}^{\mathrm{I}}_{00} = \begin{bmatrix} \Delta\mathbf{Q}^{\mathrm{I}}_{00} \\ \Delta\mathbf{M}^{\mathrm{I}}_{00} \\ \boldsymbol{\vartheta}^{\mathrm{I}}_{00} \\ \mathbf{u}^{\mathrm{I}}_{00} \end{bmatrix}, \qquad \mathbf{Z}^{\mathrm{V}}_{0\,\mathrm{f}} = \begin{bmatrix} \Delta\mathbf{Q}^{\mathrm{V}}_{0\,\mathrm{f}} \\ \Delta\mathbf{M}^{\mathrm{V}}_{0\,\mathrm{f}} \\ \boldsymbol{\vartheta}^{\mathrm{V}}_{0\,\mathrm{f}} \\ \mathbf{u}^{\mathrm{V}}_{0\,\mathrm{f}} \end{bmatrix}.$$

The boundary conditions for the rod shown in Fig. 4.9*a* are:

$$\begin{aligned} &(1) \quad \text{at} \quad \eta = 0\,, \quad \mathbf{u}^{\mathrm{I}}_{00} = \boldsymbol{\vartheta}^{\mathrm{I}}_{00} = 0\,; \\ &(2) \quad \text{at} \quad \eta = \eta_5\,, \quad \Delta\mathbf{Q}^{\mathrm{V}}_{0\,\mathrm{f}} = \Delta\mathbf{M}^{\mathrm{V}}_{0\,\mathrm{f}} = 0\,; \end{aligned}$$

here, η_5 is the dimensionless length of the fifth segment. From the boundary conditions at the right-hand end of the rod it follows that

$$a_{11}\,\Delta\mathbf{Q}^{\mathrm{I}}_{30} + a_{12}\,\Delta\mathbf{M}^{\mathrm{I}}_{00} = 0\,, \qquad a_{21}\,\Delta\mathbf{Q}^{\mathrm{I}}_{00} + a_{22}\,\Delta\mathbf{M}^{\mathrm{I}}_{00} = 0\,, \tag{4.78}$$

where a_{ij} are the elements of the matrix A.

As noted above, the frequencies can be found from the relationship

$$D = \det\begin{bmatrix} a_{11} & a_{12} \\ a_{21} & a_{22} \end{bmatrix} = 0\,. \tag{4.79}$$

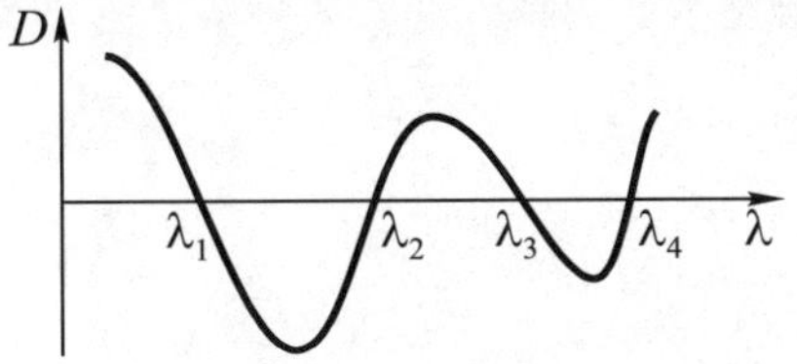

Fig. 4.10.

For conservative problems, $D = D(\lambda)$ is alternating function of λ (for example, see the determinants (4.49), (4.75) or (4.79)). The fact that the signs of these determinants are plus or minus depending on the values of initial approximation $\tilde{\lambda}_j$ will be of use when constructing numerical algorithms for determination of frequencies on the basis of determinants. The relationship between D and λ is qualitatively shown in Fig. 4.10. The frequencies correspond to the points where the graph crosses the abscissa.

Consider in detail the determination of D for a plane circular rod with intermediate supports (Fig. 4.11). The rod is constantly in the plane of drawing during vibrations. The reaction forces $\mathbf{R}_1$ and $\mathbf{R}_2$ and the inertial forces of the concentrated mass due to vibrations are shown as a dashed line. The

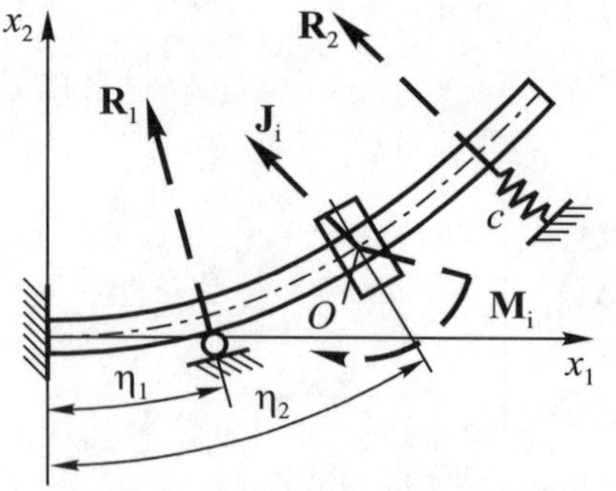

Fig. 4.11.

equations governing small vibrations of a rod in the plane of its axial line were obtained in Sect. 3.2. For a rod of constant cross section, the equations of forced small vibrations about the natural configuration are as follows:

$$\begin{aligned}
&\frac{\partial^2 u_1}{\partial \tau^2} - \frac{\partial \Delta Q_1}{\partial \eta} + æ_{30}\,\Delta Q_2 + Q_{20}\,\Delta M_3 \\
&\qquad + m\,\frac{\partial^2 u_1}{\partial \tau^2}\,\delta(\eta-\eta_2) - (\mathbf{R}_1\cdot\mathbf{e}_1)\,\delta(\eta-\eta_1) - (\mathbf{R}_2\cdot\mathbf{e}_1)\,\delta(\eta-\eta_3) = 0\,,\\
&\frac{\partial^2 u_2}{\partial \tau^2} - \frac{\partial \Delta Q_2}{\partial \eta} - æ_{30}\,\Delta Q_1 - Q_{10}\,\Delta M_3 \\
&\qquad + m\,\frac{\partial^2 u_2}{\partial \tau^2}\,\delta(\eta-\eta_2) - (\mathbf{R}_1\cdot\mathbf{e}_2)\,\delta(\eta-\eta_1) - (\mathbf{R}_2\cdot\mathbf{e}_2)\,\delta(\eta-\eta_3) = 0\,,\\
&J_{33}\,\frac{\partial^2 \vartheta_3}{\partial \tau^2} - \frac{\partial \Delta M_3}{\partial \eta} - \Delta Q_2 + J_{m0}\,\frac{\partial^2 \vartheta_3}{\partial \tau^2}\,\delta(\eta-\eta_2) = 0\,,\\
&\frac{\partial \vartheta_3}{\partial \eta} - \Delta M_3 = 0\,,\\
&\frac{\partial u_1}{\partial \eta} - æ_{30} u_2 = 0\,,\\
&\frac{\partial u_2}{\partial \eta} + æ_{30} u_1 - \vartheta_3 = 0\,;
\end{aligned} \tag{4.80}$$

here, J_{m0} is the moment of inertia of the concentrated mass about the axis that is perpendicular to the plane of drawing at the point O (Fig. 4.11).

Assuming that during vibrations the vectors $\mathbf{R}_1$ and $\mathbf{R}_2$ are perpendicular to the vectors $\mathbf{e}_1(\eta_1)$ and $\mathbf{e}_1(\eta_3)$, we have

$$\begin{aligned}
&(\mathbf{R}_1\cdot\mathbf{e}_1) = 0\,, \qquad (\mathbf{R}_1\cdot\mathbf{e}_2) = |\mathbf{R}_1| = R_{10}\,,\\
&(\mathbf{R}_2\cdot\mathbf{e}_1) = 0\,, \qquad (\mathbf{R}_2\cdot\mathbf{e}_2) = -cu_2\,.
\end{aligned}$$

Setting

$$\begin{aligned}
&\Delta Q_1 = \Delta Q_{10}\,\mathrm{e}^{\mathrm{i}\lambda\tau}\,, \qquad \Delta Q_2 = \Delta Q_{20}\,\mathrm{e}^{\mathrm{i}\lambda\tau}\,, \qquad \Delta M_3 = \Delta M_{30}\,\mathrm{e}^{\mathrm{i}\lambda\tau}\,,\\
&\vartheta_3 = \vartheta_{30}\,\mathrm{e}^{\mathrm{i}\lambda\tau}\,, \qquad u_1 = u_{10}\,\mathrm{e}^{\mathrm{i}\lambda\tau}\,, \qquad u_2 = u_{20}\,\mathrm{e}^{\mathrm{i}\lambda\tau}
\end{aligned}$$

and using (4.80), we get

$$\mathbf{Z}_0' + \mathrm{B}(\eta,\, \lambda)\, \mathbf{Z}_0 = \Delta \boldsymbol{\Phi}_0 \,, \tag{4.81}$$

where

$$\mathbf{Z}_0 = \begin{bmatrix} \Delta Q_{10} \\ \Delta Q_{20} \\ \Delta M_{30} \\ \vartheta_{30} \\ u_{10} \\ u_{20} \end{bmatrix},$$

$$\mathrm{B}(\eta,\, \lambda) = \begin{bmatrix} 0 & -æ_{30} & -Q_{20} & 0 & \lambda^2 & 0 \\ æ_{30} & 0 & Q_{10} & 0 & 0 & \lambda^2 \\ 0 & 1 & \lambda^2 J_{33} & 0 & 0 & 0 \\ 0 & 0 & -1 & 0 & 0 & 0 \\ 0 & 0 & 0 & 0 & 0 & -æ_{30} \\ 0 & 0 & 0 & -1 & æ_{30} & 0 \end{bmatrix},$$

$$\Delta \boldsymbol{\Phi}_0 = \begin{bmatrix} -m\lambda^2 u_{10}\, \delta(\eta - \eta_2) \\ -R_{10}\, \delta(\eta - \eta_1) - m\lambda^2 u_{20}\, \delta(\eta - \eta_2) + c u_{20}\, \delta(\eta - \eta_3) \\ -J_{m0}\lambda^2 \vartheta_{30}\, \delta(\eta - \eta_2) \\ 0 \\ 0 \\ 0 \end{bmatrix}.$$

The vector $\Delta \boldsymbol{\Phi}_0$ can be represented in the form

$$\Delta \boldsymbol{\Phi}_0 = \boldsymbol{\Phi}_1\, \delta(\eta - \eta_1) + \mathrm{D}^{(2)} \mathbf{Z}_0\, \delta(\eta - \eta_2) + \mathrm{D}^{(3)} \mathbf{Z}_0\, \delta(\eta - \eta_3)\,,$$

where

$$\boldsymbol{\Phi}_1 = R_{10} \begin{bmatrix} 0 \\ -1 \\ 0 \\ 0 \\ 0 \\ 0 \end{bmatrix}, \qquad \mathrm{D}^{(3)} = \begin{bmatrix} 0 & 0 & 0 & 0 & 0 & 0 \\ 0 & 0 & 0 & 0 & 0 & c \\ 0 & 0 & 0 & 0 & 0 & 0 \\ 0 & 0 & 0 & 0 & 0 & 0 \\ 0 & 0 & 0 & 0 & 0 & 0 \\ 0 & 0 & 0 & 0 & 0 & 0 \end{bmatrix},$$

$$\mathrm{D}^{(2)} = \begin{bmatrix} 0 & 0 & 0 & 0 & -m\lambda^2 & 0 \\ 0 & 0 & 0 & 0 & 0 & -m\lambda^2 \\ 0 & 0 & 0 & -J_{m0}\lambda^2 & 0 & 0 \\ 0 & 0 & 0 & 0 & 0 & 0 \\ 0 & 0 & 0 & 0 & 0 & 0 \\ 0 & 0 & 0 & 0 & 0 & 0 \end{bmatrix}.$$

The solution of (4.81) is a particular case of (4.66). Thus, we have

$$\begin{aligned} \mathbf{Z}_0 = \mathrm{K}(\eta)\, \mathbf{C} + \mathrm{G}^{(1)}(\eta,\, \eta_1)\, \boldsymbol{\Phi}_1\, H(\eta - \eta_1) + \mathrm{G}^{(2)}(\eta,\, \eta_2)\, \mathrm{D}^{(2)} \mathbf{Z}_0(\eta_2)\, H(\eta - \eta_2) \\ + \mathrm{G}^{(3)}(\eta,\, \eta_3)\, \mathrm{D}^{(3)} \mathbf{Z}_0(\eta_3)\, H(\eta - \eta_3)\,. \end{aligned} \tag{4.82}$$

In components, it reads

$$
\begin{aligned}
Z_{01} = \Delta Q_{01} = {} & \sum_{j=1}^{6} k_{1j} c_j - g_{12}^{(1)}(\eta,\, \eta_1)\, R_{10}\, H(\eta - \eta_1) \\
& - g_{11}^{(2)}(\eta,\, \eta_2)\, m\lambda^2 u_{10}(\eta_2)\, H(\eta - \eta_2) - g_{12}^{(2)}(\eta,\, \eta_2)\, m\lambda^2 u_{20}(\eta_2)\, H(\eta - \eta_2) \\
& - g_{13}^{(2)}(\eta,\, \eta_2)\, J_{m0}\lambda^2 \vartheta_{30}(\eta_2)\, H(\eta - \eta_2) + g_{12}^{(3)}(\eta,\, \eta_3)\, u_{20}(\eta_3)\, H(\eta - \eta_3)\,, \\
Z_{02} = \Delta Q_{02} = {} & \sum_{j=1}^{6} k_{2j} c_j - g_{22}^{(1)}(\eta,\, \eta_1)\, R_{10}\, H(\eta - \eta_1) \\
& - g_{21}^{(2)}(\eta,\, \eta_2)\, m\lambda^2 u_{10}(\eta_2)\, H(\eta - \eta_2) - g_{22}^{(2)}(\eta,\, \eta_2)\, m\lambda^2 u_{20}(\eta_2)\, H(\eta - \eta_2) \\
& - g_{23}^{(2)}(\eta,\, \eta_2)\, J_{m0}\lambda^2 \vartheta_{30}(\eta_2)\, H(\eta - \eta_2) + g_{22}^{(3)}(\eta,\, \eta_3)\, u_{20}(\eta_3)\, H(\eta - \eta_3)\,, \\
Z_{03} = \Delta M_{03} = {} & \sum_{j=1}^{6} k_{3j} c_j - g_{32}^{(1)}(\eta,\, \eta_1)\, R_{10}\, H(\eta - \eta_1) \\
& - g_{31}^{(2)}(\eta,\, \eta_2)\, m\lambda^2 u_{10}(\eta_2)\, H(\eta - \eta_2) - g_{32}^{(2)}(\eta,\, \eta_2)\, m\lambda^2 u_{20}(\eta_2)\, H(\eta - \eta_2) \\
& - g_{33}^{(2)}(\eta,\, \eta_2)\, J_{m0}\lambda^2 \vartheta_{30}(\eta_2)\, H(\eta - \eta_2) + g_{32}^{(3)}(\eta,\, \eta_3)\, u_{20}(\eta_3)\, H(\eta - \eta_3)\,, \\
Z_{04} = \vartheta_{30} = {} & \sum_{j=1}^{6} k_{4j} c_j - g_{42}^{(1)}(\eta,\, \eta_1)\, R_{10}\, H(\eta - \eta_1) \\
& - g_{41}^{(2)}(\eta,\, \eta_2)\, m\lambda^2 u_{10}(\eta_2)\, H(\eta - \eta_2) - g_{42}^{(2)}(\eta,\, \eta_2)\, m\lambda^2 u_{20}(\eta_2)\, H(\eta - \eta_2) \\
& - g_{43}^{(2)}(\eta,\, \eta_2)\, J_{m0}\lambda^2 \vartheta_{30}(\eta_2)\, H(\eta - \eta_2) + g_{42}^{(3)}(\eta,\, \eta_3)\, u_{20}(\eta_3)\, H(\eta - \eta_3)\,, \\
Z_{05} = u_{10} = {} & \sum_{j=1}^{6} k_{5j} c_j - g_{52}^{(1)}(\eta,\, \eta_1)\, R_{10}\, H(\eta - \eta_1) \\
& - g_{51}^{(2)}(\eta,\, \eta_2)\, m\lambda^2 u_{10}(\eta_2)\, H(\eta - \eta_2) - g_{52}^{(2)}(\eta,\, \eta_2)\, m\lambda^2 u_{20}(\eta_2)\, H(\eta - \eta_2) \\
& - g_{53}^{(2)}(\eta,\, \eta_2)\, J_{m0}\lambda^2 \vartheta_{30}(\eta_2)\, H(\eta - \eta_2) + g_{52}^{(3)}(\eta,\, \eta_3)\, u_{20}(\eta_3)\, H(\eta - \eta_3)\,, \\
Z_{06} = u_{20} = {} & \sum_{j=1}^{6} k_{6j} c_j - g_{62}^{(1)}(\eta,\, \eta_1)\, R_{10}\, H(\eta - \eta_1) \\
& - g_{61}^{(2)}(\eta,\, \eta_2)\, m\lambda^2 u_{10}(\eta_2)\, H(\eta - \eta_2) - g_{62}^{(2)}(\eta,\, \eta_2)\, m\lambda^2 u_{20}(\eta_2)\, H(\eta - \eta_2) \\
& - g_{53}^{(2)}(\eta,\, \eta_2)\, J_{m0}\lambda^2 \vartheta_{30}(\eta_2)\, H(\eta - \eta_2) + g_{62}^{(3)}(\eta,\, \eta_3)\, u_{20}(\eta_3)\, H(\eta - \eta_3)\,.
\end{aligned}
\tag{4.83}
$$

The components Z_{04}, Z_{05}, and Z_{06} are functions continuous in η while Z_{01}, Z_{02}, and Z_{03} are discontinuous functions.

Let us determine $u_{10}(\eta_2)$, $u_{20}(\eta_2)$, $\vartheta_{30}(\eta_2)$, and $u_{20}(\eta_3)$, which enter into the right-hand sides of (4.83). To do this, we put $\eta = \eta_2$ in the expressions for Z_{04}, Z_{05}, and Z_{06}. Then, we get

$$\vartheta_{30}(\eta_2) = \sum_{j=1}^{6} k_{4j}(\eta_2)\, c_j - g_{42}^{(1)}(\eta_2, \eta_1)\, R_{10}\,,$$

$$u_{10}(\eta_2) = \sum_{j=1}^{6} k_{5j}(\eta_2)\, c_j - g_{52}^{(1)}(\eta_2, \eta_1)\, R_{10}\,,$$

$$u_{20}(\eta_2) = \sum_{j=1}^{6} k_{6j}(\eta_2)\, c_j - g_{62}^{(1)}(\eta_2, \eta_1)\, R_{10}\,. \tag{4.84}$$

To find the component $u_{20}(\eta_3)$, we put $\eta = \eta_3$ in the expression for Z_{06} and thus obtain

$$u_{20}(\eta_3) = \sum_{j=1}^{6} k_{6j}(\eta_3)\, c_j - g_{62}^{(1)}(\eta_2, \eta_1)\, R_{10} - g_{61}^{(2)}(\eta_3, \eta_2)\, m\lambda^2 u_{10}(\eta_2)$$
$$- g_{62}^{(2)}(\eta_3, \eta_2)\, m\lambda^2 u_{20}(\eta_2) - g_{33}^{(2)}(\eta_3, \eta_2)\, J_{m0}\lambda^2 \vartheta_{30}(\eta_2)\,. \tag{4.85}$$

Substituting $\vartheta_{30}(\eta_2)$, $u_{10}(\eta_2)$, and $u_{20}(\eta_2)$ from (4.84) into (4.85), we can represent $u_{20}(\eta_3)$ in terms of c_j and R_{10}. Eliminating $\vartheta_{30}(\eta_2)$, $u_{10}(\eta_2)$, $u_{20}(\eta_2)$, and $u_{20}(\eta_3)$ from (4.83) and (4.84), we get the equations which contain seven unknown constants R_{10} and c_j, $i = 1, \dots, 6$. To find these constants, we can use the following seven conditions: three at each end of the rod ($\eta = 0$ and $\eta = 1$) and the condition $u_{20}(\eta_1) = 0$ at $\eta = \eta_1$.

For the rod shown in Fig. 4.11, it follows from the boundary conditions at $\eta = 0$ that $c_4 = c_5 = c_6 = 0$. Considering the conditions $u_{20}(\eta_1) = 0$ (at $\eta = \eta_1$) and $\Delta Q_{10}(1) = \Delta Q_{20}(1) = \Delta M_3(1) = 0$ (at $\eta = 1$), we get the following system of homogeneous equations:

$$k_{61}(\eta_1)\, c_1 + k_{62}(\eta_1)\, c_2 + k_{63}(\eta_1)\, c_3 = 0\,,$$
$$b_{11}(1)\, c_1 + b_{12}(1)\, c_2 + b_{13}(1)\, c_3 + b_{14}(1)\, R_{10} = 0\,,$$
$$b_{21}(1)\, c_1 + b_{22}(1)\, c_2 + b_{23}(1)\, c_3 + b_{24}(1)\, R_{10} = 0\,,$$
$$b_{31}(1)\, c_1 + b_{32}(1)\, c_2 + b_{33}(1)\, c_3 + b_{34}(1)\, R_{10} = 0\,;$$

here, the coefficients b_{ij} depend on the elements of the matrices K and $\mathrm{G}^{(i)}$.

The frequencies λ_j can be found from the condition that the determinant of this homogeneous system of equations equals zero.

4.2 Determination of Eigenvalues in Nonconservative Problems

4.2.1 Complex Eigenvalues

In the previous sections, we considered conservative systems. The inertial forces only were taken into consideration and thus the equations of small free vibrations contain second-order derivatives with respect to time. Mechanical systems governed by such equations are known as conservative. The corresponding eigenvalues λ_j are real and the corresponding functions of time are harmonic. Consequently, in the case of conservative systems, we have

$$\mathrm{e}^{\mathrm{i}\lambda_j\tau} = \cos\lambda_j\tau + \mathrm{i}\,\sin\lambda_j\tau\,.$$

The system of equations (4.1)–(4.4) governs free vibrations of a rod such that during the vibrations the total energy, the sum of the potential and kinetic energy, remains constant because forces of viscous damping are ignored.

Let us now take into consideration forces of viscous damping which are linear with the vector of velocity. We have

$$\mathbf{f}_u = \alpha_0\,\frac{\partial\mathbf{u}}{\partial\tau}\,, \qquad \mathbf{F}^{(k)}_\alpha = \alpha_m\,\frac{\partial\mathbf{u}}{\partial\tau}\,\delta(\eta-\eta_\mathrm{f})\,, \qquad \mathbf{f}_\vartheta = \alpha^{(1)}_0\,\frac{\partial\boldsymbol{\vartheta}}{\partial\tau}\,, \tag{4.86}$$

where $\mathbf{f}_u$ is a distributed force, $\mathbf{F}^{(k)}_\alpha$ is the concentrated force applied to a concentrated mass, and $\mathbf{f}_\vartheta$ is a distributed moment of forces; here, α_0, α_m, and $\alpha^{(1)}_0$ are dimensionless coefficients of forces and moments of viscous damping, respectively.

Taking into consideration only the distributed force of viscous damping $\mathbf{f}_u$, we can represent the equations of vibrations as follows:

$$n_1(\eta)\,\frac{\partial^2\mathbf{u}}{\partial\tau^2} + \alpha_0\,\frac{\partial\mathbf{u}}{\partial\tau} - \frac{\partial\Delta\mathbf{Q}}{\partial\eta} - \mathrm{A}_Q\mathrm{A}^{-1}\,\Delta\mathbf{M} - \mathrm{A}_{æ}\,\Delta\mathbf{Q} = 0\,, \tag{4.87}$$

$$\mathrm{J}\,\frac{\partial^2\boldsymbol{\vartheta}}{\partial\tau^2} - \frac{\partial\Delta\mathbf{M}}{\partial\eta} - \mathrm{A}_M\mathrm{A}^{-1}\,\Delta\mathbf{M} - \mathrm{A}_{æ}\,\Delta\mathbf{M} - \mathrm{A}_1\,\Delta\mathbf{Q} = 0\,, \tag{4.88}$$

$$\frac{\partial\boldsymbol{\vartheta}}{\partial\eta} + \mathrm{A}_{æ}\boldsymbol{\vartheta} - \mathrm{A}^{-1}\,\Delta\mathbf{M} = 0\,, \tag{4.89}$$

$$\frac{\partial\mathbf{u}}{\partial\eta} + \mathrm{A}_{æ}\mathbf{u} + \mathrm{A}_1\boldsymbol{\vartheta} = 0\,. \tag{4.90}$$

In Sect. 3.4, the equations governing small vibrations of a rod about its stationary motion were obtained. Recall that these equations (more exactly, the equation of translational motion) contain the Coriolis forces that are proportional to

$$\frac{\partial^2\boldsymbol{\vartheta}}{\partial\eta\,\partial\tau}\,.$$

Note that this term contains the first derivative with respect to time.

Owing to the forces of viscous damping, free vibrations decay and thus λ_j are to be complex numbers,

$$\lambda_j = \alpha_j \pm \mathrm{i}\,\beta_j\,.$$

If we seek the solution of (4.87)–(4.90) in the form (4.9), then, upon substitution of (4.9) into (4.87)–(4.90), the term $\alpha\,\mathrm{i}\,\lambda\mathbf{u}_0$ occurs in (4.87);

hence, $\Delta\mathbf{Q}_0$, $\Delta\mathbf{M}_0$, $\Delta\boldsymbol{\vartheta}_0$, and $\Delta\mathbf{u}_0$ are to be considered as complex functions. To solve (4.87)–(4.90), we put

$$\Delta\mathbf{Q} = \Delta\mathbf{Q}_0 e^{\lambda\tau}\,, \qquad \Delta\mathbf{M} = \Delta\mathbf{M}_0 e^{\lambda\tau}\,,$$
$$\boldsymbol{\vartheta} = \boldsymbol{\vartheta}_0 e^{\lambda\tau}\,, \qquad \mathbf{u} = \mathbf{u}_0 e^{\lambda\tau}\,, \tag{4.91}$$

where

$$\lambda = \alpha + \mathrm{i}\,\beta\,, \tag{4.92}$$
$$\Delta\mathbf{Q}_0 = \Delta\mathbf{Q}_0^{(1)} + \mathrm{i}\,\Delta\mathbf{Q}_0^{(2)}\,, \qquad \Delta\mathbf{M}_0 = \Delta\mathbf{M}_0^{(1)} + \mathrm{i}\,\Delta\mathbf{M}_0^{(2)}\,,$$
$$\boldsymbol{\vartheta}_0 = \boldsymbol{\vartheta}_0^{(1)} + \mathrm{i}\,\boldsymbol{\vartheta}_0^{(2)}\,, \qquad \mathbf{u}_0 = \mathbf{u}_0^{(1)} + \mathrm{i}\,\mathbf{u}_0^{(2)}\,. \tag{4.93}$$

Substituting (4.91)–(4.93) into (4.87)–(4.90) and separating the real and imaginary parts, we get the following eight vector equations:

$$\begin{aligned}
&\frac{\mathrm{d}\,\Delta\mathbf{Q}_0^{(1)}}{\mathrm{d}\eta} + \mathrm{A}_æ\,\Delta\mathbf{Q}_0^{(1)} + \mathrm{A}_Q\mathrm{A}^{-1}\,\Delta\mathbf{M}_0^{(1)}\\
&\qquad - \left[n_1(\alpha^2-\beta^2) + \alpha_0\alpha\right]\mathbf{u}_0^{(1)} + (2n_1\alpha\beta + \alpha_0\beta)\,\mathbf{u}_0^{(2)} = 0\,,\\
&\frac{\mathrm{d}\,\Delta\mathbf{M}_0^{(1)}}{\mathrm{d}\eta} + \mathrm{A}_1\,\Delta\mathbf{Q}_0^{(1)} + (\mathrm{A}_M\mathrm{A}^{-1} + \mathrm{A}_æ)\,\Delta\mathbf{M}_0^{(1)}\\
&\qquad - \mathrm{J}(\alpha^2-\beta^2)\,\boldsymbol{\vartheta}_0^{(1)} + \mathrm{J}2\alpha\beta\boldsymbol{\vartheta}_0^{(2)} = 0\,,\\
&\frac{\mathrm{d}\boldsymbol{\vartheta}_0^{(1)}}{\mathrm{d}\eta} + \mathrm{A}_æ\boldsymbol{\vartheta}_0^{(1)} - \mathrm{A}^{-1}\,\Delta\mathbf{M}_0^{(1)} = 0\,,\\
&\frac{\mathrm{d}\mathbf{u}_0^{(1)}}{\mathrm{d}\eta} + \mathrm{A}_æ\mathbf{u}_0^{(1)} + \mathrm{A}_1\boldsymbol{\vartheta}_0^{(1)} = 0\,,
\end{aligned} \tag{4.94}$$

$$\begin{aligned}
&\frac{\mathrm{d}\,\Delta\mathbf{Q}_0^{(2)}}{\mathrm{d}\eta} + \mathrm{A}_æ\,\Delta\mathbf{Q}_0^{(2)} + \mathrm{A}_Q\mathrm{A}^{-1}\,\Delta\mathbf{M}_0^{(2)}\\
&\qquad - \left[n_1(\alpha^2-\beta^2) + \alpha_0\alpha\right]\mathbf{u}_0^{(2)} - (2n_1\alpha\beta + \alpha_0\beta)\,\mathbf{u}_0^{(1)} = 0\,,\\
&\frac{\mathrm{d}\,\Delta\mathbf{M}_0^{(2)}}{\mathrm{d}\eta} + \mathrm{A}_1\,\Delta\mathbf{Q}_0^{(2)} + (\mathrm{A}_M\mathrm{A}^{-1} + \mathrm{A}_æ)\,\Delta\mathbf{M}_0^{(2)}\\
&\qquad - \mathrm{J}(\alpha^2-\beta^2)\,\boldsymbol{\vartheta}_0^{(2)} - \mathrm{J}2\alpha\beta\boldsymbol{\vartheta}_0^{(1)} = 0\,,\\
&\frac{\mathrm{d}\boldsymbol{\vartheta}_0^{(2)}}{\mathrm{d}\eta} + \mathrm{A}_\nu\boldsymbol{\vartheta}_0^{(2)} - \mathrm{A}^{-1}\,\Delta\mathbf{M}_0^{(2)} = 0\,,\\
&\frac{\mathrm{d}\mathbf{u}_0^{(2)}}{\mathrm{d}\eta} + \mathrm{A}_æ\mathbf{u}_0^{(2)} + \mathrm{A}_1\boldsymbol{\vartheta}_0^{(2)} = 0\,.
\end{aligned} \tag{4.95}$$

In vector–matrix form, we have

$$\mathbf{Z}_0' + \mathrm{B}(\eta,\,\alpha,\,\beta)\,\mathbf{Z}_0 = 0\,, \tag{4.96}$$

where

$$\mathbf{Z}_0 = \begin{bmatrix} \mathbf{Z}_{01} \\ \mathbf{Z}_{02} \end{bmatrix}, \qquad \mathbf{Z}_{01} = \begin{bmatrix} \Delta\mathbf{Q}_0^{(1)} \\ \Delta\mathbf{M}_0^{(1)} \\ \boldsymbol{\vartheta}_0^{(1)} \\ \mathbf{u}_0^{(1)} \end{bmatrix}, \qquad \mathbf{Z}_{02} = \begin{bmatrix} \Delta\mathbf{Q}_0^{(2)} \\ \Delta\mathbf{M}_0^{(2)} \\ \boldsymbol{\vartheta}_0^{(2)} \\ \mathbf{u}_0^{(2)} \end{bmatrix}, \tag{4.97}$$

and

$$\mathrm{B}(\eta,\, \alpha,\, \beta) =$$

$$\left[\begin{array}{cccccccc}
\mathrm{A}_{\ae} & \mathrm{A}_Q\mathrm{A}^{-1} & 0 & -\mathrm{E}\left[n_1(\alpha^2-\beta^2)-\alpha_0\alpha\right] & 0 & 0 & 0 & \mathrm{E}(2n_1\alpha+\alpha_0)\,\beta \\
\mathrm{A}_1 & (\mathrm{A}_M\mathrm{A}^{-1}+\mathrm{A}_{\ae}) & -\mathrm{J}(\alpha^2-\beta^2) & 0 & 0 & 0 & 2\alpha\beta\mathrm{J} & 0 \\
0 & -\mathrm{A}^{-1} & \mathrm{A}_{\ae} & 0 & 0 & 0 & 0 & 0 \\
0 & 0 & \mathrm{A}_1 & \mathrm{A}_{\ae} & 0 & 0 & 0 & 0 \\
0 & 0 & 0 & -\mathrm{E}(2n_1\alpha+\alpha_0)\,\beta & \mathrm{A}_{\ae} & \mathrm{A}_Q\mathrm{A}^{-1} & 0 & -\mathrm{E}\left[n_1(\alpha^2-\beta^2)-\alpha_0\alpha\right] \\
0 & 0 & -2\alpha\beta\mathrm{J} & 0 & \mathrm{A}_1 & (\mathrm{A}_M\mathrm{A}^{-1}+\mathrm{A}_{\ae}) & -\mathrm{J}(\alpha^2-\beta^2) & 0 \\
0 & 0 & 0 & 0 & \mathrm{A}^{-1} & 0 & \mathrm{A}_{\ae} & 0 \\
0 & 0 & 0 & 0 & 0 & 0 & \mathrm{A}_1 & \mathrm{A}_{\ae}
\end{array}\right].$$

Let the values of α and β be specified. Let us obtain a numerical solution of (4.96) in the form

$$\mathbf{Z}_0 = \mathrm{K}(\eta,\, \alpha,\, \beta)\,\mathbf{C}\,; \tag{4.98}$$

the function $\mathbf{Z}_0$ should satisfy the boundary conditions at $\eta = 0$ and $\eta = 1$.

For a cantilever beam, the boundary conditions are as follows:

$$(1) \quad \text{at} \quad \eta = 0\,, \quad \boldsymbol{\vartheta}_0^{(1)} = \boldsymbol{\vartheta}_0^{(2)} = \mathbf{u}_0^{(1)} = \mathbf{u}_0^{(2)} = 0\,;$$

$$(2) \quad \text{at} \quad \eta = 1\,, \quad \Delta\mathbf{Q}_0^{(1)} = \Delta\mathbf{Q}_0^{(2)} = \Delta\mathbf{M}_0^{(1)} = \Delta\mathbf{M}_0^{(2)} = 0\,.$$

Let us represent the vector $\mathbf{C}$ in the form

$$\mathbf{C} = (\mathbf{C}_{Q_1},\, \mathbf{C}_{M_1},\, \mathbf{C}_{\vartheta_1},\, \mathbf{C}_{u_1},\, \mathbf{C}_{Q_2},\, \mathbf{C}_{M_2},\, \mathbf{C}_{\vartheta_2},\, \mathbf{C}_{u_2})^{\mathrm{T}}$$

and the matrix $\mathrm{K}(\eta,\, \alpha,\, \beta)$ as

$$\mathrm{K}(\eta, \alpha, \beta) = \begin{bmatrix} \mathrm{K}_{11} & \dots & \mathrm{K}_{18} \\ \dots & \dots & \dots \\ \mathrm{K}_{81} & \dots & \mathrm{K}_{88} \end{bmatrix},$$

where K_{ij} are 3×3 matrices.

It follows from the boundary conditions at $\eta = 0$ that

$$\mathbf{C}_{\vartheta 1} = \mathbf{C}_{\vartheta 2} = \mathbf{C}_{u1} = \mathbf{C}_{u2} = 0 \,.$$

The boundary conditions at $\eta = 1$ yield the equations

$$\begin{aligned} &\mathrm{K}_{11}\mathbf{C}_{Q1} + \mathrm{K}_{12}\mathbf{C}_{M1} + \mathrm{K}_{15}\mathbf{C}_{Q2} + \mathrm{K}_{16}\mathbf{C}_{M2} = 0 \,, \\ &\mathrm{K}_{21}\mathbf{C}_{Q1} + \mathrm{K}_{22}\mathbf{C}_{M1} + \mathrm{K}_{25}\mathbf{C}_{Q2} + \mathrm{K}_{26}\mathbf{C}_{M2} = 0 \,, \\ &\mathrm{K}_{51}\mathbf{C}_{Q1} + \mathrm{K}_{52}\mathbf{C}_{M1} + \mathrm{K}_{55}\mathbf{C}_{Q2} + \mathrm{K}_{56}\mathbf{C}_{M2} = 0 \,, \\ &\mathrm{K}_{61}\mathbf{C}_{Q1} + \mathrm{K}_{62}\mathbf{C}_{M1} + \mathrm{K}_{65}\mathbf{C}_{Q2} + \mathrm{K}_{66}\mathbf{C}_{M2} = 0 \,. \end{aligned} \tag{4.99}$$

Equating the determinant of the system (4.99) to zero, we finally arrive at

$$D = \det\left[\mathrm{K}_{ij}(1, \alpha, \beta)\right] = 0 \,. \tag{4.100}$$

The real and imaginary parts of the complex-valued eigenvalues α_j and β_j can be determined from this equation.

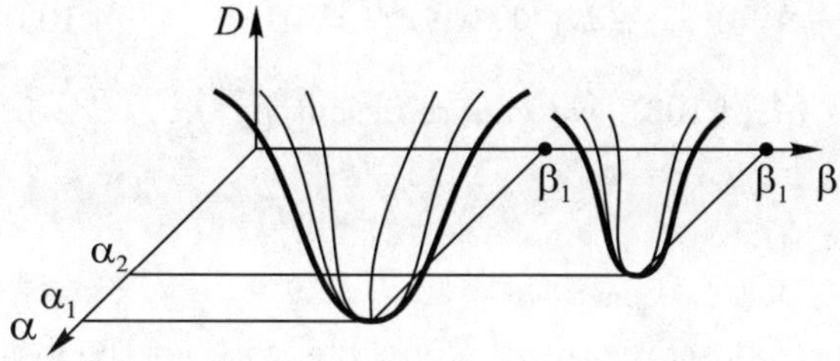

Fig. 4.12.

Recall that, for conservative problems, $D = D(\lambda)$ is alternating function of λ. Unlike conservative problems, determinants in nonconservative problems (for example, (4.100)) are of constant signs. This complicates numerical determination of the eigenvalues. The surface $D = D(\alpha, \beta)$ is qualitatively shown in Fig. 4.12. The points of contact of $D(\alpha, \beta)$ and the coordinate plane (α, β) give the sought-for eigenvalues λ_j:

$$\lambda_j = \alpha_j + \mathrm{i}\,\beta_j \,.$$

In the subsequent sections, applied problems of the dynamics of rods (both conservative and nonconservative) and examples of numerical determination of complex eigenvalues are considered.

4.3 Eigenfunctions (Eigenvectors)

4.3.1 Determination of Eigenfunctions in Conservative Problems

Having found the frequencies λ_j, we can determine the corresponding eigenfunctions $\mathbf{Z}_0^{(j)}$ from (4.14). We have

$$\frac{\mathrm{d}\mathbf{Z}_0^{(j)}}{\mathrm{d}\eta} + \mathrm{B}(\eta, \lambda_j)\,\mathbf{Z}_0^{(j)} = 0\,. \tag{4.101}$$

From this equation it follows that

$$\mathbf{Z}_0^{(j)} = \mathrm{K}(\eta, \lambda_j)\,\mathbf{C}^{(j)}\,, \tag{4.102}$$

where $\mathrm{K}(\eta, \lambda_j)$ is the fundamental matrix corresponding to λ_j.

The methods of numerical determination of the matrix $\mathrm{K}(\eta, \lambda_j)$ are discussed in Chap. 2 of Svetlitsky (2000a) and in Sect. 4.1. For example, in the case of a cantilever beam, we have $c_7^{(j)} = c_8^{(j)} = \ldots = c_{12}^{(j)} = 0$. The coefficients $c_1^{(j)} = c_2^{(j)} = \ldots = c_6^{(j)} = 0$ satisfy the relationships (4.20):

$$\begin{aligned} &k_{11}(1, \lambda_j)\,c_1^{(j)} + k_{12}(1, \lambda_j)\,c_2^{(j)} + \ldots + k_{16}(1, \lambda_j)\,c_6^{(j)} = 0\,,\\ &\ldots\\ &k_{61}(1, \lambda_j)\,c_1^{(j)} + k_{62}(1, \lambda_j)\,c_2^{(j)} + \ldots + k_{66}(1, \lambda_j)\,c_6^{(j)} = 0\,. \end{aligned} \tag{4.103}$$

Using the first five equations of (4.103), we can represent $c_1^{(j)}, \ldots, c_5^{(j)}$ in terms of $c_6^{(j)}$ as follows:

$$c_k^{(j)} = \mu_k c_6^{(j)}\,,\ k = 1, \ldots, 5\,.$$

Here, $c_6^{(j)}$ is an arbitrary factor. Consequently, it is possible, without the loss of generality, to set $c_6^{(j)} = 1$.

Solving for $\mathbf{C}^{(j)}$ from (4.103), we get the eigenvector

$$\mathbf{Z}_0^{(j)} = \mathrm{K}(\eta, \lambda_j)\,\mathbf{C}_0^{(j)}\,, \tag{4.104}$$

where

$$\mathbf{C}_0^{(j)} = (\mu_1, \mu_2, \mu_3, \mu_4, \mu_5, 1, 0, \ldots, 0)^{\mathrm{T}}\,.$$

The components $Z_{07}^{(j)}, \ldots, Z_{0,12}^{(j)}$ of the vector $\mathbf{Z}$ defined by

$$\begin{aligned} &Z_{07}^{(j)} = \vartheta_{01}^{(j)}\,, \quad Z_{08}^{(j)} = \vartheta_{02}^{(j)}\,, \quad Z_{09}^{(j)} = \vartheta_{03}^{(j)}\,,\\ &Z_{0,10}^{(j)} = u_{01}^{(j)}\,, \quad Z_{0,11}^{(j)} = u_{02}^{(j)}\,, \quad Z_{0,12}^{(j)} = u_{03}^{(j)} \end{aligned}$$

are called the *modes of vibrations of the rod.* The components $Z_{01}^{(j)}, \ldots, Z_{01}^{(j)}$ describe η-dependence of the amplitude values of the components of the vectors $\Delta\mathbf{Q}_0^{(j)}$ and $\Delta\mathbf{M}_0^{(j)}$ during the vibrations at each of the frequencies λ_j. For further analysis, it is convenient to represent the eigenvectors in the form

$$\mathbf{Z}_0^{(j)} = \begin{bmatrix} \boldsymbol{\psi}^{(j)} \\ \boldsymbol{\varphi}^{(j)} \end{bmatrix}, \tag{4.105}$$

where

$$\boldsymbol{\psi}^{(j)} = \begin{bmatrix} \Delta\mathbf{Q}_0^{(j)} \\ \Delta\mathbf{M}_0^{(j)} \end{bmatrix}, \qquad \boldsymbol{\varphi}^{(j)} = \begin{bmatrix} \boldsymbol{\vartheta}_0^{(j)} \\ \mathbf{u}_0^{(j)} \end{bmatrix}. \tag{4.106}$$

The vector $\boldsymbol{\psi}^{(j)}$ describes the stress state of the rod and the vector $\boldsymbol{\varphi}^{(j)}$ describes the deformed state of the rod. Such a representation of eigenvectors is useful in the determination of the approximate solutions of problems of dynamics of rods.

4.3.2 Orthogonality Condition for Eigenvectors

Consider two vectors $\mathbf{Z}_0^{(i)}$ and $\mathbf{Z}_0^{(j)}$ that satisfy (4.101). We have

$$\mathbf{Z}_0^{(j)\prime} + \mathrm{B}^{(j)}\mathbf{Z}_0^{(j)} = 0\,, \tag{4.107}$$

$$\mathbf{Z}_0^{(i)\prime} + \mathrm{B}^{(i)}\mathbf{Z}^{(i)} = 0\,. \tag{4.108}$$

Multiplying (4.107) by $\mathrm{E}_0\mathbf{Z}_0^{(i)}$ and (4.108) by $\mathrm{E}_0\mathbf{Z}_0^{(j)}$, where

$$\mathrm{E}_0 = \begin{bmatrix} 0 & 0 & 0 & \mathrm{E} \\ 0 & 0 & \mathrm{E} & 0 \\ 0 & \mathrm{E} & 0 & 0 \\ \mathrm{E} & 0 & 0 & 0 \end{bmatrix},$$

and summing the results, we get

$$(\mathbf{Z}_0^{(j)\prime} \cdot \mathrm{E}_0\mathbf{Z}_0^{(i)}) + (\mathbf{Z}_0^{(i)\prime} \cdot \mathrm{E}_0\mathbf{Z}_0^{(j)}) + (\mathrm{B}^{(j)}\mathbf{Z}_0^{(j)} \cdot \mathrm{E}_0\mathbf{Z}_0^{(i)}) + (\mathrm{B}^{(i)}\mathbf{Z}_0^{(i)} \cdot \mathrm{E}_0\mathbf{Z}_0^{(j)}) = 0\,. \tag{4.109}$$

The matrix E_0 satisfies the relationship $\mathrm{E}_0 = \mathrm{E}_0^{\mathrm{T}}$; hence, (4.109) can be rewritten as

$$\frac{\mathrm{d}}{\mathrm{d}\eta}(\mathrm{E}_0\mathbf{Z}_0^{(j)} \cdot \mathbf{Z}_0^{(i)}) + ([\mathrm{E}_0\mathrm{B}^{(i)} + (\mathrm{B}^{(j)})^{\mathrm{T}}\mathrm{E}_0]\,\mathbf{Z}_0^{(i)} \cdot \mathbf{Z}_0^{(j)}) = 0\,. \tag{4.110}$$

Integration of (4.110) yields

$$(\mathrm{E}_0\mathbf{Z}_0^{(j)} \cdot \mathbf{Z}_0^{(i)})\Big|_0^1 + \int_0^1 ([\mathrm{E}_0\mathrm{B}^{(i)} + (\mathrm{B}^{(j)})^{\mathrm{T}}\mathrm{E}_0]\,\mathbf{Z}_0^{(i)} \cdot \mathbf{Z}_0^{(j)})\,\mathrm{d}\eta = 0\,. \tag{4.111}$$

The first term in the left-hand side of this relationship can be written as

$$(\mathrm{E}_0\mathbf{Z}_0^{(j)} \cdot \mathbf{Z}_0^{(i)}) = (\Delta\mathbf{Q}_0^{(j)} \cdot \mathbf{u}_0^{(i)}) + (\Delta\mathbf{M}_0^{(j)} \cdot \boldsymbol{\vartheta}_0^{(i)}) \\ + (\Delta\mathbf{Q}_0^{(i)} \cdot \mathbf{u}_0^{(j)}) + (\Delta\mathbf{M}_0^{(i)} \cdot \boldsymbol{\vartheta}_0^{(j)}) . \quad (4.112)$$

In terms of boundary conditions, there are just two distinct terms in the right-hand side of (4.112), say, $(\Delta\mathbf{Q}_0^{(j)} \cdot \mathbf{u}_0^{(i)})$ and $(\Delta\mathbf{M}_0^{(j)} \cdot \boldsymbol{\vartheta}_0^{(i)})$.

In the case of homogeneous boundary conditions, these scalar products for $\eta = 0$ and $\eta = 1$ are always zero. Therefore, using (4.111), we obtain the condition for two eigenvectors to be orthogonal:

$$\int_0^1 ([\mathrm{E}_0\mathrm{B}^{(i)} + (\mathrm{B}^{(j)})^{\mathrm{T}}\mathrm{E}_0]\,\mathbf{Z}_0^{(i)} \cdot \mathbf{Z}_0^{(j)})\,\mathrm{d}\eta = 0 . \quad (4.113)$$

When deriving this orthogonality condition, we introduced the vectors $\mathrm{E}_0\mathbf{Z}^{(j)}$. As a result, the scalar products occurred. These scalar products have the dimensions of work (for example, (4.112)). Such an approach may be useful when solving equations of motion by approximate methods based on the principle of virtual displacements.

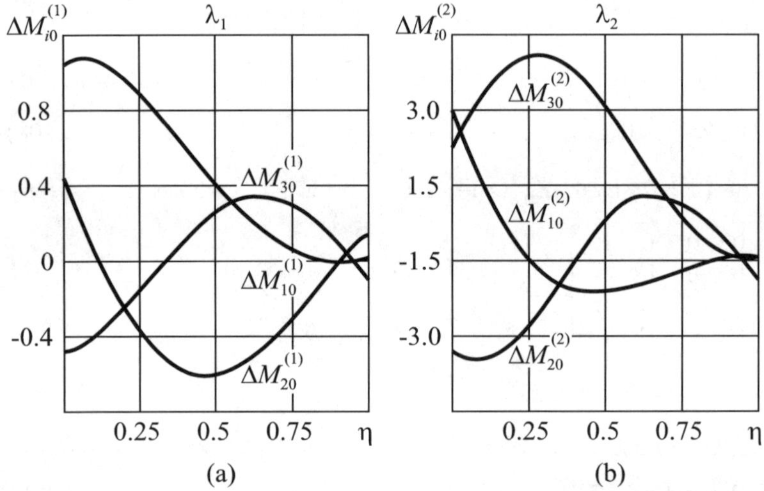

Fig. 4.13.

As an example, let us consider again the helical rod with a concentrated mass shown in Fig. 4.4*a*. We restrict our consideration to two first eigenvectors. Let the angle of helix α be 30°. The frequencies of the rod were found in Sect. 4.1 (Fig. 4.5). The components of vectors $\Delta\mathbf{Q}_0^{(1)}$ and $\Delta\mathbf{Q}_0^{(2)}$ versus η are shown in Fig. 4.13*a,b*.

The components of vectors $\Delta\mathbf{M}_0^{(1)}$ and $\Delta\mathbf{M}_0^{(2)}$ versus η are shown in Fig. 4.14*a,b*.

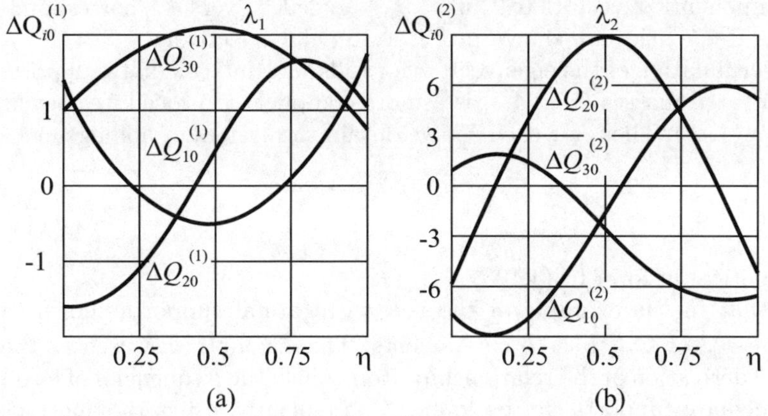

Fig. 4.14.

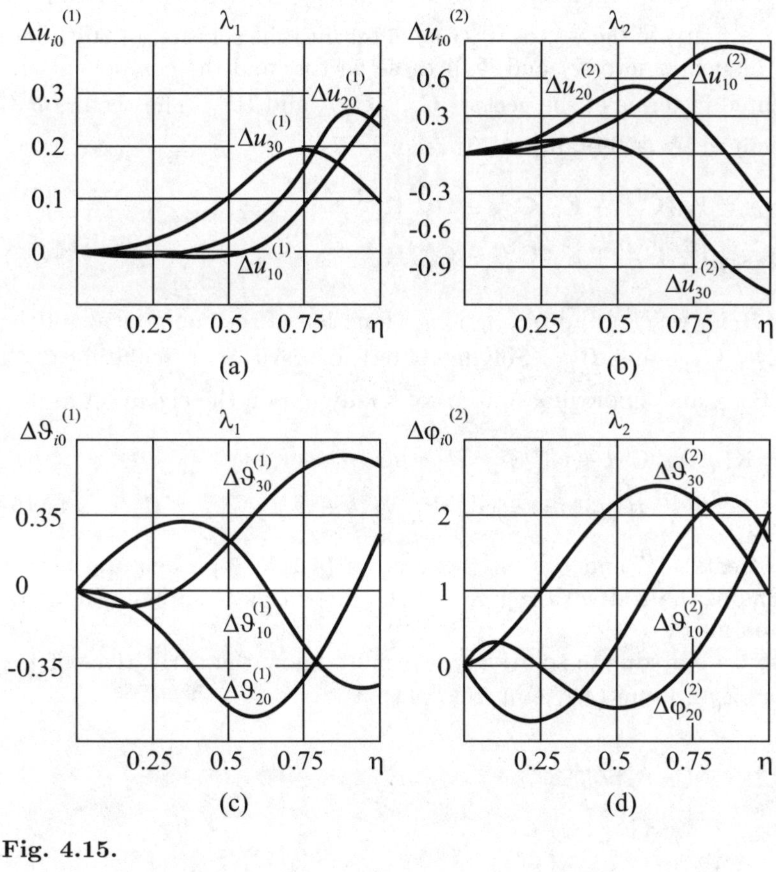

Fig. 4.15.

The components of vectors $\mathbf{u}_0^{(1)}$, $\mathbf{u}_0^{(2)}$, $\boldsymbol{\vartheta}_0^{(1)}$, and $\boldsymbol{\vartheta}_0^{(2)}$) versus η corresponding to lower frequencies are shown in Fig. 4.15*a–d*, respectively.

The determination of the eigenvectors for rods with intermediate supports and concentrated masses (Fig. 4.8) is a more complicated procedure. Let us find λ_j from (4.74). Then, for each λ_j, we obtain the system of homogeneous equations

$$\mathrm{P}(\lambda_j)\,\mathbf{C}_1^{(j)} = 0\,, \tag{4.114}$$

which is identical in form to (4.103).

Recall that in the example of the rod with lateral supports shown in Fig. 4.8, the vector $\mathbf{C}_1^{(j)}$ has 16 components. This example was used as the basis for the derivation of the relationships from which the frequencies of a rod with intermediate supports can be found. Consequently, using 15 equations of the set (4.114), we can represent 15 components of the vector $\mathbf{C}_1^{(j)}$ in terms of its 16th component. Then, we can put $\mathbf{C}_1^{(16)} = 1$. As a result, we obtain the vectors $\mathbf{C}_Q^{(j)}$, $\mathbf{C}_M^{(j)}$, and $\mathbf{R}_1^{(j)}$. The vectors $\mathbf{u}_0^{(j)}(\eta_2)$ and $\boldsymbol{\vartheta}_0^{(j)}(\eta_2)$ and the component $u_{20}^{(j)}(\eta_3)$ of the vector $\mathbf{u}_0^{(j)}(\eta_3)$ enter into the solution (4.66) (more precisely, they enter into $\boldsymbol{\Phi}_1$ and $\boldsymbol{\Phi}_2$); these vectors and the component can be represented in terms of the vectors $\mathbf{C}_Q^{(j)}$, $\mathbf{C}_M^{(j)}$, and $\mathbf{R}_1^{(j)}$. The vectors $\boldsymbol{\vartheta}_0^{(j)}$ and $\mathbf{u}_0^{(j)}$ defined by (4.69) and (4.70) read

$$\begin{aligned}
\boldsymbol{\vartheta}_0^{(j)}(\eta_2) &= \mathrm{K}_{31}\mathbf{C}_Q^{(j)} + \mathrm{K}_{32}\mathbf{C}_M^{(j)} - \mathrm{G}_{31}^{(1)}\mathbf{R}_1^{(j)}\,,\\
\mathbf{u}_0^{(j)}(\eta_2) &= \mathrm{K}_{41}\mathbf{C}_Q^{(j)} + \mathrm{K}_{42}\mathbf{C}_M^{(j)} - \mathrm{G}_{41}^{(1)}\mathbf{R}_1^{(j)}\,.
\end{aligned}$$

Using (4.71), we can find $\mathbf{u}_{20}^{(j)}(\eta_3)$ as a function of the components of the vectors $\mathbf{C}_Q^{(j)}$, $\mathbf{C}_M^{(j)}$, and $\mathbf{R}_1^{(j)}$. Solving (4.64) for each λ_j, calculating $\mathbf{C}_Q^{(j)}$, $\mathbf{C}_M^{(j)}$, and $\mathbf{R}_1^{(j)}$, and employing (4.69), we finally obtain the eigenvectors

$$\begin{aligned}
\mathbf{Z}_0^{(j)} &= \mathrm{K}(\eta,\,\lambda_j)\,\mathrm{C}^{(j)} + \mathrm{G}^{(1)}\boldsymbol{\Phi}_1^{(j)}\,H(\eta-\eta_1)\\
&\quad + \mathrm{G}^{(2)}\boldsymbol{\Phi}_2^{(j)}\,H(\eta-\eta_2) + \mathrm{G}^{(3)}\boldsymbol{\Phi}_3^{(j)}\,H(\eta-\eta_3)\,.
\end{aligned} \tag{4.115}$$

The components $u_{0\mathrm{f}}^{(j)}$ and $\vartheta_{0\mathrm{f}}^{(j)}$ of the vectors (4.115) represent the modes of vibrations of the rod with regard for the intermediate supports and the concentrated masses.

To obtain approximate solutions, one must know the derivatives of the eigenvectors $\mathbf{Z}_0^{(j)}$. From (4.101) it follows that

$$\frac{\mathrm{d}\mathbf{Z}_0^{(j)}}{\mathrm{d}\eta} = -\mathrm{B}(\eta,\,\lambda_j)\,\mathbf{Z}_0^{(j)}$$

or

$$\frac{\mathrm{d}\mathbf{Z}_0^{(j)}}{\mathrm{d}\eta} = -\mathrm{B}(\eta,\, \lambda_j)\, \mathrm{K}(\eta,\, \lambda_j)\, \mathbf{C}_0^{(j)} \,. \tag{4.116}$$

For a rod with intermediate supports, we have

$$\frac{\mathrm{d}\mathbf{Z}_0^{(j)}}{\mathrm{d}\eta} = -\mathrm{B}(\eta,\, \lambda_j)\, \Big[\mathrm{K}(\eta,\, \lambda_j)\, \mathrm{C}^{(j)} + \mathrm{G}^{(1)}\boldsymbol{\Phi}_1^{(j)}\, H(\eta - \eta_1)$$
$$+\, \mathrm{G}^{(2)}\boldsymbol{\Phi}_2^{(j)}\, H(\eta - \eta_2) + \mathrm{G}^{(3)}\boldsymbol{\Phi}_3^{(j)}\, H(\eta - \eta_3)\Big] \,. \tag{4.117}$$

4.3.3 Determination of Eigenvectors in Nonconservative Problems

Having found the eigenvalues $\lambda_j = \alpha_j + \mathrm{i}\,\beta_j$, we can insert them into (4.96) and thereby obtain the equation from which the corresponding eigenvectors $\mathbf{Z}_0^{(j)}$ can be found. We have

$$\frac{\mathrm{d}\mathbf{Z}_0^{(j)}}{\mathrm{d}\eta} + \mathrm{B}(\eta,\, \alpha_j,\, \beta_j)\, \mathbf{Z}_0^{(j)} = 0 \,. \tag{4.118}$$

Now, the values of α_j and β_j are known. From (4.118) it follows that

$$\mathbf{Z}_0^{(j)} = \mathrm{K}(\eta,\, \alpha_j,\, \beta_j)\, \mathbf{C}^{(j)} \,.$$

Consider again the example presented in Fig. 4.6. Taking into account the boundary conditions at $\eta = 1$, we get

$$\begin{aligned}
\Delta\mathbf{Q}_0^{(1)} &= \mathrm{K}_{11}\mathbf{C}_{Q_1}^{(j)} + \mathrm{K}_{12}\mathbf{C}_{M_1}^{(j)} + \mathrm{K}_{15}\mathbf{C}_{Q_2}^{(j)} + \mathrm{K}_{16}\mathbf{C}_{M_2}^{(j)} = 0\,,\\
\Delta\mathbf{M}_0^{(1)} &= \mathrm{K}_{21}\mathbf{C}_{Q_1}^{(j)} + \mathrm{K}_{22}\mathbf{C}_{M_1}^{(j)} + \mathrm{K}_{25}\mathbf{C}_{Q_2}^{(j)} + \mathrm{K}_{26}\mathbf{C}_{M_2}^{(j)} = 0\,,\\
\Delta\mathbf{Q}_0^{(2)} &= \mathrm{K}_{51}\mathbf{C}_{Q_1}^{(j)} + \mathrm{K}_{52}\mathbf{C}_{M_1}^{(j)} + \mathrm{K}_{55}\mathbf{C}_{Q_2}^{(j)} + \mathrm{K}_{56}\mathbf{C}_{M_2}^{(j)} = 0\,,\\
\Delta\mathbf{M}_0^{(2)} &= \mathrm{K}_{61}\mathbf{C}_{Q_1}^{(j)} + \mathrm{K}_{62}\mathbf{C}_{M_1}^{(j)} + \mathrm{K}_{65}\mathbf{C}_{Q_2}^{(j)} + \mathrm{K}_{66}\mathbf{C}_{M_2}^{(j)} = 0\,.
\end{aligned} \tag{4.119}$$

Using any three linear independent equations of (4.119), we can represent any three vectors in terms of the fourth one. For example, the vectors $\mathbf{C}_{Q_1}^{(j)}$, $\mathbf{C}_{Q_2}^{(j)}$, and $\mathbf{C}_{M_1}^{(j)}$ can be represented in terms of the vector $\mathbf{C}_{M_2}^{(j)}$ as follows:

$$\mathbf{C}_{Q_1}^{(j)} = \mathrm{H}_1^{(j)}\mathbf{C}_{M_2}^{(j)}\,, \qquad \mathbf{C}_{M_1}^{(j)} = \mathrm{H}_2^{(j)}\mathbf{C}_{M_2}^{(j)}\,, \qquad \mathbf{C}_{Q_2}^{(j)} = \mathrm{H}_3^{(j)}\mathbf{C}_{M_2}^{(j)}\,.$$

Let $\mathbf{C}_{M_2}^{(j)} = \mathbf{1}$, where $\mathbf{1}$ is a vector whose components are equal to unity. Since $\mathbf{C}_{\vartheta_1} = \mathbf{C}_{\vartheta_2} = \mathbf{C}_{u_1} = \mathbf{C}_{u_2} = 0$, we have

$$\mathbf{C}_0^{(j)} = (\mathrm{H}_1^{(j)}\mathbf{1},\, \mathrm{H}_2^{(j)}\mathbf{1},\, \mathbf{0},\, \mathbf{0};\;\, \mathrm{H}_3^{(j)}\mathbf{1},\, \mathbf{1},\, \mathbf{0},\, \mathbf{0})^{\mathrm{T}} = (\mathbf{C}_{01}^{(j)},\, \mathbf{C}_{02}^{(j)})^{\mathrm{T}} \,.$$

Finally, having obtained the vectors $\mathbf{C}_0^{(j)}$, we get the eigenvectors in the following form:

$$\mathbf{Z}_0^{(j)} = \mathrm{K}(\eta, \alpha_j, \beta_j)\, \mathbf{C}_0^{(j)} . \tag{4.120}$$

This formula accounts for the boundary conditions by means of the vectors $\mathbf{C}_0^{(j)}$, that is, if the boundary conditions change, then only the vectors $\mathbf{C}_0^{(j)}$ change while the matrix K remains the same. Thus, the vectors $\mathbf{Z}_0^{(j)}$ (4.120) are eigenfunctions of boundary value problems for nonconcervative systems. In view of (4.120), the vectors $\mathbf{Z}_{01}^{(j)}$ and $\mathbf{Z}_{02}^{(j)}$ (4.97) can be rewritten as

$$\mathbf{Z}_{01}^{(j)} = \tilde{\mathrm{K}}_{11}\mathbf{C}_{01}^{(j)} + \tilde{\mathrm{K}}_{12}\mathbf{C}_{02}^{(j)} , \qquad \mathbf{Z}_{02}^{(j)} = \tilde{\mathrm{K}}_{21}\mathbf{C}_{01}^{(j)} + \tilde{\mathrm{K}}_{22}\mathbf{C}_{02}^{(j)} , \tag{4.121}$$

where

$$\mathbf{C}_{01}^{(j)} = \begin{bmatrix} \mathrm{H}_1^{(j)}\mathbf{1} \\ \mathrm{H}_2^{(j)}\mathbf{1} \\ \mathbf{0} \\ \mathbf{0} \end{bmatrix} , \qquad \mathbf{C}_{02}^{(j)} = \begin{bmatrix} \mathrm{H}_3^{(j)}\mathbf{1} \\ \mathbf{1} \\ \mathbf{0} \\ \mathbf{0} \end{bmatrix} ,$$

and $\tilde{\mathrm{K}}_{ij}$ are 12×12 matrices.

The vectors $\mathbf{Z}_{01}^{(j)}$ and $\mathbf{Z}_{02}^{(j)}$ satisfy all the boundary conditions of the problem. Hence, they can be used in the approximate analysis of problems on vibrations (for instance, forced vibrations, parametric vibrations, random vibrations) based on the principle of virtual displacements. The relevant problems are considered in the subsequent sections devoted to applications of the dynamics of rods.

Using (4.118), we obtain the derivatives of the vectors $\mathbf{Z}_{01}^{(j)}$ and $\mathbf{Z}_{02}^{(j)}$:

$$\mathbf{Z}_{01}^{(j)\prime} = -\mathrm{D}_{11}\mathbf{Z}_{01}^{(j)} - \mathrm{D}_{12}\mathbf{Z}_{02}^{(j)} , \qquad \mathbf{Z}_{02}^{(j)\prime} = -\mathrm{D}_{21}\mathbf{Z}_{01}^{(j)} - \mathrm{D}_{22}\mathbf{Z}_{02}^{(j)} ; \tag{4.122}$$

here, D_{ij} are block matrices entering into the matrix D such that

$$\mathrm{D} = \begin{bmatrix} \mathrm{D}_{11} & \mathrm{D}_{12} \\ \mathrm{D}_{21} & \mathrm{D}_{22} \end{bmatrix} .$$

Eigenvectors are of extensive use in approximate methods for intricate problems of dynamics of rods.

4.4 Approximate Methods for Determination of Eigenvalues

4.4.1 Method Based on the Generalized Principle of Virtual Displacements

Some exact numerical methods for determination of eigenvalues and the corresponding eigenvectors were discussed in Sects. 4.1 and 4.2. The application of these methods implies a great deal of calculations, especially in the case of complex-valued eigenvalues because in this case any new problem should be solved anew. Consequently, approximate methods have considerable utility in design. In this section, we consider an efficient method based on the principle of virtual displacements.

Consider the linear homogeneous system of partial differential equations (4.1)–(4.4):

$$\mathbf{L}_1 = n_1(\eta)\,\frac{\partial^2 \mathbf{u}}{\partial \tau^2} - \frac{\partial \Delta \mathbf{Q}}{\partial \eta} - \mathrm{A}_Q \mathrm{A}^{-1}\,\Delta\mathbf{M} - \mathrm{A}_{æ}\,\Delta\mathbf{Q} = 0\,, \tag{4.123}$$

$$\mathbf{L}_2 = \mathrm{J}(\eta)\,\frac{\partial^2 \boldsymbol{\vartheta}}{\partial \tau^2} - \frac{\partial \Delta \mathbf{M}}{\partial \eta} - \mathrm{A}_M \mathrm{A}^{-1}\,\Delta\mathbf{M} - \mathrm{A}_{æ}\,\Delta\mathbf{M} - \mathrm{A}_1\,\Delta\mathbf{Q} = 0\,, \tag{4.124}$$

$$\mathbf{L}_3 = \frac{\partial \boldsymbol{\vartheta}}{\partial \eta} - \mathrm{A}_{æ}\boldsymbol{\vartheta} - \mathrm{A}^{-1}\,\Delta\mathbf{M} = 0\,, \tag{4.125}$$

$$\mathbf{L}_4 = \frac{\partial \mathbf{u}}{\partial \eta} + \mathrm{A}_{æ}\mathbf{u} + \mathrm{A}_1\boldsymbol{\vartheta} = 0\,. \tag{4.126}$$

Let us rewrite this system as a vector equation

$$\mathbf{L} = \mathrm{A}^{(1)}\,\frac{\partial^2 \mathbf{Z}}{\partial \tau^2} + \frac{\partial \mathbf{Z}}{\partial \eta} + \mathrm{A}^{(2)}\mathbf{Z} = 0\,, \tag{4.127}$$

where

$$\mathrm{A}^{(1)} = \begin{bmatrix} 0 & 0 & 0 & -n_1\mathrm{E} \\ 0 & 0 & -\mathrm{J} & 0 \\ 0 & 0 & 0 & 0 \\ 0 & 0 & 0 & 0 \end{bmatrix},$$

$$\mathrm{A}^{(2)} = \begin{bmatrix} \mathrm{A}_{æ} & \mathrm{A}_Q\mathrm{A}^{-1} & 0 & 0 \\ \mathrm{A}_1 & \mathrm{A}_M\mathrm{A}^{-1} + \mathrm{A}_{æ} & 0 & 0 \\ 0 & -\mathrm{A}^{-1} & \mathrm{A}_{æ} & 0 \\ 0 & 0 & \mathrm{A}_1 & \mathrm{A}_{æ} \end{bmatrix}, \qquad \mathbf{Z} = \begin{bmatrix} \Delta\mathbf{Q} \\ \Delta\mathbf{M} \\ \boldsymbol{\vartheta} \\ \mathbf{u} \end{bmatrix}.$$

We seek an approximate solution of the vector equation (4.127) in the form

$$\mathbf{Z} = \sum_{j=1}^{n} f_j(\tau)\,\mathbf{Z}_0^{(j)}(\eta)\,, \tag{4.128}$$

where $f_j(\tau)$ are unknown functions of dimensionless time τ and $\mathbf{Z}_0^{(j)}(\eta)$ are specified vectors whose components satisfy the boundary conditions of the problem. As the specified vectors $\mathbf{Z}_0^{(j)}(\eta)$, it is best to use eigenvectors of some simplified problem with the same boundary conditions.

In Sect. 4.3.1, we present the method for determination of eigenvectors for conservative problems. In this case, the eigenvectors $\mathbf{Z}_0^{(j)}$ (4.102) can be represented as

$$\mathbf{Z}_0^{(j)} = \begin{bmatrix} \boldsymbol{\psi}^{(j)} \\ \boldsymbol{\varphi}^{(j)} \end{bmatrix}, \tag{4.129}$$

where the vector $\boldsymbol{\psi}^{(j)}$ describes the variation of amplitude values of components of the vectors $\Delta\mathbf{Q}_0^{(j)}$ and $\Delta\mathbf{M}_0^{(j)}$ with η and the vector $\boldsymbol{\varphi}_0^{(j)}$ describes the variation of amplitude values of components of the vectors $\boldsymbol{\vartheta}_0^{(j)}$ and $\mathbf{u}_0^{(j)}$ (eigenmodes) when the rod vibrates at the frequency λ_j. In their turn, the vectors $\boldsymbol{\varphi}^{(j)}$ and $\boldsymbol{\psi}^{(j)}$ can be written in the form

$$\boldsymbol{\psi}^{(j)} = \begin{bmatrix} \boldsymbol{\psi}_Q^{(j)} \\ \boldsymbol{\varphi}_M^{(j)} \end{bmatrix}, \qquad \boldsymbol{\varphi}^{(j)} = \begin{bmatrix} \boldsymbol{\psi}_\vartheta^{(j)} \\ \boldsymbol{\varphi}_u^{(j)} \end{bmatrix},$$

where

$$\boldsymbol{\psi}_Q^{(j)} = \Delta\mathbf{Q}_0^{(j)}, \qquad \boldsymbol{\psi}_M^{(j)} = \Delta\mathbf{M}_0^{(j)}, \qquad \boldsymbol{\varphi}_\vartheta^{(j)} = \boldsymbol{\vartheta}_0^{(j)}, \qquad \boldsymbol{\varphi}_u^{(j)} = \mathbf{u}_0^{(j)}.$$

To obtain an approximate solution, we should first determine the eigenvectors $\mathbf{Z}_0^{(j)}(\eta)$. These vectors can be found from a simplified version of the equations, say, from the system of equations (4.5)–(4.8) for an unloaded rod of constant cross section ($n_1 = 1$) when the rotary inertia is ignored ($J = 0$). We have

$$\frac{\partial^2 \mathbf{u}}{\partial \tau^2} - \frac{\partial \Delta\mathbf{Q}}{\partial \eta} - \mathrm{A}_{æ}\, \Delta\mathbf{Q} = 0, \tag{4.130}$$

$$\frac{\partial \Delta\mathbf{M}}{\partial \eta} + \mathrm{A}_{æ}\, \Delta\mathbf{M} - \mathrm{A}_1\, \Delta\mathbf{Q} = 0, \tag{4.131}$$

$$\frac{\partial \boldsymbol{\vartheta}}{\partial \eta} + \mathrm{A}_{æ}\boldsymbol{\vartheta} - \mathrm{A}^{-1}\, \Delta\mathbf{M} = 0, \tag{4.132}$$

$$\frac{\partial \mathbf{u}}{\partial \eta} + \mathrm{A}_{æ}\mathbf{u} + \mathrm{A}_1\boldsymbol{\vartheta} = 0. \tag{4.133}$$

Solving the system of equations (4.130)–(4.133), we obtain eigenvectors $\mathbf{Z}_0^{(i)}$ (see Sects. 3.4.1 and 4.3):

$$\mathbf{Z}_0^{(i)} = \begin{bmatrix} \boldsymbol{\psi}^{(i)} \\ \boldsymbol{\varphi}^{(i)} \end{bmatrix}, \qquad \boldsymbol{\psi}^{(i)} = \begin{bmatrix} \boldsymbol{\psi}_Q^{(i)} \\ \boldsymbol{\varphi}_M^{(i)} \end{bmatrix}, \qquad \boldsymbol{\varphi}^{(i)} = \begin{bmatrix} \boldsymbol{\psi}_\vartheta^{(i)} \\ \boldsymbol{\varphi}_u^{(i)} \end{bmatrix}.$$

Restrict our consideration to two-term approximation

$$\mathbf{Z} = f^{(1)}\mathbf{Z}_0^{(1)} + f^{(2)}\mathbf{Z}_0^{(2)}. \tag{4.134}$$

Equations of motion (4.123)–(4.126) were derived by application of the d'Alembert principle. This principle allows us to reduce a problem of dynamics to a problem of statics by introducing the forces of inertia. Therefore, the vector equation (4.127) can be treated as the equilibrium equation for a rod and we can use the principle of virtual displacements stated for problems of statics.

The principle of virtual displacements, as applied to systems with distributed parameters, is stated as follows: if a mechanical system is in an equilibrium state under the action of forces including the forces of inertia, the sum of the work done by all the external and internal forces through the virtual displacements is equal to zero.

The principle of virtual displacements can also be formulated in the form of the converse statement: if the sum of the work done by all the external and internal forces through the virtual displacements is equal to zero, the mechanical system is in an equilibrium state. It is worth noting that inertial forces should be taken into consideration.

Recall that any small displacement of rod points away from the initial configuration that can be performed without breakdown of the constraints is termed *virtual* or *allowable.* Thus, virtual displacements can be interpreted as some small variations of the actual displacements that do not disturb the internal and external constraints of the system.

For space-curved rods, the virtual displacements are small displacements of the axial points $\delta \mathbf{u}$ and small angles of rotation of the attached coordinate axes $\delta \boldsymbol{\vartheta}$ that do not disturb constraints of the rod.

Equations (4.123) and (4.124) can be thought of as being of physical nature (physical relationships) while equations (4.125) and (4.126) of geometric one (geometrical relationships). The first six components of the vector $\mathbf{L}$ (4.127) have dimensions of distributed forces and distributed moments. The other six components have dimensions of linear and angular displacements.

Let the generalized displacements $\delta \mathbf{Z}_0^{(i)}$ be proportional to the corresponding eigenvectors such that

$$\delta \mathbf{Z}_0^{(1)} = \delta b_1 \mathrm{E}_0 \mathbf{Z}_0^{(1)} , \qquad \delta \mathbf{Z}_0^{(2)} = \delta b_2 \mathrm{E}_0 \mathbf{Z}_0^{(2)} , \tag{4.135}$$

where δb_1 and δb_2 are arbitrary independent coeffitients and

$$\mathrm{E}_0 = \begin{bmatrix} 0 & 0 & 0 & \mathrm{E} \\ 0 & 0 & \mathrm{E} & 0 \\ 0 & \mathrm{E} & 0 & 0 \\ \mathrm{E} & 0 & 0 & 0 \end{bmatrix} , \qquad \mathbf{Z}_0^{(i)} = \begin{bmatrix} \Delta \mathbf{Q}_0^{(i)} \\ \Delta \mathbf{M}_0^{(i)} \\ \boldsymbol{\vartheta}_0^{(i)} \\ \mathbf{u}_0^{(i)} \end{bmatrix} . \tag{4.136}$$

By a *generalized displacement* is meant not only the variations of linear displacements $\delta \mathbf{u}$ and the variations of angular displacements $\delta \boldsymbol{\vartheta}$ but also the variation of the internal forces $\delta \Delta \mathbf{Q}$ and the internal moments $\delta \Delta \mathbf{M}$.

Consequently, the generalized principle of virtual displacements includes two principles of mechanics of solids: the principle of virtual displacements and the principle of virtual change of stresses.

The matrix E_0 was introduced so as the scalar products $(\mathbf{Z} \cdot \mathrm{E}_0 \mathbf{Z}_0^{(1)})$, $(\mathbf{Z} \cdot \mathrm{E}_0 \mathbf{Z}_0^{(2)})$, etc., to have dimensions of work. Since

$$\mathrm{E}_0\mathbf{Z}_0^{(i)} = (\mathbf{u}_0^{(i)}, \boldsymbol{\vartheta}_0^{(i)}, \Delta\mathbf{M}_0^{(i)}, \Delta\mathbf{Q}_0^{(i)})^{\mathrm{T}},$$

in accordance with the principle of virtual displacements we have

$$(\mathbf{Z}\cdot\mathrm{E}_0\mathbf{Z}_0^{(i)}) = (\Delta\mathbf{Q}\cdot\mathbf{u}_0^{(i)})+(\Delta\mathbf{M}\cdot\boldsymbol{\vartheta}_0^{(i)})+(\boldsymbol{\vartheta}\cdot\Delta\mathbf{M}_0^{(i)})+(\mathbf{u}\cdot\Delta\mathbf{Q}_0^{(i)}). \quad (4.137)$$

As a result, all the terms in the right-hand side of this relationship have dimensions of work.

Substituting the approximate value of the vector $\mathbf{Z}$ (4.134) into (4.127), we get

$$\begin{aligned}\mathbf{L} &= \mathrm{A}^{(1)}\mathbf{Z}_0^{(1)}\ddot{f}_1 + \mathrm{A}^{(1)}\mathbf{Z}_0^{(2)}\ddot{f}_2 + (\mathbf{Z}_0^{(1)\prime} + \mathrm{A}^{(2)}\mathbf{Z}_0^{(1)})\,f_1 \\ &+ (\mathbf{Z}_0^{(2)\prime} + \mathrm{A}^{(2)}\mathbf{Z}_0^{(2)})\,f_2 = \boldsymbol{\delta}, \qquad \boldsymbol{\delta} \neq 0. \end{aligned} \quad (4.138)$$

Here, $\boldsymbol{\delta}$ is a residual vector function.

The work done by the inertial forces plus the work done by the external force and moment ($\Delta\mathbf{Q}$ and $\Delta\mathbf{M}$) through the virtual displacements $\delta\mathbf{u}_0$ and $\delta\boldsymbol{\vartheta}_0$ plus the work done by the virtual variation of the internal forces and moments ($\delta\Delta\mathbf{Q}_0$ and $\delta\Delta\mathbf{M}_0$) through the displacements $\mathbf{u}$ and $\boldsymbol{\vartheta}$ for the rod as a whole must equal zero.

From the mathematical point of view, the last condition is equivalent to the conditions

$$\int_0^1 (\mathbf{L}\cdot\delta\mathbf{Z}_0)\,\mathrm{d}\eta = \int_0^1 (\boldsymbol{\delta}\cdot\delta\mathbf{Z}_0)\,\mathrm{d}\eta = 0.$$

where $\delta\mathbf{Z}_0$ is the vector whose components are the virtual displacements $\delta\mathbf{u}_0$ and $\delta\boldsymbol{\vartheta}_0$ and the virtual variation of the internal force $\delta\Delta\mathbf{Q}_0$ and moment $\delta\Delta\mathbf{M}_0$.

As the vector $\delta\mathbf{Z}_0$, we can take the vector proportional to the eigenvectors $\mathbf{Z}_0^{(j)}$ of the system (4.130)–(4.133), that is, the eigenvectors (4.135):

$$\delta\mathbf{Z}_0 = \sum_{j=1}^{2} \delta\mathbf{Z}_0^{(j)} = \sum_{j=1}^{2} \delta b_j\, E_0\mathbf{Z}_0^{(j)}. \quad (4.140)$$

Consequently, in accordance to the principle of virtual displacements, the integral from the scalar product $(\boldsymbol{\delta}\cdot\delta\mathbf{Z}_0)$ has the dimensions of work for the case of dimensional values.

From (4.139) it follows that

$$\delta b_1 \int_0^1 (\boldsymbol{\delta}\cdot\mathrm{E}_0\mathbf{Z}_0^{(1)})\,\mathrm{d}\eta + \delta b_2 \int_0^1 (\boldsymbol{\delta}\cdot\mathrm{E}_0\mathbf{Z}_0^{(2)})\,\mathrm{d}\eta = 0.$$

Because δb_i are independent variables, we obtain two equations

$$\int_0^1 (\boldsymbol{\delta}\cdot\mathrm{E}_0\mathbf{Z}_0^{(1)})\,\mathrm{d}\eta = 0, \qquad \int_0^1 (\boldsymbol{\delta}\cdot\mathrm{E}_0\mathbf{Z}_0^{(2)})\,\mathrm{d}\eta = 0.$$

On rearrangement, we arrive at two equations in unknown functions f_1 and f_2:

$$\begin{aligned} &a_{11}\ddot{f}_1 + a_{12}\ddot{f}_2 + a_{11}f_1 + c_{12}f_2 = 0\,, \\ &a_{21}\ddot{f}_1 + a_{22}\ddot{f}_2 + a_{21}f_1 + c_{22}f_2 = 0\,, \end{aligned} \tag{4.141}$$

where

$$a_{11} = \int_0^1 (\mathrm{A}^{(1)}\mathbf{Z}_0^{(1)} \cdot \mathrm{E}_0\mathbf{Z}_0^{(1)})\,\mathrm{d}\eta\,, \qquad a_{12} = \int_0^1 (\mathrm{A}^{(1)}\mathbf{Z}_0^{(2)} \cdot \mathrm{E}_0\mathbf{Z}_0^{(1)})\,\mathrm{d}\eta\,,$$

$$a_{21} = \int_0^1 (\mathrm{A}^{(1)}\mathbf{Z}_0^{(1)} \cdot \mathrm{E}_0\mathbf{Z}_0^{(2)})\,\mathrm{d}\eta\,, \qquad a_{22} = \int_0^1 (\mathrm{A}^{(1)}\mathbf{Z}_0^{(2)} \cdot \mathrm{E}_0\mathbf{Z}_0^{(2)})\,\mathrm{d}\eta\,,$$

$$c_{11} = \int_0^1 ([\mathbf{Z}_0^{(1)\prime} + \mathrm{A}^{(2)}E_0\mathbf{Z}_0^{(1)}] \cdot \mathbf{Z}_0^{(1)})\,\mathrm{d}\eta\,,$$

$$c_{12} = \int_0^1 ([\mathbf{Z}_0^{(2)\prime} + \mathrm{A}^{(2)}E_0\mathbf{Z}_0^{(2)}] \cdot \mathbf{Z}_0^{(1)})\,\mathrm{d}\eta\,,$$

$$c_{21} = \int_0^1 ([\mathbf{Z}_0^{(1)\prime} + \mathrm{A}^{(2)}E_0\mathbf{Z}_0^{(1)}] \cdot \mathbf{Z}_0^{(2)})\,\mathrm{d}\eta\,,$$

$$c_{22} = \int_0^1 ([\mathbf{Z}_0^{(2)\prime} + \mathrm{A}^{(2)}E_0\mathbf{Z}_0^{(2)}] \cdot \mathbf{Z}_0^{(2)})\,\mathrm{d}\eta\,.$$

Setting $f_1 = f_{10}\mathrm{e}^{\lambda\tau}$ and $f_2 = f_{20}\mathrm{e}^{\lambda\tau}$, we obtain, by use of (4.141), the characteristic equation in λ_1 and λ_2:

$$D = \det\begin{bmatrix} a_{11}\lambda^2 + c_{11} & a_{12}\lambda^2 + c_{12} \\ a_{21}\lambda^2 + c_{21} & a_{22}\lambda^2 + c_{22} \end{bmatrix} = 0\,. \tag{4.142}$$

Consider another approach to the approximate determination of the frequencies. Let us represent the system of equations (4.123)–(4.126) as two vector equations

$$\mathbf{L}^{(1)} = \mathrm{B}^{(1)}\frac{\partial^2\mathbf{Y}}{\partial\tau^2} + \frac{\partial\mathbf{X}}{\partial\eta} + \mathrm{B}^{(2)}\mathbf{X} = 0\,, \tag{4.143}$$

$$\mathbf{L}^{(2)} = \frac{\partial\mathbf{Y}}{\partial\eta} + \mathrm{C}^{(1)}\mathbf{Y} + \mathrm{C}^{(2)}\mathbf{X} = 0\,, \tag{4.144}$$

where

$$\mathrm{B}^{(1)} = \begin{bmatrix} 0 & -n_1\mathrm{E} \\ -\mathrm{J} & 0 \end{bmatrix}, \qquad \mathbf{Y} = \begin{bmatrix} \boldsymbol{\vartheta} \\ \mathbf{u} \end{bmatrix}, \qquad \mathbf{X} = \begin{bmatrix} \Delta\mathbf{Q} \\ \Delta\mathbf{M} \end{bmatrix},$$

$$\mathrm{B}^{(2)} = \begin{bmatrix} \mathrm{A}_{æ} & \mathrm{A}_Q\mathrm{A}^{-1} \\ \mathrm{A}_1 & \mathrm{A}_M\mathrm{A}^{-1} + \mathrm{A}_{æ} \end{bmatrix},$$

$$\mathrm{C}^{(1)} = \begin{bmatrix} -\mathrm{A}_{æ} & 0 \\ \mathrm{A}_1 & \mathrm{A}_{æ} \end{bmatrix}, \qquad \mathrm{C}^{(2)} = \begin{bmatrix} 0 & -\mathrm{A}^{-1} \\ 0 & 0 \end{bmatrix}.$$

We seek a solution of the system of equations (4.143)–(4.144) in the following form:

$$\mathbf{Y} = \sum_{i=1}^{n} f^{(i)}(\tau)\,\mathbf{Y}_0^{(i)}(\eta)\,, \qquad \mathbf{X} = \sum_{i=1}^{n} \Phi^{(i)}(\tau)\,\mathbf{X}_0^{(i)}(\eta)\,; \tag{4.145}$$

here,

$$\mathbf{Y}_0^{(i)} = \boldsymbol{\varphi}^i = \begin{bmatrix} \boldsymbol{\vartheta}_0^{(i)} \\ \mathbf{u}_0^{(i)} \end{bmatrix}, \qquad \mathbf{X}_0^{(i)} = \boldsymbol{\psi}^{(i)} = \begin{bmatrix} \Delta\mathbf{Q}_0^{(i)} \\ \Delta\mathbf{M}_0^{(i)} \end{bmatrix},$$

and $f^{(i)}(\tau)$ and $\Phi^{(i)}(\tau)$ are yet unknown functions of time.

Using (4.145) in (4.143) and (4.144), we get

$$\mathbf{L}^{(1)} = \boldsymbol{\delta}_1\,, \qquad \mathbf{L}^{(2)} = \boldsymbol{\delta}_2\,, \tag{4.146}$$

where the vectors $\boldsymbol{\delta}_1$ and $\boldsymbol{\delta}_2$ are the residual functions for the vector equations (4.143) and (4.144), respectively.

Let us represent the virtual displacements in the form

$$\delta\mathbf{Y}_0^{(i)} = \delta a_i \mathrm{E}_0 \mathbf{Y}_0^{(i)}\,, \qquad \delta\mathbf{X}_0^{(i)} = \delta b_i \mathrm{E}_0 \mathbf{X}_0^{(i)}\,.$$

Considering the principle of virtual displacements for the case of two-term approximation, from (4.146) we get

$$\int_0^1 (\boldsymbol{\delta}_1 \cdot E_0 \mathbf{Y}_0^{(1)})\,\mathrm{d}\eta = 0\,, \qquad \int_0^1 (\boldsymbol{\delta}_1 \cdot E_0 \mathbf{Y}_0^{(2)})\,\mathrm{d}\eta = 0\,,$$

$$\int_0^1 (\boldsymbol{\delta}_2 \cdot E_0 \mathbf{X}_0^{(1)})\,\mathrm{d}\eta = 0\,, \qquad \int_0^1 (\boldsymbol{\delta}_2 \cdot E_0 \mathbf{X}_0^{(2)})\,\mathrm{d}\eta = 0\,.$$

On reaarangement, we obtain

$$\begin{aligned} &h_{11}\ddot{f}^{(1)} + h_{12}\ddot{f}^{(2)} + b_{11}\Phi^{(1)} + b_{12}\Phi^{(2)} = 0\,,\\ &h_{21}\ddot{f}^{(1)} + h_{22}\ddot{f}^{(2)} + b_{21}\Phi^{(1)} + b_{22}\Phi^{(2)} = 0\,, \end{aligned} \tag{4.147}$$

$$\begin{aligned} &c_{11}f^{(1)} + c_{12}f^{(2)} + d_{11}\Phi^{(1)} + d_{12}\Phi^{(2)} = 0\,,\\ &c_{21}f^{(1)} + c_{22}f^{(2)} + d_{21}\Phi^{(1)} + d_{22}\Phi^{(2)} = 0\,; \end{aligned} \tag{4.148}$$

here,

$$h_{11} = \int_0^1 (\mathrm{B}^{(1)}\mathbf{Y}_0^{(1)} \cdot \mathrm{E}_0\mathbf{Y}_0^{(1)})\,\mathrm{d}\eta\,, \qquad h_{12} = \int_0^1 (\mathrm{B}^{(1)}\mathbf{Y}_0^{(2)} \cdot \mathrm{E}_0\mathbf{Y}_0^{(1)})\,\mathrm{d}\eta\,,$$

$$h_{21} = \int_0^1 (\mathrm{B}^{(1)}\mathbf{Y}_0^{(1)} \cdot \mathrm{E}_0\mathbf{Y}_0^{(2)})\,\mathrm{d}\eta\,, \qquad h_{22} = \int_0^1 (\mathrm{B}^{(1)}\mathbf{Y}_0^{(2)} \cdot \mathrm{E}_0\mathbf{Y}_0^{(2)})\,\mathrm{d}\eta\,,$$

$$b_{11} = \int_0^1 ([\mathbf{X}_0^{(1)\prime} + \mathrm{B}^{(2)}\mathbf{X}_0^{(1)}] \cdot \mathrm{E}_0\mathbf{Y}^{(1)})\,\mathrm{d}\eta\,,$$

$$b_{12} = \int_0^1 ([\mathbf{X}_0^{(2)\prime} + \mathrm{B}^{(2)}\mathbf{X}_0^{(2)}] \cdot \mathrm{E}_0\mathbf{Y}^{(1)})\,\mathrm{d}\eta\,,$$

$$b_{21} = \int_0^1 ([\mathbf{X}_0^{(1)\prime} + \mathrm{B}^{(2)}\mathbf{X}_0^{(1)}] \cdot \mathrm{E}_0\mathbf{Y}^{(2)})\,\mathrm{d}\eta\,,$$

$$b_{22} = \int_0^1 ([\mathbf{X}_0^{(2)\prime} + \mathrm{B}^{(2)}\mathbf{X}_0^{(2)}] \cdot \mathrm{E}_0\mathbf{Y}^{(2)})\,\mathrm{d}\eta\,,$$

$$c_{11} = \int_0^1 ([\mathbf{Y}_0^{(1)\prime} + \mathrm{C}^{(1)}\mathbf{Y}_0^{(1)}] \cdot \mathrm{E}_0\mathbf{X}_0^{(1)})\,\mathrm{d}\eta\,,$$

$$c_{12} = \int_0^1 ([\mathbf{Y}_0^{(2)\prime} + \mathrm{C}^{(1)}\mathbf{Y}_0^{(2)}] \cdot \mathrm{E}_0\mathbf{X}_0^{(1)})\,\mathrm{d}\eta\,,$$

$$c_{21} = \int_0^1 ([\mathbf{Y}_0^{(1)\prime} + \mathrm{C}^{(1)}\mathbf{Y}_0^{(1)}] \cdot \mathrm{E}_0\mathbf{X}_0^{(2)})\,\mathrm{d}\eta\,,$$

$$c_{22} = \int_0^1 ([\mathbf{Y}_0^{(2)\prime} + \mathrm{C}^{(1)}\mathbf{Y}_0^{(2)}] \cdot \mathrm{E}_0\mathbf{X}_0^{(2)})\,\mathrm{d}\eta\,,$$

$$d_{11} = \int_0^1 (\mathrm{C}^{(2)}\mathbf{X}_0^{(1)} \cdot \mathrm{E}_0\mathbf{X}_0^{(1)})\,\mathrm{d}\eta\,, \qquad d_{12} = \int_0^1 (\mathrm{C}^{(2)}\mathbf{X}_0^{(2)} \cdot \mathrm{E}_0\mathbf{X}_0^{(1)})\,\mathrm{d}\eta\,,$$

$$d_{21} = \int_0^1 (\mathrm{C}^{(2)}\mathbf{X}_0^{(1)} \cdot \mathrm{E}_0\mathbf{X}_0^{(2)})\,\mathrm{d}\eta\,, \qquad d_{22} = \int_0^1 (\mathrm{C}^{(2)}\mathbf{X}_0^{(2)} \cdot \mathrm{E}_0\mathbf{X}_0^{(2)})\,\mathrm{d}\eta\,.$$

Solving for $\Phi^{(1)}$ and $\Phi^{(2)}$ from (4.148), we have

$$\Phi^{(1)} = a_{11}f^{(1)} + a_{12}f^{(2)}\,, \qquad \Phi^{(2)} = a_{21}f^{(1)} + a_{22}f^{(2)}\,.$$

Then, eliminating these functions from (4.147), we get

$$\mathrm{H}\ddot{\mathbf{f}} + \mathrm{P}\mathbf{f} = 0\,.$$

Setting $\mathbf{f} = \mathbf{f}_0 \mathrm{e}^{\mathrm{i}\lambda\tau}$, we obtain the equation

$$D = \det\left[\mathrm{P} - \lambda^2\mathrm{H}\right] = 0\,. \tag{4.149}$$

Thus, we can find the approximate values of the first two frequencies from this equation.

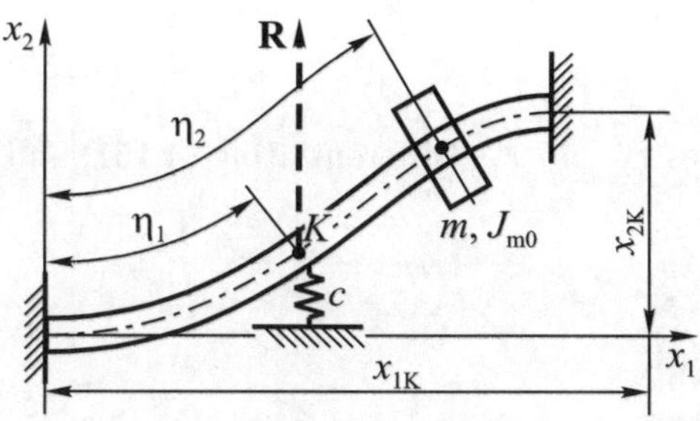

Fig. 4.16.

As an example, we determine the first two frequencies for a rod of constant cross section (Fig. 4.16). Restrict our consideration to in-plane vibrations of the rod.

First, let us find the eigenvectors for a rod with no concentrated masses and intermediate supports. To do this, we must determine some geometric characteristics of the rod.

In the natural configuration, the rod axis is defined by the equation

$$x_2 = 1 - \cos\left(\frac{\pi x_1}{x_{1\,\mathrm{f}}}\right), \tag{4.150}$$

where x_1 and x_2 are the dimensionless coordinates of the axial points of the rod. Using the method discussed in Appendix A of Svetlitsky (2000a), we can find the functions $x_j(\eta)$, $x'_j(\eta)$, and $æ_{30}(\eta)$. Differentiating (4.150) with respect to η, we get

$$x'_2 = \alpha x'_1 \sin \alpha x_1\,, \qquad \alpha = \frac{\pi}{x_{1\,\mathrm{f}}}\,. \tag{4.151}$$

The following condition for the derivatives $x_j{}'$ must be satisfied:

$$(x'_1)^2 + (x'_2)^2 = 1\,. \tag{4.152}$$

Consequently, from (4.151) and (4.152) it follows that

$$x'_1 = \frac{1}{\sqrt{1+\alpha^2 \sin^2 \alpha x_1}}\,, \qquad x'_2 = \frac{\alpha \sin \alpha x_1}{\sqrt{1+\alpha^2 \sin^2 \alpha x_1}}\,. \tag{4.153}$$

Recall that we have $x_1 = x_2 = 0$ at $\eta = 0$, and $x_1 = x_{1\,\mathrm{f}}$ and $x_2 = x_{2\,\mathrm{f}}$ at $\eta = 1$. Then, integrating (4.153), we obtain $x_1(\eta)$ and, consequently, $x_2(\eta)$, x'_1, and x'_2.

Having found x'_j, we obtain the matrix of transformation L^0 from the basis $\{\mathbf{i}_j\}$ to the basis $\{\mathbf{e}_{i0}\}$:

$$\mathrm{L}^0 = \begin{bmatrix} \cos\vartheta_{30} & \sin\vartheta_{30} \\ -\sin\vartheta_{30} & \cos\vartheta_{30} \end{bmatrix}; \tag{4.154}$$

here, $x'_1 = \cos\vartheta_{30}$ and $x'_2 = \sin\vartheta_{30}$. Because

$$æ_{30} = \sqrt{(x''_1)^2 + (x''_2)^2}\,, \tag{4.155}$$

we should know the second derivatives x''_1 and x''_1. Differentiating (4.151) and (4.152), we get

$$x'_1 x''_1 + x'_2 x''_2 = 0\,,$$
$$x''_2 = \alpha^2 \cos \alpha x_1 {x'_1}^2 + \alpha \sin \alpha x_1 x''_1\,;$$

hence,

$$x_1'' = -\frac{\alpha^2 \cos \alpha x_1}{(1+\alpha^2 \sin^2 \alpha x_1)\, 2\alpha \sin \alpha x_1}\,,$$

$$x_2'' = \frac{\alpha^2 \cos \alpha x_1}{2(1+\alpha^2 \sin^2 \alpha x_1)}\,.$$

Substituting x_j' into (4.155), after some rearrangements, we arrive at

$$\text{æ}_{30}(\eta) = \frac{\alpha^2}{2} \frac{\cos \alpha x_1}{\sqrt{1+\alpha^2 \sin^2 \alpha x_1}} = \frac{\alpha^2}{2} x_1' \cos \alpha x_1\,.$$

The equations governing free in-plane vibrations of a rod when the rotary inertia is ignored ($c = 0$ and $m = J_m = 0$) are as follows:

$$\frac{\partial^2 u_1}{\partial \tau^2} - \frac{\partial \Delta Q_1}{\partial \eta} + \text{æ}_{30}\, \Delta Q_2 = 0\,,$$
$$\frac{\partial^2 u_1}{\partial \tau^2} - \frac{\partial \Delta Q_2}{\partial \eta} - \text{æ}_{30}\, \Delta Q_1 = 0\,,$$
$$\frac{\partial \Delta M_3}{\partial \eta} + \Delta Q_2 = 0\,,$$
$$\frac{\partial \vartheta_3}{\partial \eta} - \Delta M_3 = 0\,,$$
$$\frac{\partial u_1}{\partial \eta} - \text{æ}_{30} u_2 = 0\,,$$
$$\frac{\partial u_2}{\partial \eta} + \text{æ}_{30} u_1 - \vartheta_3 = 0\,.$$

Setting

$$\Delta Q_1 = \Delta Q_{10}\, \mathrm{e}^{\mathrm{i}\lambda\tau}\,, \qquad \Delta Q_2 = \Delta Q_{20}\, \mathrm{e}^{\mathrm{i}\lambda\tau}\,, \qquad \Delta M_3 = \Delta M_{30}\, \mathrm{e}^{\mathrm{i}\lambda\tau}\,,$$
$$\vartheta_3 = \vartheta_{30}\, \mathrm{e}^{\mathrm{i}\lambda\tau}\,, \qquad \Delta u_1 = \Delta u_{10}\, \mathrm{e}^{\mathrm{i}\lambda\tau}\,, \qquad \Delta u_2 = \Delta u_{20}\, \mathrm{e}^{\mathrm{i}\lambda\tau}\,,$$

we arrive at one vector equation

$$\mathbf{Z}_0' + \mathrm{A}_0 \mathbf{Z}_0 = 0\,, \tag{4.156}$$

where

$$\mathbf{Z}_0 = \begin{bmatrix} \Delta Q_{10} \\ \Delta Q_{20} \\ \Delta M_{30} \\ \vartheta_{30} \\ u_{10} \\ u_{20} \end{bmatrix}, \qquad \mathrm{A}_0 = \begin{bmatrix} 0 & -\text{æ}_{30} & 0 & 0 & \lambda^2 & 0 \\ \text{æ}_{30} & 0 & 0 & 0 & 0 & \lambda^2 \\ 0 & 1 & 0 & 0 & 0 & 0 \\ 0 & 0 & -1 & 0 & 0 & 0 \\ 0 & 0 & 0 & 0 & 0 & -\text{æ}_{30} \\ 0 & 0 & 0 & -1 & \text{æ}_{30} & 0 \end{bmatrix}.$$

The solution of (4.156) is

$$\mathbf{Z}_0 = \mathrm{K}(\eta)\, \mathbf{C}\,, \qquad \mathrm{K}(0) = \mathrm{E}\,.$$

At $\eta = 0$, the boundary conditions are $\vartheta_{30} = u_{10} = u_{20} = 0$; thus, we have $c_4 = c_5 = c_6 = 0$.

At $\eta = 1$, the boundary conditions are $\vartheta_{30} = u_{10} = u_{20} = 0$. Therefore, we get three homogeneous equations

$$
\begin{aligned}
&k_{41}(1)\, c_1 + k_{42}(1)\, c_2 + k_{43}(1)\, c_3 = 0\,,\\
&k_{51}(1)\, c_1 + k_{52}(1)\, c_2 + k_{53}(1)\, c_3 = 0\,,\\
&k_{61}(1)\, c_1 + k_{62}(1)\, c_2 + k_{63}(1)\, c_3 = 0\,.
\end{aligned}
$$

Then, the eigenfrequencies λ_j can be found from the relationship

$$
D = \det \begin{bmatrix} k_{41}(1) & k_{42}(1) & k_{43}(1) \\ k_{51}(1) & k_{52}(1) & k_{53}(1) \\ k_{61}(1) & k_{62}(1) & k_{63}(1) \end{bmatrix} = 0\,, \tag{4.157}
$$

Using (4.157) and setting $c_3^{(j)} = 1$, we arrive at $c_1^{(j)} = \alpha_1^{(j)}$ and $c_2^{(j)} = \alpha_2^{(j)}$. Finally, the eigenfunctions (that is, the components of the vectors $\mathbf{Z}_0^{(j)}$) are

$$
\begin{aligned}
&Z_{01}^{(j)} = \Delta Q_{10}^{(j)} = k_{11}(\eta,\, \lambda_j)\, \alpha_1^{(j)} + k_{12}(\eta,\, \lambda_j)\, \alpha_2^{(j)} + k_{13}(\eta,\, \lambda_j)\,,\\
&\dots\\
&Z_{06}^{(j)} = u_{20}^{(j)} = k_{61}(\eta,\, \lambda_j)\, \alpha_1^{(j)} + k_{62}(\eta,\, \lambda_j)\, \alpha_2^{(j)} + k_{63}(\eta,\, \lambda_j)\,.
\end{aligned} \tag{4.158}
$$

Let us consider now the determination of approximate values of the frequencies for a rod with an elastic support and a concentrated mass shown in Fig. 4.16. First, let us find the components of the reaction force $\mathbf{R}$ in the attached basis. In Fig. 4.16, this force is shown as a dashed line. Assume that the force $\mathbf{R}$ remains parallel to the axis x_2 during small vibrations of the rod, that is, $\mathbf{R} = |\mathbf{R}|\mathbf{i}_2$. The force $\mathbf{R}$ depends on the vertical displacement $\mathbf{u}_K$ of the point K as follows:

$$
\mathbf{R} = -c(\mathbf{u}_K \cdot \mathbf{i}_2)\, \mathbf{i}_2\,, \qquad c = \frac{c_{\mathrm{d}} l^3}{A_{33}}\,; \tag{4.159}
$$

here, c is the dimensionless spring constant and c_{d} is the dimensional spring constant.

Using the matrix L^0 given by (4.154), we get

$$
\begin{aligned}
\mathbf{R} &= R_1 \mathbf{e}_1 + R_2 \mathbf{e}_2\\
&= -c\left[u_1(\eta_1)\, \sin^2 \vartheta_{30}(\eta_1) + u_2(\eta_1)\, \sin \vartheta_{30}(\eta_1)\, \cos \vartheta_{30}(\eta_1)\right] \mathbf{e}_1\\
&\quad - c\left[u_1(\eta_1)\, \sin \vartheta_{30}(\eta_1)\, \cos \vartheta_{30}(\eta_1) + u_2(\eta_1)\, \cos^2 \vartheta_{30}(\eta_1)\right] \mathbf{e}_2\,.
\end{aligned}
$$

The equations of small vibrations having regard to the elastic constraint and the concentrated mass read

$$\frac{\partial^2 u_1}{\partial \tau^2} - \frac{\partial \Delta Q_1}{\partial \eta} + æ_{30}\,\Delta Q_2 - R_1\,\delta(\eta-\eta_1) + n_2\,\frac{\partial^2 u_1}{\partial \tau^2}\,\delta(\eta-\eta_2) = 0\,,$$
$$\frac{\partial^2 u_2}{\partial \tau^2} - \frac{\partial \Delta Q_2}{\partial \eta} - æ_{30}\,\Delta Q_1 - R_2\,\delta(\eta-\eta_1) + n_2\,\frac{\partial^2 u_2}{\partial \tau^2}\,\delta(\eta-\eta_2) = 0\,,$$
$$J^0_{m0} = \frac{\partial^2 \vartheta_3}{\partial \tau^2}\,\delta(\eta-\eta_2) - \frac{\partial \Delta M_3}{\partial \eta} - \Delta Q_2 = 0\,, \tag{4.160}$$
$$\frac{\partial \vartheta_3}{\partial \eta} - \Delta M_3 = 0\,,$$
$$\frac{\partial u_1}{\partial \eta} - æ_{30} u_2 = 0\,,$$
$$\frac{\partial u_2}{\partial \eta} + æ_{30} u_1 - \vartheta_3 = 0\,, \tag{4.161}$$

where

$$n_2 = \frac{m}{m_0 l}\,, \qquad J^0_{m0} = \frac{J_{m0}}{m_0 l^3}\,.$$

Let us represent these equations in the form of two vector equations. We have

$$\mathbf{L}_1 = \mathrm{B}^{(1)}\,\frac{\partial^2 \mathbf{Y}}{\partial \tau^2} + \frac{\partial \mathbf{X}}{\partial \eta} + \mathrm{B}^{(2)}\mathbf{X} + \mathrm{B}^{(3)}\mathbf{Y} = 0\,, \tag{4.162}$$
$$\mathbf{L}_2 = \frac{\partial \mathbf{Y}}{\partial \eta} + \mathrm{C}^{(1)}\mathbf{Y} + \mathrm{C}^{(2)}\mathbf{X} = 0\,, \tag{4.163}$$

where

$$\mathrm{B}^{(1)} = \begin{bmatrix} 0 & -(1+n_2\,\delta_2) & 0 \\ 0 & 0 & -(1+n_2\,\delta_2) \\ -J^0_{m0}\,\delta_2 & 0 & 0 \end{bmatrix},$$
$$\mathrm{B}^{(2)} = \begin{bmatrix} 0 & -æ_{30} & 0 \\ æ_{30} & 0 & 0 \\ 0 & 1 & 0 \end{bmatrix},$$
$$\mathrm{B}^{(3)} = \begin{bmatrix} \delta_2 & c\sin^2\vartheta_{30}\,\delta_1 & c\sin\vartheta_{30}\cos\vartheta_{30}\,\delta_1 \\ 0 & c\sin\vartheta_{30}\cos\vartheta_{30}\,\delta_1 & c\cos^2\vartheta_{30}\,\delta_1 \\ 0 & 0 & 0 \end{bmatrix},$$
$$\mathrm{C}^{(1)} = \begin{bmatrix} 0 & 0 & 0 \\ 0 & 0 & -æ_{30} \\ -1 & æ_{30} & 0 \end{bmatrix}, \qquad \mathrm{C}^{(2)} = \begin{bmatrix} 0 & 0 & -1 \\ 0 & 0 & 0 \\ 0 & 0 & 0 \end{bmatrix},$$
$$\mathbf{Y} = \begin{bmatrix} \vartheta_{30} \\ u_{10} \\ u_{20} \end{bmatrix}, \qquad \mathbf{X} = \begin{bmatrix} \Delta Q_{10} \\ \Delta Q_{20} \\ \Delta M_{30} \end{bmatrix}.$$

Here, $\delta_1 = \delta(\eta-\eta_1)$ and $\delta_2 = \delta(\eta-\eta_2)$.

When solving the system of equations (4.162)–(4.164), let us restrict our consideration to the case of two-term approximation and thus set

$$\mathbf{Y} = f^{(1)}(\tau)\mathbf{Y}_0^{(1)}(\eta) + f^{(2)}\tau\mathbf{Y}_0^{(2)}(\eta)\,, \tag{4.164}$$
$$\mathbf{X} = \Phi^{(1)}(\tau)\mathbf{X}_0^{(1)}(\eta) + \Phi^{(2)}(\tau)\mathbf{X}_0^{(2)}(\eta)\,, \tag{4.165}$$

where $f^{(i)}(\tau)$ and $\Phi^{(i)}(\tau)$ are yet unknown functions of time; the components of the vectors $\mathbf{Y}_0^{(i)}$ and $\mathbf{X}_0^{(i)}$ are the eigenfunctions (4.158), that is,

$$\mathbf{Y}_0^{(i)} = \begin{bmatrix} Z_{04}^{(i)} \\ Z_{05}^{(i)} \\ Z_{06}^{(i)} \end{bmatrix} = \begin{bmatrix} \vartheta_{30}^{(i)} \\ u_{10}^{(i)} \\ u_{20}^{(i)} \end{bmatrix}, \qquad \mathbf{X}_0^{(i)} = \begin{bmatrix} Z_{01}^{(i)} \\ Z_{02}^{(i)} \\ Z_{03}^{(i)} \end{bmatrix} = \begin{bmatrix} \Delta Q_{10} \\ \Delta Q_{20} \\ \Delta M_{30} \end{bmatrix}.$$

As the generalized virtual displacements we can take the vectors

$$\delta\mathbf{Y}_0^{(i)} = \delta a_i\,\mathrm{E}_{01}\mathbf{Y}_0^{(i)}\,, \qquad \delta\mathbf{X}_0^{(i)} = \delta b_i\,\mathrm{E}_{02}\mathbf{X}_0^{(i)}\,,$$

where

$$\mathrm{E}_{01} = \begin{bmatrix} 0 & 1 & 0 \\ 0 & 0 & 1 \\ 1 & 0 & 0 \end{bmatrix}, \qquad \mathrm{E}_{02} = \begin{bmatrix} 0 & 0 & 1 \\ 1 & 0 & 0 \\ 0 & 1 & 0 \end{bmatrix}.$$

Substituting (4.164) and (4.165) into (4.162) and (4.163) and applying the principle of virtual displacements, we obtain

$$\int_0^1 (\mathbf{L}_1 \cdot \mathrm{E}_{01}\mathbf{Y}_0^{(1)})\,d\eta = 0\,, \qquad \int_0^1 (\mathbf{L}_1 \cdot \mathrm{E}_{01}\mathbf{Y}_0^{(2)})\,d\eta = 0\,,$$
$$\int_0^1 (\mathbf{L}_2 \cdot \mathrm{E}_{02}\mathbf{X}_0^{(1)})\,d\eta = 0\,, \qquad \int_0^1 (\mathbf{L}_2 \cdot \mathrm{E}_{02}\mathbf{X}_0^{(2)})\,d\eta = 0\,.$$

On rearrangement, this gives

$$\begin{aligned} &h_{11}\ddot{f}^{(1)} + h_{12}\ddot{f}^{(2)} + b_{11}\Phi^{(1)} + b_{12}\Phi^{(2)} + c_{11}f^{(1)} + c_{12}f^{(2)} = 0\,,\\ &h_{21}\ddot{f}^{(1)} + h_{22}\ddot{f}^{(2)} + b_{21}\Phi^{(1)} + b_{22}\Phi^{(2)} + c_{21}f^{(1)} + c_{22}f^{(2)} = 0\,, \end{aligned} \tag{4.166}$$
$$\begin{aligned} &d_{11}f^{(1)} + d_{12}f^{(2)} + e_{11}\Phi^{(1)} + e_{12}\Phi^{(2)} = 0\,,\\ &d_{21}f^{(1)} + d_{22}f^{(2)} + e_{21}\Phi^{(1)} + e_{22}\Phi^{(2)} = 0\,, \end{aligned} \tag{4.167}$$

where

$$h_{11} = \int_0^1 (\mathrm{B}^{(1)}\mathbf{Y}_0^{(1)} \cdot \mathrm{E}_{01}\mathbf{Y}_0^{(1)})\,d\eta\,, \qquad h_{12} = \int_0^1 (\mathrm{B}^{(1)}\mathbf{Y}_0^{(2)} \cdot \mathrm{E}_{01}\mathbf{Y}_0^{(1)})\,d\eta\,,$$
$$h_{21} = \int_0^1 (\mathrm{B}^{(1)}\mathbf{Y}_0^{(1)} \cdot \mathrm{E}_{01}\mathbf{Y}_0^{(2)})\,d\eta\,, \qquad h_{22} = \int_0^1 (\mathrm{B}^{(1)}\mathbf{Y}_0^{(2)} \cdot \mathrm{E}_{01}\mathbf{Y}_0^{(2)})\,d\eta\,,$$
$$b_{11} = \int_0^1 ([\mathbf{X}_0^{(1)\prime} + \mathrm{B}^{(2)}\mathbf{X}_0^{(1)}] \cdot \mathrm{E}_{01}\mathbf{Y}_0^{(1)})\,d\eta\,,$$

$$b_{12} = \int_0^1 ([\mathbf{X}_0^{(2)\prime} + \mathrm{B}^{(2)}\mathbf{X}_0^{(2)}] \cdot \mathrm{E}_{01}\mathbf{Y}_0^{(1)})\,\mathrm{d}\eta\,,$$

$$b_{21} = \int_0^1 ([\mathbf{X}_0^{(1)\prime} + \mathrm{B}^{(2)}\mathbf{X}_0^{(1)}] \cdot \mathrm{E}_{01}\mathbf{Y}_0^{(2)})\,\mathrm{d}\eta\,,$$

$$b_{22} = \int_0^1 ([\mathbf{X}_0^{(2)\prime} + \mathrm{B}^{(2)}\mathbf{X}_0^{(2)}] \cdot \mathrm{E}_{01}\mathbf{Y}_0^{(2)})\,\mathrm{d}\eta\,,$$

$$c_{11} = \int_0^1 (\mathrm{B}^{(3)}\mathbf{Y}_0^{(1)} \cdot \mathrm{E}_{01}\mathbf{Y}_0^{(1)})\,\mathrm{d}\eta\,, \qquad c_{12} = \int_0^1 (\mathrm{B}^{(3)}\mathbf{Y}_0^{(2)} \cdot \mathrm{E}_{01}\mathbf{Y}_0^{(1)})\,\mathrm{d}\eta\,,$$

$$c_{21} = \int_0^1 (\mathrm{B}^{(3)}\mathbf{Y}_0^{(1)} \cdot \mathrm{E}_{01}\mathbf{Y}_0^{(2)})\,\mathrm{d}\eta\,, \qquad c_{22} = \int_0^1 (\mathrm{B}^{(3)}\mathbf{Y}_0^{(2)} \cdot \mathrm{E}_{01}\mathbf{Y}_0^{(2)})\,\mathrm{d}\eta\,,$$

$$d_{11} = \int_0^1 (\mathbf{Y}_0^{(1)\prime} + \mathrm{C}^{(1)}\mathbf{Y}_0^{(1)}) \cdot \mathrm{E}_{02}\mathbf{X}_0^{(1)}\,\mathrm{d}\eta\,,$$

$$d_{12} = \int_0^1 (\mathbf{Y}_0^{(2)\prime} + \mathrm{C}^{(1)}\mathbf{Y}_0^{(2)}) \cdot \mathrm{E}_{02}\mathbf{X}_0^{(1)}\,\mathrm{d}\eta\,,$$

$$d_{21} = \int_0^1 (\mathbf{Y}_0^{(1)\prime} + \mathrm{C}^{(1)}\mathbf{Y}_0^{(1)}) \cdot \mathrm{E}_{02}\mathbf{X}_0^{(2)}\,\mathrm{d}\eta\,,$$

$$d_{22} = \int_0^1 (\mathbf{Y}_0^{(2)\prime} + \mathrm{C}^{(1)}\mathbf{Y}_0^{(2)}) \cdot \mathrm{E}_{02}\mathbf{X}_0^{(2)}\,\mathrm{d}\eta\,,$$

$$e_{11} = \int_0^1 (\mathrm{C}^{(2)}\mathbf{X}_0^{(1)} \cdot \mathrm{E}_{02}\mathbf{X}_0^{(1)})\,\mathrm{d}\eta\,, \qquad e_{12} = \int_0^1 (\mathrm{C}^{(2)}\mathbf{X}_0^{(2)} \cdot \mathrm{E}_{02}\mathbf{X}_0^{(1)})\,\mathrm{d}\eta\,,$$

$$e_{21} = \int_0^1 (\mathrm{C}^{(2)}\mathbf{X}_0^{(1)} \cdot \mathrm{E}_{02}\mathbf{X}_0^{(2)})\,\mathrm{d}\eta\,, \qquad e_{22} = \int_0^1 (\mathrm{C}^{(2)}\mathbf{X}_0^{(2)} \cdot \mathrm{E}_{02}\mathbf{X}_0^{(2)})\,\mathrm{d}\eta\,.$$

Solving for $\Phi^{(1)}$ and $\Phi^{(2)}$ from (4.167), we get

$$\Phi^{(1)} = \gamma_{11}f^{(1)} + \gamma_{12}f^{(2)}\,, \qquad \Phi^{(2)} = \gamma_{21}f^{(1)} + \gamma_{22}f^{(2)}\,.$$

Eliminating these functions from (4.166), we arrive at

$$h_{11}\ddot{f}^{(1)} + h_{12}\ddot{f}^{(2)} + g_{11}f^{(1)} + g_{12}f^{(2)} = 0\,,$$
$$h_{21}\ddot{f}^{(1)} + h_{22}\ddot{f}^{(2)} + g_{21}f^{(1)} + g_{22}f^{(2)} = 0$$

or

$$\mathrm{H}\ddot{\mathbf{f}} + \mathrm{G}\mathbf{f} = 0\,. \tag{4.168}$$

Setting $\mathbf{f} = \mathbf{f}_0\mathrm{e}^{\mathrm{i}\lambda\tau}$, we get the equation

$$D = \det\left[\mathrm{G} - \lambda^2\mathrm{H}\right] = 0\,,$$

from which λ_j can be found.

The results of calculation of the frequencies for some values of the physical parameters c, m, and J_{m0} are presented in Table 4.1. These parameters

Table 4.1.

γ	0	5	10	20
λ_1	24.00	15.10	12.13	9.18
λ_2	87.21	68.23	63.06	58.30

are proportional to γ, there minimum values are $c = 0.01$, $m = 0.01$, and $J_{m0} = 0.01$.

The method discussed in this section is a powerful tool for the determination of approximate values of the frequencies for complicated cases (rods with intermediate supports and concentrated masses). For nonconservative problems, the method allows us to find the complex eigenvalues that will be used further in the stability analysis of small vibrations.

5. Free and Forced Small Vibrations

This chapter presents theoretical foundation and computational techniques extensively used in the analysis of applied problems concerned with small free and forced vibrations of rods.

5.1 Free Vibrations

5.1.1 Initial Conditions

Let a rod with constraints be subjected to external loads. In practice, free vibrations are encountered when either one of the constraints or one of the loads is abruptly removed. Consider examples.

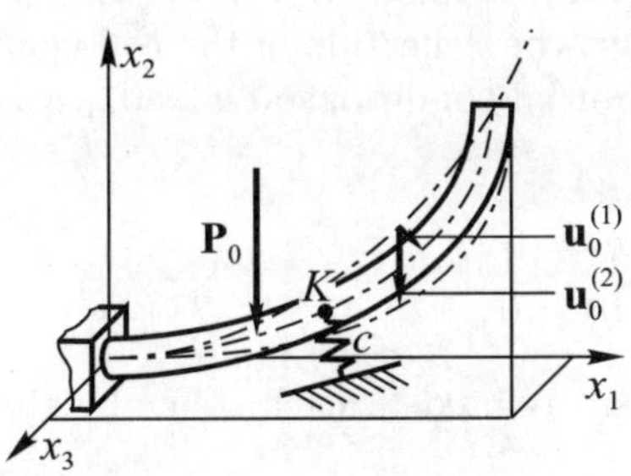

Fig. 5.1.

A loaded rod with an intermediate elastic support is shown in Fig. 5.1. When the action of the support is suddenly terminated, the rod starts to oscillate.

A rod subjected to a force $\mathbf{P}_0$ and a moment $\mathbf{T}_0$ is shown in Fig. 5.2. The dashed line indicates the rod axis in the natural configuration (that is, before application of the loads). When the force $\mathbf{P}_0$ and the moment $\mathbf{T}_0$ are abruptly removed, free vibrations of the rod occur.

A plane spiral spring (say, the executive component of a timing device) mounted on an object that is in an accelerated motion (for example, an airborne vehicle) is shown in Fig. 5.3. Due to the acceleration $\mathbf{a}$, a distributed load $\mathbf{q}$ applied to the spring appears; as a result, the shape of the spring

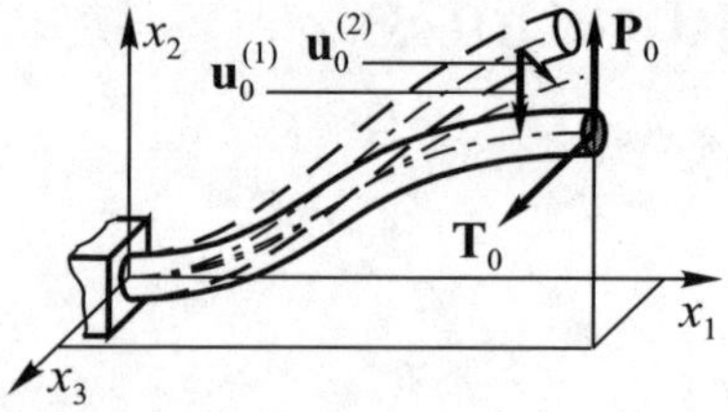

Fig. 5.2.

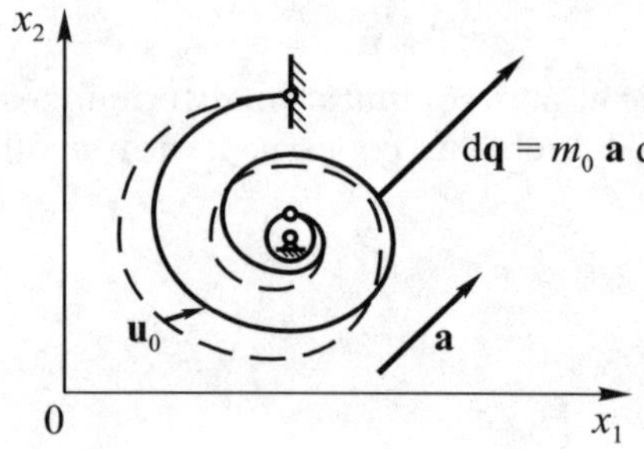

Fig. 5.3.

changes. The spring in the unloaded configuration is shown as a dashed line. If the engine of the vehicle is suddenly shut off, then the acceleration disappears and the spring breaks into free vibrations.

In these examples, the initial conditions are defined by the vectors $\mathbf{u}_0$ and $\boldsymbol{\vartheta}_0$ because the only cause for vibrations is the deflection of the rod from its natural configuration. Consequently, the initial conditions at $\tau = 0$ are as follows:

$$\mathbf{u}(0, \eta) = \mathbf{u}_0(\eta) \neq 0\,, \qquad \boldsymbol{\vartheta}(0, \eta) = \boldsymbol{\vartheta}_0(\eta) \neq 0\,,$$
$$\mathbf{v}(0, \eta) = \mathbf{v}_0(\eta) = 0\,, \qquad \boldsymbol{\omega}(0, \eta) = \boldsymbol{\omega}_0(\eta) = 0\,.$$

Note that the vectors $\mathbf{u}_0(\eta)$ and $\boldsymbol{\vartheta}_0(\eta)$ are related to each other by the equation of strain consistency

$$\frac{d\mathbf{u}_0}{d\eta} + \mathrm{A}_{æ}\mathbf{u}_0 + \mathrm{A}_1\boldsymbol{\vartheta}_0 = 0\,. \tag{5.1}$$

Let the initial conditions be functions of displacements and their first derivatives such that

$$\mathbf{u}(0, \eta) = \mathbf{u}_0(\eta) \neq 0\,, \qquad \boldsymbol{\vartheta}(0, \eta) = \boldsymbol{\vartheta}_0(\eta) \neq 0\,,$$
$$\mathbf{v}(0, \eta) = \mathbf{v}_0(\eta) \neq 0\,, \qquad \dot{\boldsymbol{\vartheta}}(0, \eta) = \boldsymbol{\omega}_0(\eta) \neq 0\,.$$

Then, the vectors $\mathbf{v}_0$ and $\boldsymbol{\omega}_0$ are related by the following equation:

$$\frac{d\boldsymbol{\vartheta}_0}{d\eta} + \mathrm{A}_{æ}\boldsymbol{\vartheta}_0 = \boldsymbol{\omega}_0 \times \mathbf{e}_1\,. \tag{5.2}$$

Initial conditions of this type occur when during forced vibrations of a rod the forces applied disappear at time τ_0. Beginning with this instant, the rod breaks into free vibrations subject to the following initial conditions:

$$\mathbf{u}(\tau_0, \eta) = \mathbf{u}_0(\eta) \neq 0\,, \qquad \boldsymbol{\vartheta}(\tau_0, \eta) = \boldsymbol{\vartheta}_0(\eta) \neq 0\,,$$
$$\mathbf{v}(\tau_0, \eta) = \mathbf{v}_0(\eta) \neq 0\,, \qquad \dot{\boldsymbol{\vartheta}}(\tau_0, \eta) = \boldsymbol{\omega}_0(\eta) \neq 0\,.$$

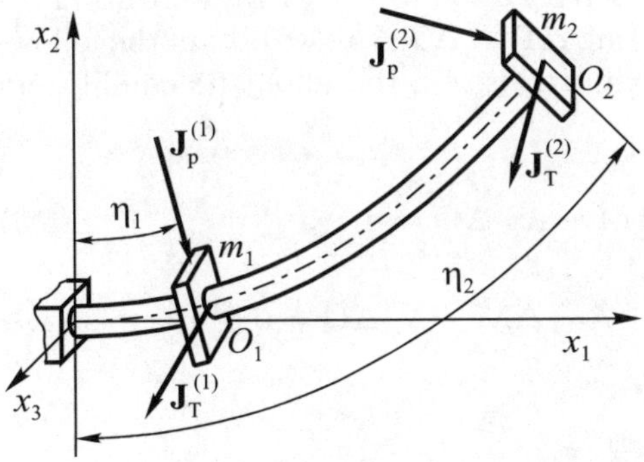

Fig. 5.4.

Consider vibrations caused by an impulsive load applied to a concentrated mass (Fig. 5.4). In this case, the following initial conditions at $\tau = 0$ hold

$$\mathbf{u}_0(\eta) = 0\,, \qquad \boldsymbol{\vartheta}_0(\eta) = 0\,,$$
$$\mathbf{v}_0(\eta_1) \neq 0\,, \qquad \boldsymbol{\omega}_0(\eta_1) \neq 0\,, \qquad \mathbf{v}_0(\eta_2) \neq 0\,, \qquad \boldsymbol{\omega}_0(\eta_2) \neq 0\,.$$

Consider in detail the initial conditions for the cases shown in Figs. 5.1–5.3. For definiteness, assume that $\mathbf{P}_0$ is a *dead* force. Under the action of the force $\mathbf{P}_0$ and due to the reaction of the elastic support, the axial points of the rod take up a new position (Fig. 5.1). This position is well defined by the vectors of linear and angular displacements of a cross section ($\mathbf{u}_0^{(1)}$ and $\boldsymbol{\vartheta}_0^{(1)}$, respectively). These vectors are related to each other by equation (5.1). In the absence of an elastic support, the vector of linear displacements of axial points is $\mathbf{u}_0^{(2)}$ and the vector of angular displacements is $\boldsymbol{\vartheta}_0^{(2)}$. Thus, at the moment the elastic support is to be removed, the initial conditions are

$$\mathbf{u}_0 = \mathbf{u}_0^{(2)} - \mathbf{u}_0^{(1)}\,, \qquad \boldsymbol{\vartheta}_0 = \boldsymbol{\vartheta}_0^{(2)} - \boldsymbol{\vartheta}_0^{(1)}\,, \qquad \mathbf{v}_0 = \boldsymbol{\omega}_0 = 0\,.$$

In the case shown in Fig. 5.3, the initial conditions identical in form to the last ones occur. If the vector of acceleration $\mathbf{a}$ lies in the plane x_1Ox_2, then the components u_{10}, u_{20}, and ϑ_{30} are the only nonzero components of the vectors of linear and angular displacements. Consequently, the initial conditions read

$$u_{30}(0,\,\eta) = 0\,, \qquad u_{10}(0,\,\eta) \neq 0\,, \qquad u_{20}(0,\,\eta) \neq 0\,,$$
$$\vartheta_{10}(0,\,\eta) = \vartheta_{20}(0,\,\eta) = 0\,, \qquad \vartheta_{30}(0,\,\eta) \neq 0\,,$$
$$\mathbf{v}_0(0,\,\eta) = \boldsymbol{\omega}_0(0,\,\eta) = 0\,.$$

5.2 Exact Numerical Solution

The equations governing small vibrations of a rod (3.11)–(3.15) were obtained in Sect. 3.1. Eliminating $\Delta\text{æ}$ and setting $\Delta\mathbf{P} = \Delta\mathbf{T} = 0$, we obtain the following equations governing small free vibrations of a rod about its equilibrium configuration:

$$n_1(\eta)\,\frac{\partial^2 \mathbf{u}}{\partial \tau^2} - \frac{\partial \Delta\mathbf{Q}}{\partial \eta} - \mathrm{A}_Q \mathrm{A}^{-1}\,\Delta\mathbf{M} - \mathrm{A}_{\text{æ}}\,\Delta\mathbf{Q} = 0\,, \tag{5.3}$$

$$\mathrm{J}(\eta)\,\frac{\partial^2 \boldsymbol{\vartheta}}{\partial \tau^2} - \frac{\partial \Delta\mathbf{M}}{\partial \eta} - (\mathrm{A}_M \mathrm{A}^{-1} + \mathrm{A}_{\text{æ}})\,\Delta\mathbf{M} - \mathrm{A}_1\,\Delta\mathbf{Q} = 0\,, \tag{5.4}$$

$$\frac{\partial \boldsymbol{\vartheta}}{\partial \eta} + \mathrm{A}_{\text{æ}}\boldsymbol{\vartheta} - \mathrm{A}^{-1}\,\Delta\mathbf{M} = 0\,, \tag{5.5}$$

$$\frac{\partial \mathbf{u}}{\partial \eta} + \mathrm{A}_{\text{æ}}\mathbf{u} + \mathrm{A}_1\boldsymbol{\vartheta} = 0\,. \tag{5.6}$$

The methods of determination of eigenvectors and eigenvalues of the system of homogeneous equations (5.3)–(5.6) were discussed in Sect. 4.1–4.3. This system can be represented as one vector equation. For $w_0 = 0$, we arrive at the special form of (4.127):

$$\mathrm{A}^{(1)}\,\frac{\partial^2 \mathbf{Z}}{\partial \tau^2} + \frac{\partial \mathbf{Z}}{\partial \eta} + \mathrm{A}^{(2)}\mathbf{Z} = 0\,. \tag{5.7}$$

Setting $\mathbf{Z} = \mathbf{Z}_0 \mathrm{e}^{\mathrm{i}\lambda\tau}$, we find λ_j and the corresponding eigenvectors $\mathbf{Z}_0^{(j)}$. Then, for each λ_j, we obtain two real value partial solutions:

$$\mathbf{Z}_1^{(j)} = \mathbf{Z}_0^{(j)} \cos\lambda_j\tau\,, \qquad \mathbf{Z}_2^{(j)} = \mathbf{Z}_0^{(j)} \sin\lambda_j\tau_0\,.$$

Consequently, the solution of equation (5.7) reads

$$\mathbf{Z}(\tau,\,\eta) = \sum_{j=1}^{\infty} (C^{(j)}\mathbf{Z}_0^{(j)} \cos\lambda_j\tau + B^{(j)}\mathbf{Z}_0^{(j)} \sin\lambda_j\tau)\,, \tag{5.8}$$

where $C^{(j)}$ and $B^{(j)}$ are arbitrary constants.

Since the vector $\mathbf{Z}$ is composed of the vectors $\Delta\mathbf{Q}$, $\Delta\mathbf{M}$, $\boldsymbol{\vartheta}$, and $\mathbf{u}$, from (5.8) it follows that

$$\Delta\mathbf{Q} = \sum_{j=1}^{\infty}(C^{(j)}\mathbf{Z}_Q^{(j)} \cos\lambda_j\tau + B^{(j)}\mathbf{Z}_Q^{(j)} \sin\lambda_j\tau)\,, \tag{5.9}$$

$$\Delta\mathbf{M} = \sum_{j=1}^{\infty}(C^{(j)}\mathbf{Z}_M^{(j)} \cos\lambda_j\tau + B^{(j)}\mathbf{Z}_M^{(j)} \sin\lambda_j\tau)\,, \tag{5.10}$$

$$\boldsymbol{\vartheta} = \sum_{j=1}^{\infty}(C^{(j)}\mathbf{Z}_\vartheta^{(j)} \cos\lambda_j\tau + B^{(j)}\mathbf{Z}_\vartheta^{(j)} \sin\lambda_j\tau)\,, \tag{5.11}$$

$$\mathbf{u} = \sum_{j=1}^{\infty}(C^{(j)}\mathbf{Z}_u^{(j)} \cos\lambda_j\tau + B^{(j)}\mathbf{Z}_u^{(j)} \sin\lambda_j\tau)\,, \tag{5.12}$$

where

$$\mathbf{Z}_Q^{(j)} = \begin{bmatrix} Z_{01}^{(j)} \\ Z_{02}^{(j)} \\ Z_{03}^{(j)} \end{bmatrix}, \qquad \mathbf{Z}_M^{(j)} = \begin{bmatrix} Z_{04}^{(j)} \\ Z_{05}^{(j)} \\ Z_{06}^{(j)} \end{bmatrix},$$

$$\mathbf{Z}_\vartheta^{(j)} = \begin{bmatrix} Z_{07}^{(j)} \\ Z_{08}^{(j)} \\ Z_{09}^{(j)} \end{bmatrix}, \qquad \mathbf{Z}_u^{(j)} = \begin{bmatrix} Z_{0,10}^{(j)} \\ Z_{0,11}^{(j)} \\ Z_{0,12}^{(j)} \end{bmatrix}; \tag{5.13}$$

here, $\mathbf{Z}_{0k}^{(j)}$ are the components of the vector $\mathbf{Z}_0^{(j)}$.

Let us now consider two sets of initial conditions.

Zero Initial Velocity In this case, we have

$$\mathbf{u}(0,\,\eta) = \mathbf{u}_0(\eta)\,, \qquad \mathbf{v}(0,\,\eta) = 0\,,$$

where $\mathbf{u}_0(\eta)$ is a specified vector.

For small vibrations, we have

$$\mathbf{v} = \frac{\partial\mathbf{u}}{\partial\tau}\,,$$

consequently, from the condition $\mathbf{v}(0,\,\eta) = 0$ it follows that $B^{(j)} = 0$.

Consider the expression for $\mathbf{u}(\eta,\,\tau)$ (5.12). For $\tau = 0$, we have $B^{(j)} = 0$ and thus

$$\mathbf{u}_0(\eta) = \sum_{j=1}^{\infty} C^{(j)}\mathbf{Z}_u^{(j)}(\eta)\,. \tag{5.14}$$

In the analysis of applied problems, the infinite summation is to be replaced by a finite one, that is,

$$\mathbf{u}_0(\eta) = \sum_{j=1}^{n} C^{(j)}\mathbf{Z}_u^{(j)}(\eta)\,. \tag{5.15}$$

Multiplying both sides of this relationship by $\mathbf{Z}_u^{(\nu)}(\eta)$ and integrating the result with respect to η between the limits $\eta = 0$ and $\eta = 1$, we get n inhomogeneous linear equations in the n unknowns $C^{(j)}$:

$$\int_0^1 (\mathbf{u}_0 \cdot \mathbf{Z}_u^{(\nu)})\, d\eta = \sum_{j=1}^{n} C^{(j)} \int_0^1 (\mathbf{Z}_u^{(j)} \cdot \mathbf{Z}_u^{(\nu)})\, d\eta\,, \qquad \nu = 1, 2, \ldots, n\,. \tag{5.16}$$

This method of determination of the constants $C^{(j)}$ can be justified in the following way. The right-hand side of (5.15) is an approximation of $\mathbf{u}_0(\eta)$, therefore,

$$\mathbf{u}_0(\eta) - \sum_{j=1}^{n} C^{(j)} \mathbf{Z}_u^{(j)}(\eta) = \boldsymbol{\delta} \neq 0\,,$$

where $\boldsymbol{\delta}$ is a residual vector function.

Integrating the square of $\boldsymbol{\delta}$, we get

$$J = \int_0^1 (\mathbf{u}_0 - \sum_{j=1}^{n} C^{(j)} \mathbf{Z}_u^{(j)})^2\, d\eta\,.$$

Then, we find $C^{(j)}$ from the minimum condition for J, that is,

$$\frac{\partial J}{\partial C^{(j)}} = 0\,,$$

consequently,

$$\int_0^1 ([\mathbf{u}_0 - \sum_{j=1}^{n} C^{(j)} \mathbf{Z}_u^{(j)}] \cdot \mathbf{Z}_u^{(\nu)})\, d\eta = 0\,, \qquad \nu = 1, \ldots, n\,.$$

This gives (5.16).

Nonzero Initial Velocity In this case, at initial moment of time, linear velocities of axial points of the rod are nonzero. Thus, we have

$$\mathbf{u}(0, \eta) = \mathbf{u}_0(\eta)\,, \qquad \mathbf{v}(0, \eta) = \mathbf{v}_0(\eta)\,,$$

that is,

$$\dot{\mathbf{u}}(0, \eta) = \mathbf{v}(0, \eta) = \mathbf{v}_0(\eta)$$

and the constants $B^{(j)}$ can be found from the relationship

$$\dot{\mathbf{u}}(0, \eta) = \mathbf{v}_0(\eta) = \sum_{j=1}^{\infty} \lambda_j B^{(j)} \mathbf{Z}_u^{(j)}\,. \tag{5.17}$$

Replacing in this relationship the infinite sum by a finite one, we get

$$\int_0^1 (\mathbf{v}\cdot\mathbf{Z}_u^{(\nu)})\, d\eta = \sum_{j=1}^{n} B^{(j)}\lambda \int_0^1 (\mathbf{Z}_u^{(j)}\cdot\mathbf{Z}_u^{(\nu)})\, d\eta\,, \qquad \nu = 1, \ldots, n\,. \quad (5.18)$$

In the case of initial conditions of general type, $C^{(j)} \neq 0$ and $B^{(j)} \neq 0$, from (5.9)–(5.12) it follows that

$$\Delta\mathbf{Q} = \sum_{j=1}^{n} (C^{(j)}\mathbf{Z}_Q^{(j)} \cos\lambda_j\tau + B^{(j)}\mathbf{Z}_Q^{(j)} \sin\lambda_j\tau)\,, \quad (5.19)$$

$$\Delta\mathbf{M} = \sum_{j=1}^{n} (C^{(j)}\mathbf{Z}_M^{(j)} \cos\lambda_j\tau + B^{(j)}\mathbf{Z}_M^{(j)} \sin\lambda_j\tau)\,, \quad (5.20)$$

$$\boldsymbol{\vartheta} = \sum_{j=1}^{n} (C^{(j)}\mathbf{Z}_\vartheta^{(j)} \cos\lambda_j\tau + B^{(j)}\mathbf{Z}_\vartheta^{(j)} \sin\lambda_j\tau)\,, \quad (5.21)$$

$$\mathbf{u} = \sum_{j=1}^{n} (C^{(j)}\mathbf{Z}_u^{(j)} \cos\lambda_j\tau + B^{(j)}\mathbf{Z}_u^{(j)} \sin\lambda_j\tau)\,. \quad (5.22)$$

Restrict our consideration to the case of $n = 2$. Then, in coordinates, equation (5.19) reads

$$\begin{aligned}
\Delta Q_1 &= C^{(1)} Z_{01}^{(1)} \cos\lambda_1\tau + C^{(2)} Z_{01}^{(2)} \cos\lambda_2\tau \\
&\qquad + B^{(1)} Z_{01}^{(1)} \sin\lambda_1\tau + B^{(2)} Z_{01}^{(2)} \sin\lambda_2\tau\,, \\
\Delta Q_2 &= C^{(1)} Z_{02}^{(1)} \cos\lambda_1\tau + C^{(2)} Z_{02}^{(2)} \cos\lambda_2\tau \\
&\qquad + B^{(1)} Z_{02}^{(1)} \sin\lambda_1\tau + B^{(2)} Z_{02}^{(2)} \sin\lambda_2\tau\,, \\
\Delta Q_3 &= C^{(1)} Z_{03}^{(1)} \cos\lambda_1\tau + C^{(2)} Z_{03}^{(2)} \cos\lambda_2\tau \\
&\qquad + B^{(1)} Z_{03}^{(1)} \sin\lambda_1\tau + B^{(2)} Z_{03}^{(2)} \sin\lambda_2\tau\,.
\end{aligned}$$

Equations (5.20)–(5.22) can be represented in similar fashion.

5.3 Approximate Solution of the Equations of Small Vibrations

Setting $w_0 = 0$ in (3.48) and taking into consideration forces of viscous damping and moments of the forces of viscous damping, we obtain the general equation of small free vibrations

$$\mathbf{L} = \mathrm{A}^{(1)} \frac{\partial^2\mathbf{Z}}{\partial\tau^2} + \mathrm{A}^{(5)} \frac{\partial\mathbf{Z}}{\partial\tau} + \mathrm{A}^{(4)} \frac{\partial\mathbf{Z}}{\partial\tau} + \frac{\partial\mathbf{Z}}{\partial\eta} + \mathrm{A}^{(2)}\mathbf{Z} = 0\,, \quad (5.23)$$

where

$$A^{(5)} = \begin{bmatrix} 0 & 0 & 0 & -\alpha_0 E \\ 0 & 0 & -\alpha_0^{(1)} E & 0 \\ 0 & 0 & 0 & 0 \\ 0 & 0 & 0 & 0 \end{bmatrix}.$$

Because of the presence of terms containing first-order time derivative in (5.23), the determination of the exact solution of this equation is cumbersome. Therefore, we shall use the principle of virtual displacements. Let us represent the vector $\mathbf{Z}$ in the form

$$\mathbf{Z} = \sum_{j=1}^{n} f^{(j)}(\tau)\mathbf{Z}_0^{(j)}(\eta), \tag{5.24}$$

where $f^{(j)}(\tau)$ are unknown functions of time and $\mathbf{Z}_0^{(j)}(\eta)$ are the eigenvectors that satisfy the boundary conditions identical to those for the vector $\mathbf{Z}$ in (5.23). In this case, the vectors $\mathbf{Z}_0^{(j)}(\eta)$ can be obtained from a simplified equation of free vibrations, say, from equation (5.7).

Let the generalized displacements be identical in form to (4.135). We have

$$\delta\mathbf{Z}^{(j)} = \delta a^{(j)} E_0 \mathbf{Z}_0^{(j)}, \qquad j = 1, \ldots, n.$$

Using the principle of virtual displacements, from (5.23) we obtain

$$\int (\mathbf{L} \cdot E_0 \mathbf{Z}_0^{(\nu)})\, d\eta = 0, \qquad \nu = 1, \ldots, n.$$

On rearrangement, this gives

$$H\ddot{\mathbf{f}} + B\dot{\mathbf{f}} + C\mathbf{f} = 0. \tag{5.25}$$

This second-order equation can be represented as two vector equations of the first order,

$$H\dot{\mathbf{Y}}_1 + B\mathbf{Y}_1 + C\mathbf{Y}_2 = 0,$$
$$\dot{\mathbf{Y}}_2 - \mathbf{Y}_1 = 0,$$

or as one vector equation,

$$\dot{\mathbf{Y}} + A\mathbf{Y} = 0, \tag{5.26}$$

where

$$\mathbf{Y} = \begin{bmatrix} \mathbf{Y}_1 \\ \mathbf{Y}_2 \end{bmatrix} = \begin{bmatrix} \dot{\mathbf{f}} \\ \mathbf{f} \end{bmatrix}, \qquad A = \begin{bmatrix} H^{-1}B & H^{-1}C \\ -E & 0 \end{bmatrix}.$$

From (5.26) it follows that

$$\mathbf{Y} = K(\tau)\mathbf{Y}_0, \qquad K(0) = E, \tag{5.27}$$

where

$$\mathbf{Y}_0 = \begin{bmatrix} \dot{\mathbf{f}}(0) \\ \mathbf{f}(0) \end{bmatrix} = \begin{bmatrix} \mathbf{C}_1 \\ \mathbf{C}_2 \end{bmatrix}, \qquad \mathrm{K}(\tau) = \begin{bmatrix} \mathrm{K}_{11} & \mathrm{K}_{12} \\ \mathrm{K}_{21} & \mathrm{K}_{22} \end{bmatrix}.$$

Thus, we arrive at

$$\dot{\mathbf{f}}(\tau) = \mathrm{K}_{11}(\tau)\,\mathbf{C}_1 + \mathrm{K}_{12}(\tau)\,\mathbf{C}_2\,, \tag{5.28}$$

$$\mathbf{f}(\tau) = \mathrm{K}_{21}(\tau)\,\mathbf{C}_1 + \mathrm{K}_{22}(\tau)\,\mathbf{C}_2\,. \tag{5.29}$$

In components, the last equation reads

$$f^{(j)} = \sum_{\nu=1}^{n} k_{j\nu}^{(21)} C_{1\nu} + \sum_{\nu=1}^{n} k_{j\nu}^{(22)} C_{2\nu}\,,$$

where $k_{j\nu}^{(21)}$ and $k_{j\nu}^{(22)}$ are the elements of the matrices K_{21} and K_{22}, respectively.

Finally, substituting the expressions for $f^{(j)}$ into (5.24), we obtain

$$\mathbf{Z} = \sum_{j=1}^{n} \left[\sum_{\nu=1}^{n} \left(C_{1\nu} k_{j\nu}^{(21)} \mathbf{Z}_0^{(j)} + C_{2\nu} k_{j\nu}^{(22)} \mathbf{Z}_0^{(j)} \right) \right]. \tag{5.30}$$

Recall that $\boldsymbol{\vartheta}_0$ and $\dot{\boldsymbol{\vartheta}}_0 = \boldsymbol{\omega}_0$ are related to $\mathbf{u}_0$ and $\mathbf{v}_0$ by equations (5.1) and (5.2). If the values $\mathbf{u}_0(\eta)$ and $\dot{\mathbf{u}}_0(\eta) = \mathbf{v}_0(\eta)$ are specified at the initial moment of time, then, we arrive at the following two equations:

$$\mathbf{u}_0(\eta) = \sum_{\nu=1}^{n} \left(C_{1\nu} \mathbf{X}_1^{(\nu)} + C_{2\nu} \mathbf{X}_2^{(\nu)} \right), \tag{5.31}$$

$$\mathbf{v}_0(\eta) = \sum_{\nu=1}^{n} \left(C_{1\nu} \mathbf{X}_3^{(\nu)} + C_{2\nu} \mathbf{X}_4^{(\nu)} \right), \tag{5.32}$$

where

$$\mathbf{X}_1^{(\nu)} = \sum_{j=1}^{n} k_{j\nu}^{(21)}(0)\,\mathbf{Z}_0^{(j)}\,, \qquad \mathbf{X}_2^{(\nu)} = \sum_{j=1}^{n} k_{j\nu}^{(22)}(0)\,\mathbf{Z}_0^{(j)}\,,$$

$$\mathbf{X}_3^{(\nu)} = \sum_{j=1}^{n} k_{j\nu}^{0(21)}(0)\,\mathbf{Z}_0^{(j)}\,, \qquad \mathbf{X}_4^{(\nu)} = \sum_{j=1}^{n} k_{j\nu}^{0(22)}(0)\,\mathbf{Z}_0^{(j)}\,.$$

Equations (5.31) and (5.32) are identical in form to (5.15) and (5.17), respectively. We can find the parameters $C_{1\nu}$ and $C_{2\nu}$ from these equations.

Multiplying both sides of (5.31) and (5.32) by $\mathbf{Z}_0^{(\rho)}$, $\rho = 1, \ldots, n$, and integrating the result between the limits $\eta = 0$ and $\eta = 1$, we obtain the following system of equations in $2n$ unknowns:

$$\int_0^1 (\mathbf{u}_0 \cdot \mathbf{Z}_0^{(\rho)})\, \mathrm{d}\eta = \sum_{\nu=1}^{n} \int_0^1 (\mathbf{X}_1^{(\nu)} \cdot \mathbf{Z}_0^{(\rho)})\, C_{1\nu}\, \mathrm{d}\eta + \sum_{\nu=1}^{n} \int_0^1 (\mathbf{X}_2^{(\nu)} \cdot \mathbf{Z}_0^{(\rho)})\, C_{2\nu}\, \mathrm{d}\eta\,, \tag{5.33}$$

$$\int_0^1 (\mathbf{v}_0 \cdot \mathbf{Z}_0^{(\rho)})\, \mathrm{d}\eta = \sum_{\nu=1}^{n} \int_0^1 (\mathbf{X}_3^{(\nu)} \cdot \mathbf{Z}_0^{(\rho)})\, C_{1\nu}\, \mathrm{d}\eta + \sum_{\nu=1}^{n} \int_0^1 (\mathbf{X}_4^{(\nu)} \cdot \mathbf{Z}_0^{(\rho)})\, C_{2\nu}\, \mathrm{d}\eta\,, \qquad \rho = 1, \ldots, n\,. \tag{5.34}$$

Consider the case $\mathbf{v}_0 = 0$ for $\tau = 0$. Then, the left-hand side of (5.34) vanishes. Having found $C_{1\nu}$ and $C_{2\nu}$ from (5.30), we obtain the vectors $\Delta\mathbf{Q}(\eta, \tau)$, $\Delta\mathbf{M}(\eta, \tau)$, $\boldsymbol{\vartheta}(\eta, \tau)$, and $\mathbf{u}(\eta, \tau)$, which are characteristics of the stress–strain state of the rod during the small vibrations.

5.4 Small Vibrations of a Rod Due to an Impulsive Loading

Consider a rod with concentrated masses (Fig. 5.4). Let us study vibrations of the rod caused by an impulsive force and an impulsive moment.

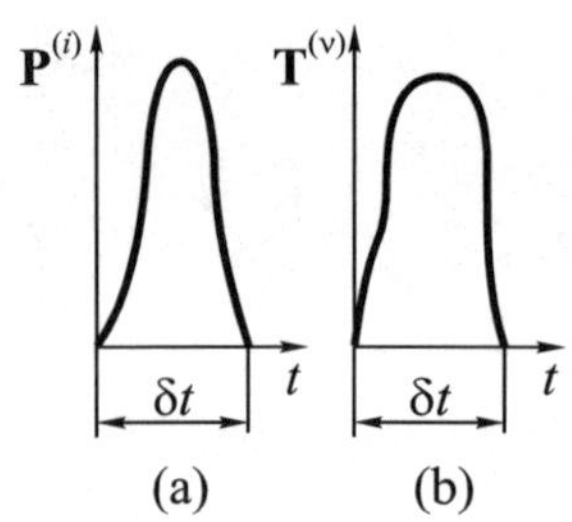

Fig. 5.5.

Impulsive loads $\mathbf{P}^{(i)}$ and $\mathbf{T}^{(\nu)}$ are shown qualitatively in Fig. 5.5*a,b*. The duration of a pulse δt is small as compared to the period of the first harmonic of the rod. Nevertheless, the integrals

$$\mathbf{J}_P^{(i)} = \int_0^{\delta t} \mathbf{P}^{(i)}\, \mathrm{d}t\,, \qquad \mathbf{J}_T^{(\nu)} = \int_0^{\delta t} \mathbf{T}^{(\nu)}\, \mathrm{d}t$$

are assumed to be finite quantities.

Once the loads have been applied, the linear and angular velocities of the concentrated masses become

$$\mathbf{v}(0,\,\eta_1) = \frac{\mathbf{J}_P^{(1)}}{m_1}\,, \qquad \boldsymbol{\omega}(0,\,\eta_1) = \mathrm{J}_{m_1}^{-1}\mathbf{J}_T^{(1)}\,,$$

$$\mathbf{v}(0,\,\eta_2) = \frac{\mathbf{J}_P^{(2)}}{m_2}\,, \qquad \boldsymbol{\omega}(0,\,\eta_2) = \mathrm{J}_{m_2}^{-1}\mathbf{J}_T^{(2)}\,,$$

where J_{m_1} and J_{m_2} are the matrices of the moments of inertia of the masses m_1 and m_2, respectively.

Let us now assume that only impulses $\mathbf{J}_P^{(1)}$ and $\mathbf{J}_P^{(2)}$ are applied to the concentrated masses. The lines of action of the forces passes permanently through the points O_1 and O_2, respectively. Hence, $\mathbf{u}(0,\,\eta) = 0$ and the first derivative of the vector $\mathbf{u}$ with respect to τ (represented in the form (5.22)) at $\tau = 0$ must satisfy the relationship

$$\dot{\mathbf{u}}\,(0,\,\eta) = \mathbf{v}(0,\,\eta_1)\,\delta(\eta{-}\eta_1){+}\mathbf{v}(0,\,\eta_2)\,\delta(\eta{-}\eta_2) = \sum_{j=1}^{n}\lambda_j B^{(j)}\mathbf{Z}_u^{(j)}\,, \tag{5.35}$$

where

$$\mathbf{v}(0,\,\eta_1) = \frac{\mathbf{J}_p^{(1)}}{m_1}\,, \qquad \mathbf{v}(0,\,\eta_2) = \frac{\mathbf{J}_P^{(2)}}{m_2}\,.$$

Recall that $\delta(\eta - \eta_1)$ and $\delta(\eta - \eta_2)$ are the Dirac delta-functions.

Multiplying both sides of (5.35) by $\mathbf{Z}_u^{(k)}$ and integrating the result, we obtain the system of equations

$$\begin{aligned}&\left(\mathbf{v}(0,\,\eta_1)\cdot\mathbf{Z}_u^{(k)}(\eta_1)\right) + \left(\mathbf{v}(0,\,\eta_2)\cdot\mathbf{Z}_u^{(k)}(\eta_2)\right)\\ &= \sum_{j=1}^{n}\lambda_j B^{(j)}\int_0^1 (\mathbf{Z}_u^{(j)}\cdot\mathbf{Z}_u^{(k)})\,\mathrm{d}\eta\,,\end{aligned} \tag{5.36}$$

from which the constants $B^{(j)}$ can be found.

Since, for $\tau = 0$, $\mathbf{u} \equiv 0$, we see that $C^{(j)} = 0$, and (5.19)–(5.22) can be solved for $\Delta\mathbf{Q}$, $\Delta\mathbf{M}$, $\boldsymbol{\vartheta}$, and $\mathbf{u}$. The last parameters are characteristics of the stress–strain state of the rod during free vibrations caused by a pulse load.

When an impulsive moment is applied to the masses at $\tau = 0$, the constants $B^{(j)}$ can be determined in a similar way.

In this case, setting $\tau = 0$ in (5.21), we arrive at

$$\boldsymbol{\omega}(0,\,\eta_1)\,\delta(\eta - \eta_1) + \boldsymbol{\omega}(0,\,\eta_2)\,\delta(\eta - \eta_2) = \sum_{j=1}^{n}\lambda_j B^{(j)}\mathbf{Z}_\nu^{(j)}\,. \tag{5.37}$$

Multiplying both sides of this equation by $\mathbf{Z}_\nu^{(k)}$ and integrating the result, we obtain

$$\left(\boldsymbol{\omega}(0,\,\eta_1)\cdot\mathbf{Z}_\nu^{(k)}(\eta_1)\right)+\left(\boldsymbol{\omega}(0,\,\eta_2)\cdot\mathbf{Z}_\nu^{(k)}(\eta_2)\right)$$
$$=\sum_{j=1}^{n}\lambda_j B^{(j)}\int_0^1(\mathbf{Z}_\nu^{(j)}\cdot\mathbf{Z}_\nu^{(k)})\,\mathrm{d}\eta\,, \tag{5.38}$$

where

$$\boldsymbol{\omega}(0,\,\eta_1)=\mathrm{J}_{m_1}^{-1}\mathbf{J}_T^{(1)}\,,\qquad \boldsymbol{\omega}(0,\,\eta_2)=\mathrm{J}_{m_2}^{-1}\mathbf{J}_T^{(2)}\,.$$

Thus, $B^{(j)}$ can be found from (5.38).

Once the constants $B^{(j)}$ are found either from (5.36) or from (5.38), we obtain the solution of (5.7).

5.5 Forced Vibrations

5.5.1 Steady-State Vibrations Under a Harmonic Excitation

First, consider the simplest case of steady-state vibrations when a rod is subjected to a harmonic force $\mathbf{P}^{(1)}$ and a harmonic moment $\mathbf{T}^{(1)}$ (Fig. 5.6) defined as

$$\mathbf{P}^{(1)}=\mathbf{P}_0^{(1)}\cos\omega_0\tau\,,\qquad \mathbf{T}^{(1)}=\mathbf{T}_0^{(1)}\cos\omega_0\tau\,,\qquad \omega_0=\frac{\omega}{p_0}\,,$$

where ω_0 is the dimensionless frequency.

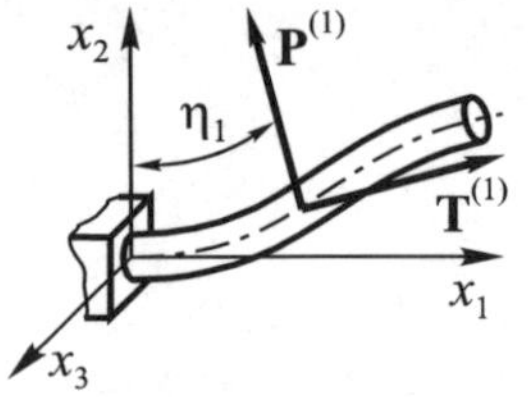

Fig. 5.6.

In the absence of damping forces, the equations governing forced vibrations of the rod about its equilibrium configuration are given by (3.17). We have

$$\mathrm{A}^{(1)}\frac{\partial^2\mathbf{Z}}{\partial\tau^2}+\frac{\partial\mathbf{Z}}{\partial\eta}+\mathrm{A}^{(2)}\mathbf{Z}=\Delta\boldsymbol{\Phi}\,, \tag{5.39}$$

where

$$\mathbf{Z} = \begin{bmatrix} \Delta\mathbf{Q} \\ \Delta\mathbf{M} \\ \boldsymbol{\vartheta} \\ \mathbf{u} \end{bmatrix}, \qquad \Delta\boldsymbol{\Phi} = \begin{bmatrix} -\Delta\mathbf{P} \\ -\Delta\mathbf{T} \\ \mathbf{0} \\ \mathbf{0} \end{bmatrix} = \begin{bmatrix} -\mathbf{P}_0^{(1)} \\ -\mathbf{T}_0^{(1)} \\ \mathbf{0} \\ \mathbf{0} \end{bmatrix} \delta(\eta - \eta_1) \cos\omega_0\tau$$

$$= -\Delta\boldsymbol{\Phi}_0\,\delta(\eta - \eta_1)\cos\omega_0\tau\,.$$

Both exact and approximate methods of solving (5.39) will be discussed further in this section. As a result, the amplitude values of vectors characterizing stress–strain state of a rod will be obtained for the case of steady-state vibrations.

5.5.2 Exact Numerical Solution

For steady-state vibrations, the solution of (5.39) can be written in the form

$$\mathbf{Z} = \mathbf{Z}_0(\eta)\cos\omega_0\tau\,. \tag{5.40}$$

In components, we have

$$\Delta\mathbf{Q} = \Delta\mathbf{Q}_0\cos\omega_0\tau\,, \qquad \Delta\mathbf{M} = \Delta\mathbf{M}_0\cos\omega_0\tau\,,$$
$$\boldsymbol{\vartheta} = \boldsymbol{\vartheta}_0\cos\omega_0\tau\,, \qquad \mathbf{u} = \mathbf{u}_0\cos\omega_0\tau\,.$$

Combining (5.40) and (5.39), we obtain

$$\frac{\mathrm{d}\mathbf{Z}_0}{\mathrm{d}\eta} + (\mathrm{A}^{(2)} - \omega_0^2\mathrm{A}^{(1)})\,\mathbf{Z}_0 = -\Delta\boldsymbol{\Phi}_0\,\delta(\eta - \eta_1)\,. \tag{5.41}$$

The solution of this inhomogeneous equation is

$$\mathbf{Z}_0 = \mathrm{K}(\eta)\,\mathbf{C} - \mathrm{G}(\eta,\,\eta_1)\,\Delta\boldsymbol{\Phi}_0\,H(\eta - \eta_1)\,, \tag{5.42}$$

where

$$\mathrm{G}(\eta,\,\eta_1) = \mathrm{K}(\eta)\,\mathrm{K}^{-1}(\eta_1)\,, \qquad \mathbf{C} = (C_Q,\,C_M,\,C_\vartheta,\,C_u)^{\mathrm{T}}\,.$$

For example, consider a rod shown in Fig. 5.5. Then, the solution (5.42) must satisfy the following boundary conditions:

(1) at $\eta = 0\,,\quad \boldsymbol{\vartheta}_0 = 0$ and $\mathbf{u}_0 = 0\,;$

(2) at $\eta = 1\,,\quad \Delta\mathbf{Q}_0 = 0$ and $\Delta\mathbf{M}_0 = 0\,.$

It follows from the boundary conditions at $\eta = 0$ that

$$\mathbf{C}_\vartheta = \mathbf{C}_u = 0\,,$$

where

$$\mathrm{K}(\eta) = \begin{bmatrix} \mathrm{K}_{11} & \mathrm{K}_{12} & \mathrm{K}_{13} & \mathrm{K}_{14} \\ \mathrm{K}_{21} & \mathrm{K}_{22} & \mathrm{K}_{23} & \mathrm{K}_{24} \\ \mathrm{K}_{31} & \mathrm{K}_{32} & \mathrm{K}_{33} & \mathrm{K}_{34} \\ \mathrm{K}_{41} & \mathrm{K}_{42} & \mathrm{K}_{43} & \mathrm{K}_{44} \end{bmatrix},$$

$$\mathrm{G}(\eta,\, \eta_1) = \begin{bmatrix} \mathrm{G}_{11} & \mathrm{G}_{12} & \mathrm{G}_{13} & \mathrm{G}_{14} \\ \mathrm{G}_{21} & \mathrm{G}_{22} & \mathrm{G}_{23} & \mathrm{G}_{24} \\ \mathrm{G}_{31} & \mathrm{G}_{32} & \mathrm{G}_{33} & \mathrm{G}_{34} \\ \mathrm{G}_{41} & \mathrm{G}_{42} & \mathrm{G}_{43} & \mathrm{G}_{44} \end{bmatrix}.$$

The boundary conditions at $\eta = 1$ yield the following equations in the unknowns $\mathbf{C}_Q$ and $\mathbf{C}_M$:

$$\begin{aligned} &\mathrm{K}_{11}(1)\,\mathbf{C}_Q + \mathrm{K}_{12}(1)\,\mathbf{C}_M = \mathrm{G}_{11}(1,\, \eta_1)\,\mathbf{P}_0^{(1)} + \mathrm{G}_{12}(1,\, \eta_1)\,\mathbf{T}_0^{(1)}\,, \\ &\mathrm{K}_{21}(1)\,\mathbf{C}_Q + \mathrm{K}_{22}(1)\,\mathbf{C}_M = \mathrm{G}_{21}(1,\, \eta_1)\,\mathbf{P}_0^{(1)} + \mathrm{G}_{22}(1,\, \eta_1)\,\mathbf{T}_0^{(1)}\,. \end{aligned} \tag{5.43}$$

Once $\mathbf{C}_Q$ and $\mathbf{C}_M$ are found, we can obtain the solution of (5.39) for steady-state forced vibrations. The arbitrary constants $\mathbf{C}_Q$ and $\mathbf{C}_M$ are linear in the vectors $\mathbf{P}_0^{(1)}$ and $\mathbf{T}_0^{(1)}$, that is,

$$\mathbf{C}_Q = \mathrm{H}_{11}\mathbf{P}_0^{(1)} + \mathrm{H}_{12}\mathbf{T}_0^{(1)}\,, \qquad \mathbf{C}_M = \mathrm{H}_{21}\mathbf{P}_0^{(1)} + \mathrm{H}_{22}\mathbf{T}_0^{(1)}\,.$$

The elements of the matrices H_{ij} are functions of the elements of the matrices $\mathrm{K}_{\nu\rho}$ and $\mathrm{G}_{\nu\rho}$.

Let us now return to the initial unknown vectors that are characteristics of the stress–strain state of the rod. Finally, we obtain the following amplitude values:

$$\begin{aligned} \Delta\mathbf{Q}_0 &= \mathrm{K}_{11}(\eta)\,\mathbf{C}_Q + \mathrm{K}_{12}(\eta)\,\mathbf{C}_M - \mathrm{G}_{11}(\eta,\, \eta_1)\,\mathbf{P}_0^{(1)}\,H(\eta - \eta_1) \\ &\quad - \mathrm{G}_{12}(\eta,\, \eta_1)\,\mathbf{T}_0^{(1)}\,H(\eta - \eta_1)\,, \\ \Delta\mathbf{M}_0 &= \mathrm{K}_{21}(\eta)\,\mathbf{C}_Q + \mathrm{K}_{22}(\eta)\,\mathbf{C}_M - \mathrm{G}_{21}(\eta,\, \eta_1)\,\mathbf{P}_0^{(1)}\,H(\eta - \eta_1) \\ &\quad - \mathrm{G}_{22}(\eta,\, \eta_1)\,\mathbf{T}_0^{(1)}\,H(\eta - \eta_1)\,, \\ \boldsymbol{\vartheta}_0 &= \mathrm{K}_{31}(\eta)\,\mathbf{C}_Q + \mathrm{K}_{32}(\eta)\,\mathbf{C}_M - \mathrm{G}_{31}(\eta,\, \eta_1)\,\mathbf{P}_0^{(1)}\,H(\eta - \eta_1) \\ &\quad - \mathrm{G}_{32}(\eta,\, \eta_1)\,\mathbf{T}_0^{(1)}\,H(\eta - \eta_1)\,, \\ \mathbf{u}_0 &= \mathrm{K}_{41}(\eta)\,\mathbf{C}_Q + \mathrm{K}_{42}(\eta)\,\mathbf{C}_M - \mathrm{G}_{41}(\eta,\, \eta_1)\,\mathbf{P}_0^{(1)}\,H(\eta - \eta_1) \\ &\quad - \mathrm{G}_{42}(\eta,\, \eta_1)\,\mathbf{T}_0^{(1)}\,H(\eta - \eta_1)\,. \end{aligned} \tag{5.44}$$

The method presented above also applies when a rod is subjected to several concentrated forces and moments. In practice, external forces and moments can be of different frequency, for example,

$$\mathbf{P}^{(1)} = \mathbf{P}_0^{(1)}\cos\omega_{01}\tau\,, \qquad \mathbf{P}^{(2)} = \mathbf{P}_0^{(2)}\cos\omega_{02}\tau\,.$$

In this case, the vector $\Delta\boldsymbol{\Phi}$ in the right-hand side of (5.39) takes the form

$$\Delta \boldsymbol{\Phi} = \Delta \boldsymbol{\Phi}_0^{(1)}\, \delta(\eta - \eta_1) \cos \omega_{01}\tau + \Delta \boldsymbol{\Phi}_0^{(2)}\, \delta(\eta - \eta_2) \cos \omega_{02}\tau \,,$$

where η_1 and η_2 are the coordinates of the points of application of the concentrated forces. The vectors $\mathbf{Z}_0^{(1)}$ and $\mathbf{Z}_0^{(2)}$ can be found from the equations

$$\begin{aligned}
&\frac{\mathrm{d}\mathbf{Z}_0^{(1)}}{\mathrm{d}\eta} + (\mathrm{A}^{(2)} - \omega_{01}^2 \mathrm{A}^{(1)})\, \mathbf{Z}_0^{(1)} = -\Delta \boldsymbol{\Phi}_0^{(1)}\, \delta(\eta - \eta_1)\,, \\
&\frac{\mathrm{d}\mathbf{Z}_0^{(2)}}{\mathrm{d}\eta} + (\mathrm{A}^{(2)} - \omega_{02}^2 \mathrm{A}^{(1)})\, \mathbf{Z}_0^{(2)} = -\Delta \boldsymbol{\Phi}_0^{(2)}\, \delta(\eta - \eta_2)\,.
\end{aligned} \tag{5.45}$$

Having determined $\mathbf{Z}_0^{(1)}$ and $\mathbf{Z}_0^{(2)}$ with regard for the boundary conditions, we find the solution of (5.39) in the form

$$\mathbf{Z} = \mathbf{Z}_0^{(1)}(\eta) \cos \omega_{01}\tau + \mathbf{Z}_0^{(2)}(\eta) \cos \omega_{02}\tau \tag{5.46}$$

or

$$\begin{aligned}
&\Delta \mathbf{Q}_0 = \Delta \mathbf{Q}_0^{(1)} \cos \omega_{01}\tau + \Delta \mathbf{Q}_0^{(2)} \cos \omega_{02}\tau \,, \\
&\Delta \mathbf{M}_0 = \Delta \mathbf{M}_0^{(1)} \cos \omega_{01}\tau + \Delta \mathbf{M}_0^{(2)} \cos \omega_{02}\tau \,, \\
&\boldsymbol{\vartheta}_0 = \boldsymbol{\vartheta}_0^{(1)} \cos \omega_{01}\tau + \boldsymbol{\vartheta}_0^{(2)} \cos \omega_{02}\tau \,, \\
&\mathbf{u}_0 = \mathbf{u}_0^{(1)} \cos \omega_{01}\tau + \mathbf{u}_0^{(2)} \cos \omega_{02}\tau \,.
\end{aligned} \tag{5.47}$$

To take into account the forces of viscous damping and their moments given by (4.86), we must add the term

$$\mathrm{A}^{(5)} \frac{\partial \mathbf{Z}}{\partial \tau}\,. \tag{5.48}$$

to the vector equation (5.39).

If we are interested in forced vibrations of a rod about its stationary motion, it is necessary to add the terms

$$\mathrm{A}^{(3)} \frac{\partial \mathbf{Z}}{\partial \tau}\,, \qquad \mathrm{E}^{(1)} \frac{\partial \mathbf{Z}}{\partial \eta}\,. \tag{5.49}$$

to the vector equation (5.39). Consequently, taking into account the terms (5.48) and (5.49), we obtain

$$\mathrm{A}^{(1)} \frac{\partial^2 \mathbf{Z}}{\partial \tau^2} + \mathrm{A}^{(6)} \frac{\partial \mathbf{Z}}{\partial \tau} + \mathrm{A}^{(2)} \mathbf{Z} + \mathrm{A}^{(4)} \frac{\partial^2 \mathbf{Z}}{\partial \eta \partial \tau} + \mathrm{E}^{(1)} \frac{\partial \mathbf{Z}}{\partial \eta} = \Delta \boldsymbol{\Phi}\,, \tag{5.50}$$

where

$$\mathrm{A}^{(6)} = \mathrm{A}^{(3)} + \mathrm{A}^{(5)}\,.$$

We seek the solution of (5.50) in the form

$$\mathbf{Z} = \mathbf{Z}_0^{(1)} \cos \omega_0\tau + \mathbf{Z}_0^{(2)} \sin \omega_0\tau \,. \tag{5.51}$$

Combining (5.51) and (5.50), we get two vector equations in the unknowns $\mathbf{Z}_0^{(1)}$ and $\mathbf{Z}_0^{(2)}$:

$$\begin{aligned}
&\mathrm{E}^{(1)}\mathbf{Z}_0^{(1)\prime} + \mathrm{A}^{(4)}\omega_0\,\mathbf{Z}_0^{(2)\prime} + (\mathrm{A}^{(2)} - \omega_0^2\mathrm{A}^{(1)})\,\mathbf{Z}_0^{(1)} + \omega_0\mathrm{A}^{(6)}\mathbf{Z}_0^{(2)}\\
&= \Delta\boldsymbol{\Phi}_0\,\delta(\eta-\eta_1)\,,\\
&-\mathrm{A}^{(4)}\omega_0\mathbf{Z}_0^{(1)\prime} + \mathrm{E}^{(1)}\mathbf{Z}_0^{(2)\prime} - \omega_0\mathrm{A}^{(6)}\mathbf{Z}_0^{(1)\prime} + (\mathrm{A}^{(2)} - \omega_0^2\mathrm{A}^{(1)})\,\mathbf{Z}_0^{(2)} = 0\,. \qquad (5.52)
\end{aligned}$$

These equations can be represented as one vector equation

$$\mathbf{Z}_0{}' + \mathrm{A}\mathbf{Z}_0 = \mathrm{A}^{(7)}\,\Delta\boldsymbol{\Phi}_{00}\,\delta(\eta-\eta_1)\,, \qquad (5.53)$$

where

$$\mathbf{Z}_0 = \begin{bmatrix} \mathbf{Z}_0^{(1)} \\ \mathbf{Z}_0^{(2)} \end{bmatrix}, \qquad \Delta\boldsymbol{\Phi}_{00} = \begin{bmatrix} \Delta\boldsymbol{\Phi}_0 \\ \mathbf{0} \end{bmatrix},$$

$$\mathrm{A} = \mathrm{A}^{(7)} \begin{bmatrix} \mathrm{A}^{(2)} - \omega_0^2\mathrm{A}^{(1)} & (\mathrm{A}^{(3)} + \mathrm{A}^{(4)})\,\omega_0 \\ -(\mathrm{A}^{(3)} + \mathrm{A}^{(4)})\,\omega_0 & \mathrm{A}^{(2)} - \omega_0^2\mathrm{A}^{(1)} \end{bmatrix},$$

$$\mathrm{A}^{(7)} = \begin{bmatrix} (\mathrm{A}^{(2)} - \omega_0^2\mathrm{A}^{(1)}) & \omega_0\mathrm{A}^{(6)} \\ -\omega_0\mathrm{A}^{(6)} & (\mathrm{A}^{(2)} - \omega_0^2\mathrm{A}^{(1)}) \end{bmatrix}^{-1}.$$

The solution of (5.53) is

$$\mathbf{Z}_0 = \mathrm{K}(\eta)\,\mathbf{C} + \mathrm{G}(\eta,\,\eta_1)\,\mathrm{A}^{(7)}\,\Delta\boldsymbol{\Phi}_{00}\,H(\eta-\eta_1)\,, \qquad (5.54)$$

where

$$\mathbf{C} = \begin{bmatrix} \mathbf{C}^{(1)} \\ \mathbf{C}^{(2)} \end{bmatrix}, \qquad \mathbf{C}^{(i)} = \begin{bmatrix} \mathbf{C}_Q^{(i)} \\ \mathbf{C}_M^{(i)} \\ \mathbf{C}_\vartheta^{(i)} \\ \mathbf{C}_u^{(i)} \end{bmatrix}, \qquad i = 1,\,2\,.$$

The vector functions $\mathbf{Z}_0^{(1)}$ and $\mathbf{Z}_0^{(2)}$ satisfy the same boundary conditions. Consequently, for the rod shown in Fig. 5.6, we have

$$\mathbf{C}_\vartheta^{(1)} = \mathbf{C}_u^{(1)} = \mathbf{C}_\vartheta^{(2)} = \mathbf{C}_u^{(2)} = 0\,.$$

The other four vectors $\mathbf{C}_Q^{(i)}$ and $\mathbf{C}_M^{(i)}$, $i = 1,\,2$, can be found from the following system of inhomogeneous equations (that is, from the boundary conditions at $\eta = 1$):

$$\begin{aligned}
&\mathrm{K}_{11}\mathbf{C}_Q^{(1)} + \mathrm{K}_{12}\mathbf{C}_M^{(1)} + \mathrm{K}_{15}\mathbf{C}_Q^{(2)} + \mathrm{K}_{16}\mathbf{C}_M^{(2)} + \mathrm{D}_{11}\mathbf{P}_0^{(1)} + \mathrm{D}_{12}\mathbf{T}_0^{(1)} = 0\,,\\
&\mathrm{K}_{21}\mathbf{C}_Q^{(1)} + \mathrm{K}_{22}\mathbf{C}_M^{(1)} + \mathrm{K}_{25}\mathbf{C}_Q^{(2)} + \mathrm{K}_{26}\mathbf{C}_M^{(2)} + \mathrm{D}_{21}\mathbf{P}_0^{(1)} + \mathrm{D}_{22}\mathbf{T}_0^{(1)} = 0\,,\\
&\mathrm{K}_{51}\mathbf{C}_Q^{(1)} + \mathrm{K}_{52}\mathbf{C}_M^{(1)} + \mathrm{K}_{55}\mathbf{C}_Q^{(2)} + \mathrm{K}_{56}\mathbf{C}_M^{(2)} + \mathrm{D}_{51}\mathbf{P}_0^{(1)} + \mathrm{D}_{52}\mathbf{T}_0^{(1)} = 0\,,\\
&\mathrm{K}_{61}\mathbf{C}_Q^{(1)} + \mathrm{K}_{62}\mathbf{C}_M^{(1)} + \mathrm{K}_{65}\mathbf{C}_Q^{(2)} + \mathrm{K}_{66}\mathbf{C}_M^{(2)} + \mathrm{D}_{61}\mathbf{P}_0^{(1)} + \mathrm{D}_{62}\mathbf{T}_0^{(1)} = 0\,; \qquad (5.55)
\end{aligned}$$

here, the matrices K_{ij} are the entries of the block matrix $K(\eta)$ and the matrices D_{ij} are the entries of the block matrix D such that, at $\eta = 1$, we have

$$D = G(\eta, \eta_1)\,E\,.$$

Once the vectors $\mathbf{C}_Q^{(i)}$ and $\mathbf{C}_M^{(i)}$ are found, we obtain the vectors $\mathbf{Z}_0^{(1)}$ and $\mathbf{Z}_0^{(2)}$ and thus the vectors $\Delta\mathbf{Q}_0^{(i)}$, $\Delta\mathbf{M}_0^{(i)}$, $\boldsymbol{\vartheta}_0^{(i)}$, and $\mathbf{u}_0^{(i)}$, $i = 1, 2$. Finally, using (5.51), we arrive at

$$\begin{aligned}
&\Delta\mathbf{Q}_0 = \Delta\mathbf{Q}_0^{(1)} \cos\omega_0\tau + \Delta\mathbf{Q}_0^{(2)} \sin\omega_0\tau\,,\\
&\Delta\mathbf{M}_0 = \Delta\mathbf{M}_0^{(1)} \cos\omega_0\tau + \Delta\mathbf{M}_0^{(2)} \sin\omega_0\tau\,,\\
&\boldsymbol{\vartheta}_0 = \boldsymbol{\vartheta}_0^{(1)} \cos\omega_0\tau + \boldsymbol{\vartheta}_0^{(2)} \sin\omega_0\tau\,,\\
&\mathbf{u}_0 = \mathbf{u}_0^{(1)} \cos\omega_0\tau + \mathbf{u}_0^{(2)} \sin\omega_0\tau\,.
\end{aligned} \tag{5.56}$$

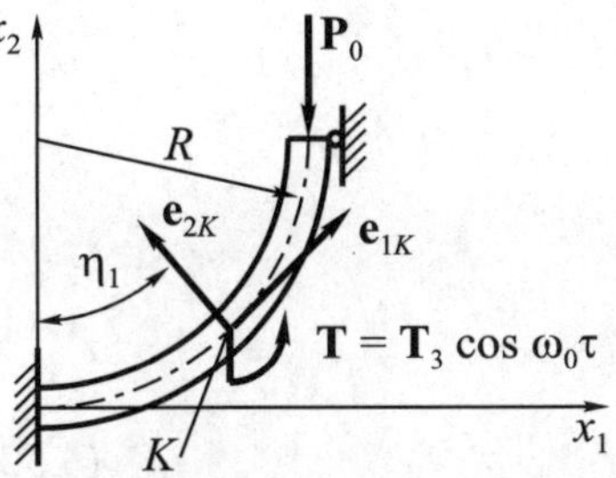

Fig. 5.7.

As an example, consider in-plane steady forced vibrations of a circular rod (Fig. 5.7). The rod is subjected to a harmonic concentrated moment

$$\mathbf{T} = T_{30} \cos\omega\tau\, \mathbf{e}_{3k}$$

and a constant force $\mathbf{P}_0$. Thus, we consider the forced vibrations about the equilibrium configuration of the rod.

With regard for the moment $\mathbf{T}$, the system of equations (3.36) (this system is a particular case of the vector equation (5.50)) reads

$$\begin{aligned}
&n_1 \frac{\partial^2 u_1}{\partial\tau^2} - \frac{\partial\Delta Q_1}{\partial\eta} + \frac{Q_{20}}{A_{33}} \Delta M_3 + æ_{30}\, \Delta Q_2 = 0\,,\\
&n_1 \frac{\partial^2 u_2}{\partial\tau^2} - \frac{\partial\Delta Q_2}{\partial\eta} - \frac{Q_{10}}{A_{33}} \Delta M_3 - æ_{30}\, \Delta Q_1 = 0\,,\\
&J_{33} \frac{\partial^2 \vartheta_3}{\partial\tau^2} - \frac{\partial\Delta M_3}{\partial\eta} - \Delta Q_2 = T_3 \cos\omega_0\tau\, \delta(\eta - \eta_1)\,,\\
&\frac{\partial\vartheta_3}{\partial\eta} - \frac{1}{A_{33}} \Delta M_3 = 0\,,
\end{aligned}$$

$$\frac{\partial u_1}{\partial \eta} - æ_{30} u_2 = 0\,,$$
$$\frac{\partial u_2}{\partial \eta} + æ_{30} u_1 - \vartheta_3 = 0\,. \tag{5.57}$$

Let the deformation of the rod caused by the force $\mathbf{P}_0$ alone be small. Then the curvature $æ_{30}$ in (5.57) can be taken as constant. For steady-state vibrations, we seek the solution of (5.57) in the form

$$\Delta Q_1 = \Delta Q_{10} \cos \omega_0 \tau\,, \qquad \Delta Q_2 = \Delta Q_{20} \cos \omega_0 \tau\,,$$
$$\Delta M_3 = \Delta M_{30} \cos \omega_0 \tau\,,$$
$$\vartheta_3 = \vartheta_{30} \cos \omega_0 \tau\,, \qquad u_1 = u_{10} \cos \omega_0 \tau\,, \qquad u_2 = u_{20} \cos \omega_0 \tau\,. \tag{5.58}$$

Substituting these expressions into (5.57), after some rearrangements, we obtain

$$\Delta Q_{10}{}' - æ_{30}\, \Delta Q_{20} - \frac{Q_{20}}{A_{33}}\, \Delta M_{30} + n_1 \omega_0^2 u_{10} = 0\,,$$
$$\Delta Q_{20}{}' + æ_{30}\, \Delta Q_{10} + \frac{Q_{10}}{A_{33}}\, \Delta M_{30} + n_1 \omega_0^2 u_{20} = 0\,,$$
$$\Delta M_{30}{}' + \Delta Q_{20} + J_{33} \omega_0^2 \vartheta_{30} = -T_{30}\, \delta(\eta - \eta_1)\,,$$
$$\vartheta_{30}{}' - \frac{1}{A_{33}}\, \Delta M_{30} = 0\,,$$
$$u_{10}' - æ_{30} u_{20} = 0\,,$$
$$u_{20}' + æ_{30} u_{10} - \vartheta_3 = 0 \tag{5.59}$$

or, in vector–matrix notation,

$$\mathbf{Z}_0{}' + \mathbf{A}\mathbf{Z}_0 = \mathbf{b}_0\, \delta(\eta - \eta_1)\,, \tag{5.60}$$

where

$$\mathbf{Z}_0 = \begin{bmatrix} \Delta Q_{10} \\ \Delta Q_{20} \\ \Delta M_{30} \\ \vartheta_{30} \\ u_{10} \\ u_{20} \end{bmatrix}, \qquad \mathbf{b}_0 = \begin{bmatrix} 0 \\ 0 \\ -T_{30} \\ 0 \\ 0 \\ 0 \end{bmatrix},$$

$$\mathrm{A} = \begin{bmatrix} 0 & -æ_{30} & -\dfrac{Q_{20}}{A_{33}} & 0 & n_1\omega_0^2 & 0 \\ æ_{20} & 0 & \dfrac{Q_{10}}{A_{33}} & 0 & 0 & n_1\omega_0^2 \\ 0 & 1 & 0 & J_{33}\omega_0^2 & 0 & 0 \\ 0 & 0 & -\dfrac{1}{A_{33}} & 0 & 0 & 0 \\ 0 & 0 & 0 & 0 & 0 & -æ_{30} \\ 0 & 0 & 0 & -1 & æ_{30} & 0 \end{bmatrix}.$$

The solution of (5.60) is

$$\mathbf{Z}_0 = \mathrm{K}(\eta)\,\mathbf{C} + \mathrm{G}(\eta,\,\eta_1)\,\mathbf{b}_0\,H(\eta-\eta_1)\,, \tag{5.61}$$

where $\mathrm{G}(\eta,\,\eta_1) = \mathrm{K}(\eta)\mathrm{K}^{-1}(\eta_1)$. Methods of determination of the fundamental matrix $\mathrm{K}(\eta)$ are discussed in Chap. 2 of Svetlitsky (2000a).

The components of the vector $\mathbf{Z}_0$ must satisfy the following boundary conditions:

$$\begin{aligned}
&(1)\quad \text{at}\quad \eta = 0\,,\quad \vartheta_{30} = u_{10} = u_{20} = 0\,;\\
&(2)\quad \text{at}\quad \eta = 1\,,\quad \Delta Q_{10} = \Delta M_{30} = 0 \quad \text{and} \quad u_{20} = 0\,.
\end{aligned}$$

It follows from the boundary conditions at $\eta = 0$ that $c_4 = c_5 = c_6 = 0$. Using the boundary conditions at $\eta = 1$, we obtain the following inhomogeneous equations in c_1, c_2, and c_3:

$$\begin{aligned}
&k_{11}(1)\,c_1 + k_{12}(1)\,c_2 + k_{13}(1)\,c_3 - g_{13}(1,\,\eta_1)\,T_{30} = 0\,,\\
&k_{31}(1)\,c_1 + k_{32}(1)\,c_2 + k_{33}(1)\,c_3 - g_{33}(1,\,\eta_1)\,T_{30} = 0\,,\\
&k_{61}(1)\,c_1 + k_{62}(1)\,c_2 + k_{63}(1)\,c_3 - g_{63}(1,\,\eta_1)\,T_{30} = 0\,.
\end{aligned} \tag{5.62}$$

Once c_1, c_2, and c_3 have been found, we obtain the amplitude values of internal forces and displacements:

$$\begin{aligned}
\Delta Q_{10} &= \sum_{j=1}^{3} k_{1j}(\eta)\,c_j - g_{13}(\eta,\,\eta_1)\,T_{30}\,H(\eta-\eta_1)\,,\\
\Delta Q_{20} &= \sum_{j=1}^{3} k_{2j}(\eta)\,c_j - g_{23}(\eta,\,\eta_1)\,T_{30}\,H(\eta-\eta_1)\,,\\
\Delta M_{30} &= \sum_{j=1}^{3} k_{3j}(\eta)\,c_j - g_{33}(\eta,\,\eta_1)\,T_{30}\,H(\eta-\eta_1)\,,\\
\vartheta_{30} &= \sum_{j=1}^{3} k_{4j}(\eta)\,c_j - g_{43}(\eta,\,\eta_1)\,T_{30}\,H(\eta-\eta_1)\,,\\
u_{10} &= \sum_{j=1}^{3} k_{5j}(\eta)\,c_j - g_{53}(\eta,\,\eta_1)\,T_{30}\,H(\eta-\eta_1)\,,\\
u_{20} &= \sum_{j=1}^{3} k_{6j}(\eta)\,c_j - g_{63}(\eta,\,\eta_1)\,T_{30}\,H(\eta-\eta_1)\,.
\end{aligned} \tag{5.63}$$

Let us consider these relationships in detail. All the functions except ΔM_{30} must be continuous in η. The internal moment ΔM_3 must have a discontinuity at the point of application of the concentrated moment. The elements of the matrix $\mathrm{G}(\eta,\,\eta_1)$ enter into the right-hand sides of (5.63). The equality $\mathrm{G}(\eta,\,\eta_1) = \mathrm{K}(\eta)\mathrm{K}^{-1}(\eta_1)$ is valid. For $\eta = \eta_1$, the matrix G is the identity matrix, that is, $G(\eta_1, \eta_1) = E$. Thus, all the elements of the matrix G except the diagonal ones equal zero, that is, $g_{ij} = 0$ for $j \neq i$ and $g_{ij} = 0$ for $j = i$. Besides, all the terms in (5.63) containing T_{30} are of the following mode: to the left of the cross section where the moment is applied, these terms are equal to zero, and to the right of this cross section, they vary monotonically. At that, all the terms in (5.63) containing T_{30} (except $g_{33}T_{30}$) are continuous functions; since $g_{33}(\eta_1, \eta_1) = 1$, the term entering into ΔM_{30} has a discontinuity at this point and the magnitude of the step is T_{30}.

5.6 Steady-state Vibrations of a Rod Under Kinematic Excitation

Let a point K on a rod axis be in a simple harmonic motion along a unit vector $\mathbf{e}_k$ (Fig. 5.8).

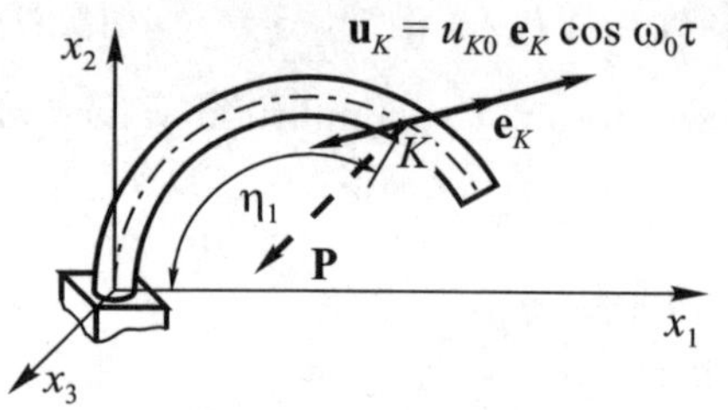

Fig. 5.8.

Denote the displacement of the point K by $\mathbf{u}_k$; then,

$$\mathbf{u}_k = u_{k0} \cos\omega_0\tau \, \mathbf{e}_k \, .$$

Let the forces of viscous damping be neglected. Suppose that the point K moves under the action of an unknown force $\mathbf{P}$ in the specified direction such that

$$\mathbf{P} = \mathbf{P}_0^{(1)} \cos\omega_0\tau \, .$$

Setting $\mathbf{T}^{(1)} = 0$, we can see that the problem under consideration coincides with the problem discussed at the beginning of Sect. 5.2. Once the vectors $\mathbf{C}_Q$ and $\mathbf{C}_M$ are found from (5.43), we can use (5.47) and arrive at

$$\mathbf{u}_0 = [\mathrm{K}_{41}(\eta)\,\mathrm{H}_{11} + \mathrm{K}_{42}(\eta)\,\mathrm{H}_{21} - \mathrm{G}_{41}(\eta, \eta_1)\,H(\eta - \eta_1)]\,\mathbf{P}_0^{(1)}$$

or

$$\mathbf{u}_0 = \mathrm{T}_u \mathbf{P}_0^{(1)} \, . \tag{5.64}$$

Thus, we obtain the relationship between $\mathbf{u}_0$ and $\mathbf{P}_0^{(1)}$.

The expressions of this type can be obtained for the other unknown vectors entering into the vector $\mathbf{Z}_0$ (5.42). Thus, we have

$$\Delta\mathbf{Q}_0 = \mathrm{T}_Q\mathbf{P}_0^{(1)}, \qquad \Delta\mathbf{M}_0 = \mathrm{T}_M\mathbf{P}_0^{(1)}, \qquad \boldsymbol{\vartheta}_0 = \mathrm{T}_\vartheta\mathbf{P}_0^{(1)}. \tag{5.65}$$

At $\eta = \eta_1$, the vector $\mathbf{u}_0$ must satisfy three additional conditions:

$$(\mathbf{u}_0 \cdot \mathbf{e}_{k1}) = u_{k1}, \qquad (\mathbf{u}_0 \cdot \mathbf{e}_{k2}) = 0, \qquad (\mathbf{u}_0 \cdot \mathbf{e}_{k3}) = 0. \tag{5.66}$$

The second and the third equations follow from the fact that the vector $\mathbf{u}_0$ is orthogonal to the plane that in its turn is perpendicular to the vector $\mathbf{e}_{k1}$.

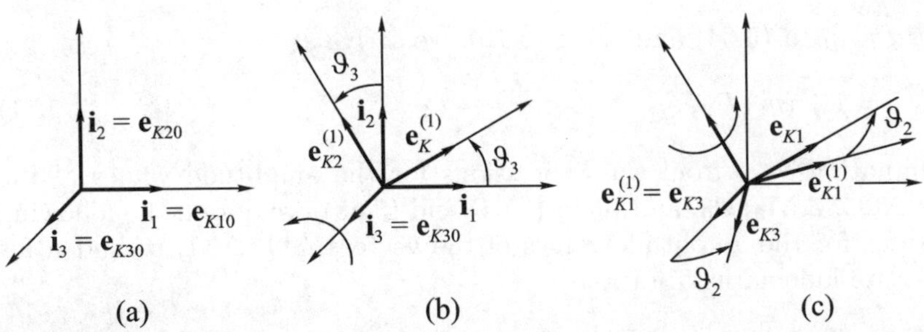

Fig. 5.9.

Consider the conditions (5.66) in detail. Let us find the matrix of transformation from the basis $\{\mathbf{i}_j\}$ to the basis $\{\mathbf{e}_{kj}\}$. Consider the case when the vector $\mathbf{e}_{k10}$ initially coincides with the vector $\mathbf{i}_1$ as shown in Fig. 5.9*a*. To bring the vector $\mathbf{e}_{k10}$ into coincidence with the vector $\mathbf{e}_k$, it is sufficient to rotate the trihedron by the angle ϑ_3 (Fig. 5.9*b*) and then by the angle ϑ_2 (Fig. 5.9*c*). Thus, the matrix of transformation reads

$$\mathrm{L} = \begin{bmatrix} \cos\vartheta_2\cos\vartheta_3 & \cos\vartheta_2\sin\vartheta_3 & -\sin\vartheta_2 \\ -\sin\vartheta_3 & \cos\vartheta_3 & 0 \\ \sin\vartheta_2\cos\vartheta_3 & \sin\vartheta_2\sin\vartheta_3 & \cos\vartheta_2 \end{bmatrix} = [l_{ij}]. \tag{5.67}$$

It follows from the relationships (5.66) that

$$\begin{aligned} &\gamma_{11}u_{01}(\eta_1) + \gamma_{12}u_{02}(\eta_1) + \gamma_{13}u_{03}(\eta_1) = u_{k0}, \\ &\gamma_{21}u_{01}(\eta_1) + \gamma_{22}u_{02}(\eta_1) + \gamma_{23}u_{03}(\eta_1) = 0, \\ &\gamma_{31}u_{01}(\eta_1) + \gamma_{32}u_{02}(\eta_1) + \gamma_{33}u_{03}(\eta_1) = 0, \end{aligned} \tag{5.68}$$

where

$$\gamma_{1\nu} = \sum_{j=1}^{3} l_{1j}(\mathbf{e}_{\nu 0} \cdot \mathbf{i}_j)\,, \qquad \gamma_{2\nu} = \sum_{j=1}^{3} l_{2j}(\mathbf{e}_{\nu 0} \cdot \mathbf{i}_j)\,,$$

$$\gamma_{3\nu} = \sum_{j=1}^{3} l_{3j}(\mathbf{e}_{\nu 0} \cdot \mathbf{i}_j)\,.$$

In vector–matrix form, the system of equations (5.68) reads

$$\Gamma \mathbf{u}_0(\eta_1) = \mathbf{a}\,, \qquad \mathbf{a} = (u_{k0},\, 0,\, 0)^{\mathrm{T}}\,, \tag{5.69}$$

therefore, we obtain

$$\mathbf{u}_0(\eta_1) = \Gamma^{-1}\mathbf{a}\,. \tag{5.70}$$

Setting $\eta = \eta_1$ in (5.64) and using (5.70), we arrive at

$$\mathbf{P}_0^{(1)} = \mathrm{T}_u^{-1}(\eta_1)\,\Gamma^{-1}\mathbf{a}\,. \tag{5.71}$$

Eliminating $\mathbf{P}_0^{(1)}$ from the expressions for the amplitude values of the vectors $\Delta\mathbf{Q}_0$, $\Delta\mathbf{M}_0$, $\boldsymbol{\vartheta}_0$, and $\mathbf{u}_0$ in (5.64) and (5.65), we obtain the following expressions for the amplitude values of the vectors $\Delta\mathbf{Q}$, $\Delta\mathbf{M}$, $\boldsymbol{\vartheta}$, and $\mathbf{u}$ for steady-state kinematic vibrations:

$$\begin{aligned}
\Delta\mathbf{Q}_0 &= \mathrm{T}_Q(\eta)\,\mathrm{T}_u^{-1}(\eta_1)\,\Gamma^{-1}(\eta_1)\,\mathbf{a}\,,\\
\Delta\mathbf{M}_0 &= \mathrm{T}_M(\eta)\,\mathrm{T}_u^{-1}(\eta_1)\,\Gamma^{-1}(\eta_1)\,\mathbf{a}\,,\\
\boldsymbol{\vartheta}_0 &= \mathrm{T}_\vartheta(\eta)\,\mathrm{T}_u^{-1}(\eta_1)\,\Gamma^{-1}(\eta_1)\,\mathbf{a}\,,\\
\mathbf{u}_0 &= \mathrm{T}_u(\eta)\,\mathrm{T}_u^{-1}(\eta_1)\,\Gamma^{-1}(\eta_1)\,\mathbf{a}\,.
\end{aligned} \tag{5.72}$$

Consider an example. Suppose that the cross section $\eta = \eta_1$ is forced to rotate such that

$$\vartheta_3(\eta_1) = \vartheta_{300}\cos\omega_0\tau\,. \tag{5.73}$$

We assume that an unknown moment,

$$\mathbf{T}_3 = T_{30}\cos\omega_0\tau\,\mathbf{e}_{k3}\,,$$

is applied to this cross section. Then, by analogy with the previous example, we can find the solution of (5.57). Using (5.62), we find the constants c_1, c_2, and c_3 as functions of T_{30}; thus, we have $c_j = \gamma_j T_{30}$, $j = 1, 2, 3$. After that, considering the expression of $\vartheta_{30}(\eta)$ for $\eta = \eta_1$ and using (5.73), we obtain the following equation in T_3:

$$\vartheta_{30}(\eta_1) = \vartheta_{300} = \left[\sum_{j=1}^{3} k_{4j}(\eta_1)\,\gamma_j\right] T_{30}\,.$$

Hence,

$$T_{30} = \frac{\vartheta_{300}}{\sum_{j=1}^{3} k_{4j}(\eta_1)\,\gamma_j} = \beta_0 \vartheta_{300}\,. \tag{5.74}$$

Finally, eliminating T_{30} from (5.63), we arrive at

$$\begin{aligned}
\Delta Q_{10} &= \left[\sum_{j=1}^{3} k_{1j}(\eta)\,\gamma_j - g_{13}(\eta,\,\eta_1)\,H(\eta-\eta_1)\right]\beta_0\vartheta_{300}\,,\\
\Delta Q_{20} &= \left[\sum_{j=1}^{3} k_{2j}(\eta)\,\gamma_j - g_{23}(\eta,\,\eta_1)\,H(\eta-\eta_1)\right]\beta_0\vartheta_{300}\,,\\
\Delta M_{30} &= \left[\sum_{j=1}^{3} k_{3j}(\eta)\,\gamma_j - g_{33}(\eta,\,\eta_1)\,H(\eta-\eta_1)\right]\beta_0\vartheta_{300}\,,\\
\vartheta_{30} &= \left[\sum_{j=1}^{3} k_{4j}(\eta)\,\gamma_j - g_{43}(\eta,\,\eta_1)\,H(\eta-\eta_1)\right]\beta_0\vartheta_{300}\,,\\
u_{10} &= \left[\sum_{j=1}^{3} k_{5j}(\eta)\,\gamma_j - g_{53}(\eta,\,\eta_1)\,H(\eta-\eta_1)\right]\beta_0\vartheta_{300}\,,\\
u_{20} &= \left[\sum_{j=1}^{3} k_{6j}(\eta)\,\gamma_j - g_{63}(\eta,\,\eta_1)\,H(\eta-\eta_1)\right]\beta_0\vartheta_{300}\,.
\end{aligned} \tag{5.75}$$

5.6.1 Approximate Solutions of (5.39) for the Case of Steady-State Vibrations

Considering two-term approximation, we seek the solution of equation (5.39) in the form

$$\mathbf{Z} = a_1 \mathbf{Z}_0^{(1)} \cos\omega_0\tau + a_2 \mathbf{Z}_0^{(2)} \cos\omega_0\tau\,, \tag{5.76}$$

where a_1 and a_2 are arbitrary constants and the vectors $\mathbf{Z}_0^{(1)}$ and $\mathbf{Z}^{(2)}$ are the eigenvectors that satisfy the boundary conditions of the problem. Recall that these vectors can be obtained as solutions of a simplified problem. For example, for the circular rod (see Fig. 5.7), these vectors can be obtained from the equations of free vibrations if the initial stress state are not taken into account (this case corresponds to the condition $\mathbf{P}_0 = 0$).

The generalized virtual displacements are

$$\delta\mathbf{Z}_0^{(1)} = \delta b_1 \mathrm{E}_0 \mathbf{Z}_0^{(1)}\,, \qquad \delta\mathbf{Z}_0^{(2)} = \delta b_2 \mathrm{E}_0 \mathbf{Z}_0^{(2)}\,.$$

Using (5.76) in (5.39), we arrive at

$$\mathbf{L}_0 = -\omega_0^2 a_1 \mathrm{A}^{(1)} \mathbf{Z}_0^{(1)} - \omega_0^2 a_2 \mathrm{A}^{(1)} \mathbf{Z}_0^{(2)} + a_1 \mathbf{Z}_0^{(1)\prime} + a_2 \mathbf{Z}_0^{(2)\prime} + a_1 \mathrm{A}^{(2)} \mathbf{Z}_0^{(1)} + a_2 \mathrm{A}^{(2)} \mathbf{Z}_0^{(2)} - \Delta\boldsymbol{\Phi}_0 \, \delta = 0 . \tag{5.77}$$

Following the approximate method discussed in Sect. 4.3, we obtain the equations

$$\int_0^1 (\mathbf{L}_0 \cdot \mathrm{E}_0 \mathbf{Z}_0^{(1)}) \, \mathrm{d}\eta = 0 , \qquad \int_0^1 (\mathbf{L}_0 \cdot \mathrm{E}_0 \mathbf{Z}_0^{(2)}) \, \mathrm{d}\eta = 0 \tag{5.78}$$

or

$$\alpha_{11} a_1 + \alpha_{12} a_2 = b_1 , \qquad \alpha_{21} a_1 + \alpha_{22} a_2 = b_2 , \tag{5.79}$$

where

$$\alpha_{11} = \int_0^1 \Big[-\omega_0^2 (\mathrm{A}^{(1)} \mathbf{Z}_0^{(1)} \cdot \mathrm{E}_0 \mathbf{Z}_0^{(1)}) + (\mathbf{Z}_0^{(1)\prime} \cdot \mathrm{E}_0 \mathbf{Z}_0^{(1)}) + (\mathrm{A}^{(2)} \mathbf{Z}_0^{(1)} \cdot \mathrm{E}_0 \mathbf{Z}_0^{(1)}) \Big] \, \mathrm{d}\eta ,$$

$$\alpha_{12} = \int_0^1 \Big[-\omega_0^2 (\mathrm{A}^{(1)} \mathbf{Z}_0^{(2)} \cdot \mathrm{E}_0 \mathbf{Z}_0^{(1)}) + (\mathbf{Z}_0^{(2)\prime} \cdot \mathrm{E}_0 \mathbf{Z}_0^{(1)}) + (\mathrm{A}^{(2)} \mathbf{Z}_0^{(2)} \cdot \mathrm{E}_0 \mathbf{Z}_0^{(1)}) \Big] \, \mathrm{d}\eta ,$$

$$\alpha_{21} = \int_0^1 \Big[-\omega_0^2 (\mathrm{A}^{(1)} \mathbf{Z}_0^{(1)} \cdot \mathrm{E}_0 \mathbf{Z}_0^{(2)}) + (\mathbf{Z}_0^{(1)\prime} \cdot \mathrm{E}_0 \mathbf{Z}_0^{(2)}) + (\mathrm{A}^{(2)} \mathbf{Z}_0^{(1)} \cdot \mathrm{E}_0 \mathbf{Z}_0^{(2)}) \Big] \, \mathrm{d}\eta ,$$

$$\alpha_{22} = \int_0^1 \Big[-\omega_0^2 (\mathrm{A}^{(1)} \mathbf{Z}_0^{(2)} \cdot \mathrm{E}_0 \mathbf{Z}_0^{(2)}) + (\mathbf{Z}_0^{(2)\prime} \cdot \mathrm{E}_0 \mathbf{Z}_0^{(2)}) + (\mathrm{A}^{(2)} \mathbf{Z}_0^{(2)} \cdot \mathrm{E}_0 \mathbf{Z}_0^{(2)}) \Big] \, \mathrm{d}\eta ,$$

$$b_1 = (\Delta\boldsymbol{\Phi}_0 \cdot \mathrm{E}_0 \mathbf{Z}_0^{(1)}(\eta_1)) , \qquad b_2 = (\Delta\boldsymbol{\Phi}_0 \cdot \mathrm{E}_0 \mathbf{Z}_0^{(2)}(\eta_1)) .$$

Once the parameters a_1 and a_2 are found, we can obtain the approximate solution (5.76) of (5.39) for the case of steady-state vibrations. The use of the approximate method is advantageous if it allows us to cut down the computation run time for a given problem (as compared to the exact numerical method). This is the case when a rod is subjected to several concentrated forces and moments.

Forces of Viscous Damping and the Coriolis Forces Consider the equations of small vibrations in which the forces of viscous damping and the Coriolis forces are taken into consideration (see (5.50)). We seek an approximate solution of (5.50) in the form

$$\mathbf{Z} = \mathbf{Z}_0^{(1)} (a_1 \cos \omega_0 \tau + b_1 \sin \omega_0 \tau) + \mathbf{Z}_0^{(2)} (a_2 \cos \omega_0 \tau + b_2 \sin \omega_0 \tau) . \tag{5.80}$$

On rearrangement, the system of equations (5.78) may be represented in the form of two relationships:

$$(I)\ \cos\omega_0\tau + (II)\ \sin\omega_0\tau = 0\,,$$
$$(III)\ \cos\omega_0\tau + (IV)\ \sin\omega_0\tau = 0\,. \tag{5.81}$$

The expressions in round brackets are functions of a_1, a_2, b_1, and b_2. In view of the functional independence of $\cos\omega_0\tau$ and $\sin\omega_0\tau$, we have

$$(I) = 0\,, \qquad (II) = 0\,, \qquad (III) = 0\,, \qquad (IV) = 0\,. \tag{5.82}$$

Periodic Forces and Moments Let us obtain an approximate solution of the equations of forced steady-state vibrations for arbitrary periodic forces and moments. The periodicity conditions are

$$\mathbf{P}^{(i)}(\tau) = \mathbf{P}^{(i)}(\tau + T)\,, \qquad \mathbf{T}^{(\nu)}(\tau) = \mathbf{T}^{(\nu)}(\tau + T)\,.$$

The forces and moments can be represented as

$$\mathbf{P}^{(i)} = \mathbf{P}_0^{(i)} F^{(i)}(\tau)\,, \qquad \mathbf{T}^{(\nu)} = \mathbf{T}_0^{(\nu)} \varPhi^{(\nu)}(\tau)\,,$$

where $F^{(i)}$ and $\varPhi^{(\nu)}$ are periodic functions.

There exist two approaches to the solution of the equations of steady-state vibrations: (1) the expansion of the solution in a Fourier series and (2) the Douffing method.

Consider the Douffing method. For the sake of simplicity, let us assume that there are one force, $\mathbf{P}^{(1)}$, and one moment, $\mathbf{T}^{(1)}$.

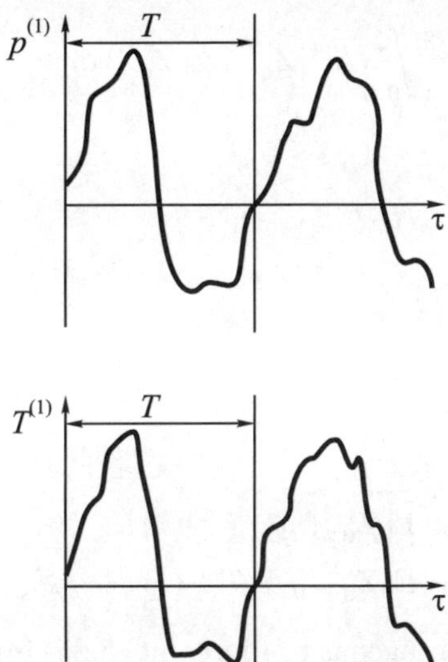

Fig. 5.10.

The force and the moment versus time are plotted in Fig. 5.10. Considering two-term approximation, we shall seek the solution of (5.39) in the form

$$\mathbf{Z} = f^{(1)}(\tau)\,\mathbf{Z}_0^{(1)}(\eta) + f^{(2)}(\tau)\,\mathbf{Z}_0^{(2)}(\eta)\,. \tag{5.83}$$

In this case, the right-hand side of (5.39) reads

$$\Delta\boldsymbol{\Phi} = \Delta\boldsymbol{\Phi}_{10}P^{(1)}(\tau)\,\delta(\eta-\eta_1) + \Delta\boldsymbol{\Phi}_{20}T^{(1)}(\tau)\,\delta(\eta-\eta_1)\,, \tag{5.84}$$

where

$$\Delta\boldsymbol{\Phi}_{10} = \begin{bmatrix} -\mathbf{P}^{(1)} \\ \mathbf{0} \\ \mathbf{0} \\ \mathbf{0} \end{bmatrix}, \qquad \Delta\boldsymbol{\Phi}_{20} = \begin{bmatrix} \mathbf{0} \\ -\mathbf{T}^{(1)} \\ \mathbf{0} \\ \mathbf{0} \end{bmatrix}.$$

Employing the principle of virtual displacements, we arrive at the following inhomogeneous differential equations in $f^{(1)}$ and $f^{(2)}$:

$$\begin{aligned} h_{11}\ddot{f}^{(1)} + h_{12}\ddot{f}^{(2)} + b_{11}f^{(1)} + b_{12}f^{(2)} &= b_1\,, \\ h_{21}\ddot{f}^{(1)} + h_{22}\ddot{f}^{(2)} + b_{21}f^{(1)} + b_{22}f^{(2)} &= b_2\,; \end{aligned} \tag{5.85}$$

here,

$$h_{11} = \int_0^1 (\mathrm{A}^{(1)}\mathbf{Z}_0^{(1)}\cdot \mathrm{E}_0\mathbf{Z}_0^{(1)})\,\mathrm{d}\eta\,, \qquad h_{12} = \int_0^1 (\mathrm{A}^{(1)}\mathbf{Z}_0^{(2)}\cdot \mathrm{E}_0\mathbf{Z}_0^{(1)})\,\mathrm{d}\eta\,,$$

$$h_{21} = \int_0^1 (\mathrm{A}^{(1)}\mathbf{Z}_0^{(1)}\cdot \mathrm{E}_0\mathbf{Z}_0^{(2)})\,\mathrm{d}\eta\,, \qquad h_{22} = \int_0^1 (\mathrm{A}^{(1)}\mathbf{Z}_0^{(2)}\cdot \mathrm{E}_0\mathbf{Z}_0^{(2)})\,\mathrm{d}\eta\,,$$

$$b_{11} = \int_0^1 ([\mathbf{Z}_0^{(1)\prime} + \mathrm{A}^{(2)}\mathbf{Z}_0^{(1)}]\cdot \mathrm{E}_0\mathbf{Z}_0^{(1)})\,\mathrm{d}\eta\,,$$

$$b_{12} = \int_0^1 ([\mathbf{Z}_0^{(2)\prime} + \mathrm{A}^{(2)}\mathbf{Z}_0^{(2)}]\cdot \mathrm{E}_0\mathbf{Z}_0^{(1)})\,\mathrm{d}\eta\,,$$

$$b_{21} = \int_0^1 ([\mathbf{Z}_0^{(1)\prime} + \mathrm{A}^{(2)}\mathbf{Z}_0^{(1)}]\cdot \mathrm{E}_0\mathbf{Z}_0^{(2)})\,\mathrm{d}\eta\,,$$

$$b_{22} = \int_0^1 ([\mathbf{Z}_0^{(2)\prime} + \mathrm{A}^{(2)}\mathbf{Z}_0^{(2)}]\cdot \mathrm{E}_0\mathbf{Z}_0^{(2)})\,\mathrm{d}\eta\,,$$

$$b_1 = (\Delta\boldsymbol{\Phi}_{10}\cdot \mathrm{E}_0\mathbf{Z}_0^{(1)}(\eta_1))\,P^{(1)}(\tau) + (\Delta\boldsymbol{\Phi}_{20}\cdot \mathrm{E}_0\mathbf{Z}_0^{(1)}(\eta_1))\,T^{(1)}(\tau)\,,$$

$$b_2 = (\Delta\boldsymbol{\Phi}_{10}\cdot \mathrm{E}_0\mathbf{Z}_0^{(2)}(\eta_1))\,P^{(1)}(\tau) + (\Delta\boldsymbol{\Phi}_{20}\cdot \mathrm{E}_0\mathbf{Z}_0^{(2)}(\eta_1))\,T^{(1)}(\tau)\,.$$

To perform a numerical analysis, it is advantageous to represent (5.85) in the form of a system of first-order equations. We have

$$\dot{\mathbf{Y}} + \mathbf{C}\mathbf{Y} = \mathbf{b}^{(1)}\,, \tag{5.86}$$

where

$$\mathrm{C} = \begin{bmatrix} 0 & \mathrm{H}^{-1}\mathrm{B} \\ -\mathrm{E} & 0 \end{bmatrix}, \qquad \mathbf{b}^{(1)} = \begin{bmatrix} \mathrm{H}^{-1}\mathbf{b} \\ \mathbf{0} \end{bmatrix},$$

$$\mathbf{Y} = \begin{bmatrix} \dot{\mathbf{f}} \\ \mathbf{f} \end{bmatrix}, \qquad \mathbf{f} = \begin{bmatrix} f_1 \\ f_2 \end{bmatrix},$$

$$\mathrm{H} = \begin{bmatrix} h_{11} & h_{12} \\ h_{21} & h_{22} \end{bmatrix}, \qquad \mathrm{B} = \begin{bmatrix} b_{11} & b_{12} \\ b_{21} & b_{22} \end{bmatrix}, \qquad \mathbf{b} = \begin{bmatrix} b_1 \\ b_2 \end{bmatrix}.$$

The solution of (5.86) can be written as

$$\mathbf{Y} = \mathrm{K}(\tau)\,\mathbf{C} + \mathbf{Y}_*(\tau)\,, \qquad \mathrm{K}(0) = \mathrm{E}\,, \tag{5.87}$$

where $\mathbf{Y}_*(\tau)$ is the partial solution of (5.86) obtained for the zero initial conditions.

For steady-state vibrations, the periodicity condition

$$\mathbf{Y}(\tau) = \mathbf{Y}(\tau + T)\,, \qquad \mathbf{Y}(0) = \mathbf{C}\,.$$

must be hold. Since, at $\tau = 0$, $\mathbf{Y}(0) = \mathbf{C}$, then $\mathbf{Y}(T) = \mathbf{Y}(0)$.

From (5.87) it follows that

$$[\mathrm{E} - \mathrm{K}(T)]\;\mathbf{Y}(0) = \mathbf{Y}_{\mathrm{i}}(T)\,. \tag{5.88}$$

Thus, $\mathbf{Y}(0)$ can be found in the form

$$\mathbf{Y}(0) = [\mathrm{E} - \mathrm{K}(T)]^{-1}\,\mathbf{Y}_{\mathrm{i}}(T)\,. \tag{5.89}$$

Having found $f^{(1)}(0)$ and $f^{(2)}(0)$, we can use them in (5.87) and thereby obtain

$$\begin{aligned} f^{(1)}(\tau) &= k_{11}(\tau)\, f^{(1)}(0) + k_{12}(\tau)\, f^{(2)}(0) + f_{\mathrm{i}\,1}(\tau)\,, \\ f^{(2)}(\tau) &= k_{21}(\tau)\, f^{(1)}(0) + k_{22}(\tau)\, f^{(2)}(0) + f_{\mathrm{i}\,2}(\tau)\,. \end{aligned} \tag{5.90}$$

Finally, we find the approximate solution of the problem, namely, the vector $\mathbf{Z}$ (5.84).

The obtained solution is valid for τ subject to $0 \le \tau \le T$. For $\tau > T$, the solution may be determined due to the periodicity condition. The solution (5.90) is valid provided the matrix

$$\mathrm{D} = [\mathrm{E} - \mathrm{K}(T)] \tag{5.91}$$

a nondegenerate matrix, that is, its determinant is nonzero.

The determinant of the matrix D vanishes when the period of a frequency of free vibrations obtained from the homogeneous equation (5.86) is equal to the period T (the case corresponds to the *resonance* mode of vibrations).

With regard for the forces of viscous damping (5.48), we arrive at the governing equation in the form

$$\mathrm{A}^{(1)} \frac{\partial^2 \mathbf{Z}}{\partial \tau^2} + \frac{\partial \mathbf{Z}}{\partial \eta} + \mathrm{A}^{(2)} \mathbf{Z} + \mathrm{A}^{(3)} \frac{\partial \mathbf{Z}}{\partial \tau} = \Delta \boldsymbol{\Phi} . \tag{5.92}$$

If we seek an approximate solution in the form (5.83), then the equation in $f^{(i)}(\tau)$ is found to be

$$\ddot{\mathbf{f}} + \mathrm{P}\dot{\mathbf{f}} + \mathrm{C}\mathbf{f} = \mathbf{b}^{(1)} . \tag{5.93}$$

In this case (that is, when the forces of viscous damping is taken into consideration), the determinant of the matrix (5.91) never vanishes.

Consequently, even in the case of resonance, the solution (5.89) remains the same.

5.7 Unsteady Forced Vibrations

Consider an approximate solution of (5.92) for the case when its right-hand side is an arbitrary function of time. If a rod is loaded as shown in Fig. 5.6, then the right-hand side of (5.92) coincides with that of (5.84) except that the functions $F^{(1)}(\tau)$ and $\Phi^{(1)}(\tau)$ are arbitrary functions of time. Considering two-term approximation, we can seek the solution to (5.92) in the form identical to (5.89).

Using the principle of virtual displacements in combination with (5.92), we obtain equation (5.93). If the rod is at rest at $\tau = 0$, then (5.93) is to be solved for the zero initial conditions. Equation (5.93) could be solved by use of either numerical or analytical methods depending on the specific type of the functions $F^{(1)}$ and $\Phi^{(1)}$.

5.8 Problems

5.1. A force **P** is applied to the end cross section of a circular rod of constant circular cross section (Fig. 5.11). The force vanishes at time $t = 0$. Study the free vibrations of the rod, namely, find the horizontal displacement of the end cross section (the point K) in the x_1-direction as a function of time. Because of circular shape of the cross section of the rod, it vibrates in the plane of drawing. Neglect the rotary inertia and the forces of viscous damping.

5.2. Consider a circular rod of constant cross section (Fig. 5.12*b*). At the instant of time $t = 0$, a concentrated constant moment **T** is suddenly applied to the rod. One of the principal axes of the cross section of the rod is perpendicular to the plane of drawing; hence, during the vibrations, the rod lies in this plane. Determine approximately the moment acting at the clamped end of the rod. Consider two-term approximation.

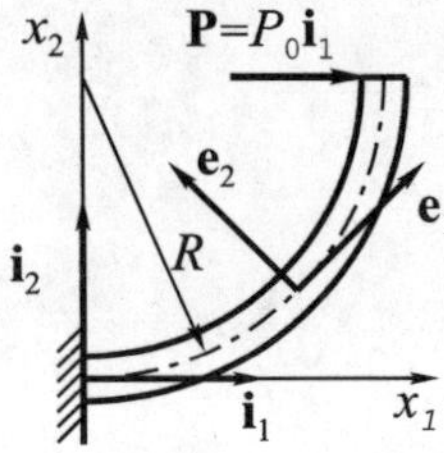

Fig. 5.11.

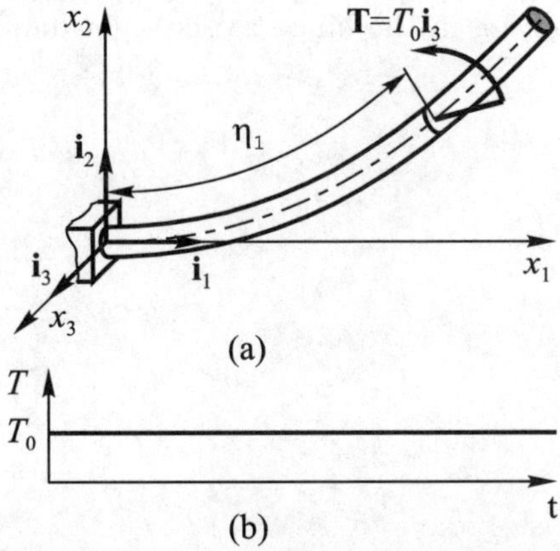

Fig. 5.12.

5.3. A force $\mathbf{P}$ applied to a rod (Fig. 5.13) disappears suddenly at time $t = 0$. Let ρ_0 be the radius of curvature of the rod axis in the natural configuration. During the vibrations the rod moves in the plane of drawing. Find the reaction at the intermediate hinge support.

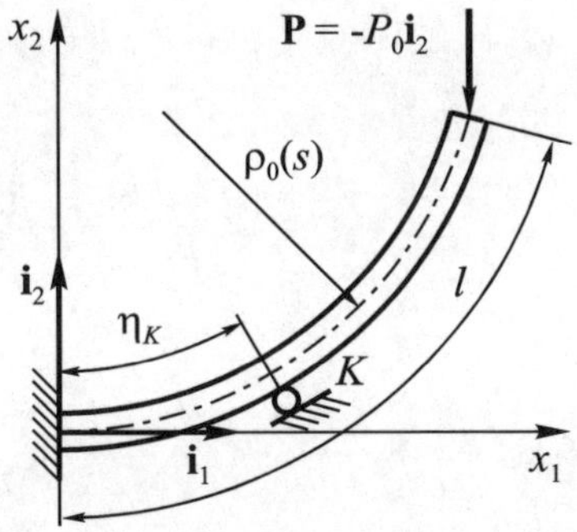

Fig. 5.13.

5.4. The right-hand end of a cylindrical spring (the point K in Fig. 5.14) is displaced by u_{x_1K} along the axis x_1. At that, the displacements of the point K along the axes x_2 and x_3 are equal zero, that is, $u_{x_2K} = u_{x_3K} = 0$.

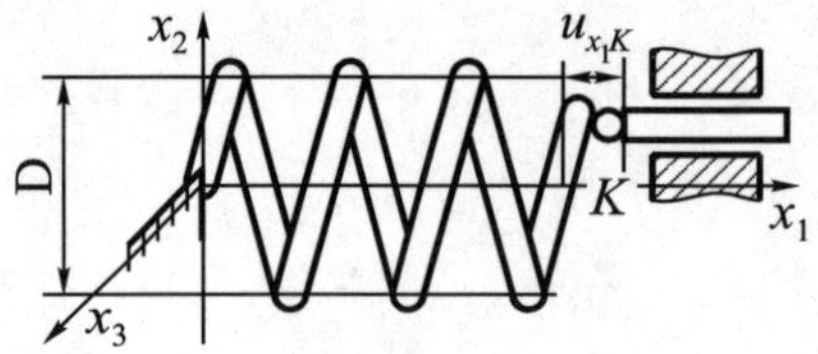

Fig. 5.14.

In other words, the spring is excited kinematically due to the hinge at the point K. Determine the reaction force acting in the hinge at the right-hand end of the spring.

6. Random Vibrations

In the previous chapters, we were concerned with problems in which the initial conditions, the values of the applied loads and the values of the prescribed displacements were assumed to be specified exactly. In practice, as a rool. The scatter of the initial data leads to the uncertainty in the determination of the stress–strain state of the rod. Thus, special computational techniques need to be elaborated to account for this effect.

Some elastic executive components of devices that are in an accelerated motion are shown in Figs. 6.1 and 6.2.

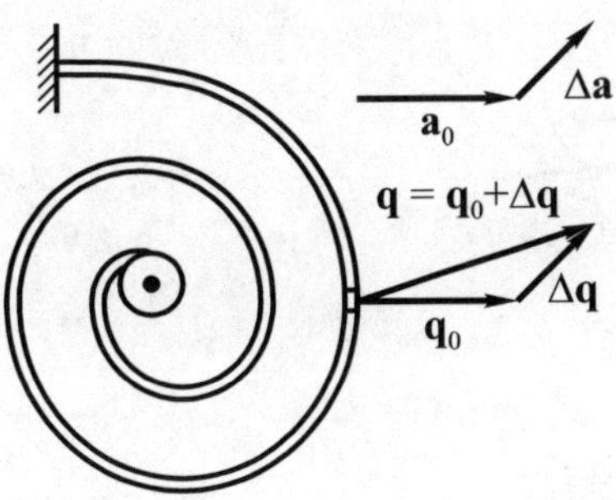

Fig. 6.1.

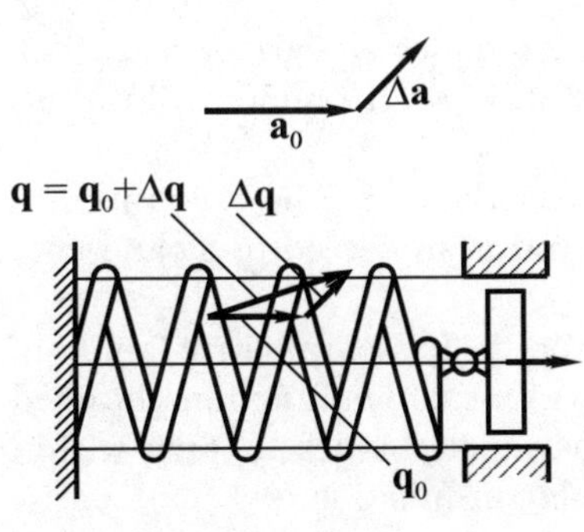

Fig. 6.2.

Additional inertial distributed forces are applied to these components. These additional inertial forces are caused by vibrations of supports of devices (more exactly, by a random component of the acceleration, $\Delta\mathbf{a}$).

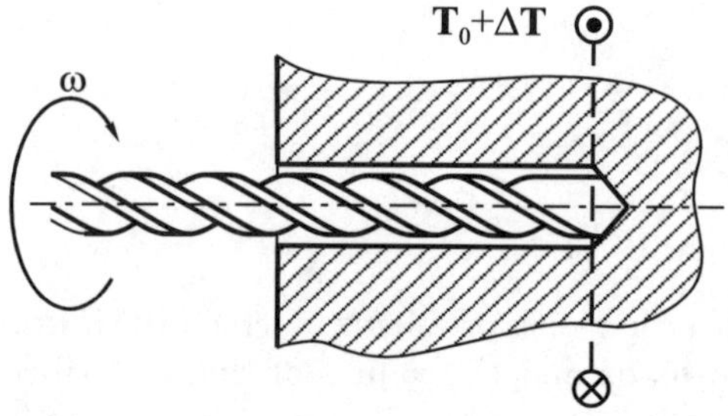

Fig. 6.3.

A drilling bit subjected to a torque is shown in Fig. 6.3. Owing to the lack of homogeneity of the metal being worked and due to random vibrations of the drill, the torque acquires a random component $\Delta\mathbf{T}$.

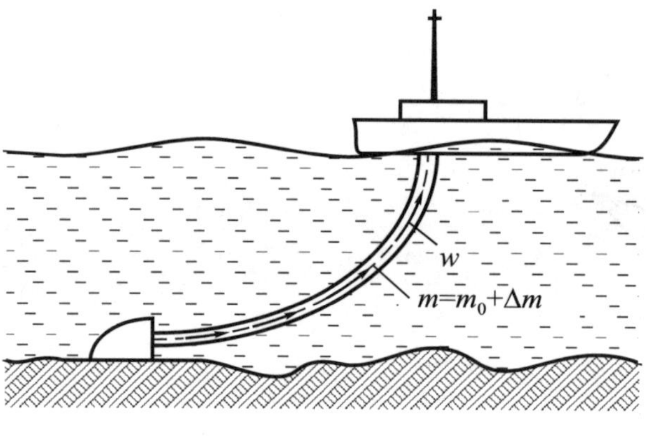

Fig. 6.4.

Figure 6.4 illustrates a pipeline designed for undersea earth excavation. The density of the seabed soil varies randomly; as a result, random vibrations of the pipeline appear.

A rod streamlined by an airflow is shown in Fig. 6.5. The flow rate has a random component $\Delta\mathbf{v}$. Consequently, the rod is subjected to a randomly changing aerodynamic force.

Recall that a random function (random process) is a parametric family of random variables. The set of numerical values of a random function obtained through an experiment (that is, when the parameter is let to vary within certain limits) is called a *realization* of the random function.

The theoretical background and computational tools for the analysis of problems related to statistical mechanics are presented in the textbooks

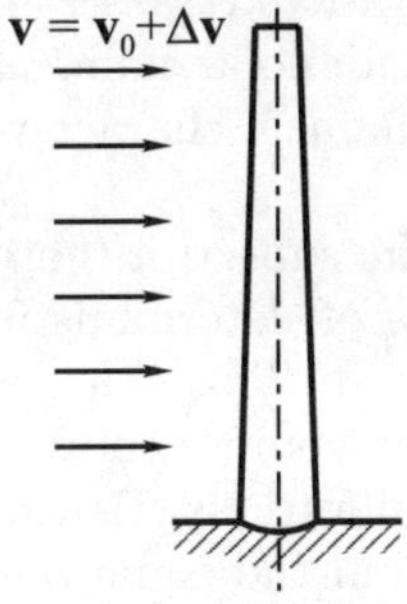

Fig. 6.5.

Bolotin (1979), Ventsel (1964), Svetlitsky (2003), and Sveshnikov (1968). Moreover, most students of technical universities have a sound training in the field of stochastic processes and are familiar with the methods of analysis of structures under random perturbation. For this reason, only a brief account of the most important points of the theory of random processes, especially of those referring to the analysis of equations for random vibrations of rods, is presented below.

6.1 Equations of Small Random Vibrations of Rods

6.1.1 The Derivation of the Equations

Consider a rod subjected to a concentrated force and a moment (Fig. 6.6). The force and the moment have random components, $\Delta\mathbf{P}$ and $\Delta\mathbf{T}$, respectively.

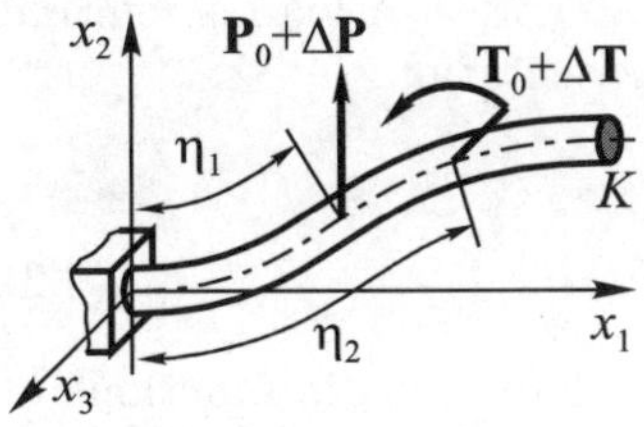

Fig. 6.6.

As an example, let a distributed random force $\Delta\mathbf{q}$ depending on $\Delta\mathbf{a}$ be applied to the rod. Then, the vector characteristics of a stress–strain state of the rod will include random components:

$$\mathbf{Q} = \mathbf{Q}_0 + \Delta\mathbf{Q} + \Delta\mathbf{Q}^{\mathrm{r}} ,$$
$$\mathbf{M} = \mathbf{M}_0 + \Delta\mathbf{M} + \Delta\mathbf{M}^{\mathrm{r}} ,$$
$$\boldsymbol{\vartheta} = \boldsymbol{\vartheta}_0 + \boldsymbol{\vartheta} + \boldsymbol{\vartheta}^{\mathrm{r}} ,$$
$$\mathbf{u} = \mathbf{u}_0 + \mathbf{u} + \mathbf{u}^{\mathrm{r}} .$$

Here, $\mathbf{Q}_0$, $\mathbf{M}_0$, $\boldsymbol{\vartheta}_0$, and $\mathbf{u}_0$ are the characteristics of the static stress–strain state, $\Delta\mathbf{Q}$, $\Delta\mathbf{M}$, $\boldsymbol{\vartheta}$, and $\mathbf{u}$ are the "deterministic" characteristics of dynamical stress–strain state, and $\Delta\mathbf{Q}^{\mathrm{r}}$, $\Delta\mathbf{M}^{\mathrm{r}}$, $\boldsymbol{\vartheta}^{\mathrm{r}}$, and $\mathbf{u}^{\mathrm{r}}$ are the dynamical characteristics of the random stress–strain state.

A few methods for determination of the vectors with the subscript 0 are presented in Svetlitsky (2000a). The methods for analysis of deterministic dynamic state of rods were discussed in the previous chapters. This chapter is concerned with the determination of the dynamic random components.

Let us restrict our consideration to the case of small vibrations. Hence, the equations of small random vibrations are separated from the equations governing the small deterministic vibrations. The former equations are

$$
\begin{aligned}
& n_1 \frac{\partial^2 \mathbf{u}^{\mathrm{r}}}{\partial \tau^2} - \frac{\partial \Delta\mathbf{Q}^{\mathrm{r}}}{\partial \eta} - \mathrm{A}_Q \mathrm{A}^{-1} \Delta\mathbf{M}^{\mathrm{r}} - \mathrm{A}_{æ}\, \Delta\mathbf{Q}^{\mathrm{r}} = \Delta\mathbf{P}^{\mathrm{r}}\,, \\
& \mathrm{J}\, \frac{\partial^2 \boldsymbol{\vartheta}^{\mathrm{r}}}{\partial \tau^2} - \frac{\partial \mathbf{M}^{\mathrm{r}}}{\partial \eta} - (\mathrm{A}_M \mathrm{A}^{-1} + \mathrm{A}_{æ})\, \Delta\mathbf{M}^{\mathrm{r}} - \mathrm{A}_1\, \Delta\mathbf{Q}^{\mathrm{r}} = \Delta\mathbf{T}^{\mathrm{r}}\,, \\
& \frac{\partial \boldsymbol{\vartheta}^{\mathrm{r}}}{\partial \eta} + \mathrm{A}_{æ} \boldsymbol{\vartheta}^{\mathrm{r}} - \mathrm{A}^{-1}\, \Delta\mathbf{M}^{\mathrm{r}} = 0\,, \\
& \frac{\partial \mathbf{u}^{\mathrm{r}}}{\partial \eta} + \mathrm{A}_{æ} \mathbf{u}^{\mathrm{r}} - \mathrm{A}^{-1} \boldsymbol{\vartheta}^{\mathrm{r}} = 0\,,
\end{aligned}
\tag{6.1}
$$

where

$$
\Delta\mathbf{P}^{\mathrm{r}} = \Delta\mathbf{q}^{\mathrm{r}} + \sum_{i=1}^{n} \Delta\mathbf{P}^{\mathrm{r}\,(i)}\, \delta(\eta - \eta_i)\,,
$$

$$
\Delta\mathbf{T}^{\mathrm{r}} = \Delta\boldsymbol{\mu}^{\mathrm{r}} + \sum_{\nu=1}^{\rho} \Delta\mathbf{T}^{\mathrm{r}\,(\nu)}\, \delta(\eta - \eta_\nu)\,.
$$

Random, time-dependent loads fall into two categories. A load is referred to the first category if it allows the following representation:

$$
\begin{aligned}
\Delta\mathbf{q}^{\mathrm{r}} &= \Delta\mathbf{q}_0^{\mathrm{r}} f_q(\tau)\,, & \Delta\mathbf{P}^{\mathrm{r}\,(i)} &= \Delta\mathbf{P}_0^{\mathrm{r}\,(i)} f_P(\tau)\,, \\
\Delta\boldsymbol{\mu}^{\mathrm{r}} &= \Delta\boldsymbol{\mu}_0^{\mathrm{r}} f_\mu(\tau)\,, & \Delta\mathbf{T}^{\mathrm{r}\,(\nu)} &= \Delta\mathbf{T}_0^{\mathrm{r}\,(\nu)} f_T(\tau)\,;
\end{aligned}
\tag{6.2}
$$

here, f_q, f_P, f_μ, and f_T are knowing in advance, deterministic functions of time; the components of the vectors

$$
\Delta\mathbf{q}_0^{\mathrm{r}}\,, \qquad \Delta\mathbf{P}_0^{\mathrm{r}(i)}\,, \qquad \Delta\boldsymbol{\mu}_0^{\mathrm{r}}\,, \qquad \Delta\mathbf{T}_0^{\mathrm{r}(\nu)}
$$

are random quantities. For example, if the functions $f(\tau)$ are harmonic, then $\Delta\mathbf{q}^{\mathrm{r}}$reads $\Delta\mathbf{q}^{\mathrm{r}}$, reads

$$
\Delta\mathbf{q}^{\mathrm{r}} = \Delta\mathbf{q}_0^{\mathrm{r}} \cos\omega_0(\tau)\,. \tag{6.3}
$$

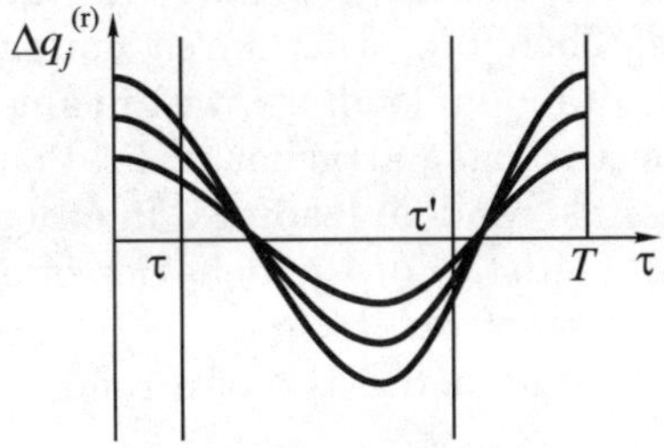

Fig. 6.7.

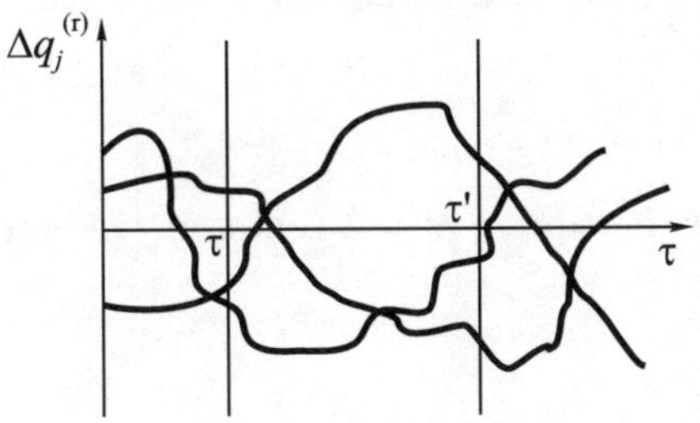

Fig. 6.8.

A selection of realizations of Δq_j^{r} defined by the last relationship are shown in Fig. 6.7. We see that distinct realizations of the random load are similar. This is true not only for the load (6.3) but also for the general case (6.2).

The second category consists of loads $\Delta \mathbf{q}^{\mathrm{r}}$, $\Delta \mathbf{P}^{\mathrm{r}(i)}$, $\Delta \boldsymbol{\mu}^{\mathrm{r}}$, and $\Delta \mathbf{T}^{\mathrm{r}(\nu)}$ that are random both in amplitude and in time. An example is shown in Fig. 6.8.

Equations (6.1) govern small random vibrations of a rod about the deterministic equilibrium configuration.

Only deterministic equilibrium configurations will be considered throughout the rest of the book and therefore will be simply referred to as *equilibrium configurations.*

To solve the equations of random vibrations (6.1), one must know all the probabilistic characteristics of the input random variables (that is, each component of the input vectors $\Delta \mathbf{q}^{\mathrm{r}}$, $\Delta \boldsymbol{\mu}^{\mathrm{r}}$, $\Delta \mathbf{P}^{\mathrm{r}(i)}$, and $\Delta \mathbf{T}^{\mathrm{r}(\nu)}$). These characteristics are: mathematical expectations (expected values), auto-correlation functions, cross-correlation functions, and a spectral density functions. The analysis of a problem implies the determination of these probabilistic characteristics for the components of the vectors $\Delta \mathbf{Q}^{\mathrm{r}}$, $\Delta \mathbf{M}^{\mathrm{r}}$, $\boldsymbol{\vartheta}^{\mathrm{r}}$, and $\mathbf{u}^{\mathrm{r}}$.

6.2 Basic Notions of the Statistical Dynamics

6.2.1 Random Variables (Processes) and Their Deterministic Characteristics

Three realizations of the random variable $\Delta q_j^{\mathrm{r}}(\tau)$ are shown in Figs. 6.7 and 6.8. Within the framework of statistic approach the analysis of random functions, we should study not each function and its properties, but the properties of the set of functions as a whole. Therefore, when investigating random vibrations of a rod under the action of random loadings, we are able to study the motion of the rod not with respect to one particular realization of the load but with respect to a range of possible random loadings. In other words, our goal is to obtain the general characteristics of the behavior of a rod under a set of random loads not under a distinct realization.

The following nonrandom functions[1] are the characteristics of a random variable X.

(1) The *mathematical expectation* (or expected value) of the random variable X is a function defined as

$$m_x(\tau) = M[X(\tau)]\,.$$

(2) The mathematical expectation of the square of the centered random variable

$$X^\circ(\tau) = X(\tau) - m_x(\tau)$$

is called the *variance* of the random variable X, that is,

$$D_x(\tau) = M[X^\circ(\tau)^2]\,.$$

(3) The *density function* $f(x,\, \tau)$, which now a time-dependent quantity.

(4) The *correlation function* (the *covariance function*) is a measure of the dependence between two realizations at instances of time τ and $\hat{\tau}$ of the centered random process $X(\tau)$ (see Figs. 6.7 and 6.8)

$$K_x(\tau,\, \hat{\tau}) = M[X^\circ(\tau)\, X^\circ(\hat{\tau})]\,,$$

where

$$X^\circ(\tau) = X(\tau) - m_x(\tau)\,, \qquad X^\circ(\hat{\tau}) = X(\hat{\tau}) - m_x(\hat{\tau})\,.$$

(5) The spectral *density function* is usually denoted by $S_x(\omega)$, where ω is a continuous frequency.

The characteristics of a stationary random variable are time independent, that is,

$$m_x = \text{const}\,, \qquad D_x = \text{const}\,,$$

and the correlation function is a function of one variable $\hat{\tau} - \tau = \tau_1$, that is,

$$K_x(\hat{\tau},\, \tau) = K_x(\hat{\tau} - \tau) = K_x(\tau_1)\,.$$

The variance of a stationary random variable reads

---

[1] It should be noted that the terminology and definitions used in this chapter correlate with the terminology and definitions introduced in Svetlitsky (2003).

$$D_x = K_x(0) = \text{const} .$$

The correlation function of a stationary random process and the spectral density function are related to each other by the Wiener–Khintchin formulas

$$K_x(\tau_1) = \int_{-\infty}^{\infty} S_x(\omega) \cos \omega\tau_1 \, \mathrm{d}\omega , \tag{6.4}$$

$$S_x(\omega) = \frac{1}{2\pi} \int_{-\infty}^{\infty} K_x(\tau_1) \cos \omega\tau_1 \, \mathrm{d}\tau_1 . \tag{6.5}$$

Setting $\tau_1 = 0$ in (6.4), we obtain another useful relationship

$$K_x(0) = D_x = \int_{-\infty}^{\infty} S_x(\omega) \, \mathrm{d}\omega . \tag{6.6}$$

The derivation of (6.4) will be given below.

6.2.2 Probabilistic Characteristics of Solutions of Linear Inhomogeneous Differential Equations

Consider a linear equation of the second order

$$\ddot{y} + a_1(\tau)\,\dot{y} + a_2(\tau)\,\dot{y} = X(\tau) , \tag{6.7}$$

where $X(\tau)$ is a random variable whose probabilistic characteristics $m_x(\tau)$, $K_x(\tau, \hat{\tau})$, and $D_x(\tau)$ are known.

The solution of (6.7) is of the form

$$y(\tau) = c_1 k_{11}(\tau) + c_2 k_{12}(\tau) + \int_0^\tau g_{11}(\tau, \zeta)\, X(\zeta) \, \mathrm{d}\zeta ,$$

$$\dot{y}(\tau) = c_1 k_{21}(\tau) + c_2 k_{22}(\tau) + \int_0^\tau g_{21}(\tau, \zeta)\, X(\zeta) \, \mathrm{d}\zeta .$$

Assuming that, for $\tau = 0$, $y(0) = \dot{y}(0) = 0$, we arrive at

$$y(\tau) = \int_0^\tau g_{11} X \, \mathrm{d}\zeta , \qquad \dot{y}(\tau) = \int_0^\tau g_{21} X \, \mathrm{d}\zeta . \tag{6.8}$$

The mathematical expectation and their time derivatives are as follows:

$$m_y(\tau) = M[Y(\tau)] = M\left[\int_0^\tau g_{11} X \, \mathrm{d}\zeta\right] ,$$

$$m_{\dot{y}}(\tau) = M[\dot{Y}(\tau)] = M\left[\int_0^\tau g_{21} X \, \mathrm{d}\zeta\right] . \tag{6.9}$$

It is known from the theory of stochastic processes that

$$M\left[\int F\,\mathrm{d}\zeta\right] = \int M[F]\,\mathrm{d}\zeta\,.$$

The functions g_{11} and g_{21} are nonrandom; therefore, we have

$$M[g_{i1}X] = g_{i1}M[X] = g_{i1}m_x\,.$$

Consequently, it follows from (6.9) that

$$m_y(\tau) = \int_0^\tau g_{11}m_x(\zeta)\,\mathrm{d}\zeta\,, \qquad m_{\dot{y}}(\tau) = \int_0^\tau g_{21}m_x(\zeta)\,\mathrm{d}\zeta\,. \tag{6.10}$$

Let us obtain the correlation function for the solution and its first derivative. The centered functions Y and Y° read

$$Y^\circ(\tau) = Y - m_y = \int_0^\tau g_{11}X^\circ\,\mathrm{d}\zeta\,,$$

$$\dot{Y}^\circ(\tau) = \dot{Y} - m_{\dot{y}} = \int_0^\tau g_{21}X^\circ\,\mathrm{d}\zeta\,.$$

The correlation function is

$$K_y(\tau,\,\hat{\tau}) = M\left[\int_0^\tau g_{11}X^\circ(\zeta)\,\mathrm{d}\zeta \int_0^{\hat{\tau}} g_{11}X^\circ(\xi)\,\mathrm{d}\xi\right]$$

or

$$K_y = \int_0^\tau \int_0^{\hat{\tau}} g_{11}(\tau,\,\zeta)\,g_{11}(\hat{\tau},\,\xi)\,M[X^\circ(\zeta)\,X^\circ(\xi)]\,\mathrm{d}\zeta\,\mathrm{d}\xi\,.$$

Since

$$M[X^\circ(\zeta)\,X^\circ(\xi)] = K_x(\zeta,\,\xi)\,,$$

we finally arrive at

$$K_y = \int_0^\tau \int_0^{\hat{\tau}} g_{11}(\tau,\,\zeta)\,g_{11}(\hat{\tau},\,\xi)\,K_x(\zeta,\,\xi)\,\mathrm{d}\zeta\,\mathrm{d}\xi\,. \tag{6.11}$$

Similarly, the correlation function for the first derivative is as follows:

$$K_{\dot{y}} = \int_0^\tau \int_0^{\hat{\tau}} g_{21}(\tau,\,\zeta)\,g_{21}(\hat{\tau},\,\xi)\,K_x(\zeta,\,\xi)\,\mathrm{d}\zeta\,\mathrm{d}\xi\,. \tag{6.12}$$

Suppose that the random variable $X(\tau)$ is stationary; then, $X(\tau)$ and a stationary solution $y(\tau)$ can be represented by the Fourier integrals

$$X(t) = \int_{-\infty}^{\infty} X_0(\omega)\,\mathrm{e}^{\mathrm{i}\omega t}\,\mathrm{d}\omega\,, \tag{6.13}$$

$$Y(t) = \int_{-\infty}^{\infty} Y_0(\omega)\,\mathrm{e}^{\mathrm{i}\omega t}\,\mathrm{d}\omega\,. \tag{6.14}$$

It is worth noting that in the case of stationary random loads, a stationary solution exists provided the motion of the rod is governed by a differential equation with constant coefficients, that is, time-independent coefficients. Thus, equation (6.7) has a stationary solution provided $a_1 = \text{const}$ and $a_2 = \text{const}$.

The correlation function for the random variable $X^\circ(t)$ is

$$K_x[X^\circ(t)\, X^{\circ *}(\hat{t}\,)] = M\left[\int_{-\infty}^{\infty}\int_{-\infty}^{\infty} \mathrm{e}^{\mathrm{i}\omega t - \mathrm{i}\,\hat{\omega}\hat{t}} X_0(\omega)\, X_0^*(\hat{\omega})\, \mathrm{d}\omega\, \mathrm{d}\hat{\omega}\right]$$

or

$$K_x = \int_{-\infty}^{\infty}\int_{-\infty}^{\infty} \mathrm{e}^{\mathrm{i}\omega t - \mathrm{i}\,\hat{\omega}\hat{t}} M[X_0(\omega)\, X_0^*(\hat{\omega})]\, \mathrm{d}\omega\, \mathrm{d}\hat{\omega}\,. \tag{6.15}$$

For a stationary random variable, the correlation function is a function of the difference of time instants $\tau = t - \hat{t}$; in this case, we have

$$M[X_0(\omega)\, X_0^*(\hat{\omega})] = S_x(\hat{\omega})\, \delta(\hat{\omega} - \omega)\,. \tag{6.16}$$

Using (6.16) in (6.15) and integrating the result with respect to $\hat{\omega}$, we get

$$K_x(\tau) = \int_{-\infty}^{\infty} S_x(\omega)\, \mathrm{e}^{\mathrm{i}\omega\tau}\, \mathrm{d}\omega\,, \qquad \tau = t - \hat{t}\,. \tag{6.17}$$

The inverse transform reads

$$S_x(\omega) = \frac{1}{2\pi}\int_{-\infty}^{\infty} K_x(\tau)\, \mathrm{e}^{-\mathrm{i}\omega\tau}\, \mathrm{d}\tau\,. \tag{6.18}$$

The relationships (6.17) and (6.18) are known as the Wiener–Khintchin formulas.

Substituting (6.13) and (6.14) into (6.6) (for $a_1 = a_2 = \text{const}$), we obtain

$$Y_0(\omega) = W(\mathrm{i}\,\omega)\, X_0(\omega)\,, \tag{6.19}$$

where

$$W(\mathrm{i}\,\omega) = \frac{1}{(\mathrm{i}\,\omega)^2 + a_1\,\mathrm{i}\,\omega + a_2}\,.$$

Using (6.18) in (6.14), we arrive at

$$y(t) = \int_{-\infty}^{\infty} W(\mathrm{i}\,\omega)\, X_0(\omega)\, \mathrm{e}^{\mathrm{i}\omega t}\, \mathrm{d}\omega\,.$$

Similarly, for time $\hat{t}$, we have

$$y(\hat{t}\,) = \int_{-\infty}^{\infty} W(\mathrm{i}\,\hat{\omega})\, X_0(\hat{\omega})\, \mathrm{e}^{\mathrm{i}\hat{\omega}\hat{t}}\, \mathrm{d}\hat{\omega}\,.$$

Therefore, the correlation function $K_y[y(t)y(\hat{t}\,)]$ takes the form

$$K_y(t, \hat{t}) = \int_{-\infty}^{\infty}\int_{-\infty}^{\infty} W(\mathrm{i}\,\hat{\omega})\, W^*(\mathrm{i}\,\omega)\, M[X_0(\omega)\, X_0^*(\hat{\omega})]\, \mathrm{e}^{\mathrm{i}\omega t - \mathrm{i}\,\hat{\omega}\hat{t}}\, \mathrm{d}\omega\, \mathrm{d}\hat{\omega}.$$

In view of (6.16), this gives

$$K_y(\tau) = \int_{-\infty}^{\infty} |W(\mathrm{i}\,\omega)|^2 S_x(\omega)\, \mathrm{d}\omega\,. \tag{6.20}$$

By use of (6.14), we obtain

$$K_y(\tau) = \int_{-\infty}^{\infty} S_y(\omega)\, \mathrm{e}^{\mathrm{i}\omega\tau}\, \mathrm{d}\omega\,. \tag{6.21}$$

This expression is identical in form to (6.17).

It follows from (6.20) and (6.21) that

$$S_y(\omega) = |W(\mathrm{i}\,\omega)|^2 S_x(\omega)\,. \tag{6.22}$$

The differential equation in one unknown (6.7) is of frequent occurrence in the analysis of mechanical systems with one degree of freedom. Such an equation also arises when the principle of virtual displacements is used for determination of a one-term approximate solution of the equations of small vibrations. In the case of a mechanical system with several degrees of freedom, the calculation becomes cumbersome. A detailed discussion of methods for solution of systems of linear differential equations can be found in Gusev, Svetlitsky (1984), Svetlitsky (2003), and Sveshnikov (1968). Additional information on the analysis of problems related to the statistical dynamics can be found in the section devoted to applied problems.

6.2.3 Estimation of Durability with Consideration for Random Stresses

The reader may ask the question: What is the practical advantage of studying random vibrations of rods? As it has been repeatedly mentioned above, the mechanics of rods is presented in this book as a theoretical background and computational tools for the analysis of structures and structural members that can be treated as rods. The analysis of such structures under real loadings involves the determination of their stress–strain state.

When we consider static loads, we determine a static stress–strain state. Knowing the static stress–strain state of a structure, we are able to determine the limiting values of stresses and therefore find out whether the element still ensures safe operation. For example, the real acceleration of an object with an elastic structural members mounted on it (say, an executive component of a gauge) (see Figs. 6.1 and 6.2) must not cause a plastic deformation of the elastic member. Otherwise, the residual deformation may lead to inaccuracy in the gauge readings.

Similarly, in the case of dynamic loads, the determination of stresses that result in plastic deformation is of primary interest because these plastic deformations are inadmissible. If a rod, for instance, the valve spring of an engine, is in a stationary mode of oscillatory motion, then, to estimate the durability of the rod, it is essential to know the amplitude values of the stresses.

The situation is more difficult when dealing with rods exerted by random loads. Actually, the stress–strain state caused by random loads cannot be determined exactly. Anyway, it is clear that random static or dynamic loadings lead to a random stress-strain state of a structural element; therefore, in the case of random loadings, we have no opportunity to determine stresses in a rod uniquely. It is clear that the serviceability of a structure may, on occasion, strongly depend on its stress–strain state. Consequently, it is essential to be able to estimate random stress–strain states of structural elements. For example, even if the water surface is still, the non-homogeneity of the seabed inevitably causes random vibrations of the pipeline by which the soil is taken up (Fig. 6.4). Consequently, dynamic stresses in the pipeline are of random nature and the life duration of the pipeline is to be estimated with consideration for random components of stresses.

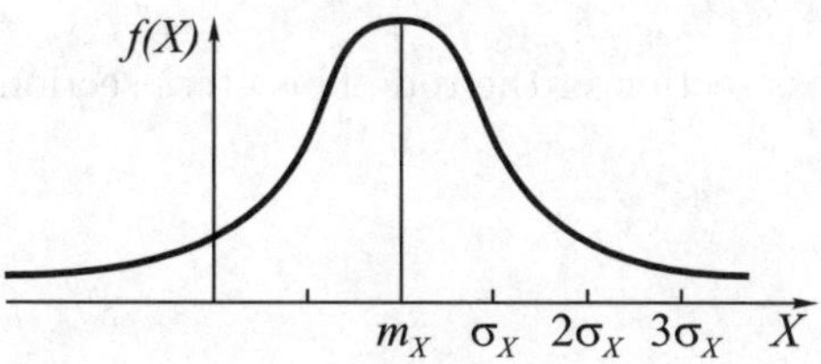

Fig. 6.9.

To estimate the influence of random components of stresses and displacements on the serviceability of a structure, one must employ some relationships for obtaining non-random quantitative characteristics of stress–strain state of a structure. (For instance, this is the case when the life duration estimation of a structure in stationary random oscillatory motion is determined by a conventional method presented in Feodosiev (1999); the last method requires the knowledge of the extreme values of the stresses.) One such a relationship is the formula for the maximum value of a random variable that follows a normal distribution (a Gaussian distribution) (Fig. 6.9), namely,

$$x_{\max} = m_x + 3\sigma_x \, , \qquad \sigma_x^2 = D_x \, , \tag{6.23}$$

where m_x is the mathematical expectation of the random variable and σ_x is the standard deviation (the root-mean-square deviation).

For example, solving the equations of small random vibrations of a rod, we obtain $\Delta M_2^{\mathrm{r}}(\eta, \tau)$ and $\Delta M_3^{\mathrm{r}}(\eta, \tau)$. Assuming these quantities possess a normal distribution, we find their maximum values to be

$$\max(\Delta M_j^{\mathrm{r}}) = m_{AM_j}(\eta, \tau) + 3\sigma_{\Delta M_j}(\eta, \tau), \qquad j = 2, 3.$$

Knowing the limiting possible values of the moments, we may determine the corresponding maximum values of the normal stresses in a rod:

$$\max(\Delta M_j^{\mathrm{r}}) = m_{AM_j}(\eta, \tau) + 3\sigma_{\Delta M_j}(\eta, \tau), \qquad j = 2, 3.$$

It is worth noting that the normal stresses σ depend linearly on the bending moments.

It should be pointed out that in some graduate-level textbooks (see Feodosiev (1999) and Ventsel (1964)) the letter σ stands both for standard deviation and normal stress.

In the case of stationary vibrations, the amplitude values of the maximum stresses do not depend on time. The total normal stress at in a rod is

$$\sigma_n(\eta) = \sigma_0(\eta) + \sigma_{\max}^{\mathrm{r}}(\eta),$$

where $\sigma_0(\eta)$ is the normal stress caused by the static loads. If we know σ_n as a function of η, then we can find the cross section of the rod at which σ_n has a maximum

$$\max \sigma_n(\eta^*) = \max(\sigma_0(\eta^*) + \sigma_{\max}^{\mathrm{r}}(\eta^*)),$$

where η^* is the coordinate of the cross section of the rod. This cross section of the rod is called a *weak section.*

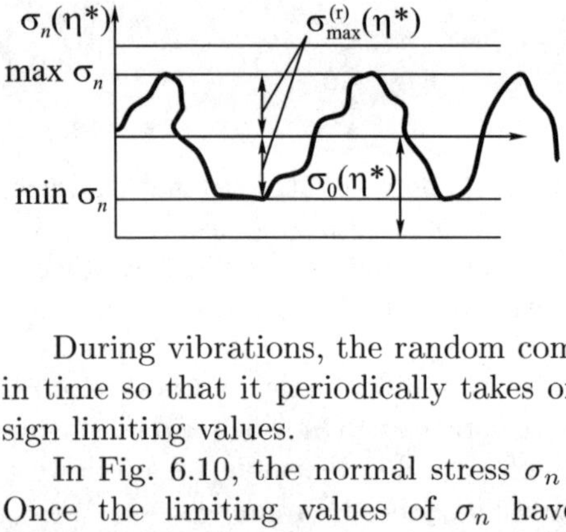

Fig. 6.10.

During vibrations, the random component of the normal stress σ_{r} varies in time so that it periodically takes on equal in magnitude but different in sign limiting values.

In Fig. 6.10, the normal stress σ_n is qualitatively plotted against time. Once the limiting values of σ_n have been found, we may determine a safety factor for the fatigue strength by using the well-known formulas (Feodosiev (1999)).

6.3 Stationary Random Vibrations

6.3.1 Probabilistic Characteristics of a Solution

Introducing terms accounting for forces of viscous damping into the first equation of the system (6.1), we can represent the system in the form of the vector equation (5.50) as follows:

$$\mathrm{A}^{(1)}\frac{\partial^2 \mathbf{Z}^{\mathrm{r}}}{\partial \tau^2}+\frac{\partial \mathbf{Z}^{\mathrm{r}}}{\partial \eta}+\mathrm{A}^{(2)}\mathbf{Z}^{\mathrm{r}}+\mathrm{A}^{(3)}\frac{\partial \mathbf{Z}^{\mathrm{r}}}{\partial \tau}=\Delta\boldsymbol{\Phi}^{\mathrm{r}}\,; \tag{6.24}$$

here,

$$\Delta\boldsymbol{\Phi}^{\mathrm{r}}=(\Delta\mathbf{P}^{\mathrm{r}}\,\delta(\eta-\eta_1),\,\Delta\mathbf{T}^{\mathrm{r}}\,\delta(\eta-\eta_2),\,0,\,0)^{\mathrm{T}}\,.$$

We seek an approximate solution of (6.24) in the form

$$\mathbf{Z}^{\mathrm{r}}=\sum_{i=1}^{n}\mathbf{Z}_0^{(i)}(\eta)\,f^{(i)}(\tau)\,.$$

Considering only two-term approximation, we get

$$\mathbf{Z}^{\mathrm{r}}=\mathbf{Z}_0^{(1)}\,f^{(1)}+\mathbf{Z}_0^{(2)}\,f^{(2)}\,. \tag{6.25}$$

By the principle of virtual displacements, from (6.24) it follows that

$$\begin{aligned}
&h_{11}\ddot{f}^{(1)}+h_{12}\ddot{f}^{(2)}+b_{11}\dot{f}^{(1)}+b_{12}\dot{f}^{(2)}+c_{11}f^{(1)}+c_{12}f^{(2)}\\
&\quad=\sum_{j=1}^{3}\left[\Delta P_j(\tau)\,u_{0j}^{(1)}(\eta_1)+\Delta T_j(\tau)\,\vartheta_{0j}^{(1)}(\eta_2)\right],\\
&h_{21}\ddot{f}^{(1)}+h_{22}\ddot{f}^{(2)}+b_{21}\dot{f}^{(1)}+b_{22}\dot{f}^{(2)}+c_{21}f^{(1)}+c_{22}f^{(2)}\\
&\quad=\sum_{j=1}^{3}\left[\Delta P_j(\tau)\,u_{0j}^{(2)}(\eta_1)+\Delta T_j(\tau)\,\vartheta_{0j}^{(2)}(\eta_2)\right].
\end{aligned}$$

This system of two equations in two unknown functions $f^{(1)}$ and $f^{(2)}$ can be represented as a vector equation:

$$\mathrm{H}\ddot{\mathbf{f}}+\mathrm{B}\dot{\mathbf{f}}+\mathrm{C}\mathbf{f}=\mathrm{D}^{(1)}\,\Delta\mathbf{P}^{\mathrm{r}}+\mathrm{D}^{(2)}\,\Delta\mathbf{T}^{\mathrm{r}}\,; \tag{6.26}$$

here,

$$\mathrm{D}^{(1)}=\begin{bmatrix} u_{01}^{(1)} & u_{02}^{(1)} & u_{03}^{(1)}\\ u_{01}^{(2)} & u_{02}^{(2)} & u_{03}^{(2)}\end{bmatrix},\qquad \mathrm{D}^{(2)}=\begin{bmatrix} \vartheta_{01}^{(1)} & \vartheta_{02}^{(1)} & \vartheta_{03}^{(1)}\\ \vartheta_{01}^{(2)} & \vartheta_{02}^{(2)} & \vartheta_{03}^{(2)}\end{bmatrix}.$$

Now we will discuss a general method for determination of the spectral density functions and the cross-spectral densities of the components of the

n-dimensional vector $\mathbf{f}$ when the spectral density functions of the stationary random perturbations in the right-hand side of (6.26) are given.

By definition, the cross-correlation function of two centered random processes, $f^{(j)}$ and $f^{(\nu)}$, is as follows (see Svetlitsky (2003)):

$$K(\tau,\,\hat{\tau}) = M[f^{(j)}(\tau)f^{(\nu)}(\hat{\tau})]\,. \tag{6.27}$$

Using the Fourier transform (6.14), we obtain

$$\begin{aligned} f^{(j)} &= \int_{-\infty}^{\infty} f_0^{(j)}(\omega)\,\mathrm{e}^{\mathrm{i}\omega\tau}\,\mathrm{d}\omega\,,\\ f^{(\nu)} &= \int_{-\infty}^{\infty} f_0^{(\nu)}(\hat{\omega})\,\mathrm{e}^{-\mathrm{i}\hat{\omega}\hat{\tau}}\,\mathrm{d}\hat{\omega}\,. \end{aligned} \tag{6.28}$$

Substituting these expressions into (6.27), we arrive at

$$K_{f^{(j)}f^{(\nu)}} = \int_{-\infty}^{\infty}\int_{-\infty}^{\infty} M[f_0^{(j)}(\omega)f_0^{(\nu)}(\hat{\omega})]\,\mathrm{e}^{\mathrm{i}\,(\omega\tau-\hat{\omega}\hat{\tau})}\,\mathrm{d}\omega\,\mathrm{d}\hat{\omega}\,. \tag{6.29}$$

In the case of stationary vibrations, the correlation function $K_{f^{(j)}}$ and the cross-correlation function $K_{f^{(j)}f^{(\nu)}}$ are functions of one variable, the difference of instants of time $\tau - \hat{\tau} = \tau_1$. The cross-correlation functions (6.29) depend on τ_1 if and only if

$$M[f_0^{(j)}(\omega)f_0^{(\nu)}(\hat{\omega})] = S_{f^{(j)}f^{(\nu)}}(\hat{\omega})\,\delta(\hat{\omega}-\omega)\,, \tag{6.30}$$

where $S_{f^{(j)}f^{(\nu)}}(\hat{\omega})$ is a function of $\hat{\omega}$. This function is called the *cross-spectral density of two stationary random processes.*

Substituting (6.30) into (6.29) and integrating with respect to $\hat{\omega}$, we get

$$K_{f^{(j)}f^{(\nu)}}(\tau,\,\hat{\tau}) = \int_{-\infty}^{\infty} S_{f^{(j)}f^{(\nu)}}(\omega)\,\mathrm{e}^{\mathrm{i}\omega\tau}\,\mathrm{d}\omega\,. \tag{6.31}$$

Let us represent the correlation function $K_{f^{(j)}f^{(\nu)}}$ in terms of spectral density functions of the components of the vectors $\Delta\mathbf{P}$ and $\Delta\mathbf{T}$. Employing the Fourier transform, from (6.28) we get

$$\begin{aligned} \mathbf{f}_0(\mathrm{i}\,\omega) &= W(\mathrm{i}\,\omega)[\mathrm{D}^{(1)}\Delta\mathbf{P}_0(\omega) + \mathrm{D}^{(2)}\Delta\mathbf{T}_0(\omega)]\,,\\ W(\mathrm{i}\,\omega) &= [(\mathrm{i}\,\omega)^2\mathrm{H} + \mathrm{i}\,\omega\mathrm{B} + \mathrm{C}]^{-1} \end{aligned}$$

or

$$\mathbf{f}_0 = W^{(P)}(\mathrm{i}\,\omega)\,\Delta\mathbf{P}_0 + W^{(T)}(\mathrm{i}\,\omega)\,\Delta\mathbf{T}_0\,. \tag{6.32}$$

The components of the vectors $\mathbf{u}_0^{(1)}$, $\mathbf{u}_0^{(2)}$, $\boldsymbol{\vartheta}_0^{(1)}$, and $\boldsymbol{\vartheta}_0^{(2)}$ for $\eta = \eta_1$ enter into the matrix $\mathrm{D}^{(1)}$, and these components for $\eta = \eta_2$ enter into the matrix $\mathrm{D}^{(2)}$ (Fig. 6.6).

In components, we have

$$f_0^{(j)}(\omega) = \sum_{\rho=1}^{n_1} W_{j\rho}^{(P)} P_\rho + \sum_{\rho=1}^{n_1} W_{j\rho} T_\rho , \qquad n_1 \le n , \tag{6.33}$$

where P_ρ and T_ρ are the components of the vectors $\mathbf{\Delta P}_0$ and $\mathbf{\Delta T}_0$. Therefore,

$$\begin{aligned} M[f_0^{(j)}(\omega) f_0^{(\nu)*}(\hat{\omega})] &= \sum_{\rho=1}^{n_1}\sum_{k=1}^{n_1} W_{j\rho}^{(P)} W_{\nu k}^{(P)*} M[P_\rho(\omega) P_k^*(\hat{\omega})] \\ &+ \sum_{\rho=1}^{n_1}\sum_{k=1}^{n_1} W_{j\rho}^{(P)} W_{\nu k}^{(T)*} M[P_\rho(\omega) T_k^*(\hat{\omega})] \\ &+ \sum_{\rho=1}^{n_1}\sum_{k=1}^{n_1} W_{j\rho}^{(T)}(\omega) W_{\nu k}^{(P)*} M[T_\rho(\omega) P_k^*(\hat{\omega})] \\ &+ \sum_{\rho=1}^{n_1}\sum_{k=1}^{n_1} W^{(T)}(\omega) W^{(T)*}(\hat{\omega}) M[T_\rho(\omega) T_k^*(\hat{\omega})] . \end{aligned} \tag{6.34}$$

Since

$$K_{f^{(j)}f^{(\nu)}}(\tau, \hat{\tau}) = \int_{-\infty}^{\infty}\int_{-\infty}^{\infty} M[f_0^{(j)}(\omega) f_0^{(\nu)*}(\hat{\omega})] \, \mathrm{e}^{\mathrm{i}\,(\omega\tau - \hat{\omega}\hat{\tau})} \, \mathrm{d}\omega \, \mathrm{d}\hat{\omega} , \tag{6.35}$$

we obtain, eliminating $M[f_0^{(j)} f_0^{(\nu)*}]$ by use of (6.30), the following relationship:

$$\begin{aligned} K_{f^{(j)}f^{(\nu)}} = \int_{-\infty}^{\infty}\int_{-\infty}^{\infty} \Bigg\{ \sum_{\rho=1}^{n_1}\sum_{k=1}^{n_1} \Big(& W_{j\rho}^{(P)} W_{\nu k}^{(P)} M[P_\rho P_k^*] \\ & + W_{j\rho}^{(P)} W_{\nu k}^{(T)*} M[P_\rho T_k^*] + W_{j\rho}^{(T)} W_{\nu k}^{(P)*} M[T_\rho P_k^*] \\ & + W_{j\rho}^{(T)} W_{\nu k}^{(T)*} M[T_\rho T_k^*] \Big) \Bigg\} \, \mathrm{e}^{\mathrm{i}\,(\omega\tau - \hat{\omega}\hat{\tau})} \, \mathrm{d}\omega \, \mathrm{d}\hat{\omega} . \end{aligned} \tag{6.36}$$

For $K_{f^{(j)}f^{(\nu)}}$ to be a function of the difference of instants of time, the following equalities must hold:

$$\begin{aligned} M[P_\rho P_k^*] &= S_{P_\rho P_k}(\hat{\omega}) \, \delta(\hat{\omega} - \omega) , \\ M[P_\rho T_k^*] &= S_{P_\rho T_k}(\hat{\omega}) \, \delta(\hat{\omega} - \omega) , \\ M[T_\rho P_k^*] &= S_{T_\rho P_k}(\hat{\omega}) \, \delta(\hat{\omega} - \omega) , \\ M[T_\rho T_k^*] &= S_{T_\rho T_k}(\hat{\omega}) \, \delta(\hat{\omega} - \omega) . \end{aligned} \tag{6.37}$$

Here, $S_{P_\rho P_k}$, $S_{P_\rho T_k}$,$S_{T_\rho P_k}$, and $S_{T_\rho T_k}$ are the cross-spectral densities of the components of the vectors $\mathbf{\Delta P}$ and $\mathbf{\Delta T}$.
The equalities (6.37) are identical to (6.30).

Substituting (6.37) into the expression of $K_{f^{(j)}f^{(\nu)}}$ and integrating with respect to $\hat{\omega}$, we obtain the cross-correlation function $K_{f^{(j)}f^{(\nu)}}$ for stationary dependent components of the vector $\mathbf{f}$:

$$K_{f^{(j)}f^{(\nu)}}(\tau_1) = \int_{-\infty}^{\infty} \left\{ \sum_{\rho=1}^{n_1} \sum_{k=1}^{n_1} \left(W_{j\rho}^{(P)}(\omega) W_{\nu k}^{(P)*} S_{P_\rho P_k} \right. \right.$$

$$+ W_{j\rho}^{(P)} W_{\nu k}^{(T)*} S_{P_\rho T_k} + W_{j\rho}^{(T)} W_{\nu k}^{(P)*} S_{T_\rho P_k}$$

$$\left. \left. + \; W_{j\rho}^{(T)} W_{\nu k}^{(T)*} S_{T_\rho T_k} \right) \right\} \mathrm{e}^{\mathrm{i}\omega\tau_1} \, \mathrm{d}\omega \, . \quad (6.38)$$

Thus, we have obtained another expression for that cross-correlation function. Therefore, equating the right-hand sides of the expressions (6.31) and (6.38), we get the identity

$$\int_{-\infty}^{\infty} \left[S_{f^{(j)}f^{(\nu)}} - \sum_{\rho=1}^{n_1} \sum_{k=1}^{n_1} \left(W_{j\rho}^{(P)} W_{\nu k}^{(P)*} S_{P_\rho P_k} + W_{j\rho}^{(P)} W_{\nu k}^{(T)*} S_{P_\rho T_k} \right. \right.$$

$$\left. \left. + \; W_{j\rho}^{(T)} W_{\nu k}^{(P)*} S_{T_\rho P_k} + W_{j\rho}^{(T)} W_{\nu k}^{(T)*} S_{T_\rho T_k} \right) \right] \mathrm{e}^{\mathrm{i}\omega\tau_1} \, \mathrm{d}\omega \equiv 0 \, .$$

From this identity it follows that

$$S_{f^{(j)}f^{(\nu)}}(\omega) = \sum_{\rho=1}^{n_1} \sum_{k=1}^{n_1} \left(W_{j\rho}^{(P)} W_{\nu k}^{(P)*} S_{P_\rho P_k} + W_{j\rho}^{(P)} W_{\nu k}^{(T)*} S_{P_\rho T_k} \right.$$

$$\left. + \; W_{j\rho}^{(T)} W_{\nu k}^{(P)*} S_{T_\rho P_k} + W_{j\rho}^{(T)} W_{\nu k}^{(T)*} S_{T_\rho T_k} \right) . \quad (6.39)$$

If the components of the vectors $\mathbf{\Delta P}$ and $\mathbf{\Delta T}$ are independent random variables, then we should put in the last formula

$$S_{P_\rho T_k} = S_{T_\rho P_k} = 0 \, .$$

The result is

$$S_{f^{(j)}f^{(\nu)}} = \sum_{\rho=1}^{n_1} \sum_{k=1}^{n_1} \left(W_{j\rho}^{(P)} W_{\nu k}^{(P)*} S_{P_\rho P_k} + W_{j\rho}^{(T)} W_{\nu k}^{(T)*} S_{T_\rho T_k} \right) . \quad (6.40)$$

If, in addition, the components of each of the vectors $\mathbf{\Delta P}$ and $\mathbf{\Delta T}$ are independent, then, using (6.39), we get

$$S_{f^{(j)}f^{(\nu)}} = \sum_{\rho=1}^{n_1} \left(W_{j\rho}^{(P)} W_{\nu k}^{(P)*} S_{P_\rho} + W_{j\rho}^{(T)} W_{\nu k}^{(T)*} S_{T_\rho} \right) , \quad (6.41)$$

where

$$S_{P_\rho} = S_{P_\rho P_\rho} \, , \qquad S_{T_\rho} = S_{T_\rho T_\rho} \, .$$

Return to the analysis of the vector equation (6.24). Let us obtain a two-term approximate solution of (6.24). By use of (6.26), we obtain the following expression for the vector $\mathbf{f}$ within the frequency domain:

$$\mathbf{f}_0(\omega) = W(\mathrm{i}\,\omega)\,\mathrm{D}^{(1)}\,\Delta\mathbf{P}_0(\omega) + W(\mathrm{i}\,\omega)\,\mathrm{D}^{(2)}\,\Delta\mathbf{T}_0(\omega)\,;$$

here,

$$W(\mathrm{i}\,\omega) = \left[(\mathrm{i}\,\omega)^2\mathrm{H} + \mathrm{i}\,\omega\mathrm{B} + \mathrm{C}\right]^{-1}\,.$$

The components of the vector $\mathbf{f}$ read

$$f_0^{(1)}(\omega) = \sum_{k=1}^{3} u_{0k}^{(1)}(W_{11}\,P_k) + \sum_{k=1}^{3} u_{0k}^{(2)}(W_{12}\,P_k)$$
$$+ \sum_{k=1}^{3} \vartheta_{0k}^{(1)}(W_{11}\,T_k) + \sum_{k=1}^{3} \vartheta_{0k}^{(2)}(W_{12}\,T_k)\,,$$
$$f_0^{(2)}(\omega) = \sum_{k=1}^{3} u_{0k}^{(1)}(W_{21}\,P_k) + \sum_{k=1}^{3} u_{0k}^{(2)}(W_{22}\,P_k)$$
$$+ \sum_{k=1}^{3} \vartheta_{0k}^{(1)}(W_{21}\,T_k) + \sum_{k=1}^{3} \vartheta_{0k}^{(2)}(W_{22}\,T_k)\,. \tag{6.42}$$

We assume that the spectral density functions S_{Pk} and S_{Tk} of the components of the vectors $\mathbf{P}_j$ and $\mathbf{T}_j$ are given a priori. Let us restrict our consideration to the case when the vector components are independent random variables; (hence, their cross-spectral functions equal zero). It is also possible a case when the components are somehow interrelated, say, the vectors $\Delta\mathbf{P}$ and $\Delta\mathbf{T}$ are of constant directions but their magnitudes may vary randomly. We have

$$P_j = |\Delta\mathbf{P}|\,(\mathbf{e}_P\cdot\mathbf{e}_j)\,,\qquad T_j = |\Delta\mathbf{T}|\,(\mathbf{e}_T\cdot\mathbf{e}_j)\,,$$

where $\mathbf{e}_P$ and $\mathbf{e}_T$ are the unit vectors directed along the vectors $\Delta\mathbf{P}$ and $\Delta\mathbf{T}$, respectively.

Let us now obtain the spectral density function and the cross-spectral density for the components $f^{(1)}$ and $f^{(2)}$ of the vector $\mathbf{f}$ when the components of the vectors $\Delta\mathbf{P}$ and $\Delta\mathbf{T}$ are independent random variables. In this case, we have

$$K_{T_\rho P_k} = K_{P_\rho T_k} = K_{P_\rho P_k} = K_{T_\rho T_k} = 0\,,$$

Using the relationship (6.36), we get

$$S_{f^{(1)}}(\omega) = \sum_{k=1}^{3}\beta_k^{(1)}(\mathrm{i}\,\omega)\,S_{P_k}(\omega) + \sum_{k=1}^{3}\gamma_k^{(1)}(\mathrm{i}\,\omega)\,S_{T_k}(\omega)\,,$$
$$S_{f^{(2)}}(\omega) = \sum_{k=1}^{3}\beta_k^{(2)}(\mathrm{i}\,\omega)\,S_{P_k}(\omega) + \sum_{k=1}^{3}\gamma_k^{(2)}(\mathrm{i}\,\omega)\,S_{T_k}(\omega)\,, \tag{6.43}$$

where

$$\begin{aligned}
\beta_k^{(1)}(\mathrm{i}\,\omega) &= (u_{0k}^{(1)})^2 W_{11} W_{11}^* + u_{0k}^{(1)} u_{0k}^{(2)} W_{11} W_{12}^* + u_{0k}^{(2)} u_{0k}^{(1)} W_{12} W_{11}^* \\
&\quad + (u_{0k}^{(2)})^2 W_{12} W_{12}^* , \\
\gamma_k^{(1)}(\mathrm{i}\,\omega) &= (\vartheta_{0k}^{(1)})^2 W_{11} W_{11}^* + \vartheta_{0k}^{(1)} \vartheta_{0k}^{(2)} W_{11} W_{12}^* + \vartheta_{0k}^{(2)} \vartheta_{0k}^{(1)} W_{12} W_{11}^* \\
&\quad + (\vartheta_{0k}^{(2)})^2 W_{12} W_{12}^* , \\
\beta_k^{(2)}(\mathrm{i}\,\omega) &= (u_{0k}^{(1)})^2 W_{21} W_{21}^* + u_{0k}^{(1)} u_{0k}^{(2)} W_{21} W_{22}^* + u_{0k}^{(2)} u_{0k}^{(1)} W_{22} W_{21}^* \\
&\quad + (u_{0k}^{(2)})^2 W_{22} W_{22}^* , \\
\gamma_k^{(2)}(\mathrm{i}\,\omega) &= (\vartheta_{0k}^{(1)})^2 W_{21} W_{21}^* + \vartheta_{0k}^{(1)} \vartheta_{0k}^{(2)} W_{21} W_{22}^* + \vartheta_{0k}^{(2)} \vartheta_{0k}^{(1)} W_{22} W_{21}^* \\
&\quad + (\vartheta_{0k}^{(2)})^2 W_{22} W_{22}^* .
\end{aligned} \tag{6.44}$$

The cross-spectral densities are

$$\begin{aligned}
S_{f^{(1)} f^{(2)}} &= \sum_{k=1}^{3} \beta_k^{(3)} S_{P_k} + \sum_{k=1}^{3} \gamma_k^{(3)} S_{T_k} , \\
S_{f^{(2)} f^{(1)}} &= \sum_{k=1}^{3} \beta_k^{(4)} S_{P_k} + \sum_{k=1}^{3} \gamma_k^{(4)} S_{T_k} ,
\end{aligned} \tag{6.45}$$

where

$$\begin{aligned}
\beta_k^{(3)} &= (u_{0k}^{(1)})^2 W_{21}^* W_{11} + u_{0k}^{(1)} u_{0k}^{(2)} W_{21}^* W_{12} \\
&\qquad + u_{0k}^{(2)} u_{0k}^{(1)} W_{22}^* W_{11} + (u_{0k}^{(2)})^2 W_{22}^* W_{12} , \\
\gamma_k^{(3)} &= (\vartheta_{0k}^{(1)})^2 W_{21}^* W_{11} + \vartheta_{0k}^{(1)} \vartheta_{0k}^{(2)} W_{21}^* W_{12} \\
&\qquad + \vartheta_{0k}^{(2)} \vartheta_{0k}^{(1)} W_{22}^* W_{11} + (\vartheta_{0k}^{(2)})^2 W_{22}^* W_{12} , \\
\beta_k^{(4)} &= (u_{0k}^{(1)})^2 W_{11}^* W_{21} + u_{0k}^{(1)} u_{0k}^{(2)} W_{12}^* W_{21} \\
&\qquad + u_{0k}^{(2)} u_{0k}^{(1)} W_{11}^* W_{22} + (u_{0k}^{(2)})^2 W_{12}^* W_{22} , \\
\gamma_k^{(4)} &= (\vartheta_{0k}^{(1)})^2 W_{11}^* W_{21} + \vartheta_{0k}^{(1)} \vartheta_{0k}^{(2)} W_{12}^* W_{21} \\
&\qquad + \vartheta_{0k}^{(2)} \vartheta_{0k}^{(1)} W_{11}^* W_{22} + (\vartheta_{0k}^{(2)})^2 W_{12}^* W_{22} .
\end{aligned} \tag{6.46}$$

The expressions for the spectral density functions $S_{f^{(i)}}(\omega)$ (6.27) and the cross-spectral densities $S_{f^{(i)} f^{(k)}}(\omega)$ (6.29) can be obtained by use of the Wiener–Khintchin formulas (6.17) and (6.18). Recall that the Wiener–Khintchin formulas relate correlation functions and cross-correlation functions to spectral density functions.

Let us examine in greater detail one of the coefficients $\beta_k^{(j)}$ and $\gamma_k^{(j)}$, say, $\beta_k^{(1)}$. Consider the products of W_{ij} and $W_{k\nu}^*$ in the right-hand side of (6.44). The product of functions with the same indices equals to the square of the absolute value of this function, for example, $(W_{11} W_{11}^*) = |W_{11}|^2$. The square of the absolute value of a complex-valued function is a real-valued function. The product of functions with different indices is a complex-valued function, for example,

$$(W_{11}W_{12}^*) = (W_{11}^{(1)} + \mathrm{i}\, W_{11}^{(2)})\,(W_{12}^{(1)} - \mathrm{i}\, W_{12}^{(2)})$$
$$= W_{11}^{(1)}W_{12}^{(1)} + W_{11}^{(2)}W_{12}^{(2)} + \mathrm{i}\,(W_{11}^{(2)}W_{12}^{(1)} - W_{11}^{(1)}W_{12}^{(2)})\,, \qquad (6.47)$$

where $W_{11}^{(1)}$ and $W_{12}^{(1)}$ are the real parts and $W_{11}^{(2)}$ and $W_{12}^{(2)}$ are the imaginary parts. The real and imaginary parts are even and odd functions of ω, respectively, that is,

$$W_{11}^{(1)}(\omega) = W_{11}^{(1)}(-\omega)\,, \qquad W_{11}^{(2)}(\omega) = -W_{11}^{(2)}(-\omega)\,.$$

The product of two odd functions is an even function. Hence, the real part of the right-hand side of (6.31) is an even function and the imaginary part is an odd one. These facts will be used in the determination of the variances of the components of the vector **Z** (6.25).

The components of the vector $\mathbf{Z}^{\mathrm{r}}$ (6.25) are

$$Z_j^{\mathrm{r}}(\eta,\tau) = Z_{0j}^{(1)}(\eta)\, f^{(1)}(\tau) + Z_{0j}^{(2)}(\eta)\, f^{(2)}(\tau)\,.$$

Applying the Fourier transform to both sides of this equation, we get

$$Z_j^{\mathrm{r}}(\eta,\omega) = Z_{0j}^{(1)}(\eta)\, f^{(1)}(\omega) + Z_{0j}^{(2)}(\eta)\, f^{(2)}(\omega)\,.$$

By analogy with the derivation of (6.45), we obtain the spectral density functions of the components of the vector Z_j^{r}:

$$S_{z_j}(\eta,\omega) = (Z_{0j}^{(1)})^2 S_{f^{(1)}} + Z_{0j}^{(1)} Z_{0j}^{(2)} (S_{f^{(1)}f^{(2)}} + S_{f^{(2)}f^{(1)}}) + (Z_{0j}^{(2)})^2 S_{f^{(2)}}\,; \qquad (6.48)$$

here, $S_{f^{(1)}}$, $S_{f^{(2)}}$ and $S_{f^{(1)}f^{(2)}}$ are defined by (6.42) and (6.45), respectively. Using (6.47), we can represent the coefficients $\beta_k^{(j)}$ and $\gamma_k^{(j)}$, $k = 1, 2, 3$, $j = 1, 2, 3, 4$, as follows:

$$\beta_k^{(j)} = \beta_k^{(j)(1)} + i\beta_k^{(j)(2)}\,, \qquad \gamma_k^{(j)} = \gamma_k^{(j)(1)} + i\gamma_k^{(j)(2)}\,;$$

here, $\beta_k^{(j)(1)}$ and $\gamma_k^{(j)(1)}$ are even functions of ω and $\beta_k^{(j)(2)}$ and $\gamma_k^{(j)(2)}$ are odd functions of ω.

Hence, we obtain

$$S_{f^{(1)}}(\mathrm{i}\,\omega) = \sum_{k=1}^{3}(\beta_k^{(1)(1)} S_{P_k} + \gamma_k^{(1)(1)} S_{T_k}) + \mathrm{i}\sum_{k=1}^{3}(\beta_k^{(1)(2)} S_{P_k} + \gamma_k^{(1)(2)} S_{T_k})\,,$$

$$S_{f^{(2)}}(\mathrm{i}\,\omega) = \sum_{k=1}^{3}(\beta_k^{(2)(1)} S_{P_k} + \gamma_k^{(2)(1)} S_{T_k})$$

$$+\,\mathrm{i}\sum_{k=1}^{3}(\beta_k^{(2)(2)}S_{P_k}+\gamma_k^{(2)(2)}S_{T_k})\,, \tag{6.49}$$

$$S_{f^{(1)}f^{(2)}}(\mathrm{i}\,\omega)=\sum_{k=1}^{3}(\beta_k^{(3)(1)}S_{P_k}+\gamma_k^{(3)(1)}S_{T_k})$$

$$+\,\mathrm{i}\sum_{k=1}^{3}(\beta_k^{(3)(2)}S_{P_k}+\gamma_k^{(3)(2)}S_{T_k})\,,$$

$$S_{f^{(2)}f^{(1)}}(\mathrm{i}\,\omega)=\sum_{k=1}^{3}(\beta_k^{(4)(1)}S_{P_k}+\gamma_k^{(4)(1)}S_{T_k})$$

$$+\,\mathrm{i}\sum_{k=1}^{3}(\beta_k^{(4)(2)}S_{P_k}+\gamma_k^{(4)(2)}S_{T_k})\,. \tag{6.50}$$

The relationships (6.49) can be rearranged as follows:

$$S_{f^{(1)}}=S_{f^{(1)}}^{(1)}+\mathrm{i}\,S_{f^{(1)}}^{(2)}\,,\qquad S_{f^{(2)}}=S_{f^{(2)}}^{(1)}+\mathrm{i}\,S_{f^{(2)}}^{(2)}\,, \tag{6.51}$$

$$S_{f^{(1)}f^{(2)}}=S_{f^{(1)}f^{(2)}}^{(1)}+\mathrm{i}\,S_{f^{(1)}f^{(2)}}^{(2)}\,,$$

$$S_{f^{(2)}f^{(1)}}=S_{f^{(2)}f^{(1)}}^{(1)}+\mathrm{i}\,S_{f^{(2)}f^{(1)}}^{(2)}\,. \tag{6.52}$$

Moreover,

$$S_{f^{(1)}}^{(1)}=S_{f^{(2)}}^{(1)}\,,\qquad S_{f^{(1)}}^{(2)}=-S_{f^{(2)}}^{(2)}\,,$$

$$S_{f^{(1)}f^{(2)}}^{(1)}=S_{f^{(2)}f^{(1)}}^{(1)}\,,\qquad S_{f^{(1)}f^{(2)}}^{(2)}=-S_{f^{(2)}f^{(1)}}^{(2)}\,. \tag{6.53}$$

The sum of the cross-spectral densities in (6.48), by using (6.51) and (6.53), takes the form

$$S_{f^{(1)}f^{(2)}}+S_{f^{(2)}f^{(1)}}=2S_{f^{(1)}f^{(2)}}^{(1)}\,.$$

From this relationship it is obviously seen that the sum is an even function of ω.

The cross-spectral densities of the components of the vector $\mathbf{Z}^{\mathrm{r}}$ are

$$S_{z_j z_\nu}=Z_{0j}^{(1)}Z_{0\nu}^{(1)}S_{f^{(1)}}+Z_{0j}^{(1)}Z_{0\nu}^{(2)}(S_{f^{(1)}f^{(2)}}+S_{f^{(2)}f^{(1)}})+Z_{0j}^{(2)}Z_{0\nu}^{(2)}S_{f^{(2)}}\,.$$

The variance of Z_j reads

$$D_{z_j}=\int_{-\infty}^{\infty}S_{z_j}(\eta,\,\omega)\,\mathrm{d}\omega$$

or

$$D_{z_j} = (Z_{0j}^{(1)}(\eta))^2 \int_{-\infty}^{\infty} S_{f^{(1)}}\,\mathrm{d}\omega + 2Z_{0j}^{(1)}(\eta)\,Z_{0j}^{(2)}(\eta) \int_{-\infty}^{\infty} S_{f^{(1)}f^{(2)}}^{(1)}\,\mathrm{d}\omega + (Z_{0j}^{(2)}(\eta))^2 \int_{-\infty}^{\infty} S_{f^{(2)}}\,\mathrm{d}\omega\,. \tag{6.54}$$

The integrals of an odd function are equal to zero; hence, the last relationship yields

$$D_{z_j}(\eta) = (Z_{0j}^{(1)}(\eta))^2 \sum_{k=1}^{3} \int_{-\infty}^{\infty} (\beta_k^{(1)(1)} S_{P_k} + \gamma_k^{(1)(1)} S_{T_k})\,\mathrm{d}\omega + Z_{0j}^{(1)}(\eta)\,Z_{0j}^{(2)}(\eta) \sum_{k=1}^{3} \int_{-\infty}^{\infty} (\beta_k^{(4)(1)} S_{P_k} + \gamma_k^{(4)(1)} S_{T_k})\,\mathrm{d}\omega + (Z_{0j}^{(2)}(\eta))^2 \sum_{k=1}^{3} \int_{-\infty}^{\infty} (\beta_k^{(2)(1)} S_{P_k} + \gamma_k^{(2)(1)} S_{T_k})\,\mathrm{d}\omega\,. \tag{6.55}$$

The standard deviation is

$$\sigma_{z_j}(\eta) = \sqrt{D_{z_j}(\eta)}\,. \tag{6.56}$$

Let us next consider special cases of the equations obtained.

Absence of Moments For $\Delta\mathbf{P} \neq 0$, $\Delta\mathbf{T} = 0$, in the case of one-term approximation,

$$\mathbf{Z} = \mathbf{Z}_0^{(1)} f^{(1)},$$

the equation in $f^{(1)}(\tau)$ is as follows:

$$h_{11}\ddot{f}^{(1)} + b_{11}\dot{f}^{(1)} + c_{11} f^{(1)} = \sum_{j=1}^{3} u_{0j}^{(1)}(\eta_1)\,\Delta P_j(\tau)\,.$$

The frequency function is

$$W(\mathrm{i}\,\omega) = W_{11}(\mathrm{i}\,\omega) = \frac{1}{-\omega^2 h_{11} + \mathrm{i}\,\omega b_{11} + c_{11}} = W_{11}^{(1)}(\omega) + \mathrm{i}\,W_{11}^{(2)}(\omega)\,,$$

where

$$W_{11}^{(1)}(\omega) = \frac{c_{11} - \omega^2 h_{11}}{(c_{11} - \omega^2 h_{11})^2 + \omega^2 b_{11}^2}\,,$$

$$W_{11}^{(2)}(\omega) = -\frac{\omega b_{11}}{(c_{11} - \omega^2 h_{11})^2 + \omega^2 b_{11}^2}\,.$$

The Fourier transform of $f^{(1)}$ reads

$$f^{(1)}(\omega) = \sum_{k=1}^{3} u_{0k}^{(1)}(\eta_1) W_{11}(\omega)\, \Delta P_k(\omega)\,.$$

Therefore, from (6.43) it follows that

$$S_{f^{(1)}}(\omega) = \sum_{k=1}^{3} (u_{0k}^{(1)})^2 |W_{11}|^2 S_{P_k}(\omega)\,.$$

The variance of $Z_j(\eta,\, \omega)$ can be written as

$$D_{z_j}(\eta) = (Z_{0j}^{(1)}(\eta))^2 \sum_{k=1}^{3} \int_{-\infty}^{\infty} (u_{0k}^{(1)}(\eta_1))^2\, |W_{11}|^2 S_{P_k}(\omega)\, \mathrm{d}\omega\,.$$

Absence of Forces For $\Delta \mathbf{P} = 0$ and $\Delta \mathbf{T} \neq 0$, we have

$$D_{z_j}(\eta) = (Z_{0j}^{(1)}(\eta))^2 \sum_{k=1}^{3} \int_{-\infty}^{\infty} (\vartheta_{0k}^{(1)}(\eta_2))^2\, |W_{11}|^2 S_{T_k}(\omega)\, \mathrm{d}\omega\,.$$

The standard deviations are

$$\sigma_{z_j}(\eta) = Z_{0j}^{(1)}(\eta) \sqrt{\sum_{k=1}^{3} (u_{0k}^{(1)}(\eta_1))^2 \int_{-\infty}^{\infty} |W_{11}|^2 S_{P_k}(\omega)\, \mathrm{d}\omega}\,,$$

$$\sigma_{z_j}(\eta) = Z_{0j}^{(1)}(\eta) \sqrt{\sum_{k=1}^{3} (\vartheta_{0k}^{(1)}(\eta_2))^2 \int_{-\infty}^{\infty} |W_{11}|^2 S_{T_k}(\omega)\, \mathrm{d}\omega}\,.$$

In some special cases, the integrals in the relationships for D_{z_j} and σ_{z_j} can be found analytically (see Appendix of Svetlitsky (2003)). If this is not the case, the infinite limits of integration should be replaced by some finite numbers (large enough to ensure the accuracy required), and thus the integrals can be evaluated numerically. These large but finite values should be taking with regard to the conditions of accuracy. By means of expressions for variances and standard deviations (6.39) and (6.40), we can obtain the variance and the standard deviation of Z_j for each cross section of the rod. Consequently, we can find the cross sections of the rod at which the functions Z_j attain their maximum values. Then, using the three sigma rule, we can estimate this maximum (for $m_{z_j} = 0$) as follows:

$$\max Z_j(\eta_{*j}) = Z_{j0}(\eta_{*j}) + 3\sigma_{z_j}(\eta_{*j})\,, \tag{6.57}$$

where the quantity $Z_{j0}(\eta_{*j})$ corresponds to the static load and η_{*j} is the coordinate of the cross section of the rod at which the right-hand side of (6.57) attains a maximum.

For example, for the bending moments M_2 and M_3, we have

$$\max Z_5(\eta_{*5}) = \max M_2(\eta_{*5}) = M_{20}(\eta_{*5}) + 3\sigma_{\Delta_{M_2}}(\eta_{*5}),$$
$$\max Z_6(\eta_{*6}) = \max M_3(\eta_{*6}) = M_{30}(\eta_{*6}) + 3\sigma_{\Delta_{M_3}}(\eta_{*6}).$$

The maximum normal stresses at these cross sections caused by bending moments acting separately are

$$\max\sigma = \frac{\max Z_5(\eta_{*5})}{W_2}, \qquad \max\sigma = \frac{\max Z_6(\eta_{*6})}{W_3},$$

where W_2 and W_3 are the moduli of the cross sections.

To estimate the durability, we must calculate the total normal stress at the cross section of the rod caused by M_2 and M_3. For example, in the case of a rod of rectangular cross section, the amplitude values of the normal stresses are

$$\max\sigma(\eta_{*5}) = \frac{\max Z_5(\eta_{*5})}{W_2} + \frac{Z_6(\eta_5)}{W_3},$$
$$\max\sigma(\eta_{*6}) = \frac{\max Z_6(\eta_{*6})}{W_3} + \frac{Z_5(\eta_6)}{W_2},$$

where

$$Z_6(\eta_5) = M_{30}(\eta_5) + 3\sigma_{\Delta_{M_3}}(\eta_5), \qquad Z_5(\eta_6) = M_{20}(\eta_6) + 3\sigma_{\Delta_{M_2}}(\eta_6).$$

In Fig. 6.10, the normal stress is qualitatively plotted against time. The minimum values of the normal stresses at these cross sections of the rod are

$$\min\sigma(\eta_5) = \frac{1}{W_2}(M_{20} - 3\sigma_{\Delta_{M_2}}) + \frac{1}{W_3}(M_{30} - 3\sigma_{\Delta_{M_3}}),$$
$$\min\sigma(\eta_6) = \frac{1}{W_3}(M_{30} - 3\sigma_{\Delta_{M_3}}) + \frac{1}{W_2}(M_{20} - 3\sigma_{\Delta_{M_2}}).$$

Once $\max\sigma(\eta_5)$, $\min\sigma(\eta_5)$, $\max\sigma(\eta_6)$, and $\min\sigma(\eta_6)$ have been determined, we can, using the well-known formulas, find the safety factor for the fatigue strength.

6.4 Nonstationary Random Vibrations

6.4.1 The Solution of the Equations for Nonstationary Vibrations

In the previous section, we dealt with random forces and random vibrations produced by these forces. It was assumed that the probabilistic characteristics of the forces (mathematical expectation, variance, etc.) and the components of the vector $\mathbf{Z}^{\mathrm{r}}(\eta, \tau)$ were time independent. Such vibrations may occur in a system whose transient response time is small as compared to the characteristic time of operating mode. Besides, such stationary vibrations occur if and only if the equations for vibrations are equations with constant coefficients and the loads applied to the rod may be represented in the form of stationary random variables.

In actual practice, of some interest is the transient oscillatory process that starts when loads are applied and lasts until the system is in the steady state mode (stationary mode). For example, consider a rod exerted by a suddenly applied force of random magnitude and direction. This arises the question: What is the mode of motion of a rod after the force is applied? The motion of the rod cannot be thought of as stationary, even though the applied force may be represented in the form of a stationary random variable. In the general case, the loads applied to a rod are of arbitrary character. These loads may be represented in the form of nonstationary random variables.

In this case, the probabilistic characteristics of the solutions of the equations for vibrations of a rod are time dependent. This is true even for equations with constant coefficients. This seriously complicates the analysis because we cannot employ the spectral theory.

Consider the equations of small random vibrations of a rod (6.24). We seek an approximate solution of these equations in the form (6.25). As a result, for a rod subjected to a force ΔP and a moment ΔT (Fig. 6.6), we obtain a system of equations identical in form to (6.26):

$$\mathrm{H}\ddot{\mathbf{f}} + \mathrm{B}\mathbf{f} + \mathrm{C}\mathbf{f} = \mathrm{D}^{(1)}\,\Delta\mathbf{P} + \mathrm{D}^{(2)}\,\Delta\mathbf{T}\,. \tag{6.58}$$

Setting

$$\dot{\mathbf{f}}(\tau) = \mathbf{Y}_1(\tau)\,, \qquad \mathbf{f}(\tau) = \mathbf{Y}_2(\tau)\,, \tag{6.58}$$

we can represented the vector equation of the second order (6.58) as a system of two vector equations of the first order

$$\dot{\mathbf{Y}}_2 - \mathbf{Y}_1 = 0 \tag{6.59}$$

or

$$\dot{\mathbf{Y}}(\tau) + \mathrm{A}\mathbf{Y}(\tau) = \mathrm{B}^{(1)}\,\Delta\mathbf{F}^{(1)}(\tau) + \mathrm{B}^{(2)}\,\Delta\mathbf{F}^{(2)}(\tau)\,, \tag{6.60}$$

where

$$\mathbf{Y} = \begin{bmatrix} \mathbf{Y}_1 \\ \mathbf{Y}_2 \end{bmatrix} = \begin{bmatrix} \dot{\mathbf{f}} \\ \mathbf{f} \end{bmatrix}, \qquad \mathrm{A} = \begin{bmatrix} \mathrm{H}^{-1}\mathrm{B} & \mathrm{H}^{-1}\mathrm{C} \\ -\mathrm{E} & 0 \end{bmatrix},$$

$$\Delta\mathbf{F}^{(1)} = \begin{bmatrix} \Delta\mathbf{P} \\ \mathbf{0} \end{bmatrix}, \qquad \Delta\mathbf{F}^{(2)} = \begin{bmatrix} \Delta\mathbf{T} \\ \mathbf{0} \end{bmatrix},$$

$$\mathrm{B}^{(1)} = \begin{bmatrix} \mathrm{H}^{-1}\mathrm{D}^{(1)} & 0 \\ 0 & 0 \end{bmatrix}, \qquad \mathrm{B}^{(2)} = \begin{bmatrix} \mathrm{H}^{-1}\mathrm{D}^{(2)} & 0 \\ 0 & 0 \end{bmatrix}.$$

Restrict our consideration to the case of zero initial data at $\tau = 0$. Then, the solution of (6.60) reads

$$\mathbf{Y}(\tau) = \int_0^{\tau} \mathrm{G}(\tau,\, \hat{\tau})\, \mathrm{B}^{(1)}\, \boldsymbol{\Delta}\mathbf{F}^{(1)}(\hat{\tau})\, \mathrm{d}\hat{\tau} + \int_0^{\tau} \mathrm{G}(\tau,\, \hat{\tau})\, \mathrm{B}^{(2)}\, \boldsymbol{\Delta}\mathbf{F}^{(2)}(\hat{\tau})\, \mathrm{d}\hat{\tau}\,, \tag{6.61}$$

where $\mathrm{C}(\tau,\, \hat{\tau})$ is a Green matrix such that

$$\mathrm{G}(\tau,\, \hat{\tau}) = \mathrm{K}(\tau)\, \mathrm{K}^{-1}(\hat{\tau})\,;$$

here, $\mathrm{K}(\tau)$ is the fundamental matrix of solutions of a homogeneous equation corresponding to (6.60) (this homogeneous equation is the same as (6.60) except that its right-hand side is zero). If the elements of the matrix A are time-independent functions, then

$$\mathrm{G}(\tau,\, \hat{\tau}) = \mathrm{K}(\tau - \hat{\tau})\,.$$

When studying non-stationary random processes, in the general case, it is necessary to determine the probabilistic characteristics of the solution (that is, the components of the state vector $\mathbf{Z}(\eta,\, \tau)$) at time $\tau = \tau_{\mathrm{f}}$, that is,

$$\mathbf{Y}(\tau)\big|_{\tau=\tau_{\mathrm{f}}} = \mathbf{Y}(\tau_{\mathrm{f}})\,.$$

Assume now that the following probabilistic characteristics of the vectors $\boldsymbol{\Delta}\mathbf{P}$ and $\boldsymbol{\Delta}\mathbf{T}$, which enter into the expressions for $\boldsymbol{\Delta}\mathbf{F}^{(1)}$ and $\boldsymbol{\Delta}\mathbf{F}^{(2)}$, are specified, that is, assume that the mathematical expectations $\mathbf{m}_P$ and $\mathbf{m}_T$, the auto-correlation function $\mathrm{K}_{P_{\mathrm{K}}P_i}$ and the cross-correlation function $\mathrm{K}_{T_{\mathrm{K}}T_j}$ are specified. Let us obtain these probabilistic characteristics for the components of the vector $Y(\tau_{\mathrm{K}})$. For simplicity, we assume that the vectors $\boldsymbol{\Delta}\mathbf{P}$ and $\boldsymbol{\Delta}\mathbf{T}$ are independent random variables.

The mathematical expectation of the vector $\mathbf{Y}(\tau_{\mathrm{f}})$b is

$$\mathbf{m}_{\mathbf{Y}} = M[\mathbf{Y}(\tau_{\mathrm{f}})] = \int_0^{\tau_{\mathrm{f}}} \mathrm{GB}^{(1)}\mathbf{m}_{F^{(1)}}\, \mathrm{d}\tau^{(2)} + \int_0^{\tau_{\mathrm{f}}} \mathrm{GB}^{(2)}\mathbf{m}_{F^{(2)}}\, \mathrm{d}\tau^{(2)}\,, \tag{6.62}$$

where $\mathbf{m}_{F^{(1)}}$ and $\mathbf{m}_{F^{(2)}}$ are the mathematical expectations of the vectors $\boldsymbol{\Delta}\mathbf{F}^{(1)}$ and $\boldsymbol{\Delta}\mathbf{F}^{(2)}$ defined as

$$\mathbf{m}_{F^{(1)}} = \begin{bmatrix} \mathbf{m}_P \\ \mathbf{0} \end{bmatrix}, \qquad \mathbf{m}_{F^{(2)}} = \begin{bmatrix} \mathbf{m}_T \\ \mathbf{0} \end{bmatrix}.$$

In components, equation (6.62) reads

$$m_{y_1} = m_{\dot{f}^{(1)}} = \sum_{j=1}^{n} \int_0^{\tau_f} p_{1j}^{(1)} m_{P_j} \, d\tau^{(2)} + \sum_{j=1}^{n} \int_0^{\tau_f} p_{1j}^{(2)} m_{T_j} \, d\tau^{(2)} ,$$

$$m_{y_2} = m_{\dot{f}^{(2)}} = \sum_{j=1}^{n} \int_0^{\tau_f} p_{2j}^{(1)} m_{P_j} \, d\tau^{(2)} + \sum_{j=1}^{n} \int_0^{\tau_f} p_{2j}^{(2)} m_{T_j} \, d\tau^{(2)} ,$$

$$m_{y_3} = m_{f^{(1)}} = \sum_{j=1}^{n} \int_0^{\tau_f} p_{3j}^{(1)} m_{P_j} \, d\tau^{(2)} + \sum_{j=1}^{n} \int_0^{\tau_f} p_{3j}^{(2)} m_{T_j} \, d\tau^{(2)} ,$$

$$m_{y_4} = m_{f^{(2)}} = \sum_{j=1}^{n} \int_0^{\tau_f} p_{4j}^{(1)} m_{P_j} \, d\tau^{(2)} + \sum_{j=1}^{n} \int_0^{\tau_f} p_{4j}^{(2)} m_{T_j} \, d\tau^{(2)} ,$$

where the parameters $p_{ij}^{(1)}$ and $p_{ij}^{(2)}$ are the elements of the matrices

$$\mathrm{P}^{(1)} = \mathrm{GB}^{(1)} , \qquad \mathrm{P}^{(2)} = \mathrm{GB}^{(2)} ,$$

respectively.

To determine the correlation functions, consider the components of the vector $\mathbf{Y}(\tau_f)$:

$$Y_j(\tau_f) = \int_0^{\tau_f} \sum_{k=1}^{3} p_{1k}^{(1)} \, \Delta P_k \, d\tau^{(2)} + \int_0^{\tau_f} \sum_{k=1}^{3} p_{1k}^{(2)} \, \Delta T_k \, d\tau^{(2)} . \tag{6.63}$$

The correlation functions are

$$f_{y_j y_\nu}(\tau_f, \hat{\tau}_f) = M[Y_j^\circ(\tau_f) \, Y_j^\circ(\hat{\tau}_f)] , \tag{6.64}$$

where Y_j° are centered random variables. Substituting (6.63) into (6.64) and varying the upper limits of integration, after some rearrangements, we arrive at

$$\begin{aligned} K_{y_j y_\nu}(\tau_f, \tau_f^{(1)}) &= \sum_{k=1}^{3} \sum_{i=1}^{3} \int_0^{\tau_f} \int_0^{\tau_f^{(1)}} p_{jk}^{(1)}(\tau_f, \tau^{(2)}) \, p_{\nu i}^{(1)}(\tau_f^{(1)}, \tau^{(2)(1)}) \\ &\qquad \times K_{P_k P_i}(\tau^{(2)}, \tau^{(2)(1)}) \, d\tau^{(2)} \, d\tau^{(2)(1)} \\ &+ \sum_{k=1}^{3} \sum_{i=1}^{3} \int_0^{\tau_f} \int_0^{\tau_f^{(1)}} p_{jk}^{(2)}(\tau_f, \tau^{(2)}) \, p_{\nu i}^{(2)}(\tau_f^{(1)}, \tau^{(2)(1)}) \\ &\qquad \times K_{T_k T_i}(\tau^{(2)}, \tau^{(2)(1)}) \, d\tau^{(2)} \, d\tau^{(2)(1)} , \end{aligned} \tag{6.65}$$

$$\begin{aligned} K_{y_j y_j}(\tau_f, \tau_f^{(1)}) &= \sum_{k=1}^{3} \sum_{i=1}^{3} \int_0^{\tau_f} \int_0^{\tau_f^{(1)}} p_{jk}^{(1)}(\tau_f, \tau^{(2)}) \, p_{ji}^{(1)}(\tau_f^{(1)}, \tau^{(2)(1)}) \\ &\qquad \times K_{P_k P_i}(\tau^{(2)}, \tau^{(2)(1)}) \, d\tau^{(2)} \, d\tau^{(2)(1)} \\ &+ \sum_{k=1}^{3} \sum_{i=1}^{3} \int_0^{\tau_f} \int_0^{\tau_f^{(1)}} p_{jk}^{(2)}(\tau_f, \tau^{(2)}) \, p_{ji}^{(2)}(\tau_f^{(1)}, \tau^{(2)(1)}) \\ &\qquad \times K_{T_k T_i}(\tau^{(2)}, \tau^{(2)(1)}) \, d\tau^{(2)} \, d\tau^{(2)(1)} . \end{aligned} \tag{6.66}$$

Integrating and then putting $\tau_{\mathrm{f}} = \tau_{\mathrm{f}}^{(1)}$, we obtain the variance of the components of the vector $\mathbf{Y}(\tau_{\mathrm{f}})$:

$$D_{y_j}(\tau_{\mathrm{f}}) = K_{y_j y_j}(\tau_{\mathrm{f}}, \tau_{\mathrm{f}}^{(1)})\Big|_{\tau_{\mathrm{f}}^{(1)}=\tau_{\mathrm{f}}} .$$

Setting $\tau_{\mathrm{f}} = \tau_{\mathrm{f}}^{(1)}$ in (6.65), we obtain the cross-correlation function for the components of the vector $\mathrm{Y}(\tau_{\mathrm{f}})$. If the components of the vectors $\Delta\mathbf{P}$ and $\Delta\mathbf{T}$ are independent random variables, then

$$K_{P_k P_i} = \begin{cases} 0, & \text{for} \quad k \neq i, \\ K_{P_k}, & \text{for} \quad k = i, \end{cases} \qquad K_{T_k T_i} = \begin{cases} 0, & \text{for} \quad k \neq i, \\ K_{T_k}, & \text{for} \quad k = i. \end{cases}$$

Using (6.50) and (6.51), for $j \neq \nu$, we get

$$\begin{aligned} K_{y_j y_\nu} = {} & \sum_{k=1}^{3} \int_0^{\tau_{\mathrm{f}}} \int_0^{\tau_{\mathrm{f}}^{(1)}} p_{jk}^{(1)}(\tau_{\mathrm{f}}, \tau^{(2)})\, p_{\nu k}^{(1)}(\tau_{\mathrm{f}}^{(1)}, \tau^{(2)(1)}) \\ & \times K_{P_k}(\tau^{(2)}, \tau^{(2)(1)})\, \mathrm{d}\tau^{(2)}\, \mathrm{d}\tau^{(2)(1)} \\ & + \sum_{k=1}^{3} \int_0^{\tau_{\mathrm{f}}} \int_0^{\tau_{\mathrm{f}}^{(1)}} p_{jk}^{(2)}(\tau_{\mathrm{f}}, \tau^{(2)})\, p_{\nu k}^{(2)}(\tau_{\mathrm{f}}^{(1)}, \tau^{(2)(1)}) \\ & \times K_{T_k}(\tau^{(2)}, \tau^{(2)(1)})\, \mathrm{d}\tau^{(2)}\, \mathrm{d}\tau^{(2)(1)} , \end{aligned} \tag{6.67}$$

$$\begin{aligned} K_{y_j y_j} = {} & \sum_{k=1}^{3} \int_0^{\tau_{\mathrm{f}}} \int_0^{\tau_{\mathrm{f}}^{(1)}} p_{jk}^{(1)}(\tau_{\mathrm{f}}, \tau^{(2)})\, p_{jk}^{(1)}(\tau_{\mathrm{f}}^{(1)}, \tau^{(2)(1)}) \\ & \times K_{P_k}(\tau^{(2)}, \tau^{(2)(1)})\, \mathrm{d}\tau^{(2)}\, \mathrm{d}\tau^{(2)(1)} \\ & + \sum_{k=1}^{3} \int_0^{\tau_{\mathrm{f}}} \int_0^{\tau_{\mathrm{f}}^{(1)}} p_{jk}^{(2)}(\tau_{\mathrm{f}}, \tau^{(2)})\, p_{jk}^{(2)}(\tau_{\mathrm{f}}^{(1)}, \tau^{(2)(1)}) \\ & \times K_{T_k}(\tau^{(2)}, \tau^{(2)(1)})\, \mathrm{d}\tau^{(2)}\, \mathrm{d}\tau^{(2)(1)} . \end{aligned} \tag{6.68}$$

Once the probabilistic characteristics of the components of the vector $\mathbf{Y}(\tau_{\mathrm{f}})$ have been found, the probabilistic characteristics of the components of the vectors $\mathbf{f}$ and $\dot{\mathbf{f}}$ are as follows:

$$\begin{aligned} & K_{y_1 y_2} = K_{\dot{f}^{(1)} \dot{f}^{(2)}} , && K_{y_2 y_1} = K_{\dot{f}^{(2)} \dot{f}^{(1)}} , && K_{y_3 y_4} = K_{f^{(1)} f^{(2)}} , \\ & K_{y_4 y_3} = K_{f^{(2)} f^{(1)}} , && K_{y_1 y_3} = K_{\dot{f}^{(1)} f^{(1)}} , && K_{y_1 y_4} = K_{\dot{f}^{(1)} f^{(2)}} , \\ & K_{y_2 y_3} = K_{\dot{f}^{(2)} f^{(1)}} , && K_{y_2 y_4} = K_{\dot{f}^{(2)} f^{(2)}} , && \\ & K_{y_3 y_3} = K_{f^{(1)} f^{(1)}} , && K_{y_4 y_4} = K_{f^{(2)} f^{(2)}} . && \end{aligned}$$

Then, we can find the probabilistic characteristics of the components of the vector $\mathbf{Z}^{\mathrm{r}}(\eta, \tau)$ (see (6.25)) defined as

$$Z_j^{\mathrm{r}}(\eta, \tau) = Z_{0j}^{(1)} f^{(1)}(\tau) + Z_{0j}^{(2)} f^{(2)}(\tau) , \tag{6.69}$$

and the probabilistic characteristics of the time derivatives of the components

$$\dot{Z}^{\mathrm{r}}_j(\eta,\, \tau) = Z^{(1)}_{0j}\, \dot{f}^{(1)}(\tau) + Z^{(2)}_{0j}\, \dot{f}^{(2)}(\tau)\,. \tag{6.70}$$

The correlation moment and the variance of the components Z^{r}_j at time τ_{f} are as follows:

$$K_{z_j z_j} = D_{z_j}(\tau_{\mathrm{f}}) = (Z^{(1)}_{0j})^2 M[f^{(1)}(\tau_{\mathrm{f}})\, f^{(1)}(\tau_{\mathrm{f}})]$$
$$+ 2 Z^{(1)}_{0j} Z^{(2)}_{0j} M[f^{(1)}(\tau_{\mathrm{f}})\, f^{(2)}(\tau_{\mathrm{f}})] + (Z^{(2)}_{0j})^2 M[f^{(2)}(\tau_{\mathrm{f}})\, f^{(2)}(\tau_{\mathrm{f}})]\,,$$
$$D_{\dot{z}_j}(\tau_{\mathrm{f}}) = (Z^{(1)}_{0j})^2 M[f^{(1)}(\tau_{\mathrm{f}})\, f^{(1)}(\tau_{\mathrm{f}})]$$
$$+ 2 Z^{(1)}_{0j} Z^{(2)}_{0j} M[f^{(1)}(\tau_{\mathrm{f}})\, f^{(2)}(\tau_{\mathrm{f}})] + (Z^{(2)}_{0j})^2 M[f^{(2)}(\tau_{\mathrm{f}})\, f^{(2)}(\tau_{\mathrm{f}})]\,.$$

The correlation function and the variance of the derivatives $\dot{Z}^{\mathrm{r}}_j$ can be determined in a similar way.

In compact notation, the variances of the components Z^{r}_j and their derivatives $\dot{Z}^{\mathrm{r}}_j$ can be written as

$$D_{z_j}(\tau_{\mathrm{f}}) = (Z^{(1)}_{0j})^2 K_{y_3 y_3} + 2 Z^{(1)}_{0j} Z^{(2)}_{0j} K_{y_3 y_4} + (Z^{(2)}_{0j})^2 K_{y_4 y_4}\,,$$
$$D_{\dot{z}_j}(\tau_{\mathrm{f}}) = (Z^{(1)}_{0j})^2 K_{y_1 y_1} + 2 Z^{(1)}_{0j} Z^{(2)}_{0j} K_{y_1 y_2} + (Z^{(2)}_{0j})^2 K_{y_2 y_2}\,, \tag{6.71}$$

where $K_{y_i y_j}$ can be found from the formulas (6.52) and (6.53) for $\tau_{\mathrm{f}} = \tau^{(1)}_{\mathrm{f}}$.

Consider an example. A rod shown in Fig. 6.6 is suddenly exerted by a random force and a random moment with zero mathematical expectations, that is, $m_{P_j} = m_{T_j} = 0$. Let us find a locus of the point K at a given time.

From (6.54) it follows that the components of the displacement vector of the point K are

$$u^{\mathrm{r}}_{10} = Z^{\mathrm{r}}_{10}(\eta_K,\, \tau)\,, \qquad u^{\mathrm{r}}_{20} = Z^{\mathrm{r}}_{11}(\eta_K,\, \tau)\,, \qquad u^{\mathrm{r}}_{30} = Z^{\mathrm{r}}_{12}(\eta_K,\, \tau)\,.$$

Using (6.56), we obtain their standard deviations

$$\sigma_{z10} = \sqrt{D_{z10}}\,, \qquad \sigma_{z11} = \sqrt{D_{z11}}\,, \qquad \sigma_{z12} = \sqrt{D_{z12}}\,.$$

Applying the three sigma rule, we get

$$\begin{matrix}\max\\ \min\end{matrix}\; u^{\mathrm{r}}_{j0} = \pm 3\sigma_{z,9+j}\,, \qquad j = 1,\, 2,\, 3\,.$$

Recall that we consider arbitrary realizations of the random force and the random moment. We see that the displacement vector of K (if transferred parallel to itself so that its initial point coincides with the origin of the Cartesian reference frame) lies within an ellipsoid whose semi-axes are $\max u^{\mathrm{r}}_{j0}$, that is, the components of the vector $\mathbf{u}^{\mathrm{r}}_{\mathrm{K}}(\eta_{\mathrm{K}},\, \tau_{\mathrm{K}})$ satisfy the inequality

$$\left(\frac{u^{\mathrm{r}}_{10}}{3\sigma_{z_{10}}}\right)^2 + \left(\frac{u^{\mathrm{r}}_{20}}{3\sigma_{z_{11}}}\right)^2 + \left(\frac{u^{\mathrm{r}}_{30}}{3\sigma_{z_{12}}}\right)^2 \le 1\,.$$

In conclusion, it should be remarked once again that the action of random loads on structures, mechanisms, and gauges often lead to their failure and misoperation. That is why random loads must be considered in the design.

7. Straight Rods

The previous chapters of the book are devoted to the theoretical basics of the dynamics of rods. In these chapters, various methods for derivation of the equations of motion for space-curved rods subject to time-dependent loads have been presented. The following types of loads have been considered: distributed, concentrated, dead, depending on linear and angular displacements and on their first derivatives with respect to independent parameters. The equations of small vibrations of space-curved rods were derived and the techniques of determination of the frequencies and the vibration modes were presented.

As a rule, the general equations of motion as well as the equations of small vibrations are complicated while the equations encountered in applications are special cases of the general equations. Hence, it makes sense to consider in detail some special cases as well as the problems they arise from.

In many cases, the analysis of applied problems concerned with the behavior of elastic structural elements can be performed within the framework of the theory of straight rods. At first glance it would seem that the straight rod as a physical model is very simple. Nevertheless, even with such a simplification, the problems remain difficult for studying and are of practical interest.

7.1 Equations of Small Vibrations

7.1.1 Equations Referred to the Attached Axes

The equations governing small vibrations of rods of variable cross section that are straight in the natural configuration can be obtained by setting

$$
\begin{aligned}
&æ_{20} = æ_{30} = 0\,, \qquad Q_{20} = Q_{30} = 0\,, \qquad Q_{10} \neq 0\,,\\
&M_{20} = M_{30} = 0\,, \qquad M_{10} \neq 0\,, \qquad u_1 \approx 0
\end{aligned}
$$

in the general equations (3. 11)–(3. 15). Then, disregarding the initial twisting (that is, setting $æ_{10} = 0$), we obtain the equations referred to the attached coordinate frame

$$n_1(\eta)\,\frac{\partial^2 \mathbf{u}}{\partial \tau^2} - \frac{\partial \Delta \mathbf{Q}}{\partial \eta} + Q_{10}\mathrm{A}_1\mathrm{A}^{-1}\,\Delta \mathbf{M} = \Delta \mathbf{P}\,, \tag{7.1}$$

$$J\,\frac{\partial^2 \boldsymbol{\vartheta}}{\partial \tau^2} - \frac{\partial \Delta \mathbf{M}}{\partial \eta} + M_{10}\mathrm{A}_1\mathrm{A}^{-1}\,\Delta \mathbf{M} - \mathrm{A}_1\,\Delta \mathbf{Q} = \Delta \mathbf{T}\,, \tag{7.2}$$

$$\frac{\partial \boldsymbol{\vartheta}}{\partial \eta} - \mathrm{A}^{-1}\,\Delta \mathbf{M} = 0\,, \tag{7.3}$$

$$\frac{\partial \mathbf{u}}{\partial \eta} + \mathrm{A}_1 \boldsymbol{\vartheta} = 0\,. \tag{7.4}$$

In components, considering the attached axes, we obtain

$$\begin{aligned}
&n_1\,\frac{\partial^2 u_1}{\partial \tau^2} - \frac{\partial \Delta Q_1}{\partial \eta} = \Delta P_1\,,\\
&n_1\,\frac{\partial^2 u_2}{\partial \tau^2} - \frac{\partial \Delta Q_2}{\partial \eta} - \frac{Q_{10}}{A_{33}}\,\Delta M_3 = \Delta P_2\,,\\
&n_1\,\frac{\partial^2 u_3}{\partial \tau^2} - \frac{\partial \Delta Q_3}{\partial \eta} + \frac{Q_{10}}{A_{22}}\,\Delta M_2 = \Delta P_3\,,
\end{aligned} \tag{7.5}$$

$$\begin{aligned}
&J_{11}\,\frac{\partial^2 \vartheta_1}{\partial \tau^2} - \frac{\partial \Delta M_1}{\partial \eta} = \Delta T_1\,,\\
&J_{22}\,\frac{\partial^2 \vartheta_2}{\partial \tau^2} - \frac{\partial \Delta M_2}{\partial \eta} - \frac{M_{10}}{A_{33}}\,\Delta M_3 + \Delta Q_3 = \Delta T_2\,,\\
&J_{33}\,\frac{\partial^2 \vartheta_3}{\partial \tau^2} - \frac{\partial \Delta M_3}{\partial \eta} + \frac{M_{10}}{A_{22}}\,\Delta M_2 - \Delta Q_2 = \Delta T_3\,,
\end{aligned} \tag{7.6}$$

$$\begin{aligned}
&\frac{\partial \vartheta_1}{\partial \eta} - \frac{\Delta M_1}{A_{11}} = 0\,,\\
&\frac{\partial \vartheta_2}{\partial \eta} - \frac{\Delta M_2}{A_{22}} = 0\,,\\
&\frac{\partial \vartheta_3}{\partial \eta} - \frac{\Delta M_3}{A_{33}} = 0\,,
\end{aligned} \tag{7.7}$$

$$\begin{aligned}
&\frac{\partial u_1}{\partial \eta} = 0\,,\\
&\frac{\partial u_2}{\partial \eta} - \vartheta_3 = 0\,,\\
&\frac{\partial u_3}{\partial \eta} + \vartheta_2 = 0\,,
\end{aligned} \tag{7.8}$$

$$\Delta M_j = \mathrm{A}_{jj}\Delta æ_j\,, \qquad j = 1,\,2,\,3\,. \tag{7.9}$$

Some equations of the system of equations (7.5)–(7.9) are independent from the others. Thus, the following equations of this set may be separated:

$$n_1\,\frac{\partial^2 u_1}{\partial \tau^2} - \frac{\partial \Delta Q_1}{\partial \eta} = \Delta P_1\,, \tag{7.10}$$

$$J_{11}\,\frac{\partial^2\vartheta_1}{\partial\tau^2} - \frac{\partial\Delta M_1}{\partial\eta} = \Delta T_1\,, \tag{7.11}$$

$$\frac{\partial\vartheta_1}{\partial\eta} - \frac{\Delta M_1}{A_{11}} = 0\,, \tag{7.12}$$

$$\frac{\partial u_1}{\partial\eta} = 0\,. \tag{7.13}$$

It follows from the last equation that $u_1 = 0$. In more expanded notation, the forces and moments that appear in these equations read

$$\Delta P_j = \Delta q_j(\eta,\,\tau) + \sum_{i=1}^{n} \Delta P_j^{(i)}(\tau)\,\delta(\eta-\eta_i)\,,$$

$$\Delta T_j = \Delta \mu_j(\eta,\,\tau) + \sum_{\nu=1}^{\rho} \Delta T_j^{(\nu)}(\tau)\,\delta(\eta-\eta_\nu)\,, \qquad j = 2,\,3\,. \tag{7.14}$$

In practice, loads applied to the rod may depend on displacements and on the first derivatives of the displacements of the rod axis points and of the angles of rotation of the cross sections of the rod with respect to η and τ. As an example, we assume that ΔP_j and ΔT_j depend on u_k and ϑ_k, $k = 2,\,3$, only, that is,

$$\Delta P_j = \Delta P_j(u_2,\,u_3,\,\vartheta_2,\,\vartheta_3)\,, \qquad \Delta T_j = \Delta T_j(u_2,\,u_3,\,\vartheta_2,\,\vartheta_3)\,. \tag{7.15}$$

then equations (7.5)–(7.9) (except (7.10)–(7.13), which are uncoupled from the others) are coupled. Let $M_{10} = 0$. Then, if either the forces ΔP_j and moments ΔT_j do not depend on u_k and ϑ_k and on their derivatives with respect to η and τ or they do depend on u_k and ϑ_k as

$$\Delta P_2 = \Delta P_2(u_2,\,\vartheta_3)\,, \qquad \Delta P_3 = \Delta P_3(u_3,\,\vartheta_2)\,, \tag{7.16}$$

then the set of equations (7.5)–(7.9) (except (7.10)–(7.13)) decomposes into two sets of equations.

(1) The equations governing vibrations of the rod in the plane of the vectors $(\mathbf{e}_1,\,\mathbf{e}_2)$ (see Fig. 7.1)

$$n_1\,\frac{\partial^2 u_2}{\partial\tau^2} - \frac{\partial\Delta Q_2}{\partial\eta} - \frac{Q_{10}}{A_{33}}\,\Delta M_3 = \Delta P_2(u_2,\,\vartheta_3)\,, \tag{7.17}$$

$$J_{33}\,\frac{\partial^2\vartheta_3}{\partial\tau^2} - \frac{\partial\Delta M_3}{\partial\eta} - \Delta Q_2 = \Delta T_3(u_2,\,\vartheta_3)\,, \tag{7.18}$$

$$\frac{\partial\vartheta_3}{\partial\eta} - \frac{\Delta M_3}{A_{33}} = 0\,, \tag{7.19}$$

$$\frac{\partial u_2}{\partial\eta} - \vartheta_3 = 0\,, \qquad \Delta M_3 = A_{33}\,\Delta æ_3\,. \tag{7.20}$$

(2) The equations governing vibrations of the rod in the plane of the vectors $\mathbf{e}_1$, $\mathbf{e}_3$

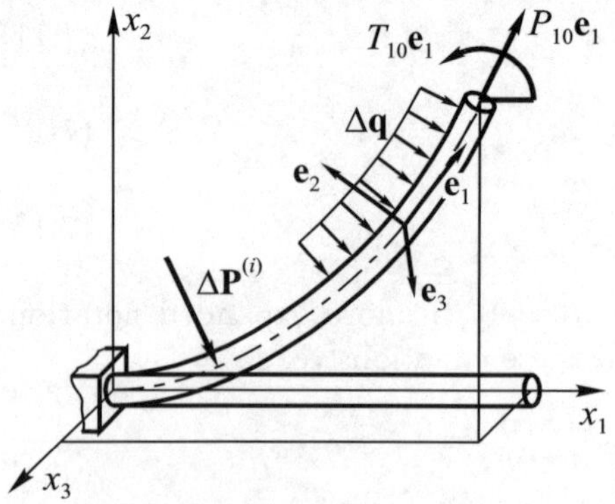

Fig. 7.1.

$$n_1 \frac{\partial^2 u_3}{\partial \tau^2} - \frac{\partial \Delta Q_3}{\partial \eta} + \frac{Q_{10}}{A_{22}} \Delta M_2 = \Delta P_3(u_3, \vartheta_2)\,, \tag{7.21}$$

$$J_{22} \frac{\partial^2 \vartheta_2}{\partial \tau^2} - \frac{\partial \Delta M_2}{\partial \eta} + \Delta Q_3 = \Delta T_2(u_3, \vartheta_2)\,, \tag{7.22}$$

$$\frac{\partial \vartheta_2}{\partial \eta} - \frac{\Delta M_2}{A_{22}} = 0\,, \tag{7.23}$$

$$\frac{\partial u_3}{\partial \eta} + \vartheta_2 = 0\,, \qquad \Delta M_2 = A_{22} \Delta æ_2\,. \tag{7.24}$$

The expression (7.14) for the forces ΔP_j contains the distributed force $\Delta q_j(\eta, \tau)$. Consider a rod on a linearly elastic foundation. Suppose that the rod is in an oscillatory motion; then, with regard for the force of viscous damping, the distributed force acting on the rod reads

$$\Delta q_j = -k u_j(\eta, \tau) - \alpha \frac{\partial u_j(\eta, \tau)}{\partial \tau}\,, \tag{7.25}$$

where

$$k = \frac{k_0 l^4}{A_{33}^{(0)}}\,;$$

here, k is the dimensionless stiffness coefficient of the foundation and α is a dimensionless coefficient of the force of viscous damping.

By analogy with (6.24), we can represent each of the sets (7.17)–(7.20) and (7.21)–(7.24) in vector–matrix form.

(1) Equations (7.17)–(7.20), for the case of Δq_j defined by (7.25), become

$$\mathrm{A}^{(1)} \frac{\partial^2 \mathbf{Z}}{\partial \tau^2} + \frac{\partial \mathbf{Z}}{\partial \eta} + \mathrm{A}^{(2)} \mathbf{Z} + \mathrm{A}^{(3)} \frac{\partial \mathbf{Z}}{\partial \tau} = \Delta \boldsymbol{\Phi}\,; \tag{7.26}$$

here,

$$\mathbf{Z} = \begin{bmatrix} \Delta Q_2 \\ \Delta M_3 \\ \vartheta_3 \\ u_2 \end{bmatrix}, \qquad \mathrm{A}^{(1)} = \begin{bmatrix} 0 & 0 & 0 & -n_1 \\ 0 & 0 & -J_{33} & 0 \\ 0 & 0 & 0 & 0 \\ 0 & 0 & 0 & 0 \end{bmatrix},$$

$$\mathrm{A}^{(2)} = \begin{bmatrix} 0 & \dfrac{Q_{10}}{A_{33}} & 0 & -k \\ 1 & 0 & 0 & 0 \\ 0 & -\dfrac{1}{A_{33}} & 0 & 0 \\ 0 & 0 & -1 & 0 \end{bmatrix},$$

$$\mathrm{A}^{(3)} = \begin{bmatrix} 0 & 0 & 0 & -\alpha \\ 0 & 0 & 0 & 0 \\ 0 & 0 & 0 & 0 \\ 0 & 0 & 0 & 0 \end{bmatrix}, \qquad \Delta\boldsymbol{\Phi} = \begin{bmatrix} -\Delta P_2 \\ -\Delta T_3 \\ 0 \\ 0 \end{bmatrix},$$

$$\Delta P_2 = \sum_{i=1}^{n} \Delta P_2^{(i)}\, \delta(\eta - \eta_i)\,.$$

(2) Equations (7.21)–(7.24) can be rearranged as

$$\mathrm{A}_1^{(1)} \frac{\partial^2 \mathbf{Z}_1}{\partial \tau^2} + \frac{\partial \mathbf{Z}_1}{\partial \eta} + \mathrm{A}_1^{(2)} \mathbf{Z}_1 + \mathrm{A}_1^{(3)} \frac{\partial \mathbf{Z}_1}{\partial \tau} = \Delta\boldsymbol{\Phi}_1\,; \tag{7.27}$$

here,

$$\mathbf{Z}_1 = \begin{bmatrix} \Delta Q_3 \\ \Delta M_2 \\ \vartheta_2 \\ u_3 \end{bmatrix}, \qquad \mathrm{A}_1^{(1)} = \begin{bmatrix} 0 & 0 & 0 & -n_1 \\ 0 & 0 & -J_{22} & 0 \\ 0 & 0 & 0 & 0 \\ 0 & 0 & 0 & 0 \end{bmatrix},$$

$$\mathrm{A}_1^{(2)} = \begin{bmatrix} 0 & -\dfrac{Q_{10}}{A_{22}} & 0 & -k \\ -1 & 0 & 0 & 0 \\ 0 & -\dfrac{1}{A_{22}} & 0 & 0 \\ 0 & 0 & 1 & 0 \end{bmatrix},$$

$$\mathrm{A}_1^{(3)} = \begin{bmatrix} 0 & 0 & 0 & -\alpha \\ 0 & 0 & 0 & 0 \\ 0 & 0 & 0 & 0 \\ 0 & 0 & 0 & 0 \end{bmatrix}, \qquad \Delta\boldsymbol{\Phi}_1 = \begin{bmatrix} -\Delta P_3 \\ -\Delta T_2 \\ 0 \\ 0 \end{bmatrix},$$

$$\Delta P_3 = \sum_{j=1}^{n} \Delta P_3^{(j)}\, \delta(\eta - \eta_j)\,.$$

7.1.2 Equations Referred to the Cartesian Frame

The equations can be obtained by specification of the set of equations (3.27)–(3.31). Setting

$$
\text{æ}_{j0} = 0\,, \qquad \mathrm{L}^0 = \mathrm{E}\,,
$$
$$
\mathrm{A}_L^{(1)} = \mathrm{A}_1 = \begin{bmatrix} 0 & 0 & 0 \\ 0 & 0 & -1 \\ 0 & 1 & 0 \end{bmatrix}, \qquad \mathrm{A}_L = \mathrm{A}_1 = \begin{bmatrix} 0 & 0 & 0 \\ 0 & 0 & -1 \\ 0 & 1 & 0 \end{bmatrix},
$$
$$
\mathrm{A}_{LQ} = \begin{bmatrix} 0 & 0 & 0 \\ 0 & -Q_{x_1 0} & 0 \\ 0 & 0 & -Q_{x_1 0} \end{bmatrix},
$$

we have

$$
\begin{aligned}
& n_1 \frac{\partial^2 \mathbf{u}_x}{\partial \tau^2} - \frac{\partial \Delta \mathbf{Q}_x}{\partial \eta} = \Delta \mathbf{P}_x\,, \\
& J \frac{\partial^2 \boldsymbol{\vartheta}}{\partial \tau^2} - \frac{\partial \Delta \mathbf{M}_x}{\partial \eta} - \mathrm{A}_1\, \Delta \mathbf{Q}_x + M_{10} \mathrm{A}_1 \mathrm{A}^{-1}\, \Delta \mathbf{M}_x - \mathrm{A}_{LQ} \boldsymbol{\vartheta} = \Delta \mathbf{T}_x\,, \\
& \frac{\partial \boldsymbol{\vartheta}}{\partial \eta} - \mathrm{A}^{-1}\, \Delta \mathbf{M}_x = 0\,, \\
& \frac{\partial \mathbf{u}_x}{\partial \eta} + \mathrm{A}_1 \boldsymbol{\vartheta} = 0\,.
\end{aligned} \tag{7.28}
$$

In components (a special case of (3.44)–(3.48)), for $u_{x_1} \approx 0$, we get

$$
\begin{aligned}
& n_1 \frac{\partial^2 u_{x_j}}{\partial \tau^2} - \frac{\partial \Delta Q_{x_j}}{\partial \eta} = \Delta P_{x_j}\,, \qquad j = 2,\, 3\,, \\
& J_{11} \frac{\partial^2 \vartheta_1}{\partial \tau^2} - \frac{\partial \Delta M_{x_1}}{\partial \eta} = \Delta T_{x_1}\,, \\
& J_{22} \frac{\partial^2 \vartheta_2}{\partial \tau^2} - \frac{\partial \Delta M_{x_2}}{\partial \eta} - \frac{M_{10}}{A_{33}}\, \Delta M_{x_3} + \Delta Q_{x_3} + Q_{x_1 0} \vartheta_2 = \Delta T_{x_2}\,, \\
& J_{33} \frac{\partial^2 \vartheta_3}{\partial \tau^2} - \frac{\partial \Delta M_{x_3}}{\partial \eta} + \frac{M_{10}}{A_{22}}\, \Delta M_{x_2} - \Delta Q_{x_2} + Q_{x_2 0} \vartheta_3 = \Delta T_{x_3}\,, \\
& \frac{\partial \vartheta_j}{\partial \eta} - \frac{\Delta M_{x_j}}{A_{jj}} = 0\,, \qquad j = 1,\, 2,\, 3\,, \\
& \frac{\partial u_{x_1}}{\partial \eta} = 0\,, \\
& \frac{\partial u_{x_2}}{\partial \eta} - \vartheta_3 = 0\,, \\
& \frac{\partial u_{x_3}}{\partial \eta} + \vartheta_2 = 0\,, \\
& \Delta M_{x_j} = A_{jj}\, \Delta \text{æ}_j\,, \qquad j = 1,\, 2,\, 3\,.
\end{aligned} \tag{7.29}
$$

This set of equations includes the equations of torsional vibrations, which are identical in form to (7.11) and (7.12):

$$J_{11}\,\frac{\partial^2\vartheta_1}{\partial\tau^2} - \frac{\partial\Delta M_1}{\partial\eta} = \Delta T_{x_1}\,,$$
$$\frac{\partial\vartheta_1}{\partial\eta} - \frac{\Delta M_{x_1}}{A_{11}} = 0\,. \tag{7.30}$$

In the general case, after excluding (7.30), the rest of the set of equations (7.29) is coupled. In terms of the assumptions similar to (7.16), we get, for $M_{10} = 0$, the following independent systems of equations.

(1) The equations governing vibrations of the rod in the plane x_1Ox_2 (see Fig. 7.1):

$$n_1\,\frac{\partial^2 u_{x_2}}{\partial\tau^2} - \frac{\partial\Delta Q_{x_2}}{\partial\eta} = \Delta P_{x_2}\,,$$
$$J_{33}\,\frac{\partial^2\vartheta_3}{\partial\tau^2} - \frac{\partial\Delta M_{x_3}}{\partial\eta} - \Delta Q_{x_2} + Q_{x_10}\vartheta_3 = \Delta T_{x_3}\,,$$
$$\frac{\partial\vartheta_3}{\partial\eta} - \frac{\Delta M_{x_3}}{A_{33}} = 0\,,$$
$$\frac{\partial u_{x_2}}{\partial\eta} - \vartheta_3 = 0\,, \qquad \Delta M_{x_3} = A_{33}\,\Delta æ_3\,. \tag{7.31}$$

(2) The equations governing vibrations of the rod in the plane x_1Ox_3:

$$n_1\,\frac{\partial^2 u_{x_3}}{\partial\tau^2} - \frac{\partial\Delta Q_{x_3}}{\partial\eta} = \Delta P_{x_3}\,,$$
$$J_{22}\,\frac{\partial^2\vartheta_2}{\partial\tau^2} - \frac{\partial\Delta M_{x_2}}{\partial\eta} + \Delta Q_{x_3} + Q_{x_10}\vartheta_2 = \Delta T_{x_2}\,,$$
$$\frac{\partial\vartheta_2}{\partial\eta} - \frac{\Delta M_{x_2}}{A_{22}} = 0\,,$$
$$\frac{\partial u_{x_3}}{\partial\eta} + \vartheta_2 = 0\,, \qquad \Delta M_{x_2} = A_{22}\,\Delta æ_2\,. \tag{7.32}$$

For a rod on an elastic foundation (with regard for the force of viscous damping), we have

$$\Delta q_{x_j} = -k u_{x_j} - \alpha\,\frac{\partial u_{x_j}}{\partial\tau}\,. \tag{7.33}$$

By analogy with (7.26) and (7.27), the sets of equations (7.31) and (7.32) can be written in vector-matrix form.

For a rod of constant cross section (A_{22} = const, $A_{33} = 1$, $n_1 = 1$), we get upon elimination of the unknowns from (7.29) the following equations in u_{x_2} and u_{x_3}:

$$\frac{\partial^4 u_{x2}}{\partial\eta^4} - J_{33}\,\frac{\partial^4 u_{x2}}{\partial\eta^2\,\partial\tau^2} - \frac{M_{10}}{A_{22}}\,\frac{\partial^3 u_{x3}}{\partial\eta^3} - Q_{x10}\,\frac{\partial^2 u_{x2}}{\partial\eta^2} + \frac{\partial^2 u_{x2}}{\partial\tau^2}$$
$$= \Delta P_{x_2} - \frac{\partial\Delta T_{x3}}{\partial\eta}, \qquad (7.34)$$

$$A_{22}\,\frac{\partial^4 u_{x3}}{\partial\eta^4} - J_{22}\,\frac{\partial^4 u_{x3}}{\partial\eta^2\,\partial\tau^2} + M_{10}\,\frac{\partial^3 u_{x2}}{\partial\eta^3} - Q_{x10}\,\frac{\partial^2 u_{x3}}{\partial\eta^2} + \frac{\partial^2 u_{x3}}{\partial\tau^2}$$
$$= \Delta P_{x_3} - \frac{\partial\Delta T_{x2}}{\partial\eta}, \qquad (7.35)$$

where

$$\Delta P_{x_j} = \Delta q_{x_j} + \sum_{i=1}^{n} \Delta P_{x_j}^{(i)}\,\delta(\eta-\eta_i)\,,$$

$$\Delta T_{x_j} = \Delta\mu_{x_j} + \sum_{\nu=1}^{\rho} \Delta T_{x_j}^{(\nu)}\,\delta(\eta-\eta_\nu)\,, \qquad j = 2,\,3\,.$$

Similarly, equations to be solved for u_2 and u_3 can be obtained from the systems of equations (7.17)–(7.20) and (7.21)–(7.24), respectively.

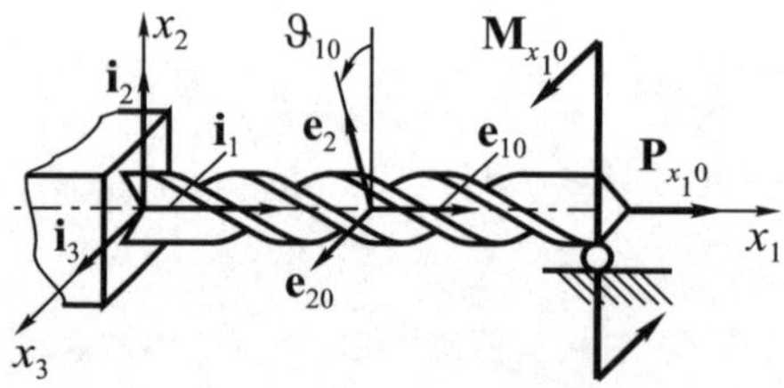

Fig. 7.2.

7.1.3 Equations of Small Vibrations for Naturally Twisted Rods

The equilibrium equations for loaded naturally twisted rods (Fig. 7.2) when deflections from the equilibrium position are small were derived in Sect. 4.3 of the first part of the book (Svetlitsky (2000a)). For a rod of constant cross section these are equations (4.124)–(4.127) (see Svetlitsky (2000a)), and for a rod subject solely to distributed forces these are equations (4.128)–(4.134) (see Svetlitsky (2000a)). Let us introduce the forces of inertia

$$-\frac{\partial^2 u_2}{\partial\tau^2}\,, \qquad -\frac{\partial^2 u_3}{\partial\tau^2}\,, \qquad -J_{22}\,\frac{\partial^2\vartheta_2}{\partial\tau^2}\,, \qquad -J_{33}\,\frac{\partial^2\vartheta_3}{\partial\tau^2}$$

into (4.129), (4.130), (4.132) and (4.133) (Svetlitsky (2000a)). Let us also set

$$Q_2 = \Delta Q_2\,, \qquad Q_3 = \Delta Q_3\,, \qquad M_2 = \Delta M_2\,, \qquad M_3 = \Delta M_3\,,$$

where ΔQ_2, ΔQ_3, ΔM_2, and ΔM_3 are dynamic loads.
As a result, we obtain the following equations of small vibrations for straight naturally twisted rods:

$$\begin{aligned}
&\frac{\partial^2 u_2}{\partial \tau^2} - \frac{\partial \Delta Q_2}{\partial \eta} + æ_{10}\,\Delta Q_3 + Q_{10}\,\Delta M_3 = \Delta P_2\,,\\
&\frac{\partial^2 u_3}{\partial \tau^2} - \frac{\partial \Delta Q_3}{\partial \eta} - æ_{10}\,\Delta Q_2 - \frac{Q_{10}}{A_{22}}\,\Delta M_3 = \Delta P_3\,,\\
&J_{22}\,\frac{\partial^2 \vartheta_2}{\partial \tau^2} - \frac{\partial \Delta M_2}{\partial \eta} + æ_{10}^{(1)}\,\Delta M_3 + \Delta Q_3 = \Delta T_2\,,\\
&\frac{\partial^2 \vartheta_3}{\partial \tau^2} - \frac{\partial \Delta M_3}{\partial \eta} - æ_{10}^{(1)}\,\Delta M_2 - \Delta Q_2 = \Delta T_3\,,\\
&\frac{\partial \vartheta_2}{\partial \eta} - æ_{10}\vartheta_3 - \frac{\Delta M_2}{A_{22}} = 0\,,\\
&\frac{\partial \vartheta_3}{\partial \eta} + æ_{10}\vartheta_2 - \Delta M_3 = 0\,,\\
&\frac{\partial u_2}{\partial \eta} - æ_{10}u_3 - \vartheta_3 = 0\,,\\
&\frac{\partial u_3}{\partial \eta} + æ_{10}u_2 + \vartheta_2 = 0\,,
\end{aligned} \tag{7.36}$$

where

$$æ_{10} = \frac{\partial \vartheta_{10}}{\partial \eta}\,, \qquad æ_{10}^{(1)} = \frac{\partial \vartheta_{10}}{\partial \eta} - \frac{M_{10}}{A_{11}}\,;$$

here, Q_{10} is a compressive force and ϑ_{10} is the angle which describes the rotation of the principal axes of the cross section relative to the natural axes (see Fig. 7.2).

7.1.4 Equations of Flexural-and-Torsional Vibrations

In the previous sections, we dealt with rods such that the centroid of a cross section coincides with its flexural center. A cross section of a rod with its centroid O_1 and flexural center O_2 is shown in Fig. 7.3*a*. Qualitatively, an airplane wing or a turbine blade has the cross section of the same configuration.

Recall the meaning of the concept *flexural center*. Consider a cross section of a rod and the forces acting in the plane of this cross section. For any fixed point in the plane, these system of forces can be replaced by a single, resultant, force, **Q**, and a single, resultant, moment, **M**, about this point of reference. The magnitude and the direction of the resultant force **Q** does not depend on the reference point but the moment **M** does. The moment **M** depends on the normal and shear stresses at the cross section of the rod

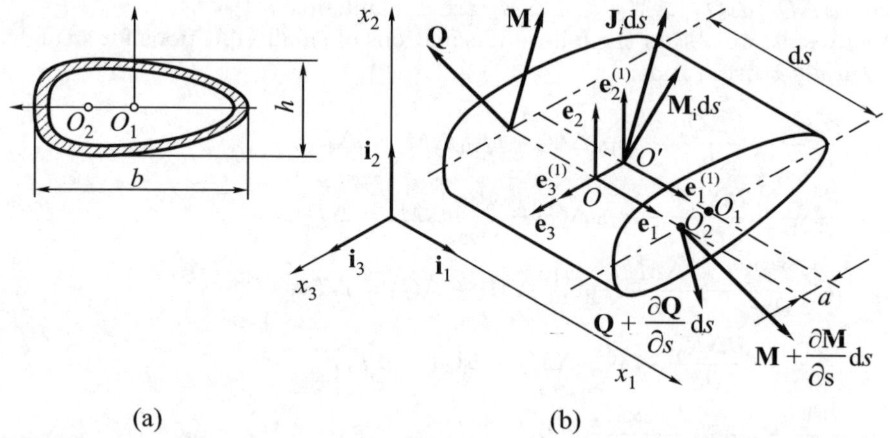

Fig. 7.3.

including the shear stresses caused by the shear forces Q_2 and Q_3. Considering any other point, say, the point Q_1, we see that the moments differ by $\mathbf{a} \times \mathbf{Q}$

The point at which the moment caused by the shear stresses equals zero is called the *flexural center* (Feodosiev (1999)) or the *center of elasticity* (Fung (1959)). In what follows, we shall restrict our consideration to the following case: *the cross section has an axis of symmetry and the points O_1 and O_2 belong to this axis.*

Let the coordinate frame attached to the rod axis (the bases $\{\mathbf{e}_j\}$) be associated with the line of flexural centers. Then, as in the case when the points O_2 and O_1 coincide, the vectors $\mathbf{Q}$ and $\mathbf{M}$ are independent.

In accordance with the methods of mechanics, deriving the equations of motion for a rod element, we may use the following basic theorems: the theorem of motion of the center of mass of mechanical system (in the case under consideration, the motion of a rod element) and the theorem of motion a mechanical system relative to its center of mass. The d'Alembert principle also may be used (this principle was used earlier when we have derived the equations of motion of a rod). Thus, using the d'Alembert principle, we obtain the following equations of motion for an element of a rod:

$$\mathbf{J}_{\mathrm{i}}\, ds + \frac{\partial \mathbf{Q}}{\partial s}\, ds + \mathbf{q}\, ds = 0\,, \tag{7.37}$$

$$\mathbf{M}_{\mathrm{i}}\, ds + \left(\mathbf{M} + \frac{\partial \mathbf{M}}{\partial s}\, ds\right) - \mathbf{M} + \left(\frac{ds}{2}\,\mathbf{e}_1 + \mathbf{a}\right) \times \left(\mathbf{Q} + \frac{\partial \mathbf{Q}}{\partial s}\, ds\right)$$
$$- \left(\frac{ds}{2}\,\mathbf{e}_1 + \mathbf{a}\right) \times \mathbf{Q} + \boldsymbol{\mu} = 0\,; \tag{7.38}$$

here,

$$\mathbf{J}_\mathrm{i} = -m_0 \frac{\partial^2 \mathbf{u}_0}{\partial t^2}, \qquad \mathbf{M}_\mathrm{i} = -\mathbf{J} \frac{\partial^2 \boldsymbol{\vartheta}}{\partial t^2}$$

are the inertial force and the moment of inertia of the element of the rod about the principal central axes of the element.

In (7.38), the total moment is calculated relative to the centroid O' of the element. On rearrangement, we get

$$\mathbf{b}_1 = m_0 \frac{\partial^2 \mathbf{u}_{O'}}{\partial t^2} - \frac{\partial \mathbf{Q}}{\partial s} = 0, \tag{7.39}$$

$$\mathbf{b}_2 = \mathbf{J} \frac{\partial^2 \boldsymbol{\vartheta}}{\partial t^2} - \frac{\partial \mathbf{M}}{\partial s} - (\mathbf{e}_1 \times \mathbf{Q}) - \left(\mathbf{a} \times \frac{\partial \mathbf{Q}}{\partial s} \right) = 0. \tag{7.40}$$

Here, the vector $\mathbf{a}$ is the vector from the point O_1 to the point O_2.

Hereafter, the equations are assumed to be in dimensionless notation. In the general case, the rod may be of varying cross section; hence, m_0, $\mathbf{J}$, and $|\mathbf{a}|$ may depend on the coordinate s.

Recall that we consider vibrations of the rod about its stress-free natural configuration when $\mathbf{Q}_0 = \mathbf{M}_0 = 0$. Consequently, $\mathbf{Q} = \Delta\mathbf{Q}$ and $\mathbf{M} = \Delta\mathbf{M}$. Therefore, the equations of small vibrations (7.39) and (7.40) become

$$\mathbf{b}_1 = n_1 \frac{\partial^2 \mathbf{u}_{O'}}{\partial \tau^2} - \frac{\partial \Delta\mathbf{Q}}{\partial \eta} = 0, \tag{7.41}$$

$$\mathbf{b}_2 = \mathbf{J} \frac{\partial^2 \boldsymbol{\vartheta}}{\partial \tau^2} - \frac{\partial \Delta\mathbf{M}}{\partial \eta} - \mathrm{A}_1 \, \Delta\mathbf{Q} - \left(\mathbf{a} \times \frac{\partial \Delta\mathbf{Q}}{\partial \eta} \right) = 0. \tag{7.42}$$

The derivatives in these equations may be thought of as local derivatives because the terms $\Delta æ \times \Delta\mathbf{Q}$, $\Delta æ \times \Delta\mathbf{M}$, and $\boldsymbol{\omega} \times \mathbf{u}_{0'}$ can be neglected.

The vector $\mathbf{u}_{O'}$ in (7.39) is the vector of displacements of the centroids of the cross sections of the rod. The previously obtained relationship between the moment $\Delta\mathbf{M}$ and the increment of the curvature $\Delta æ$, $\mathbf{M} = \mathrm{A}\, \Delta æ$, where $\Delta M_i = A_{ii} \, \Delta æ_i$, remain the same in the frame $\{\mathbf{e}_j\}$ associated with the line of flexural centers of the cross sections of the rod. Consequently, to obtain the equations in scalar form, it is necessary that the components ΔM_j enter into the equations; this is true when the vector equations (7.39) and (7.40) are represented in terms of projections onto the axes associated with the line of flexural centers. The velocity of the points that lie on the line of flexural centers is

$$\mathbf{v} = \mathbf{v}_{O'} + \boldsymbol{\omega} \times \mathbf{a},$$

where $\mathbf{v}_{O'}$ is the velocity of the centroid of the rod element, $\mathbf{v}$ is the velocity of the point O (Fig. 7. 3*b*), $\boldsymbol{\omega}$ is the angular velocity of the element. For small displacements and angles of rotation, we have

$$\frac{\partial \mathbf{u}_{O'}}{\partial \tau} = \frac{\partial \mathbf{u}}{\partial \tau} - \frac{\partial \boldsymbol{\vartheta}}{\partial \tau} \times \mathbf{a}, \qquad \mathbf{a} = a\mathbf{e}_3 .$$

Since

$$\frac{\partial \boldsymbol{\vartheta}}{\partial \tau} \times \mathbf{a} = \begin{vmatrix} \mathbf{e}_1 & \mathbf{e}_2 & \mathbf{e}_3 \\ \dot{\vartheta}_1 & \dot{\vartheta}_2 & \dot{\vartheta}_3 \\ 0 & 0 & a \end{vmatrix} = \dot{\vartheta}_2\, a\, \mathbf{e}_1 - \dot{\vartheta}_1\, a\, \mathbf{e}_2\,,$$

we arrive at

$$\frac{\partial \mathbf{u}_{O'}}{\partial \tau} = \frac{\partial \mathbf{u}}{\partial \tau} - \frac{\partial \vartheta_2}{\partial \tau}\, a\mathbf{e}_1 + \frac{\partial \vartheta_1}{\partial \tau}\, a\mathbf{e}_2\,. \tag{7.43}$$

Upon elimination of the vector

$$\frac{\partial \Delta \mathbf{Q}}{\partial \eta}$$

from (7.42) by using (7.41) and taking into consideration that $\mathbf{a} = a\,\mathbf{e}_3$, we get

$$\mathbf{b}_2 = \mathrm{J}\,\frac{\partial^2 \boldsymbol{\vartheta}}{\partial \tau^2} - \frac{\partial \Delta \mathbf{M}}{\partial \eta} - \mathrm{A}\,\Delta \mathbf{Q} - \left(\mathbf{a} \times n_1\, \frac{\partial^2 \mathbf{u}_{O'}}{\partial \tau^2}\right) = 0\,. \tag{7.44}$$

Using (7.43), we eliminate the vector $\mathbf{u}_{O'}$ from (7.41) and (7.44) and thus obtain

$$\mathbf{b}_1 = n_1\, \frac{\partial^2 \mathbf{u}}{\partial \tau^2} - n_1 a \left(\frac{\partial^2 \vartheta_2}{\partial \tau^2}\, \mathbf{e}_1 - \frac{\partial^2 \vartheta_1}{\partial \tau^2}\, \mathbf{e}_2\right) - \frac{\partial\, \Delta \mathbf{Q}}{\partial \eta} = 0\,, \tag{7.45}$$

$$\begin{aligned} \mathbf{b}_2 = \mathrm{J}\,\frac{\partial^2 \boldsymbol{\vartheta}}{\partial \tau^2} &- \frac{\partial\, \Delta \mathbf{M}}{\partial \eta} - \mathrm{A}_1\, \Delta \mathbf{Q} - a n_1 \left(\mathbf{e}_3 \times \frac{\partial^2 \mathbf{u}}{\partial \tau^2}\right) \\ &+ a^2 n_1 \left[\mathbf{e}_3 \times \left(\frac{\partial^2 \vartheta_2}{\partial \tau^2}\, \mathbf{e}_1 - \frac{\partial^2 \vartheta_1}{\partial \tau^2}\, \mathbf{e}_2\right)\right] = 0\,. \end{aligned} \tag{7.46}$$

On rearrangement, the last equation takes the form

$$\begin{aligned} \mathbf{b}_2 = \mathrm{J}\,\frac{\partial^2 \boldsymbol{\vartheta}}{\partial \tau^2} &- \frac{\partial \Delta \mathbf{M}}{\partial \eta} + a n_1\, \frac{\partial^2 u_2}{\partial \tau^2}\, \mathbf{e}_1 - a n_1\, \frac{\partial^2 u_1}{\partial \tau^2}\, \mathbf{e}_2 \\ &+ a^2 n_1\, \frac{\partial^2 \vartheta_2}{\partial \tau^2}\, \mathbf{e}_2 + a^2 n_1\, \frac{\partial^2 \vartheta_1}{\partial \tau^2}\, \mathbf{e}_1 = 0\,. \end{aligned} \tag{7.47}$$

Since the derivatives are local ones, we get

$$\frac{\partial^2 \boldsymbol{\vartheta}}{\partial \tau^2} = \sum_{j=1}^{3} \frac{\partial^2 \vartheta_j}{\partial \tau^2}\, \mathbf{e}_j\,, \qquad \frac{\partial \Delta \mathbf{M}}{\partial \eta} = \sum_{j=1}^{3} \frac{\partial \Delta M_j}{\partial \eta}\, \mathbf{e}_j\,.$$

Considering the scalar products $\mathbf{b}_1 \cdot \mathbf{e}_j = 0$ and $\mathbf{b}_2 \cdot \mathbf{e}_j = 0$ ($j = 1,\, 2,\, 3$), we obtain, on rearrangement, the following scalar equations:

$$
\begin{aligned}
& n_1 \frac{\partial^2 u_1}{\partial \tau^2} - a n_1 \frac{\partial^2 \vartheta_2}{\partial \tau^2} - \frac{\partial \Delta Q_1}{\partial \eta} = 0 \,, \\
& n_1 \frac{\partial^2 u_2}{\partial \tau^2} + a n_1 \frac{\partial^2 \vartheta_1}{\partial \tau^2} - \frac{\partial \Delta Q_2}{\partial \eta} = 0 \,, \\
& n_1 \frac{\partial^2 u_3}{\partial \tau^2} - \frac{\partial \Delta Q_3}{\partial \eta} = 0 \,, \\
& J_{11}^{(1)} \frac{\partial^2 \vartheta_1}{\partial \tau^2} - \frac{\partial \Delta M_1}{\partial \eta} + a n_1 \frac{\partial^2 u_2}{\partial \tau^2} = 0 \,, \\
& J_{22}^{(1)} \frac{\partial^2 \vartheta_2}{\partial \tau^2} - \frac{\partial \Delta M_2}{\partial \eta} - a n_1 \frac{\partial^2 u_1}{\partial \tau^2} + \Delta Q_3 = 0 \,, \\
& J_{33}^{(1)} \frac{\partial^2 \vartheta_3}{\partial \tau^2} - \frac{\partial \Delta M_3}{\partial \eta} - \Delta Q_2 = 0 \,, \\
& \frac{\partial \vartheta_1}{\partial \eta} - \frac{\Delta M_1}{A_{11}} = 0 \,, \\
& \frac{\partial \vartheta_2}{\partial \eta} - \frac{\Delta M_2}{A_{22}} = 0 \,, \\
& \frac{\partial \vartheta_3}{\partial \eta} - \frac{\Delta M_3}{A_{33}} = 0 \,, \\
& \frac{\partial u_1}{\partial \eta} = 0 \,, \\
& \frac{\partial u_2}{\partial \eta} - \vartheta_3 = 0 \,, \\
& \frac{\partial u_3}{\partial \eta} + \vartheta_2 = 0 \,;
\end{aligned}
\tag{7.48}
$$

here,

$$
J_{11}^{(1)} = J_{11} + a^2 n_1 \,, \qquad J_{22}^{(1)} = J_{22} + a^2 n_1 \,.
$$

For small vibrations of straight rods, we have $u_1 = 0$. On rearrangement, (7.48) yields

$$
\frac{\partial^2}{\partial \eta^2}\left(A_{33} \frac{\partial^2 u_2}{\partial \eta^2}\right) - \frac{\partial}{\partial \eta}\left(J_{33} \frac{\partial^3 u_2}{\partial \eta \, \partial \tau^2}\right) + n_1 \frac{\partial^2 u_2}{\partial \tau^2} + a n_1 \frac{\partial^2 \vartheta_1}{\partial \tau^2} = 0 \,,
$$

$$
\frac{\partial}{\partial \eta}\left(A_{11} \frac{\partial \vartheta_1}{\partial \eta}\right) - J_{11}^{(1)} \frac{\partial^2 \vartheta_1}{\partial \tau^2} - a n_1 \frac{\partial^2 u_2}{\partial \tau^2} = 0 \,, \tag{7.49}
$$

$$
\frac{\partial^2}{\partial \eta^2}\left(A_{22} \frac{\partial^2 u_3}{\partial \eta^2}\right) - \frac{\partial}{\partial \eta}\left(J_{22}^{(1)} \frac{\partial^2 u_3}{\partial \eta \, \partial \tau^2}\right) + n_1 \frac{\partial^2 u_3}{\partial \tau^2} = 0 \,, \tag{7.50}
$$

$$
a n_1 \frac{\partial^2 \vartheta_2}{\partial \tau^2} + \frac{\partial \Delta Q_1}{\partial \eta} = 0 \,. \tag{7.51}
$$

The system of equations (7.49) allows us to study flexural–and–torsional vibrations of a rod of variable cross section. Equation (7.50) describes flexural vibrations of the rod within the plane x_1Ox_3. In the case of small vibrations, (7.50) is uncoupled from (7.49). Recall that each cross section of the rod under consideration has an axis of symmetry, and the points O_1 and O_2 (the centroid and the flexural center) lie on this axis. If a cross section has no axes of symmetry, then the vector **a** has two nonzero components with respect to the coordinate axes associated with the centroid of the rod element. Hence, we have three equations governing flexural–and–torsional vibrations of the rod.

By use of (7.51), we can determine the axial load ΔQ_1. Since

$$\vartheta_2 = -\frac{\partial u_3}{\partial \eta},$$

we get

$$\frac{\partial \Delta Q_1}{\partial \eta} = an_1 \frac{\partial^3 u_3}{\partial \eta\, \partial \tau^2}$$

or

$$\Delta Q_1 = \int_0^{\eta} an_1 \frac{\partial^3 u_3}{\partial \eta\, \partial \tau^2}\, \mathrm{d}\eta + c(\tau)\,. \tag{7.52}$$

An arbitrary function of time c can be found through the condition $\Delta Q_1 = 0$ at $\eta = 1$. Therefore,

$$c(\tau) = -\int_0^{1} an_1 \frac{\partial^3 u_3}{\partial \eta\, \partial \tau^2}\, \mathrm{d}\eta\,.$$

The displacement u_3 in (7.52) can be found from (7.50).

7.1.5 Equations of Vibrations with Regard for Rotary Inertia and Shear

Consider a rod of variable cross section subjected to a dead load q (Fig. 7.4*a*). A rod element $\mathrm{d}x_1$ is shown in Fig. 7.4*b*.
Due to a shear strain, the end cross sections of the rod element are rotated by an additional angle γ_{m}. Hence, from Fig. 7.4*a* it is evident that the total angle of rotation for this element is

$$\theta = \gamma_{\mathrm{m}} + \frac{\partial u_{x_2}}{\partial s}\,. \tag{7.53}$$

To set up a relationship between the shear force ΔQ_2 and the angle γ_{m}, we should represent the shear strain energy of the rod element in terms of the work done by the force Q_2 (Fig. 7.4*b*). In the case of static loading, we have

$$\frac{1}{2}\Delta Q_2 \gamma_{\mathrm{m}}\, \mathrm{d}s = \frac{1}{2}\int_F \tau\gamma\, \mathrm{d}s\, \mathrm{d}F\,, \tag{7.54}$$

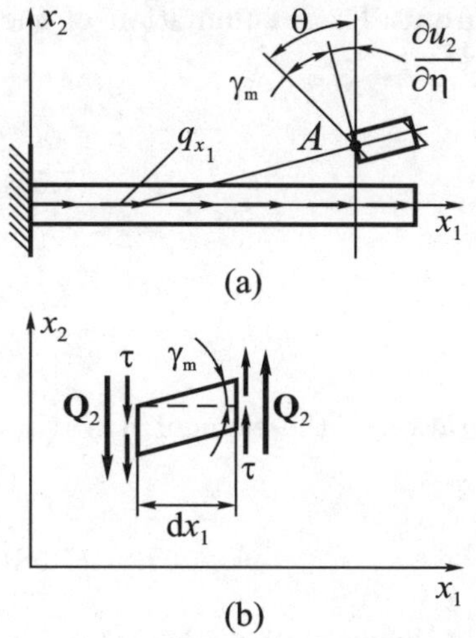

Fig. 7.4.

where F is the cross-sectional area. We have (see Feodosiev (1999))

$$\tau = \frac{\Delta Q_2 \, S^*_{x_3}}{J_{x_3} b}\,, \qquad \gamma = \frac{\tau}{G}\,.$$

Using these relationships in (7.54), we get

$$\Delta Q_{x_2} = \frac{G}{k_2}\gamma_{\mathrm{m}} = \frac{G}{k_2}\left(\theta - \frac{\partial u_{x_2}}{\partial s}\right), \tag{7.55}$$

where

$$k_2 = \int_F \frac{S^{*2}_{x_3}\,\mathrm{d}F}{b^2 J^2_{x_3}}\,. \tag{7.55}$$

For the subsequent discussion it is convenient to introduce a dimensionless coefficient $k_1(s) = k_2 F(0)$.

Equation of Translational Motion of the Rod Element. Summation of the forces in the x_2-direction yields the equation of translational motion of the rod element

$$m_0 \frac{\partial^2 u_{x_2}}{\partial t^2} - \frac{\partial \Delta Q_{x_2}}{\partial s} - \frac{\partial}{\partial s}\left(Q_{x_1}\frac{\partial u_{x_2}}{\partial s}\right) = 0\,, \tag{7.56}$$

Eliminating ΔQ_{x_2}, we get

$$m_0 \frac{\partial^2 u_{x_2}}{\partial t^2} - \frac{GF}{k_1}\left(\frac{\partial \theta}{\partial s} - \frac{\partial^2 u_{x_2}}{\partial s^2}\right) - \frac{\partial}{\partial s}\left(Q_{x_1}\frac{\partial u_{x_2}}{\partial s}\right) = 0\,.$$

Equation of Rotation of the Rod Element. From summation of the moments about the point A (Fig. 7.4a), we get

$$\frac{\partial \Delta M_{x3}}{\partial s}\,\mathrm{d}s + \mathrm{d}M_{\mathrm{i}} + \Delta Q_{x2}\,\mathrm{d}s = 0\,, \tag{7.57}$$

where

$$\mathrm{d}M_{\mathrm{i}} = -\mathrm{d}J\,\frac{\partial^2\theta}{\partial t^2} = -J_{x3}\rho\,\frac{\partial^2\theta}{\partial t^2}\,\mathrm{d}s$$

is the moment of inertial forces due to the rotation of the element. On rearrangement, (7.57) gives

$$\frac{\partial \Delta M_{x3}}{\partial s} - \rho J_{x3}\,\frac{\partial^2\theta}{\partial t^2} + \Delta Q_{x2} = 0\,. \tag{7.58}$$

The moment ΔM_{x3} is linear in the change of the absolute angle of rotation of cross section θ, that is,

$$\Delta M_{x3} = A_{33}\,\frac{\partial\theta}{\partial s}\,, \tag{7.59}$$

Therefore, (7.58) takes the form

$$\frac{\partial}{\partial s}\left(A_{33}\,\frac{\partial\theta}{\partial s}\right) - \rho J_{x3}\,\frac{\partial^2\theta}{\partial t^2} + \frac{GF}{k_1}\left(\theta - \frac{\partial u_{x2}}{\partial s}\right) = 0\,. \tag{7.60}$$

Thus, we have developed the system of two equations (7.56) and (7.60) in the unknowns θ and u_{x2}.

Equations of Vibrations of a Straight Rod with Regard for Rotary Inertia and Shear. For a rod of variable cross section it is sensible not to eliminate ΔQ_{x2} and ΔM_{x3} from the equations of motion (7.56) and (7.58). We have the system of four equations (7.56), (7.58), (7.55) and (7.59) in four unknowns u_{x2}, θ, ΔM_{x3}, and ΔQ_{x2}. In nondimensional form, these equations read

$$n_1\,\frac{\partial^2 u_{x2}}{\partial\tau^2} - a_1\,\frac{\partial\Delta Q_{x2}}{\partial\eta} - a_2\,\frac{\partial\theta}{\partial\eta} - a_3\,\Delta Q_{x2} - a_4\theta = 0\,, \tag{7.61}$$

$$J_{33}\,\frac{\partial^2\theta}{\partial\tau^2} - \frac{\partial\Delta M_{x3}}{\partial\eta} - \Delta Q_{x2} = 0\,, \tag{7.62}$$

$$\frac{\partial\theta}{\partial\eta} - \frac{\Delta M_{x3}}{A_{33}} = 0\,, \tag{7.63}$$

$$\frac{\partial u_{x2}}{\partial\eta} + k_1 n_2 n_3\,\Delta Q_{x2} - \theta = 0\,, \tag{7.64}$$

where

$$a_1 = 1 - Q_{x_1} k_1 n_2 n_3 \,, \qquad a_2 = Q_{x_1} \,,$$

$$a_3 = \frac{\partial Q_{x_1}}{\partial \eta} k_1 - Q_{x_1} \frac{\partial k_1}{\partial \eta} \,, \qquad a_4 = \frac{\partial Q_{x_1}}{\partial \eta} \,,$$

$$n_2 = \frac{E}{G} \,, \qquad n_3 = \frac{J_{x_3}(0)}{l^2 F(0)} \,.$$

For example, for the rod shown in Fig. 7.4a, the nondimensional axial load is

$$Q_{x_1} = q_{x_1}(1 - \eta) \,.$$

A dimensional distributed load can be written as $(q_{x_1})_{\mathrm{d}} = n q_{x_1}$, where $q_{x_1} = 7.87 \cdot A_{33}/l^3$ is the critical load for a cantilever beam. Hence, the dimensionless distributed load reads

$$q_{x_1} = \gamma n \,, \qquad \gamma = 7.87 \,, \qquad n < 1 \,.$$

Using (7.61)–(7.64), we obtain a vector equation

$$\mathrm{A}^{(1)} \frac{\partial^2 \mathbf{Z}}{\partial \tau^2} + \mathrm{E}^{(1)} \frac{\partial \mathbf{Z}}{\partial \eta} + \mathrm{A}^{(2)} \mathbf{Z} = 0 \,, \tag{7.65}$$

where

$$\mathbf{Z} = \begin{bmatrix} \Delta Q_{x_2} \\ \Delta M_{x_3} \\ \theta \\ u_{x_2} \end{bmatrix}, \qquad \mathrm{A}^{(1)} = \begin{bmatrix} 0 & 0 & 0 & n_1 \\ 0 & 0 & -J_{33} & 0 \\ 0 & 0 & 0 & 0 \\ 0 & 0 & 0 & 0 \end{bmatrix},$$

$$\mathrm{E}^{(1)} = \begin{bmatrix} a_1 & 0 & a_2 & 0 \\ 0 & 1 & 0 & 0 \\ 0 & 0 & 1 & 0 \\ 0 & 0 & 0 & 1 \end{bmatrix}, \qquad \mathrm{A}^{(2)} = \begin{bmatrix} a_3 & 0 & a_4 & 0 \\ 1 & 0 & 0 & 0 \\ 0 & -\dfrac{1}{A_{33}} & 0 & 0 \\ k_1 n_2 n_3 & 0 & -1 & 0 \end{bmatrix}.$$

The solutions to the system of equations (7.61)–(7.64) must satisfy the following boundary conditions.

For a cantilever beam:

(1) at $\eta = 0$, $\theta = 0$ and $u_{x_2} = 0$;

(2) at $\eta = 1$, $\Delta Q_{x_2}(1) = \dfrac{1}{k_1 n_2 n_3} \left(\theta - \dfrac{\partial u_{x_2}}{\partial \eta} \right) \Big|_{\eta=1} = 0$

and $\theta'(1) = 0$.

For a rod with simply supported ends:

(1) at $\eta = 0$, $u_{x_2}(0) = 0$ and $\theta'(0) = 0$;

(2) at $\eta = 1$, $u_{x_2}(1) = 0$ and $\theta'(1) = 0$.

The equations of small vibrations for straight rods obtained in this section may be used in a number of applications.

7.2 Determination of Frequencies and Vibration Modes

7.2.1 Solution to the Equations of Free Vibrations

Consider the simplest case: free vibrations of a straight rod of constant cross section. Using (7.34) and setting $u_{x_2} = u$, we get

$$\frac{\partial^4 u}{\partial \eta^4} + \frac{\partial^2 u}{\partial \tau^2} = 0\,. \tag{7.66}$$

Setting $u = u_0(\eta)\,\mathrm{e}^{\mathrm{i}\,\lambda\tau}$, we obtain

$$u_0^{\mathrm{IV}} - \lambda^2 u_0 = 0\,. \tag{7.67}$$

Substituting $u_0 = a\,\mathrm{e}^{m\eta}$, we arrive at the following characteristic equation to be solved for m:

$$m^4 - \lambda^2 = 0\,;$$

hence,

$$m_{1,2} = \pm\sqrt{\lambda} = \pm\lambda_0\,, \qquad m_{3,4} = \pm\mathrm{i}\,\sqrt{\lambda} = +\mathrm{i}\,\lambda_0\,, \qquad \lambda_0 = \sqrt{\lambda}\,.$$

For each root m_j there exists a partial solution

$$u_{0j} = k_j^{(1)} = \mathrm{e}^{m_j\eta}$$

or

$$u_{01} = k_1^{(1)} = \mathrm{e}^{\lambda_0\eta}\,, \qquad u_{02} = k_2^{(1)} = \mathrm{e}^{-\lambda_0\eta}\,,$$
$$u_{03} = k_3^{(1)} = \cos\lambda_0\eta\,, \qquad u_{04} = k_4^{(1)} = \sin\lambda_0\eta\,.$$

As a result, the solution to the homogeneous equation (7.67) is seen to be

$$u_0 = c_1k_1^{(1)} + c_2k_2^{(1)} + c_3k_3^{(1)} + c_4k_4^{(1)}\,. \tag{7.68}$$

Differentiation of this relationship yields

$$\begin{aligned}
u_{x_20} &= u_0 = c_1k_1^{(1)} + c_2k_2^{(1)} + c_3k_3^{(1)} + c_4k_4^{(1)}\,,\\
\vartheta_{30} &= u_0' = c_1k_1^{(1)\prime} + c_2k_2^{(1)\prime} + c_3k_3^{(1)\prime} + c_4k_4^{(1)\prime}\,,\\
\Delta M_{x_30} &= u_0'' = c_1k_1^{(1)\prime\prime} + c_2k_2^{(1)\prime\prime} + c_3k_3^{(1)\prime\prime} + c_4k_4^{(1)\prime\prime}\,,\\
\Delta Q_{x_20} &= -u_0''' = -c_1k_1^{(1)\prime\prime\prime} - c_2k_2^{(1)\prime\prime\prime} - c_3k_3^{(1)\prime\prime\prime} - c_4k_4^{(1)\prime\prime\prime}\,.
\end{aligned} \tag{7.69}$$

The functions $k_j^{(1)}$ and their derivatives can be organized into a fundamental matrix

$$\mathrm{K}^{(1)}(\eta) = \begin{bmatrix} -k_1^{(1)'''} & -k_2^{(1)'''} & -k_3^{(1)'''} & -k_4^{(1)'''} \\ k_1^{(1)''} & k_2^{(1)''} & k_3^{(1)''} & k_4^{(1)''} \\ -k_1^{(1)'} & k_2^{(1)'} & k_3^{(1)'} & k_4^{(1)'} \\ k_1^{(1)} & k_2^{(1)} & k_3^{(1)} & k_4^{(1)} \end{bmatrix}$$

$$= \begin{bmatrix} -\lambda_0^3\,\mathrm{e}^{\lambda_0\eta} & \lambda_0^3\,\mathrm{e}^{-\lambda_0\eta} & -\lambda_0^3\sin\lambda_0\eta & \lambda_0^3\cos\lambda_0\eta \\ \lambda_0^2\,\mathrm{e}^{\lambda_0\eta} & \lambda_0^2\,\mathrm{e}^{-\lambda_0\eta} & -\lambda_0^2\cos\lambda_0\eta & -\lambda_0^2\sin\lambda_0\eta \\ \lambda_0\,\mathrm{e}^{\lambda_0\eta} & -\lambda_0\,\mathrm{e}^{-\lambda_0\eta} & -\lambda_0\sin\lambda_0\eta & \lambda_0\cos\lambda_0\eta \\ \mathrm{e}^{\lambda_0\eta} & \mathrm{e}^{-\lambda_0\eta} & \cos\lambda_0\eta & \sin\lambda_0\eta \end{bmatrix}. \quad (7.70)$$

The arrangement of the rows and columns in the fundamental matrix (7.70) is dictated by the arrangement of the components of the vector

$$\mathbf{Z} = \mathbf{Z}\left(\Delta Q_{x_2},\, \Delta M_{x_3},\, \vartheta_3,\, u_{x_2}\right).$$

For the fundamental matrix $\mathrm{K}^{(1)}(\eta)$ the condition $\mathrm{K}^{(1)}(0) = \mathrm{E}$ does not hold. There is no harm in that but it is convenient to perform manipulations and computations when the condition under consideration holds. Thus, it is convenient to deal with the fundamental matrix

$$\mathrm{K}(\eta) = \mathrm{K}^{(1)}(\eta)(\mathrm{K}^{(1)}(0))^{-1},$$

for which $\mathrm{K}(0) = \mathrm{E}$. Here, $(\mathrm{K}^{(1)}(0))^{-1}$ is the matrix inverse to the matrix $\mathrm{K}^{(1)}(\eta)$ at $\eta = 0$.

Any linear combination of the partial solutions k_j is a solution to (7.67). Hence, by choosing the combination appropriately, we can simplify further computations, particularly, the determination of the arbitrary constants c_j. It is a common practice to take as the members of such combinations the Krylov functions defined as

$$K_1 = \frac{1}{2}(\cosh\lambda_0\eta + \cos\lambda_0\eta), \qquad K_2 = \frac{1}{2}(\sinh\lambda_0\eta + \sin\lambda_0\eta),$$
$$K_3 = \frac{1}{2}(\cosh\lambda_0\eta - \cos\lambda_0\eta), \qquad K_4 = \frac{1}{2}(\sinh\lambda_0\eta - \sin\lambda_0\eta). \quad (7.71)$$

Now, we seek a solution to (7.67) in the form

$$u_0 = \sum_{j=1}^{4} c_j K_j. \quad (7.72)$$

The derivatives of the Krylov functions are

$$\begin{aligned} K_1' &= \lambda_0 K_4, & K_1'' &= \lambda_0^2 K_3, & K_1''' &= \lambda_0^3 K_2, \\ K_2' &= \lambda_0 K_1, & K_2'' &= \lambda_0^2 K_4, & K_2''' &= \lambda_0^3 K_3, \\ K_3' &= \lambda_0 K_2, & K_3'' &= \lambda_0^2 K_1, & K_3''' &= \lambda_0^3 K_4, \\ K_4' &= \lambda_0 K_3, & K_4'' &= \lambda_0^2 K_2, & K_4''' &= \lambda_0^3 K_1. \end{aligned} \quad (7.73)$$

7.2.2 Determination of Eigenvalues (Eigenfrequencies)

The solution (7.72) is to satisfy the four boundary-condition equations. For illustrative purposes, consider a rod shown in Fig. 7.5*a*.

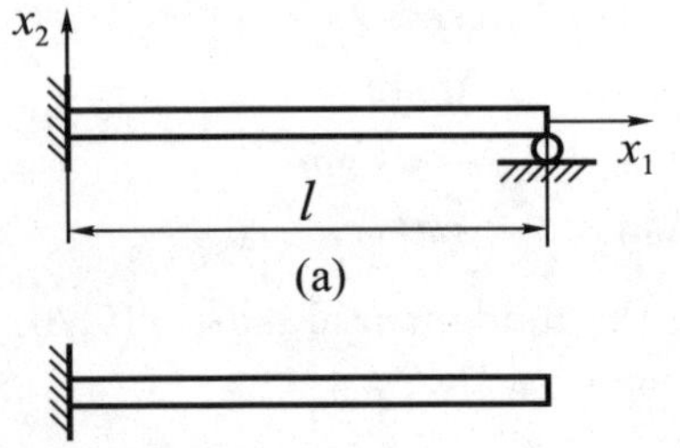

Fig. 7.5.

The boundary conditions are as follows:

$$\begin{aligned}
&(1)\quad \text{at}\quad \eta = 0\,, \quad u_0 = u_{x_2 0} = 0 \quad \text{and} \quad u_0' = \vartheta_{30} = 0\,;\\
&(2)\quad \text{at}\quad \eta = 1\,, \quad u_0 = u_{x_2 0} = 0 \quad \text{and} \quad u_0'' = \Delta M_{30} = 0\,.
\end{aligned}$$

From the conditions at $\eta = 0$ and in view of the expressions for the derivatives, it follows that $c_1 = c_2 = 0$. The boundary conditions at $\eta = 1$ yield two equations

$$\begin{aligned}
&c_3 \mathrm{K}_3(1,\, \lambda_0) + c_4 \mathrm{K}_4(1,\, \lambda_0) = 0\,,\\
&c_3 \mathrm{K}_1(1,\, \lambda_0) + c_4 \mathrm{K}_2(1,\, \lambda_0) = 0\,.
\end{aligned} \tag{7.74}$$

Equating the determinant of these equations to zero,

$$D = \det \begin{bmatrix} \mathrm{K}_3(1,\, \lambda_0) & \mathrm{K}_4(1,\, \lambda_0) \\ \mathrm{K}_1(1,\, \lambda_0) & \mathrm{K}_2(1,\, \lambda_0) \end{bmatrix} = 0\,, \tag{7.75}$$

and using (7.73), we obtain the equation to be solved for the frequencies

$$\cosh \lambda_0 \sin \lambda_0 - \sinh \lambda_0 \cos \lambda_0 = 0\,. \tag{7.76}$$

This gives

$$\lambda_{01} = 3.926\,, \qquad \lambda_{02} = 7.068\,, \qquad \lambda_{0n} = \frac{(4n+1)\,\pi}{4}\,, \quad n > 2\,.$$

The dimensional frequencies of the vibrations of the rod are

$$p_j = (\lambda_{0j})^2 \sqrt{\frac{A_{33}}{m_0 l^4}}\,.$$

For a cantilever beam (Fig. 7. 5*b*), we have

$$D = \det \begin{bmatrix} \mathrm{K}_1 & \mathrm{K}_2 \\ \mathrm{K}_4 & \mathrm{K}_1 \end{bmatrix} = 0$$

and

$$\cos\lambda_0 \cosh\lambda_0 + 1 = 0 \,. \tag{7.77}$$

The equation has the following roots:

$$\lambda_{01} = 1.875\,, \qquad \lambda_{02} = 4.694\,, \qquad \lambda_{0n} = \frac{(2n-1)\,\pi}{2}\,, \quad n > 2\,.$$

7.2.3 Determination of Eigenfunctions (Eigenmodes)

We will illustrate the determination of the eigenfunctions by the example of the rods shown in Figs. 7.5*a, b*. Having found λ_{0j} for the rod in Fig. 7.5*a*, we use (7.74) to find

$$c_3^{(j)} = -\frac{\mathrm{K}_4(1,\,\lambda_{0j})}{\mathrm{K}_3(1,\,\lambda_{0j})}\,c_4^{(j)}\,.$$

The arbitrary constant $c_4^{(j)}$ can be set equal to unity.

Since $c_1^{(j)} = c_2^{(j)} = 0$, it follows from (7.72) that the eigenfunctions are

$$\varphi^{(j)}(\eta) = u_0^{(j)}(\eta) = \mathrm{K}_2(\eta,\,\lambda_{0j}) - \frac{\mathrm{K}_4(1,\,\lambda_{0j})}{\mathrm{K}_2(1,\,\lambda_{0j})}\,\mathrm{K}_3(\eta,\,\lambda_{0j})\,. \tag{7.78}$$

For the case shown in Fig. 7.5*b*, the eigenfunctions are as follows:

$$\varphi^{(j)}(\eta) = u_0^{(j)}(\eta) = \mathrm{K}_2(\eta,\,\lambda_{0j}) - \frac{\mathrm{K}_2(1,\,\lambda_{0j})}{\mathrm{K}_1(1,\,\lambda_{0j})}\,\mathrm{K}_1(\eta,\,\lambda_{0j})\,. \tag{7.79}$$

For the jth frequency, the function $\varphi^{(j)}(\eta)$ describes the variation of the amplitude values with the coordinate η. The derivatives of the functions $\varphi^{(j)}(\eta)$ characterize the variation of the amplitude values of the angle of inclination of the tangent line to the rod axis, $\vartheta_{30}^{(j)}(\eta)$, the bending moment, $\Delta M_{x30}^{(j)}(\eta)$, and of the shear force $\Delta Q_{x20}^{(j)}(\eta)$ corresponding to each frequency λ_{0j}. The eigenfunctions (7.66) have been obtained for the simplest case, namely, for transversal vibrations of a rod of constant cross section. Nevertheless, these eigenfunctions can be used for approximate analysis of more complicated equations of transversal vibrations, for example, for rods of variable cross section loaded by concentrated dynamic forces, for rods under the action of liquid or air flows, etc.

7.2.4 Numerical Determination of Frequencies and Vibration Modes

To find the frequencies and the vibration modes numerically, it would be convenient to represent the equations of vibrations in the form of equations (7.5)–(7.9) (for $\Delta P_j = \Delta T_j = 0$). Ignoring the initial twisting of the rod, we can rewrite the system of equations (7.5)–(7.9) for free vibrations as one vector equation

$$\mathrm{A}^{(1)} \frac{\partial^2 \mathbf{Z}}{\partial \tau^2} + \frac{\partial \mathbf{Z}}{\partial \eta} + \mathrm{A}^{(2)} \mathbf{Z} = 0 , \qquad (7.80)$$

where

$$\mathbf{Z} = \begin{bmatrix} \Delta Q_2 \\ \Delta Q_3 \\ \Delta M_2 \\ \Delta M_3 \\ \vartheta_2 \\ \vartheta_3 \\ u_2 \\ u_3 \end{bmatrix}, \quad \mathrm{A}^{(1)} = \begin{bmatrix} 0 & 0 & 0 & 0 & 0 & 0 & -n_1 & 0 \\ 0 & 0 & 0 & 0 & 0 & 0 & 0 & -n_1 \\ 0 & 0 & 0 & 0 & -J_{22} & 0 & 0 & 0 \\ 0 & 0 & 0 & 0 & 0 & -J_{33} & 0 & 0 \\ 0 & 0 & 0 & 0 & 0 & 0 & 0 & 0 \\ 0 & 0 & 0 & 0 & 0 & 0 & 0 & 0 \\ 0 & 0 & 0 & 0 & 0 & 0 & 0 & 0 \\ 0 & 0 & 0 & 0 & 0 & 0 & 0 & 0 \end{bmatrix},$$

$$\mathrm{A}^{(2)} = \begin{bmatrix} 0 & 0 & 0 & \frac{Q_{10}}{A_{33}} & 0 & 0 & 0 & 0 \\ 0 & 0 & -\frac{Q_{10}}{A_{33}} & 0 & 0 & 0 & 0 & 0 \\ 0 & -1 & 0 & \frac{M_{10}}{A_{33}} & 0 & 0 & 0 & 0 \\ 1 & 0 & -\frac{M_{10}}{A_{33}} & 0 & 0 & 0 & 0 & 0 \\ 0 & 0 & -\frac{1}{A_{22}} & 0 & 0 & 0 & 0 & 0 \\ 0 & 0 & 0 & -\frac{1}{A_{33}} & 0 & 0 & 0 & 0 \\ 0 & 0 & 0 & 0 & 0 & -1 & 0 & 0 \\ 0 & 0 & 0 & 0 & 1 & 0 & 0 & 0 \end{bmatrix}.$$

Equation (7.80) describes the most general case of forced vibrations of a straight rod in two mutually orthogonal planes. The method of numerical determination of the frequencies for equations identical in form to (7.80) was discussed in Chapter 4. Setting

$$\mathbf{Z} = \mathbf{Z}_0 \, \mathrm{e}^{\mathrm{i}\,\lambda \eta} ,$$

we get

$$\mathbf{Z}_0' + \mathrm{B}(\eta,\, \lambda)\, \mathbf{Z}_0 = 0 , \qquad (7.81)$$

where

$$\mathrm{B}(\eta, \lambda) = \mathrm{A}^{(2)} - \lambda^2 \mathrm{A}^{(1)} .$$

The only difference between (7.81) and (4.14) is that the matrix B in (7.81) is of dimensions 8×8 and this matrix in (4.14) is of dimensions 12×12. Such a difference is of little importance and thus the method presented in Sect. 4.1 can be applied to (7.81).

The equations governing independent vibrations of a rod in two mutually orthogonal planes were obtained in Sect. 7.1 (equations (7.26) and (7.27)).

To determine the frequencies, one should put

$$\Delta \boldsymbol{\Phi} = \Delta \boldsymbol{\Phi}_1 = 0 , \qquad \mathrm{A}^{(3)} \frac{\partial \mathbf{Z}}{\partial \tau} = 0 , \qquad \mathrm{A}_1^{(3)} \frac{\partial \mathbf{Z}_1}{\partial \tau} = 0$$

in these equations. As a result, we arrive at

$$\mathrm{A}^{(1)} \frac{\partial^2 \mathbf{Z}}{\partial \tau^2} + \frac{\partial \mathbf{Z}}{\partial \eta} + \mathrm{A}^{(2)} \mathbf{Z} = 0 ,$$

$$\mathrm{A}_1^{(1)} \frac{\partial^2 \mathbf{Z}_1}{\partial \tau^2} + \frac{\partial \mathbf{Z}_1}{\partial \eta} + \mathrm{A}_1^{(2)} \mathbf{Z}_1 = 0 ,$$

where the matrices $\mathrm{A}^{(1)}$, $\mathrm{A}^{(2)}$, $\mathrm{A}_1^{(1)}$, and $\mathrm{A}_1^{(2)}$ correspond to the matrices in (7.26) and (7.27).

Setting

$$\mathbf{Z} = \mathbf{Z}_0 \, \mathrm{e}^{\mathrm{i} \lambda \tau} , \qquad \mathbf{Z}_1 = \mathbf{Z}_{01} \, \mathrm{e}^{\mathrm{i} \lambda \tau} ,$$

we get

$$\mathbf{Z}_0' + \mathrm{B}(\eta, \lambda) \, \mathbf{Z}_0 = 0 , \tag{7.82}$$

$$\mathbf{Z}_{01}' + \mathrm{B}_1(\eta, \lambda) \, \mathbf{Z}_{01} = 0 , \tag{7.83}$$

where

$$\mathrm{B}(\eta, \lambda) = \mathrm{A}^{(2)} - \lambda^2 \mathrm{A}^{(1)} , \qquad \mathrm{B}_1(\eta, \lambda) = \mathrm{A}_1^{(2)} - \lambda^2 \mathrm{A}_1^{(1)} .$$

In the last equations, the matrices B and B_1 are of dimensions 4×4.

Having found the eigenvalues λ_j, we can proceed, employing the method given in Sect. 4.1, with the determination of the eigenfunctions for each of the equations (7.81), (7.82) and (7.83).

Consider some examples to illustrate the determination of the frequencies of straight rods.

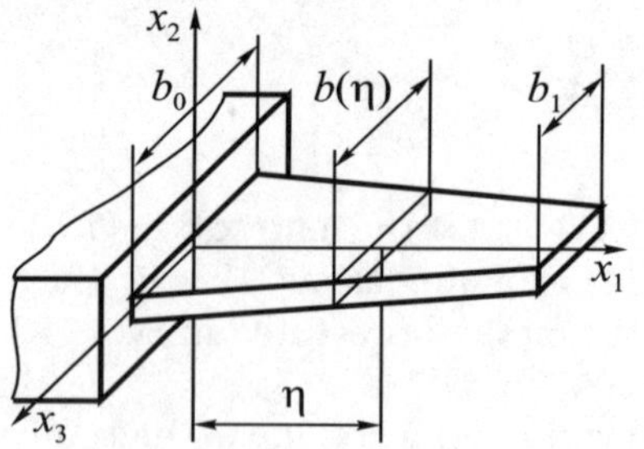

Fig. 7.6.

Determination of the Frequencies of a Rod of Variable Cross Section. A rod of variable cross section is shown in Fig. 7.6. We ignore the rotary inertia. Consider the following modes of variation of the rod width:

$$(1)\quad b(\eta) = b_0\, \mathrm{e}^{-\alpha_1 \eta}\,, \qquad \alpha_1 = \ln \frac{b_1}{b_0}\,,$$

$$(2)\quad b(\eta) = \left[1 - \left(1 - \frac{b_1}{b_0}\right)\eta\right] b_0\,.$$

For these cases, the dimensionless coefficients n and A are

$$(1)\quad n_1(\eta) = \mathrm{e}^{-\alpha_1 \eta}\,, \qquad A_{33}(\eta) = \mathrm{e}^{-\alpha_1 \eta}\,,$$

$$(2)\quad n_1(\eta) = 1 - \left(1 - \frac{b_1}{b_0}\right)\eta\,, \qquad A_{33}(\eta) = 1 - \left(1 - \frac{b_1}{b_0}\right)\eta\,.$$

Then, we can set

$$Q_{10} = 0\,, \qquad a = 0\,, \qquad k = 0\,, \qquad \Delta \boldsymbol{\Phi} = 0$$

in (7.26). By setting $\mathbf{Z} = \mathbf{Z}_0\, \mathrm{e}^{\mathrm{i}\,\lambda\tau}$, we arrive at

$$\mathbf{Z}_0' + \mathrm{B}(\eta,\, \lambda)\, \mathbf{Z}_0 = 0\,, \tag{7.84}$$

where

$$\mathrm{B}(\eta,\, \lambda) = \begin{bmatrix} 0 & 0 & 0 & n_1(\eta)\,\lambda^2 \\ 1 & 0 & 0 & 0 \\ 0 & -\dfrac{1}{A_{33}(\eta)} & 0 & 0 \\ 0 & 0 & -1 & 0 \end{bmatrix}.$$

By solving (7.84) numerically, we get the following relationship between b_1/b_0 and λ_{0j}.

The results corresponding to the first case are presented in Table 7.1.
The results corresponding to the second case are presented in Table 7.2.

Table 7.1.

b_1/b_0	1	0.8	0.6	0.4
λ_{01}	3.516	3.761	4.102	4.591
λ_{02}	22.035	22.511	23.056	24.031
λ_{03}	61.696	62.451	63.011	63.951

Table 7.2.

b_1/b_0	1	0.8	0.6	0.4
λ_{01}	3.516	3.672	4.098	4.585
λ_{02}	22.035	22.502	23.119	24.022
λ_{03}	61.697	62.153	62.777	63.754

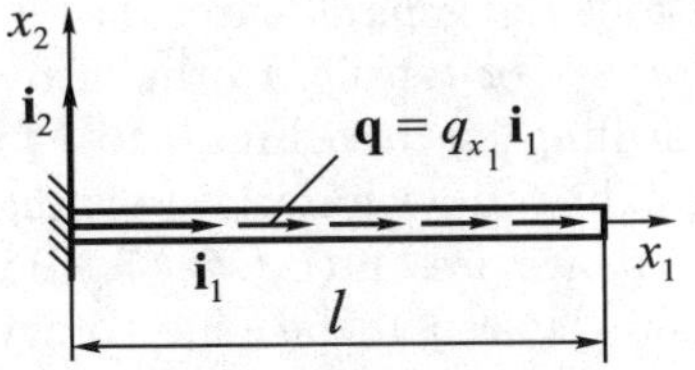

Fig. 7.7.

Determination of the Frequencies of a Cantilever Beam of Constant Cross Section Subjected to a Dead Distributed Load. A cantilever beam of constant cross section ($n_1 = A_{33} = 1$) is shown in Fig. 7.7. Let the beam be loaded by a dead distributed force $\mathbf{q}$.

Let us use equations (7.31) referred to the Cartesian axes; hence, we have $\Delta q_{x_1} = \Delta q_{x_2} = 0$. The axial force is

$$Q_{10} = q^0_{x_1}(1-\eta) = \gamma n(1-\eta)\,,$$

where

$$\gamma = 7.87\,, \qquad n = \frac{q_{x1}}{q_{x*}}\,;$$

here, q_{x*} is the critical distributed load for the cantilever beam.

In the example under consideration, the matrix $\mathrm{B}(\eta\,, \lambda)$ reads

$$\mathrm{B}(\eta,\,\lambda) = \begin{bmatrix} 0 & 0 & 0 & \lambda^2 \\ 1 & 0 & -\gamma n(1-\eta) & 0 \\ 0 & -1 & 0 & 0 \\ 0 & 0 & -1 & 0 \end{bmatrix}.$$

The values of λ_{0j} for some values of n are shown in Table 7.3.

Table 7.3.

n	-0.8	-0.4	0	0.4	0.8
λ_{01}	1.5740	2.7254	3.5161	4.1570	4.7104
λ_{02}	20.7654	21.4098	22.0341	20.6410	23.2308
λ_{03}	60.4132	61.0593	61.6971	62.3280	62.9507

Determination of the Frequencies of a Straight Rod of Constant Cross Section with Initial Twisting Subjected to a Concentrated Axial Force and a Concentrated Torque. Consider a straight rod of constant cross section with initial twisting subjected to a concentrated axial force P_{x_10} and a concentrated torque $M_{x_{10}}$ applied at $\eta = 1$ as shown in Fig. 7.2.

The rod is simply supported at $\eta = 1$. Consider the general case, that is, the axial force $Q_{10} = P_{x_10}$ can be either compressive or tensile. In the latter case, we have the problem of vibrations of a drilling bit (see Fig. 0.10). In Sect. 7.1, equations of small vibrations of a straight rod with initial twisting subjected to an axial force Q_{10} and a moment M_{10} are presented (see (7.36)). On rearrangement, the matrix $\mathrm{B}(\eta\,,\lambda)$ for (7.36) is seen to be (the rotary inertia of the rod element is ignored)

$$\mathrm{B}(\eta,\,\lambda) = \begin{bmatrix} 0 & -æ_{10} & 0 & -Q_{10} & 0 & 0 & \lambda^2 & 0 \\ æ_{10} & 0 & 0 & \dfrac{Q_{10}}{A_{22}} & 0 & 0 & 0 & \lambda^2 \\ 0 & -1 & 0 & -æ_{10}^{(1)} & 0 & 0 & 0 & 0 \\ 1 & 0 & æ_{10}^{(1)} & 0 & 0 & 0 & 0 & 0 \\ 0 & 0 & -\dfrac{1}{A_{33}} & 0 & 0 & -æ_{10} & 0 & 0 \\ 0 & 0 & 0 & -1 & æ_{10} & 0 & 0 & 0 \\ 0 & 0 & 0 & 0 & 0 & -1 & 0 & -æ_{10} \\ 0 & 0 & 0 & 0 & 1 & 0 & æ_{10} & 0 \end{bmatrix}, \tag{7.85}$$

where

$$æ_{10} = \frac{\partial \vartheta_{10}}{\partial \eta}, \qquad æ_{10}^{(1)} = \frac{\partial \vartheta_{10}}{\partial \eta} - \frac{M_{10}}{A_{11}},$$

$$Q_{10} = \frac{(Q_{10})_{\mathrm{d}}}{Q_{1\,\mathrm{cr}}}, \qquad Q_{1\,\mathrm{cr}} = \frac{20A_{33}(0)}{l^2}.$$

For numerical evaluation, we put $\vartheta_{10} = \vartheta_{100}\eta$, where ϑ_{100} is the constant angle of mutual rotation of the end cross sections of the rod.

By solving numerically (7.84) with the matrix (7.85) for certain values of Q_{10} and M_{10}, we find the dimensionless frequencies λ_{0j}, $j = 1\,,2\,,3\,,4$, as

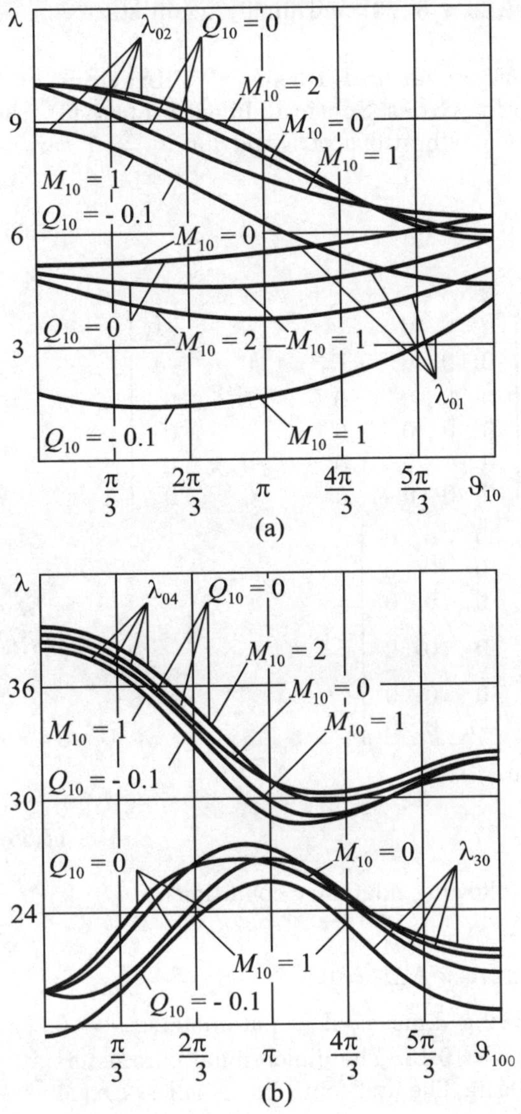
λ
λ02
Q10 = 0
M10 = 2
M10 = 0
M10 = 1
M10 = 1
Q10 = - 0.1
M10 = 0
Q10 = 0
M10 = 2
M10 = 1
λ01
Q10 = - 0.1
M10 = 1
9
6
3
π/3
2π/3
π
4π/3
5π/3
ϑ10
(a)
λ
λ04
Q10 = 0
M10 = 2
M10 = 0
M10 = 1
M10 = 1
Q10 = - 0.1
36
30
24
Q10 = 0
M10 = 0
λ30
M10 = 1
Q10 = - 0.1
π/3
2π/3
π
4π/3
5π/3
ϑ100
(b)

Fig. 7.8.

functions of ϑ_{100} (during the computation, the fundamental matrix $\mathrm{K}(\eta)$ is subject to corrections according to the procedure presented in Sect. 4.1). The graphs of λ_{01} and λ_{02} are given in Fig. 7.8*a* and the graphs of λ_{03} and λ_{04} are shown in Fig. 7.8*b*. It is seen that the first two frequencies come closer as ϑ_{100} increases while the third and fourth frequencies start getting closer ($\vartheta_{100} < \pi$) then move apart ($\vartheta_{100} < 5\pi/3$) and finally again start coming closer ($\vartheta_{100} > 5\pi/3$).

Frequencies and Modes of Flexural-and-Torsional Vibrations of a Cantilever Beam of Constant Cross Section. In view of (7.48), the equations of flexural-and-torsional vibrations of a rod for $n_1 = 1$ can be written as

$$\mathrm{A}^{(1)}\frac{\partial^2 \mathbf{Z}}{\partial \tau^2} + \frac{\partial \mathbf{Z}}{\partial \eta} + \mathrm{A}^{(2)}\mathbf{Z} = 0\,, \tag{7.86}$$

where

$$\mathbf{Z} = \begin{bmatrix} \Delta Q_2 \\ \Delta M_1 \\ \Delta M_3 \\ \vartheta_1 \\ \vartheta_3 \\ u_2 \end{bmatrix}, \quad \mathrm{A}^{(1)} = \begin{bmatrix} 0 & 0 & 0 & -a & 0 & -1 \\ 0 & 0 & 0 & -J_{11}^{(1)} & 0 & -a \\ 0 & 0 & 0 & 0 & -J_{33}^{(1)} & 0 \\ 0 & 0 & 0 & 0 & 0 & 0 \\ 0 & 0 & 0 & 0 & 0 & 0 \\ 0 & 0 & 0 & 0 & 0 & 0 \end{bmatrix},$$

$$\mathrm{A}^{(2)} = \begin{bmatrix} 0 & 0 & 0 & 0 & 0 & 0 \\ 0 & 0 & 0 & 0 & 0 & 0 \\ -1 & 0 & 0 & 0 & 0 & 0 \\ 0 & -\dfrac{1}{A_{11}} & 0 & 0 & 0 & 0 \\ 0 & 0 & -1 & 0 & 0 & 0 \\ 0 & 0 & 0 & -1 & 0 & 0 \end{bmatrix}. \tag{7.87}$$

Setting $\mathbf{Z} = \mathbf{Z}_0\,\mathrm{e}^{\mathrm{i}\,\lambda\tau}$, we arrive at

$$\frac{\mathrm{d}\mathbf{Z}_0}{\mathrm{d}\eta} + (\mathrm{A}^{(2)} - \lambda^2\mathrm{A}^{(1)})\,\mathbf{Z}_0 = 0\,. \tag{7.88}$$

The vector $\mathbf{Z}_0$ must satisfy the following boundary conditions:

(1) at $\eta = 0, \quad u_2 = \vartheta_1 = \vartheta_3 = 0;$

(2) at $\eta = 1, \quad \Delta Q_2 = \Delta M_1 = \Delta M_3 = 0.$

For simulation, we use following dimensionless parameters: $a = 0.065$, $J_{11}^{(1)} = 0.02$, $J_{33}^{(1)} = 0.02$, and $A_{11} = 0.15$. The dimensional parameters are: $l = 2.0\,\mathrm{m}$, $b = 0.3\,\mathrm{m}$, and $h = 0.06\,\mathrm{m}$. The first four dimensionless frequencies obtained are

$$\lambda_{01} = 3.192\,, \qquad \lambda_{02} = 5.553\,, \qquad \lambda_{03} = 13.051\,, \qquad \lambda_{04} = 21.162\,.$$

The graphs of $u_{20}(\eta)$ and $\vartheta_{10}(\eta)$ for the first three frequencies are shown in Fig. 7.9*a*–*d*.

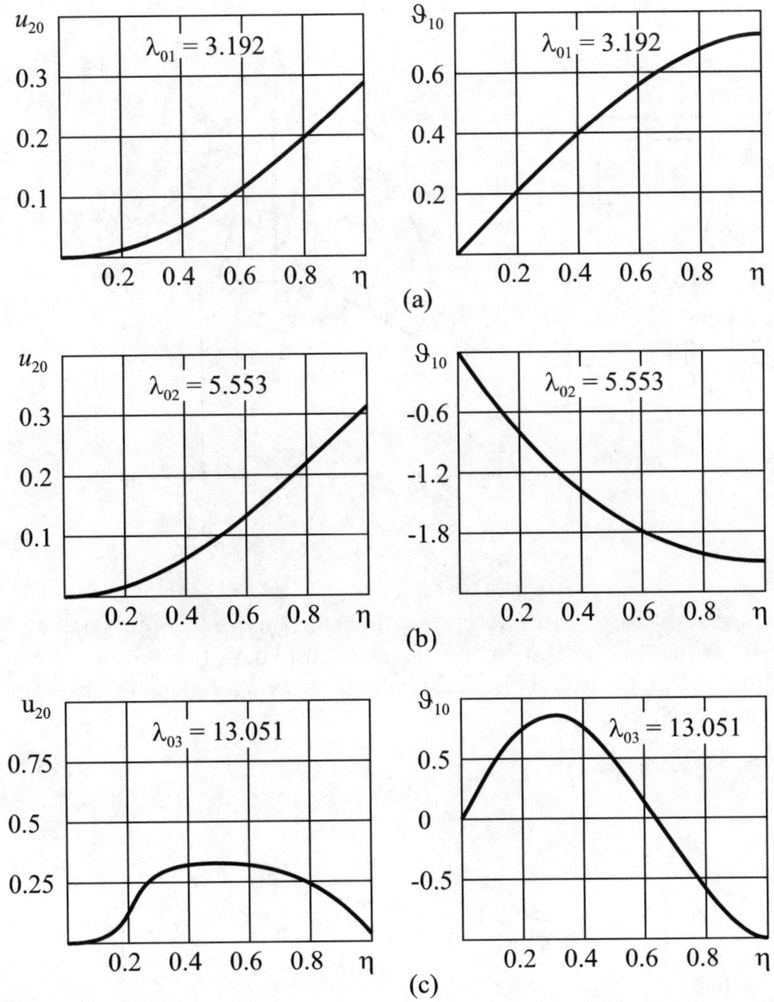

u20
λ01 = 3.192
0.3
0.2
0.1
0.2 0.4 0.6 0.8 η
ϑ10
λ01 = 3.192
0.6
0.4
0.2
0.2 0.4 0.6 0.8 η
(a)
u20
λ02 = 5.553
0.3
0.2
0.1
0.2 0.4 0.6 0.8 η
ϑ10
λ02 = 5.553
-0.6
-1.2
-1.8
0.2 0.4 0.6 0.8 η
(b)
u20
λ03 = 13.051
0.75
0.5
0.25
0.2 0.4 0.6 0.8 η
ϑ10
λ03 = 13.051
0.5
0
-0.5
0.2 0.4 0.6 0.8 η
(c)

Fig. 7.9.

7.2.5 The Method of Initial Parameters

This method was discussed in Sect. 4.1. To illustrate how the method works let us consider a straight rod comprised of three segments *I*, *II*, and *III* (Fig. 7.10*a*).

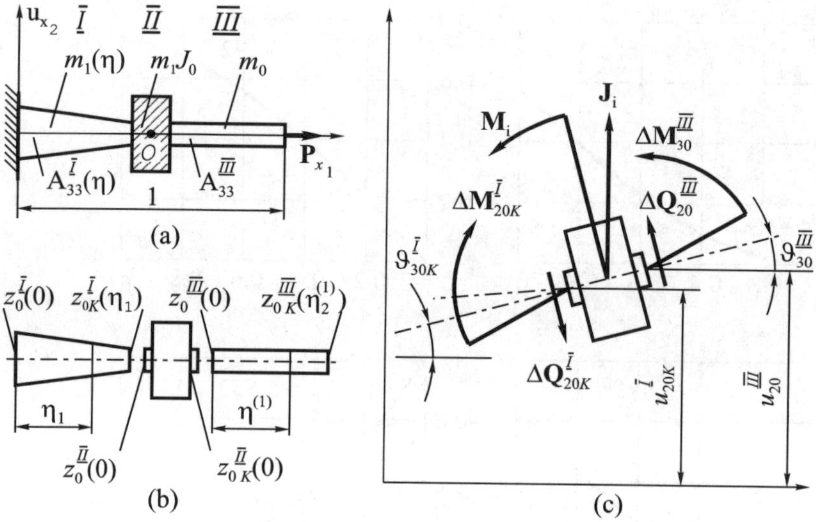

Fig. 7.10.

The rod is subjected to a tensile force P_{x_1}; hence, $Q_{10} = P_{x_1}$. Our goal is to study the vibrations of the rod in the plane of drawing. Therefore, we can use the system of equations (7.17)–(7.20), which, on rearrangement, gives the vector equation (7.82):

$$\mathbf{Z}_0' + \mathrm{B}(\eta,\, \lambda)\, \mathbf{Z}_0 = 0\,, \tag{7.89}$$

where

$$\mathbf{Z}_0 = \begin{bmatrix} \Delta Q_2 \\ \Delta M_3 \\ \vartheta_3 \\ u_2 \end{bmatrix}, \qquad \mathrm{B}(\eta,\, \lambda) = \begin{bmatrix} 0 & \dfrac{Q_{10}}{A_{33}} & 0 & n_1\lambda^2 \\ 1 & 0 & \lambda^2 J_{33} & 0 \\ 0 & -\dfrac{1}{A_{33}} & 0 & 0 \\ 0 & 0 & -1 & 0 \end{bmatrix}.$$

Equation (7.89) is valid both for rods of variable cross section (the segment *I* in Fig. 7.10*b*) and constant cross section (the segment *II*). For the segment *I*, the parameters n_1, A_{33}, and J_{33} depend on the coordinate η, and for the segment *II*, these parameters are independent of η. The dimensionless parameters for the segment *III* are as follows:

$$\tilde{n}_1^{\mathrm{III}} = \frac{m_0}{m_1(0)}\,, \qquad \tilde{A}_{33}^{\mathrm{III}} = \frac{A_{33}^{\mathrm{III}}}{A_{33}^{\mathrm{III}}(0)}\,, \qquad \tilde{J}_{33}^{\mathrm{III}} = \frac{J_{x_0}^{\mathrm{III}}}{F_1(0)\,l^2}\,.$$

The amplitude values of the components of the vectors $\mathbf{Z}_0^{(j)}$ can be found from the system of equations

$$\frac{\mathrm{d}\mathbf{Z}_0^{\mathrm{I}}}{\mathrm{d}\eta} + \mathrm{B}^{\mathrm{I}}(\eta,\,\lambda)\,\mathbf{Z}_0^{\mathrm{I}} = 0\,, \qquad 0 \le \eta \le \eta_1\,, \tag{7.90}$$

$$\frac{\mathrm{d}\mathbf{Z}_0^{\mathrm{III}}}{\mathrm{d}\eta} + \mathrm{B}^{\mathrm{III}}(\eta,\,\lambda)\,\mathbf{Z}_0^{\mathrm{III}} = 0\,, \qquad 0 \le \eta \le \eta_2^{(1)}\,, \tag{7.91}$$

where $\eta_2^{(1)} = 1 - \eta_1$ and

$$\mathrm{B}^{\mathrm{I}} = \begin{bmatrix} 0 & \dfrac{Q_{10}}{A_{33}} & 0 & n_1\lambda^2 \\ 1 & 0 & \lambda^2 J_{33} & 0 \\ 0 & -\dfrac{1}{A_{33}} & 0 & 0 \\ 0 & 0 & -1 & 0 \end{bmatrix},$$

$$\mathrm{B}^{\mathrm{III}} = \begin{bmatrix} 0 & \dfrac{Q_{10}}{A_{33}^{\mathrm{III}}} & 0 & n_1^{\mathrm{III}}\lambda^2 \\ 1 & 0 & \lambda^2 J_{33}^{\mathrm{III}} & 0 \\ 0 & -\dfrac{1}{A_{33}^{\mathrm{III}}} & 0 & 0 \\ 0 & 0 & -1 & 0 \end{bmatrix}.$$

For each of equations (7.90) and (7.91), we have

$$\mathbf{Z}_0^{\mathrm{I}}(\eta) = \mathrm{K}^{\mathrm{I}}(\eta)\,\mathbf{Z}_{00}^{\mathrm{I}}(0)\,, \qquad 0 \le \eta \le \eta_1\,, \tag{7.92}$$

$$\mathbf{Z}_0^{\mathrm{III}}(\eta) = \mathrm{K}^{\mathrm{III}}(\eta)\,\mathbf{Z}_{00}^{\mathrm{III}}(0)\,, \qquad 0 \le \eta \le \eta_2^{(1)}\,, \tag{7.93}$$

where $\mathbf{Z}_{00}^{\mathrm{I}} = \mathbf{Z}_0^{\mathrm{I}}(0)$ and $\mathbf{Z}_{00}^{\mathrm{III}} = \mathbf{Z}_0^{\mathrm{III}}(0)$ (Fig. 7.10*b*). It follows from (7.92) and (7.93) that

$$\mathbf{Z}_{0\mathrm{K}}^{\mathrm{I}}(\eta_1) = \mathrm{K}^{\mathrm{I}}(\eta_1)\,\mathbf{Z}_0^{\mathrm{I}}(0)\,, \tag{7.94}$$

$$\mathbf{Z}_{0\mathrm{K}}^{\mathrm{III}}(\eta_2^{(1)}) = \mathrm{K}^{\mathrm{III}}(\eta_2^{(1)})\,\mathbf{Z}_0^{\mathrm{III}}(0)\,, \tag{7.95}$$

where $\mathrm{K}^{\mathrm{I}}(\eta_1)$ and $\mathrm{K}^{\mathrm{III}}(\eta_2^{(1)})$ are the transition matrices for segments *I* and *III*. Recall that the segment *I* is of variable cross section and the segment *III* is of constant cross section.

Let us obtain the transition matrix for the segment *II*. Assuming that the dimensions of the concentrated mass are small as compared to the rod length and taking into consideration the forces of inertia, we get (Fig. 7.10*c*)

$$\Delta Q_{20}^{\mathrm{III}} = \frac{\Delta Q_{20K}^{\mathrm{I}}}{A_{33}^{0}} - \beta_1 \lambda_0^2 u_{20K}^{\mathrm{I}} \,,$$
$$\Delta M_{30}^{\mathrm{III}} = \frac{\Delta M_{30K}^{\mathrm{I}}}{J_{33}^{0}} - \beta_2 \lambda_0^2 \vartheta_{30K}^{\mathrm{I}} \,,$$
$$\vartheta_{30}^{\mathrm{III}} = \vartheta_{30K}^{\mathrm{I}} \,,$$
$$u_{20}^{\mathrm{III}} = u_{20K}^{\mathrm{I}} \,, \tag{7.96}$$

where

$$\beta_1 = \frac{m}{m_1(0) l \tilde{A}_{33}} \,, \qquad \beta_2 = \frac{J_0}{m_1(0) l^3 \tilde{A}_{33}} \,.$$

In vector–matrix notation, the system of equations (7.96) reads

$$\mathbf{Z}_0^{\mathrm{III}}(0) = \mathrm{K}^{\mathrm{II}} \mathbf{Z}_{0\mathrm{K}}^{\mathrm{I}}(\eta_1) \,, \tag{7.97}$$

where K^{II} is the transition matrix for the segment bearing the concentrated mass,

$$\mathrm{K}^{\mathrm{II}} = \begin{bmatrix} \frac{1}{A_{33}} & 0 & 0 & -\beta_1 \lambda_0^2 \\ 0 & \frac{1}{A_{33}} & -\beta_2 \lambda_0^2 & 0 \\ 0 & 0 & 1 & 0 \\ 0 & 0 & 0 & 1 \end{bmatrix} .$$

Now let us obtain a relationship between the vector $\mathbf{Z}_{00}^{\mathrm{I}}$ and the vector $\mathbf{Z}_{0K}^{\mathrm{III}}$. Using (7.94), (7.95) and (7.97), we get

$$\mathbf{Z}_{0K}^{\mathrm{III}} = \mathrm{K}^{\mathrm{III}} \mathrm{K}^{\mathrm{II}} \mathrm{K}^{\mathrm{I}} \mathbf{Z}_0^{\mathrm{I}} = \mathrm{K} \mathbf{Z}_0^{\mathrm{I}} \,. \tag{7.98}$$

In the example under consideration, the components of the vectors $\mathbf{Z}_0^{\mathrm{I}}$ and $\mathbf{Z}_{0K}^{\mathrm{III}}$ must satisfy the following boundary conditions:

(1) at $\eta = 0 \,, \quad u_{20}^{\mathrm{I}} = \vartheta_{30}^{\mathrm{I}} = 0 \,;$

(2) at $\eta^{(1)} = \eta_2^{(1)} \,, \quad \Delta Q_{20K}^{\mathrm{III}} = \Delta M_{30K}^{\mathrm{III}} = 0 \,.$

It follows from the boundary conditions at $\eta_2^{(1)}$ that

$$k_{11}\, \Delta Q_{20}^{\mathrm{I}} + k_{12}\, \Delta M_{30}^{\mathrm{I}} = 0 \,,$$
$$k_{12}\, \Delta Q_{20}^{\mathrm{I}} + k_{22}\, \Delta M_{30}^{\mathrm{I}} = 0 \,, \tag{7.99}$$

where k_{ij} are the elements of the matrix K. The dimensionless frequencies can be found through the condition $D = 0$, where D is the determinant of the coefficient of (7.99).

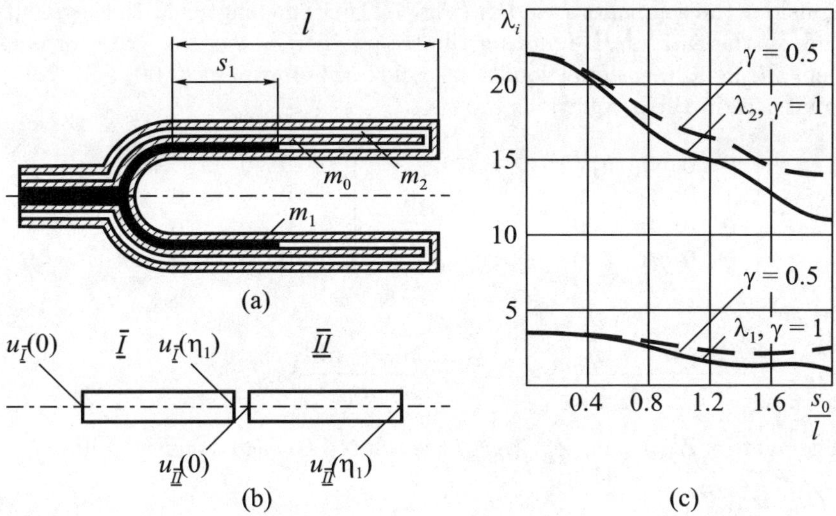

Fig. 7.11.

As an example, determine the frequencies of a tuning fork shown in Fig. 7.11*a*. These frequencies can be obtained as follows. The prongs of the fork have cavities (channels) filled with two liquids (for example, water and mercury). Let the masses of liquids be m_1 and m_2. The natural frequencies of a prong depend on the lengths of the liquid columns. To find the feasible range of change for, say, the first frequency p_1, $p_{11} \le p_1 \le p_{12}$, it is sufficient to consider the following limiting cases (Fig. 7.11*a*):

(1) $s_1 = 0$ and (2) $s_1 = 2l$, where s_1 is the length of the mercury column. We have

$$p_{11} = 3.516\sqrt{\frac{EJ_x}{l^4(m_0 + 2m_1)}}\,, \qquad p_{12} = 3.516\sqrt{\frac{EJ_x}{l^4(m_0 + 2m_2)}}\,,$$

where m_0 is the mass per unit length of the rod.

The ratio between the frequencies is

$$\beta_0 = \frac{p_{11}}{p_{12}} = \sqrt{\frac{m_0 + 2m_2}{m_0 + 2m_1}} = \sqrt{\frac{\rho_0 + 2\rho_2\,\dfrac{F}{F_0}}{\rho_0 + 2\rho_1\,\dfrac{F}{F_0}}}\,,$$

where ρ_0, ρ_1, and ρ_2 are the densities of the rod, mercury and water, respectively, F is the cross-sectional area of the channel, and F_0 is the cross-sectional area of the rod. For $F/F_0 = 2$, the ratio of frequencies is 2.28. This means that by varying the length of the liquid columns, we can obtain the first frequency more than twice as large as its natural frequency.

Consider the segments I and II (Fig. 7.11b). The lengths of the segments depend on the ratio s_1/l. Ignoring the rotary inertia, we can write for each segment the nondimensional equations of flexural vibrations (7.90) and (7.91). The matrices in these equations are

$$\mathrm{B}^{\mathrm{I}} = \begin{bmatrix} 0 & 0 & 0 & n_1^{(1)}\lambda^2 \\ 1 & 0 & 0 & 0 \\ 0 & 0 & 0 & 0 \\ 0 & 0 & 0 & 0 \end{bmatrix}, \qquad \mathrm{B}^{\mathrm{II}} = \begin{bmatrix} 0 & 0 & 0 & n_1^{(2)}\lambda^2 \\ 0 & 0 & 0 & 0 \\ 0 & 0 & 0 & 0 \\ 0 & 0 & 0 & 0 \end{bmatrix},$$

where

$$n_1^{(1)} = \frac{m_0 + m_1 + m_2}{m_0}, \qquad n_1^{(2)} = \frac{m_0 + 2m_1}{m_0}.$$

The vectors $\mathbf{Z}_0^{\mathrm{I}}(0)$ and $\mathbf{Z}_{0K}^{\mathrm{II}}(\eta_2^{(1)})$ are related to each other as follows:

$$\mathbf{Z}_{0K}^{\mathrm{II}} = \mathrm{K}^{\mathrm{I}}\mathbf{Z}_0^{\mathrm{I}}.$$

In view of the boundary conditions, the equations to be solved for the frequencies λ_j are

$$k_{33}k_{44} - k_{34}k_{43} = 0. \tag{7.100}$$

By solving the equation numerically, we get the graphs of the first two dimensionless frequencies λ_1 and λ_2 versus s_1/l for two values of $\gamma = F/F_0$ (Fig. 7.11c).

In Sect. 4.1, a method that implies no matrix multiplication was discussed. The method is based on the use of the generalized functions $H(\eta)$ and $\delta(\eta)$. The method can be used for the analysis of flexural vibrations especially when the rod consists of many segments of different elastic properties. For such problems, because of a great many of matrix multiplications, the use of the traditional method of initial parameters entails severe computational problems.

7.3 Small Vibrations About a Stationary Motion

7.3.1 Equations of Small Vibrations of a Straight Rod which is in a Longitudinal Motion

The general nonlinear equations of motion for a space-curved rod (Fig. 2.4) which is forced to rotate with angular velocity $\boldsymbol{\omega}_0$ and move along its axis with velocity $\mathbf{w}_0$ are presented in Sect. 2.1. The equations governing a stationary motion (the shape of the rod axis remains the same) are presented in Sect. 2.4. The equations of small vibrations of about a stationary motion are presented in Sect. 3.4. These equations apply when the rod axis is a spatial curve. The equations of small vibrations of the rods about rectilinear motion (for example, a belt transmission (Fig. 0.5)) can be obtained by direct specialization of the general equations. Nevertheless, to outline some peculiarities, it would be relevant to derive the equations of small vibrations of a rod in rectilinear stationary motion anew.

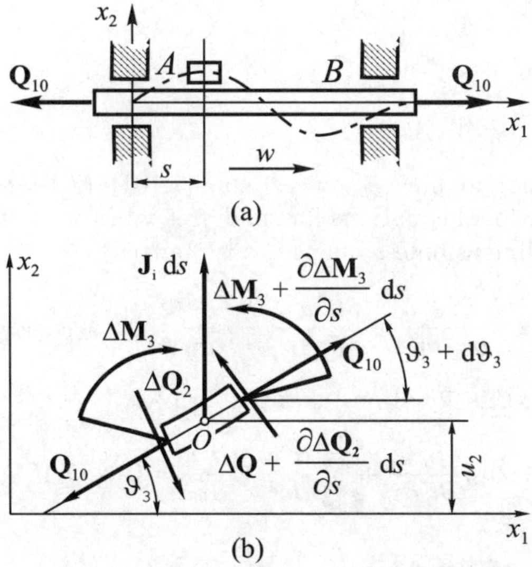

Fig. 7.12.

First, consider the simplest case: a rod is in an in-plane oscillatory motion (Fig. 7.12*a*).

Many engineering components can be described by the equations obtained for such an ideal system; for instance, vibrations of saw bands, belt transmissions, coilers, etc. Assume that the rotary inertia and the inertia of longitudinal motion can be neglected. We will use the Eulerian variables. In terms of these variables, the inertia of an element of the moving rod (Fig. 7.12*b*) reads

$$J_{\mathrm{i}} = -m_0 \frac{\mathrm{d}^2 u}{\mathrm{d}t^2} = -m_0 \left(\frac{\partial^2 u}{\partial t^2} + 2w \frac{\partial^2 u}{\partial s\, \partial t} + w^2 \frac{\partial^2 u}{\partial s^2} + \frac{\partial w}{\partial t} \frac{\partial u}{\partial s} \right). \quad (7.101)$$

Summation of forces in the x_2-direction gives

$$J_{\mathrm{i}} + \frac{\partial \Delta Q_2}{\partial s} + Q_{10} \frac{\mathrm{d}\vartheta_3}{\mathrm{d}s} = 0\,,$$

or

$$m_0 \frac{\mathrm{d}^2 u}{\mathrm{d}t^2} - \frac{\partial \Delta Q_2}{\partial s} - Q_{10} \frac{\partial^2 u_2}{\partial s^2} = 0\,. \quad (7.102)$$

From summation of moments about the center of gravity (the point O in Fig. 7.12*b*), we get

$$\frac{\partial \Delta M_3}{\partial s} + \Delta Q_2 = 0\,, \qquad \Delta M_3 = A_{33}\,\frac{\partial \vartheta_3}{\partial s}\,. \tag{7.103}$$

Now we introduce the following dimensionless quantities:

$$\eta = \frac{s}{l}\,, \qquad \tilde{u} = \frac{u}{l}\,, \qquad \Delta\tilde{Q}_2 = \frac{\Delta Q_2 l^2}{A_{33}}\,, \qquad \tilde{Q}_{10} = \frac{Q_{10} l^2}{A_{33}}\,,$$

$$\Delta\tilde{M}_3 = \frac{\Delta M_3 l}{A_{33}}\,, \qquad t = \frac{\tau}{p_0}\,, \qquad p_0 = \sqrt{\frac{A_{33}}{m_0 l^4}}\,,$$

$$\tilde{w} = \frac{w}{p_0 l}\,, \qquad \tilde{A}_{33} = \frac{A_{33}}{A_{33}(0)}\,.$$

For a rod of constant cross section we have $\tilde{A}_{33} = 1$. Using (7.101)–(7.103), we obtain the following nondimensional equations for small free vibrations of a rod (a tilde in the notation of dimensionless quantities is omitted):

$$\frac{\partial^4 u}{\partial \eta^4} - \frac{\partial}{\partial \eta}\left(Q_1\,\frac{\partial u}{\partial \eta}\right) + \frac{\partial^2 u}{\partial \tau^2} + 2w\,\frac{\partial^2 u}{\partial \tau\,\partial \eta} + w^2\,\frac{\partial^2 u}{\partial \eta^2} + \frac{\partial w}{\partial \tau}\,\frac{\partial u}{\partial \eta} = 0\,. \tag{7.104}$$

In particular, for $w = w_0 = \text{const}$, this equation takes the form

$$L(u) = \frac{\partial^4 u}{\partial \eta^4} - \frac{\partial}{\partial \eta}\left(Q_1\,\frac{\partial u}{\partial \eta}\right) + 2w_0\,\frac{\partial^2 u}{\partial \eta\,\partial \tau} + w_0^2\,\frac{\partial^2 u}{\partial \eta^2} + \frac{\partial^2 u}{\partial \tau^2} = 0\,. \tag{7.105}$$

7.3.2 Frequencies and Modes of Vibrations

The key feature of equation (7.105) is that due to the Coriolis force, there is a term containing time derivative of odd order. This fact complicates the determination of the eigenvalues of the boundary value problem. The presence of odd-order derivatives suggests that the problem may be non-conservative and consequently the eigenvalues are complex numbers. Sometimes it can be easily found out whether a given system is conservative or not. For example, in the case shown in Fig. 7.12*b*, the problem is conservative because the input, in unit time, of the kinetic energy $(m_0 w^2/2)$ at the cross section A equals the kinetic energy of elements of the rod leaving the cross section B. Therefore, the total energy of the segment AB of the vibrating rod remains the same. Recall that the eigenvalues λ are real numbers in the case of a conservative problem. We seek a solution to (7.105) in the form $u = u_0(\eta)\,\mathrm{e}^{\mathrm{i}\,\lambda\tau}$. Substituting this into (7.105), we get

$$\frac{\mathrm{d}^4 u_0}{\mathrm{d}\eta^4} + 2w_0\,\mathrm{i}\,\lambda\,\frac{\mathrm{d}u_0}{\mathrm{d}\eta} - (Q_{10} - w_0^2)\,\frac{\mathrm{d}^2 u_0}{\mathrm{d}\eta_2} - \lambda^2 u_0 = 0\,. \tag{7.106}$$

The function u_0 is a complex function; hence, we can represent it in the form

$$u_0 = u_0^{(1)} + \mathrm{i}\,u_0^{(2)}\,. \tag{7.107}$$

The real and imaginary parts of the left-hand side of (7.106) are zero, that is,

$$u_0^{(1)\mathrm{IV}} - 2w_0\lambda u_0^{(2)\prime} - (Q_{10} - w_0^2)\, u_0^{(1)\prime\prime} - \lambda^2 u_0^{(1)} = 0\,,$$
$$u_0^{(2)\mathrm{IV}} + 2w_0\lambda u_0^{(1)\prime} - (Q_{10} - w_0^2)\, u_0^{(2)\prime\prime} - \lambda^2 u_0^{(2)} = 0\,. \qquad (7.108)$$

Recall that the superscript IV denotes the fourth derivative with respect to η.

These equations can be written as a set of eight equations of the first order. In its turn, this set can be represented as follows:

$$\mathbf{Z}' + \mathrm{B}\mathbf{Z} = 0\,, \qquad (7.109)$$

where

$$\mathbf{Z} = (\Delta Q_2^{(1)},\, \Delta Q_2^{(2)},\, \Delta M_3^{(1)},\, \Delta M_3^{(2)},\, \vartheta_3^{(1)},\, \vartheta_3^{(2)},\, u_0^{(1)},\, u_0^{(2)})^{\mathrm{T}}$$

and

$$\mathrm{B} = \begin{bmatrix} 0 & 0 & -(Q_{10} - w_0^2) & 0 & 0 & -2w_0\lambda & -\lambda^2 & 0 \\ 0 & 0 & 0 & -(Q_{10} - w_0^2) & 2w_0\lambda & 0 & 0 & -\lambda^2 \\ 1 & 0 & 0 & 0 & 0 & 0 & 0 & 0 \\ 0 & 1 & 0 & 0 & 0 & 0 & 0 & 0 \\ 0 & 0 & -1 & 0 & 0 & 0 & 0 & 0 \\ 0 & 0 & 0 & -1 & 0 & 0 & 0 & 0 \\ 0 & 0 & 0 & 0 & -1 & 0 & 0 & 0 \\ 0 & 0 & 0 & 0 & 0 & -1 & 0 & 0 \end{bmatrix}.$$

The components of the vector $\mathbf{Z}$ are complex functions, which can be represented in the form (7.107).

For fixed values of λ and w_0, we get the solution to (7.109) in the form

$$\mathbf{Z} = \mathrm{K}(\eta)\mathbf{C}\,. \qquad (7.110)$$

This solution must satisfy the following boundary conditions:

(1) at $\eta = 0$, $u_0^{(1)} = u_0^{(2)} = 0$ and $\vartheta_3^{(1)} = \vartheta_3^{(2)} = 0$;

(2) at $\eta = 1$, $u_0^{(1)} = u_0^{(2)} = 0$ and $\vartheta_3^{(1)} = \vartheta_3^{(2)} = 0$.

The boundary conditions at $\eta = 0$ give for the components of the vector $\mathbf{C}$: $c_5 = c_6 = c_7 = c_8 = 0$. The boundary conditions at $\eta = 1$ lead to the system of four homogeneous equations

$$k_{11}c_1 + k_{12}c_2 + \ldots + k_{14}c_4 = 0\,,$$
$$\ldots$$
$$k_{41}c_1 + k_{42}c_2 + \ldots + k_{44}c_4 = 0\,. \qquad (7.111)$$

For a fixed Q_{10}, the dimensionless frequencies can be found from the equation

$$D = \det \begin{bmatrix} k_{15}(1, \lambda) & \cdots & k_{18}(1, \lambda) \\ \cdots & \cdots & \cdots \\ k_{45}(1, \lambda) & \cdots & k_{48}(1, \lambda) \end{bmatrix} = 0\,. \tag{7.112}$$

When finding frequencies for an equation that does not contain odd-order time derivatives (for example, (7.66)), the sign of the determinant, say (7.75), depends on λ ($\lambda = \lambda_0^2$). In this case, the sought-for λ_j can be easily found numerically. Qualitatively, the graph of D versus λ is shown in Fig. 4.10 (Sect. 4.1). For an equation of small vibrations with odd-order time derivatives, the determinant D given by (7.112) does not change its sign under change of λ. For this case, the surface D is qualitatively shown in Fig. 4.12. Such a relationship between D and λ seriously complicates the determination of eigenvalues λ_k. For each λ_j found from (7.111), we represent c_1, c_2, and c_3 in terms of c_4 as follows:

$$c_1^{(k)} = \gamma_{1k} c_4^{(k)}\,, \qquad c_2^{(k)} = \gamma_{2k} c_4^{(k)}\,, \qquad c_3^{(k)} = \gamma_{3k} c_4^{(k)}\,, \quad k = 1, 2, 3, \ldots\,.$$

Denote the mode of vibration by φ_k. This mode is given by the real part of u_0. Then, setting $c_4^{(k)} = 1$, we arrive at

$$u_0^{(1)(k)} = \varphi^{(k)} = k_{71}(\lambda_k, \eta)\,\gamma_{1k} + k_{72}(\lambda_k, \eta)\,\gamma_{2k} + k_{73}(\lambda_k, \eta)\,\gamma_{3k} + k_{74}(\lambda_k, \eta)\,. \tag{7.113}$$

7.3.3 Approximate Calculation of the Frequencies

To calculate approximately the frequencies of vibrations for a moving rod, one can use the principle of virtual displacements. We seek a solution to (7.105) in the form

$$u = \sum_{j=1}^{n} f^{(j)}(\tau)\varphi^{(j)}(\eta)\,, \tag{7.114}$$

where the functions $\varphi^{(j)}(\eta)$ are assumed to satisfy the boundary conditions of the problem. As such functions, we can take the functions that satisfy equation (7.66),

$$\frac{\partial^4 u}{\partial \eta^4} + \frac{\partial^2 u}{\partial \tau^2} = 0\,.$$

Let the virtual displacements be of the form

$$\delta u = \sum_{j=1}^{n} \delta A_j \varphi^{(j)}(\eta)\,.$$

According to the principle of virtual displacements, we obtain the system of equations

$$\int_0^1 L(u)\varphi^{(j)}\,\mathrm{d}\eta = 0\,, \qquad j = 1, 2, \ldots, n\,,$$

or, in expanded form,

$$a_{ii}\ddot{f}^{(i)} + \sum_{j=1}^{n} b_{ij}\dot{f}^{(j)} + \sum_{j=1}^{n} c_{ij}f^{(j)} = 0\,, \tag{7.115}$$

where

$$a_{ii} = \int_0^1 (\varphi^{(j)})^2\,\mathrm{d}\eta\,, \qquad b_{ij} = 2w_0\int_0^1 \varphi^{(j)\prime}\varphi^{(i)}\,\mathrm{d}\eta\,,$$

$$c_{ij} = w_0^2\int_0^1 \varphi^{(j)\prime\prime}\varphi^{(i)}\,\mathrm{d}\eta + \int_0^1 \varphi^{(j)\mathrm{IV}}\varphi^{(i)}\,\mathrm{d}\eta\,.$$

Integrating by parts, one can easily show that

$$b_{ij} = -b_{ji}\,, \qquad b_{ii} = 0\,, \qquad c_{ij} = c_{ji}\,.$$

The system of equations (7.115) can be represented as follows:

$$\ddot{\mathbf{f}} + \mathrm{B}\dot{\mathbf{f}} + \mathrm{C}\mathbf{f} = 0\,, \tag{7.116}$$

where

$$\mathbf{f} = \begin{bmatrix} f^{(1)} \\ f^{(2)} \\ \vdots \\ f^{(n)} \end{bmatrix}, \qquad \mathrm{B} = \left[\frac{2w_0 b_{ij}}{a_{ii}}\right], \qquad \mathrm{C} = \left[\frac{c_{ij}}{a_{ii}}\right].$$

We seek a solution to (7.116) in the form $\mathbf{f} = \mathbf{f}_0\,\mathrm{e}^{\mathrm{i}\lambda\tau}$. The corresponding characteristic equation is

$$\det\left[-\lambda^2\mathrm{E} + \mathrm{i}\,\mathrm{B}\lambda + \mathrm{C}\right] = 0\,.$$

The matrix $[-\lambda^2\mathrm{E} + \mathrm{i}\,\mathrm{B}\lambda + \mathrm{C}]$ is a Hermitian matrix. Its eigenvalues, μ_j, are real; hence, λ_j are also real. For example, setting $n = 2$ in (7.114), we see that the two dimensionless frequencies satisfy the equation

$$D = \det\begin{bmatrix} \dfrac{c_{11}}{a_{11}} - \lambda^2 & \dfrac{c_{12}}{a_{11}} - \mathrm{i}\,\dfrac{b_{12}}{a_{11}}\,\lambda \\ \dfrac{c_{12}}{a_{22}} + \mathrm{i}\,\dfrac{b_{12}}{a_{22}}\,\lambda & \dfrac{c_{22}}{a_{22}} - \lambda^2 \end{bmatrix} = 0\,;$$

hence,

$$(c_{11} - a_{11}\lambda^2)\,(c_{22} - a_{22}\lambda^2) - (c_{12} - \mathrm{i}\,b_{12}\lambda)\,(c_{12} + \mathrm{i}\,b_{12}\lambda) = 0\,,$$

This gives

$$\lambda_{1,\,2} = \sqrt{\frac{a_{22}c_{11} + a_{11}c_{22} + b_{12} \mp \sqrt{d}}{2a_{11}a_{22}}}\,,$$

where

$$d = a_{22}c_{11} + a_{11}c_{22} + b_{12}^2 - 4a_{11}a_{22}(c_{11}c_{22} - c_{12}^2)\,.$$

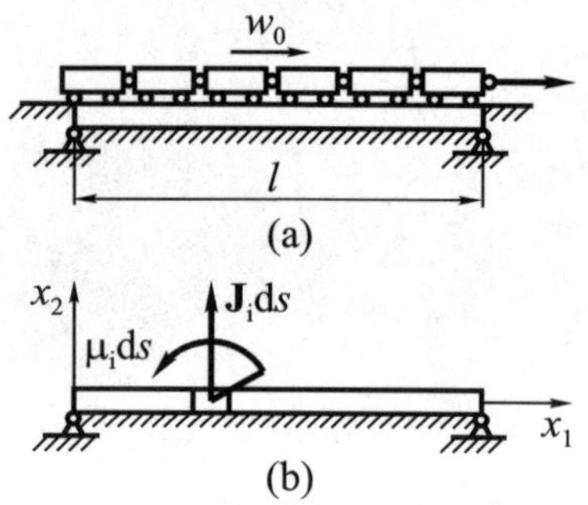

Fig. 7.13.

As an example, consider vibrations of a railway track (Fig. 7.13a). Such a track can be treated as a rod on an elastic foundation. There is a train of infinite length moving along the track. A rod of zero bending stiffness can be used as a model of the train. Such an idealization is justified if the wheel base for each car of the train is much less than the length of the rod l. The load on the track due to the train may be thought of as being continuously distributed. Each car of the train acts upon the track with two contact forces, which can be replaced by the resultant force, J_{i}, and moment μ_{i} (Fig. 7.13b)

$$J_{\mathrm{i}} = -m\,\frac{\mathrm{d}^2 u_2}{\mathrm{d}t^2}\,, \qquad \mu_{\mathrm{i}} = -J_0\,\frac{\mathrm{d}^2 \vartheta_3}{\mathrm{d}t^2}\,. \tag{7.117}$$

here, m and J_0 are the mass of the car and the moment of inertia of the car relative to the axis perpendicular to the plane of drawing at the center of gravity. Denote the velocity of the train by w_0. Using the Eulerian variables, we can represent (7.117) in nondimensional form as follows:

$$J_{\mathrm{i}} = -n_{11}\left(\frac{\partial^2 u_2}{\partial \tau^2} + 2w_0\,\frac{\partial^2 u_2}{\partial \tau\,\partial \eta} + w_0^2\,\frac{\partial^2 u_2}{\partial \eta^2}\right),$$

$$\mu_{\mathrm{i}} = -J_0^0\left(\frac{\partial^3 u_2}{\partial \eta\,\partial \tau^2} + 2w_0\,\frac{\partial^3 u_2}{\partial \tau\,\partial \eta^2} + w_0^2\,\frac{\partial^3 u_2}{\partial \eta^3}\right), \qquad \vartheta_3 = \frac{\partial u_2}{\partial \eta}\,,$$

where

$$n_{11} = \frac{m}{m_0 l}\,, \qquad J_0^0 = \frac{J_0}{m_0 l^3}\,.$$

It is worth noting that n_{11} and J_0^0 are dimensionless coefficients while w_0 is the dimensionless velocity.

Using (7.101)–(7.103) and ignoring the rotary inertia, we obtain the following equations of vibrations for a rod subjected to a traveling load:

$$L = (1+n_{11})\,\frac{\partial^2 u_2}{\partial \tau^2} + (1 - J_0^0 w_0^2)\,\frac{\partial^4 u_2}{\partial \eta^4} + 2 w_0 n_{11}\,\frac{\partial^2 u_2}{\partial \tau\,\partial \eta}$$
$$- 2 w_0 J_0^0\,\frac{\partial^4 u_2}{\partial \tau\,\partial \eta^3} - J_0^0\,\frac{\partial^4 u_2}{\partial \eta^2\,\partial \tau^2} + n_{11} w_0^2\,\frac{\partial^2 u_2}{\partial \eta^2} + k u_2 = 0\,. \qquad (7.118)$$

Consider the virtual displacements in the form

$$\delta u_2 = \sum_1^n \delta a_{(i)} u_2^{(i)}\,,$$

where the functions $u_2^{(i)}$ satisfy the boundary conditions of the problem (for example, for a rod with simply supported ends, we have $u_2^{(i)} = \sin(i\,\pi\eta)$). Then, the two-term approximate solution to (7.118) reads

$$u_2 = f^{(1)} u_2^{(1)} + f^{(2)} u_2^{(2)}\,.$$

Using the principle of virtual displacements, we get

$$\int_0^1 L(u_2^{(1)}, u_2^{(2)})\, u_2^{(1)}\, \mathrm{d}\eta = 0\,, \qquad \int_0^1 L(u_2^{(1)}, u_2^{(2)})\, u_2^{(2)}\, \mathrm{d}\eta = 0\,.$$

On rearrangement, this gives

$$a_{11}\ddot f_1 + a_{12}\ddot f_2 + b_{11}\dot f_1 + b_{12}\dot f_2 + c_{11} f_1 + c_{12} f_2 = 0\,,$$
$$a_{21}\ddot f_1 + a_{22}\ddot f_2 + b_{21}\dot f_1 + b_{22}\dot f_2 + c_{21} f_1 + c_{22} f_2 = 0\,,$$

where

$$a_{11} = \frac{1}{2}\left[(1+n_{11}) + \pi^2 J_0^0\right]\,, \qquad a_{12} = 0\,,$$

$$a_{21} = 0\,, \qquad a_{22} = \frac{1}{2}\left[(1+n_{11}) + 4\pi^2 J_0^0\right]\,,$$

$$b_{11} = 0\,, \qquad b_{12} = -\frac{8}{3}\, w_0 (n_{11} + 4\pi^2 J_0^0)\,,$$

$$b_{21} = \frac{8}{3}\, w_0 (n_{11} + \pi^2 J_0^0)\,, \qquad b_{22} = 0\,,$$

$$c_{11} = \frac{1}{2}\left[(1 - w_0^2 J_0^0)\,\pi^4 - n_{11}\pi^2 w_0^2 + k\right]\,, \qquad c_{12} = 0\,,$$

$$c_{21} = 0\,, \qquad c_{22} = \frac{1}{2}\left[16\pi^2 (1 - w_0^2 J_0^0) - n_{11}\, 4\pi^2 w_0^2 + k\right]\,.$$

Setting $f_1 = f_{10}\,e^{\lambda\tau}$ and $f_2 = f_{20}\,e^{\lambda\tau}$ and disregarding the Coriolis force, we obtain the characteristic equation

$$a_0\lambda^4 + a_2\lambda^2 + a_4 = 0\,,$$

where

$$a_0 = a_{11}a_{22}\,, \qquad a_2 = a_{11}c_{22} + a_{22}c_{11}\,, \qquad a_4 = c_{11}c_{22}\,.$$

Finally, we arrive at

$$\lambda_{1,2} = \pm\mathrm{i}\sqrt{\frac{c_{11}}{a_{11}}}\,, \qquad \lambda_{3,4} = \pm\mathrm{i}\sqrt{\frac{c_{22}}{a_{22}}}\,,$$

and thus the frequencies of vibrations are

$$\beta_1 = \sqrt{\frac{c_{11}}{a_{11}}}\,, \qquad \beta_2 = \sqrt{\frac{c_{22}}{a_{22}}}\,.$$

7.4 Free Vibrations of a Rotating Rod

7.4.1 Equations of Free Vibrations

The vector equations (3.38)–(3.40) governing small vibrations of a rotating rod of circular cross section were obtained in Sect.3.3. For $w_0 = 0$ and $\omega_0 \neq 0$, we obtained equations (3.77). These equations were written in terms of projections in a coordinate frame attached to the rod.

Fig. 7.14.

Now we derive, on the basis of those equations, the equations of small flexural vibrations of a straight rod (Fig. 7.14). To do this, we put

$$\Delta T_1 = \Delta T_2 = 0\,, \qquad æ_{10} \approx 0\,, \qquad æ_{20} = æ_{30} = 0\,,$$
$$Q_{10} \neq 0\,, \qquad Q_{20} = Q_{30} = 0\,, \qquad M_{10} \neq 0\,, \qquad M_{20} = M_{30} = 0\,,$$
$$A_{22} = A_{33}\,, \qquad J_{22} = J_{33}\,.$$

in (3.38)–(3.40) and (3.70). As a result, we get

$$n_1 \frac{\partial^2 u_2}{\partial \tau^2} - \frac{\partial \Delta Q_2}{\partial \eta} - \frac{Q_{10}}{A_{33}} \Delta M_3 = 0\,,$$
$$n_1 \frac{\partial^2 u_3}{\partial \tau^2} - \frac{\partial \Delta Q_3}{\partial \eta} + \frac{Q_{10}}{A_{33}} \Delta M_2 = 0\,, \tag{7.119}$$
$$J_{22} \frac{\partial \vartheta_2}{\partial \tau^2} + J_{11}\omega_{10} \frac{\partial \vartheta_3}{\partial \tau} - \frac{\partial \Delta M_2}{\partial \eta} - \frac{M_{10}}{A_{33}} \Delta M_3 + \Delta Q_3 = 0\,,$$
$$J_{33} \frac{\partial^2 \vartheta_3}{\partial \tau^2} - J_{11}\omega_{10} \frac{\partial \vartheta_2}{\partial \tau} - \frac{\partial \Delta M_3}{\partial \eta} + \frac{M_{10}}{A_{33}} \Delta M_2 - \Delta Q_2 = 0\,, \tag{7.120}$$
$$\frac{\partial \vartheta_2}{\partial \eta} - \frac{\Delta M_2}{A_{33}} = 0\,,$$
$$\frac{\partial \vartheta_3}{\partial \eta} - \frac{\Delta M_3}{A_{33}} = 0\,, \tag{7.121}$$
$$\frac{\partial u_2}{\partial \eta} - \vartheta_3 = 0\,,$$
$$\frac{\partial u_3}{\partial \eta} + \vartheta_2 = 0\,. \tag{7.122}$$

Recall that the equations obtained in Sect. 2.1 and the equations of small vibrations of a rod rotating about its axis (Sect. 3.3) were developed with respect to the coordinate frame attached to the axial line of the conduit containing the rotating rod; at that, we supposed that the conduit is of zero inertia of rotation. Equations (7.120) contain the distributed gyroscopic moments

$$J_{11}\omega_{10} \frac{\partial \vartheta_j}{\partial \tau}\,, \quad j = 1, 2\,.$$

7.4.2 Exact Numerical Method for Determination of Frequencies

The system of equations (7.119)–(7.120) can be represented in the form of one vector equation

$$\mathbf{L} = \mathrm{A}^{(1)} \frac{\partial^2 \mathbf{Z}}{\partial \tau^2} + \frac{\partial \mathbf{Z}}{\partial \eta} + \mathrm{A}^{(2)} \mathbf{Z} + \mathrm{A}^{(3)} \frac{\partial \mathbf{Z}}{\partial \tau} = 0\,, \tag{7.123}$$

where

$$\mathbf{Z} = \begin{bmatrix} \Delta Q_2 \\ \Delta Q_3 \\ \Delta M_2 \\ \Delta M_3 \\ \vartheta_2 \\ \vartheta_3 \\ u_2 \\ u_3 \end{bmatrix}, \quad \mathrm{A}^{(1)} = \begin{bmatrix} 0 & 0 & 0 & 0 & 0 & 0 & -n_1 & 0 \\ 0 & 0 & 0 & 0 & 0 & 0 & 0 & -n_1 \\ 0 & 0 & 0 & 0 & -J_{22} & 0 & 0 & 0 \\ 0 & 0 & 0 & 0 & 0 & -J_{33} & 0 & 0 \\ 0 & 0 & 0 & 0 & 0 & 0 & 0 & 0 \\ 0 & 0 & 0 & 0 & 0 & 0 & 0 & 0 \\ 0 & 0 & 0 & 0 & 0 & 0 & 0 & 0 \end{bmatrix},$$

$$
\mathrm{A}^{(2)} = \begin{bmatrix}
0 & 0 & 0 & -\dfrac{Q_{10}}{A_{33}} & 0 & 0 & 0 & 0 \\
0 & 0 & -\dfrac{Q_{10}}{A_{33}} & 0 & 0 & 0 & 0 & \\
0 & -1 & 0 & \dfrac{M_{10}}{A_{33}} & 0 & 0 & 0 & 0 \\
1 & 0 & -\dfrac{M_{10}}{A_{33}} & 0 & 0 & 0 & 0 & \\
0 & 0 & -\dfrac{1}{A_{33}} & 0 & 0 & 0 & 0 & 0 \\
0 & 0 & 0 & -\dfrac{1}{A_{33}} & 0 & 0 & 0 & 0 \\
0 & 0 & 0 & 0 & 0 & -1 & 0 & 0 \\
0 & 0 & 0 & 0 & 1 & 0 & 0 & 0
\end{bmatrix},
$$

$$
\mathrm{A}^{(3)} = \begin{bmatrix}
0 & 0 & 0 & 0 & 0 & 0 & 0 & 0 \\
0 & 0 & 0 & 0 & 0 & 0 & 0 & 0 \\
0 & 0 & 0 & 0 & 0 & -\omega_{10} J_{11} & 0 & 0 \\
0 & 0 & 0 & 0 & \omega_{10} J_{11} & 0 & 0 & 0 \\
0 & 0 & 0 & 0 & 0 & 0 & 0 & 0 \\
0 & 0 & 0 & 0 & 0 & 0 & 0 & 0 \\
0 & 0 & 0 & 0 & 0 & 0 & 0 & 0 \\
0 & 0 & 0 & 0 & 0 & 0 & 0 & 0
\end{bmatrix}.
$$

Let

$$
\mathbf{Z} = \mathbf{Z}_0 \, \mathrm{e}^{\mathrm{i}\,\lambda\tau}, \qquad \mathbf{Z}_0 = \mathbf{Z}_{01} + \mathrm{i}\,\mathbf{Z}_{02},
$$

then (7.123) gives

$$
\mathbf{Z}_0' + \mathrm{B}(\eta, \lambda, \omega_0)\, \mathbf{Z}_0 = 0, \tag{7.124}
$$

where $\mathrm{B} = \mathrm{A}^{(2)} - \lambda^2 \mathrm{A}^{(1)} + \mathrm{i}\,\lambda \mathrm{A}^{(3)}$.

Due to the distributed gyroscopic moments, the determination of the eigenvalues by means of an exact numerical method becomes complicated. This is explained by the fact that the matrix B in (7.124) has complex elements.

For the real and imaginary parts of (7.124), we get

$$
\begin{aligned}
&\mathbf{Z}_{01}' + (\mathrm{A}^{(2)} - \lambda^2 \mathrm{A}^{(1)})\, \mathbf{Z}_{01} - \lambda \mathrm{A}^{(3)} \mathbf{Z}_{02} = 0, \\
&\mathbf{Z}_{02}' + (\mathrm{A}^{(2)} - \lambda^2 \mathrm{A}^{(1)})\, \mathbf{Z}_{02} + \lambda \mathrm{A}^{(3)} \mathbf{Z}_{01} = 0
\end{aligned}
$$

or

$$
\mathbf{Y}_0' + \mathrm{A}\mathbf{Y}_0 = 0, \tag{7.125}
$$

where

$$\mathbf{Z}_{01} = (\Delta Q_2^{(1)}, \Delta Q_3^{(1)}, \Delta M_2^{(1)}, \Delta M_3^{(1)}, \vartheta_2^{(1)}, \vartheta_3^{(1)}, u_2^{(1)}, u_3^{(1)})^{\mathrm{T}},$$

$$\mathbf{Z}_{02} = (\Delta Q_2^{(2)}, \Delta Q_3^{(2)}, \Delta M_2^{(2)}, \Delta M_3^{(2)}, \vartheta_2^{(2)}, \vartheta_3^{(2)}, u_2^{(2)}, u_3^{(2)})^{\mathrm{T}},$$

$$\mathbf{Y}_0 = \begin{bmatrix} \mathbf{Z}_{01} \\ \mathbf{Z}_{02} \end{bmatrix}, \qquad \mathrm{A}^{(1)} = \begin{bmatrix} \mathrm{A}^{(2)} - \lambda^2 \mathrm{A}^{(1)} & -\lambda \mathrm{A}^{(3)} \\ \lambda \mathrm{A}^{(3)} & \mathrm{A}^{(2)} - \lambda^2 \mathrm{A}^{(1)} \end{bmatrix}.$$

A solution $\mathbf{Y}_0 = \mathbf{KC}$ to (7.125) must satisfy the following 16 boundary conditions:

$$\begin{aligned}
&(1) \quad \text{at} \quad \eta = 0, \quad \vartheta_2^{(1)} = \vartheta_2^{(2)} = 0, \quad \vartheta_3^{(1)} = \vartheta_3^{(2)} = 0, \\
&\qquad\qquad\qquad\quad u_2^{(1)} = u_2^{(2)} = 0, \quad u_3^{(1)} = u_3^{(2)} = 0; \\
&(2) \quad \text{at} \quad \eta = 1, \quad \vartheta_2^{(1)} = \vartheta_2^{(2)} = 0, \quad \vartheta_3^{(1)} = \vartheta_3^{(2)} = 0, \\
&\qquad\qquad\qquad\quad u_2^{(1)} = u_2^{(2)} = 0, \quad u_3^{(1)} = u_3^{(2)} = 0.
\end{aligned}$$

It follows from the boundary conditions at $\eta = 0$ that

$$c_5 = c_6 = c_7 = c_8 = 0, \qquad c_{13} = c_{14} = c_{15} = c_{16} = 0.$$

The conditions at $\eta = 1$ give the system of eight equations:

$$\begin{aligned}
&k_{51}c_1 + \ldots + k_{54}c_4 + k_{59}c_9 + \ldots + k_{512}c_{12} = 0, \\
&\ldots \\
&k_{81}c_1 + \ldots + k_{84}c_4 + k_{89}c_9 + \ldots + k_{812}c_{12} = 0, \\
&k_{131}c_1 + \ldots + k_{134}c_4 + k_{139}c_9 + \ldots + k_{1312}c_{12} = 0, \\
&\ldots \\
&k_{161}c_1 + \ldots + k_{164}c_4 + k_{169}c_9 + \ldots + k_{1612}c_{12} = 0.
\end{aligned} \tag{7.126}$$

The frequencies of vibrations can be found as functions of Q_{10}, M_{10}, and ω_0 through the condition $D = 0$, where D is the determinant of the matrix of the system of equations (7.126).

By use of the system of equations (7.119)–(7.122), we obtain the following equations in the displacements u_2 and u_3:

$$\begin{aligned}
&\frac{\partial^2}{\partial\eta^2}\left(A_{33}\frac{\partial^2 u_2}{\partial\eta^2}\right) - Q_{10}\frac{\partial^2 u_2}{\partial\eta^2} + M_{10}\frac{\partial^3 u_3}{\partial\eta^3} - \frac{\partial}{\partial\eta}\left(J_{33}\frac{\partial^3 u_2}{\partial\eta\,\partial\tau^2}\right) \\
&\qquad - \omega_{10}\frac{\partial}{\partial\eta}\left(J_{11}\frac{\partial^2 u_3}{\partial\tau\,\partial\eta}\right) + n_1\frac{\partial^2 u_2}{\partial\tau^2} = 0,
\end{aligned} \tag{7.127}$$

$$\begin{aligned}
&\frac{\partial^2}{\partial\eta^2}\left(A_{33}\frac{\partial^2 u_3}{\partial\eta^2}\right) - Q_{10}\frac{\partial^2 u_3}{\partial\eta^2} - M_{10}\frac{\partial^3 u_2}{\partial\eta^3} - \frac{\partial}{\partial\eta}\left(J_{22}\frac{\partial^3 u_3}{\partial\eta\,\partial\tau^2}\right) \\
&\qquad + \omega_{10}\frac{\partial}{\partial\eta}\left(J_{11}\frac{\partial^2 u_2}{\partial\tau\,\partial\eta}\right) + n_1\frac{\partial^2 u_3}{\partial\tau^2} = 0.
\end{aligned} \tag{7.128}$$

Let us find the frequencies for a special case, namely, for a rotating rod of constant cross section with simply supported ends ($M_{10} = 0$). In the case under consideration, we have $n_1 = 1$ and $A_{33} = 1$. Using (7.127) and (7.128), we get the following system of equations:

$$\frac{\partial^4 u_2}{\partial \eta^4} - Q_{10}\frac{\partial^2 u_2}{\partial \eta^2} - J_{33}\frac{\partial^4 u_2}{\partial \eta^2\,\partial \tau^2} - J_{11}\omega_{10}\frac{\partial^3 u_3}{\partial \tau\,\partial \eta^2} + \frac{\partial^2 u_2}{\partial \tau^2} = 0\,, \quad (7.129)$$

$$\frac{\partial^4 u_3}{\partial \eta^4} - Q_{10}\frac{\partial^2 u_3}{\partial \eta^2} - J_{33}\frac{\partial^4 u_3}{\partial \eta^2\,\partial \tau^2} + J_{11}\omega_{10}\frac{\partial^3 u_2}{\partial \eta^2\,\partial \tau} + \frac{\partial^2 u_3}{\partial \tau^2} = 0\,. \quad (7.130)$$

Taking the product of (7.130) by the imaginary unit and adding the result to (7.129), we arrive at

$$\frac{\partial^4 u}{\partial \eta^4} - Q_{10}\frac{\partial^2 u}{\partial \eta^2} + \mathrm{i}\,\omega_{10}J_{11}\frac{\partial^3 u}{\partial \tau\,\partial \eta^2} - J_{33}\frac{\partial^4 u}{\partial \eta^2\,\partial \tau^2} + \frac{\partial^2 u}{\partial \tau^2} = 0\,, \quad (7.131)$$

where $u = u_2 + \mathrm{i}\,u_3$.

We seek a solution to (7.131) in the form $u = u_0(\eta)\,\mathrm{e}^{\mathrm{i}\,\lambda\tau}$. Hence,

$$u_0^{\mathrm{IV}} - a u_0'' - \lambda^2 u_0 = 0\,, \quad (7.132)$$

where

$$a = Q_{10} - J_{33}\lambda^2 + \omega_{10}J_{11}\lambda\,.$$

Consider now the solution to (7.132) in the form $u_0 = A\,\mathrm{e}^{k\eta}$. The coefficient k can be determined from the characteristic equation

$$k^4 - ak^2 - \lambda^2 = 0\,.$$

The roots of the equation are

$$k_{1,\,2} = \pm\sqrt{\frac{a + \sqrt{a^2 + 4\lambda^2}}{2}}\,, \qquad k_{3,\,4} = \pm\sqrt{\frac{a - \sqrt{a^2 + 4\lambda^2}}{2}}\,. \quad (7.133)$$

Finally, the solution to (7.132) is seen to be

$$u_0(\eta) = c_1\,\mathrm{e}^{k_1\eta} + c_2\,\mathrm{e}^{k_2\eta} + c_3\,\mathrm{e}^{k_3\eta} + c_4\,\mathrm{e}^{k_4\eta}\,. \quad (7.134)$$

For a rod with simply supported ends, the following boundary conditions must hold:

(1) at $\eta = 0$, $u_0(0) = u_0''(0) = 0$;

(2) at $\eta = 1$, $u_0(1) = u_0''(1) = 0$.

These conditions yield a system of homogeneous equations whose determinant reads

$$D = \det \begin{bmatrix} 1 & 1 & 1 & 1 \\ \mathrm{e}^{k_1} & \mathrm{e}^{k_2} & \mathrm{e}^{k_3} & \mathrm{e}^{k_4} \\ k_1^2 & k_2^2 & k_3^2 & k_4^2 \\ k_1^2\,\mathrm{e}^{k_1} & k_2^2\,\mathrm{e}^{k_2} & k_3^2\,\mathrm{e}^{k_3} & k_4^2\,\mathrm{e}^{k_4} \end{bmatrix} . \tag{7.135}$$

Since $k_1^2 = k_2^2$ and $k_3^2 = k_4^2$, the determinant can be rearranged as follows:

$$D = \det \begin{bmatrix} 1 & 1 & 1 & 1 \\ \mathrm{e}^{k_1} & \mathrm{e}^{k_2} & \mathrm{e}^{k_3} & \mathrm{e}^{k_4} \\ 0 & 0 & k_4^2 - k_1^2 & k_4^2 - k_1^2 \\ 0 & 0 & (k_4^2 - k_1^2)\,\mathrm{e}^{k_3} & (k_4^2 - k_1^2)\,\mathrm{e}^{k_4} \end{bmatrix} .$$

Therefore,

$$D = (\mathrm{e}^{k_2} - \mathrm{e}^{k_1})\,(k_4^2 - k_1^2)^2(\mathrm{e}^{k_4} - \mathrm{e}^{k_3})\,.$$

The eigenvalues λ_j can be found through the condition $D = 0$. Since $k_4^2 \neq k_1^2$, the determinant may become zero if $\mathrm{e}^{k_2 - k_1} = 1$ and $\mathrm{e}^{k_4 - k_3} = 1$ or $k_2 - k_1 = \mathrm{i}\,2\pi j$ and $k_4 - k_3 = \mathrm{i}\,2\pi j$, where i is the imaginary unit. Using (7.123), we finally arrive at

$$\lambda_{1,2}^{(j)} = \frac{\pi^2 j^2 (J_{11}\omega_{10}) \pm \sqrt{d}}{2(1 + \pi^2 j^2 J_{33})}\,,$$

where

$$d = \pi^4 j^4 (J_{11}\omega_{10})^2 + 4(1 + \pi^2 j^2 J_{33})\,\pi^2 j^2 (\pi^2 j^2 + Q_{10})\,.$$

The obtained analytic formulas for the frequencies allow us to estimate the role of the parameters J_{ii}, J_{33}, ω_{10}, and Q_{10}.

7.5 Free and Forced Vibrations

7.5.1 Free Vibrations of Rods of Constant Cross Section

In this subsection, we consider the method of solution of the simplest equation, that is, the equation of free vibrations of a straight rod of constant cross section. This method, along with the methods discussed in the previous chapters, provides a foundation for the analysis of more complicated problems. (See the problems for self-study at the end of the chapter.)

Consider a rod shown in Fig. 7.15. The segment $0.5 < \eta < 1$ of the rod is subjected to a distributed load. The load vanishes at $t = 0$, and the rod breaks into vibrations in the plane x_1Ox_2. Consider the simplest situation, that is, the equations of vibrations of a straight rod of constant cross section when the rotary inertia is neglected. These equations (when damping forces are ignored) were obtained in Sect. 7.1:

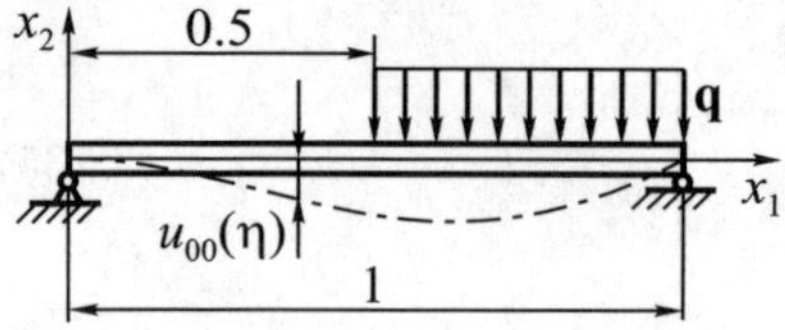

Fig. 7.15.

$$\frac{\partial^2 u}{\partial \tau^2} + \frac{\partial^4 u}{\partial \eta^4} = 0\,, \qquad u = u_{x_2}\,. \tag{7.136}$$

Recall that we sought the solution in the form

$$u = u_0(\eta)\,\mathrm{e}^{\mathrm{i}\,\lambda\tau}\,. \tag{7.137}$$

In Sect. 7.2, a method for determination of the eigenvalues λ_j and the corresponding eigenfunctions $\varphi^{(j)}(\eta)$ was presented. In view of (7.137), each λ_j gives two partial solutions

$$u_1^{(j)} = \varphi^{(j)}(\eta)\cos\lambda_j\tau\,, \qquad u_2^{(j)} = \varphi^{(j)}(\eta)\sin\lambda_j\tau\,. \tag{7.138}$$

Hence, the general solution to (7.137) is

$$u = \sum_{j=1}^{n}(c_1^{(j)}\cos\lambda_j\tau + c_2^{(j)}\sin\lambda_j\tau)\,\varphi^{(j)}(\eta)\,. \tag{7.139}$$

At the initial instant of time, the mode of the rod deflection due to the static load (Fig. 7.15) is defined by the equilibrium equation

$$\frac{\mathrm{d}^4 u_{00}}{\mathrm{d}\eta^4} + q\,H(\eta - 0.5) = 0\,. \tag{7.140}$$

Integration of this equation yields

$$u_{00} = -\frac{q}{24}(\eta - 0.5)^4\,H(\eta - 0.5) + \frac{c_1\eta^3}{6} + \frac{c_2\eta^2}{2} + c_3\eta + c_4\,.$$

Arbitrary constants c_i are to be calculated through the boundary conditions:

$$\begin{aligned} &(1) \quad \text{at} \quad \eta = 0\,, \quad u_{00} = u''_{00} = 0\,;\\ &(2) \quad \text{at} \quad \eta = 1\,, \quad u_{00} = u''_{00} = 0\,. \end{aligned}$$

Finally, we arrive at

$$u_{00} = -\frac{q}{24}(\eta - 0.5)^4\,H(\eta - 0.5) + \frac{q}{12}\,0.25\eta^3 - \frac{q}{24}\,0.4375\eta\,.$$

If before the vibration, the velocities of the rod axis points were zero, then the constants $c_2^{(j)}$ in (7.139) must equal zero. Hence, equation (7.139), which describes the shape of the rod axis, becomes

$$u = \sum_{j=1}^{n} c_1^{(j)} \varphi^{(j)}(\eta) \cos \lambda_j \tau \,. \tag{7.141}$$

For $\tau = 0$, the following condition must hold:

$$u_{00}(\eta) = \sum_{j=1}^{n} c_1^{(j)} \varphi^{(j)}(\eta) \,. \tag{7.142}$$

Multiplying this relationship by $\varphi^{(k)(\eta)}$ and integrating the result between the limits $\eta = 0$ and $\eta = 1$, we obtain the equation, from which $c_1^{(j)}$ can be found:

$$\int_0^1 u_{00} \varphi^{(k)} \, \mathrm{d}\eta = \sum_{j=1}^{n} c_1^{(j)} \int_0^1 \varphi^{(j)} \varphi^{(k)} \, \mathrm{d}\eta \,. \tag{7.143}$$

Using the condition of orthogonality for the eigenfunctions $\varphi^{(j)}$, we arrive at

$$c_1^{(j)} = \frac{\displaystyle\int_0^1 u_{00} \varphi^{(j)} \, \mathrm{d}\eta}{\displaystyle\int_0^1 (\varphi^{(j)})^2 \, \mathrm{d}\eta} \,. \tag{7.144}$$

Having found $c_1^{(j)}$, we can determine, by use of (7.142), the deflection of the rod during the vibrations.

With regard for the forces of viscous damping, (7.136) becomes

$$\frac{\partial^2 u}{\partial \tau^2} + \alpha \frac{\partial u}{\partial \tau} + \frac{\partial^4 u}{\partial \eta^4} = 0 \,. \tag{7.145}$$

The equation can be solved by setting $u = u_0(\eta) \, f(\tau)$. Substituting this into (7.145), we get

$$-\frac{u_0^{\mathrm{IV}}}{u_0} = \frac{\ddot{f}}{f} + \alpha \frac{\dot{f}}{f} = -\lambda^2 \,. \tag{7.146}$$

Taking into account (7.140), we arrive at

$$u_0^{\mathrm{IV}} - \lambda^2 u_0 = 0 \,, \qquad \ddot{f} + \alpha \dot{f} + \lambda^2 f = 0 \,. \tag{7.147}$$

The first equation, the equation in $u_0(\eta)$, can be used to find the eigenvalues λ_j and eigenfunctions $\varphi^{(j)}(\eta)$. For each λ_j, it follows from the equation in $f(\tau)$ that

$$f_j = \mathrm{e}^{-\alpha\tau/2} (c_1^{(j)} \cos p_j \tau + c_2^{(j)} \sin p_j \tau) \,. \tag{7.148}$$

A partial solution to (7.145) reads

$$u^{(j)} = \mathrm{e}^{-\alpha\tau/2}(c_1^{(j)} \cos p_j\tau + c_2^{(j)} \sin p_j\tau)\,\varphi^{(j)}(\eta)\,, \tag{7.149}$$

where

$$p_j = \frac{\sqrt{4\lambda_j^2 - \alpha^2}}{2}\,.$$

The general solution to (7.145) is

$$u = \sum_{j=1}^{n} \mathrm{e}^{-\alpha\tau/2}(c_1^{(j)} \cos p_j\tau + c_2^{(j)} \sin p_j\tau)\,\varphi^{(j)}(\eta)\,. \tag{7.150}$$

For the rod shown in Fig. 7.15, the constants $c_1^{(j)}$ are to be determined through (7.144) and the conditions $c_2^{(j)} = 0$. The displacements of the rod axis points (with the forces of viscous damping taken into consideration) are given by

$$u = \sum_{j=1}^{n} \mathrm{e}^{-\alpha\tau/2}\, c_1^{(j)} \cos p_j\tau \cdot \varphi^{(j)}(\eta)\,. \tag{7.151}$$

In more general cases, besides initial deflection $u_{00}(\eta)$, the initial velocities of the rod axis points, $\dot{u}_{00}(\eta)$, could also be specified.

7.5.2 Free Vibrations of Rods of Variable Cross Section

The equations of free vibrations of a rod of variable cross section (both with rotary inertia taken into consideration or ignored) can usually be solved only numerically. The relevant methods, which apply in the most general case of a space-curved rod, were given in Chap. 5. We shall use these methods to solve (7.17)–(7.20), which describe vibrations of a straight rod of variable cross section. The rotary inertia will be taken into consideration. These equations can be represented as one vector equation, which is a special case of (7.26) for $Q_{10} = \alpha = k = 0$ and $\mathbf{\Delta}\boldsymbol{\Phi} = 0$,

$$\mathrm{A}^{(1)} \frac{\partial^2 \mathbf{Z}}{\partial\tau^2} + \frac{\partial \mathbf{Z}}{\partial\eta} + \mathrm{A}^{(2)}\mathbf{Z} = 0\,. \tag{7.152}$$

Setting $\mathbf{Z} = \mathbf{Z}_0\,\mathrm{e}^{\mathrm{i}\,\lambda\tau}$ and upon determination of λ_j, we get two partial solutions

$$\mathbf{Z}_1^{(j)} = \mathbf{Z}_0^{(j)} \cos\lambda_j\tau\,, \qquad \mathbf{Z}_2^{(j)} = \mathbf{Z}_0^{(j)} \sin\lambda_j\tau\,.$$

The general solution to (7.152) is

$$\mathbf{Z} = \sum_{j=1}^{n} (c_1^{(j)} \cos\lambda_j\tau + c_2^{(j)} \sin\lambda_j\tau)\,\mathbf{Z}_0^{(j)}\,. \tag{7.153}$$

In the general case, the boundary conditions for $\tau = 0$ are as follows:

$$Z_4(0,\,\eta) = u_{x_2}(0,\,\eta) = u_{00}(\eta)\,,$$
$$\dot{Z}_4(0,\,\eta) = \dot{u}_{x_2}(0,\,\eta) = v_{00}(\eta)\,.$$

It follows from (7.153) that

$$u_{00}(\eta) = \sum_{j=1}^{n} c_1^{(j)} Z_{04}^{(j)}\,, \qquad v_{00}(\eta)_j = \sum_{j=1}^{n} c_2^{(j)} \lambda_j Z_{04}^{(j)}\,.$$

Arbitrary constants $c_1^{(j)}$ and $c_2^{(j)}$ can be found from the equations

$$\int_0^1 \mathbf{u}_{00}\mathbf{Z}_{04}^{(k)}\,\mathrm{d}\eta = \sum_{j=1}^{n}\int_0^1 c_1^{(j)}\mathbf{Z}_{04}^{(j)}\mathbf{Z}_{04}^{(k)}\,\mathrm{d}\eta\,,$$
$$\int_0^1 \mathbf{v}_{00}\mathbf{Z}_{04}^{(k)}\,\mathrm{d}\eta = \sum_{j=1}^{n}\int_0^1 c_2^{(j)}\mathbf{Z}_{04}^{(j)}\mathbf{Z}_{04}^{(k)}\,\mathrm{d}\eta\,, \qquad k = 1,\,2,\,\ldots,\,n\,. \qquad (7.154)$$

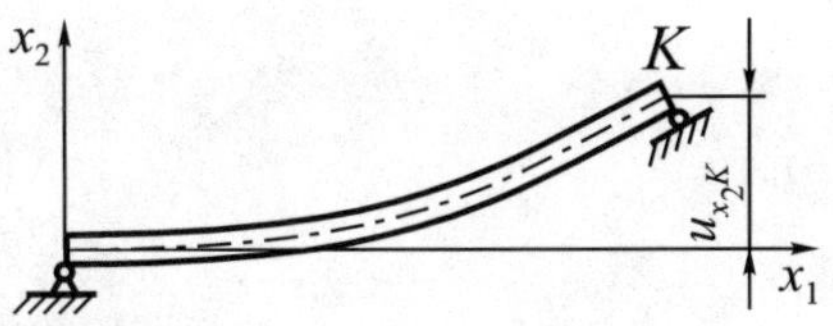

Fig. 7.16.

Consider an example presented in Fig. 7.16. The rod is simply supported at the right-hand end. The vertical component of the displacement of the point K, u_{x_2K}, is assumed to be known. The solution to the corresponding problem of statics gives

$$u_{00} = u_{x_2K}\left(\frac{3}{2}\,\eta^2 - \eta^3\right)\,.$$

Suppose that the hinged support at the right-hand end disappears at $\tau = 0$. Our task is to find the displacement of the point K as a function of time. The initial conditions are as follows:

$$Z_4(0,\,\eta) = u_{00}\,, \qquad \dot{Z}_4(0,\,\eta) = 0\,.$$

Hence, (7.153) gives

$$Z_4(\tau,\,\eta) = u_{x_2}(\tau,\,\eta) = \sum_{j=1}^{n} c_1^{(j)} Z_{04}^{(j)} \cos\lambda_j\tau$$

The constants $c_2^{(j)}$ are zero.

At $\tau = 0$, we have

$$u_{00} = u_{x_2K}\left(\frac{3}{2}\eta^2 - \eta^3\right) = \sum_{j=1}^{n} c_1^{(j)} Z_{04}^{(j)} .$$

In accordance with (7.154), the arbitrary constants are to be found from the relationship

$$\int_0^1 u_{x_2K}\left(\frac{3}{2}\eta^2 - \eta^3\right) Z_{04}^{(k)}\,\mathrm{d}\eta = \sum_{j=1}^{n}\int_0^1 c_1^{(j)} Z_{04}^{(j)} Z_{04}^{(k)}\,\mathrm{d}\eta . \tag{7.155}$$

7.5.3 Forced Steady-State Vibrations

To illustrate how the equation of forced steady-state vibrations of a rod can be solved analytically, consider an example shown in Fig. 7.17.

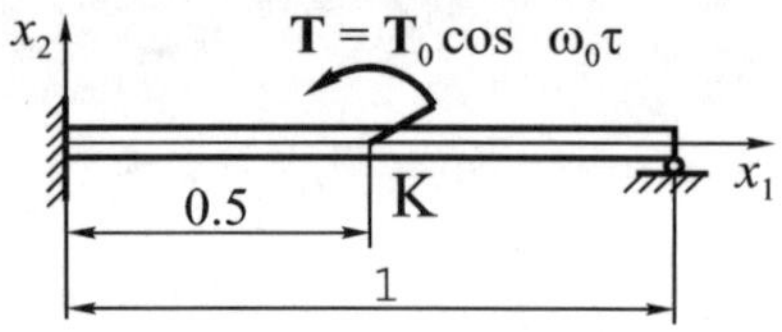

Fig. 7.17.

A *concentrated harmonic moment* is applied at the cross section K. Ignoring damping forces, we can rewrite the equations of forced vibrations for a rod of constant cross section in the form

$$\frac{\partial^2 u}{\partial \tau^2} + \frac{\partial^4 u}{\partial \eta^4} - T_0 \cos\omega_0\tau\,\delta'(\eta - 0.5) = 0 . \tag{7.156}$$

Setting $u = u_0(\eta)\cos\omega_0\tau$, we get

$$u_0^{\mathrm{IV}} - \omega_0^2 u_0 = T_0\,\delta'(\eta - 0.5) . \tag{7.157}$$

The general solution to this equation is

$$u_0(\eta) = c_1K_1(\eta) + c_2K_2(\eta) + c_3K_3(\eta) + c_4K_4(\eta) + \frac{K_3\left[\sqrt{\omega_0}(\eta - \eta_K)\right]}{\omega_0} T_0\,H(\eta - 0.5) . \tag{7.158}$$

Once the constants c_j have been determined, it is easy to obtain the exact solution to (7.156). For the rod shown in Fig. 7.17, the boundary conditions are as follows:

(1) at $\eta = 0$, $u_0 = u_0' = 0$ and $c_1 = c_2 = 0$;

(2) at $\eta = 1$, $u_0 = u_0'' = 0$.

Then, we have

$$K_3(1)\,c_3 + K_4(1)\,c_4 = -\frac{K_3(0.5)}{\omega_0}\,T_0\,,$$
$$K_1(1)\,c_3 + K_2(1)\,c_4 = -\frac{K_1(0.5)}{\omega_0}\,T_0\,.$$

This gives

$$c_3 = \frac{[K_1(0.5)\,K_4(1) - K_3(0.5)\,K_2(1)]\,T_0}{[K_3(1)\,K_2(1) - K_1(1)\,K_4(1)]\,\omega_0}\,,$$
$$c_4 = \frac{[K_1(1)\,K_3(0.5) - K_3(1)\,K_1(0.5)]\,T_0}{[K_3(1)\,K_2(1) - K_1(1)\,K_4(1)]\,\omega_0}\,.$$

Finally, we arrive at

$$u_0(\eta) = c_3 K_3(\eta,\,\omega_0) + c_4 K_4(\eta,\,\omega_0) + \frac{K_3\,[\omega_0(\eta-\eta_K)]}{\omega_0}\,T_0\,H(\eta-0.5)\,. \tag{7.159}$$

Differentiating this relationship, we find the amplitude values of the moment, $\Delta M_{x30}(\eta)$, and shear force, $\Delta Q_{x20}(\eta)$. The solution (7.159) is valid under the following condition:

$$\Delta = K_3(1)\,K_2(1) - K_1(1)\,K_4(1) \neq 0\,.$$

The denominator vanishes when ω_0 is an eigenfrequency of the rod (the case of resonance).

Forced Steady-State Vibrations Due to Non-Harmonic Periodic Forces. The graph of a non-harmonic concentrated moment periodic in time is shown in Fig. 7.18.

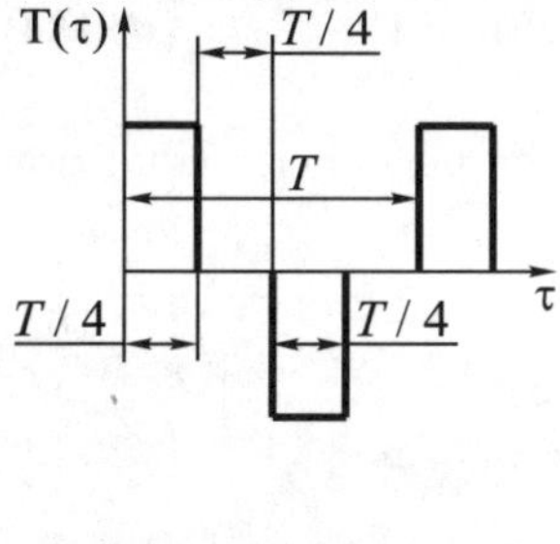

Fig. 7.18.

The moment $T(\tau)$ can be expanded in a Fourier series as follows:

$$T(\tau) = \sum_{j=0}^{n}(c_j \sin j\omega_0\tau + b_j \cos j\omega_0\tau)\,, \qquad \omega_0 = \frac{2\pi}{T}\,, \tag{7.160}$$

After that, we can find the solution to (7.156) for each term on the right-hand side of (7.160) and then sum up the solutions obtained.

Consider the Douffing method that does not use the representation (7.160). To illustrate how the method works, consider a rod in Fig. 7.17. Suppose that $T(\tau)$ varies with time as shown in Fig. 7.18.

Let

$$u = \sum_{j=1}^{n} f^{(j)}(\tau)\,\varphi^{(j)}(\eta)\,. \tag{7.161}$$

In accordance with the principle of virtual displacements, from (7.156) we get the system of equations

$$\sum_{j=1}^{n} \int_0^1 \left[\ddot{f}^{(j)}\varphi^{(j)} + f^{(j)}(\varphi^{(j)})^{\mathrm{IV}} - T(\tau)\right] \varphi^{(k)}\,\mathrm{d}\eta = 0\,, \quad k = 1, 2, \ldots, n\,.$$

On rearrangement, we obtain the set of equations identical in form to (5.85). In vector–matrix form, this set reads

$$\ddot{\mathbf{f}} + \mathrm{H}^{-1}\,\mathrm{B}\,\mathbf{f} = \mathrm{H}^{-1}\,\mathbf{b}\,. \tag{7.162}$$

By setting $\dot{\mathbf{f}} = \mathbf{Y}_1$ and $\mathbf{f} = \mathbf{Y}_2$ (see Sect. 5.2), this equation can be represented as a system of first-order differential equations

$$\dot{\mathbf{Y}} + \mathrm{C}\mathbf{Y} = \mathbf{b}^{(1)}\,, \tag{7.163}$$

where

$$\mathbf{Y} = \begin{bmatrix} \mathbf{Y}_1 \\ \mathbf{Y}_2 \end{bmatrix}, \qquad \mathrm{C} = \begin{bmatrix} 0 & \mathrm{H}^{-1}\mathrm{B} \\ -\mathrm{E} & 0 \end{bmatrix}, \qquad \mathbf{b}^{(1)} = \begin{bmatrix} \mathrm{H}^{-1}\mathbf{b} \\ \mathbf{0} \end{bmatrix}.$$

The solution to (7.163) is

$$\mathbf{Y} = \mathrm{K}(\tau)\,\mathbf{Y}_0 + \int_0^{\tau} \mathrm{G}(\tau,\,\tau_1)\,\mathbf{b}^{(1)}(\tau_1)\,\mathrm{d}\tau_1\,, \qquad \mathrm{K}(0) = \mathrm{E}\,. \tag{7.164}$$

For $\tau = 0$, we have $\mathbf{Y}_0 = \mathbf{Y}(0)$. It follows from the periodicity condition for the solution that

$$\mathbf{Y}(0) = \mathrm{K}(T)\,\mathbf{Y}(0) + \mathbf{Y}_{\mathrm{p}}\,, \tag{7.165}$$

where

$$\mathbf{Y}_{\mathrm{p}} = \int_0^{T} \mathrm{G}(T,\,\tau_1)\,\mathbf{b}^{(1)}(\tau_1)\,\mathrm{d}\tau_1\,.$$

Using (7.165), we get

$$\mathbf{Y}(0) = [\mathrm{E} - \mathrm{K}(T)]^{-1}\,\mathbf{Y}_{\mathrm{p}}\,.$$

Finally, the solution to (7.156) on the interval $0 \le \tau \le T$ takes the form

$$\mathbf{Y}(\tau) = \mathrm{K}(\tau)\,[\mathrm{E} - \mathrm{K}(T)]^{-1}\,\mathbf{Y}_{\mathrm{p}} + \mathbf{Y}_{\mathrm{p}}(\tau)\,. \tag{7.166}$$

This solution is also valid on all succeeding intervals $(\nu - 1)T \le \tau \le \nu T$, where ν is the number of an interval. When solving the equation numerically, the partial solution $\mathbf{Y}_{\mathrm{p}}$ of the inhomogeneous equation can be obtained without resort to the Green matrix $\mathrm{G}(\tau, \tau_1)$. This approach simplifies the process of solution. To do this, it is sufficient to solve (7.163) under the zero initial conditions and obtain the vector $\mathbf{Y}_{\mathrm{p}}(T)$, which is used when determining the vector $\mathbf{Y}(0)$. Note that the components of the vector $\mathbf{Y}(0)$ may not coincide with the amplitude values of the corresponding components of the vector $\mathbf{Y}(\tau)$.

7.5.4 Forced Unsteady Vibrations

A rod subjected to a varying in time force P is shown in Fig. 7.19*b*.

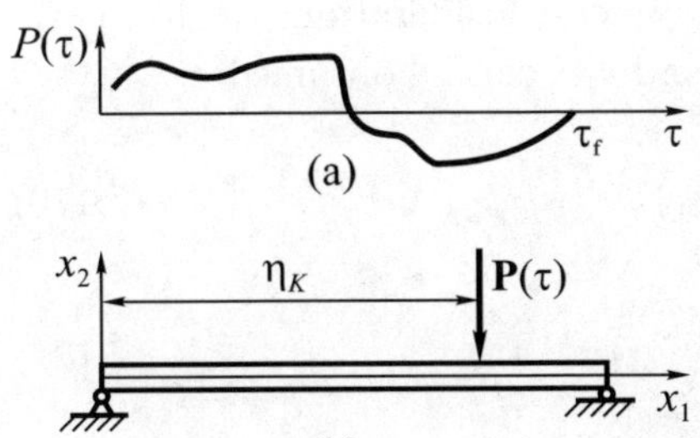

Fig. 7.19.

In the general case, since the force P acts during a finite time interval $0 \le \tau \le \tau_k$, the rod vibrations would be unsteady. The equations of small vibrations for a rod of constant cross section are

$$\frac{\partial^2 u}{\partial \tau^2} + \frac{\partial^4 u}{\partial \eta^4} - P(\tau)\,\delta(\eta - \eta_k) = 0\,. \tag{7.167}$$

To use the principle of virtual displacements, we put

$$u = \sum_{j=1}^{n} f^{(j)}(\tau)\,\varphi^{(j)}(\eta)\,.$$

Then, we get the system of equations identical in form to (7.163),

$$\dot{\mathbf{Y}} + \mathrm{B}^{(1)}\mathbf{Y} = \mathbf{b}^{(1)}\,. \tag{7.168}$$

For the zero initial conditions, the solution to this equation is $\mathbf{Y}(\tau) = \mathbf{Y}_{\mathrm{p}}(\tau)$. The components of the vector $\mathbf{Y}(\tau)$ are the functions $f^{(j)}$ and their first derivatives; hence, having found the vector $\mathbf{Y}(\tau)$, we obtain the displacements of the rod axis points, $u(\eta, \tau)$, and the velocities of the points:

$$\frac{\partial u}{\partial \tau} = \sum_{j=1}^{n} \dot{f}^{(j)} \varphi^{(j)} . \tag{7.169}$$

When $\tau > \tau_{\mathrm{f}}$, the rod is in free oscillatory motion. The initial conditions for this motion are as follows:

$$u(\eta, \tau_{\mathrm{f}}) = \sum_{j=1}^{n} f^{(j)}(\tau_{\mathrm{f}})\, \varphi^{(j)}(\eta) ,$$

$$\dot{u}(\eta, \tau_{\mathrm{f}}) = \sum_{j=1}^{n} \dot{f}^{(j)}(\tau_{\mathrm{f}})\, \varphi^{(j)}(\eta) .$$

The solution (7.139) to equation (7.136) was obtained for the case where damping forces were ignored. Recall that equation (7.136) governs free vibrations. In the case of free vibrations under consideration, we have two equations from which the constants $c_1^{(j)}$ and $c_2^{(j)}$ can be obtained

$$u(\eta, \tau_{\mathrm{f}}) = \sum_{j=1}^{n} c_1^{(j)} \varphi^{(j)}(\eta) = \sum_{j=1}^{n} f^{(j)}(\tau_{\mathrm{f}})\, \varphi^{(j)}(\eta) , \tag{7.170}$$

$$\dot{u}(\eta, \tau_{\mathrm{f}}) = \sum_{j=1}^{n} \lambda_j c_2^{(j)} \varphi^{(j)}(\eta) = \sum_{j=1}^{n} \dot{f}^{(j)}(\tau_{\mathrm{f}})\, \varphi^{(j)}(\eta) . \tag{7.171}$$

Therefore,

$$c_1^{(j)} = f^{(j)}(\tau_{\mathrm{f}}) , \qquad c_2^{(j)} = \frac{\dot{f}^{(j)}(\tau_{\mathrm{f}})}{\lambda_j} .$$

Finally, we arrive at

$$u(\eta, \tau) = \sum_{j=1}^{n} \left[f^{(j)}(\tau_{\mathrm{f}}) \cos \lambda_j \tau + \frac{\dot{f}^{(j)}(\tau_{\mathrm{f}})}{\lambda_j} \sin \lambda_j \tau \right] \varphi^{(j)}(\eta) . \tag{7.172}$$

Recall that this formula corresponds to the free vibrations of the rod when the load ceases its action.

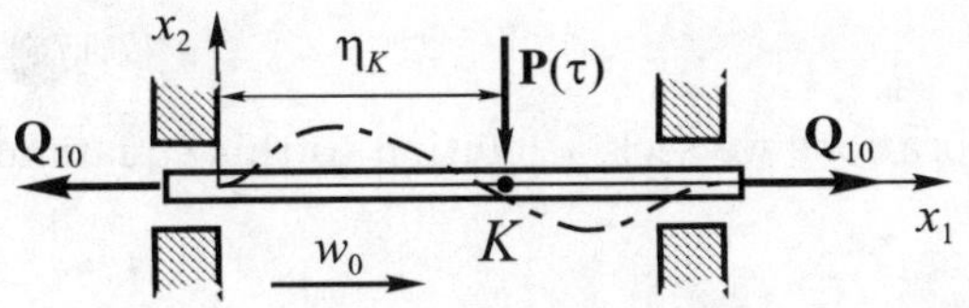

Fig. 7.20.

7.5.5 Forced Vibrations About a Stationary Motion

Consider a rod in rectilinear, steady state motion (Fig. 7.20). The equations of small vibrations of a rod are as follows:

$$L(u) = \frac{\partial^4 u}{\partial \eta^4} + 2w_0 \frac{\partial^2 u}{\partial \eta \, \partial \tau} - (Q_{10} - w_0^2) \frac{\partial^2 u}{\partial \eta^2} + \frac{\partial^2 u}{\partial \tau^2} - P_{x_2} = 0 \,. \quad (7.173)$$

This equation is a special case of (7.105) for $Q_1 = Q_{10} = \text{const}$.

Consider steady state vibrations of a rod under the action of a harmonically changing force

$$P_{x_2}(t) = P_0 \cos \omega_0 \tau \cdot \delta(\eta - \eta_K) \,. \quad (7.174)$$

When using the principle of virtual displacements, we should seek a solution in the form

$$u = \sum_{j=1}^{n} f^{(j)}(\tau) \, \varphi^{(j)}(\eta) \,.$$

The following system of equations results

$$\int_0^1 L(u) \, \varphi^{(k)} \, \mathrm{d}\eta = 0 \,, \qquad k = 1, 2, \ldots, n \,,$$

On rearrangement, we have

$$h_{kk} \ddot{f}^{(k)} + \sum_{j=1}^{n} (2w_0 b_{kj} + \alpha h_{kj} \, \delta_{kj}) \, \dot{f}^{(j)} + \sum_{j=1}^{n} \left[\lambda_j^2 \, \delta_{kj} - (Q_{10} - w_0^2) \, c_{kj} \right] f^{(j)}$$
$$- P_0 \varphi^{(k)}(\eta_k) \cos \omega_0 \tau = 0 \,, \quad (7.175)$$

where

$$b_{kj} = -b_{jk} = \int_0^1 \varphi^{(j)\prime} \varphi^{(k)} \, \mathrm{d}\eta \,, \qquad h_{kj} = \int_0^1 \varphi^{(j)} \varphi^{(k)} \, \mathrm{d}\eta \,,$$

$$c_{kj} = \int_0^1 \varphi^{(j)\prime\prime} \varphi^{(k)} \, \mathrm{d}\eta \,.$$

Recall that δ_{kj} is the Kronecker delta.

Because of the orthogonality condition for the functions $\varphi^{(j)}$, we see that $h_{kj} = 0$ for $k \neq j$. In vector–matrix form, (7.175) reads

$$\ddot{\mathbf{f}} + \mathrm{B}\dot{\mathbf{f}} + \mathrm{C}\mathbf{f} = \mathbf{b}_0 \cos\omega_0\tau \,. \tag{7.176}$$

In the case of steady state vibrations we seek a solution to this equation in the form

$$\mathbf{f} = \mathbf{f}_{10}\cos\omega_0\tau + \mathbf{f}_{20}\sin\omega_0\tau \,.$$

On rearrangement, the following equations to be solved for the vectors $\mathbf{f}_{10}$ and $\mathbf{f}_{20}$ result

$$\begin{aligned}
&(\mathrm{C} - \omega_0^2\mathrm{E})\,\mathbf{f}_{10} + \omega_0\mathrm{B}\mathbf{f}_{20} = \mathbf{b}_0 \,,\\
&-\omega_0\mathrm{B}\mathbf{f}_{10} + (\mathrm{C} - \omega_0^2\mathrm{E})\,\mathbf{f}_{20} = 0 \,.
\end{aligned}$$

Having found $\mathbf{f}_{10}$ and $\mathbf{f}_{20}$, we get the components of the vector $\mathbf{f}$:

$$f^{(j)}(\tau) = f_{10}^{(j)}\cos\omega_0\tau + f_{20}^{(j)}\sin\omega_0\tau \,,$$

Thus, the solution to (7.173) is

$$u(\eta,\,\tau) = \sum_{j=1}^{n}\left(f_{10}^{(j)}\cos\omega_0\tau + f_{20}^{(j)}\sin\omega_0\tau\right)\varphi^{(j)}(\eta) \tag{7.177}$$

or

$$u(\eta,\,\tau) = \sum_{j=1}^{n} u_0^{(j)}(\eta)\cos(\omega_0\tau + \tilde{\varphi}_j) \,, \tag{7.178}$$

where

$$u_0^{(j)} = \varphi^{(j)}\sqrt{(f_{10}^{(j)})^2 + (f_{20}^{(j)})^2} \,, \qquad \tan\tilde{\varphi}_j = \frac{f_{20}^{(j)}}{f_{10}^{(j)}} \,.$$

7.5.6 Forced Vibrations Due to Kinematic Excitation

To illustrate vibrations due to kinematic excitations, consider a rod in Fig. 7.21*a*. One cross section of the rod, $\eta = \eta_K$, is forced to oscillate harmonically in a prearranged manner. We do not consider here the piece of equipment that maintains the vibrations of the cross section K and thus suppose that the vibrations occur due to a periodic force $P(\tau)$ as shown in Fig. 7.21*b*.

Thus, we have a problem of forced vibrations of a rod subjected to a concentrated periodic force. A similar problem, vibrations due to a concentrated moment $T(\tau)$, was considered above.

The equation of small vibrations for the rod is

$$\frac{\partial^2 u}{\partial\tau^2} + \frac{\partial^4 u}{\partial\eta^4} - P_0\cos\omega_0\tau\,\delta(\eta - \eta_k) = 0 \,. \tag{7.179}$$

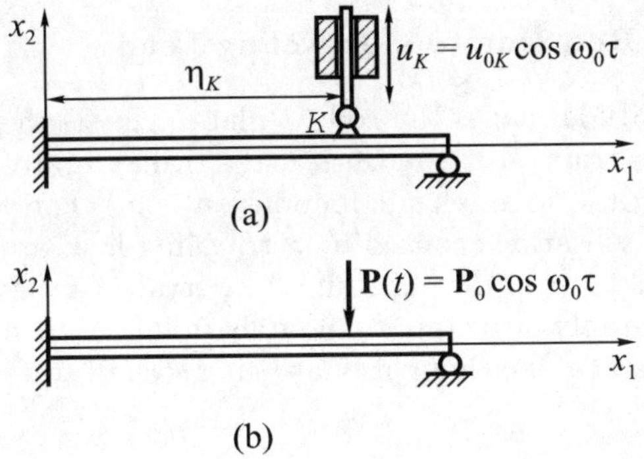

Fig. 7.21.

Setting $u = u_0(\eta)\cos\omega_0\tau$, we get

$$u_0^{\mathrm{IV}} - \omega_0^2 u_0 = P_0\,\delta(\eta - \eta_k)\,. \tag{7.180}$$

The solution to this equation is identical in form to (7.158):

$$u_0 = c_1K_1 + c_2K_2 + c_3K_3 + c_4K_4 + \frac{K_4\left[\sqrt{\omega_0}\,(\eta - \eta_K)\right]}{(\sqrt{\omega_0})^3}\,P_0\,H(\eta - \eta_K)\,. \tag{7.181}$$

It follows from the boundary conditions at $\eta = 0$ that $c_1 = c_2 = 0$. The other unknowns, c_3, c_4, and P_0, can be found from the three following conditions.

(1) At $\eta = \eta_K$, we have $u_0 = u_{0K}$. Consequently,

$$c_3K_3(\eta_K) + c_4K_4(\eta_K) = u_{0K}\,. \tag{7.182}$$

(2) At $\eta = 1$, we have $u'' = 0$ and $u''' = 0$. Consequently,

$$u_0'' = c_3K_1(1) + c_4K_2(1) + \frac{K_2\left[\sqrt{\omega_0}\,(1 - \eta_K)\right]}{\sqrt{\omega_0}}\,P_0 = 0\,. \tag{7.183}$$

(3) At $\eta = 1$, we also have $u''' = 0$. Consequently,

$$u_0''' = c_3K_4(1) + c_4K_1(1) + K_1\left[\sqrt{\omega_0}\,(1 - \eta_K)\right]P_0 = 0\,. \tag{7.184}$$

It follows from the system of equations (7.182)–(7.183) that $c_3 = \beta_1 u_{0K}$, $c_4 = \beta_2 u_{0K}$, and $P_0 = \beta_3 u_{0K}$. Then, the amplitude values of displacements are seen to be

$$u_0(\eta) = u_{0K}\left[\beta_1K_1(\eta) + \beta_2K_4(\eta)\right] + \frac{K_4\left[\sqrt{\omega_0},\,(\eta - \eta_K)\right]}{(\sqrt{\omega_0})^3}\,\beta_3u_{0K}\,H(\eta - \eta_K)\,. \tag{7.185}$$

7.5.7 Vibrations of a Straight Rod Due to a Traveling Load

A traveling load of constant magnitude makes the rod oscillate. This is the reason why a bridge vibrates when a train passes across it and a train vibrates when it traveling along a railway track on an elastic foundation (soil). For a rod of finite length (a bridge), the vibrations caused by a traveling load are unsteady because the load acts on the rod during a finite interval of time. When the length of a rod is considerably large (approximately infinite), then the vibrations due to a traveling load can be thought of as being steady-state vibrations.

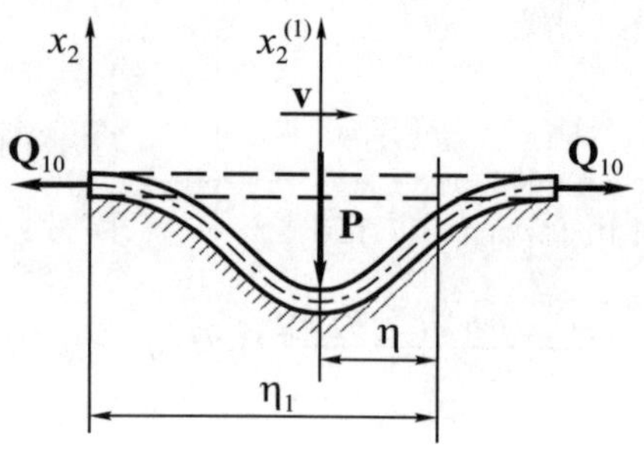

Fig. 7.22.

In the general case, when dealing with an elastic system under the action of a traveling load, it is essential to know both the mass of the moving element and the mass of the whole elastic system. At the same time, when a weight moves steadily along an infinite rod on an elastic, solid foundation and the deflection of the rod due to the weight remains the same (Fig. 7.22), the mass of the weight is of no importance because there is no acceleration in the x_2-direction.

The equation of forced flexural vibrations of a rod of constant cross section on an elastic foundation (when the damping forces, rotary inertia, and the shear inertia are ignored) is as follows:

$$\frac{\partial^4 u}{\partial \eta_1^4} + \frac{\partial^2 u}{\partial \tau^2} - Q_{10}\,\frac{\partial^2 u}{\partial \eta_1^2} + 4\alpha_1^4 u = -P_0\,\delta\left(\eta_1 - \frac{v\tau}{lp_0}\right), \qquad 4\alpha_1^4 = k_0\,\frac{l^2}{A_{33}}; \tag{7.186}$$

here, l is a characteristic length (for example, the distance the load moves in a unit time, $l = v1$); the dimensionless coordinate η ranges from 0 to ∞; k_0 is the dimensional stiffness coefficient of the foundation.

Refer the rod vibrations to a frame of reference with the origin at P moving with the steady speed v (Fig. 7.22). Let us introduce a new variable

$$\eta = \eta_1 - \frac{v\tau}{lp_0}$$

Then, equation (7.186) reads

$$\frac{\mathrm{d}^4 u}{\mathrm{d}\eta^4} - Q_{10}\,\frac{\mathrm{d}^2 u}{\mathrm{d}\eta^2} + \left(\frac{v}{l p_0}\right)^2 \frac{\mathrm{d}^2 u}{\mathrm{d}\eta^2} + 4\alpha_1^4 u = -P_0\,\delta(\eta)\,. \tag{7.187}$$

Note that all the terms containing partial derivatives with respect to time vanish because the motion of the rod relative to the moving frame is steady-state (see Biderman (1980)).
Then, the corresponding characteristic equation reads

$$\lambda^4 + 4\nu^2\lambda^2 + 4\alpha_1^4 = 0\,, \tag{7.188}$$

where

$$4\nu^2 = \frac{v^2}{p_0^2 l^2} - Q_{10}\,.$$

It has the following roots:

$$\lambda_{1,\,2,\,3,\,4} = \pm\sqrt{2\left(-\nu^2 \pm \sqrt{\nu^4 - \alpha_1^4}\right)}\,.$$

For $\nu^2 > 0$ and $\nu^4 < \alpha_1^4$, the roots are complex numbers. We have

$$\lambda_{1-4} = \pm\gamma \pm \mathrm{i}\,\beta \qquad \gamma = \sqrt{\alpha_1^2 - \nu^2}\,, \qquad \beta = \sqrt{\alpha_1^2 + \nu^2}\,.$$

The solution to (7.187) is

$$u = \mathrm{e}^{\gamma\eta}(c_1 \cos\beta\eta + c_2 \sin\beta\eta) + \mathrm{e}^{-\gamma\eta}(c_3 \cos\beta\eta + c_4 \sin\beta\eta) + \Phi(\eta)\,, \tag{7.189}$$

where $\Phi(\eta)$ is the partial solution to the inhomogeneous equation (7.187).

The origin of the moving frame is chosen so that the load is zero both for $\eta > 0$ and $\eta < 0$; hence, $\Phi(\eta) = 0$ and the rod deflection can be found upon use of the homogeneous equation. The solution (7.189) increases without bound as $\eta \to +\infty$ or $\eta \to -\infty$. This, of course, is in conflict with the physical meaning of the solution. Thus, omitting the terms responsible for the unlimited growth, we get the following formulas for the deflections to the left and to the right of the origin:

$$\begin{aligned} u_- &= \mathrm{e}^{\gamma\eta}(c_1 \cos\beta\eta + c_2 \sin\beta\eta)\,, \qquad && \eta < 0\,, \\ u_+ &= \mathrm{e}^{-\gamma\eta}(c_3 \cos\beta\eta + c_4 \sin\beta\eta)\,, && \eta > 0\,. \end{aligned}$$

Since the function u and its first three derivatives are continuous at the origin, we have

$$\begin{aligned} u_-(0) &= u_+(0)\,, \qquad & u'_-(0) &= u'_+(0)\,, \\ u''_-(0) &= u''_+(0)\,, & u'''_-(0) &= u'''_+(0) + P_0\,. \end{aligned} \tag{7.190}$$

Calculating the constants c_j through these conditions, we obtain the following formulas for the deflection:

$$u_- = -\frac{P_0\,\mathrm{e}^{\gamma\eta}}{8\alpha_1^2\gamma}\left(\cos\beta\eta - \frac{\gamma}{\beta}\sin\beta\eta\right),$$
$$u_+ = -\frac{P_0\,\mathrm{e}^{-\gamma\eta}}{8\alpha_1^2\gamma}\left(\cos\beta\eta + \frac{\gamma}{\beta}\sin\beta\eta\right). \tag{7.191}$$

The dimensionless deflection at the point of application of the force P_0 is

$$|u_\mp(0)| = \frac{P_0}{8\alpha_1^2\gamma}.$$

For $\gamma = 0$ it tends to infinity. This situation occurs when the dimensionless velocity of the load is

$$v_* = \sqrt{4\alpha_1^2 + Q_{10}}\,. \tag{7.192}$$

In dimensional form, it reads

$$(v_*)_\mathrm{d} = \sqrt{\frac{A_{33}}{m_0}}\sqrt{\frac{Q_{10}}{A_{33}} + 2\sqrt{\frac{k}{A_{33}}}},$$

where k is the stiffness coefficient of the elastic foundation.

This result clearly shows that the critical velocity increases with the value of the tensile force Q_{10}.

Now let $\nu^2 < 0$ and $\nu^4 - \alpha_1^4 < 0$. We have

$$\gamma = \sqrt{\alpha_1^2 + \nu^2}\,, \qquad \beta = \sqrt{\alpha_1^2 - \nu^2}\,.$$

Hence, the deflection at the origin is a bounded quantity. Thus, in this case, there is no critical velocity.

Let us find out the physical sense of the critical velocity. The homogeneous equation (7.186) has the solution

$$u = c\sin\frac{\pi}{\lambda}(\eta_1 - \tilde{c}\tau)\,, \tag{7.193}$$

where λ is the nondimentional half-wave length and $\tilde{c}$ is the dimensionless phase velocity of the moving wave. Substituting (7.193) into (7.186), we get

$$\tilde{c} = \sqrt{\frac{\pi^2}{\lambda^2} + Q_{10} + 4\alpha_1^4\frac{\lambda^2}{\pi^2}}\,. \tag{7.194}$$

It follows from (7.194) that the phase velocity of the wave depends on its length.

Let us now find the minimum value of $\tilde{c}$. The condition

$$\frac{\mathrm{d}\tilde{c}}{\mathrm{d}\lambda} = 0$$

gives

$$\lambda_* = \frac{\pi}{\sqrt{2}\,\alpha_1}$$

and then

$$\tilde{c}_{\min} = \sqrt{4\alpha_1^2 + Q_{10}}\,,$$

which coincides with the formula for the critical velocity (7.192).

Consider a steady-state motion of a rod with rotary inertia taken into consideration. In this case, the equation of flexural vibrations of the rod (7.186) is

$$\frac{\partial^4 u}{\partial \eta_1^4} + \frac{\partial^2 u}{\partial \tau^2} - Q_{10}\,\frac{\partial^2 u}{\partial \eta_1^2} - J_{33}\,\frac{\partial^4 u}{\partial \eta_1^2\,\partial \tau^2} + 4\alpha_1^4 u = P_0\,\delta(\eta_1 - v\tau)\,, \tag{7.195}$$

where v is the dimensionless velocity.

The equation of steady-state vibrations in the moving frame of reference is as follows:

$$(1 - J_{33}v^2)\,\frac{\mathrm{d}^4 u}{\mathrm{d}\eta^4} + (v^2 - Q_{10})\,\frac{\mathrm{d}^2 u}{\partial \eta^2} + 4\alpha_1^4 u = -P_0\,\delta(\eta)\,. \tag{7.196}$$

The characteristic equation reads

$$\lambda^4 + 4\nu_1^2\lambda^2 + 4\alpha_{11}^4 = 0\,,$$

where

$$\nu_1^2 = \frac{\nu^2}{1 - J_{33}v^2}\,, \qquad \alpha_{11}^4 = \frac{\alpha_1^4}{1 - J_{33}v^2}\,.$$

The roots of the characteristic equation are

$$\lambda_{1-4} = \pm\gamma_1 \pm \mathrm{i}\,\beta_1\,, \qquad \gamma_1 = \sqrt{\alpha_{11}^2 - \nu_1^2}\,, \qquad \beta_1 = \sqrt{\alpha_{11}^2 + \nu_1^2}\,.$$

The further solving of (7.196) is the same as that of (7.187). As before, we should find the critical velocity through the condition $\gamma_1 = 0$, which upon rearrangement gives

$$v_{**}^4 - (2Q_{10} - 16\alpha_1^4 J_{33})\,v_{**}^2 + Q_{10}^2 - 16\alpha_1^4 = 0\,. \tag{7.197}$$

Consequently, the critical velocity is

$$v_{**}^2 = Q_{10} - 8\alpha_1^4 J_{33} \pm \sqrt{-16Q_{10}\alpha_1^4 J_{33} + 64J_{33}^2\alpha_1^8 + 16\alpha_1^4}\,. \tag{7.198}$$

Since, for $J_{33} = 0$, the relationship (7.198) is to coincide with (7.192), we should take the plus sign in (7.198). For $Q_{10} = 0$, we have

$$v_{**}^2 = \sqrt{16\,\alpha_1^4 + 64 J_{33}^2 \alpha_1^8} + 8\,\alpha_1^4 J_{33}\,. \tag{7.199}$$

Usually, the dimensionless coefficient J_{33} is a small quantity. Hence, expanding the square root in (7.198) in a series and omitting higher-order terms, we arrive at the following expression for the critical velocity:

$$v_{**}^2 = v_*^2 - J_{33}(8\,\alpha_1^4 + 2\,\alpha_1^2 Q_{10} + 8\,\alpha_1^6 J_{33})\,, \tag{7.200}$$

where

$$v_*^2 = 4\alpha_1^2 + Q_{10}\,.$$

This formula specifies the relationship between the axial force Q_{10} and the rotary inertia on one hand and the critical velocity on the other hand.

7.6 Random Vibrations

7.6.1 Probability Characteristics

In Chap. 6, random vibrations of a space-curved rod were considered. The formulas of Chap. 6 may be simplified for the case of straight rods. At the same time, it is more convenient to obtain the relationships for a straight rod from (7.167).

As an example, consider stationary random vibrations of a rod shown in Fig. 7.19*b*. The force P is a stationary random function with specified characteristics. In particular, we know its spectral density $S_P(\omega)$.

Let us take into consideration the force of viscous damping. Then, we get

$$\frac{\partial^2 u}{\partial \tau^2} + \alpha\,\frac{\partial u}{\partial \tau} + \frac{\partial^4 u}{\partial \eta^4} - P(\tau)\,\delta(\eta - \eta_k) = 0\,. \tag{7.201}$$

Let u be given by (7.161) with $n = 2$, that is, let us restrict our consideration to the case of two-term approximation in the relationship (7.161). Then, considering (7.167) and using the principle of virtual displacements, we obtain the following equations in $f^{(1)}(\tau)$ and $f^{(2)}(\tau)$:

$$\begin{aligned} &\ddot f^{(1)} + \alpha \dot f^{(1)} + \lambda_1^2 f^{(1)} = b_1 P(\tau)\,,\\ &\ddot f^{(2)} + \alpha \dot f^{(2)} + \lambda_2^2 f^{(2)} = b_2 P(\tau)\,; \end{aligned} \tag{7.202}$$

here,

$$b_1 = \frac{\varphi^{(1)}(\eta_k)}{\displaystyle\int_0^1 (\varphi^{(1)})^2\,\mathrm{d}\eta}\,, \qquad b_2 = \frac{\varphi^{(2)}(\eta_k)}{\displaystyle\int_0^1 (\varphi^{(2)})^2\,\mathrm{d}\eta}\,.$$

In the example under consideration, equations (7.202) are uncoupled. The spectral density of the input force $S_P(\omega)$ and the spectral density of the output $S_{f^{(1)}}(\omega)$ and $S_{f^{(2)}}(\omega)$ are related to each other by (6.22) in which $W_j(\mathrm{i}\,\omega)$ are

$$W_1(\mathrm{i}\,\omega) = \frac{b_1}{(\mathrm{i}\,\omega)^2 + \alpha\,\mathrm{i}\,\omega + \lambda_1^2}\,, \qquad W_2(\mathrm{i}\,\omega) = \frac{b_2}{(\mathrm{i}\,\omega)^2 + \alpha\,\mathrm{i}\,\omega + \lambda_2^2}\,.$$

Let us obtain the spectral density $S_u(\eta\,,\omega)$ for the deflection

$$u = f^{(1)}(\tau)\,\varphi^{(1)}(\eta) + f^{(2)}(\tau)\,\varphi^{(2)}(\eta)\,. \tag{7.203}$$

Using the Wiener-Khintchin formula, we get

$$K_u(\eta,\,\tau,\,\hat{\tau}) = \int_{-\infty}^{\infty} \mathrm{e}^{\mathrm{i}\,\omega\tau_2} S_u(\eta,\,\omega)\,\mathrm{d}\omega\,, \qquad \tau_2 = \tau - \hat{\tau}\,. \tag{7.204}$$

For

$$K_u = M\left[u(\tau,\,\eta)\,u^*(\hat{\tau},\,\eta)\right]\,,$$

we have

$$K_u = M\left[\left(f^{(1)}(\tau)\,\varphi^{(1)} + f^{(2)}(\tau)\,\varphi^{(2)}\right)\left(f^{(1)*}(\hat{\tau})\,\varphi^{(1)} + f^{(2)*}(\hat{\tau})\,\varphi^{(2)}\right)\right]$$

or

$$\begin{aligned} K_u(\eta,\,\tau,\,\hat{\tau}) = {} & (\varphi^{(1)})^2 K_{f^{(1)}}(\tau,\,\hat{\tau}) + \varphi^{(1)}\varphi^{(2)} K_{f^{(1)}f^{(2)}}(\tau,\,\hat{\tau}) \\ & + \varphi^{(1)}\varphi^{(2)} K_{f^{(1)}f^{(2)}}(\tau,\,\hat{\tau}) + (\varphi^{(2)})^2 K_{f^{(2)}}(\tau,\,\hat{\tau})\,. \end{aligned} \tag{7.205}$$

The correlation functions on the right-hand side of the equation are to be represented in terms of the spectral densities $S_{f^{(i)}}(\omega)$ and $S_{f^{(i)}f^{(k)}}(\omega)$, which, on their turn, depend on $S_P(\omega)$. The relationship between the spectral density of input and the spectral density of the output $S_P(\omega)$ was discussed in Sect. 6.2 (see (6.22)).

In the problem under consideration, we have

$$S_{f^{(1)}}(\omega) = |W_1(\mathrm{i}\,\omega)|^2 S_P(\omega)\,. \tag{7.206}$$

Similarly, we get the following expressions of the spectral density $S_{f^{(2)}}(\omega)$ and of the cross-spectral densities $S_{f^{(1)}f^{(2)}}$ and $S_{f^{(2)}f^{(1)}}$:

$$S_{f^{(2)}}(\omega) = |W_2(\mathrm{i}\,\omega)|^2 S_P(\omega)\,, \tag{7.207}$$

$$S_{f^{(1)}f^{(2)}} = W_1(\mathrm{i}\,\omega)\,W_2^*(\mathrm{i}\,\omega)\,S_P(\omega)\,, \tag{7.208}$$

$$S_{f^{(2)}f^{(1)}} = W_2(\mathrm{i}\,\omega)\,W_1^*(\mathrm{i}\,\omega)\,S_P(\omega)\,. \tag{7.209}$$

Substituting the expressions for $K_{f^{(1)}}$, $K_{f^{(2)}}$, $K_{f^{(1)}f^{(2)}}$, and $K_{f^{(2)}f^{(1)}}$ into (7.205) and using the Weiner-Khintchin formula, we get

$$K_u(\eta,\,\tau,\,\hat{\tau}) = \int_{-\infty}^{\infty} \mathrm{e}^{\mathrm{i}\,\omega\tau_2} \Big[(\varphi^{(1)})^2 S_{f^{(1)}} + \varphi^{(1)}\varphi^{(2)} S_{f^{(1)}f^{(2)}} + \varphi^{(1)}\varphi^{(2)} S_{f^{(2)}f^{(1)}} + (\varphi^{(2)})^2 S_{f^{(2)}}\Big]\,\mathrm{d}\omega\,. \quad (7.210)$$

On rearrangement, we arrive at

$$S_u(\eta,\,\omega) = (\varphi^{(1)})^2 S_{f^{(1)}} + \varphi^{(1)}\varphi^{(2)}(S_{f^{(1)}f^{(2)}} + S_{f^{(2)}f^{(1)}}) + (\varphi^{(2)})^2 S_{f^{(2)}}\,. \quad (7.211)$$

Substituting (7.206)–(7.209) into this equation, we obtain

$$S_u(\eta,\,\omega) = \Big[(\varphi^{(1)})^2|W_1|^2 + (\varphi^{(2)})^2|W_2|^2 + \varphi^{(1)}\varphi^{(2)}(W_1W_2^* + W_2W_1^*)\Big]\, S_P\,.$$

Let us represent the functions $W_1(\mathrm{i}\,\omega)$ and $W_2(\mathrm{i}\,\omega)$ in the form

$$W_1(\mathrm{i}\,\omega) = W_1^{(1)}(\omega) + \mathrm{i}\,W_1^{(2)}(\omega)\,,$$
$$W_2(\mathrm{i}\,\omega) = W_2^{(1)}(\omega) + \mathrm{i}\,W_2^{(2)}(\omega)\,.$$

Then, on rearrangement, we arrive at

$$W_1W_2^* + W_2W_1^* = 2(W_1^{(1)}W_2^{(1)} + W_2^{(2)}W_1^{(2)})\,.$$

Finally, we obtain

$$S_u(\eta,\,\omega) = \Big[(\varphi^{(1)})^2|W_1|^2 + (\varphi^{(2)})^2|W_2|^2 + 2\varphi^{(1)}\varphi^{(2)}(W_1^{(1)}W_2^{(1)} + W_2^{(2)}W_1^{(2)})\Big]\, S_P\,. \quad (7.212)$$

Consequently, we find the amplitude values of the variance of the displacements of the rod axis points to be

$$D_u(\eta) = \int_{-\infty}^{\infty} S_u(\eta,\,\omega)\,\mathrm{d}\omega\,. \quad (7.213)$$

Similarly, the spectral densities of the amplitude values of the moment $\Delta M_{x30} = u''$ and of the shear force $\Delta Q_{x20} = u'''$ are as follows:

$$S_{\Delta M}(\eta,\,\omega) = \Big[(\varphi^{(1)''})^2|W_1|^2 + (\varphi^{(2)''})^2|W_2|^2 + 2\varphi^{(1)''}\varphi^{(2)''}(W_1^{(1)}W_2^{(1)} + W_2^{(2)}W_1^{(2)})\Big]\, S_P\,, \quad (7.214)$$

$$S_{\Delta Q}(\eta,\,\omega) = \Big[(\varphi^{(1)'''})^2|W_1|^2 + (\varphi^{(2)'''})^2|W_2|^2 + 2\varphi^{(1)'''}\varphi^{(2)'''}(W_1^{(1)}W_2^{(1)} + W_2^{(2)}W_1^{(2)})\Big]\, S_P\,. \quad (7.215)$$

The variances of the amplitude values of the moment and of the shear force are as follows:

$$D_{\Delta M}(\eta) = \int_{-\infty}^{\infty} S_{\Delta M}\, \mathrm{d}\omega\,, \qquad D_{\Delta Q}(\eta) = \int_{-\infty}^{\infty} S_{\Delta Q}\, \mathrm{d}\omega\,. \tag{7.216}$$

Suppose that the amplitude values of u, ΔM_{x30}, and ΔQ_{x20} are normally distributed with zero mathematical expectation; then, amplitude values are

$$\begin{aligned} &\max u(\eta) = 3\sigma_u(\eta)\,, \qquad \max \Delta M_{x30}(\eta) = 3\sigma_{\Delta M}(\eta)\,,\\ &\max \Delta Q_{x20}(\eta) = 3\sigma_{\Delta Q}(\eta)\,, \end{aligned} \tag{7.217}$$

These characteristics of the stress strain state of a rod under the action of random forces are of great importance in design.

7.7 Parametric Vibrations

7.7.1 Equations of Parametric Vibrations of a Straight Rod

Straight rods subjected to a periodic axial force, $P(\tau)$, and a periodic torque, $M(\tau)$, are shown in Fig. 7.23*a, b*. Recall that the force and the moment appear in the equations of small vibrations, say, (7.34) and (7.35). Hence, these equations are equations with periodic coefficients.

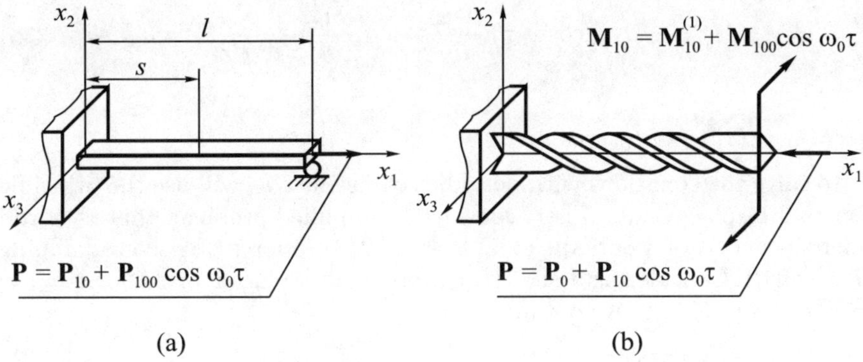

Fig. 7.23.

The rod in Fig. 7.23*b* (a drilling bit) is forced to oscillate along its axis (the high-frequency drilling or vibrodrilling). Due to the axial vibration with the specified frequency ω, the cutting force and the cutting moment acquire an additional component of periodic nature.

There are some monographs and textbooks that provide a detailed discussion of parametric vibrations in mechanical systems (for instance, see

Bolotin (1956) and Svetlitsky (1982)). That is why, in this section, only general methods for some simple cases of parametric vibrations of straight rods are considered. It is worth noting that these methods may be used with profit when studying parametric vibrations of space-curved rods.

Consider some examples of equations of parametric vibrations of straight rods. First, consider a rod that oscillates in the plane of drawing (Fig. 7.23*a*). With regard for the force of viscous damping, it follows from (7.34) that

$$L_1 = \frac{\partial^4 u_{x2}}{\partial \eta^4} - J_{33}\frac{\partial^4 u_{x2}}{\partial \eta^2\,\partial \tau^2} - (P_{10} + P_{100}\cos\omega_0\tau)\frac{\partial^2 u_{x2}}{\partial \eta^2} + \alpha\frac{\partial u_{x2}}{\partial \tau} + \frac{\partial^2 u_{x2}}{\partial \tau^2} = 0\,, \tag{7.218}$$

where

$$Q_{x_1 0} = P_{10} + P_{100}\cos\omega_0\tau\,.$$

For the rod shown in Fig. 7.23*b*, the coupled equations (7.34) and (7.35) take the form

$$L_1 = \frac{\partial^4 u_{x2}}{\partial \eta^4} - J_{33}\frac{\partial^4 u_{x2}}{\partial \eta^2\,\partial \tau^2} - (P_{10} + P_{100}\cos\omega_0\tau)\frac{\partial^2 u_{x2}}{\partial \eta^2} + \frac{M_{10}^{(1)} + M_{100}\cos\omega_0\tau}{A_{22}}\frac{\partial^3 u_{x3}}{\partial \eta^3} + \alpha\frac{\partial u_{x2}}{\partial \tau} + \frac{\partial^2 u_{x2}}{\partial \tau^2} = 0\,, \tag{7.219}$$

$$L_2 = A_{22}\frac{\partial^4 u_{x3}}{\partial \eta^4} - J_{22}\frac{\partial^4 u_{x3}}{\partial \eta^2\,\partial \tau^2} - (P_{10} + P_{100}\cos\omega_0\tau)\frac{\partial^2 u_{x3}}{\partial \eta^2} - (M_{10}^{(1)} + M_{100}\cos\omega_0\tau)\frac{\partial^3 u_{x2}}{\partial \eta^3} + \alpha\frac{\partial u_{x3}}{\partial \tau} + \frac{\partial^2 u_{x3}}{\partial \tau^2} = 0\,, \tag{7.220}$$

where

$$M_{10} = M_{10}^{(1)} + M_{100}\cos\omega_0\tau\,.$$

To solve the equations with periodic coefficients, we will use the principle of virtual displacements. First, consider a simplified problem, that is, a rod under the action of a periodic axial force (7.218) (a force of viscous damping is neglected). Considering a one-term approximate solution of the form $u_{x_2} = \phi^{(1)} f(\tau)$, from (7.128) we obtain

$$\ddot{f} + (a_1 - a_2\cos\omega_0\tau)\,f = 0\,, \tag{7.221}$$

where

$$a_1 = \frac{1}{a_0}\int_0^1 \left[\left(A_{33}\varphi^{(1)\prime\prime}\right)^{\prime\prime}\varphi^{(1)} - P_{10}\varphi^{(1)\prime\prime}\varphi^{(1)}\right] d\eta\,,$$

$$a_2 = \frac{P_{100}}{a_0}\int_0^1 \varphi^{(1)\prime\prime}\varphi^{(1)}\, d\eta\,,$$

$$a_0 = \int_0^1 \left[n_1(\varphi^{(1)})^2 - J_{33}\varphi^{(1)\prime\prime}\varphi^{(1)}\right] d\eta\,.$$

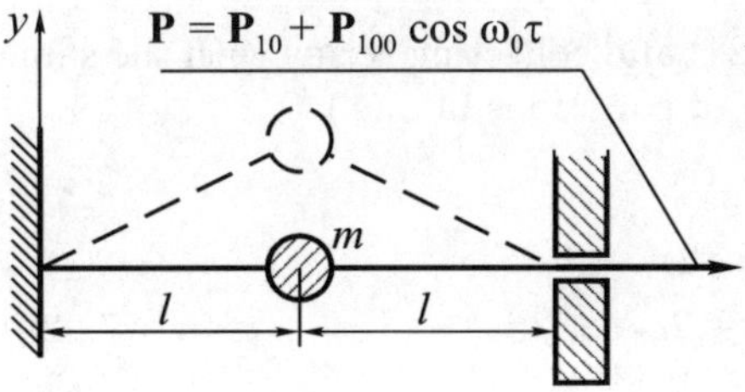

Fig. 7.24.

Equation (7.221) is identical to the equation of vibrations of a single-degree-of-freedom system, say, like that in Fig. 7.24. If the force due to the gravity is neglected, then the equation of small vibrations of a point mass m attached to a string (see Fig. 7.24) is

$$\ddot{y} + \left(\frac{2P_{10}}{ml} + \frac{2P_{100}}{ml} \cos \omega_0 \tau \right) y = 0 . \tag{7.222}$$

7.7.2 Methods for Solution to Equations of Parametric Vibrations of Single-Degree-of-Freedom Systems

The Method of Small Parameter. Let

$$\omega_0 \tau = 2\tau_1 . \tag{7.223}$$

Then, (7.221) and (7.222) become

$$\ddot{f} + (a + 2q \cos 2\tau_1) f = 0 . \tag{7.224}$$

This equation is known as the Mathieu equation. Treating the coefficient q as a small parameter, we seek a solution to (7.224) in the form

$$y_m = \cos m\tau + qc_1(\tau) + q^2 c_2(\tau) + q^3 c_3(\tau) + \dots , \tag{7.225}$$
$$a_m = m^2 + q\alpha_1 + q^2 \alpha_2 + q^3 \alpha_3 + \dots \tag{7.226}$$

and

$$y_m^{(1)} = \sin m\tau + qc_1^{(1)}(\tau) + q^2 c_2^{(1)}(\tau) + q^3 c_3^{(1)}(\tau) + \dots , \tag{7.227}$$
$$a_m^{(1)} = m^2 + q\alpha_1^{(1)} + q^2 \alpha_2^{(1)} + q^3 \alpha_3^{(1)} + \dots , \tag{7.228}$$

It follows from (7.225) and (7.227) that y_m and $y_m^{(1)}$ (for $q = 0$ and $a = m^2$) are the exact partial solutions to (7.224). Periodic solutions correspond to specific values of the parameters a and q; hence, we can subdivide all the solutions to equation (7.224) into two categories: stable solutions and unstable ones. Let us find the periodic solutions.

The parameter m is not necessarily integer; hence, there exist infinitely many solutions given by the series (7.225) and (7.227). Let us obtain partial solutions for some integer values of m, say, for $m = 1$ and $m = 2$. Substituting the expressions for y_m and a_m into (7.224) and collecting terms with the same powers q_k, we get the following system of equations in $c_i(\tau)$:

$$\begin{aligned} &\ddot{c}_1 + c_1 = (1-\alpha_1)\cos\tau + \cos 3\tau\,,\\ &\ddot{c}_2 + c_2 = -\alpha_1 c_1 - \alpha_2\cos\tau + 2c_1\cos 2\tau\,,\\ &\ddot{c}_3 + c_3 = -\alpha_1 c_2 - \alpha_2 c_1 - \alpha_3\cos\tau + 2c_2\cos 2\tau\,,\\ &\dots \end{aligned} \tag{7.229}$$

For the solutions to these equations to be bounded, the arbitrary constants α_i are to be as follows:

$$\alpha_1 = 1\,, \qquad \alpha_2 = -\frac{1}{8}\,, \qquad \alpha_3 = -\frac{1}{64}\,.$$

The result is

$$\begin{aligned} y_1 &= \cos\tau - \frac{1}{8}q\cos 3\tau + \frac{1}{64}q^2\left(-\cos 3\tau + \frac{1}{3}\cos 5\tau\right)\\ &\quad - \frac{1}{512}q^3\left(\frac{1}{3}\cos 3\tau - \frac{4}{9}\cos 5\tau + \frac{1}{18}\cos 7\tau\right) + \dots\,,\\ a_1 &= 1 + q - \frac{1}{8}q^2 - \frac{1}{64}q^3 + \dots\,. \end{aligned}$$

Similarly, we get

$$\begin{aligned} y_1^{(1)} &= \sin\tau - \frac{1}{8}q\sin 3\tau + \frac{1}{64}q^2\left(-\sin 3\tau + \frac{1}{3}\sin 5\tau\right)\\ &\quad - \frac{1}{512}q^3\left(\frac{1}{3}\sin 3\tau - \frac{4}{9}\sin 5\tau + \frac{1}{18}\sin 7\tau\right) + \dots\,,\\ a_1^{(1)} &= 1 - q - \frac{1}{8}q^2 + \frac{1}{64}q^3 + \dots\,. \end{aligned}$$

The accepted notation for the functions y_1 and $y_1^{(1)}$ is $y_1 = \mathrm{ce}_1(\tau,\, q)$ and $y_1^{(1)} = \mathrm{se}_1(\tau,\, q)$. The functions $\mathrm{ce}_1(\tau,\, q)$ and $\mathrm{se}_1(\tau,\, q)$ are known as the first-order Mathieu functions of the first kind. The period of $\mathrm{ce}_1(\tau,\, q)$ and $\mathrm{se}_1(\tau,\, q)$ is 2π.

The second-order functions of the first kind correspond to the value $m = 2$. On rearrangement, we arrive at

$$\begin{aligned} \mathrm{ce}_2(\tau,\, q) &= \cos 2\tau - \frac{1}{8}q\left(\frac{2}{3}\cos 4\tau - 2\right) + \frac{1}{384}q^2\cos 6\tau\\ &\quad - \frac{1}{512}q^3\left(\frac{1}{45}\cos 8\tau + \frac{43}{27}\cos 4\tau + \frac{40}{3}\right) + \dots\,,\\ a_2 &= 4 + \frac{5}{12}q^2 - \frac{763}{13824}q^4 + \dots\,,\\ \mathrm{se}_2(\tau,\, q) &= \sin 2\tau - \frac{1}{12}q\sin 4\tau + \frac{1}{384}q^2\sin 6\tau \end{aligned}$$

$$- \frac{1}{512} q^3 \left(\frac{1}{45} \sin 8\tau - \frac{5}{27} \sin 4\tau \right) + \dots ,$$

$$a_2^{(1)} = 4 - \frac{1}{12} q^2 + \frac{5}{13824} q^4 + \dots .$$

The functions $\mathrm{ce}_2(\tau, q)$ and $\mathrm{se}_2(\tau, q)$ are periodic with period π.

For the third-order functions of the first kind, the parameters a_3 and $a_3^{(1)}$ are

$$a_3 = 9 + \frac{1}{16} q^2 + \frac{1}{64} q^3 + \dots ,$$

$$a_3^{(1)} = 9 + \frac{1}{16} q^2 - \frac{1}{64} q^3 + \dots .$$

For the particular case of $m = 0$, it can be shown that the parameters a_0 and $a_0^{(1)}$ are

$$a_0 = a_0^{(1)} = -\frac{1}{2} q^2 + \frac{7}{128} q^4 - \frac{29}{2304} q^6 + \dots .$$

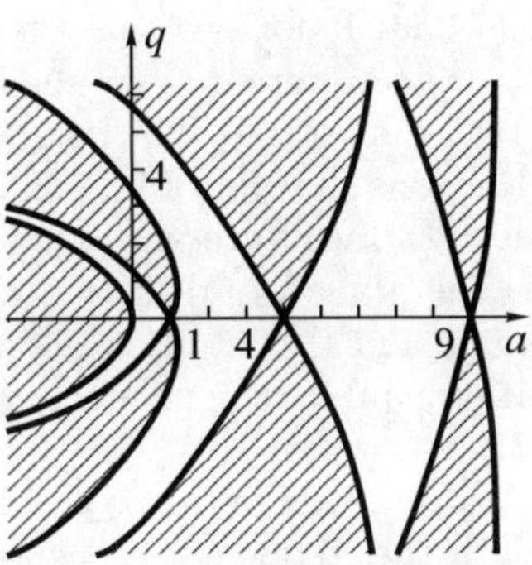

Fig. 7.25.

Relationships between $a_i(q)$ and $a_i^{(1)}(q)$, on the one hand, and q, on the other hand, for which periodic solutions exist are called the eigenvalues of the Mathieu functions of the first kind. The eigenvalues a_i and $a_i^{(1)}$ are of great importance in the approximate solutions obtains because the curves $a_i(q)$ and $a_i^{(1)}(q)$ divide the plane $(a\,, q)$ into regions of stable solutions and regions of unstable ones (which are further referred as *stable* and *unstable* regions) (Fig. 7.25). It is yet unknown which regions provide stable solutions and which regions provide unstable ones. To clarify the situation, consider approximate non-periodic solutions to (7.224) for a fractional number $m = \nu$. We have

$$\begin{aligned}
&\mathrm{ce}_\nu(\tau,\, q) = \cos \nu\tau + q c_1 + q^2 c_2 + \dots\,, \\
&a_\nu = \nu^2 + q\alpha_1 + q^2\alpha_2 + \dots\,,
\end{aligned} \tag{7.230}$$

$$\begin{aligned}
&\mathrm{se}_\nu(\tau,\, q) = \sin \nu\tau + q c_1^{(1)} + q^2 c_2^{(1)} + \dots\,, \\
&a_\nu = \nu^2 + q\alpha_1^{(1)} + q^2\alpha_2^{(1)} + \dots\,.
\end{aligned} \tag{7.231}$$

Substituting these into (7.224), we obtain

$$\begin{aligned}
\mathrm{ce}_\nu(\tau,\, q) &= \cos \nu\tau - \frac{1}{4}\, q \left(\frac{\cos(\nu+2)\,\tau}{\nu+1} - \frac{\cos(\nu-2)\,\tau}{\nu-1} \right) \\
&+ \frac{1}{32}\, q^2 \left(\frac{\cos(\nu+4)\,\tau}{(\nu+1)(\nu+2)} + \frac{\cos(\nu-4)\,\tau}{(\nu-1)(\nu-2)} \right) - \dots\,,
\end{aligned} \tag{7.232}$$

$$\begin{aligned}
\mathrm{se}_\nu(\tau,\, q) &= \sin \nu\tau - \frac{1}{4}\, q \left(\frac{\sin(\nu+2)\,\tau}{\nu+1} - \frac{\sin(\nu-2)\,\tau}{\nu-1} \right) \\
&+ \frac{1}{32}\, q^2 \left(\frac{\sin(\nu+4)\,\tau}{(\nu+1)(\nu+2)} + \frac{\sin(\nu-4)\,\tau}{(\nu-1)(\nu-2)} \right) - \dots\,,
\end{aligned} \tag{7.233}$$

$$a_\nu = a_\nu^{(1)} = \nu^2 + \frac{q^2}{2(\nu^2-1)} + \frac{(5\nu^2+7)}{32(\nu^2-1)^3(\nu^2-4)}\, q^4 + \dots\,. \tag{7.234}$$

From the approximate solutions (7.232) and (7.233) it follows that the solutions corresponding to the fractional number ν are non-periodic bounded functions of time. Hence, these solutions can be treated as stable. The corresponding curves $a_\nu = a_\nu(q)$ lie in the non-hatched regions shown in Fig. 7.25. Thus, considering the functions of fractional values of ν, we have an opportunity to separate stable and unstable regions on the plane $(a,\, q)$. Namely, the hatched regions correspond to unstable solutions and the blank regions correspond to stable solutions. The plane $(a,\, q)$ with unstable regions marked in is called the Ince–Strutt diagram (see Fig. 7.25).

Consider the unstable region that corresponds to $a = 1$. If we put $a_2 = 0$ in (7.221), then the equation of free vibrations with the frequency $p_1^2 = a_1$ results. Upon change of time (7.223), we get

$$a = \frac{4p_1^2}{\omega^2}\,.$$

Hence, we have $a = 1$ provided $\omega = 2p_1$ (that is, when the parameter ω equals the doubled frequency of free vibrations of the system). The unstable region in the Ince–Strutt diagram for $a = 1$ is called the region of principal parametric resonance. For the regions corresponding to $a = 4$, we have $\omega = p_1$. The Ince–Strutt diagram clearly shows why parametric vibrations may constitute a real danger. The matter is that the values of ω for which the unstable vibrations (parametric resonance) occur form an interval of nonzero length unlike the case of conventional resonance when unstable vibrations occur at isolated value of the frequency ω.

The advantage of the method for solution to the Mathieu equation presented above is that the values of a and q corresponding to unstable solutions were obtained without actual determination of the solutions themselves. At the same time, the weaknesses of the method should also be pointed out. First, the method is unsuitable for systems of equations (though many of mechanical systems can be approximated by (7.224)). Second, it is required that the parameter q is small because the series (7.225)–(7.228) may diverge for large q. Third, the method is hardly applied when terms responsible for viscous damping are included in the equation.

The Rayleigh Method. With regard for the force of viscous damping, the equation of parametric vibrations takes the form

$$\ddot{f} + n\dot{f} + (a_1 - a_2 \cos \omega_0 \tau)\, f = 0\,, \tag{7.235}$$

where

$$n = \frac{1}{a_0} \int_0^1 (\varphi^{(1)})^2 \,\mathrm{d}\eta\,.$$

In accordance with the Rayleigh method, we seek the solution to (7.235) in the form

$$f = \sum_{j=1,3,5} \left(A_j \sin \frac{j\omega_0 \tau}{2} + B_j \cos \frac{j\omega_0 \tau}{2} \right). \tag{7.236}$$

The period of this solution is $2T$ that is twice as much as the period of the parameter. For illustrative purposes, consider only a monomial approximation

$$f = A_1 \sin \frac{\omega_0}{2}\tau + B_1 \cos \frac{\omega_0}{2}\tau\,. \tag{7.237}$$

Substituting (7.237) into (7.235) and retaining only the terms containing $\cos(\omega_0\tau/2)$ and $\sin(\omega_0\tau/2)$, we obtain the following system of equations:

$$\begin{aligned} &\left[a_1 - \frac{a_2}{2} - \left(\frac{\omega_0}{2}\right)^2\right] A_1 - n\left(\frac{\omega_0}{2}\right) B_1 = 0\,, \\ &n\left(\frac{\omega_0}{2}\right) A_1 + \left[a_1 + \frac{a_2}{2} - \left(\frac{\omega_0}{2}\right)^2\right] B_1 = 0\,. \end{aligned} \tag{7.238}$$

First, let us determine the unstable region in the absence of the force of viscous damping ($n = 0$). The boundaries of the regions (see Fig. 7.26) can be defined by use of (7.238) (for $A_1 \neq 0$ and $B_1 \neq 0$):

$$\begin{aligned} &a_1 - \frac{a_2}{2} - \left(\frac{\omega_0}{2}\right)^2 = 0\,, \\ &a_1 - \frac{a_2}{2} + \left(\frac{\omega_0}{2}\right)^2 = 0\,. \end{aligned}$$

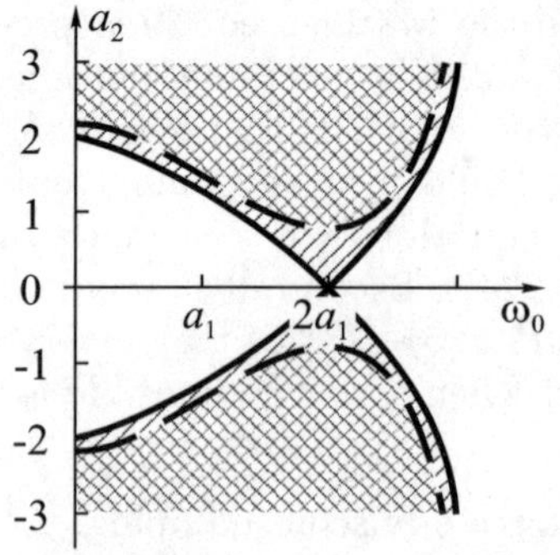

Fig. 7.26.

As expected, the unstable region originates at the value $\omega_0 = 2\sqrt{a_1}$, which is equal to the doubled natural frequency.

Let us now find the regions for the case when the force of viscous damping is taken into consideration.

Equating the determinant of the system of equations (7.238) to zero, we get

$$\left[a_1 - \frac{a_2}{2} - \left(\frac{\omega_0}{2}\right)^2\right]\left[a_1 + \frac{a_2}{2} - \left(\frac{\omega_0}{2}\right)^2\right] + n^2\left(\frac{\omega_0}{2}\right)^2 = 0\,. \qquad (7.239)$$

This equation sets up a correspondence between critical values of parameters. (The parameters a_1 and a_2 are critical when they correspond to the boundaries of the unstable regions.) Solving it for a_2, we get

$$a_2 = \pm 2\sqrt{n^2\left(\frac{\omega_0}{2}\right)^2 + \left[a_1 - \left(\frac{\omega_0}{2}\right)^2\right]^2}\,.$$

The dashed line in Fig. 7.26 shows the boundaries of the unstable regions for $n = 0.5$. The presence of dissipative forces may substantially shrink the unstable regions. The region corresponding to $\omega_0 = 2\sqrt{a_1}$ is called the principal unstable region. This region can be determined more accurately by considering an approximate solution of the form

$$f = A_1 \sin\frac{\omega_0}{2}\tau + B_1\cos\frac{\omega_0}{2}\tau + A_2\sin\frac{3}{2}\omega_0\tau + B_2\cos\frac{3}{2}\omega_0\tau\,.$$

The following system of homogeneous equations results:

$$\left[a_1 - \frac{a_2}{2} - \left(\frac{\omega_0}{2}\right)^2\right]A_1 - \frac{n\omega_0}{2}B_1 + \frac{\omega_0}{2}A_2 = 0\,,$$

$$\frac{n\omega_0}{2}A_1 + \left[a_1 + \frac{a_2}{2} - \left(\frac{\omega_0}{2}\right)^2\right]B_1 + \frac{a_2}{2}B_2 = 0\,,$$

$$\frac{a_2}{2}A_1 + \left[a_1 - \left(\frac{3}{2}\omega_0\right)^2\right]A_2 - \frac{3n\omega_0}{2}B_2 = 0\,,$$

$$\frac{a_2}{2}\,B_1 + \frac{3n\omega_0}{2}\,A_2 + B_2 = 0\,. \tag{7.240}$$

Equating the determinant of this system of equations to zero, we obtain

$$D = \det\begin{bmatrix} a_1 - \dfrac{a_2}{2} - \dfrac{\omega_0^2}{4} & -\dfrac{n\omega_0}{2} & \dfrac{a_2}{2} & 0 \\ \dfrac{n\omega_0}{2} & a_1 + \dfrac{a_2}{2} - \dfrac{\omega_0^2}{4} & 0 & \dfrac{a_2}{2} \\ \dfrac{a_2}{2} & 0 & a_1 - \dfrac{9\omega_0^2}{4} & -\dfrac{3n\omega_0}{2} \\ 0 & \dfrac{a_2}{2} & \dfrac{3n\omega_0}{2} & 1 \end{bmatrix} = 0\,.$$

Expanding the determinant, we obtain the relationship $a_2 = a_2(\omega_0, n)$. This relationship defines more accurately the boundaries of the principal unstable region and provides a crude approximation for the boundaries of the next unstable region, for which $\omega_0 = 2\sqrt{a_1}/3$. Then, using a six-term approximate solution, we can find the boundaries of the first two unstable regions more accurately and obtain a crude approximation for the unstable region for $\omega_0 = 2\sqrt{a_1}/5$.

To determine unstable regions whose boundaries correspond to periodic solutions with period T, we should seek a series solution to (7.235):

$$f = A_0 + \sum_{j=2,\,4,\,6}\left(A_j \sin\frac{\omega_0\tau}{2} + B_j \cos\frac{\omega_0\tau}{2}\right)\,.$$

Considering the first approximation ($j = 2$), we get the following system of equations in A_0, A_2, and B_2:

$$\begin{aligned} &A_0 + \frac{1}{2}\,a_2 B_2 = 0\,,\\ &(a_1 - \omega_0^2)\,A_2 - n\omega_0 B_2 = 0\,,\\ &n\omega_0 A_2 - (a_1 - \omega_0^2)\,B_2 + a_2 A_0 = 0\,. \end{aligned} \tag{7.241}$$

Equating the determinant of the coefficients to zero gives

$$(a_1 - \omega_0^2)\left(a_1 - \omega_0^2 - \frac{1}{2}\,a_2^2\right) + n^2\omega_0^2 = 0\,. \tag{7.242}$$

The unstable region corresponding to the equation of the boundary in terms of the coordinates a_1 and a_2 (for $n = 0$) is shown in Fig. 7.27. The region originates at the point $\omega_0 = \sqrt{a_1}$.

To find the boundaries of the region more accurately, we should consider solutions of higher-order approximation. For the second approximation, it follows from (7.235) that

$$f = A_0 + A_2 \sin\omega_0\tau + B_2 \cos\omega_0\tau + A_4 \sin 2\omega_0\tau + B_4 \cos 2\omega_0\tau\,.$$

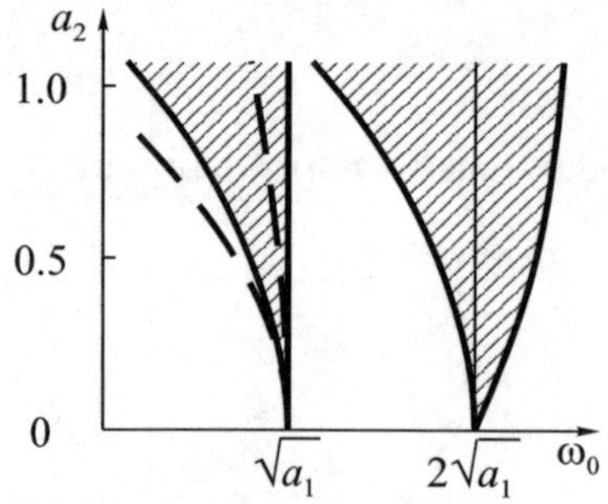

Fig. 7.27.

This gives

$$
\begin{aligned}
&A_0 + \frac{1}{2}\,a_2 B_2 = 0\,,\\
&(a_1 - \omega_0^2)\,A_2 - n\omega_0 A_2 + \frac{a_2}{2}\,A_4 = 0\,,\\
&n\omega_0 A_2 + (a_1 - \omega_0^2)\,B_2 + a_2 A_0 + \frac{a_2}{2}\,B_4 = 0\,,\\
&\frac{a_2}{2}\,A_2 + (a_1 - 4\omega_0^2)\,A_4 - 2n\omega_0 B_4 = 0\,,\\
&\frac{a_2}{2}\,B_2 + 2n\omega_0 A_4 + (a_1 - 4\omega_0^2)\,B_4 = 0\,.
\end{aligned}
\tag{7.243}
$$

By equating the determinant of this system of equations to zero, we obtain a relationship between the parameters, which for $n = 0$ reads

$$
\begin{aligned}
&(a_1 - \omega_0^2)\left[\left(a_1 - \omega_0^2 - \frac{a_2^2}{2}\right)(a_1 - 4\omega_0^2)^2 - \frac{a_2^2}{2}(a_1 - 4\omega_0^2)\right]\\
&\qquad + \frac{a_2^2}{4}\left[\frac{a_2^2}{4} - \left(a_1 - \omega_0^2 - \frac{a_2^2}{2}\right)(a_1 - 4\omega_0^2)\right] = 0\,.
\end{aligned}
\tag{7.244}
$$

The refined boundaries obtained from this relationship are shown in Fig. 7.27 as a dashed line. From this figure it is seen that in the case of the second approximation the boundaries of the regions intersect when the parameter a_2 is large. For some values of the coefficients n and a_j in the relationship (7.235), unstable regions may differ substantially from those obtained on the bases of the Mathieu equation. Note that the Rayleigh method determines the unstable regions just approximately. It is interesting then to compare the approximate solutions to the exact ones. To determine the unstable regions exactly, one should solve the homogeneous equation (7.235). A method for exact numerical determination of unstable regions can be found in the literature (for instance, see Svetlitsky (1982)).

The Rayleigh Method for Systems of Equations with Periodic Coefficients. Consider an approximate solution to (7.218) of the form

$$u = \sum_{j=1}^{k} \varphi^{(j)}(\eta)\, f^{(j)}(\tau)\,,$$

Then, making manipulations as in the case of application of the principle of virtual displacements, we get the system of equations in the functions $f^{(j)}$ which can be written as a vector equation

$$\ddot{\mathbf{f}} + \mathrm{A}^{(1)}\dot{\mathbf{f}} + \mathrm{A}^{(2)}(t)\,\mathbf{f} = 0\,, \tag{7.245}$$

where $\mathrm{A}^{(1)}$ is a matrix with constant elements and the elements of the matrix $\mathrm{A}^{(2)}$ are periodic functions. For example, suppose that the elements of the matrix $\mathrm{A}^{(2)}$ are

$$a_{ij}^{(2)} = a_{ij0}^{(2)} + b_{ij}^{(2)} \cos\omega_0\tau\,; \tag{7.246}$$

then, the matrix $\mathrm{A}^{(2)}$ can be written as

$$\mathrm{A}^{(2)} = \mathrm{A}_0^{(2)} + \mathrm{B}^{(2)}\cos\omega_0\tau\,. \tag{7.247}$$

Denote the period of variation of the coefficients of the equation by

$$T = \frac{2\pi}{\omega_0}\,.$$

We seek a $2T$-periodic solution to (7.245) in the form

$$\mathbf{f} = \sum_{j=1,3,5} \left(\mathbf{a}_j \sin\frac{j\omega_0\tau}{2} + \mathbf{b}_j\,\frac{j\omega_0\tau}{2} \right)\,. \tag{7.248}$$

Then, we substitute (7.247) and (7.248) into (7.245). Using the relationships

$$\cos\omega_0\tau\,\sin\frac{j\omega_0\tau}{2} = \frac{1}{2}\left[\sin\left(\frac{j}{2}-1\right)\omega_0\tau + \sin\left(\frac{j}{2}+1\right)\omega_0\tau\right],$$

$$\sin\omega_0\tau\,\sin\frac{j\omega_0\tau}{2} = \frac{1}{2}\left[\cos\left(1-\frac{j}{2}\right)\omega_0\tau - \cos\left(1-\frac{j}{2}\right)\omega_0\tau\right],$$

$$\cos\omega_0\tau\,\cos\frac{j\omega_0\tau}{2} = \frac{1}{2}\left[\cos\left(1-\frac{j}{2}\right)\omega_0\tau + \cos\left(1+\frac{j}{2}\right)\omega_0\tau\right],$$

$$\cos\omega_0\tau\,\sin\frac{j\omega_0\tau}{2} = \frac{1}{2}\left[\sin\left(\frac{j}{2}-1\right)\omega_0\tau + \sin\left(\frac{j}{2}+1\right)\omega_0\tau\right],$$

and combining terms with the same multipliers $\sin(k\omega_0\tau/2)$ or $\cos(k\omega_0\tau/2)$, we get

$$(I)_1 \sin\frac{\omega_0}{2}\,\tau + (I)_2 \cos\frac{\omega_0}{2}\,\tau + (II)_1 \cos\frac{3}{2}\,\omega_0\tau + (II)_2 \sin\frac{3}{2}\,\omega_0\tau$$
$$+ (III)_1 \sin\frac{5}{2}\,\omega_0\tau + (III)_2 \cos\frac{5}{2}\,\omega_0\tau + \ldots = 0\,.$$

Equating the expressions in brackets to zero, we obtain the system of algebraic equations to be solved for the unknown vectors $\mathbf{a}_j$ and $\mathbf{b}_j$:

$$(I)_1 = 0\,, \qquad (I)_2 = 0\,,$$
$$(II)_1 = 0\,, \qquad (II)_2 = 0\,, \qquad \ldots .$$

For example, for a one-term approximate solution, $j = 1$, this system is

$$\left[\mathrm{A}_0^{(2)} - \left(\frac{\omega_0}{2}\right)^2 \mathrm{E} - \frac{1}{2}\mathrm{B}^{(2)}\right]\mathbf{a}_1 - \frac{\omega_0}{2}\mathrm{A}^{(1)}\mathbf{b}_1 = 0\,,$$
$$\frac{\omega_0}{2}\mathrm{A}^{(1)}\mathbf{a}_1 + \left[\mathrm{A}_0^{(2)} - \left(\frac{\omega_0}{2}\right)^2 \mathrm{E} + \frac{1}{2}\mathrm{B}^{(2)}\right]\mathbf{b}_1 = 0\,. \qquad (7.249)$$

For the system to have a nonzero solution, the determinant must vanish, that is,

$$D = \det\begin{bmatrix} \mathrm{A}_0^{(2)} - \frac{\omega_0^2}{4}\mathrm{E} - \frac{1}{2}\mathrm{B}^{(2)} & -\frac{\omega_0}{2}\mathrm{A}^{(1)} \\ \frac{\omega_0}{2}\mathrm{A}^{(1)} & \mathrm{A}_0^{(2)} - \frac{\omega_0^2}{4}\mathrm{E} + \frac{1}{2}\mathrm{B}^{(2)} \end{bmatrix} = 0\,. \qquad (7.250)$$

This equation allows us to obtain an equation relating the critical values of the parameters of the system and thus to find the boundaries of the principal unstable regions lying in the vicinity of the frequencies $\omega_0 = 2p_k$, where p_k are the natural frequencies of the rod.

Consider a special case when the elements of the matrix $\mathrm{A}^{(1)}$ are zero (forces of viscous damping are ignored). Using (7.250), we arrive at

$$D_1 = \det\left[\mathrm{A}_0^{(2)} - \frac{\omega_0^2}{4}\mathrm{E} - \frac{1}{2}\mathrm{B}^{(2)}\right] = 0\,,$$
$$D_2 = \det\left[\mathrm{A}_0^{(2)} - \frac{\omega_0^2}{4}\mathrm{E} + \frac{1}{2}\mathrm{B}^{(2)}\right] = 0\,.$$

The boundaries of the unstable regions can be found from these conditions.

Solving for $b_{11}^{(2)}$ (an element of the matrix $\mathrm{B}^{(2)}$), we get the unstable regions in the plane $(b_{11}^{(2)}, \omega_0)$. Qualitatively, these regions are shown in Fig. 7.28*a*. In the figure, p_i are dimensionless frequencies. The unstable regions for the case when a force of viscous damping is taken into consideration are shown in Fig. 7.28 as a dashed line. The unstable regions that correspond to a T-periodic solution

$$\mathbf{f} = \mathbf{a}_0 + \sum_{j=2,\,4,\,6}\left(\mathbf{a}_j \sin\frac{j\omega_0\tau}{2} + \mathbf{b}_j \cos\frac{j\omega_0\tau}{2}\right)$$

can be determined similarly.

Considering a one-term approximate solution, $j = 2$, we arrive at the following system of equations in the vectors $\mathbf{a}_0$, $\mathbf{a}_2$, and $\mathbf{b}_2$:

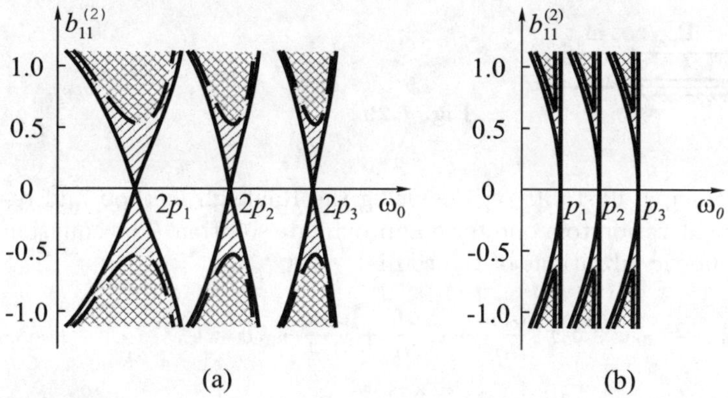

Fig. 7.28.

$$
\begin{aligned}
&\mathrm{A}_0^{(2)}\mathbf{a}_0 + \frac{1}{2}\,\mathrm{B}^{(2)}\mathbf{b}_2 = 0\,,\\
&\mathrm{B}^{(2)}\mathbf{a}_0 + \mathrm{A}^{(1)}\omega_0\mathbf{a}_2 + (\mathrm{A}_0^{(2)} - \omega_0^{(2)}\mathrm{E})\,\mathbf{b}_2 = 0\,,\\
&(\mathrm{A}_0^{(2)} - \omega_0^{(2)}\mathrm{E})\,\mathbf{a}_2 - \mathrm{A}^{(1)}\omega_0\mathbf{b}_2 = 0\,.
\end{aligned}
$$

Eliminating the vector $\mathbf{a}_0$, we get the system of two equations

$$
\begin{aligned}
&\mathrm{A}^{(1)}\omega_0\mathbf{a}_2 + \left[\mathrm{A}_0^{(2)} - \omega_0^2\mathrm{E} - \frac{1}{2}\,\mathrm{B}^{(2)}(\mathrm{A}_0^{(2)})^{-1}\mathrm{B}^{(2)}\right]\,\mathbf{b}_2 = 0\,,\\
&(\mathrm{A}_0^{(2)} - \omega_0^{(2)}\mathrm{E})\,\mathbf{a}_2 - \mathrm{A}^{(1)}\omega_0\mathbf{b}_2 = 0\,.
\end{aligned}
$$

The boundaries of the unstable regions can be found through the condition

$$
D = \det\begin{bmatrix} \mathrm{A}^{(1)}\omega_0 & \mathrm{A}_0^{(2)} - \omega_0^2\mathrm{E} - \frac{1}{2}\,\mathrm{B}^{(2)}(\mathrm{A}_0^{(2)})^{-1}\mathrm{B}^{(2)} \\ \mathrm{A}_0^{(3)} - \omega_0^2\mathrm{E} & -\mathrm{A}^{(1)}\omega_0 \end{bmatrix} = 0\,. \tag{7.251}
$$

Let us ignore the force of viscous damping. Then, the boundaries of the unstable regions are to be found from the equations

$$
D_1 = \det\left[\mathrm{A}_0^{(2)} - \omega_0^2\mathrm{E} - \frac{1}{2}\,\mathrm{B}^{(2)}(\mathrm{A}_0^{(2)})^{-1}\mathrm{B}^{(2)}\right] = 0\,, \tag{7.252}
$$

$$
D_2 = \det\left[\mathrm{A}_0^{(2)} - \omega_0^2\mathrm{E}\right] = 0\,. \tag{7.253}
$$

For the case of a free parameter $b_{11}^{(2)}$, these regions are shown qualitatively in Fig. 7.28*b*.

As an example, consider a rod of constant cross section lying on an elastic foundation (Fig. 7.29). An axial oscillatory force is applied to the rod. Let

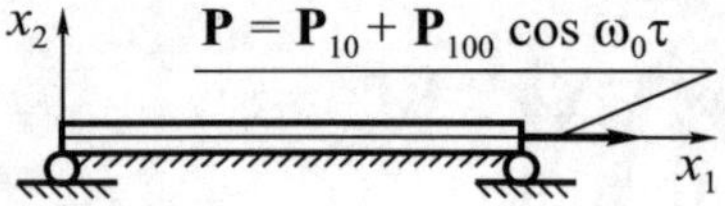

Fig. 7.29.

us obtain the principal unstable regions using the Rayleigh method and restricting our consideration to a one-term approximate solution. The equation of flexural parametric vibrations of the rod is

$$\frac{\partial^4 u}{\partial \eta^4} - (P_{10} + P_{100}\cos\omega_0\tau)\frac{\partial^2 u}{\partial \eta^2} + \alpha\frac{\partial u}{\partial \tau} + \frac{\partial^2 u}{\partial \tau^2} + ku = 0\,.$$

Setting $u = f(\tau)\sin\pi\eta$, we obtain the following equation in f:

$$\ddot{f} + \left[k + \pi^4 + \pi^2(P_{10} + P_{100}\cos\omega_0\tau)\right] f = 0\,.$$

This equation can be represented as

$$\ddot{f} + \alpha\dot{f} + (a_1 + a_2\cos\omega_0\tau)\, f = 0\,, \tag{7.254}$$

where

$$a_1 = k + \pi^4 + \pi^2 P_{10}\,, \qquad a_2 = \pi^2 P_{100}\,.$$

In accordance with the Rayleigh method, we put

$$f = A_1\cos\frac{\omega_0}{2}\tau + B_1\sin\frac{\omega_0}{2}\tau\,. \tag{7.255}$$

Substituting this into (7.254), we get two equations

$$\left[a_1 - \left(\frac{\omega_0}{2}\right)^2 + \frac{a_2}{2}\right]A_1 + \frac{\alpha\omega_0}{2}B_1 = 0\,,$$

$$-\frac{\alpha\omega_0}{2}A_1 + \left[a_1 - \left(\frac{\omega_0}{2}\right)^2 - \frac{a_2}{2}\right]B_1 = 0\,. \tag{7.256}$$

The boundaries of the principal unstable region are determined through the condition that the determinant of the system of equations (7.256) is zero,

$$\left[a_1 - \left(\frac{\omega_0}{2}\right)^2 + \frac{a_2}{2}\right]\left[a_1 - \left(\frac{\omega_0}{2}\right)^2 - \frac{a_2}{2}\right] + \frac{\alpha^2\omega_0^2}{4} = 0\,. \tag{7.257}$$

For $\alpha = 0$, this equation gives the limiting boundaries of the unstable region

$$a_2 = 2\left[-a_1 + \left(\frac{\omega_0}{2}\right)^2\right], \qquad a_2 = 2\left[a_1 - \left(\frac{\omega_0}{2}\right)^2\right].$$

The unstable region is shown by hatching in Fig. 7.30.

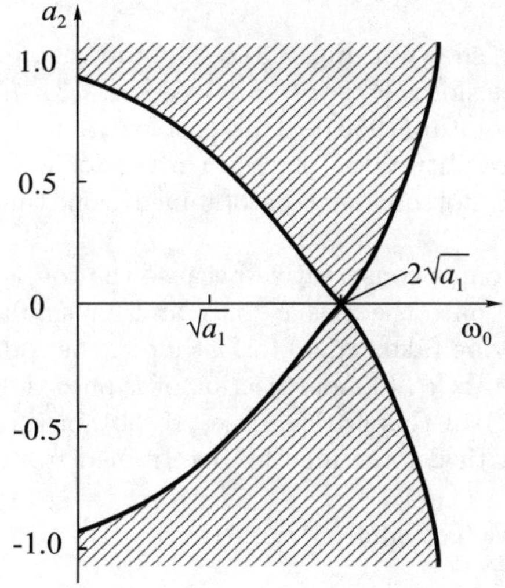

Fig. 7.30.

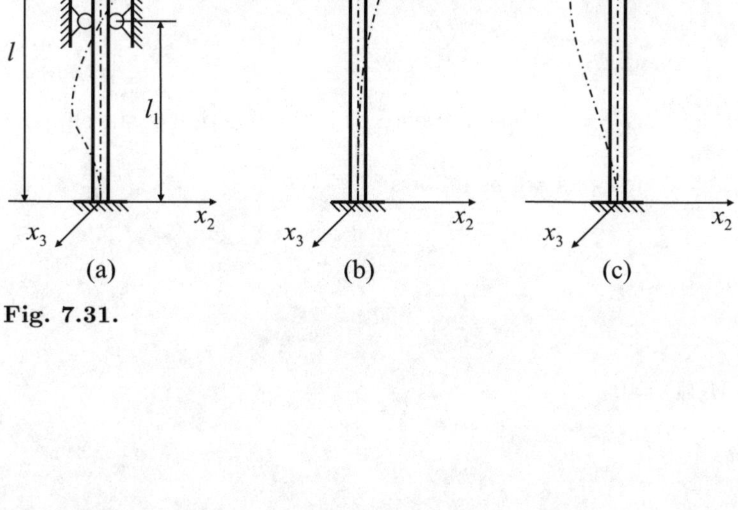

Fig. 7.31.

7.8 Stability of Straight Rods Loaded by Follower Forces

7.8.1 Stability of a Rod with a Lateral Support

A straight rod of circular cross section is shown in Fig. 7.31*a*. The rod is supported by a hinge and loaded by a follower force **P**. For $l_1 = 0$ (see Fig. 7.31*b*), we have the classical Reut problem (Bolotin (1956)). This problem is nonconservative because dynamic loss of stability takes place when the force P achieves some critical value. The critical force cannot be obtained from the equations of equilibrium.

For $l_1 = l$ (see Fig. 7.31*c*), the problem is conservative because the rod is loaded by the projection of the force P onto the axis x_1 only and for small values of the angle α this projection may be taken equal to P, that is, the rod is loaded by the dead force $P_{x_1} = P\,\cos\alpha \approx P$. (The projection of P onto the axis x_2 is compensated by the support.) In this critical case, we obtain the classical Euler problem and thus the critical force may be determined from the equations of equilibrium.

The following question arises: Up to what values of $l_1^{(1)}$ $(0 < l_1 \leq l_1^{(1)})$ the dynamic loss of stability may take place?

To answer this question, let us consider small vibrations of a rod taking into account a local linkage (a pin-type support) in the plane x_1Ox_2 (see Fig. 7.31*a*). The equations of small vibrations of a straight rod of constant cross section can be obtained as a particular case of the system of equations (7.36). Assuming for simplicity $\Delta Q_1 = Q_1$, $\Delta Q_2 = Q_2$, and $\Delta M_3 = M_3$, we arrive at

$$\begin{aligned}
&\frac{\partial^2 u_2}{\partial \tau^2} - \frac{\partial Q_2}{\partial \eta} - Q_{10} M_3 - \Delta P_2 = 0\,,\\
&\frac{\partial M_3}{\partial \eta} - Q_2 = 0\,,\\
&\frac{\partial \vartheta_3}{\partial \eta} - M_3 = 0\,,\\
&\frac{\partial u_2}{\partial \eta} - \vartheta_3 = 0\,,
\end{aligned} \tag{7.258}$$

where $Q_{10} = P$ and $\Delta P_2 = R\,\delta(\eta - \eta_1)$; here, R is the reaction of the support due to vibrations.

The boundary conditions are as follows.

(1) At $\eta = 0$,

$$\vartheta_3(0) = u_2(0) = 0\,.$$

(2) At $\eta = 1$,

$$Q_2(1) = M_3(1) = 0\,.$$

In the vector-matrix form, the system of equations (7.258) reads

$$\mathrm{A}^{(1)} \frac{\partial^2 \mathbf{Z}}{\partial \tau^2} + \frac{\partial \mathbf{Z}}{\partial \eta} + \mathrm{A}^{(2)} \mathbf{Z} = \mathbf{b}\,, \tag{7.259}$$

where

$$\mathbf{Z} = (Q_2,\, M_3,\, \vartheta_3,\, u_2)^{\mathrm{T}}\,, \qquad \mathbf{b} = (-R\,\delta(\eta - \eta_1),\, 0,\, 0,\, 0)^{\mathrm{T}}$$

$$\mathrm{A}^{(1)} = \begin{bmatrix} 0 & 0 & 0 & -1 \\ 0 & 0 & 0 & 0 \\ 0 & 0 & 0 & 0 \\ 0 & 0 & 0 & 0 \end{bmatrix}, \qquad \mathrm{A}^{(2)} = \begin{bmatrix} 0 & Q_{10} & 0 & 0 \\ -1 & 0 & 0 & 0 \\ 0 & -1 & 0 & 0 \\ 0 & 0 & -1 & 0 \end{bmatrix}.$$

To determine the complex eigenvalues, let us use the representation

$$\mathbf{Z} = (\mathbf{Z}_0^{(1)} + \mathrm{i}\,\mathbf{Z}_0^{(2)})\,\mathrm{e}^{\lambda\tau}\,, \quad \mathbf{R} = (\mathbf{R}_{01} + \mathrm{i}\,\mathbf{R}_{02})\,\mathrm{e}^{\lambda\tau}\,, \quad \lambda = \alpha + \mathrm{i}\,\beta\,, \tag{7.260}$$

where

$$\mathbf{Z}_0^{(j)} = (Q_{20}^{(j)},\, M_{30}^{(j)},\, \vartheta_{30}^{(j)},\, u_{20}^{(j)})^{\mathrm{T}}\,, \qquad j = 1,\, 2\,.$$

Substituting $\mathbf{Z}$ given by (7.260) into the vector equation (7.259), we get

$$\begin{aligned} &\mathrm{A}^{(1)}\,(\mathbf{Z}_0^{(1)} + \mathrm{i}\,\mathbf{Z}_0^{(2)})\,(\alpha^2 + 2\,\mathrm{i}\,\alpha\beta - \beta^2) + \frac{\partial \mathbf{Z}_0^{(1)}}{\partial \eta} + \mathrm{i}\,\frac{\partial \mathbf{Z}_0^{(2)}}{\partial \eta} \\ &+ \mathrm{A}^{(2)}\,(\mathbf{Z}_0^{(1)} + \mathrm{i}\,\mathbf{Z}_0^{(2)}) = \mathbf{R}_{01}\,\delta(\eta - \eta_1) + \mathrm{i}\,\mathbf{R}_{02}\,\delta(\eta - \eta_1)\,, \end{aligned} \tag{7.261}$$

where

$$\mathbf{R}_{01} = (-R_{01},\, 0,\, 0,\, 0)^{\mathrm{T}}\,, \qquad \mathbf{R}_{02} = (-R_{02},\, 0,\, 0,\, 0)^{\mathrm{T}}\,.$$

Then, separating the real and imaginary parts, we obtain two equations

$$\begin{aligned} &\mathrm{A}^{(1)}\mathbf{Z}_0^{(1)}\,(\alpha^2 - \beta^2) - \mathrm{A}^{(1)}\mathbf{Z}_0^{(2)}\,2\alpha\beta + \frac{\partial \mathbf{Z}_0^{(1)}}{\partial \eta} + \mathrm{A}^{(2)}\mathbf{Z}_0^{(1)} = \mathbf{R}_{01}\,\delta(\eta - \eta_1)\,, \\ &\mathrm{A}^{(1)}\mathbf{Z}_0^{(1)}\,2\alpha\beta + \mathrm{A}^{(1)}\mathbf{Z}_0^{(2)}\,(\alpha^2 - \beta^2) + \frac{\partial \mathbf{Z}_0^{(2)}}{\partial \eta} + \mathrm{A}^{(2)}\mathbf{Z}_0^{(2)} = \mathbf{R}_{02}\,\delta(\eta - \eta_1)\,, \end{aligned} \tag{7.262}$$

or

$$\frac{d\mathbf{Y}}{d\eta} + \mathrm{D}\,\mathbf{Y} = \mathbf{R}_0\,, \tag{7.263}$$

where

$$\mathbf{Y} = \left(\mathbf{Z}_0^{(1)},\, \mathbf{Z}_0^{(2)}\right)^{\mathrm{T}}, \qquad \mathbf{R}_0 = (\mathbf{R}_{01},\, \mathbf{R}_{02})^{\mathrm{T}}\,,$$

$$\mathrm{D} = \begin{bmatrix} (\alpha^2 - \beta^2)\,\mathrm{A}^{(1)} + \mathrm{A}^{(2)} & -2\alpha\beta\mathrm{A}^{(1)} \\ 2\alpha\beta\mathrm{A}^{(1)} & (\alpha^2 - \beta^2)\,\mathrm{A}^{(1)} + \mathrm{A}^{(2)} \end{bmatrix}.$$

The solution to the vector equation (7.263) is

$$\mathbf{Y} = \mathrm{K}(\eta,\, \alpha,\, \beta)\, \mathbf{C} + \int_0^{\eta} \mathrm{G}(\eta,\, \tilde{\eta})\, \mathbf{R}_0\, \delta(\eta - \tilde{\eta})\, \mathrm{d}\tilde{\eta}$$
$$= \mathrm{K}\mathbf{C} + \mathrm{G}(\eta,\, \eta_1)\, \mathbf{R}_0\, H(\eta - \eta_1)\,, \qquad \mathrm{K}(0) = \mathrm{E}\,. \tag{7.264}$$

From (7.7) it follows that

$$\mathbf{Z}_0^{(1)} = \mathrm{K}_{11}\mathbf{C}_1 + \mathrm{K}_{12}\mathbf{C}_2 + G_{11}\mathbf{R}_{01}\, H(\eta - \eta_1) + G_{12}\mathbf{R}_{02}\, H(\eta - \eta_1)\,,$$
$$\mathbf{Z}_0^{(2)} = \mathrm{K}_{21}\mathbf{C}_1 + \mathrm{K}_{22}\mathbf{C}_2 + G_{21}\mathbf{R}_{01}\, H(\eta - \eta_1) + G_{22}\mathbf{R}_{02}\, H(\eta - \eta_1)\,, \tag{7.265}$$

where K_{ij} and G_{ij} are block matrices.

At $\eta = 0$, the following boundary conditions must hold:

$$\vartheta_{30}^{(1)}(0) = u_{20}^{(1)}(0) = 0\,, \qquad \vartheta_{30}^{(2)}(0) = u_{20}^{(2)}(0) = 0\,.$$

Therefore,

$$\mathbf{C}_1 = (c_1^{(1)},\, c_2^{(1)},\, 0,\, 0)^{\mathrm{T}}\,, \qquad \mathbf{C}_2 = (c_1^{(2)},\, c_2^{(2)},\, 0,\, 0)^{\mathrm{T}}\,.$$

At $\eta = 1$, the following boundary conditions must hold:

$$Q_{20}^{(1)}(1) = M_{30}^{(1)}(1) = 0\,, \qquad Q_{20}^{(2)}(1) = M_{30}^{(2)}(1) = 0\,,$$

consequently, using (7.265), we obtain four equations:

$$k_{11}c_1^{(1)} + k_{12}c_2^{(1)} + k_{15}c_1^{(2)} + k_{16}c_2^{(2)} - g_{11}(1,\, \eta_1)\, R_{01} - g(1,\, \eta_1)\, R_{02} = 0\,,$$
$$k_{21}c_1^{(1)} + k_{22}c_2^{(1)} + k_{25}c_1^{(2)} + k_{26}c_2^{(2)} - g_{21}R_{01} + g_{22}R_{02} = 0\,,$$
$$k_{51}c_1^{(1)} + k_{52}c_2^{(1)} + k_{55}c_1^{(2)} + k_{56}c_2^{(2)} - g_{51}R_{01} - g_{55}R_{02} = 0\,,$$
$$k_{61}c_1^{(1)} + k_{62}c_2^{(1)} + k_{65}c_1^{(2)} + k_{66}c_2^{(2)} - gR_{01} - g_{65}R_{02} = 0\,. \tag{7.266}$$

This system of equations contains six unknown parameters, namely, $c_1^{(1)}$, $c_2^{(1)}$, $c_1^{(2)}$, $c_2^{(2)}$, R_{01}, and R_{02}. Two additional equations may be obtained from the following conditions:

$$u_{20}^{(1)}(\eta_1) = 0\,, \qquad u_{20}^{(2)}(\eta_1) = 0\,.$$

At $\eta = \eta_1$, from (7.8) it follows that

$$u_{20}^{(1)} = k_{41}(\eta_1)c_1^{(1)} + k_{42}(\eta_1)c_2^{(1)} + k_{45}(\eta_1)c_1^{(2)} + k_{46}(\eta_1)c_2^{(2)} = 0\,,$$
$$u_{20}^{(2)} = k_{81}(\eta_1)c_1^{(1)} + k_{82}(\eta_1)c_2^{(1)} + k_{85}(\eta_1)c_1^{(2)} + k_{86}(\eta_1)c_2^{(2)} = 0\,. \tag{7.267}$$

We consider the cross section with coordinate $\eta = \eta_1 - \varepsilon$, where ε is an infinitisimal quantity. Hence, $H(\eta_1 - \varepsilon - \eta_1) = 0$. The complex eigenvalues λ_j ($\lambda_j = \alpha_j + \mathrm{i}\,\beta_j$) can be found from the relationship

$$D(P_k,\, \alpha_j,\, \beta_j) = 0\,, \tag{7.268}$$

where D is the determinant of the system of equations (7.266)–(7.267) and P_k is the value of the load $(0 < P_k \le P_*)$.

For each value of the force P_k, we fix $\alpha^{(\rho)}$ and $\beta^{(\rho)}$, $\rho = 1, 2, 3, \ldots$ and then obtain such their values α_j and β_j that the determinante D is equal to zero. As a result, we get the complex conjugate eigenvalues $\lambda_j = \alpha_j \pm \mathrm{i}\,\beta_j$.

The graphs of the real and imaginary parts of the two first eigenvalues (α_1, α_2, β_1, and β_2) versus P for some values of η_1 are shown in Figs. 7.32, 7.33, and 7.34*a*.

If viscous damping is neglected, the critical forces P_* (for various values of η_1) for which a dynamic loss of stability takes place correspond to the points at which the frequencies β_1 and β_2 coincide (see Figs. *a*–*e*). For $P > P_*$, the complex eigenvalues have nonzero real parts. For $P > P_*$, we have $\lambda_1 = \alpha_1 \pm \mathrm{i}\,\beta_{1,2}$ and $\lambda_2 = -\alpha_2 \pm \mathrm{i}\,\beta_{1,2}$ such that, $|\alpha_1| = |\alpha_2|$. If there is no viscous damping, the real parts of the complex eigenvalues are zero for under-critical values of P (that is, for $P < P_*$) and $\lambda_1 = \pm\mathrm{i}\,\beta_1$ and $\lambda_2 = \pm\mathrm{i}\,\beta_2$.

In the absence of viscous damping, for subcritical values of P, the rod, after being deflected from the vertical, will execute (undamped) vibrations with finite amplitude of the points of its axis. In the presence of forces of viscous damping, the vibrations at $P < P_*$ will decay because the real parts of the eigenvalues are negative (partial solutions include the multipliers $\mathrm{e}^{-\alpha_j \tau}$).

When the force P exceeds the critical force just slightly (Figs. *a*–*e*), the rod deflected from its vertical position will vibrate with growing amplitide because one of the partial solutions includes the multiplier $\mathrm{e}^{\alpha_j \tau}$.

One interesting peculiarity of the problem under consideration is that when the parameter η_1 varies from 0 to 0.5, dynamic loss of stability (the flutter) takes place.

A theoretically possible critical force P_{**} for which the first frequency is zero is shown in Figs. 7.33*b* and 7.34*a*. This force corresponds to the static loss of stability. We see that $P_{**} > P_*$, consequently, as the follower force P continuously increases, the dynamic loss of stability occurs before the static loss of stability.

In greater details, the behavior of α_j and β_j for $\eta_1 = 0.5$ is represented in Figs. 7.34*b* and 7.34*c*.

The graphs of α_j and β_j, $j = 1, 2$, for $\eta_1 = 0.501$ are shown in Fig. 7.35*a*. From this figure it follows that the loss of stability of the rod takes place when the first frequency becomes zero. This case corresponds to the static loss of stability (the divergence). The graphs of α_j and β_j for $\eta_1 > 0.5$ shown in Fig. 7.36 and 7.37 indicate that the static loss of stability takes place for $\eta_1 > 0.5$ $(0.5 < \eta_1 < 1)$.

The stable and unstable regions in the plane (P, η_1) for a rod loaded by a follower force are shown in Fig. 7.38. Regions I_1 and I_2 are stable and regions II_1 and II_2 are unstable; region II_1 corresponds to the dynamic instability and region II_2 corresponds to the static instability.

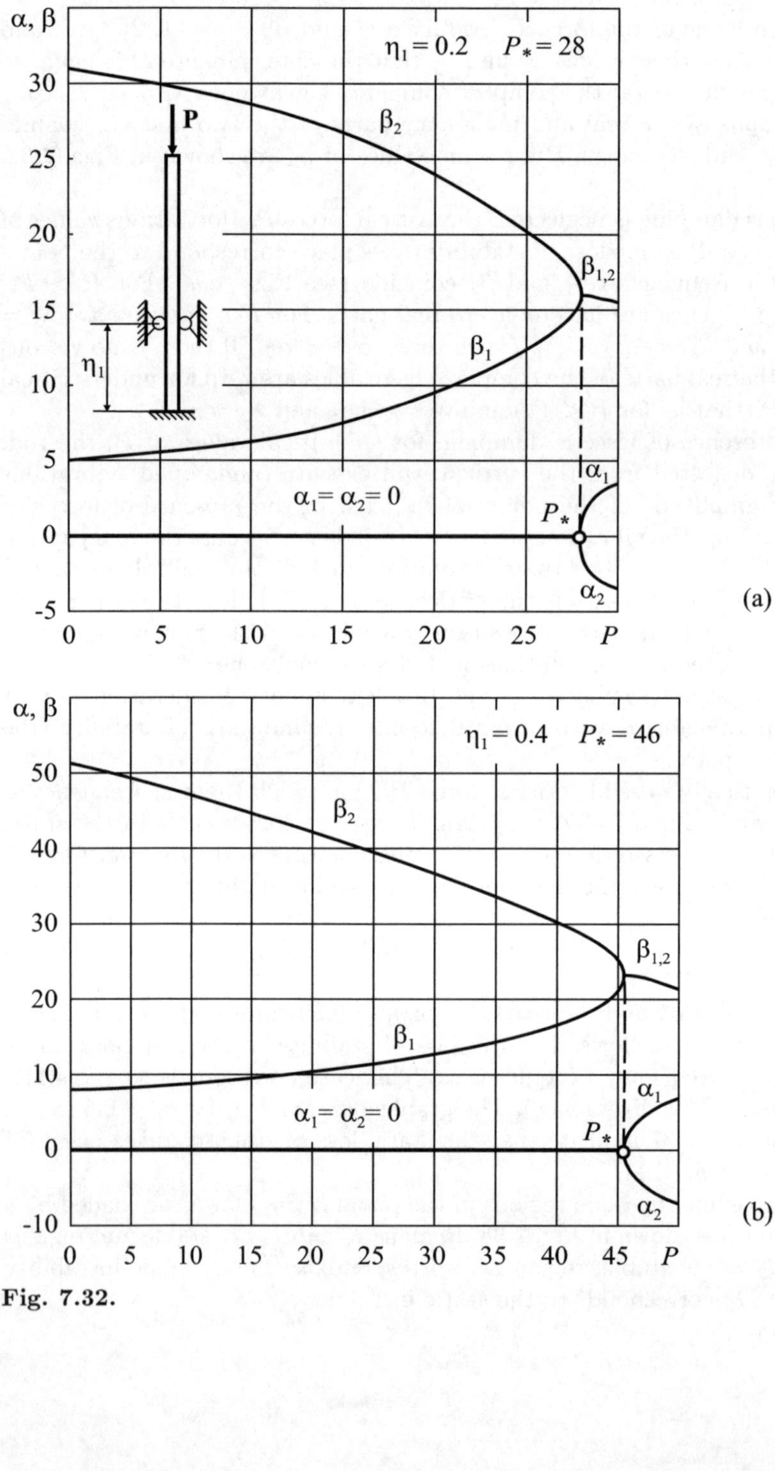
α, β
η₁ = 0.2 P* = 28
P
β₂
β₁
β₁,₂
η₁
α₁ = α₂ = 0
α₁
α₂
P*
(a)
α, β
η₁ = 0.4 P* = 46
β₂
β₁
β₁,₂
α₁ = α₂ = 0
α₁
α₂
P*
(b)

Fig. 7.32.

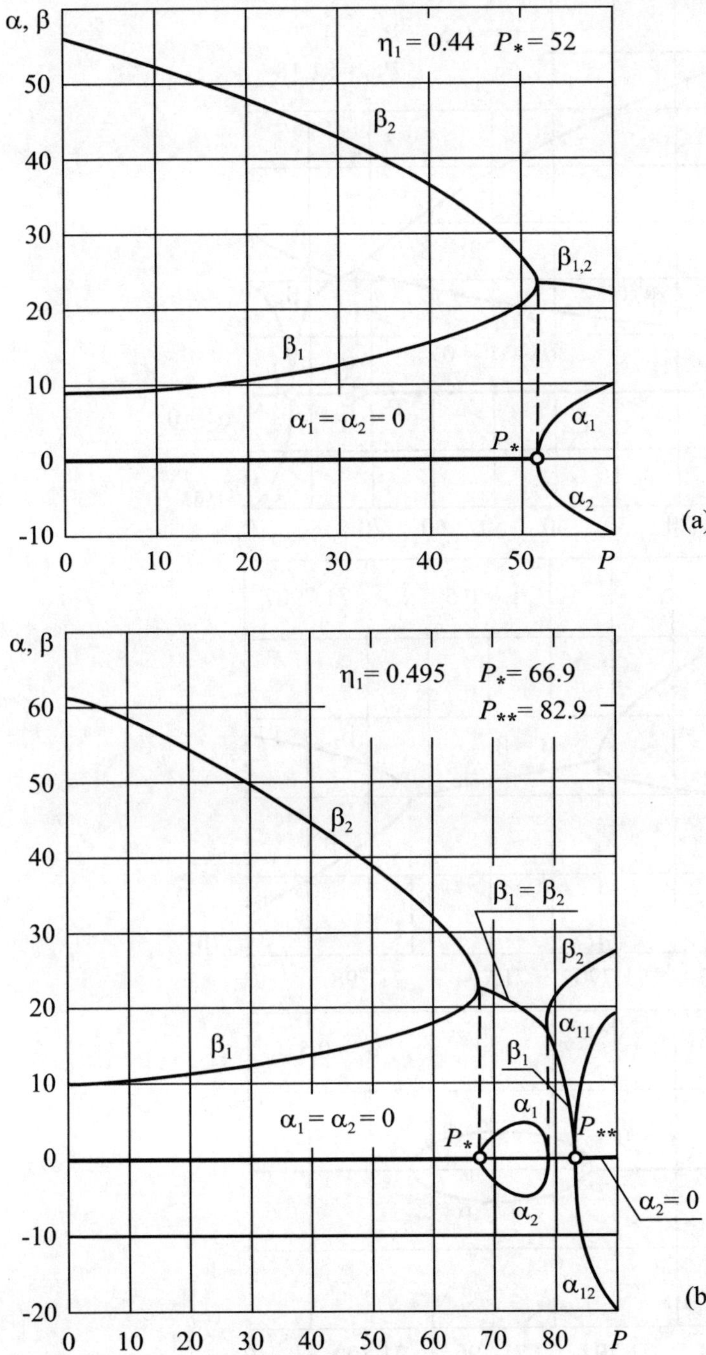
α, β
η₁ = 0.44 P* = 52
β₂
β1,2
β₁
α₁ = α₂ = 0
α₁
P*
α₂
(a)
P
α, β
η₁ = 0.495 P* = 66.9
P** = 82.9
β₂
β₁ = β₂
β₂
α11
β₁
α₁
P**
α₁ = α₂ = 0
P*
α₂
α₂ = 0
α12
(b)
P

Fig. 7.33.

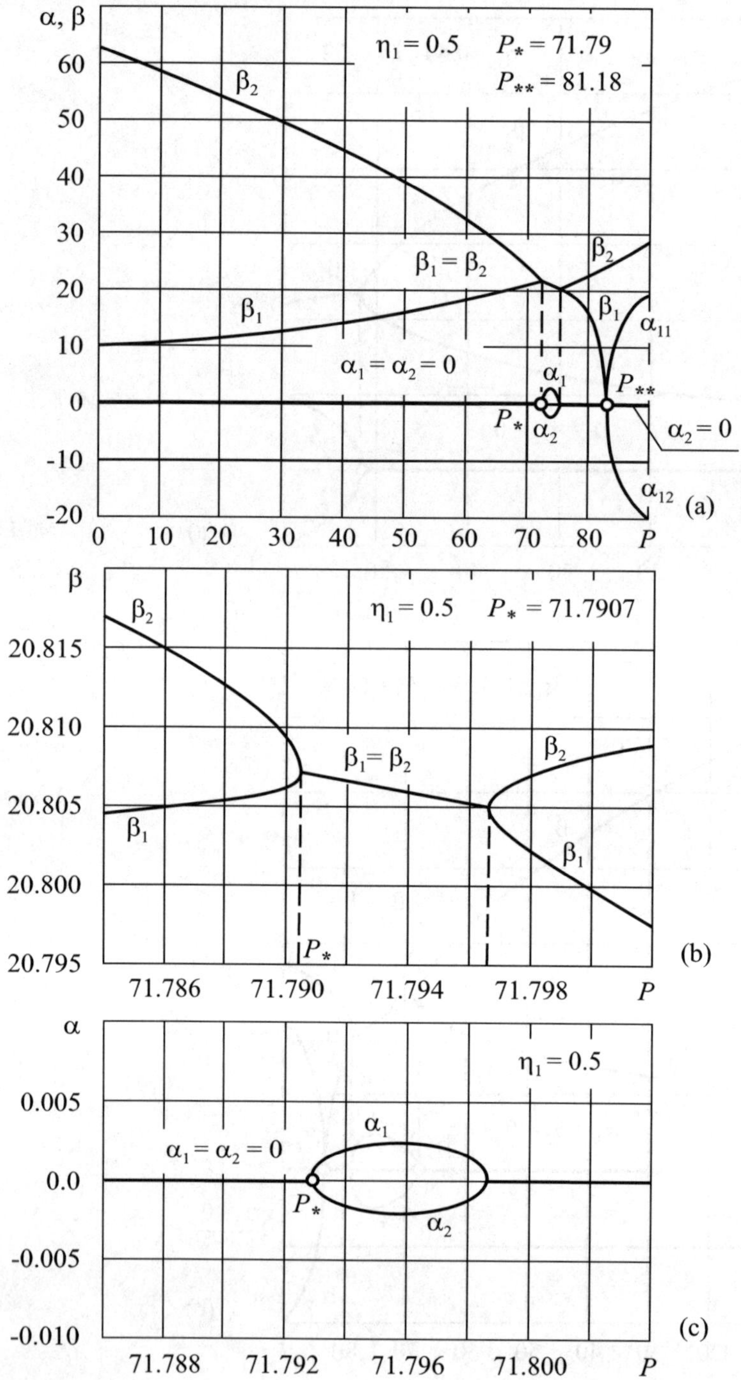
α, β
η₁ = 0.5
$P_* = 71.79$
$P_{**} = 81.18$
β₂
β₁ = β₂
β₁
α₁ = α₂ = 0
α₁
α₂
P_*
P_{**}
α₁₁
α₁₂
α₂ = 0
60
50
40
30
20
10
0
-10
-20
0
10
20
30
40
50
60
70
80
P
(a)
β
η₁ = 0.5
$P_* = 71.7907$
β₂
β₁ = β₂
β₁
P_*
20.815
20.810
20.805
20.800
20.795
71.786
71.790
71.794
71.798
P
(b)
α
η₁ = 0.5
α₁
α₁ = α₂ = 0
α₂
P_*
0.005
0.0
-0.005
-0.010
71.788
71.792
71.796
71.800
P
(c)

Fig. 7.34.

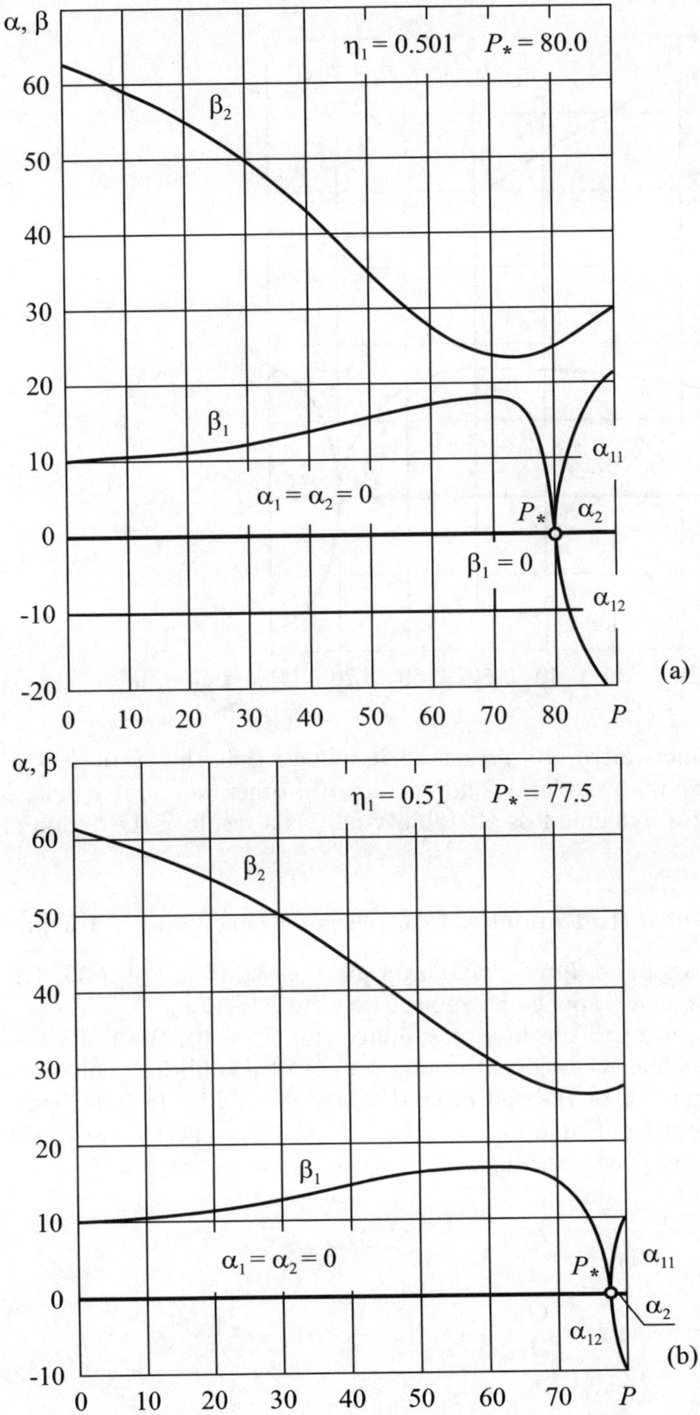
α, β
η₁ = 0.501 P* = 80.0
β₂
β₁
α₁ = α₂ = 0
β₁ = 0
P*
α₁₁
α₂
α₁₂
(a)
P
α, β
η₁ = 0.51 P* = 77.5
β₂
β₁
α₁ = α₂ = 0
P*
α₁₁
α₂
α₁₂
(b)
P

Fig. 7.35.

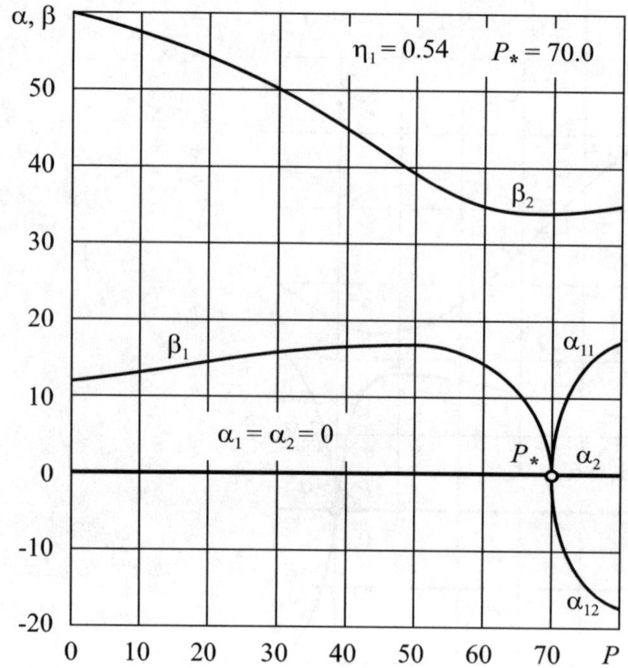

Fig. 7.36.

From the numerical results presented it follows that the static loss of stability may take place under a follower force. In other words, it is not to be supposed that a dynamic loss of stability only is a result of the action of follower forces.

7.8.2 Stability of a Rod under a Compressive Force and a Torque

A rod loaded by a follower force P and a torque T is shown in Fig. 7.39. The force and the torque are applied at an arbitrary cross section.

In this problem, after the loss of stability (for $T > 0$), the rod's axis is a space curved line. Consequently, the system of equilibrium equations governing the behavior of the rod after the loss of stability includes eight equations. This system of equations may be obtained as a special case of the system of equations (7.36). Setting $\Delta Q_1 = Q_1$, $\Delta Q_2 = Q_2$, and $\Delta M_3 = M_3$ gives

$$\frac{\partial^2 u_2}{\partial \tau^2} - \frac{\partial Q_2}{\partial \eta} + æ_{10} Q_3 - \frac{Q_{10}}{A_{33}} M_3 = 0\,,$$

$$\frac{\partial^2 u_3}{\partial \tau^2} - \frac{\partial Q_3}{\partial \eta} - æ_{10} Q_2 + \frac{Q_{10}}{A_{22}} M_2 = 0\,,$$

$$\frac{\partial M_2}{\partial \eta} - \left(æ_{10} - \frac{M_{10}}{A_{33}} \right) M_3 - Q_3 - T\vartheta_3\, \delta(\eta - \eta_1) = 0\,,$$

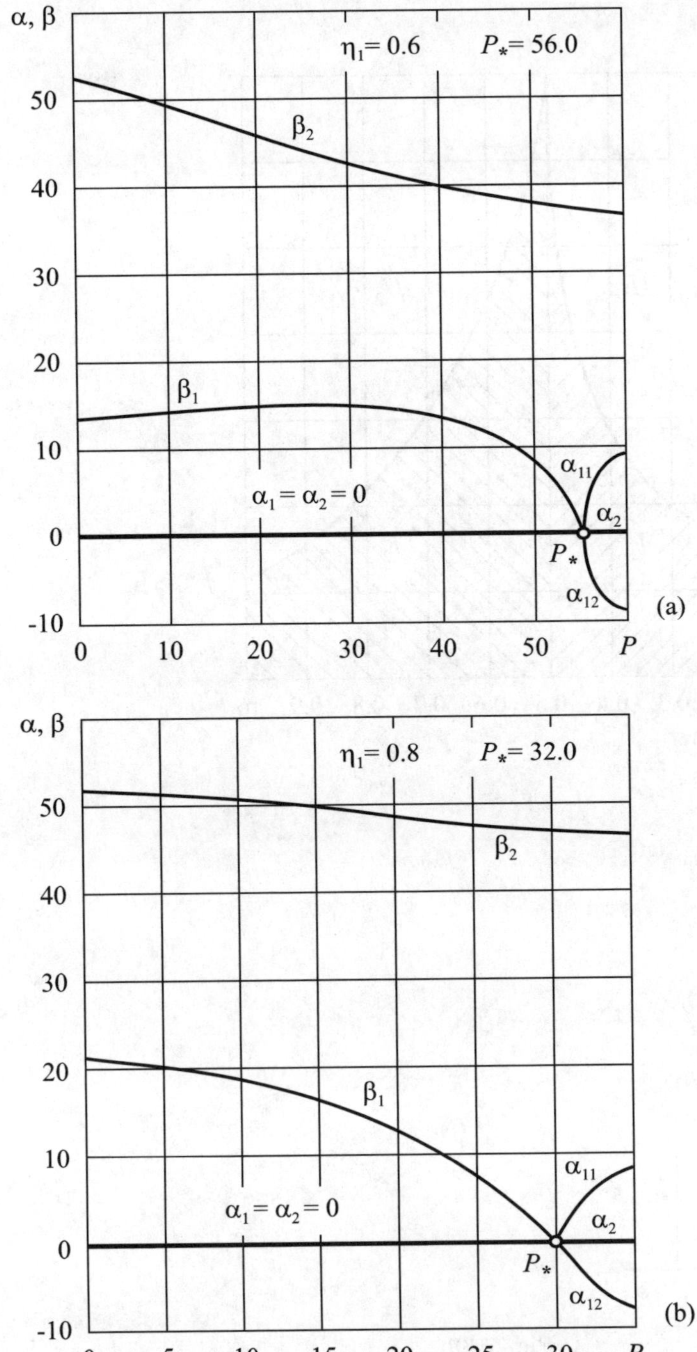
α, β
η₁= 0.6
P*= 56.0
β₂
β₁
α₁₁
α₁ = α₂ = 0
α₂
P*
α₁₂
(a)
P
α, β
η₁= 0.8
P*= 32.0
β₂
β₁
α₁₁
α₁ = α₂ = 0
α₂
P*
α₁₂
(b)
P

Fig. 7.37.

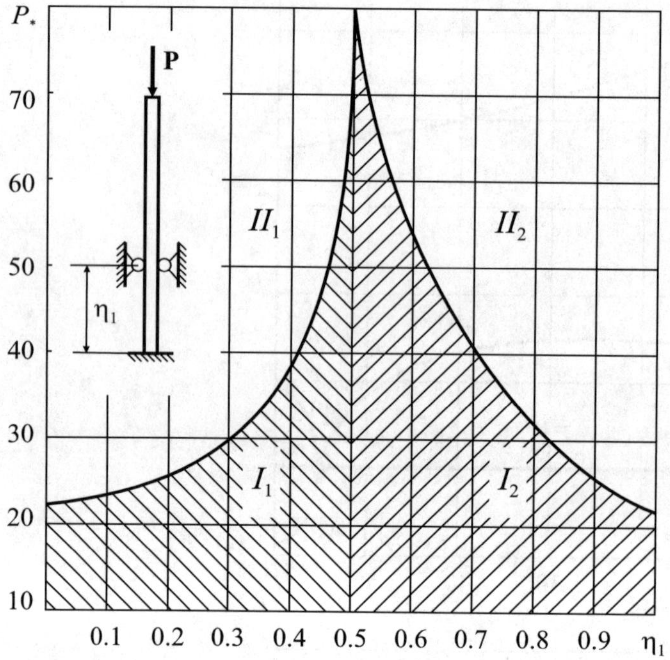

Fig. 7.38.

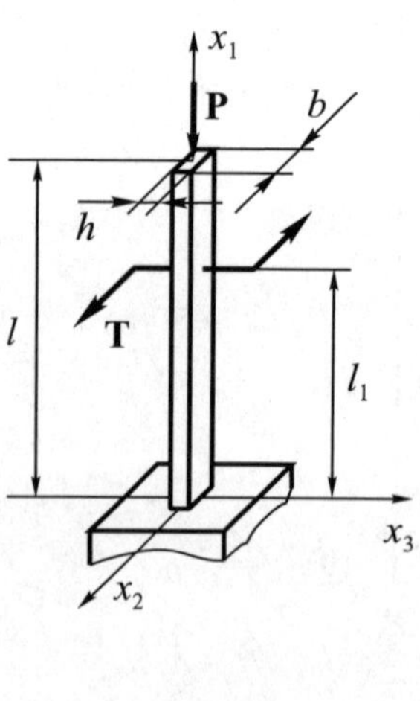

Fig. 7.39.

$$
\begin{aligned}
&\frac{\partial M_3}{\partial \eta} + \left(æ_{10} - \frac{M_{10}}{A_{33}} \right) M_2 + Q_2 + T\vartheta_2\, \delta(\eta - \eta_1) = 0\,, \\
&\frac{\partial \vartheta_2}{\partial \eta} - æ_{10}\vartheta_3 - \frac{M_2}{A_{22}} = 0\,, \\
&\frac{\partial \vartheta_3}{\partial \eta} + æ_{10}\vartheta_2 - \frac{M_3}{A_{33}} = 0\,, \\
&\frac{\partial u_2}{\partial \eta} - æ_{10}u_3 - \vartheta_3 = 0\,, \\
&\frac{\partial u_3}{\partial \eta} + æ_{10}u_2 + \vartheta_2 = 0\,,
\end{aligned}
\tag{7.269}
$$

where

$$
\begin{aligned}
&Q_{10} = -P\,, \qquad M_{10} = T\,[1 - H(\eta - \eta_1)]\,, \\
&æ_{10} = \frac{T}{A_{11}}\,[1 - H(\eta - \eta_1)]\,.
\end{aligned}
\tag{7.270}
$$

In vector–matrix form, we have

$$
\mathrm{A}^{(1)} \frac{\partial^2 \mathbf{Z}}{\partial \tau^2} + \frac{\partial \mathbf{Z}}{\partial \eta} + \mathrm{A}^{(2)} \bar{\mathbf{Z}} = \mathbf{b}\,,
\tag{7.271}
$$

where

$$
\begin{aligned}
&\bar{Z} = (Q_2,\, Q_3,\, M_2,\, M_3,\, \vartheta_2,\, \vartheta_3,\, u_2,\, u_3)^{\mathrm{T}}\,, \\
&\bar{b} = \mathrm{D}\,\mathbf{Z}\,\delta(\eta - \eta_1)\,,
\end{aligned}
$$

$$
\mathrm{A}^{(1)} = \begin{bmatrix}
0 & 0 & 0 & 0 & 0 & 0 & -1 & 0 \\
0 & 0 & 0 & 0 & 0 & 0 & 0 & -1 \\
0 & 0 & 0 & 0 & 0 & 0 & 0 & 0 \\
0 & 0 & 0 & 0 & 0 & 0 & 0 & 0 \\
0 & 0 & 0 & 0 & 0 & 0 & 0 & 0 \\
0 & 0 & 0 & 0 & 0 & 0 & 0 & 0 \\
0 & 0 & 0 & 0 & 0 & 0 & 0 & 0 \\
0 & 0 & 0 & 0 & 0 & 0 & 0 & 0
\end{bmatrix},
$$

$$
\mathrm{A}^{(2)} = \begin{bmatrix}
0 & -æ_{10} & 0 & \dfrac{Q_{10}}{A_{33}} & 0 & 0 & 0 & 0 \\
æ_{10} & 0 & -\dfrac{Q_{10}}{A_{22}} & 0 & 0 & 0 & 0 & 0 \\
0 & -1 & 0 & æ_{10} & 0 & 0 & 0 & 0 \\
1 & 0 & æ_{10} - \dfrac{M_{10}}{A_{22}} & 0 & 0 & 0 & 0 & 0 \\
0 & 0 & -\dfrac{1}{A_{22}} & 0 & 0 & -æ_{10} & 0 & 0 \\
0 & 0 & 0 & -\dfrac{1}{A_{33}} & æ_{10} & 0 & 0 & 0 \\
0 & 0 & 0 & 0 & 0 & -1 & 0 & -æ_{10} \\
0 & 0 & 0 & 0 & 1 & 0 & æ_{10} & 0
\end{bmatrix},
$$

$$D = \begin{bmatrix} 0 & 0 & 0 & 0 & 0 & 0 & 0 & 0 \\ 0 & 0 & 0 & 0 & 0 & 0 & 0 & 0 \\ 0 & 0 & 0 & 0 & 0 & T & 0 & 0 \\ 0 & 0 & 0 & 0 & -T & 0 & 0 & 0 \\ 0 & 0 & 0 & 0 & 0 & 0 & 0 & 0 \\ 0 & 0 & 0 & 0 & 0 & 0 & 0 & 0 \\ 0 & 0 & 0 & 0 & 0 & 0 & 0 & 0 \\ 0 & 0 & 0 & 0 & 0 & 0 & 0 & 0 \end{bmatrix} .$$

We seek the solution to the vector equation (7.271) in the form

$$\mathbf{Z} = [\mathbf{Z}_1(\eta) + \mathrm{i}\,\mathbf{Z}_2(\eta)]\, \mathrm{e}^{(\alpha+\mathrm{i}\,\beta)\,\tau} . \tag{7.272}$$

Then, substituting (7.272) into (7.271) and separating real and imaginary parts, we obtain

$$\frac{d\mathbf{Z}_1}{d\eta} + \left[A^{(2)} - (\beta^2 - \alpha^2)\, A^{(1)}\right] \mathbf{Z}_1 - 2\alpha\beta A^{(1)} \mathbf{Z}_2 = D\, \mathbf{Z}_1\, \delta(\eta - \eta_1) , \tag{7.273}$$

$$\frac{d\mathbf{Z}_2}{d\eta} + \left[A^{(2)} - (\beta^2 - \alpha^2)\, A^{(1)}\right] \mathbf{Z}_2 + 2\alpha\beta A^{(1)} \mathbf{Z}_1 = D\, \mathbf{Z}_2\, \delta(\eta - \eta_1) . \tag{7.274}$$

Let

$$\mathbf{Z}_0 = (\mathbf{Z}_1, \mathbf{Z}_2)^{\mathrm{T}} = (Z_{0,1}, Z_{0,2}, \ldots, Z_{0,16})^{\mathrm{T}} ;$$

we get

$$\frac{d\mathbf{Z}_0}{d\eta} + H\, \mathbf{Z}_0 = D^{(1)}\, \mathbf{Z}_0 , \tag{7.275}$$

where

$$H = \begin{bmatrix} A^{(2)} - (\beta^2 - \alpha^2)\, A^{(1)} & -2\alpha\beta A^{(1)} \\ 2\alpha\beta A^{(1)} & A^{(2)} - (\beta^2 - \alpha^2)\, A^{(1)} \end{bmatrix} ,$$

$$D^{(1)} = \begin{bmatrix} D & 0 \\ 0 & D \end{bmatrix} .$$

We seek the solution to the inhomogeneous differential equation (7.275) in the form

$$\mathbf{Z}_0(\eta) = K(\eta)\, \mathbf{C} + \int_0^{\eta} G(\eta, \tilde{\eta})\, D^{(1)}\, \mathbf{Z}_0(\tilde{\eta})\, \delta(\tilde{\eta} - \eta_1)\, d\tilde{\eta} , \tag{7.276}$$

where

$$G(\eta, \tilde{\eta}) = K(\eta) \cdot K^{(-1)}(\tilde{\eta}) , \qquad K(0) = E ;$$

here, $G(\eta, \tilde{\eta})$ is the Green matrix. Then, using the properties of the δ-function, we get

$$\mathbf{Z}_0(\eta) = \mathrm{K}(\eta)\,\mathbf{C} + \mathrm{G}(\eta,\, \eta_1)\,\mathrm{D}^{(1)}\,\mathbf{Z}_0(\eta_1)\,H(\eta - \eta_1)\,. \tag{7.277}$$

The boundary conditions are as follows.

(1) At $\eta = 0$,

$$Z_{0,5}(0) = Z_{0,6}(0) = Z_{0,7}(0) = Z_{0,8}(0) = 0\,,$$
$$Z_{0,13}(0) = Z_{0,14}(0) = Z_{0,15}(0) = Z_{0,16}(0) = 0\,.$$

(2) At $\eta = 1$,

$$Z_{0,1}(1) = Z_{0,2}(1) = Z_{0,3}(1) = Z_{0,4}(1) = 0\,,$$
$$Z_{0,9}(1) = Z_{0,10}(1) = Z_{0,11}(1) = Z_{0,12}(1) = 0\,.$$

From the boundary conditions at $\eta = 0$ it follows that

$$c_5 = c_6 = c_7 = c_8 = 0\,, \qquad c_{13} = c_{14} = c_{15} = c_{16} = 0\,.$$

Using the boundary conditions at $\eta = 1$ and the solution (7.277), we get

$$\mathbf{Z}_0(1) = \mathrm{K}(1)\,\mathbf{C} + \mathrm{G}(1,\, \eta_1)\,\mathrm{D}^{(1)}\,\mathbf{Z}_0(\eta_1)\,H(\eta - \eta_1)\,, \tag{7.278}$$

where

$$\mathbf{C} = (c_1,\, \ldots,\, c_4,\, 0,\, 0,\, 0,\, 0,\, c_9,\, \ldots,\, c_{12},\, 0,\, 0,\, 0,\, 0)^{\mathrm{T}}\,.$$

Therefore, taking into consideration the relationship

$$\mathbf{Z}_0(\eta_1) = \mathrm{K}(1)\,\mathbf{C}\,,$$

we thereby obtain

$$\mathbf{Z}_0(1) = \left[\mathrm{K}(1) + \mathrm{G}(1,\, \eta_1)\,\mathrm{D}^{(1)}\,\mathrm{K}(\eta_1)\right]\,\mathbf{C}\,. \tag{7.279}$$

Employing the boundary conditions at $\eta = 1$, we finally arrive at the following system of eight homogeneous equations:

$$\begin{bmatrix} k_{11}(1) & \ldots & k_{14}(1) & k_{19}(1) & \ldots & k_{1.12}(1) \\ \ldots & & & & & \\ k_{41}(1) & \ldots & k_{44}(1) & k_{49}(1) & \ldots & k_{4.12}(1) \\ k_{91}(1) & \ldots & k_{94}(1) & k_{99}(1) & \ldots & k_{9.12}(1) \\ \ldots & & & & & \\ k_{12.1}(1) & \ldots & k_{12.4}(1) & k_{12.9}(1) & \ldots & k_{12.12}(1) \end{bmatrix} \begin{bmatrix} c_1 \\ \vdots \\ c_4 \\ c_9 \\ \vdots \\ c_{12} \end{bmatrix}$$
$$= \mathrm{K}^{(1)}(1)\,\mathbf{C}^{(1)} = 0\,. \tag{7.280}$$

For the existence of non-trivial solutions to (7.280), the determinant $D^{(2)}$ of the matrix $\mathrm{K}^{(1)}(1)$ must be equal to zero:

$$D^{(2)} = \det \mathrm{K}^{(1)}(1) = 0\,. \tag{7.281}$$

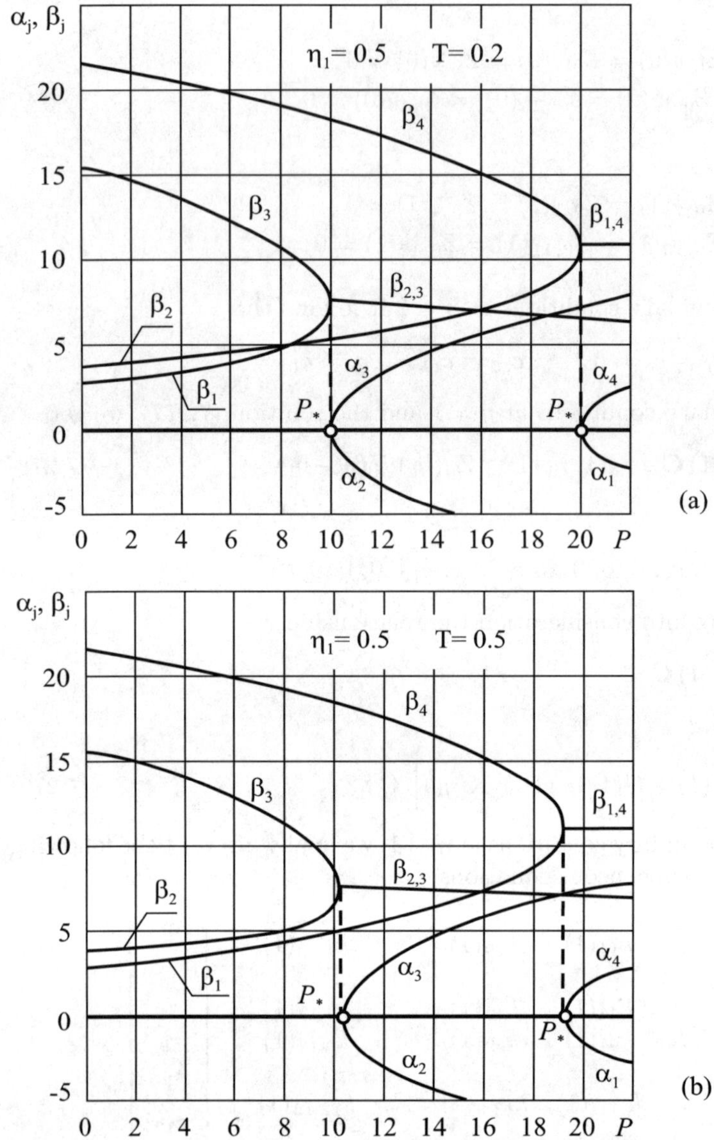

αj, βj
η1= 0.5 T= 0.2
β4
β3
β1,4
β2,3
β2
β1
α3
α4
P*
P*
α2
α1
(a)
0 2 4 6 8 10 12 14 16 18 20 P
20
15
10
5
0
-5
αj, βj
η1= 0.5 T= 0.5
β4
β3
β1,4
β2,3
β2
β1
α3
α4
P*
P*
α2
α1
(b)
0 2 4 6 8 10 12 14 16 18 20 P

Fig. 7.40.

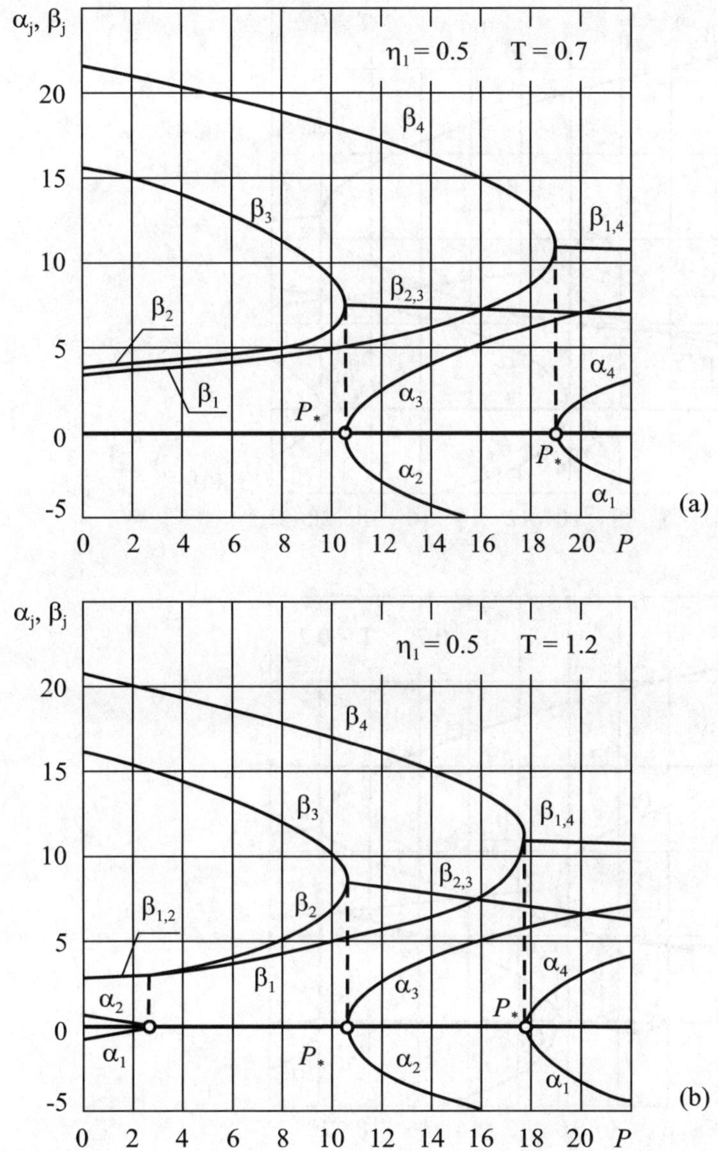

αj, βj
η1 = 0.5 T = 0.7
β4
β3
β1,4
β2,3
β2
β1
α4
α3
P*
α2
α1
(a)
P
αj, βj
η1 = 0.5 T = 1.2
β4
β3
β1,4
β2,3
β2
β1,2
β1
α4
α3
α2
α1
P*
(b)
P

Fig. 7.41.

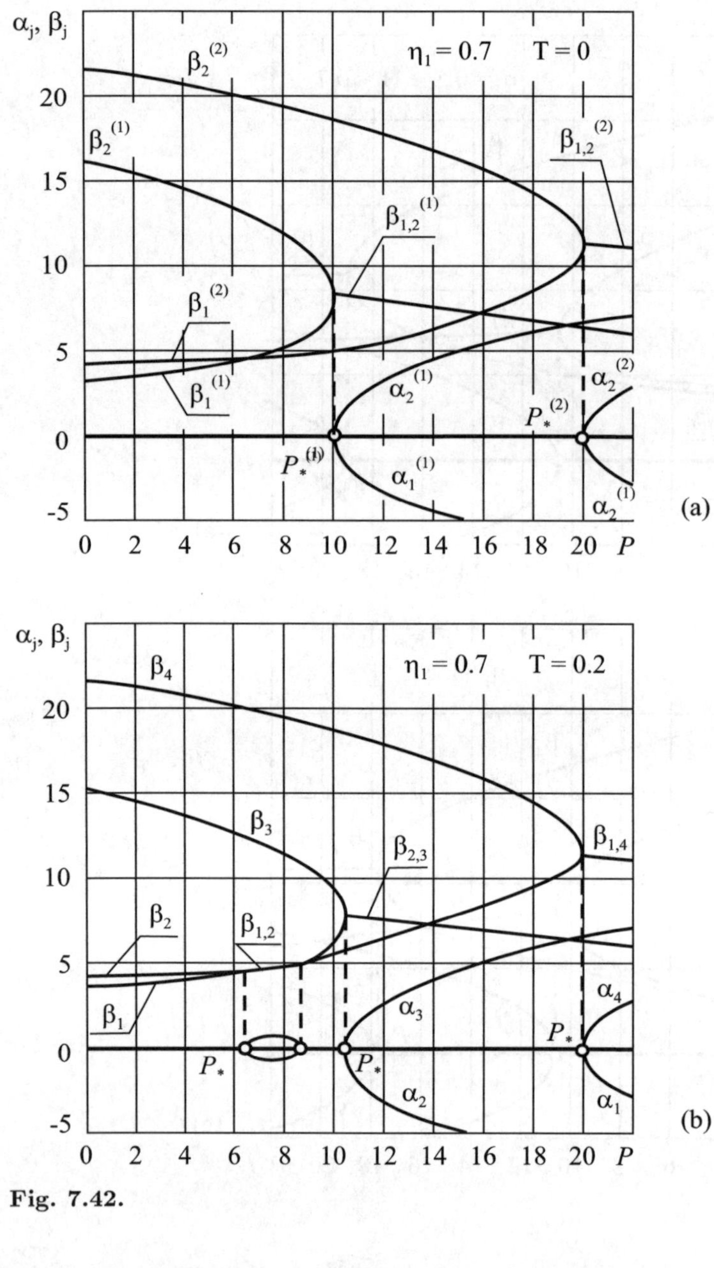

Fig. 7.42.

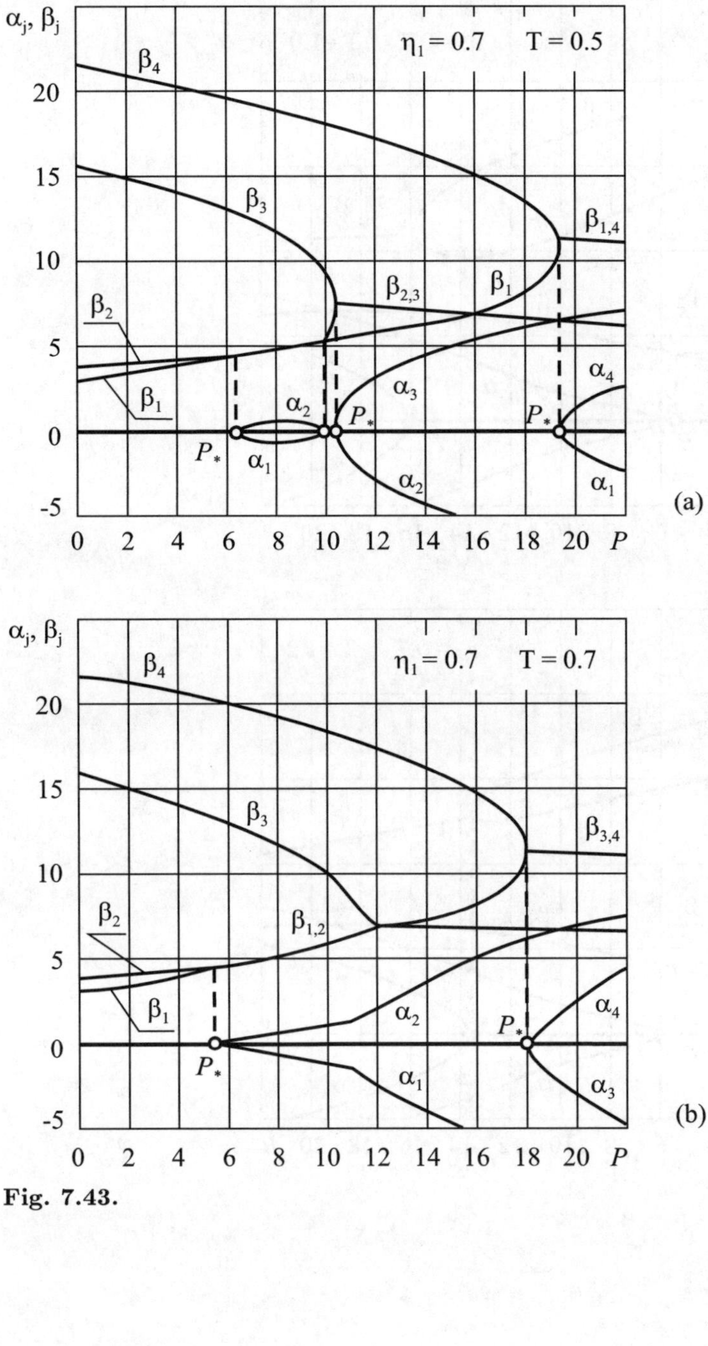
αj, βj
η1 = 0.7
T = 0.5
β4
β3
β1,4
β2,3
β1
β2
β1
α2
α3
α4
P*
P*
P*
α1
α2
α1
(a)
P
αj, βj
η1 = 0.7
T = 0.7
β4
β3
β3,4
β1,2
β2
β1
α2
α4
P*
P*
α1
α3
(b)
P

Fig. 7.43.

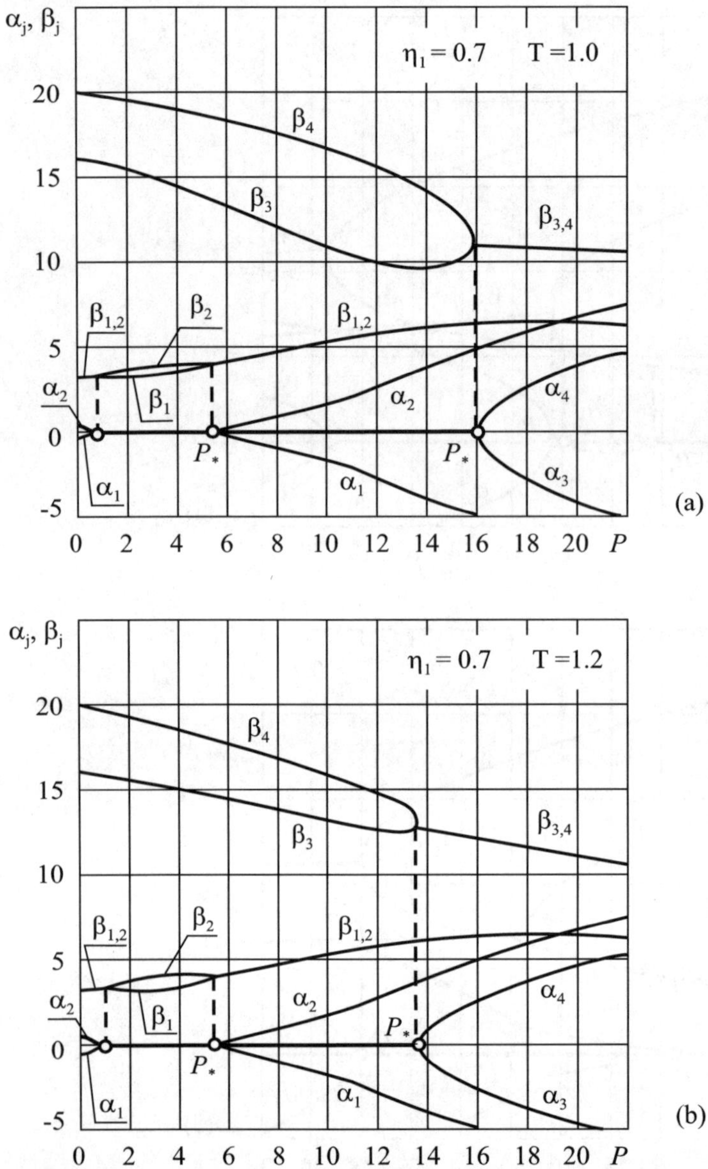
αj, βj
η1 = 0.7 T =1.0
20
15
10
5
0
-5
β4
β3
β3,4
β1,2
β2
β1,2
α2
β1
α2
α4
P*
P*
α1
α1
α3
0 2 4 6 8 10 12 14 16 18 20 P
(a)
αj, βj
η1 = 0.7 T =1.2
β4
β3
β3,4
β1,2
β2
β1,2
α2
β1
α2
α4
P*
P*
α1
α1
α3
(b)

Fig. 7.44.

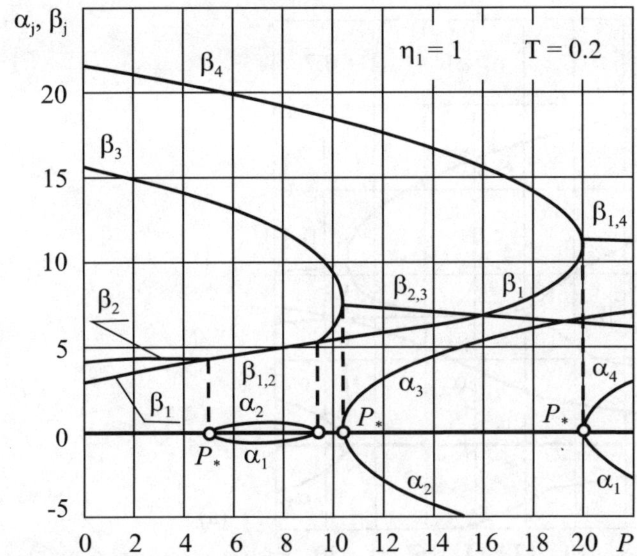

α_j, β_j
$\eta_1 = 1$
T = 0.2
β_4
β_3
$\beta_{1,4}$
$\beta_{2,3}$
β_1
β_2
$\beta_{1,2}$
α_3
α_4
α_2
β_1
P_*
α_1
α_2
α_1
20
15
10
5
0
-5
0 2 4 6 8 10 12 14 16 18 20 P

(a)

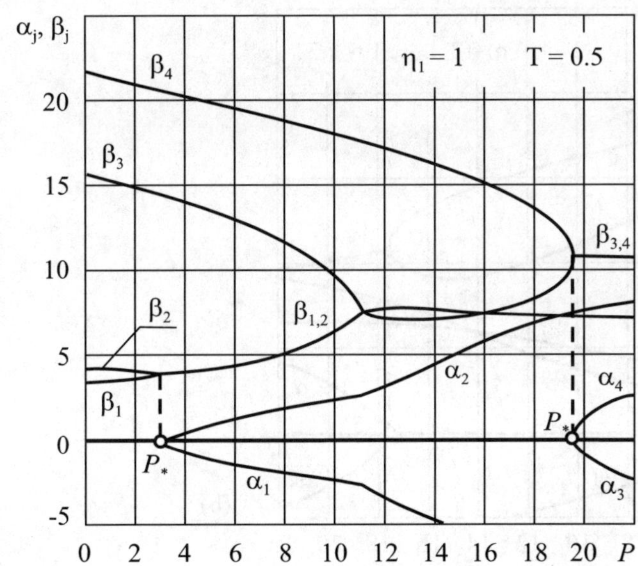

α_j, β_j
$\eta_1 = 1$
T = 0.5
β_4
β_3
$\beta_{3,4}$
β_2
$\beta_{1,2}$
α_2
α_4
β_1
P_*
α_1
α_3
20
15
10
5
0
-5
0 2 4 6 8 10 12 14 16 18 20 P

(b)

Fig. 7.45.

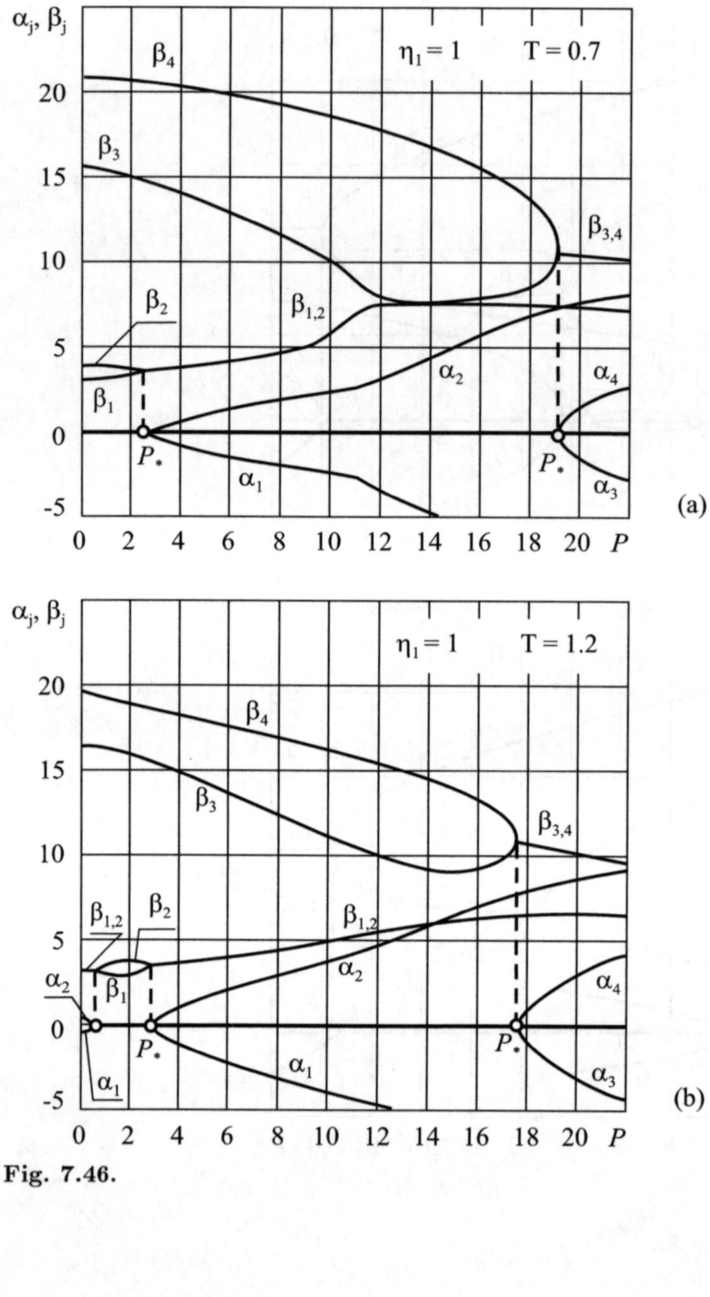
α_j, β_j
$\eta_1 = 1$
T = 0.7
β_4
β_3
$\beta_{3,4}$
β_2
$\beta_{1,2}$
β_1
α_2
α_4
P_*
P_*
α_1
α_3
(a)
α_j, β_j
$\eta_1 = 1$
T = 1.2
β_4
β_3
$\beta_{3,4}$
$\beta_{1,2}$
β_2
$\beta_{1,2}$
α_2
β_1
α_2
α_4
P_*
P_*
α_1
α_1
α_3
(b)
P

Fig. 7.46.

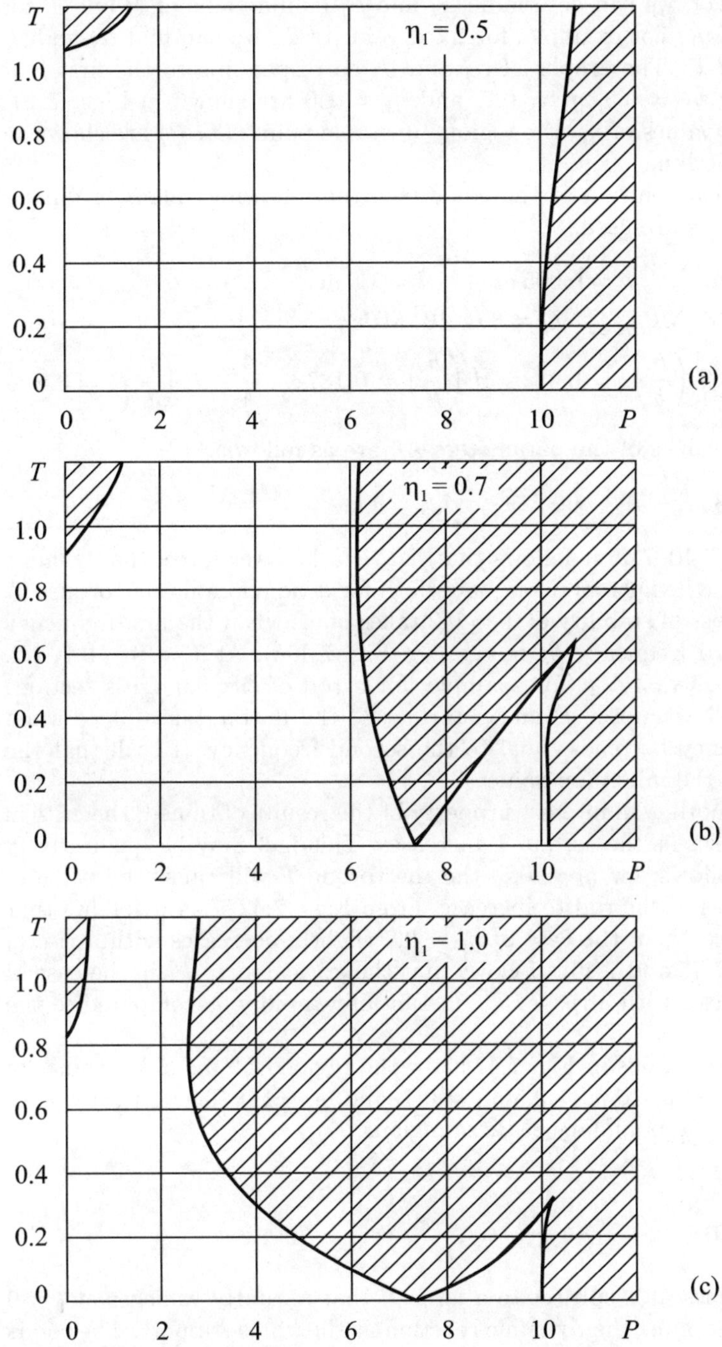

Fig. 7.47.

Using (7.281), we can determine α_j and β_j as functions of P and T for some values of η_1. For example, for fixed value of T, we can find α_j and β_j as functions of P. The graphs of α_j and β_j corresponding to the first four eigenvalues for $\eta_1 = 0.5$, $\eta_1 = 0.7$, and $\eta_1 = 1.0$ are shown in Figs. 7.40–7.46. For some values of η_1, the region "unstable pairs" $(T,\, P)$ are shown in Fig. 7.47 by hatching.

For numerical solution of the equations, the following values of dimensional parameters are used:

$$b = 0.007\,\mathrm{m}\,, \qquad h = 0.005\,\mathrm{m}\,, \qquad l = 1.0\,\mathrm{m}\,,$$
$$E = 2.0 \cdot 10^5\,\mathrm{MPa}\,, \qquad G = 8.0 \cdot 10^4\,\mathrm{MPa}\,,$$
$$\frac{A_{11}}{A_{33}} = 4.8\,\beta \left(\frac{h}{b}\right)^2 = 0.48\,, \quad \beta \left(\frac{b}{h}\right) = 0.187\,, \quad \frac{A_{22}}{A_{33}} = \left(\frac{h}{b}\right)^2 = 0.5\,.$$

Dimensionless values of the parameters A_{ii} are as follows:

$$A_{11} = 0.48\,, \qquad A_{22} = 0.5\,, \qquad A_{33} = 1\,.$$

From Figs. 7.40–7.46 it follows that if P is a follower force, the dynamic loss of stability (the flutter) takes place both for dead and follower torques T. The dynamic loss of stability at $T = 0.2$ takes place when the first frequency β_1 and the third frequency β_3 merge (see Fig. 7.40*a*). At $T = 0.5$, 0.7, 1.2, for a rod with $A_{22} = A_{33}$ (for example, for a rod of circular cross section) loaded by a follower force P and a torque T, the flutter takes place when the first frequency becomes equal to the second frequency. (Recall that the critical force is the minimum value of P_*.)

It is worth noting an interest property of the results obtained: the critical force P_* increases as the torque T increases. This fact may be qualitatively explained as follows: for $\eta_1 \le 0.5$, the the torque T will cause the value of bending stiffness of the rod to increase. From Figs. 7.42–7.44 it follows that the critical force P_* in the case of $\eta_1 = 0.7$ slightly decreases with a rise of T. From Figs. 7.45 and 7.46 it follows that the critical force P_* in the case of $\eta_1 = 1.0$ decreases with a rise of T substantially rapidly as compared to the case of $\eta_1 = 0.7$.

The hatched regions in Fig. 7.47 show the region of "unstable" values of $(T,\, P)$ corresponding to $\eta_1 = 0.5$, $\eta_1 = 0.7$, and $\eta_1 = 1.0$. The critical values, $(T_*,\, P_*)$, lie on the boundary of these regions.

7.9 Problems

7.1. A point mass m is subject to a force P that abruptly vanishes at $t = 0$ (Fig. 7.48). Determine the dynamic reaction at the hinge support. The rod is of constant cross section. Neglect the rotary inertia and the force of viscous damping.

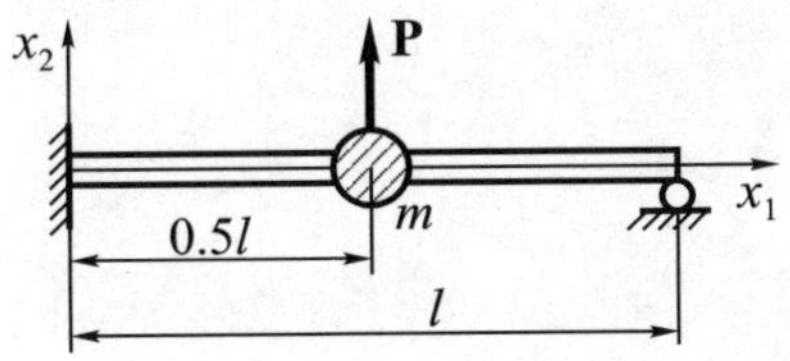

Fig. 7.48.

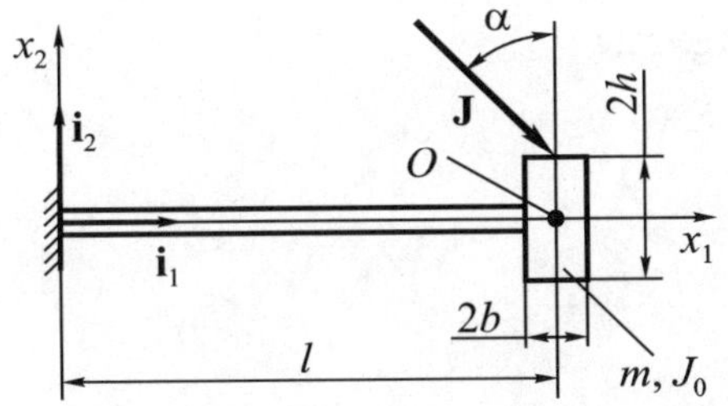

Fig. 7.49.

7.2. A point mass m is under the action of a impulsive force **J** (Fig. 7.49). After the pulse ceases, the mass breaks into vibrations. Find the angle of rotation of the mass in the plane of drawing.

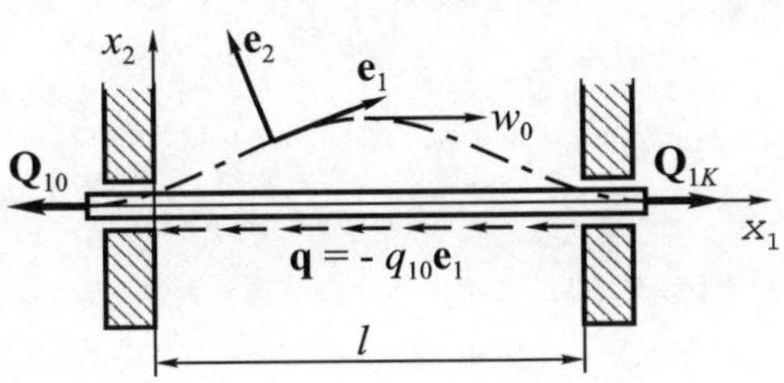

Fig. 7.50.

7.3. A rod moving in a viscous medium at a speed w_0 is shown in Fig. 7.50. Suppose that the rod is under the action of a follower uniformly distributed force $\mathbf{q}_1 = -q_{10}\mathbf{e}_1$, where $q_{10} = \beta\omega_0^2$. Find the first two eigenvalues. Use two-term approximation. Neglect the rotary inertia and the normal components of the force of viscous damping.

7.4. The lateral support at the cross section K vanishes at $t = 0$ (Fig. 7.51). As a result, the rod is set in free vibrations. Find the dynamic reaction of the hinge support at $s = l$.

7.5. Find the amplitude values of the steady-state vibrations of the point K (Fig. 7.52).

7.6. The end cross section K of a rod is forced to rotate periodically (Fig. 7.53). Find the amplitude values of the moment at the clamped end, $s = 0$, during the steady-state vibrations.

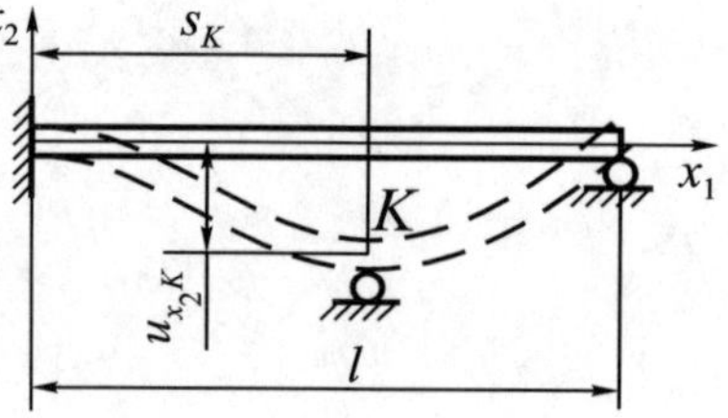

Fig. 7.51.

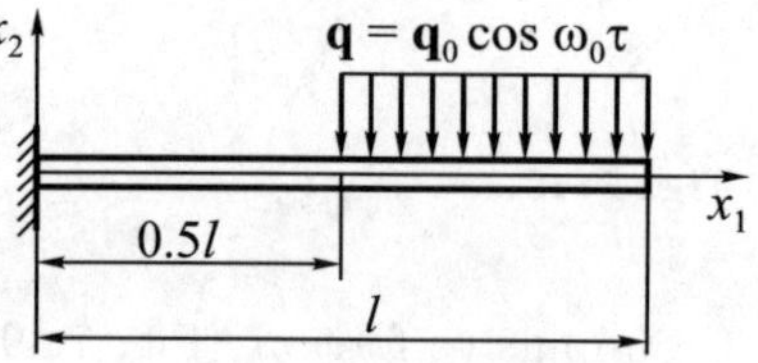

Fig. 7.52.

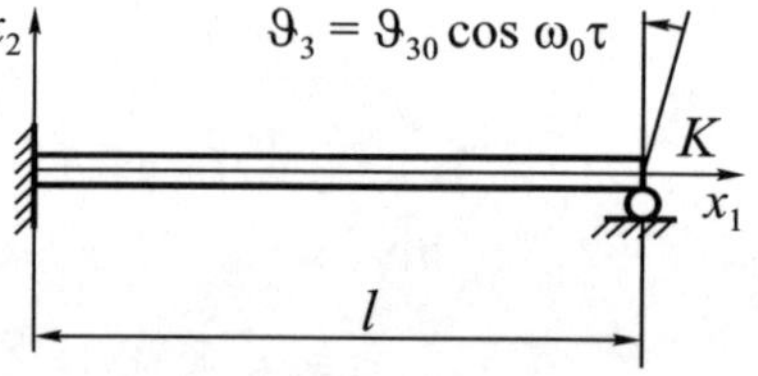

Fig. 7.53.

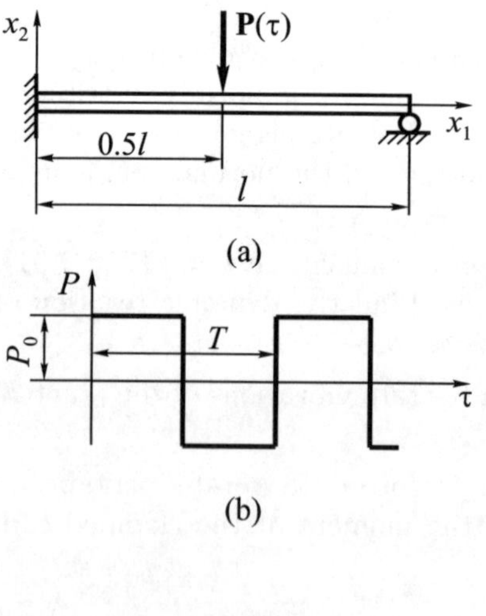

Fig. 7.54.

7.7. A rod of constant cross section shown in Fig. 7.54*a* is subjected to a periodic force $P(t)$ (Fig. 7.54*b*). Use the Douffing method to solve the equation of forced steady-state vibrations. You may also use the principle of virtual displacements considering two-term approximation.

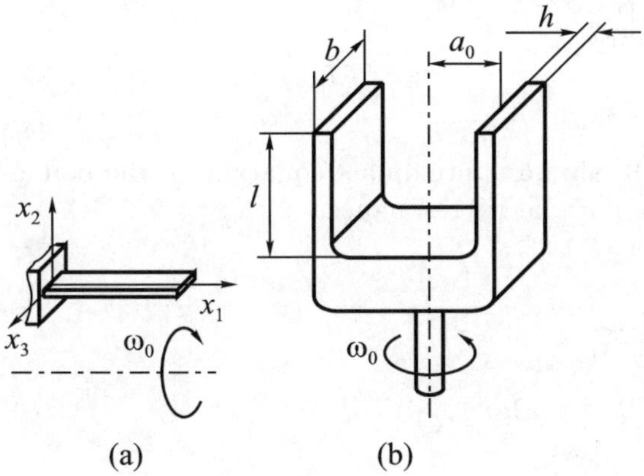

Fig. 7.55.

7.8. A gyrotron (a vibrating gyroscope) is shown in Fig. 7.55*a*. A segment of the gyroscope can be treated as a cantilever rod (Fig. 7.55*b*) whose moment of inertia about the x_2-axis is much larger than that with respect to the x_3-axis. In this condition, it can be assumed that during vibrations the points of the segment's axis remain in the plane x_1Ox_2. Use an approximate method to calculate the first two frequencies as a function of the angular velocity of the gyroscope ω_0.

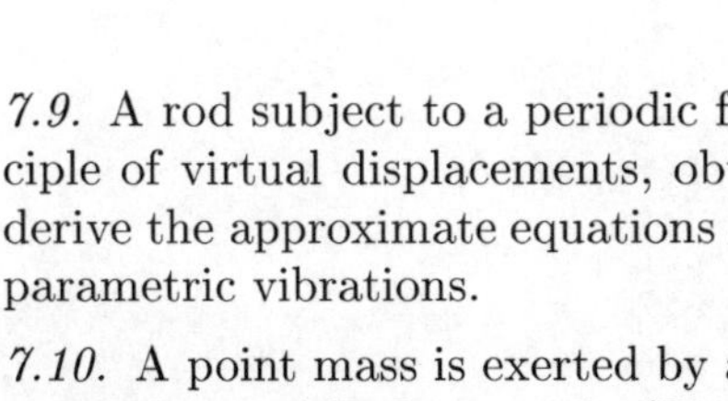

Fig. 7.56.

7.9. A rod subject to a periodic force is shown in Fig. 7.56. Using the principle of virtual displacements, obtain a one-term approximate solution and derive the approximate equations of the boundaries of the principal region of parametric vibrations.

7.10. A point mass is exerted by a periodic force whose line of action makes an angle α with the x_2-axis (Fig. 7.57). Using the principle of virtual displacements, solve the equations of vibrations of the rod (confine yourself to

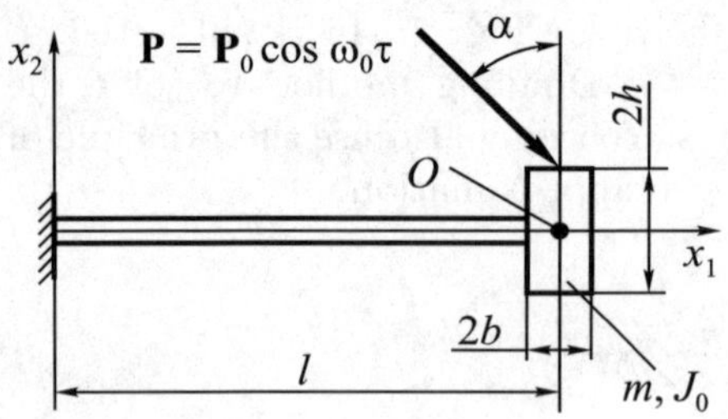

Fig. 7.57.

a one-term approximation). Obtain an approximate equation for the boundaries of the principal region of parametric resonance.

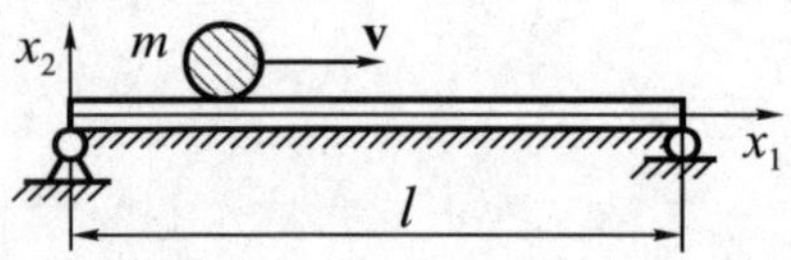

Fig. 7.58.

7.11. A point mass m moves along a rod on an elastic foundation with velocity v (Fig. 7.58). Using a two-term approximation, obtain the angle of rotation of the cross section K at the instant the mass rolls off the rod.

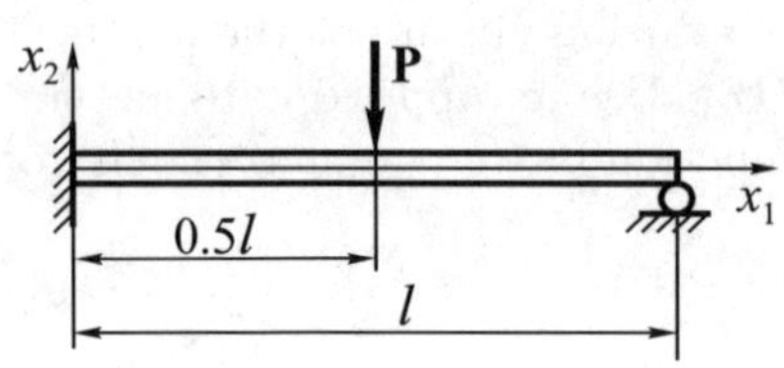

Fig. 7.59.

7.12. A rod is subjected to a stationary random concentrated force $\mathbf{P}$ (Fig. 7.59) whose probability characteristics are as follows: $m_P = 0$ and $S_P(\omega)$. Find the maximum possible magnitude of the reaction at the hinge $s = l$. Assume that the reaction is normally distributed. Use a one-term approximation.

8. Dynamics of Rods Interacting with Airflow or Liquid Flow

Elastic structural elements (rods, plates, shells) interacting with airflow or liquid flow are of frequent occurrence in applications. The main peculiarity is that the force on an elastic element due to the surrounding flow depends on the configuration of the deformed element. To perform the analysis of such a problem, both the equations related to mechanics of solid deformable body and the equations of aerohydrodynamics are needed. The branch of mechanics that involves mechanics of solid deformable body and aerohydrodynamics is called the theory of aerohydroelasticity.

In the classical problems of aerohydroelasticity, it is assumed that the aerohydrodynamic forces are independent of the shape of the body in the flow. In applications, this simplification is not always justified (for example, in the case of nonlinear problems of mechanics of rods); the aerohydrodynamic forces may considerably change when rod-like elements deform. If the interrelation between the force on a body and the body's shape is taken into consideration, then the very character of this force changes: the force should be treated as a feedback force and the corresponding problems of dynamics of elastic structural elements become nonconservative.

In what follows, the forces of interaction between a body and the ambient airflow or liquid flow are simply termed *aerodynamic forces* and the corresponding problem is called a *problem of aeroelasticity.* In particular, the theory of aeroelasticity covers the analysis of the interaction of rods with the ambient airflow or liquid flow. Examples are shown in Figs. 0.10-0.16.

Patterns of the flow about a rod of circular cross section are shown in Fig. 8.1*a, b* and for a rod of non-circular cross section in Fig. 8.1*c, d*. The points O and O_1 are the center of gravity and the flexural center of the cross sections.

The major problems of aerohydroelasticity of rod-like structural elements are as follows.

(1) Determination of the stress–strain state of an oscillating rod-like element interacting with a flow. This is important to ensure normal operation of the element.

(2) Determination of the eigenvalues and eigenvectors as a function of the flow velocity (direction and magnitude), the geometry of the rod axis, the shape of the cross section, and the boundary conditions.

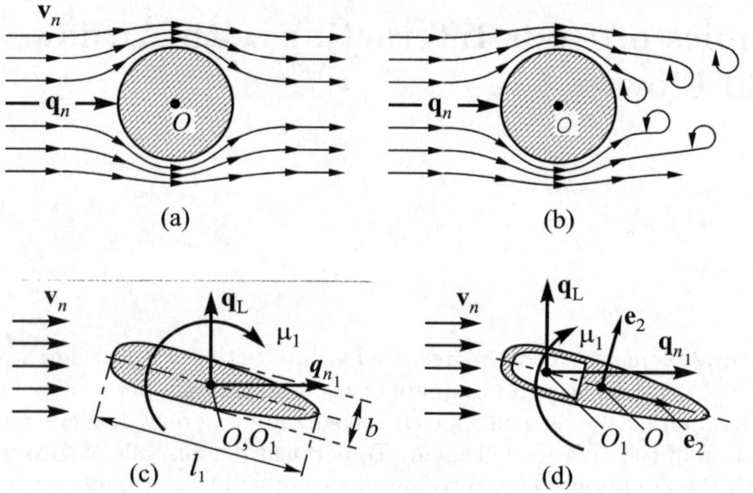

Fig. 8.1.

Knowing the complex eigenvalues, we can study the stability of the rod in the flow, namely, the static stability (*divergency*) and the dynamic stability (*flutter*) and find the corresponding critical values of the flow velocity.

(3) The study of forced oscillations of a rod that may occur due to one of the following reasons: shedding of the Kármán vortices, a kinematic excitation, short-time gusts of wind.

To solve these problems (as well as other problems of aerohydroelasticity) it is essential to know the aerodynamic forces, the determination of which entails a great amount of experimental investigation. Generally, for an arbitrary direction of the flow velocity, the aerodynamic forces can hardly be determined (even if the force is assumed to be independent of the change in the body's shape). Needless to say that the problem gets more complicated when the force is assumed to be dependent on the deformation of the body.

Consider a wire interacting with a steady-state flow of air (Fig. 8.2). To determine the static stress–strain state in the wire, we must know the relationship between the distributed aerodynamic force on the wire and the shape of its axis. In its turn, the shape of the axis depends on the magnitude and direction of the vector $\mathbf{v}_0$.

The relationship involves the angle between the flow velocity $\mathbf{v}_0$ and the vector $\mathbf{e}_1$, and this angle is of arbitrary value. This relationship can hardly be obtained experimentally. Even if we found the force for a particular set of boundary conditions and a particular $\mathbf{v}_0$, we would not be able to use this result for calculation of the force for some other boundary conditions and another direction of the vector $\mathbf{v}_0$. In other words, an experiment is to be performed for each particular case.

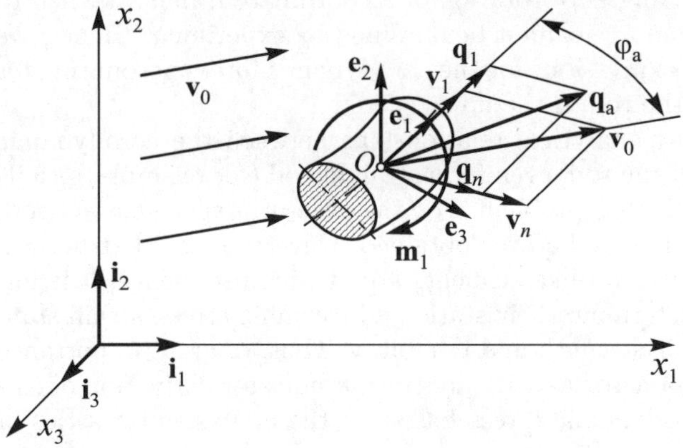

Fig. 8.2.

To obtain a numerical solution for a given applied problem, one must first derive the explicit formulas for the distributed aerodynamic forces. These formulas are supposed to account for the direction of the velocity of the steady-state flow and deviations of the rod from its equilibrium configuration (for $\mathbf{v}_0 = 0$).

A method for derivation of the formulas for the projections of the distributed aerodynamic forces in the Cartesian frame was presented in Chapter 6 of Svetlitsky (2000a).

There the most general case was considered: the vector of the flow velocity $\mathbf{v}$ made an arbitrary angle with the vector $\mathbf{e}_1$ tangent to the rod axis, the equilibrium configuration of the rod was given by nonlinear equations. The flow about the rod was assumed to be steady state and vortex-free. Besides, the local aerodynamic force, $\mathbf{q}_a$, (Fig. 8.2) was represented as a sum of two forces, $\mathbf{q}_n$ and $\mathbf{q}_1$: the drag force $\mathbf{q}_n$ lies in the plane orthogonal to the vector $\mathbf{e}_1$. The drag force was supposed to be a function of the square of the projection of the vector $\mathbf{v}_0$ on this plane, $\mathbf{v}_n$. The force $\mathbf{q}_1$ was directed along the tangent to the rod axis. Its magnitude was assumed to be a function of the square of the projection of the vector $\mathbf{v}$ on the tangent vector $\mathbf{e}_1$, $\mathbf{v}_1$. Such a representation of the distributed aerodynamic force $\mathbf{q}_a$ on a rod element implies that the forces $\mathbf{q}_n$ and $\mathbf{q}_1$ do not influence each other. As a result, the calculated value of the load on the rod exceeds the experimentally obtained value. This means that we analyze a structure for stresses and strains under the most unfavorable conditions assuming that it is subject to "static" loading.

From here on, we will use $\mathbf{q}_n$ for rods of circular cross section (Fig. 8.1*a, b*) and $\mathbf{q}_{n_1}$ for rods of noncircular cross section (Fig. 8.1*c, d*).

The problems of dynamics of rods in the flow become more complicated if the aerodynamic force on the rod is assumed to depend not only on the static stress–strain state of the rod but also on the velocities of points of the rod axis and on the angles of rotation of a coordinate frame attached to the rod. In this case, the force cannot be obtained by experiment. Hence, we need explicit analytical expressions for the aerodynamic forces accounting for dynamical behavior of the rod in the flow.

The derivation of the analytical relationships between the aerodynamic forces and the motion of the rod, even for a straight rod (for example, airfoils and blades), is a complicated problem. For the case of a space-curved rod, these relationships can hardly be ever obtained. Nevertheless, in structural applications, space-curved rod-like elements are of wide use; hence, it is important to be able to determine their static and dynamic stress–strain state in order to estimate their strength and reliability. Thus, of great importance for nonlinear problems of aeroelasticity are the methods for derivation of empirical formulas for aerodynamic forces based on the analysis of a series of measurements made in particular cases.

A method for determination of the static aerodynamic forces in the case of an immovable space-curved rod subject to a steady-state flow was discussed in Chapter 6 of Svetlitsky (2000a).

This method can be generalized to the case when the rod is in motion. To calculate approximately the drag $\mathbf{q}_n$ and $\mathbf{q}_{n_1}$ (Fig. 8.1*a–d*), the lift $\mathbf{q}_{\mathrm{L}}$ (Fig. 8.1*c, d*), and the aerodynamic moment $\boldsymbol{\mu}_1$ (Fig. 8.1*c, d*) on an element of the rod, it can be assumed that the flow about the element is quasi-stationary and plane. This is known as the plane-sections hypothesis in aerohydrodynamics. In terms of this hypothesis, the flow around each element of the rod is stationary and characterized by the velocity at infinity, $\mathbf{v}_0$. The velocity of the plane flow, $\mathbf{v}_n$, is equal to the projection of the velocity vector $\mathbf{v}_0$ on the plane orthogonal to the vector $\mathbf{e}_1$. In accordance with the plane-sections hypothesis, the flow about a small element of the rod is assumed to be plane; hence, the local velocity of the flow, $\mathbf{v}_n$, is orthogonal to the vector $\mathbf{e}_1$ while the velocity component $\mathbf{v}_1$ of the flow produces little effect on the drag $\mathbf{q}_{n_1}$, the lift $\mathbf{q}_{\mathrm{L}}$, and the moment $\boldsymbol{\mu}_1$.

In practice, for some values of the Reynolds number, Re, the shedding of Kármán vortices occurs (Fig. 8.1*b*). Experimentally, it has been shown that for a rod of circular cross section there is no flow separation for $\mathrm{Re} < 10$ (Devnin (1975) and Fung (1959)). With increasing Re, two stationary vortices appear at a distance from the rod. As Re increases further, the vortices expand. Finally, for $\mathrm{Re} > 40$, they are carried downstream and two new vortices appear at their former positions. After a while, these new vortices abandon the rod, etc. This takes place up to $\mathrm{Re} \approx 150$. The laminar vortex wake appears downstream of the rod. The vortices of the wake are periodic both in time and space (Fig. 8.1*b*). The period of the vortex separation is governed by the Strouchal number, Sh (Devnin (1975), Fung (1959)). The separating vortices result, particularly, in dynamic forces orthogonal to the vector of the flow velocity. This effect is of great importance for structural applications. These periodic forces may lead to oscillations of the rod with considerably large amplitude. This phenomenon is called a *wind resonance* (Feodosiev (1999), Fung (1959)). Consider an elastically supported absolutely rigid straight cylindrical rod whose axis is perpendicular to the flow velocity. In this case (see Devnin (1975) for experimental data), the rod tends to oscillate in synchrony with the shedding of vortices and these oscillations occur in the plane orthogonal to the vector of the flow velocity.

A Kármán vortex wake and the accompanying force occur not only in the case of circular rods but also for most high-drag bodies (that is, bodies of poor aerohydrodynamic performance (Fung (1959))).

For a further discussion of the flow about a straight cylindrical rod and for experimental investigations on the subject, see Grafskii, Kazakevich (1983), Devnin (1975), and Fung (1959). The influence of the Kármán forces on oscillations of space-curved rods is discussed in Sect. 8.1.

8.1 Aerodynamic Forces on a Rod Moving in a Flow

To study the dynamical interaction of an elastic rod with the surrounding flow and to be able to perform a numerical analysis of applied problems, it is essential to know the formulas for the aerodynamic forces, say, for their projections either in an immovable or in a movable frame of reference. These projections appear in the equations of motion of the rod. The formulas can be obtained by the method we used when the rod was assumed to be immovable, namely, we can put that the flow about the rod is quasi-stationary.

Employing the approximate expressions for the aerodynamic forces obtained by this method, we can derive both linear and nonlinear equations of motion of the rod and therefore investigate free and forced oscillations of rods of both circular and non-circular cross section (Fig. 8.1).

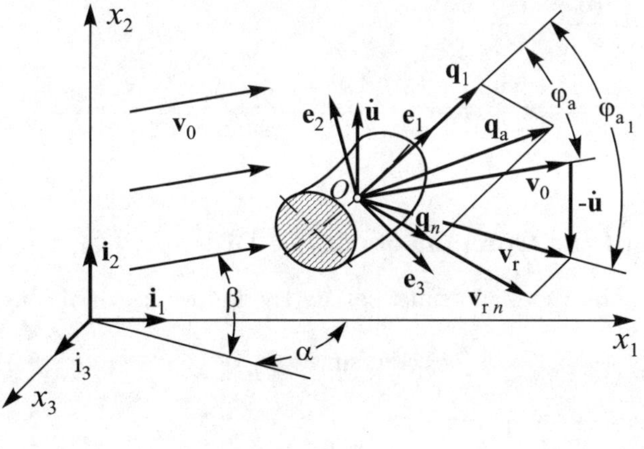

Fig. 8.3.

The general case when the vector $\mathbf{v}_0$ is arbitrarily directed is shown in Fig. 8.3. In the case of statics (see Svetlitsky (2000a)), the aerodynamic force was obtained for a special case, the vector $\mathbf{v}_0$ is parallel to the plane x_1Ox_3 (that is, $\beta = 0$).

Unlike the case of an immovable rod, the aerodynamic force on a moving rod depends on the relative velocity of the flow, which is given by

$$\begin{aligned}&\mathbf{v}_\mathrm{r} = \mathbf{v}_0 - \dot{\mathbf{u}}(\eta,\, \tau)\,,\\ &\mathbf{v}_0 = v_0\mathbf{e}_0 = v_0(\cos\alpha\cos\beta\,\mathbf{i}_1 + \sin\beta\,\mathbf{i}_2 + \sin\alpha\cos\beta\,\mathbf{i}_3)\,;\end{aligned} \tag{8.1}$$

here, $\dot{\mathbf{u}}(\eta,\, \tau)$ is the vector of the velocity of the rod axis points. The vector $\mathbf{v}_\mathrm{r}$ reads

$$\begin{aligned}\mathbf{v}_\mathrm{r} = v_0\mathbf{e}_\mathrm{r}^{(1)} = v_0[(&\cos\alpha\cos\beta - \beta_0\dot{u}_{x_1})\,\mathbf{i}_1\\ &+ (\sin\beta - \beta_0\dot{u}_{x_2})\,\mathbf{i}_2 + (\sin\alpha\cos\beta - \beta_0\dot{u}_{x_3})\,\mathbf{i}_3]\,,\end{aligned} \tag{8.2}$$

where $\dot{u}_{x_j}$ are nondimensional velocities and $\beta_0 = l\,p_0/v_0$.

8.1.1 Aerodynamic Forces Acting on Rods of Circular Cross Section

Recall that an assumption was made that the normal component is a function of the square of the projection of the relative velocity $\mathbf{v}_\mathrm{r}$ on the plane defined by the vectors $\mathbf{e}_2$ and $\mathbf{e}_3$; the tangent component is a function of the square of the projection of $\mathbf{v}_\mathrm{r}$ on the tangent to the rod axis. In terms of the assumption, we can write the following nondimensional relations:

$$|\mathbf{q}_n| = \frac{1}{2}\,\frac{c_n\rho d|\mathbf{v}_{\mathrm{r}n}|^2}{A_{33}}\,l^3\,, \qquad |\mathbf{q}_\mathrm{L}| = \frac{1}{2}\,\frac{c_1\rho d|\mathbf{v}_{\mathrm{r}\,1}|^2}{A_{33}}\,l^3\,. \tag{8.3}$$

When the rod is in motion, the dynamical angle φ_{a_1} is the angle between the vectors $\mathbf{e}_1$ and $\mathbf{v}_\mathrm{r}$ (Fig. 8.3), therefore,

$$\begin{aligned}\cos\varphi_{a_1} &= \frac{(\mathbf{v}_\mathrm{r}\cdot\mathbf{e}_1)}{|\mathbf{v}_\mathrm{r}|} = \frac{(\mathbf{e}_\mathrm{r}^{(1)}\cdot\mathbf{e}_1)}{|\mathbf{e}_\mathrm{r}^{(1)}|}\\ &= \frac{(\cos\alpha\cos\beta-\beta_0\dot{u}_{x_1})\,x_1'+(\sin\beta-\beta_0\dot{u}_{x_2})\,x_2'+(\sin\alpha\cos\beta-\beta_0\dot{u}_{x_3})\,x_3'}{\sqrt{(\cos\alpha\cos\beta - \beta_0\dot{u}_{x_1})^2 + (\sin\beta - \beta_0\dot{u}_{x_2})^2 + (\sin\alpha\cos\beta - \beta_0\dot{u}_{x_3})^2}}\,.\end{aligned} \tag{8.4}$$

8.1.2 Determination of the Components of the Drag

As in the case of statics, the vector $\mathbf{q}_n$ must satisfy the following conditions:

$$(|\mathbf{q}_n|)^2 = q_{nx_1}^2 + q_{nx_2}^2 + q_{nx_3}^2 = \left(\frac{1}{2}\,c_n\rho d v_\mathrm{r}^2\sin^2\varphi_{a_1}\right)^2, \tag{8.5}$$

$$\mathbf{q}_n\cdot\mathbf{e}_1 = q_{nx_1}x_1' + q_{nx_2}x_2' + q_{nx_3}x_3' = 0\,, \tag{8.6}$$

$$\mathbf{q}_n[\mathbf{q}_\mathrm{r}\times\mathbf{e}_1] = \mathbf{q}_n\cdot\mathbf{c} = q_{nx_1}c_1 + q_{nx_2}c_2 + q_{nx_3}c_3 = 0\,; \tag{8.7}$$

here,

$$c_1 = v_{rx_2} x_3' - v_{rx_3} x_2' \, ,$$
$$c_2 = v_{rx_3} x_1' - v_{rx_1} x_3' \, ,$$
$$c_3 = v_{rx_1} x_2' - v_{rx_2} x_1' \, , \tag{8.8}$$

where

$$v_{rx_1} = v_0(\cos\alpha\cos\beta - \beta_0 \dot{u}_{x_1}) \, ,$$
$$v_{rx_2} = v_0(\sin\beta - \beta_0 \dot{u}_{x_2}) \, ,$$
$$v_{rx_3} = v_0(\sin\alpha\cos\beta - \beta_0 \dot{u}_{x_3}) \, .$$

Solving for q_{nx_1} and q_{nx_2} from (8.6) and (8.7), we get these quantities as functions of q_{nx_3}:

$$q_{nx_1} = -q_{nx_3} \frac{\Delta_1}{\Delta_3} \, , \qquad q_{nx_2} = -q_{nx_3} \frac{\Delta_2}{\Delta_3} \, ; \tag{8.9}$$

here,

$$\Delta_1 = c_2 x_3' - c_3 x_2' \, , \qquad \Delta_2 = c_3 x_1' - c_1 x_3' \, , \qquad \Delta_3 = c_2 x_1' - c_1 x_2' \, . \tag{8.10}$$

It follows from (8.5) that

$$q_{nx_3} = \frac{q_{n0} \left(\frac{v_r}{v_0}\right)^2 \sin^2 \varphi_{a_1} \cdot \Delta_3}{\sqrt{\Delta_1^2 + \Delta_2^2 + \Delta_3^2}} \, , \qquad q_{n0} = \frac{c_n \rho d v_0^2 l^3}{2A_{33}} \, . \tag{8.11}$$

Then, from (8.9) it follows that

$$q_{nx_1} = -q_{n0} \left(\frac{v_r}{v_0}\right)^2 \sin^2 \varphi_{a_1} \cdot \frac{\Delta_1}{D} \, ,$$
$$q_{nx_2} = -q_{n0} \left(\frac{v_r}{v_0}\right)^2 \sin^2 \varphi_{a_1} \cdot \frac{\Delta_2}{D} \, , \tag{8.12}$$

where

$$D = \sqrt{\Delta_1^2 + \Delta_2^2 + \Delta_3^2} \, . \tag{8.13}$$

Let us rearrange the formula for D. Consider the vector $\boldsymbol{\Delta}$ with components Δ_i given by (8.11). This vector can be represented as the vector product of the vectors $\mathbf{c}$ and $\mathbf{e}_1$:

$$\boldsymbol{\Delta} = \mathbf{c} \times \mathbf{e}_1 \, ; \tag{8.14}$$

here, the components of the vector $\mathbf{c}$ are c_1, c_2, and c_3 from (8.8). Hence, $D = |\boldsymbol{\Delta}|$ and

$$D^2 = \sum_{i=1}^{3} \Delta_i^2 \,.$$

Then, substituting the components of the vector $\mathbf{c}$ (8.8) into (8.4), we get $\boldsymbol{\Delta} = ((\mathbf{v}_r \times \mathbf{e}_1) \times \mathbf{e}_1)$. Using the cab-bac relation for the triple vector product, we arrive at

$$\boldsymbol{\Delta} = -\mathbf{v}_\mathrm{r}(\mathbf{e}_1 \cdot \mathbf{e}_1) + \mathbf{e}_1(\mathbf{v}_\mathrm{r} \cdot \mathbf{e}_1) = -\mathbf{v}_\mathrm{r} + \mathbf{e}_1(\mathbf{v}_\mathrm{r} \cdot \mathbf{e}_1) \,. \tag{8.15}$$

Since $\mathbf{v}_\mathrm{r} = \mathbf{v}_{\mathrm{r}n} + \mathbf{e}_1\,(\mathbf{v}_\mathrm{r} \cdot \mathbf{e}_1)$, it follows from (8.15) that

$$\boldsymbol{\Delta} = -\mathbf{v}_{\mathrm{r}n} = -\mathbf{v}_\mathrm{r} \sin\varphi_{a_1} \,; \tag{8.16}$$

therefore,

$$D^2 = v_\mathrm{r}^2 \sin^2\varphi_{a_1} \,. \tag{8.17}$$

Upon direct calculation of Δ_i, a similar result was obtained for the case of statics. Finally, the components of the drag on the moving rod are

$$\begin{aligned} q_{nx_1} &= -q_{n0}\,\frac{v_\mathrm{r}}{v_0}\,\frac{\Delta_1}{v_0}\,\sin\varphi_{a_1} \,, \\ q_{nx_2} &= -q_{n0}\,\frac{v_\mathrm{r}}{v_0}\,\frac{\Delta_2}{v_0}\,\sin\varphi_{a_1} \,, \\ q_{nx_3} &= q_{n0}\,\frac{v_\mathrm{r}}{v_0}\,\frac{\Delta_3}{v_0}\,\sin\varphi_{a_1} \,. \end{aligned} \tag{8.18}$$

In more expanded notation, we have

$$\begin{aligned} q_{nx_1} &= q_{n0}\,\frac{v_\mathrm{r}}{v_0}\,(\cos\alpha\cos\beta - x_1'\cos\varphi_{a1} + \Delta_{x_1})\,\sin\varphi_{a_1} \,, \\ q_{nx_2} &= q_{n0}\,\frac{v_\mathrm{r}}{v_0}\,(\sin\beta - x_2'\cos\varphi_{a1} + \Delta_{x_2})\,\sin\varphi_{a_1} \,, \\ q_{nx_3} &= q_{n0}\,\frac{v_\mathrm{r}}{v_0}\,(\sin\alpha\cos\beta - x_3'\cos\varphi_{a1} + \Delta_{x_3})\,\sin\varphi_{a_1} \,, \end{aligned} \tag{8.19}$$

where

$$\begin{aligned} \Delta_{x_1} &= \beta_0[\dot{u}_{x_1}(1 - x_1^{2'}) - \dot{u}_{x_2}x_1'x_2' - \dot{u}_{x_3}x_1'x_3'] \,, \\ \Delta_{x_2} &= \beta_0[-\dot{u}_{x_1}x_1'x_2' + \dot{u}_{x_2}(1 - x_2^{2'}) - \dot{u}_{x_3}x_1'x_3'] \,, \\ \Delta_{x_3} &= \beta_0[-\dot{u}_{x_1}x_1'x_3' - \dot{u}_{x_2}x_2'x_3' + \dot{u}_{x_3}(1 - x_3^{2'})] \,. \end{aligned} \tag{8.20}$$

For $\dot{\mathbf{u}} = 0$, the angle φ_{α_1} coincides with the angle φ_α introduced in Chap. 4 of Svetlitsky (2000a). When studying the motion of a rod relative to its equilibrium configuration, it is sensible to write the equations in terms of deviations; hence, only that component of the drag that is due to the motion of the rod is contained in the equations. To do this, it is sufficient to subtract the corresponding static components $q_{n\,x_j0}$ from (8.18) and (8.19), that is, to employ the projections $q_{nx_j}^{(1)} = q_{nx_j} - q_{nx_j0}$ in the attached axes.

8.1.3 Determination of the Tangential Force

In view of

$$\mathbf{q}_1 = |\mathbf{q}_1|\,\mathbf{e}_1 = q_{10}\left(\frac{v_{\mathrm{r}}}{v_0}\right)^2 \cos^2\varphi_{a_1}\cdot\mathbf{e}_1\,, \tag{8.21}$$

the projections of $\mathbf{q}_1$ in an immovable frame of reference are

$$\begin{aligned} q_{1x_1} &= q_{10}\left(\frac{v_{\mathrm{r}}}{v_0}\right)^2 x_1'\cos^2\varphi_{a_1}\,,\\ q_{1x_2} &= q_{10}\left(\frac{v_{\mathrm{r}}}{v_0}\right)^2 x_2'\cos^2\varphi_{a_1}\,,\\ q_{1x_3} &= q_{10}\left(\frac{v_{\mathrm{r}}}{v_0}\right)^2 x_3'\cos^2\varphi_{a_1}\,, \end{aligned} \tag{8.22}$$

where

$$q_{10} = \frac{c_1\rho d v_0^2 l^3}{2A_{33}}\,\mathrm{sgn}(\cos\varphi_{a_1})\,.$$

To study nonlinear oscillations about the equilibrium configuration, it is reasonable to subtract the static component q_{1x_j0} from (8.22). As a result, we have

$$q_{1x_j}^{(1)} = q_{1x_j} - q_{1x_j0}\,. \tag{8.23}$$

8.1.4 Determination of components of the Kármán Forces

Due to the shedding of vortices, a rod of circular cross section is acted upon by the periodic distributed Kármán force

$$\mathbf{q}_{\mathrm{K}} = \frac{1}{2}c_{\mathrm{K}}\rho d v_{\mathrm{r}n}^2\frac{l^3}{A_{33}}\sin\omega_0\tau\cdot\mathbf{e}_{\mathrm{K}} = q_{\mathrm{K}0}\sin\omega_0\tau\cdot\mathbf{e}_{\mathrm{K}}\,, \tag{8.24}$$

where $\omega_0 = \omega/p_0$ is the nondimensional frequency of vortex separation, c_{K} is an aerodynamic coefficient; the vector $\mathbf{e}_{\mathrm{K}}$ is orthogonal to the vector $\mathbf{v}_{\mathrm{r}n}$. It is assumed that in the case of a steady-state flow (in (8.24) $\omega_0 = \mathrm{const}$), the Kármán forces are harmonic. One peculiarity of the Kármán forces acting on an elastically supported rod is of special note. For a rigidly fixed rod, the nondimensional frequencies of vortex separation depends on the Strouhal number $\mathrm{Sh} = \omega d/v_0$ (if $\mathbf{v}_{\mathrm{r}n} = \mathbf{v}_0$). A large body of experimental research (for reference, see Devnin (1975) and Feodosiev (1999)) indicate that for $40 < \mathrm{Re} < 10^5$ (under-crisis flow), the average value of the Strouhal number for a rigidly fixed rod of circular cross section is 0.2 and the frequency of vortex separation ω depends linearly on the flow velocity v_0. When a rod is elastically

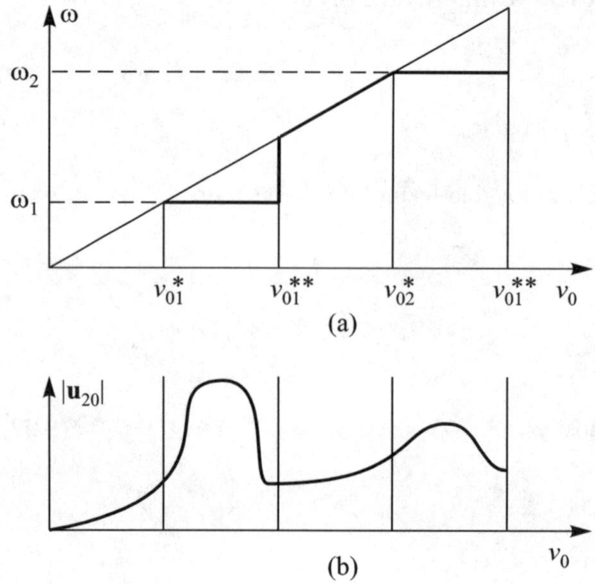

Fig. 8.4.

supported in the plane orthogonal to the flow velocity, the frequency ω is no longer linear in v_0.

For a movable straight rod, the graph of the frequency of vortex separation ω is qualitatively shown in Fig. 8.4*a*. It can be seen that for certain values of the flow velocity, the frequency of the vortex separation is correlated with the natural frequency of the rod. If the rod is not absolutely rigid, then, as the flow velocity increases, this phenomenon occurs for the other natural frequencies. On the intervals of the flow velocity for which the synchronization occurs, the amplitude of transverse oscillations of the rod considerably increases (the wind resonance) as shown in Fig. 8.4*b*. In this figure, the oscillation amplitude is denoted by $|\mathbf{u}_{20}|$. Such amplitudes are bounded in magnitude due to dissipative forces and other nonlinear effects; however, they could become considerably large and therefore need to be taken into consideration.

The flow velocity $\mathbf{v}_{0i}$ (Fig. 8.4*a*) for which the synchronization between the frequency of the vortex shedding and the natural frequency of the rod occurs can be found from the condition

$$\beta_i = \omega = \frac{0.2v_{0i}}{d}, \tag{8.25}$$

where β_i is the rod natural frequency.

The flow velocity $\mathbf{v}_{0i}$ for which the synchronization is disturbed can be found experimentally. For a rod of circular cross section, experimental studies show that the interval of the flow velocity for which the synchronization occurs (Fig. 8.4*a*) is approximately as follows:

$$|v_{0i}^{**} - v_{0i}^{*}| = 0.25 v_{0i}^{**} \,. \tag{8.26}$$

Using the aerodynamic plane-sections hypothesis, we can extend the results for straight rods to the case of space-curved rods, that is, we may assume that each small element of a space-curved rod is acted upon by a local distributed Kármán force $\mathbf{q}_\mathrm{K}$ (8.24). Since $\mathbf{v}_{\mathrm{r}n}(\eta, \tau)$ varies along the rod, we conclude that the local frequency of vortex shedding depends on the nondimensional coordinate η:

$$\omega_0(\eta, \tau) = \mathrm{Sh}\, \frac{v_{\mathrm{r}n}(\eta, \tau)}{dp_0} \qquad \left(\omega_0(\eta, \tau) = \frac{\omega}{p_0}\right).$$

In this general case (when the local frequency of vortex shedding depends on the nondimensional coordinate η), the analysis of rod oscillations becomes extremely complicated. Indeed, along the rod there could be encountered segments on which the vortex shedding frequency varies with η as well as segments on which there is no shedding at all (say, a segment where $\mathrm{Re} < 10$). The primary interest is the case of oscillation when the synchronization occurs. In this case, the whole rod, including the segments from which vortex shedding has occured, oscillates with the same natural frequency. This case is mostly dangerous (the wind resonance). However, the case is comparatively trivial for mathematical treatment because the vortex shedding frequency $\omega_0 = \mathrm{const}$ and therefore the formulas (8.24) for $\mathbf{q}_\mathrm{K}$ can be applied to the whole rod.

Note that on the intervals of variation of the flow velocity (8.26), the nondimensional frequency ω_0 from (8.24) remains constant while the amplitude value $q_{\mathrm{K}0}$ varies. This is important when determining the amplitudes of the steady-state oscillations of a rod in a flow with velocity subject to $v_{0i}^{*} < v_{0i} < v_{0i}^{**}$.

Let us consider the most simple expression for the magnitude of the Kármán force $|\mathbf{q}_\mathrm{K}|$ when the aerodynamic coefficients c_K are assumed to be constant for Re varying within a certain interval. Statistical investigations have shown that the mean-square value of the coefficient c_K for $10^3 < \mathrm{Re} < 10^5$ changes insignificantly (see Fung (1959)). Therefore, the assumption that c_k are constant is justified.

The distributed Kármán forces must satisfy the following relations:

$$(\mathbf{q}_\mathrm{K} \cdot \mathbf{e}_1) = 0 \,, \qquad (\mathbf{q}_\mathrm{K} \cdot \mathbf{v}_{\mathrm{r}n}) = 0 \,. \tag{8.27}$$

The projections of the force $\mathbf{q}_\mathrm{K}$ in the immovable coordinate frame can be found from the equations

$$\begin{aligned}
&q_{\mathrm{K}x_1}^2 + q_{\mathrm{K}x_2}^2 + q_{\mathrm{K}x_3}^2 = q_{\mathrm{K}0}^2 \left(\frac{v_\mathrm{r}}{v_0}\right)^4 \sin^2 \omega_0 \tau \sin^4 \varphi_{a_1} \,, \\
&q_{\mathrm{K}x_1}(v_{\mathrm{r}n})\, x_1 + q_{\mathrm{K}x_2}(v_{\mathrm{r}n})\, x_2 + q_{\mathrm{K}x_3}(v_{\mathrm{r}n})\, x_3 = 0 \,, \\
&q_{\mathrm{K}x_1} x_1' + q_{\mathrm{K}x_2} x_2' + q_{\mathrm{K}x_3} x_3' = 0 \,,
\end{aligned} \tag{8.28}$$

where

$$q_{\mathrm{K}0} = \frac{c_{\mathrm{K}} \rho d v_0^2 l^3}{2A_{33}} .$$

On rearrangement, from the system of equations (8.28) we get

$$\begin{aligned} q_{\mathrm{K}x_1} &= -q_{\mathrm{K}0} \frac{v_{\mathrm{r}}}{v_0} \frac{\Delta_1}{v_0} \sin\varphi_{a1} \sin\omega_0\tau , \\ q_{\mathrm{K}x_2} &= -q_{\mathrm{K}0} \frac{v_{\mathrm{r}}}{v_0} \frac{\Delta_2}{v_0} \sin\varphi_{a1} \sin\omega_0\tau , \\ q_{\mathrm{K}x_3} &= q_{\mathrm{K}0} \frac{v_{\mathrm{r}}}{v_0} \frac{\Delta_3}{v_0} \sin\varphi_{a1} \sin\omega_0\tau , \end{aligned} \tag{8.29}$$

where

$$\begin{aligned} \frac{\Delta_1}{v_0} &= (\sin\alpha\cos\beta - \beta_0 \dot{u}_{x_3})\, x_2' - (\sin\beta - \beta_0 \dot{u}_{x_2})\, x_3' , \\ \frac{\Delta_2}{v_0} &= (\sin\alpha\cos\beta - \beta_0 \dot{u}_{x_3})\, x_1' - (\cos\alpha\cos\beta - \beta_0 \dot{u}_{x_1})\, x_3' , \\ \frac{\Delta_3}{v_0} &= (\cos\alpha\cos\beta - \beta_0 \dot{u}_{x_1})\, x_2' - (\sin\beta - \beta_0 \dot{u}_{x_2})\, x_1' . \end{aligned}$$

Using the analytical expressions for the projections of the aerodynamic force in the immovable frame (8.29), we can find the projections of the force in the coordinate frame attached to the rod axis. It follows from the first equation of (8.27) that

$$\mathbf{q}_{\mathrm{K}} = q_{\mathrm{K}2}\mathbf{e}_2 + q_{\mathrm{K}3}\mathbf{e}_3 , \tag{8.30}$$

where

$$q_{\mathrm{K}2} = (\mathbf{q}_{\mathrm{K}} \cdot \mathbf{e}_2) = \Big(\Big(\sum_{j=1}^{3} q_{\mathrm{K}x_j} \mathbf{l}_j\Big) \cdot \mathbf{e}_2\Big) = \sum_{j=1}^{3} q_{\mathrm{K}x_j} l_{2j}^{(1)} ,$$

$$q_{\mathrm{K}3} = (\mathbf{q}_{\mathrm{K}} \cdot \mathbf{e}_3) = \Big(\Big(\sum_{j=1}^{3} q_{\mathrm{K}x_j} \mathbf{l}_j\Big) \cdot \mathbf{e}_3\Big) = \sum_{j=1}^{3} q_{\mathrm{K}x_j} l_{3j}^{(1)} ;$$

here, $l_{2j}^{(1)}$ and $l_{3j}^{(1)}$ are the elements of the matrix $\mathrm{L}^{(1)}$, which is the matrix of transformation from the basis $\{\mathbf{i}_j\}$ to the basis $\{\mathbf{e}_j\}$.

8.1.5 Lift and Twisting Moment Acting on a Rod of Non-Circular Cross Section

A rod of non-circular axisymmetric cross section immersed in a flow is subjected to a lift, $\mathbf{q}_{\mathrm{L}}$, and a moment, $\boldsymbol{\mu}_1$, (Fig. 8.1*c, d*). A rod cross section whose gravity center, O, and flexural center, O_1, coincide is shown in Fig. 8.1*c*.

Consider a rod's cross section shown in Fig. 8.1*d*. The equations of motion and the equilibrium equations become simplified when referred to the flexural center, thus it is sensible to refer the force $\mathbf{q}_{\mathrm{L}}$ and the moment $\boldsymbol{\mu}_1$ to the point O_1. Therefore, the relative velocity is

$$\mathbf{v}_{\mathrm{r}} = \mathbf{v}_0 - \dot{\mathbf{u}}^{(1)} ,$$

where $\dot{\mathbf{u}}^{(1)}$ is the velocity of the point O_1. As it has been shown experimentally, the magnitude of the lifting force $\mathbf{q}_{\mathrm{L}}$ in nondimensional notation is

$$|\mathbf{q}_{\mathrm{L}}| = \frac{c_{\mathrm{L}}(\alpha_a)\,\rho b v_{\mathrm{r}n}^2 l^3}{2A_{33}} = q_{\mathrm{L}00} c_{\mathrm{L}} \left(\frac{v_{\mathrm{r}}}{v_0}\right)^2 \sin^2\varphi_{a_1} , \tag{8.31}$$

where

$$q_{\mathrm{L}00} = \frac{\rho b l^3 v_0^2}{2A_{33}} , \qquad v_{\mathrm{r}n} = |\mathbf{v}_{\mathrm{r}n}| \sin\varphi_{a_1} ,$$

and the magnitude of the twisting moment $\boldsymbol{\mu}_1$ is

$$|\boldsymbol{\mu}_1| = \frac{c_m(\alpha_a)\,\rho b v_{\mathrm{r}n}^2 l_1 l^2}{2A_{33}} = \mu_{100} c_m \left(\frac{v_{\mathrm{r}}}{v_0}\right)^2 \sin^2\varphi_{a_1} , \tag{8.32}$$

where

$$\mu_{100} = \frac{\rho b l_1 l^2 v_0^2}{2A_{33}} ;$$

here, α_a is the local angle of attack (the angle between the velocity vector $\mathbf{v}_{\mathrm{r}n}$ and the vector $\mathbf{e}_3^{(1)}$ directed along the axis of symmetry of the cross section (Fig. 8.5)).

An aerodynamic force lying in the plane orthogonal to the vector $\mathbf{e}_1^{(1)}$ is shown in Fig. 8.6.

Components of the Lift $\mathbf{q}_{\mathrm{L}}$. Let us find the projections of the force $\mathbf{q}_{\mathrm{L}}$ and moment $\boldsymbol{\mu}_1$ in the immovable Cartesian frame of reference x_i. The vector $\mathbf{q}_{\mathrm{L}}$ must satisfy the following conditions:

$$(\mathbf{q}_{\mathrm{L}} \cdot \mathbf{e}_1^{(1)}) = 0 , \qquad (\mathbf{q}_{\mathrm{L}} \cdot \mathbf{v}_{\mathrm{r}n}) = 0 .$$

In components, these conditions together with (8.31) give

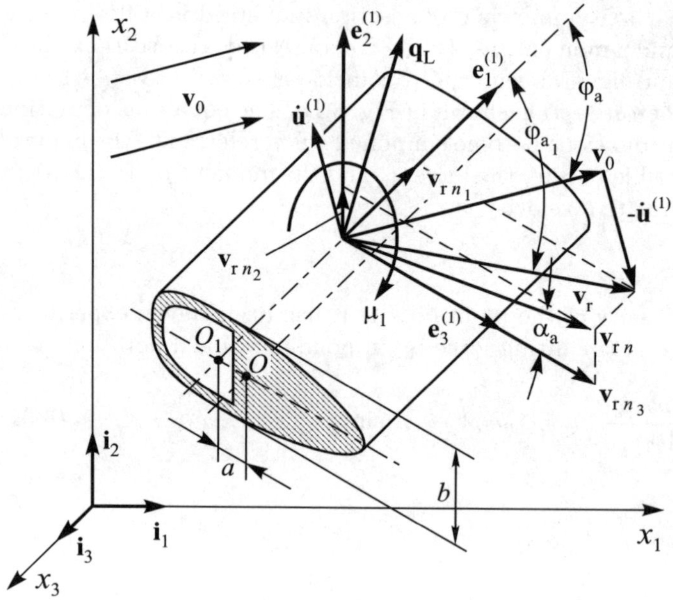

x2
e2(1)
qL
e1(1)
φa
v0
u̇(1)
φa1
v0
vrn1
-u̇(1)
vrn2
vr
μ1
e3(1)
αa
vrn
O1
O
vrn3
i2
a
b
i1
x1
i3
x3

Fig. 8.5.

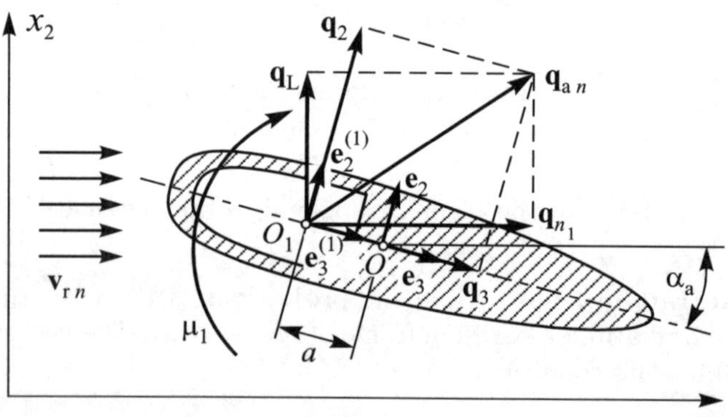

x2
q2
qL
qan
e2(1)
e2
qn1
O1
e3(1)
O
e3
q3
αa
vrn
μ1
a
x1

Fig. 8.6.

$$q_{\mathrm{L}x_1} x_1^{(1)\prime} + q_{\mathrm{L}x_2} x_2^{(1)\prime} + q_{\mathrm{L}x_3} x_3^{(1)\prime} = 0\,,$$

$$q_{\mathrm{L}x_1}(\mathbf{v}_{\mathrm{r}n})\, x_1 + q_{\mathrm{L}x_2}(\mathbf{v}_{\mathrm{r}n})\, x_2 + q_{\mathrm{L}x_3}(\mathbf{v}_{\mathrm{r}n})\, x_3 = 0\,,$$

$$q_{\mathrm{L}x_1}^2 + q_{\mathrm{L}x_2}^2 + q_{\mathrm{L}x_3}^2 = q_{\mathrm{L}00}^2 c_{\mathrm{L}}^2 \left(\frac{v_{\mathrm{r}}}{v_0}\right)^4 \sin^4 \varphi_{a_1}\,, \tag{8.33}$$

where $x_j^{(1)}$ are the coordinates of the points on the line through the flexural centers of the cross sections. Then,

$$q_{\mathrm{L}x_1} = -q_{\mathrm{L}00} c_{\mathrm{L}} \frac{v_{\mathrm{r}}}{v_0} \frac{\Delta_1}{v_0} \sin \varphi_{a1}\,,$$

$$q_{\mathrm{L}x_2} = -q_{\mathrm{L}00} c_{\mathrm{L}} \frac{v_{\mathrm{r}}}{v_0} \frac{\Delta_2}{v_0} \sin \varphi_{a1}\,,$$

$$q_{\mathrm{L}x_3} = q_{\mathrm{L}00} c_{\mathrm{L}} \frac{v_{\mathrm{r}}}{v_0} \frac{\Delta_3}{v_0} \sin \varphi_{a1}\,, \tag{8.34}$$

where

$$\frac{\Delta_1}{v_0} = \frac{1}{v_0} \left[(\mathbf{v}_{\mathrm{r}n})_{x_2} x_3^{(1)\prime} - (\mathbf{v}_{\mathrm{r}n})_{x_3} x_2^{(1)\prime} \right],$$

$$\frac{\Delta_2}{v_0} = \frac{1}{v_0} \left[(\mathbf{v}_{\mathrm{r}n})_{x_3} x_1^{(1)\prime} - (\mathbf{v}_{\mathrm{r}n})_{x_2} x_3^{(1)\prime} \right],$$

$$\frac{\Delta_3}{v_0} = \frac{1}{v_0} \left[(\mathbf{v}_{\mathrm{r}n})_{x_2} x_1^{(1)\prime} - (\mathbf{v}_{\mathrm{r}n})_{x_1} x_2^{(1)\prime} \right]. \tag{8.35}$$

Let us now find the projections $(\mathbf{v}_{\mathrm{r}n})_{x_2}$ and $(\mathbf{v}_{\mathrm{r}n})_{x_3}$ entering in (8.35). First, we should find the projections of the vector $\mathbf{v}_{\mathrm{r}n}$ in the attached frame. We have

$$v_{\mathrm{r}n2} = (\mathbf{v}_{\mathrm{r}n} \cdot \mathbf{e}_2^{(1)}) = (\mathbf{v}_{\mathrm{r}} \cdot \mathbf{e}_2^{(1)}) = v_{\mathrm{r}x_1} l_{21}^{(1)} + v_{\mathrm{r}x_2} l_{22}^{(1)} + v_{\mathrm{r}x_3} l_{23}^{(1)}\,,$$

$$v_{\mathrm{r}n3} = (\mathbf{v}_{\mathrm{r}n} \cdot \mathbf{e}_3^{(1)}) = (\mathbf{v}_{\mathrm{r}} \cdot \mathbf{e}_3^{(1)}) = v_{\mathrm{r}x_1} l_{31}^{(1)} + v_{\mathrm{r}x_2} l_{32}^{(1)} + v_{\mathrm{r}x_3} l_{33}^{(1)}\,.$$

Since

$$\mathbf{v}_{\mathrm{r}} = \mathbf{v}_{\mathrm{r}n} + \mathbf{v}_{\mathrm{r}1} = \mathbf{v}_{\mathrm{r}n} + |\mathbf{v}_{\mathrm{r}1}|\, \mathbf{e}_1\,,$$

we get the coordinates of the vector $\mathbf{v}_{\mathrm{r}n}$ in the attached basis

$$\mathbf{v}_{\mathrm{r}n} = (\mathbf{v}_{\mathrm{r}} \cdot \mathbf{e}_2^{(1)})\, \mathbf{e}_2^{(1)} + (\mathbf{v}_{\mathrm{r}} \cdot \mathbf{e}_3^{(1)})\, \mathbf{e}_3^{(1)}\,, \tag{8.36}$$

Therefore, the Cartesian coordinates of the vector $\mathbf{v}_{\mathrm{r}n}$ are $(\mathbf{v}_{\mathrm{r}n})_{x_j} = (\mathbf{v}_{\mathrm{r}n} \cdot \mathbf{i}_j)$. In components, we have

$$(\mathbf{v}_{\mathrm{r}n})_{x_1} = (\mathbf{v}_{\mathrm{r}} \cdot \mathbf{e}_2^{(1)})\, l_{21}^{(1)} + (\mathbf{v}_{\mathrm{r}} \cdot \mathbf{e}_3^{(1)})\, l_{31}^{(1)}\,,$$

$$(\mathbf{v}_{\mathrm{r}n})_{x_2} = (\mathbf{v}_{\mathrm{r}} \cdot \mathbf{e}_2^{(1)})\, l_{22}^{(1)} + (\mathbf{v}_{\mathrm{r}} \cdot \mathbf{e}_3^{(1)})\, l_{32}^{(1)}\,,$$

$$(\mathbf{v}_{\mathrm{r}n})_{x_3} = (\mathbf{v}_{\mathrm{r}} \cdot \mathbf{e}_2^{(1)})\, l_{23}^{(1)} + (\mathbf{v}_{\mathrm{r}} \cdot \mathbf{e}_3^{(1)})\, l_{33}^{(1)}\,.$$

Finally, on rearrangement, the components of the lift are found to be

$$\begin{aligned}
q_{\mathrm{L}x_1} &= q_{\mathrm{L}00} c_{\mathrm{L}} \frac{v_{\mathrm{r}}}{v_0} \Big[(\sin\alpha \cos\beta - \beta_0 \dot{u}_{x_3}^{(1)})\, x_2^{(1)\prime} \\
&\qquad - (\sin\beta - \beta_0 \dot{u}_{x_2}^{(1)})\, x_3^{(1)\prime} \Big] \sin\varphi_{a_1} \operatorname{sgn}\alpha_a \,, \\
q_{\mathrm{L}x_2} &= q_{\mathrm{L}00} c_{\mathrm{L}} \frac{v_{\mathrm{r}}}{v_0} \Big[(\sin\alpha \cos\beta - \beta_0 \dot{u}_{x_3}^{(1)})\, x_1^{(1)\prime} \\
&\qquad - (\cos\alpha \cos\beta - \beta_0 \dot{u}_{x_1}^{(1)})\, x_3^{(1)\prime} \Big] \sin\varphi_{a_1} \operatorname{sgn}\alpha_a \,, \\
q_{\mathrm{L}x_3} &= q_{\mathrm{L}00} c_{\mathrm{L}} \frac{v_{\mathrm{r}}}{v_0} \Big[(\cos\alpha \cos\beta - \beta_0 \dot{u}_{x_1}^{(1)})\, x_2^{(1)\prime} \\
&\qquad - (\sin\beta - \beta_0 \dot{u}_{x_2}^{(1)})\, x_1^{(1)\prime} \Big] \sin\varphi_{a_1} \operatorname{sgn}\alpha_a \,,
\end{aligned} \tag{8.37}$$

The direction of the vector $\mathbf{q}_{\mathrm{L}}$ depends on the angle of attack α_a by means of the function $\operatorname{sgn}\alpha_a$.

Let us derive the components of the vector $\mathbf{q}_{\mathrm{L}}$ in the attached axes. The condition $(\mathbf{q}_{\mathrm{L}} \cdot \mathbf{e}_1^{(1)}) = 0$ was used for determining the components of the vector $\mathbf{q}_{\mathrm{L}}$ in the Cartesian axes. From this condition it follows that

$$\mathbf{q}_{\mathrm{L}} = q_{\mathrm{L}2} \mathbf{e}_2^{(1)} + q_{\mathrm{L}3} \mathbf{e}_3^{(1)} \,, \tag{8.38}$$

where

$$\begin{aligned}
q_{\mathrm{L}2} &= (\mathbf{q}_{\mathrm{L}} \cdot \mathbf{e}_2^{(1)}) = \Big(\Big(\sum_{j=1}^{3} q_{\mathrm{L}x_j} \mathbf{i}_j \Big) \cdot \mathbf{e}_2^{(1)} \Big) = \sum_{j=1}^{3} q_{\mathrm{L}x_j} l_{2j}^{(1)} \,, \\
q_{\mathrm{L}3} &= (\mathbf{q}_{\mathrm{L}} \cdot \mathbf{e}_3^{(1)}) = \Big(\Big(\sum_{j=1}^{3} q_{\mathrm{L}x_j} \mathbf{i}_j \Big) \cdot \mathbf{e}_3^{(1)} \Big) = \sum_{j=1}^{3} q_{\mathrm{L}x_j} l_{3j}^{(1)} \,.
\end{aligned}$$

Cartesian Components of the Aerodynamic Moment. In the Cartesian frame of reference, the vector $\mathbf{e}_1^{(1)}$ reads

$$\mathbf{e}_1^{(1)} = \sum_{j=1}^{3} x_j^{(1)\prime} \mathbf{i}_j \,,$$

Therefore,

$$\boldsymbol{\mu}_1 = \mu_1 \mathbf{e}_1^{(1)} = \sum_{j=1}^{3} \mu_{100} c_m \left(\frac{v_{\mathrm{r}}}{v_0} \right)^2 x_j^{(1)\prime} \sin^2\varphi_{a_1} \operatorname{sgn}\alpha_a \, \mathbf{i}_j \,, \tag{8.39}$$

The Cartesian components of the vector $\boldsymbol{\mu}_1$ are

$$\mu_{1x_j} = \mu_{100} c_m \left(\frac{v_{\mathrm{r}}}{v_0} \right)^2 x_j^{(1)\prime} \sin^2\varphi_{a_1} \operatorname{sgn}\alpha_a \,, \tag{8.40}$$

8.1.6 Angle of Attack α_a

For a rod of non-circular cross section (Fig. 8.5), the aerodynamic coefficients c_{n_1}, $c_{\rm L}$, and c_m depend on the angle of attack α_a. If the rod cross section has an axis of symmetry, then the angle of attack is the angle between the unit base vector $\mathbf{e}_3^{(1)}$ directed along the axis of symmetry and the local normal component $\mathbf{v}_{\mathrm{r}n}$ orthogonal to $\mathbf{e}_1$ (Fig. 8.5). The vector $\mathbf{v}_{\mathrm{r}n}$ belongs to the plane defined by the vectors $\mathbf{e}_2^{(1)}$ and $\mathbf{e}_3^{(1)}$, therefore,

$$\sin\alpha_a = \frac{(\mathbf{v}_{\mathrm{r}n}\cdot\mathbf{e}_2^{(1)})}{|\mathbf{v}_{\mathrm{r}n}|} = \frac{(\mathbf{v}_{\mathrm{r}n}\cdot\mathbf{e}_2^{(1)})}{\sqrt{(\mathbf{v}_{\mathrm{r}n}\cdot\mathbf{e}_2^{(1)})^2+(\mathbf{v}_{\mathrm{r}n}\cdot\mathbf{e}_3^{(1)})^2}}\,. \tag{8.41}$$

Since

$$(\mathbf{v}_{\mathrm{r}n}\cdot\mathbf{e}_2^{(1)}) = (\mathbf{v}_{\mathrm{r}}\cdot\mathbf{e}_2^{(1)})\,, \qquad (\mathbf{v}_{\mathrm{r}n}\cdot\mathbf{e}_3^{(1)}) = (\mathbf{v}_{\mathrm{r}}\cdot\mathbf{e}_3^{(1)})\,,$$

we have

$$\sin\alpha_a = \frac{(\mathbf{v}_{\mathrm{r}}\cdot\mathbf{e}_2^{(1)})}{\sqrt{(\mathbf{v}_{\mathrm{r}}\cdot\mathbf{e}_2^{(1)})^2+(\mathbf{v}_{\mathrm{r}}\cdot\mathbf{e}_3^{(1)})^2}}\,,$$

where

$$\begin{aligned}\mathbf{v}_{\mathrm{r}} = v_0[(\cos\alpha\cos\beta-\beta_0\dot{u}_{x_1}^{(1)})\,\mathbf{i}_1+(\sin\beta-\beta_0\dot{u}_{x_2}^{(1)})\,\mathbf{i}_2 \\ +(\sin\alpha\cos\beta-\beta_0\dot{u}_{x_3}^{(1)})\,\mathbf{i}_3] = v_0\mathbf{e}_{\mathrm{r}}^{(1)}\,,\end{aligned}$$

$$|\mathbf{e}_{\mathrm{r}}^{(1)}|\neq 1\,. \tag{8.42}$$

If $\dot{u}_{x_j}^{(1)}=0$, then $\mathbf{e}_{\mathrm{r}}^{(1)}=\mathbf{e}_v$ and $|\mathbf{e}_v|=1$. Finally, we get

$$\sin\alpha_a = \frac{(\mathbf{e}_{\mathrm{r}}^{(1)}\cdot\mathbf{e}_2^{(1)})}{\sqrt{(\mathbf{e}_{\mathrm{r}}^{(1)}\cdot\mathbf{e}_2^{(1)})^2+(\mathbf{e}_{\mathrm{r}}^{(1)}\cdot\mathbf{e}_3^{(1)})^2}}\,. \tag{8.43}$$

During oscillations the angle of attack varies with time and thus can be considered an additional unknown parameter. Hence, one more relationship between the angle of attack α_a, and the components $\dot{u}_{x_j}^{(1)}$ of the velocity vector $\dot{\mathbf{u}}^{(1)}$ of the flexural center, on the one hand, and the angles ϑ_j, on the other hand, is required. In the general case, the angle of attack α_a also depends on the static state of the rod, that is,

$$\alpha_a(\eta,\tau) = \alpha_a^{(0)}(\eta)+\Delta\alpha_a(\eta,\tau)\,, \tag{8.44}$$

where $\alpha_a^{(0)}=\alpha_{a00}+\alpha_{a0}$ is the static angle of attack; here, $\alpha_{a00}(\eta)$ is the natural angle of attack, $\alpha_{a0}(\eta)$ is the angle that results from static deformation of the rod due to the flow (Fig. 8.7), and $\Delta\alpha_a(\eta,\tau)$ is the dynamical component of the angle of attack. The dynamical component may be a finite

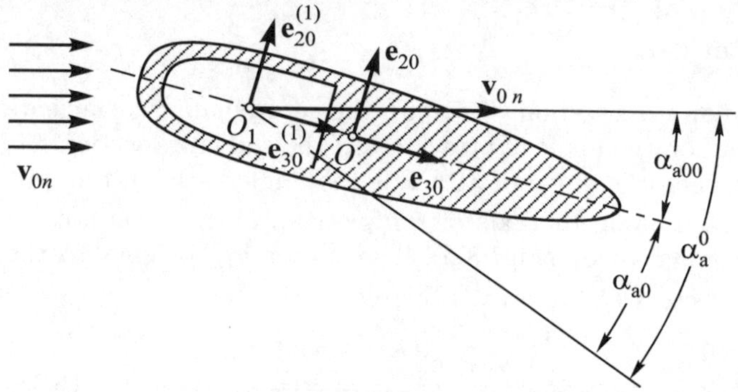

Fig. 8.7.

quantity. The angle $\alpha_a^{(0)}(\eta)$ can be found by solving the equilibrium equations of the rod (Svetlitsky (2000a)).

Let us rearrange the relationship (8.43). We have

$$\mathbf{e}_2^{(1)} = \sum_{j=1}^{3} l_{2j}^{(1)} \mathbf{i}_j \,, \qquad \mathbf{e}_3^{(1)} = \sum_{j=1}^{3} l_{3j}^{(1)} \mathbf{i}_j \,,$$

where $l_{2j}^{(1)}$ and $l_{3j}^{(1)}$ are the elements of the matrix $\mathrm{L}^{(1)} = \mathrm{L}\mathrm{L}^0$ (see (A.44), (A.55), and (A.57) in Appendix A in Svetlitsky (2000a)). Then, (8.43) takes the form

$$\sin\alpha_a = \frac{\sum_{j=1}^{3} e_{0x_j}^{(1)} \cdot l_{2j}^{(1)}}{\sqrt{\left(\sum_{j=1}^{3} e_{0x_j}^{(1)} \cdot l_{2j}^{(1)}\right)^2 + \left(\sum_{j=1}^{3} e_{0x_j}^{(1)} \cdot l_{3j}^{(1)}\right)^2}} \,, \tag{8.45}$$

where

$$e_{0x_1}^{(1)} = e_{\nu x_1} - \beta_0 \dot{u}_{x_1}^{(1)} \,, \qquad e_{0x_2}^{(1)} = e_{\nu x_2} - \beta_0 \dot{u}_{x_2}^{(1)} \,, \qquad e_{0x_3}^{(1)} = e_{\nu x_3} - \beta_0 \dot{u}_{x_3}^{(1)} \,,$$

$$e_{\nu x_1} = \cos\alpha\cos\beta \,, \qquad e_{\nu x_2} = \sin\beta \,, \qquad e_{\nu x_3} = \sin\alpha\cos\beta \,.$$

The elements $l_{ij}^{(1)}$ of the matrix $\mathrm{L}^{(1)}$ satisfy the conditions

$$l_{ij}^{(1)} l_{kj}^{(1)} = \delta_{ik} \,, \qquad l_{jk}^{(1)} l_{ji}^{(1)} = \delta_{ki} \,.$$

For example,

$$\sum_{j=1}^{3} l_{j1}^{(1)^2} = 1\,, \quad \sum_{j=1}^{3} l_{j2}^{(1)^2} = 1\,, \quad \sum_{j=1}^{3} l_{j3}^{(1)^2} = 1\,.$$

$$l_{11}^{(1)} l_{13}^{(1)} + l_{21}^{(1)} l_{23}^{(1)} + l_{31}^{(1)} l_{33}^{(1)} = 0\,,$$

$$l_{11}^{(1)} l_{12}^{(1)} + l_{21}^{(1)} l_{22}^{(1)} + l_{31}^{(1)} l_{32}^{(1)} = 0\,,$$

$$l_{12}^{(1)} l_{13}^{(1)} + l_{22}^{(1)} l_{23}^{(1)} + l_{32}^{(1)} l_{33}^{(1)} = 0\,. \tag{8.46}$$

Using these relations in (8.45), we get

$$\sin\alpha_a = \frac{\sum_{j=1}^{3} e_{0x_j}^{(1)} \cdot l_{2j}^{(1)}}{\sqrt{|\mathbf{e}_{\mathrm{r}}^{(1)}|^2 - (\mathbf{e}_{\mathrm{r}}^{(1)} \cdot \mathbf{e}_1^{(1)})^2}} = \frac{\sum_{j=1}^{3} e_{0x_j}^{(1)} \cdot l_{2j}^{(1)}}{|\mathbf{e}_{\mathrm{r}}^{(1)}|\sqrt{1 - \cos^2\varphi_{a_1}}}\,. \tag{8.47}$$

In a more expanded notation, $l_{2j}^{(1)}$ read

$$\begin{aligned}
l_{21}^{(1)} &= -l_{11}^{(0)} \sin\vartheta_3 + l_{21}^{(0)} \cos\vartheta_1 \cos\vartheta_3 + l_{31}^{(0)} \cos\vartheta_3 \sin\vartheta_1\,,\\
l_{22}^{(1)} &= -l_{12}^{(0)} \sin\vartheta_3 + l_{22}^{(0)} \cos\vartheta_1 \cos\vartheta_3 + l_{32}^{(0)} \cos\vartheta_3 \sin\vartheta_1\,,\\
l_{23}^{(1)} &= -l_{13}^{(0)} \sin\vartheta_3 + l_{23}^{(0)} \cos\vartheta_1 \cos\vartheta_3 + l_{33}^{(0)} \cos\vartheta_3 \sin\vartheta_1\,,
\end{aligned}$$

and thus

$$L^{(0)} = [l_{ij}^{(0)}] = \begin{bmatrix} \cos\vartheta_2^0 \cos\vartheta_3^0 & \cos\vartheta_2^0 \sin\vartheta_3^0 \cos\vartheta_1^0 + \cos\vartheta_1^0 \cos\vartheta_3^0 & \cos\vartheta_2^0 \sin\vartheta_3^0 \sin\vartheta_1^0 - \sin\vartheta_2^0 \cos\vartheta_1^0 \\ -\sin\vartheta_3^0 & \cos\vartheta_1^0 \cos\vartheta_3^0 & \cos\vartheta_3^0 \sin\vartheta_1^0 \\ \sin\vartheta_2^0 \cos\vartheta_3^0 & \sin\vartheta_2^0 \sin\vartheta_3^0 \cos\vartheta_1^0 - \cos\vartheta_3^0 \sin\vartheta_1^0 & \sin\vartheta_2^0 \sin\vartheta_3^0 \sin\vartheta_1^0 + \cos\vartheta_2^0 \cos\vartheta_1^0 \end{bmatrix};$$

here, ϑ_j^0 are some known angles, which describe the location of the frame $\{\mathbf{e}_{j0}^{(1)}\}$ associated with the rod static configuration relative to the immovable frame $\{\mathbf{i}_j\}$.

When solving the problem numerically, the angle of attack α_a (8.47) is to be evaluated at time τ_{k-1} and then, using the graphs of $c_{n_1}(\alpha_a)$, $c_{\mathrm{L}}(\alpha_a)$, and $c_m(\alpha_a)$ obtained experimentally and the equations of motion, the state of the system can be determined at time τ_k, where $\tau_k = \tau_{k-1} + \Delta\tau$.

8.2 Aerodynamic Forces on a Rod of Circular Cross Section Under Small Oscillations

In the case of small oscillations of a rod, the aerodynamic forces can be written as

$$\mathbf{q}_n = \mathbf{q}_{n0} + \Delta\mathbf{q}_n\,, \qquad \mathbf{q}_1 = \mathbf{q}_{10} + \Delta\mathbf{q}_1\,,$$

where the vectors with the subscript 0 are the force and the moment applied to the rod in the equilibrium configuration, the small quantities $\Delta\mathbf{q}_n$ and $\Delta\mathbf{q}_1$ are the vectors of the aerodynamic forces due to small oscillations of the rod. For small $\dot{u}_{x_j}$, the relative velocity reads

$$v_{\mathrm{r}} = v_0[1 - \beta_0(\cos\alpha\cos\beta\dot{u}_{x_1} + \sin\beta\dot{u}_{x_2} + \sin\alpha\cos\beta\dot{u}_{x_3})]\,. \tag{8.48}$$

8.2.1 Determination of the Components of the Vector $\Delta\mathbf{q}_n$

Expanding the expression for q_{nx_j} (8.19) in a series, using (8.48), and retaining only linear terms, we get the relation

$$\begin{aligned}\Delta q_{nx_i} = (a_{i1}^{(1)}u'_{x_1} + a_{i2}^{(1)}u'_{x_2} + a_{i3}^{(1)}u'_{x_3}) + (a_{i1}^{(2)}\dot{u}_{x_1} + a_{i2}^{(2)}\dot{u}_{x_2} + a_{i3}^{(2)}\dot{u}_{x_3})\\ + q_{n0}\beta_0(c_{i1}\dot{u}_{x_1} + c_{i2}\dot{u}_{x_2} + c_{i3}\dot{u}_{x_3})\sin\varphi_{a0}\,,\end{aligned}$$

where

$$\begin{aligned}\sin\varphi_{a0} &= \sqrt{1-\cos^2\varphi_{a0}}\,,\\ \cos\varphi_{a0} &= x'_{10}\cos\alpha\cos\beta + x'_{20}\sin\beta + x'_{30}\sin\alpha\cos\beta\,.\end{aligned}$$

In vector notation, we have

$$\Delta\mathbf{q}_n = \mathrm{A}^{(1)}\mathbf{u}'_x + \mathrm{A}^{(2)}\dot{\mathbf{u}}_x\,. \tag{8.49}$$

The elements of the matrix $\mathrm{A}^{(1)}$ read

$$\begin{aligned}a_{11}^{(1)} &= \frac{q_{n0}}{\sin\varphi_{a0}}[\cos\alpha\cos\beta(x'_{10}\cos 2\varphi_{a0} - \cos\varphi_{a0}\cos\alpha\cos\beta)\\ &\qquad - \cos\varphi_{a0}\sin^2\varphi_{a0}]\,,\\ a_{12}^{(1)} &= \frac{q_{n0}}{\sin\varphi_{a0}}[\sin\beta(x'_{10}\cos 2\varphi_{a0} - \cos\varphi_{a0}\cos\alpha\cos\beta)]\,,\\ a_{13}^{(1)} &= \frac{q_{n0}}{\sin\varphi_{a0}}[\sin\alpha\cos\beta(x'_{10}\cos 2\varphi_{a0} - \cos\varphi_{a0}\cos\alpha\cos\beta)]\,,\\ a_{21}^{(1)} &= \frac{q_{n0}}{\sin\varphi_{a0}}[\cos\alpha\cos\beta(x'_{20}\cos 2\varphi_{a0} - \cos\varphi_{a0}\sin\beta)]\,,\\ a_{22}^{(1)} &= \frac{q_{n0}}{\sin\varphi_{a0}}[\sin\beta(x'_{20}\cos 2\varphi_{a0} - \cos\varphi_{a0}\sin\beta) - \cos\varphi_{a0}\sin^2\varphi_{a0}]\,,\\ a_{23}^{(1)} &= \frac{q_{n0}}{\sin\varphi_{a0}}[\sin\alpha\cos\beta(x'_{20}\cos 2\varphi_{a0} - \cos\varphi_{ao}\sin\beta)]\,,\end{aligned}$$

$$a_{31}^{(1)} = \frac{q_{n0}}{\sin \varphi_{a0}} \left[\cos\alpha \cos\beta (x'_{30} \cos 2\varphi_{a0} - \cos\varphi_{a0} \sin\alpha \cos\beta)\right],$$

$$a_{32}^{(1)} = \frac{q_{n0}}{\sin \varphi_{a0}} \left[\sin\beta (x'_{30} \cos 2\varphi_{a0} - \cos\varphi_{a0} \sin\alpha \cos\beta)\right],$$

$$a_{33}^{(1)} = \frac{q_{n0}}{\sin \varphi_{a0}} \left[\sin\alpha \cos\beta (x'_{30} \cos 2\varphi_{a0} - \cos\varphi_{a0} \sin\alpha \cos\beta) - \cos\varphi_{a0} \sin^2 \varphi_{a0}\right].$$

The elements of the matrix $A^{(2)}$ are as follows:

$$a_{11}^{(2)} = \frac{q_{n0}\beta_0}{\sin \varphi_{a0}} \left[({x'_{10}}^2 - 1) \sin^2 \varphi_{a0} - (x'_{10} \cos\varphi_{a0} - \cos\alpha \cos\beta)^2\right],$$

$$a_{12}^{(2)} = \frac{q_{n0}\beta_0}{\sin \varphi_{a0}} \left[x'_{10} x'_{20} \sin^2 \varphi_{a0} - (x'_{10} \cos\varphi_{a0} - \cos\alpha \cos\beta)(x'_{20} \cos\varphi_{a0} - \sin\beta)\right],$$

$$a_{13}^{(2)} = \frac{q_{n0}\beta_0}{\sin \varphi_{a0}} \left[x'_{10} x'_{30} \sin^2 \varphi_{a0} - (x'_{10} \cos\varphi_{a0} - \cos\alpha \cos\beta)(x'_{30} \cos\varphi_{a0} - \sin\alpha \cos\beta)\right],$$

$$a_{21}^{(2)} = a_{12}^{(21)},$$

$$a_{22}^{(2)} = \frac{q_{n0}\beta_0}{\sin \varphi_{a0}} \left[({x'_{20}}^2 - 1) \sin^2 \varphi_{a0} - (x'_{20} \cos\varphi_{a0} - \sin\beta)^2\right],$$

$$a_{23}^{(2)} = \frac{q_{n0}\beta_0}{\sin \varphi_{a0}} \left[x'_{20} x'_{30} \sin^2 \varphi_{a0} - (x'_{20} \cos\varphi_{a0} - \sin\beta)(x'_{30} \cos\varphi_{a0} - \sin\alpha \cos\beta)\right],$$

$$a_{31}^{(2)} = a_{13}^{(21)}, \qquad a_{32}^{(2)} = a_{23}^{(21)},$$

$$a_{33}^{(2)} = \frac{q_{n0}\beta_0}{\sin \varphi_{a0}} \left[({x'_{30}}^2 - 1) \sin^2 \varphi_{a0} - (x'_{30} \cos\varphi_{a0} - \sin\alpha \cos\beta)^2\right].$$

8.2.2 The Components of the Vector $\Delta \mathbf{q}_1$ for a Rod of Constant Cross Section

We have

$$\Delta q_{1x_i} = (a_{i1}^{(3)} u'_{x_1} + a_{i2}^{(3)} u'_{x_2} + a_{i3}^{(3)} u'_{x_3}) + (a_{i1}^{(4)} \dot{u}_{x_1} + a_{i2}^{(4)} \dot{u}_{x_2} + a_{i3}^{(4)} \dot{u}_{x_3}).$$

In vector notation, we have

$$\Delta \mathbf{q}_1 = A^{(3)} \mathbf{u}'_x + A^{(4)} \dot{\mathbf{u}}_x. \tag{8.50}$$

The elements of the matrix $A^{(3)}$ are

$$a_{11}^{(3)} = q_{10} \cos\varphi_{a0} (2x'_{10} \cos\alpha \cos\beta + \cos\varphi_{a0}),$$

$$a_{12}^{(3)} = 2q_{10} \cos\varphi_{a0} x'_{10} \sin\beta,$$

$$a_{13}^{(3)} = 2q_{10} \cos\varphi_{a0} x'_{10} \sin\alpha \cos\beta,$$

$$
\begin{aligned}
a_{21}^{(3)} &= 2q_{10}\cos\varphi_{a0}x'_{20}\cos\alpha\cos\beta\,,\\
a_{22}^{(3)} &= q_{10}\cos\varphi_{a0}\left(2x'_{20}\sin\beta+\cos\varphi_{a0}\right),\\
a_{23}^{(3)} &= 2q_{10}\cos\varphi_{a0}x'_{20}\sin\alpha\cos\beta\,,\\
a_{31}^{(3)} &= 2q_{10}\cos\varphi_{a0}x'_{30}\cos\alpha\cos\beta\,,\\
a_{32}^{(3)} &= 2q_{10}\cos\varphi_{a0}x'_{30}\sin\beta\,,\\
a_{33}^{(3)} &= q_{10}\cos\varphi_{a0}\left(2x'_{30}\sin\alpha\cos\beta+\cos\varphi_{a0}\right).
\end{aligned}
$$

The elements of the matrix $\mathrm{A}^{(4)}$ read

$$
\begin{aligned}
a_{11}^{(4)} &= -2q_{10}\beta_0{x'_{10}}^2\cos\varphi_{a0}\,,\\
a_{12}^{(4)} &= -2q_{10}\beta_0x'_{10}x'_{20}\cos\varphi_{a0}\,,\\
a_{13}^{(4)} &= -2q_{10}\beta_0x'_{10}x'_{30}\cos\varphi_{a0}\,,\\
a_{21}^{(4)} &= a_{12}^{(4)}\,,\\
a_{22}^{(4)} &= -2q_{10}\beta_0{x'_{20}}^2\cos\varphi_{a0}\,,\\
a_{23}^{(4)} &= -2q_{10}\beta_0x'_{20}x'_{30}\cos\varphi_{a0}\,,\\
a_{31}^{(4)} &= a_{13}^{(4)}\,,\qquad a_{32}^{(4)} = a_{23}^{(4)}\,,\\
a_{33}^{(4)} &= -2q_{10}\beta_0{x'_{30}}^2\cos\varphi_{a0}\,.
\end{aligned}
$$

Sometimes, when studying oscillations of a rod in a flow, it is more advantageous to use the vector $\boldsymbol{\vartheta}$ instead of the vector $\mathbf{u}'_x$. Since

$$
\begin{aligned}
\mathbf{u}'_x &= \mathbf{e}_1-\mathbf{e}_{10}=\sum_{j=1}^{3}\left(l_{1j}^{(1)}\mathbf{i}_j-l_{1j}^{(0)}\mathbf{i}_j\right)\\
&= \left(\vartheta_3 l_{21}^{(0)}-\vartheta_2 l_{31}^{(0)}\right)\mathbf{i}_1+\left(\vartheta_3 l_{22}^{(0)}-\vartheta_2 l_{32}^{(0)}\right)\mathbf{i}_2+\left(\vartheta_3 l_{23}^{(0)}-\vartheta_2 l_{33}^{(0)}\right)\mathbf{i}_3\,,
\end{aligned}
$$

we have

$$
\mathbf{u}'_x = \mathrm{A}_\vartheta\boldsymbol{\vartheta}\,,
$$

where

$$
\mathrm{A}_\vartheta = \begin{bmatrix} 0 & -l_{31}^{(0)} & l_{21}^{(0)} \\ 0 & -l_{32}^{(0)} & l_{22}^{(0)} \\ 0 & -l_{33}^{(0)} & l_{23}^{(0)} \end{bmatrix}.
$$

As a result, the vectors $\Delta\mathbf{q}_n$ and $\Delta\mathbf{q}_1$ referred to the immovable frame can be written as

$$
\begin{aligned}
\Delta\mathbf{q}_n &= \mathrm{A}^{(1)}\mathrm{A}_\vartheta\boldsymbol{\vartheta}+\mathrm{A}^{(2)}\dot{\mathbf{u}}_x\,,\\
\Delta\mathbf{q}_1 &= \mathrm{A}^{(3)}\mathrm{A}_\vartheta\boldsymbol{\vartheta}+\mathrm{A}^{(4)}\dot{\mathbf{u}}_x\,.
\end{aligned}
\tag{8.51}
$$

8.3 Aerodynamic Forces Due to Small Oscillations for a Rod of Non-Circular Symmetrical Cross Section

8.3.1 Dynamic Component of the Angle of Attack $\Delta\alpha_a$ of a Rod Under Vibrations

Consider a rod of non-circular cross section such that its the center of gravity (the point O) do not coincide with the flexural center (the point O_1) (Figs. 8.5, 8.6). Ignoring the influence of the small angle of attack $\Delta\alpha_a$ on the aerodynamic coefficient c_1, we can write the other aerodynamic coefficients ($c_{n_1}(\alpha_a)$, $c_{\mathrm{L}}(\alpha_a)$, and $c_m(\alpha_a)$) as follows:

$$
\begin{aligned}
c_{n_1}(\alpha_a) &= c_{n_1 0}(\alpha_a^0) + \left.\frac{\partial c_{n_1}}{\partial \alpha_a}\right|_{\alpha_a=\alpha_a^0} \Delta\alpha_a(\tau,\eta)\,,\\
c_{\mathrm{L}}(\alpha_a) &= c_{\mathrm{L}0}(\alpha_a^0) + \left.\frac{\partial c_{\mathrm{L}}}{\partial \alpha_a}\right|_{\alpha_a=\alpha_a^0} \Delta\alpha_a(\tau,\eta)\,,\\
c_m(\alpha_a) &= c_{m0}(\alpha_a^0) + \left.\frac{\partial c_m}{\partial \alpha_a}\right|_{\alpha_a=\alpha_a^0} \Delta\alpha_a(\tau,\eta)\,,
\end{aligned} \tag{8.52}
$$

where

$$
c_1(\alpha_a) \approx c_1(\alpha_a^0)\,, \qquad \alpha_a = \alpha_a^0 + \Delta\alpha_a\,.
$$

The angle of attack α_a^0 for the static configuration of the rod (the angle between the vectors $\mathbf{v}_{0n}$ and $\mathbf{e}_{30}^{(1)}$) (see (Fig. 8.7)) can be found from the condition

$$
\sin\alpha_a^0 = \frac{(\mathbf{v}_{0n}\cdot\mathbf{e}_{20}^{(1)})}{|\mathbf{v}_{0n}|} = \frac{(\mathbf{e}_v\cdot\mathbf{e}_{20}^{(1)})}{\sin\varphi_{a0}}\,,
$$

or the condition

$$
\sin\alpha_a^{(0)} = \frac{l_{21}^{(1)(0)}\cos\alpha\cos\beta + l_{22}^{(1)(0)}\sin\beta + l_{23}^{(1)(0)}\sin\alpha\cos\beta}{\sin\varphi_{a0}}\,, \tag{8.53}
$$

where φ_{a0} is the angle between the vectors $\mathbf{v}_0$ and $\mathbf{e}_{10}^{(1)}$; here, $l_{2j}^{(1)(0)}$ are the elements of the matrix

$$
\mathrm{L}^{(1)(0)} = \mathrm{L}(\vartheta_{j0})\cdot\mathrm{L}^0(\vartheta_{j00})\,;
$$

in addition, we have

$$
\begin{aligned}
\cos\varphi_{a0} &= x_1^{(1)'}\cos\alpha\cos\beta + x_2^{(1)'}\sin\beta + x_3^{(1)'}\sin\alpha\cos\beta\,,\\
x_j^{(1)'} &= l_{1j}^{(1)(0)}\,,
\end{aligned}
$$

where ϑ_{j0} are the angles of rotation of the coordinate frame attached to the rod axis with the origin at O_1 (these angles describe the rotation corresponding to the static loading of the rod by the flow).

Using the equilibrium equations for a rod of non-circular cross section in a flow, we can determine the static stress–strain state of the rod and the angle of attack $\alpha_a^0(\eta)$, which affects the aerodynamic coefficients (8.52).

In Sect. 8.1.3, the general relation (8.47) for $\sin\alpha_a$ was obtained. The relation is valid if $\Delta\alpha_a$ is a finite quantity. For small oscillations of a rod in a flow, we can assume that $\dot{u}_{xj}^{(1)}$, $u_{xj}^{(1)'}$, ϑ_j and the angle of attack $\Delta\alpha_a$ are small. The parameters $l_{2j}^{(1)}$ and $l_{1j}^{(1)}$ in (8.47) (recall that $l_{1j}^{(1)}$ appear in the expressions for $\cos\varphi_{a_1}$) are the elements of the matrix $\mathrm{L}^{(1)}$, which for small angles φ_j reads (see Svetlitsky (2000a), page 363)

$$\mathrm{L}^{(1)} = \Delta\mathrm{L}\mathrm{L}^{(0)} = \begin{bmatrix} 1 & \vartheta_3 & -\vartheta_2 \\ -\vartheta_3 & 1 & \vartheta_1 \\ \vartheta_2 & -\vartheta_1 & 1 \end{bmatrix} \mathrm{L}^{(0)}, \tag{8.54}$$

therefore,

$$\begin{aligned} l_{11}^{(1)} &= l_{11}^{(0)} + \vartheta_3 l_{21}^{(0)} - \vartheta_2 l_{31}^{(0)}, & l_{21}^{(1)} &= l_{21}^{(0)} + \vartheta_1 l_{31}^{(0)} - \vartheta_3 l_{11}^{(0)}, \\ l_{12}^{(1)} &= l_{12}^{(0)} + \vartheta_3 l_{22}^{(0)} - \vartheta_2 l_{32}^{(0)}, & l_{22}^{(1)} &= l_{22}^{(0)} + \vartheta_1 l_{32}^{(0)} - \vartheta_3 l_{12}^{(0)}, \\ l_{13}^{(1)} &= l_{13}^{(0)} + \vartheta_3 l_{23}^{(0)} - \vartheta_2 l_{33}^{(0)}, & l_{23}^{(1)} &= l_{23}^{(0)} + \vartheta_1 l_{33}^{(0)} - \vartheta_3 l_{13}^{(0)}. \end{aligned} \tag{8.55}$$

By using (8.55), the nominator of (8.47) can be rearranged as

$$\begin{gathered} \sum_{j=1}^{3} e_{0x_j}^{(1)}(l_{2j}^{(0)}+\Delta l_{2j}) = (\mathbf{e}_v\cdot\mathbf{e}_{20}^{(1)})+(\mathbf{e}_v\cdot\mathbf{e}_{30}^{(1)})\,\vartheta_1-(\mathbf{e}_v\cdot\mathbf{e}_{10}^{(1)})\,\vartheta_3-\beta_0(\mathbf{e}_{20}^{(1)}\cdot\dot{\mathbf{u}}_x^{(1)}), \\ (\mathbf{e}_v\cdot\mathbf{e}_{k0}^{(1)}) = \sum_{j=1}^{3} e_{vx_j} l_{kj}^{(0)}, \qquad k = 1,\, 2,\, 3\,. \end{gathered} \tag{8.56}$$

The denominator of (8.47) contains the summands $|\mathbf{e}_{\mathrm{r}}^{(1)}|^2$ and $(\mathbf{e}_{\mathrm{r}}^{(1)}\cdot\mathbf{e}_1^{(1)})^2$. The first summand to an accuracy of the second-order terms is

$$\begin{aligned} |\mathbf{e}_{\mathrm{r}}^{(1)}|^2 &= \sum_{j=1}^{3}(e_{0x_j}^{(1)})^2 \\ &= (e_{vx_1} - \beta_0\dot{u}_{x_1}^{(1)})^2 + (e_{vx_2} - \beta_0\dot{u}_{x_2}^{(1)})^2 + (e_{vx_3} - \beta_0\dot{u}_{x_3}^{(1)})^2 \\ &= 1 - 2\beta_0(\mathbf{e}_v\cdot\dot{\mathbf{u}}_x^{(1)}), \qquad \sum_{j=1}^{3} e_{vx_j}^2 = 1\,. \end{aligned} \tag{8.57}$$

The second summand, $(\mathbf{e}_{\mathrm{r}}^{(1)}\cdot\mathbf{e}_1^{(1)})^2$, can be written as

$$(\mathbf{e}_{\mathrm{r}}^{(1)} \cdot \mathbf{e}_1^{(1)})^2 = [(e_{vx_1} - \beta_0 \dot{u}_{x_1}^{(1)})\, l_{11}^{(1)} + (e_{vx_2} - \beta_0 \dot{u}_{x_2}^{(1)})\, l_{12}^{(1)} + (e_{vx_3} - \beta_0 \dot{u}_{x_3}^{(1)})\, l_{13}^{(1)}]^2 ,$$

$$l_{1j}^{(1)} = l_{1j}^{(0)} + \Delta l_{1j} ,$$

or

$$(\mathbf{e}_{\mathrm{r}}^{(1)} \cdot \mathbf{e}_1^{(1)})^2 = [(\mathbf{e}_v \cdot \mathbf{e}_{10}^{(1)}) + (\mathbf{e}_v \cdot \mathbf{e}_{20}^{(1)})\, \vartheta_3 - (\mathbf{e}_v \cdot \mathbf{e}_{30}^{(1)})\, \vartheta_2 - \beta_0 (\mathbf{e}_{10}^{(1)} \cdot \dot{\mathbf{u}}_x^{(1)})]^2 .$$

Retaining only the terms that are linear in small quantities ϑ_j and $\dot{u}_{x_j}^{(1)}$, we get

$$(\mathbf{e}_{\mathrm{r}}^{(1)} \cdot \mathbf{e}_1^{(1)})^2 = (\mathbf{e}_v \cdot \mathbf{e}_{10}^{(1)})^2 + 2\, (\mathbf{e}_v \cdot \mathbf{e}_{10}^{(1)})\, [(\mathbf{e}_v \cdot \mathbf{e}_{20}^{(1)})\, \vartheta_3 - (\mathbf{e}_v \cdot \mathbf{e}_{30}^{(1)})\, \vartheta_2 - \beta_0 (\mathbf{e}_{10} \cdot \dot{\mathbf{u}}_x^{(1)})] ,$$

$$(\mathbf{e}_v \cdot \mathbf{e}_{10}^{(1)}) = \cos \varphi_{a0} . \tag{8.58}$$

Consequently, we have

$$|\mathbf{e}_{\mathrm{r}}^{(1)}|^2 - (\mathbf{e}_{\mathrm{r}}^{(1)} \cdot \mathbf{e}_1^{(1)})^2 = 1 - \cos^2 \varphi_{a0} + 2\, (\Delta_1 - \Delta_2 + \Delta_3) , \tag{8.59}$$

where

$$\Delta_1 = \beta_0 \left([(\mathbf{e}_v \cdot \mathbf{e}_{10}^{(1)})\, \mathbf{e}_{10}^{(1)} - \mathbf{e}_v] \cdot \dot{\mathbf{u}}_x^{(1)} \right) ,$$
$$\Delta_2 = (\mathbf{e}_v \cdot \mathbf{e}_{10}^{(1)})\, (\mathbf{e}_v \cdot \mathbf{e}_{20}^{(1)})\, \vartheta_3 ,$$
$$\Delta_3 = (\mathbf{e}_v \cdot \mathbf{e}_{10}^{(1)})\, (\mathbf{e}_v \cdot \mathbf{e}_{30}^{(1)})\, \vartheta_2 ,$$
$$(\mathbf{e}_v \cdot \mathbf{e}_{10}^{(1)}) = \cos \varphi_{a0} .$$

The expression (8.59) stands in the denominator of (8.47) under the root sign; hence, by expanding in a series and taking only the linear terms, we get

$$\frac{1}{\sqrt{|\mathbf{e}_{\mathrm{r}}^{(1)}|^2 - (\mathbf{e}_{\mathrm{r}}^{(1)} \cdot \mathbf{e}_1^{(1)})^2}} = \frac{1}{\sqrt{1 - \cos^2 \varphi_{a0}}} \left(1 - \frac{\Delta_1 - \Delta_2 + \Delta_3}{1 - \cos^2 \varphi_{a0}} \right) .$$

Finally, in view of (8.56), we arrive at

$$\sin \alpha_a = \frac{1}{\sin \varphi_{a0}} [(\mathbf{e}_v \cdot \mathbf{e}_{20}^{(1)}) + (\mathbf{e}_v \cdot \mathbf{e}_{30}^{(1)})\, \vartheta_1 - (\mathbf{e}_v \cdot \mathbf{e}_{10}^{(1)})\, \vartheta_3 - \beta_0 (\mathbf{e}_{20}^{(1)} \cdot \dot{\mathbf{u}}_x^{(1)})] \left[1 - \frac{\Delta_1 - \Delta_2 + \Delta_3}{\sin^2 \varphi_{a0}} \right] = \sin \alpha_a^0 + \Delta , \tag{8.60}$$

where

$$\Delta = \sum_{j=1}^{3} \tilde{d}_j^{(1)} \vartheta_j + \sum_{j=1}^{3} \tilde{d}_j^{(2)} \dot{u}_{xj} ,$$

$$\tilde{d}_1^{(1)} = \frac{(\mathbf{e}_v \cdot \mathbf{e}_{30}^{(1)})}{\sin \varphi_{a0}} ,$$

$$\tilde{d}_2^{(1)} = -\frac{(\mathbf{e}_v \cdot \mathbf{e}_{20}^{(1)}) (\mathbf{e}_v \cdot \mathbf{e}_{10}^{(1)}) (\mathbf{e}_v \cdot \mathbf{e}_{30}^{(1)})}{\sin^3 \varphi_{a0}} ,$$

$$\tilde{d}_3^{(1)} = -\left[\frac{(\mathbf{e}_v \cdot \mathbf{e}_{10}^{(1)})}{\sin \varphi_{a0}} - \frac{(\mathbf{e}_v \cdot \mathbf{e}_{10}^{(1)}) (\mathbf{e}_v \cdot \mathbf{e}_{20}^{(1)}) (\mathbf{e}_v \cdot \mathbf{e}_{20}^{(1)})}{\sin^3 \varphi_{a0}}\right] ,$$

$$\tilde{d}_j^{(2)} = \left(\left\{-\frac{\beta_0 \mathbf{e}_{20}^{(1)}}{\sin \varphi_{a0}} + \frac{\beta_0 (\mathbf{e}_v \cdot \mathbf{e}_{20}^{(1)}) [(\mathbf{e}_v \cdot \mathbf{e}_{10}^{(1)}) \mathbf{e}_{10}^{(1)} - \mathbf{e}_v]}{\sin^3 \varphi_{a0}}\right\} \cdot \mathbf{i}_j\right) .$$

For small values of the angle of attack $\Delta\alpha_a$, we get

$$\sin(\alpha_a^0 + \Delta\alpha_a) = \sin \alpha_a^0 + \cos \alpha_a^0 \cdot \Delta\alpha_a , \tag{8.61}$$

therefore,

$$\Delta\alpha_a = \frac{\Delta}{\cos \alpha_a^0} = (\mathbf{D}^{(1)} \cdot \boldsymbol{\vartheta}) + (\mathbf{D}^{(2)} \cdot \dot{\mathbf{u}}_x^{(1)}) , \qquad \Delta = \Delta\alpha_a \cos \alpha_a^0 . \tag{8.62}$$

The components of the vectors $\mathbf{D}^{(1)}$ and $\mathbf{D}^{(2)}$ are

$$d_1^{(1)} = \frac{(\mathbf{e}_v \cdot \mathbf{e}_{30}^{(1)})}{\sin \varphi_{a0} \cos \alpha_a^0} ,$$

$$d_2^{(1)} = -\frac{(\mathbf{e}_v \cdot \mathbf{e}_{20}^{(1)}) (\mathbf{e}_v \cdot \mathbf{e}_{10}^{(1)}) (\mathbf{e}_v \cdot \mathbf{e}_{30}^{(1)})}{\sin^3 \varphi_{a0} \cos \alpha_a^0} ,$$

$$d_3^{(1)} = \frac{(\mathbf{e}_v \cdot \mathbf{e}_{10}^{(1)}) (\mathbf{e}_v \cdot \mathbf{e}_{20}^{(1)}) (\mathbf{e}_v \cdot \mathbf{e}_{20}^{(1)})}{\sin^3 \varphi_{a0} \cos \alpha_a^0} - \frac{(\mathbf{e}_v \cdot \mathbf{e}_{10}^{(1)})}{\sin \varphi_{a0} \cos \alpha_a^0} ,$$

$$d_1^{(2)} = -\frac{d_0[(\mathbf{e}_v \cdot \mathbf{e}_{10}^{(1)}) l_{11}^{(0)} - \cos \alpha \cos \beta]}{\cos \alpha_a^0} - \frac{\beta_0 l_{21}^{(0)}}{\sin \varphi_{a0} \cos \alpha_a^0} ,$$

$$d_2^{(2)} = -\frac{d_0}{\cos \alpha_a^0} [(\mathbf{e}_v \cdot \mathbf{e}_{10}^{(1)}) l_{12}^{(0)} - \sin \beta] - \frac{\beta_0 l_{22}^{(0)}}{\sin \varphi_{a0} \cos \alpha_a^0} ,$$

$$d_3^{(2)} = -\frac{d_0}{\cos \alpha_a^0} [(\mathbf{e}_v \cdot \mathbf{e}_{10}^{(1)}) l_{13}^{(0)} - \sin \alpha \cos \beta] - \frac{\beta_0 l_{23}^{(0)}}{\sin \varphi_{a0} \cos \alpha_a^0} , \tag{8.63}$$

where

$$d_0 = \frac{\beta_0 (\mathbf{e}_v \cdot \mathbf{e}_{20}^{(1)})}{\sin^3 \varphi_{a0}} .$$

The coefficients c_{n_1}, c_{L}, and c_m entering into the expressions for the aerodynamic forces q_1, q_{n_1} and q_{L} and the moment μ_1 depend on the angle of

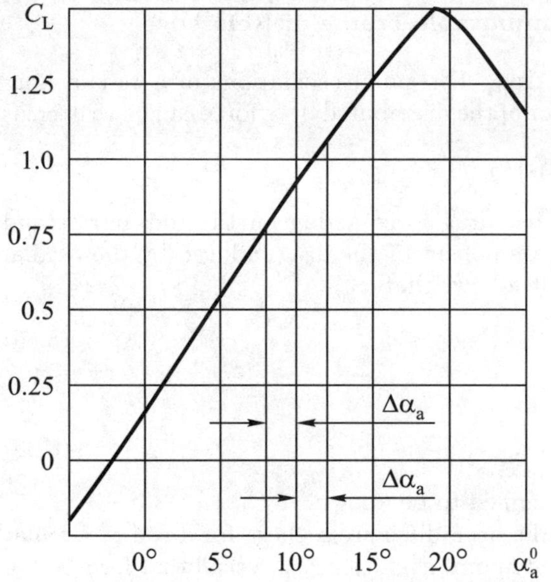

Fig. 8.8.

attack α_a (8.52). These coefficients should be determined experimentally. For example, the graphs of the lift coefficient c_L for some rods of unsmooth cross sectional profile are given in Fung (1959). For smooth, aircraft-wing type profiles, the plots of c_L versus the angle of attack α_a^0 can be found in Prandtl, Tietjens (1931) and Mises (1945). As an example, one such a plot from Mises (1945) is shown in Fig. 8.8. The other graphs of $c_L(\alpha_a)$ from Fung (1959) and Prandtl, Tietjens (1931) look qualitatively the same, that is, for a rather big range of change of the angle of attack the coefficient depends on the angle of attack almost linearly. In Fig. 8.8, this range is $-5° \leq \alpha_a \leq 20°$.

For example, consider small oscillations of a rod about the equilibrium configuration. Let the angle of attack at a given section be equal to 10°. Then, according to Fig. 8.8, the lift coefficient $c_L(\alpha_a)$, with account for the dynamical angle of attack $+\Delta\alpha_a$, is

$$c_L(\alpha_a) = 0.90 + 0.068\Delta\alpha_a \,,$$
$$\left.\frac{dc_L}{d\alpha_a}\right|_{\alpha_a=\alpha_a^0=10°} \approx 0.068 \,.$$

Given experimentally obtained relationships between c_L, c_m, and the angle of attack α_a^0, we can find similarly the aerodynamic coefficients that account for the dynamical angle of attack $\Delta\alpha_a$.

8.3.2 Projections of the Aerodynamic Forces $\Delta \mathbf{q}_{n_1}$ and $\Delta \mathbf{q}_{\mathrm{L}}$ and the Moment $\Delta \mu_1$ in the Immovable Frame of Reference

Projections of the Drag $\Delta \mathbf{q}_{n_1}$ For small oscillations of a rod of non-circular cross section, the vector of the distributed drag force can be written as

$$\mathbf{q}_{n_1}(\eta, \tau) = \mathbf{q}_{n_1 0}(\eta) + \Delta \mathbf{q}_{n_1}(\eta, \tau), \tag{8.64}$$

where $\mathbf{q}_{n_1 0}(\eta)$ is the distributed drag force acting on the rod at rest and $\Delta \mathbf{q}_{n_1}(\eta, \tau)$ is the dynamical component of the drag induced by the oscillations of the rod. From (8.64) it follows that

$$\Delta \mathbf{q}_{n_1} = \mathbf{q}_{n_1} - \mathbf{q}_{n_1 0}. \tag{8.65}$$

In components, this reads

$$\Delta q_{n_1 x_j} = q_{n_1 x_j} - q_{n_1 x_j 0}. \tag{8.66}$$

The components $q_{n_1 x_j 0}$ are assumed to be known.

The components $q_{n_1 x_j}$ of the drag differ from those for a rod of circular cross section (8.19) in the aerodynamic coefficient c_{n_1}; recall that for a rod of non-circular cross section this coefficient depends on the angle of attack $\Delta\alpha_a$.

Thus, for a small angle of attack, the relationships (8.19) yields

$$q_{n_1 x_j} = q_{n_1 00}\left(c_{n_1 0} + \left.\frac{\partial c_{n_1}}{\partial \alpha_a}\right|_{\alpha_a = \alpha_a^0} \Delta\alpha_a\right) \frac{v_{\mathrm{r}}}{v_0} \times \left(e_{v x_j} - l_{1j}^{(1)} \cos\varphi_{a_1} + \Delta_{x_j}\right) \sin\varphi_{a_1},$$

$$q_{n_1 00} = \frac{\rho b v_0^2 l^3}{2 A_{33}}. \tag{8.67}$$

For small oscillations, it follows from (8.4) that

$$\cos\varphi_{a_1} = \frac{\cos\varphi_{a0} + (\mathbf{e}_v \cdot \mathbf{e}_{20}^{(1)})\,\vartheta_3 - (\mathbf{e}_v \cdot \mathbf{e}_{30}^{(1)})\,\vartheta_2 - \beta_0(\mathbf{e}_{10}^{(1)} \cdot \dot{\mathbf{u}}_x^{(1)})}{1 - \beta_0(\mathbf{e}_v \cdot \dot{\mathbf{u}}_x^{(1)})}$$

or

$$\cos\varphi_{a_1} = \cos\varphi_{a_0} + \Delta_1^{(1)} + \Delta_2^{(1)} - \Delta_3^{(1)} = \cos\varphi_{a0} + \Delta_4,$$

where

$$\Delta_4 = \Delta_1^{(1)} + \Delta_2^{(1)} - \Delta_3^{(1)}.$$

Hence,

$$\sin\varphi_{a_1} = \sqrt{1 - \cos^2\varphi_{a_1}} = \sin\varphi_{a0} - \cot\varphi_{a0} \cdot \Delta_4,$$

where

$$\Delta_1^{(1)} = \beta_0 \left([\cos\varphi_{a0}\, \mathbf{e}_v - \mathbf{e}_{10}^{(1)}] \cdot \dot{\mathbf{u}}_x \right),$$
$$\Delta_2^{(1)} = (\mathbf{e}_v \cdot \mathbf{e}_{20}^{(1)})\, \vartheta_3,$$
$$\Delta_3^{(1)} = (\mathbf{e}_v \cdot \mathbf{e}_{30}^{(1)})\, \vartheta_3.$$

The terms Δ_{x_j} entering into (8.67) can be found from (8.20) with $x'_j = x'_{j0}$.

Using (8.67), on rearrangement we get

$$q_{n_1 x_j} = q_{n_1 x_j 0} + \Delta q_{n_1 x_j}, \tag{8.68}$$

where

$$\Delta q_{n_1 x_j} = \sum_{k=1}^{3} n_{jk}^{(1)} \vartheta_k + \sum_{k=1}^{3} n_{jk}^{(2)} \dot{u}_{x_k}^{(1)} \tag{8.69}$$

or, in the matrix–vector notation, we have

$$\Delta \mathbf{q}_{n_1} = \mathrm{N}^{(1)} \boldsymbol{\vartheta} + \mathrm{N}^{(2)} \dot{\mathbf{u}}_x^{(1)}. \tag{8.70}$$

The elements of the matrices $\mathrm{N}^{(1)}$ and $\mathrm{N}^{(2)}$ are given in Appendix A.1.

Projections of the Lift $\Delta \mathbf{q}_\mathrm{L}$ The projections of the aerodynamic lift were obtained in Sect. 8.1.2 (see the relationships (8.37)).

Consider small oscillations of the rod. For example, the component $q_{\mathrm{L}x_1}$ reads

$$q_{\mathrm{L}x_1} = q_{\mathrm{L}0} \left(c_{\mathrm{L}0} + \frac{\partial c_\mathrm{L}}{\partial \alpha_a} \Delta\alpha_a \right) \left(1 - \beta_0 (\mathbf{e} \cdot \dot{\mathbf{u}}_x^{(1)}) \right) \times \left(e_{vx_3} l_{12}^0 - e_{vx_2} l_{13}^{(0)} + \Delta_5 \right) (\sin\varphi_{a0} - \cot\varphi_{a0} \cdot \Delta_4), \tag{8.71}$$

where

$$\Delta_5 = (e_{vx_2} l_{33}^{(0)} - e_{vx_3} l_{32}^{(0)})\, \vartheta_2 + (e_{vx_3} l_{22}^{(0)} - e_{vx_2} l_{23}^{(0)})\, \vartheta_3 + \beta_0 l_{13}^{(0)} \dot{u}_{x_2}^{(1)} - \beta_0 l_{12}^{(0)} \dot{u}_{x_3}^{(1)}.$$

On rearrangement, from (8.71) it follows that

$$q_{\mathrm{L}x_j} = q_{\mathrm{L}x_j 0} + \Delta q_{\mathrm{L}x_j},$$

where

$$\Delta q_{\mathrm{L}x_j} = \sum_{k=1}^{3} p_{jk}^{(1)} \vartheta_k + \sum_{k=1}^{3} p_{jk}^{(2)} \dot{u}_{x_k}^{(1)}.$$

Finally, we arrive at

$$\Delta \mathbf{q}_\mathrm{L} = \mathrm{P}^{(1)} \boldsymbol{\vartheta} + \mathrm{P}^{(2)} \dot{\mathbf{u}}_x^{(1)}. \tag{8.72}$$

The elements of the matrices $\mathrm{P}^{(1)}$ and $\mathrm{P}^{(2)}$ are given in Appendix A.2.

Projections of the Moment $\Delta\boldsymbol{\mu}_1$ in the Immovable Frame of Reference During Small Oscillations. The aerodynamic moment $\boldsymbol{\mu}_1$ given by (8.39) can be written as

$$\boldsymbol{\mu}_1 = \mu_{10}^{(1)}\left(c_{m_0} + \left.\frac{\partial c_m}{\partial \alpha_a}\right|_{\alpha_a=\alpha_a^0} \Delta\alpha_a\right)\left(1 - 2\beta_0(\mathbf{e}_v \cdot \dot{\mathbf{u}}_x^{(1)})\right) \sin^2\varphi_{a_1} \cdot \mathbf{e}_1^{(1)}, \tag{8.73}$$

where

$$\mu_{10}^{(1)} = \mu_{100}\,\mathrm{sgn}\,\alpha_a\,.$$

With respect to the Cartesian frame, using (8.40), we can write

$$\mu_{1x_j} = \mu_{10}^{(1)}\left(c_{m_0} + \frac{\partial c_m}{\partial \alpha_a}\Delta\alpha_a\right)\left(1 - 2\beta_0(\mathbf{e}_v \cdot \dot{\mathbf{u}}_x^{(1)})\right)\dot{x}_j^{(1)} \sin^2\varphi_{a_1}\,,$$
$$j = 1,\,2,\,3\,. \tag{8.74}$$

For small oscillations, we have

$$\sin^2\varphi_{a_1} = \sin^2\varphi_{a0} - 2\cos\varphi_{a0}\cdot\Delta_4\,,$$
$$\dot{x}_j^{(1)} = l_{1j}^{(0)} + \vartheta_3 l_{2j}^{(0)} - \vartheta_2 l_{3j}^{(0)}\,. \tag{8.75}$$

Substituting (8.75) and (8.62) into (8.74) and then retaining only the terms linear in the small quantities ϑ_j and u_{x_j}, we obtain

$$\mu_{1x_j} = \mu_{1x_j0} + \Delta\mu_{1x_j}\,, \tag{8.76}$$

where

$$\mu_{1x_j0} = \mu_{10}^{(1)} c_{m_0} l_{1j}^{(0)} \sin^2\varphi_{a0}\,,$$
$$\Delta\mu_{1x_j} = \mu_{10}^{(1)}\Big[-\gamma_1 l_{1j}^{(0)}\Delta_4 - \gamma_2 l_{1j}^{(0)}(\mathbf{e}_v\cdot\dot{\mathbf{u}}_x^{(1)})$$
$$+\gamma_3 l_{1j}^{(0)}((\mathbf{D}^{(1)}\cdot\boldsymbol{\vartheta}) + (\mathbf{D}^{(2)}\cdot\dot{\mathbf{u}}_x^{(1)})) + \gamma_4(l_{2j}^{(0)}\vartheta_3 - l_{3j}\vartheta_2)\Big]\,;$$

here,

$$\gamma_1 = 2c_{m0}\cos\varphi_{a0}\,, \qquad \gamma_2 = 2\beta_0 c_{m0}\sin^2\varphi_{a0}\,,$$
$$\gamma_3 = \frac{\partial c_m}{\partial\alpha_a}\sin^2\varphi_{a0}\,, \qquad \gamma_4 = c_{m0}\sin^2\varphi_{a0}\,.$$

It is assumed that the sign of $c_m(\alpha_a)$ coincides with the sign of α_a.

The expression for $\Delta\mu_{1x_j}$ can be rearranged as

$$\Delta\mu_{1x_j} = \sum_{k=1}^{3} m_{jk}^{(1)}\vartheta_k + \sum_{k=1}^{3} m_{jk}^{(2)}\dot{u}_{x_k}^{(1)}\,,$$

In vector notation, this reads

$$\Delta\boldsymbol{\mu}_1 = \mathrm{M}^{(1)}\boldsymbol{\vartheta} + \mathrm{M}^{(2)}\dot{\mathbf{u}}_x^{(1)}\,, \tag{8.77}$$

The elements of the matrices $\mathrm{M}^{(1)}$ and $\mathrm{M}^{(2)}$ are given in Appendix A.3.

8.3.3 The Aerodynamic Forces $\Delta\mathbf{q}_n$, $\Delta\mathbf{q}_{n_1}$, $\Delta\mathbf{q}_{\mathrm{L}}$, and $\Delta\mathbf{q}_1$ and the Moment $\Delta\boldsymbol{\mu}_1$ in the Coordinate Frame Attached to the Rod Axis

The Vectors $\Delta\mathbf{q}_n$, $\Delta\mathbf{q}_{n_1}$, and $\Delta\mathbf{q}_{\mathrm{L}}$ The Cartesian components of the vectors $\Delta\mathbf{q}_n$ and $\Delta\mathbf{q}_1$ (8.51), $\Delta\mathbf{q}_{n_1}$ (8.70), $\Delta\mathbf{q}_{\mathrm{L}}$ (8.72) and the moment $\Delta\boldsymbol{\mu}_1$ (8.77) have been found. To set up the equations of small oscillations in the coordinate frame attached to the rod axis, we need to know the projections of the vectors in this frame. To evaluate the projections of $\Delta\mathbf{q}_n$, $\Delta\mathbf{q}_{n_1}$, and $\Delta\mathbf{q}_{\mathrm{L}}$ in the frame attached to the rod we can use the matrix of transformation from the basis $\{\mathbf{i}_j\}$ to the basis $\{\mathbf{e}_j^{(1)}\}$, $\mathrm{L}^{(1)}$. For small angles ϑ_j, we may assume that $\mathrm{L}^{(1)} = \mathrm{L}^{(0)}$. The result is

$$\Delta\mathbf{q}_{ne} = \mathrm{L}^{(0)}\Delta\mathbf{q}_n\,, \qquad \Delta\mathbf{q}_{n_1e} = \mathrm{L}^{(0)}\Delta\mathbf{q}_{n_1}\,, \qquad \Delta\mathbf{q}_{\mathrm{L}e} = \mathrm{L}^{(0)}\Delta\mathbf{q}_{\mathrm{L}}\,,$$
$$\dot{\mathbf{u}}_x^{(1)} = \mathrm{L}^{(0)\mathrm{T}}\dot{\mathbf{u}}^{(1)}\,,$$

or

$$\Delta\mathbf{q}_{ne} = \mathrm{L}^{(0)}\mathrm{A}^{(1)}\mathrm{A}_{\vartheta}\boldsymbol{\vartheta} + \mathrm{L}^{(0)}\mathrm{A}^{(2)}\mathrm{L}^{(0)\mathrm{T}}\dot{\mathbf{u}}^{(1)} = \sum_{j=1}^{3}\Delta q_{nj}\mathbf{e}_j\,,$$

$$\Delta\mathbf{q}_{n_1e} = \mathrm{L}^{(0)}\mathrm{N}^{(1)}\boldsymbol{\vartheta} + \mathrm{L}^{(0)}\mathrm{N}^{(2)}\mathrm{L}^{(0)\mathrm{T}}\dot{\mathbf{u}}^{(1)} = \sum_{j=1}^{3}\Delta q_{n_1j}\mathbf{e}_j\,,$$

$$\Delta\mathbf{q}_{\mathrm{L}e} = \mathrm{L}^{(0)}\mathrm{P}^{(1)}\boldsymbol{\vartheta} + \mathrm{L}^{(0)}\mathrm{P}^{(2)}\mathrm{L}^{(0)\mathrm{T}}\dot{\mathbf{u}}^{(1)} = \sum_{j=1}^{3}\Delta q_{\mathrm{L}j}\mathbf{e}_j\,. \tag{8.78}$$

The Vector $\Delta\mathbf{q}_1$ For small oscillations, the vector $\Delta\mathbf{q}_1$ (8.21) takes the form

$$\mathbf{q}_1 = q_{10}\left(\frac{v_{\mathrm{r}}}{v_0}\right)^2 \cos^2\varphi_{a_1}\mathbf{e}_1^{(1)} = q_{10}\left[1 - 2\beta_0(\mathbf{e}_v\cdot\dot{\mathbf{u}}_x^{(1)})\right]$$
$$\times\left[\cos^2\varphi_{a0} + 2\,(\varDelta_1 + \varDelta_2 - \varDelta_3)\cos\varphi_{a0}\right]\mathbf{e}_1^{(1)} = \mathbf{q}_{10} + \Delta q_1\mathbf{e}_1^{(1)}\,,$$

where

$$\mathbf{q}_{10} = q_{10}\cos^2\varphi_{a0}\cdot\mathbf{e}_1^{(1)}\,,$$
$$\Delta q_1 = -2q_{10}(\mathbf{e}_v\cdot\mathbf{e}_{30}^{(1)})\,\vartheta_2\cos\varphi_{a0}$$
$$+\,2q_{10}(\mathbf{e}_v\cdot\mathbf{e}_{20}^{(1)})\,\vartheta_3\cos\varphi_{a0} - 2q_{10}\beta_0(\mathbf{e}_{10}^{(1)}\cdot\dot{\mathbf{u}}_x^{(1)})\cos\varphi_{a0}\,,$$
$$q_{10} = q_{100}c_{10}\,. \tag{8.79}$$

The components of the vector $\Delta\mathbf{q}_1$ in the attached frame are

$$\Delta q_1 \neq 0\,, \qquad \Delta q_2 = \Delta q_3 = 0\,.$$

Using the vector $\dot{\mathbf{u}}^{(1)}$ in (8.79), we can represent Δq_1 as

$$\Delta q_1 = \sum_{j=1}^{3} t_j^{(1)} \vartheta_j + \sum_{j=1}^{3} t_j^{(2)} \dot{u}_j^{(1)}, \tag{8.80}$$

where

$$t_1^{(1)} = 0\,, \qquad t_2^{(1)} = -2q_{10}(\mathbf{e}_v \cdot \mathbf{e}_{30}^{(1)}) \cos\varphi_{a0}\,,$$
$$t_3^{(1)} = 2q_{10}(\mathbf{e}_v \cdot \mathbf{e}_{20}^{(1)}) \cos\varphi_{a0}\,, \qquad t_j^{(2)} = -2q_{10} l_{j1}^{(0)} \cos\varphi_{a0}\,.$$

The Aerodynamic Moment $\Delta\boldsymbol{\mu}_1$ For small oscillations, the aerodynamic moment $\Delta\boldsymbol{\mu}_1$ given by (8.73) can be represented as follows:

$$\begin{aligned}\boldsymbol{\mu}_1 = \mu_{10}^{(1)}\left(c_{m0} + \frac{\partial c_m}{\partial \alpha_a}\,\Delta\alpha_a\right)\,(1 - 2\beta_0(\mathbf{e}_v \cdot \dot{\mathbf{u}}_x))\\ \times\left(\sin^2\varphi_{a0} - 2\cos\varphi_{a0}\cdot\varDelta_4\right)\,\mathbf{e}_1^{(1)} = \boldsymbol{\mu}_{10} + \Delta\mu_1\mathbf{e}_1^{(1)}\,,\end{aligned} \tag{8.81}$$

here,

$$\begin{aligned}\Delta\mu_1 &= \mu_{10}^{(1)}\Big[-2c_{m0}\beta_0(\mathbf{e}_v\cdot\dot{\mathbf{u}}_x^{(1)})\,\sin^2\varphi_{a0} - 2c_{mo}\cos\varphi_{a0}\cdot\varDelta_4\\ &\quad+\frac{\partial c_m}{\partial\alpha_a}\left((\mathbf{D}^{(1)}\cdot\boldsymbol{\vartheta}) + (\mathbf{D}^{(2)}\cdot\dot{\mathbf{u}}_x^{(1)})\right)\sin^2\varphi_{a0}\Big]\,,\\ \Delta\mu_2 &= \Delta\mu_3 = 0\,.\end{aligned} \tag{8.82}$$

On rearrangement, we arrive at

$$\Delta\mu_1 = \sum_{j=1}^{3} k_j^{(1)}\vartheta_j + \sum_{j=1}^{3} k_j^{(2)}\dot{u}_j^{(1)}\,. \tag{8.83}$$

Note that the expressions for the aerodynamic forces $\Delta\mathbf{q}_{n_1}(\tau,\,\eta)$, $\Delta\mathbf{q}_{\mathrm{L}}(\tau,\,\eta)$, and $\Delta\mathbf{q}_1(\tau,\,\eta)$ and the moment $\Delta\boldsymbol{\mu}_1(\tau,\,\eta)$ developed in this chapter are just approximate. The analytical relationships obtained for these forces and moments remain valid for arbitrary orientation of the space-curved rod relative to the flow. This allows us to perform numerical analysis of rod oscillations in the flow and, in particular, the computation of the critical velocity of the flow at which either the static loss of stability (the divergency) or the dynamical loss of stability (the flutter) occurs.

8.4 Equations of Small Oscillations of a Rod in an Airflow or Liquid Flow

8.4.1 Equations of Small Oscillations of a Space-Curved Rod

The equations of small oscillations of a rod about its equilibrium configuration were obtained in Chap. 3 (see equations (3.11)–(3.15)). The rod was assumed

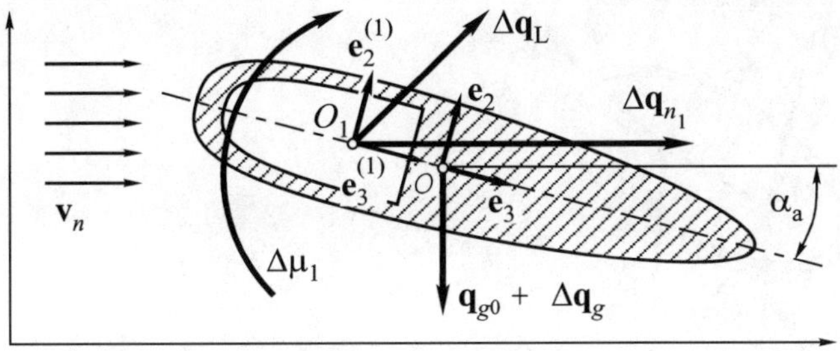

Fig. 8.9.

to be of varying cross section and acted on by arbitrary forces. Recall that when deriving the equations, we considered only rods whose center of gravity (the point O) and the flexural center (the point O_1) coincided (Fig. 8.5).

Apply equations (3.11)– (3.15) to a rod of constant cross section (in this case, we have $n_1 = 1$ and $A_{33} = 1$). First, let us represent the equations in the form that allows a more convenient generalization to the case where the center of gravity and the flexural center do not coincide (see Fig. 8.9). We have

$$\frac{\partial \Delta\mathbf{Q}}{\partial \eta} + \mathrm{A}_{æ}\Delta\mathbf{Q} + \mathrm{A}_Q\mathrm{A}^{-1}\Delta\mathbf{M} + \Delta\mathbf{q}_J = -\Delta\mathbf{P}\,, \tag{8.84}$$

$$\frac{\partial \Delta\mathbf{M}}{\partial \eta} + (\mathrm{A}_M\mathrm{A}^{-1} + \mathrm{A}_{æ})\,\Delta\mathbf{M} + \mathrm{A}_1\Delta\mathbf{Q} + \Delta\boldsymbol{\mu}_J = -\Delta\mathbf{T}\,, \tag{8.85}$$

$$\frac{\partial \boldsymbol{\vartheta}}{\partial \eta} + \mathrm{A}_{æ}\boldsymbol{\vartheta} - \mathrm{A}^{-1}\Delta\mathbf{M} = 0\,, \tag{8.86}$$

$$\frac{\partial \mathbf{u}^{(1)}}{\partial \eta} + \mathrm{A}_{æ}\mathbf{u}^{(1)} + \mathrm{A}_1\boldsymbol{\vartheta} = 0\,, \qquad \Delta\mathbf{M} = \mathrm{A}\Delta æ\,, \tag{8.87}$$

where $\Delta\mathbf{q}_J$ and $\Delta\boldsymbol{\mu}_J$ are the distributed inertial force and moment. The terms $\Delta\mathbf{Q}$ and $\Delta\mathbf{M}$ in (8.84) and (8.85) are the increments to the vectors $\mathbf{Q}$ and $\mathbf{M}$ with respect to the principal axes of the cross section with center O_1.

The vector $\mathbf{u}^{(1)}$ in (8.87) is the vector of displacements of the points on the line through the flexural centers of the cross sections (Fig. 8.10). The elements of the matrix depend on æ, which is the curvature vector of the line through the centers of gravity.

For small oscillations, the distributed force of inertia $\Delta\mathbf{q}_J$ is

$$\Delta\mathbf{q}_J = -\frac{\partial^2 \mathbf{u}}{\partial \tau^2}\,. \tag{8.88}$$

Let us obtain the expression for $\Delta\mathbf{q}_J$ referred to the flexural center of a cross section.

Two configurations of a cross section are shown in Fig. 8.10. The cross section in equilibrium is marked by '0' and the cross section in motion is marked by '1'. It follows from Fig. 8.10 that

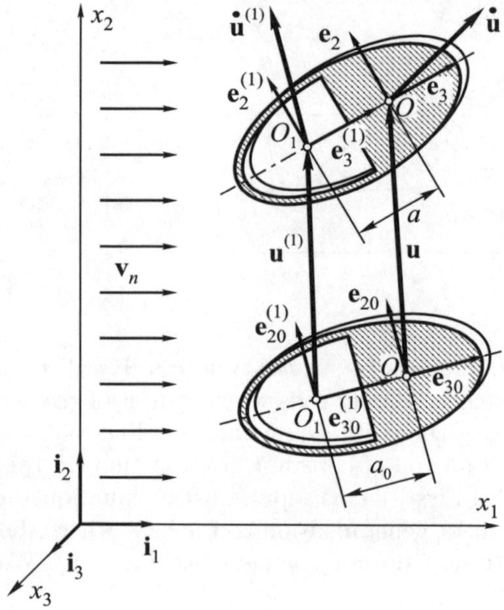

Fig. 8.10.

$$\mathbf{u}^{(1)} + \mathbf{a} = \mathbf{a}_0 + \mathbf{u}$$

or

$$\mathbf{u} = \mathbf{u}^{(1)} + (\mathbf{a} - \mathbf{a}_0)\,. \tag{8.89}$$

The velocity of the point O reads

$$\frac{\partial \mathbf{u}}{\partial \tau} = \frac{\partial \mathbf{u}^{(1)}}{\partial \tau} + \frac{\partial \mathbf{a}}{\partial \tau}\,.$$

Using local derivative notation, we get

$$\frac{\partial \mathbf{u}}{\partial \tau} = \frac{\tilde{\partial} \mathbf{u}^{(1)}}{\partial \tau} + \boldsymbol{\omega} \times \mathbf{u}^{(1)} + \frac{\tilde{\partial} \mathbf{a}}{\partial \tau} + \boldsymbol{\omega} \times \mathbf{a}\,.$$

Putting $\boldsymbol{\omega} \times \mathbf{u}^{(1)} \approx 0$ for the case of small oscillations, we obtain

$$\frac{\partial \mathbf{u}}{\partial \tau} = \frac{\tilde{\partial} \mathbf{u}^{(1)}}{\partial \tau} + \boldsymbol{\omega} \times \mathbf{a}\,.$$

The acceleration of the center of gravity of the rod element is

$$\frac{\partial^2 \mathbf{u}}{\partial \tau^2} = \frac{\tilde{\partial}^2 \mathbf{u}^{(1)}}{\partial \tau^2} + \frac{\tilde{\partial} \boldsymbol{\omega}}{\partial \tau} \times \mathbf{a}\,. \tag{8.90}$$

We have

$$\boldsymbol{\omega} \times \frac{\partial \mathbf{a}}{\partial \tau} = \boldsymbol{\omega} \times \left(\frac{\tilde{\partial} \mathbf{a}}{\partial \tau} + \boldsymbol{\omega} \times \mathbf{a}\right) = \boldsymbol{\omega} \times (\boldsymbol{\omega} \times \mathbf{a}) \approx 0\,, \qquad \frac{\tilde{\partial} \mathbf{a}}{\partial \tau} = 0\,.$$

For small oscillations, we can write

$$\boldsymbol{\omega} = \frac{\partial \boldsymbol{\vartheta}}{\partial \tau}\,, \qquad \boldsymbol{\vartheta} = \sum_{j=1}^{3} \vartheta_j \mathbf{e}_j\,,$$

Finally, omitting the tilde in the local derivative notation, we arrive at

$$\Delta \mathbf{q}_J = -\frac{\partial^2 \mathbf{u}}{\partial \tau^2} = -\left(\frac{\partial^2 \mathbf{u}^{(1)}}{\partial \tau^2} + \frac{\partial^2 \boldsymbol{\vartheta}}{\partial \tau^2} \times \mathbf{a}\right). \tag{8.91}$$

Then, equation (8.84) takes the form

$$\frac{\partial^2 \mathbf{u}^{(1)}}{\partial \tau^2} + \frac{\partial^2 \boldsymbol{\vartheta}}{\partial \tau^2} \times \mathbf{a} - \frac{\partial \Delta \mathbf{Q}}{\partial \eta} - \mathrm{A}_{æ} \Delta \mathbf{Q} - \mathrm{A}_Q \mathrm{A}^{-1} \Delta \mathbf{M} + \Delta \mathbf{P} = 0\,. \tag{8.92}$$

A rod subjected to a flow is acted on by the distributed aerodynamic forces and the force due to gravity. Assume that there are no concentrated forces and moments applied to the rod, and the rod has no point masses attached to it. In this case (Fig. 8.9), we have

$$\Delta \mathbf{P} = \Delta \mathbf{q}_{n_1} + \Delta \mathbf{q}_{\mathrm{L}} + \Delta \mathbf{q}_{\mathrm{l}} + \Delta \mathbf{q}_g\,, \tag{8.93}$$

where $\Delta \mathbf{q}_g$ is the increment to the vector of the distributed force of gravity with respect to the coordinate frame attached to the rod axis.

The nondimensional distributed force of gravity (Fig. 8.9) reads

$$\mathbf{q}_g = -q_q \mathbf{i}_2\,, \qquad q_g = \frac{m_0 l^3}{A_{33}}\, g\,,$$

where g is the acceleration due to gravity.

In the attached frame, we have

$$\mathbf{q}_g = q_g \left(l_{12}^{(1)} \mathbf{e}_1 + l_{22}^{(1)} \mathbf{e}_2 + l_{32}^{(1)} \mathbf{e}_3\right).$$

For small oscillations, we get

$$\begin{aligned}
l_{12}^{(1)} &= l_{12}^{(0)} + \vartheta_3 l_{22}^{(0)} - \vartheta_2 l_{32}^{(0)}\,,\\
l_{22}^{(1)} &= l_{22}^{(0)} + \vartheta_1 l_{32}^{(0)} - \vartheta_3 l_{12}^{(0)}\,,\\
l_{32}^{(1)} &= l_{32}^{(0)} + \vartheta_2 l_{12}^{(0)} - \vartheta_1 l_{22}^{(0)}\,,
\end{aligned}$$

Therefore,

$$\mathbf{q}_g = \mathbf{q}_{g\,0} + \Delta \mathbf{q}_g\,,$$

where

$$\Delta \mathbf{q}_g = -q_g \, \mathrm{L}^{(3)} \boldsymbol{\vartheta} \, ; \tag{8.94}$$

here,

$$\mathrm{L}^{(3)} = \begin{bmatrix} 0 & -l_{32}^{(0)} & l_{22}^{(0)} \\ l_{32}^{(0)} & 0 & -l_{12}^{(0)} \\ -l_{22}^{(0)} & l_{12}^{(0)} & 0 \end{bmatrix} .$$

The vectors in the right-hand side of (8.93) are referred to the coordinate frame attached to the rod axis. These vectors were discussed in Sect. 8.3.3.

The vector $\mathbf{a}$ (see Fig. 8.10) is

$$\mathbf{a} = a \, \mathbf{e}_3^{(1)} \, ,$$

where $a = a^{(1)}/l$ is a nondimensional quantity; here, $a^{(1)}$ is the dimensional distance between the points O and O_1. For a rod of varying cross section, the parameter a depends on η.

Consequently, we have

$$\frac{\partial^2 \boldsymbol{\vartheta}}{\partial \tau^2} \times a \mathbf{e}_3^{(1)} = a \, \frac{\partial^2 \vartheta_2}{\partial \tau^2} \, \mathbf{e}_1^{(1)} - a \, \frac{\partial^2 \vartheta_1}{\partial \tau^2} \, \mathbf{e}_2^{(1)} \, .$$

The following equation results

$$\frac{\partial^2 \mathbf{u}^{(1)}}{\partial \tau^2} + a \, \frac{\partial^2 \vartheta_2}{\partial \tau^2} \, \mathbf{e}_1^{(1)} - a \, \frac{\partial^2 \vartheta_1}{\partial \tau^2} \, \mathbf{e}_2^{(1)} - \frac{\partial \Delta \mathbf{Q}}{\partial \eta} - \mathrm{A}_{æ} \Delta \mathbf{Q} - \mathrm{A}_Q \mathrm{A}^{-1} \Delta \mathbf{M}$$
$$+ \Delta \mathbf{q}_{n_1 e} + \Delta \mathbf{q}_{\mathrm{L}e} + \Delta \mathbf{q}_{1e} + \Delta \mathbf{q}_{g\,e} = 0 \, . \tag{8.95}$$

The distributed inertial moment about the flexural center $\Delta \boldsymbol{\mu}_J$ enters into (8.85). This distributed inertial moment reads

$$\Delta \boldsymbol{\mu}_J = -J \, \frac{\partial^2 \boldsymbol{\vartheta}}{\partial \tau^2} - \mathbf{a} \times \frac{\partial^2 \mathbf{u}}{\partial \tau^2} \, .$$

Using (8.91), we get

$$\Delta \boldsymbol{\mu}_J = -J \, \frac{\partial^2 \boldsymbol{\vartheta}}{\partial \tau^2} - \mathbf{a} \times \left(\frac{\partial^2 \mathbf{u}^{(1)}}{\partial \tau^2} + \frac{\partial^2 \boldsymbol{\vartheta}}{\partial \tau^2} \times \mathbf{a} \right) . \tag{8.96}$$

On rearrangement, we obtain

$$\Delta \boldsymbol{\mu}_J = -J \, \frac{\partial^2 \boldsymbol{\vartheta}}{\partial \tau^2} + a \, \frac{\partial^2 u_2^{(1)}}{\partial \tau^2} \, \mathbf{e}_1^{(1)} - a \, \frac{\partial^2 u_1^{(1)}}{\partial \tau^2} \, \mathbf{e}_2^{(1)} - a^2 \frac{\partial^2 \vartheta_1}{\partial \tau^2} \, \mathbf{e}_1^{(1)} - a^2 \frac{\partial^2 \vartheta_2}{\partial \tau^2} \, \mathbf{e}_2^{(1)} \, . \tag{8.97}$$

The moment $\Delta \mathbf{T}$ about the flexural center (see Fig. 8.9) is

$$\Delta \mathbf{T} = -\Delta \boldsymbol{\mu}_1 - \mathbf{a} \times \Delta \mathbf{q}_g \, . \tag{8.98}$$

As a result, we obtain equation (8.85) referred to the flexural center of the rod cross section

$$J\,\frac{\partial^2 \boldsymbol{\vartheta}}{\partial \tau^2} - a\,\frac{\partial^2 u_2^{(1)}}{\partial \tau^2}\,\mathbf{e}_1^{(1)} + a\,\frac{\partial^2 u_1^{(1)}}{\partial \tau^2}\,\mathbf{e}_2^{(1)} + a^2\frac{\partial^2 \vartheta_1}{\partial \tau^2}\,\mathbf{e}_1^{(1)} + a^2\frac{\partial^2 \vartheta_2}{\partial \tau^2}\,\mathbf{e}_2^{(1)} - \frac{\partial \Delta\mathbf{M}}{\partial \eta}$$
$$- \left(\mathrm{A}_M \mathrm{A}^{-1} + \mathrm{A}_{æ}\right)\Delta\mathbf{M} - \mathrm{A}_1 \Delta\mathbf{Q} + \Delta\boldsymbol{\mu}_1 + \mathbf{a}\times\Delta\mathbf{q}_g = 0\,, \qquad (8.99)$$

where

$$\mathbf{a}\times\Delta\mathbf{q}_g = -a q_g\,\mathrm{L}_g \boldsymbol{\vartheta}\,.$$

Equations (8.95) and (8.99) govern small oscillations of a rod in a flow about its equilibrium configuration. (Recall that the elements of the matrices $\mathrm{A}_{æ}$, A_Q, and A_M depend on the static stress–strain state of the rod.) The equilibrium configuration of a rod in a flow may differ significantly from the equilibrium configuration of the rod in a quiescent medium. The static stress–strain state of a rod is described by the nonlinear equilibrium equations discussed in Svetlitsky (2000a).

The vector equations (8.95), (8.99), (8.85), and (8.87) constitute the system of equations that governs small oscillations of a space-curved rod of non-circular symmetrical cross section when the center of gravity and the flexural center do not coincide. Using the system of equations, one can study both static and dynamic stability of the rod in the flow and determine the critical flow velocities.

For stability analysis, in view of the fact that the mechanical system is not conservative, it is sensible to put

$$\Delta\mathbf{Q} = \left(\Delta\mathbf{Q}_0^{(1)} + \mathrm{i}\,\Delta\mathbf{Q}_0^{(2)}\right)\,\mathrm{e}^{(\alpha+\mathrm{i}\,\beta)\,\tau}\,,$$
$$\Delta\mathbf{M} = \left(\Delta\mathbf{M}_0^{(1)} + \mathrm{i}\,\Delta\mathbf{M}_0^{(2)}\right)\,\mathrm{e}^{(\alpha+\mathrm{i}\,\beta)\,\tau}\,,$$
$$\boldsymbol{\vartheta} = \left(\boldsymbol{\vartheta}_0^{(1)} + \mathrm{i}\,\boldsymbol{\vartheta}_0^{(2)}\right)\,\mathrm{e}^{(\alpha+\mathrm{i}\,\beta)\,\tau}\,,$$
$$\mathbf{u}^{(1)} = \left(\mathbf{u}_0^{(1)} + \mathrm{i}\,\mathbf{u}_0^{(2)}\right)\,\mathrm{e}^{(\alpha+\mathrm{i}\,\beta)\,\tau}\,.$$

As a result, we get the equations similar to (4.94)–(4.96) with the aerodynamic forces $\Delta\mathbf{q}_{\mathrm{L}}$, $\Delta\mathbf{q}_{n_1}$, and $\Delta\mathbf{q}_1$ and the aerodynamic moment $\Delta\boldsymbol{\mu}_1$ taken into consideration (all these quantities depend on $\boldsymbol{\vartheta}$ and $\dot{\mathbf{u}}^{(1)}$). For example, we have

$$\Delta\mathbf{q}_{\mathrm{L}e} = \mathrm{L}^{(0)}\mathrm{P}^{(1)}\left(\boldsymbol{\vartheta}_0^{(1)} + \mathrm{i}\,\boldsymbol{\vartheta}_0^{(2)}\right)\,\mathrm{e}^{(\alpha+\mathrm{i}\,\beta)\,\tau}$$
$$+\,\mathrm{L}^{(0)}\mathrm{P}^{(2)}\mathrm{L}^{(0)\mathrm{T}}\left(\mathbf{u}_0^{(1)} + \mathrm{i}\,\mathbf{u}_0^{(2)}\right)\,(\alpha + \mathrm{i}\,\beta)\,\mathrm{e}^{(\alpha+\mathrm{i}\,\beta)\,\tau}\,.$$

The complex eigenvalues $\lambda_j = \alpha_j + \mathrm{i}\,\beta_j$ can be found by using the method discussed in Chap. 4 or in Sect. 9.3. The flow velocity for which the real part of the first eigenvalue α_1 becomes zero is the critical velocity. If at the same time the imaginary part β_1 differs from zero, then, with further increase in the flow velocity, the dynamic loss of stability (the flutter) occurs.

8.4.2 Small Oscillations of a Straight Rod in a Flow

In Sect. 7.1.4, the equations of free oscillations of a rod referred to the center of gravity were obtained. The equations of small oscillations of a rod in a flow referred to the flexural center of the rod cross section were derived in Sect. 8.4.1. It is conventional to refer the aerodynamic forces to the flexural center of a cross section.

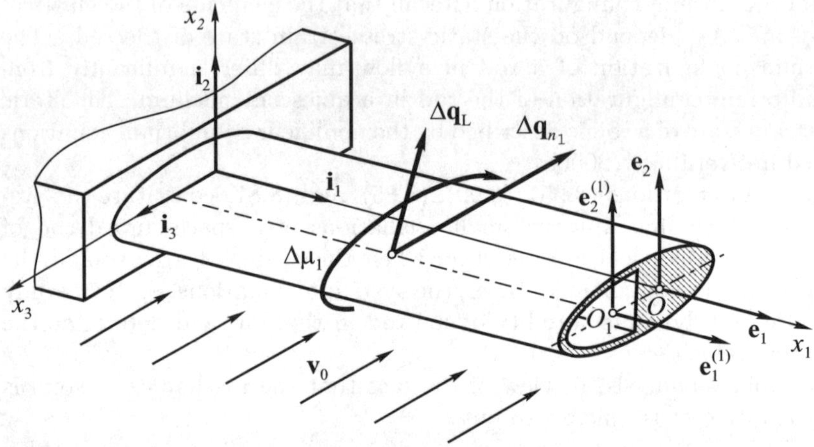

Fig. 8.11.

For a straight rod in a stationary flow (Fig. 8.11), we have $\alpha_a^0 = 0$. Then, we can put

$$\begin{aligned}
&Q_{10} = Q_{20} = 0\,, \qquad Q_{30} \neq 0\,,\\
&M_{10} = 0\,, \qquad M_{20} \neq 0\,, \qquad M_{30} = 0\,,\\
&æ_{10} = 0\,, \qquad æ_{20} \neq 0\,, \qquad æ_{30} = 0\,.
\end{aligned}$$

In this case, the matrices $\mathrm{A}_æ$, A_Q, and A_M read

$$\mathrm{A}_æ = \begin{bmatrix} 0 & 0 & æ_{20} \\ 0 & 0 & 0 \\ -æ_{20} & 0 & 0 \end{bmatrix}, \qquad \mathrm{A}_Q = \begin{bmatrix} 0 & Q_{30} & 0 \\ -Q_{30} & 0 & 0 \\ 0 & 0 & 0 \end{bmatrix},$$

$$\mathrm{A}_M = \begin{bmatrix} 0 & 0 & -M_{20} \\ 0 & 0 & 0 \\ M_{20} & 0 & 0 \end{bmatrix},$$

Using the relationships (8.95), (8.99), (8.86), and (8.87), we get the following equations in the coordinate frame attached to the rod axis:

$$
\begin{aligned}
&\frac{\partial^2 u_1^{(1)}}{\partial \tau^2} + a\,\frac{\partial^2 \vartheta_2}{\partial \tau^2} - \frac{\partial \Delta Q_1}{\partial \eta} - æ_{20}\Delta Q_3 - \frac{Q_{30}}{A_{22}}\Delta M_2 + \Delta q_{a1} + \Delta q_{g\,1} = 0\,,\\
&\frac{\partial^2 u_2^{(1)}}{\partial \tau^2} - a\,\frac{\partial^2 \vartheta_1}{\partial \tau^2} - \frac{\partial \Delta Q_2}{\partial \eta} + \frac{Q_{30}}{A_{11}}\Delta M_1 + \Delta q_{a2} + \Delta q_{g\,3} = 0\,,\\
&\frac{\partial^2 u_3^{(1)}}{\partial \tau^2} - \frac{\partial \Delta Q_3}{\partial \eta} + æ_{20}\Delta Q_1 + \Delta q_{a3} + \Delta q_{g\,3} = 0\,,
\end{aligned}
\tag{8.100}
$$

$$
\begin{aligned}
&(J_{11} + a^2)\,\frac{\partial^2 \vartheta_1}{\partial \tau^2} - a\,\frac{\partial^2 u_2^{(1)}}{\partial \tau^2} - \frac{\partial \Delta M_1}{\partial \eta}\\
&\qquad\qquad + \left(\frac{M_{20}}{A_{33}} - æ_{20}\right)\Delta M_3 + \Delta \mu_1 - a\Delta q_{g\,2} = 0\,,\\
&(J_{22} + a^2)\,\frac{\partial^2 \vartheta_2}{\partial \tau^2} + a\,\frac{\partial^2 u_1^{(1)}}{\partial \tau^2} - \frac{\partial \Delta M_2}{\partial \eta} + \Delta Q_3 + a\Delta q_{g\,1} = 0\,,\\
&J_{33}\,\frac{\partial^2 \vartheta_3}{\partial \tau^2} - \frac{\partial \Delta M_3}{\partial \eta} - \left(\frac{M_{20}}{A_{11}} - æ_{20}\right)\Delta M_1 - \Delta Q_2 = 0\,,
\end{aligned}
\tag{8.101}
$$

$$
\begin{aligned}
&\frac{\partial \vartheta_1}{\partial \eta} + æ_{20}\vartheta_3 - \frac{\Delta M_1}{A_{11}} = 0\,,\\
&\frac{\partial \vartheta_2}{\partial \eta} - \frac{\Delta M_2}{A_{22}} = 0\,,\\
&\frac{\partial \vartheta_3}{\partial \eta} + æ_{20}\vartheta_1 - \Delta M_3 = 0\,,
\end{aligned}
\tag{8.102}
$$

$$
\begin{aligned}
&\frac{\partial u_1^{(1)}}{\partial \eta} + æ_{20}u_3^{(1)} = 0\,,\\
&\frac{\partial u_2^{(1)}}{\partial \eta} - \vartheta_3 = 0\,,\\
&\frac{\partial u_3^{(1)}}{\partial \eta} + \vartheta_2 = 0\,,
\end{aligned}
\tag{8.103}
$$

where

$$
\begin{aligned}
&\Delta q_{a1} = \Delta q_{n_1 1} + \Delta q_{\mathrm{L}1} + \Delta q_1\,, \qquad \Delta q_{a2} = \Delta q_{n_1 2} + \Delta q_{\mathrm{L}2}\,,\\
&\Delta q_{a3} = \Delta q_{n_1 3} + \Delta q_{\mathrm{L}3}\,, \qquad \Delta q_{g\,j} = -q_g\,(\mathrm{L}^{(3)}\boldsymbol{\vartheta}\cdot \mathbf{e}_j)\,.
\end{aligned}
$$

This set of equations allows us the study of flexural-and-torsional oscillations of a straight rod when the bending stiffnesses A_{22} and A_{33} are of the same order of magnitude.

In the problem under consideration, it is possible to estimate the influence of the bending stiffness A_{22} on the critical velocities of the divergency and the flutter. Traditionally, when examining the stability of an aircraft wing (blade) (see Ventsel (1964) and Feodosiev (1999)), the stiffness A_{22} is assumed to be much larger than the stiffness A_{33}, that is, the bending of the rod in the plane x_1Ox_3 can be neglected (Fig. 8.11). Hence, when the condition is no longer valid, the classical equations of flexural-and-torsional oscillations, in which $A_{22} = \infty$, cannot be used to estimate the influence of the stiffness A_{22} on the values of the critical velocities. On the basis of (8.100)–(8.103), equations with $A_{22} = \infty$ will be derived in the next section.

8.4.3 Flexural-and-Torsional Oscillations of a Model of an Aircraft Wing

In what follows, we derive the equations of flexural-and-torsional oscillations of an idealized wing model. The equations help better understand some physical features of oscillations of real wings exerted by airflow. During the derivation, we assume that $A_{22} \gg A_{33}$ (that is, $A_{22} = \infty$) and $\alpha_a^0 = 0$. The gravity force is neglected ($q_g = \Delta q_g = 0$)). Therefore, $u_3^{(1)} = 0$, $\vartheta_2 = 0$, $æ_{20} = 0$, $Q_{30} \neq 0$, and $M_{20} \neq 0$. Moreover, we assume that the aerodynamic forces Δq_{aj} and the moment $\Delta\mu_1$ depend only on ϑ_1 and $\dot{u}_2^{(1)}$, that is,

$$\Delta q_{aj} = h_j^{(1)}\vartheta_1 + h_j^{(2)}\dot{u}_2^{(1)}, \qquad \Delta\mu_1 = a_1\vartheta_1 + a_2\dot{u}_2^{(1)}. \tag{8.104}$$

These formulas may be obtained from (8.78), (8.80) and (8.83) by retaining the terms that depend on ϑ_1 or $\dot{u}_2^{(1)}$ and omitting all the other terms.

Using (8.100)–(8.103), we get

$$\frac{\partial \Delta Q_1}{\partial \eta} - h_1^{(1)}\vartheta_1 - h_1^{(2)}\dot{u}_2^{(1)} = 0, \tag{8.105}$$

$$\frac{\partial^2 u_2^{(1)}}{\partial \tau^2} - a\frac{\partial^2 \vartheta_1}{\partial \tau^2} - \frac{\partial \Delta Q_2}{\partial \eta} + \frac{Q_{30}}{A_{11}}\Delta M_1 + h_2^{(1)}\vartheta_1 + h_2^{(2)}\dot{u}_2^{(1)} = 0, \tag{8.106}$$

$$\frac{\partial \Delta Q_3}{\partial \eta} - h_3^{(1)}\vartheta_1 - h_3^{(2)}\dot{u}_2^{(1)} = 0, \tag{8.107}$$

$$(J_{11} + a^2)\frac{\partial^2 \vartheta_1}{\partial \tau^2} - a\frac{\partial^2 u_2^{(1)}}{\partial \tau^2} - \frac{\partial \Delta M_1}{\partial \eta} + M_{20}\Delta M_3 + a_1\vartheta_1 + a_2\frac{\partial u_2^{(1)}}{\partial \tau} = 0, \tag{8.108}$$

$$a\frac{\partial^2 u_1^{(1)}}{\partial \tau^2} - \frac{\partial \Delta M_2}{\partial \eta} + \Delta Q_3 = 0, \tag{8.109}$$

$$J_{33}\frac{\partial^2 \vartheta_3}{\partial \tau^2} - \frac{\partial \Delta M_3}{\partial \eta} - \frac{M_{20}}{A_{11}}\Delta M_1 - \Delta Q_2 = 0, \tag{8.110}$$

$$\frac{\partial \vartheta_1}{\partial \eta} - \frac{\partial \Delta M_1}{A_{11}} = 0, \tag{8.111}$$

$$\frac{\partial \vartheta_3}{\partial \eta} - \Delta M_3 = 0\,, \tag{8.112}$$

$$\frac{\partial u_2^{(1)}}{\partial \eta} - \vartheta_3 = 0\,, \tag{8.113}$$

$$\frac{\partial u_1^{(1)}}{\partial \eta} = 0\,,$$

$$\frac{\partial u_3^{(1)}}{\partial \eta} = 0\,, \qquad u_1^{(1)} = u_3^{(1)} \equiv 0\,. \tag{8.114}$$

To obtain the equations of interest governing flexural-and-torsional oscillations of a straight rod (the model of a wing), consider the relationships (8.106), (8.108), and (8.100)-(8.113). Eliminating successively the quantities ϑ_3, ΔM_3, ΔM_1, and ΔQ_2 from these equations, we get the equations which in nondimensional form read

$$\begin{aligned}
&\frac{\partial^4 u_2^{(1)}}{\partial \eta^4} - J_{33}\frac{\partial^4 u_2^{(1)}}{\partial \eta^2 \partial \tau^2} + \frac{\partial^2 u_2^{(1)}}{\partial \tau^2} - a\,\frac{\partial^2 \vartheta_1}{\partial \tau^2} \\
&\quad + \frac{\partial}{\partial \eta}\left(M_{20}\frac{\partial \vartheta_1}{\partial \eta}\right) + Q_{30}\frac{\partial \vartheta_1}{\partial \eta} + h_2^{(1)}\vartheta_1 + h_2^{(2)}\,\frac{\partial u_2^{(1)}}{\partial \tau} = 0\,, \\
&A_{11}\frac{\partial^2 \vartheta_1}{\partial \eta^2} - \left(J_{11} + a^2\right)\frac{\partial^2 \vartheta_1}{\partial \tau^2} + a\,\frac{\partial^2 u_2^{(1)}}{\partial \tau^2} - M_{20}\frac{\partial \vartheta_3}{\partial \eta} \\
&\quad + a_1\vartheta_1 + a_2\,\frac{\partial u_2^{(1)}}{\partial \tau} = 0\,.
\end{aligned} \tag{8.115}$$

Unlike the conventional equations, equations (8.115) account for the rotary inertia of the rod element about the axis orthogonal to the plane x_1Ox_2 (see Fig. 8.11),

$$J_{33}\,\frac{\partial^4 u_2^{(1)}}{\partial \eta^2\,\partial \tau^2}\,,$$

and the static stress–strain state of the rod (described by the quantities Q_{30} and M_{20}) due to the drag $\mathbf{q}_{n_1 0}$ (recall that for $\alpha_a^0 = 0$, $\mathbf{q}_{\mathrm{L}0} = 0$). Having found ϑ_1 and $u_2^{(1)}$ from (8.115) and using (8.105), (8.107), and (8.109), we can determine ΔQ_1, ΔQ_3, and ΔM_2.

Substituting

$$\vartheta_1 = (\vartheta_{10} + \mathrm{i}\,\vartheta_{20})\,\mathrm{e}^{(\alpha + \mathrm{i}\,\beta)\,\tau}\,, \qquad u_2^{(1)} = (u_{20}^{(1)} + \mathrm{i}\,u_{20}^{(2)})\,\mathrm{e}^{(\alpha + \mathrm{i}\,\beta)\,\tau}$$

into the system of equations (8.115), we can find the complex eigenvalues $\lambda_j = \alpha_j + \mathrm{i}\,\beta_j$ as functions of the flow speed v_0. (Recall that the coefficients $h_2^{(1)}$, $h_2^{(2)}$, a_1, and a_2 depend on the flow velocity.)

8.5 Nonstationary Vibrations of "Rigid" Wires of Power Lines under Action of Aerodynamic Forces

A wire between two poles of a power transmission line under the action of an airflow is shown in Fig. 8.12*a*. The vector of flow velocity is parallel to the plane x_1Ox_3. Let us consider two main cases of loading.

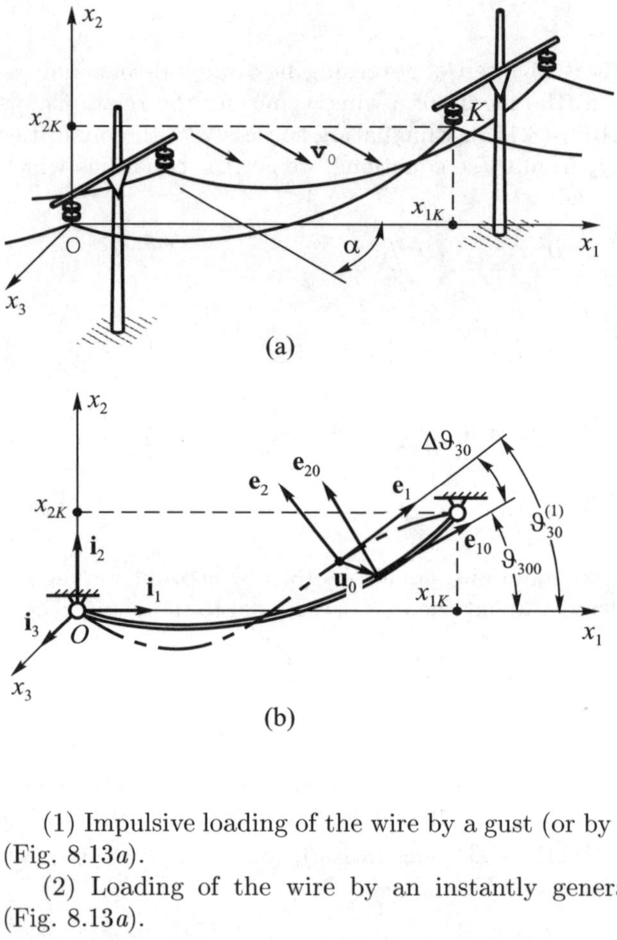

Fig. 8.12.

(1) Impulsive loading of the wire by a gust (or by a transient shock wave) (Fig. 8.13*a*).

(2) Loading of the wire by an instantly generated stationary airflow (Fig. 8.13*a*).

The duration time of the impulsive loading, Δt, is taken to be small as compared to the period of oscillations corresponding to the first eigenvalue.

In course of designing power transmission lines, the problems of strength and reliability of wires under dynamic loading are of pronounced importance.

Conventionally, a wire is considered as an absolutely flexible rod with zero bending and torsional stiffness coefficients. When analyzing strength of wires under static loading, such an assumption is acceptable. When analyzing

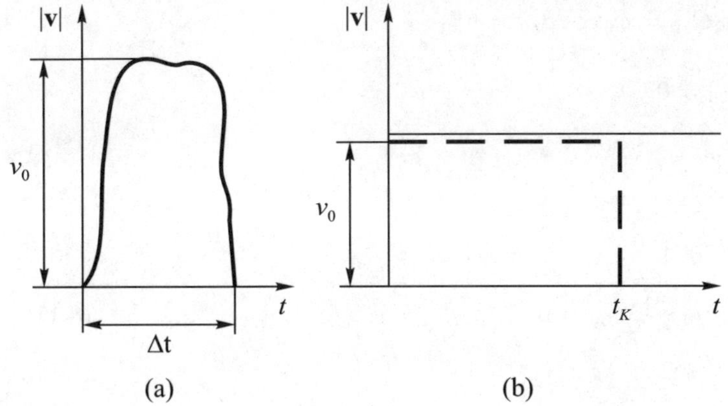

Fig. 8.13.

strength of wires under dynamic loading, such an idealized model may result in incorrect estimates of strength and reliability of structures. Consequently, determination of the dynamic stress–strain state of wires with considering real stiff parameters is one of the main problems in power transmission lines design.

When "stiff" wires vibrate, stresses in cross sections vary periodically due to bending and torsional moments. These may result in generation of zones of plastic deformations in the wire and in fatigue failure of the wire. Using the simplified mathematical model of a wire (as an absolutely flexible rod), we have no opportunity to obtain information on stresses generated by the internal moments; hence, in this case we have no chance to estimate practical reliability of wires.

8.5.1 Determination of Static Stress–Strain State of a "Stiff" Wire

Two coordinate frames, $\{\mathbf{e}_{j0}\}$ and $\{\mathbf{e}_j\}$, are shown in Fig. 12*b*. The coordinate frame $\{\mathbf{e}_{j0}\}$ is attached to the axial line of an absolutely flexible rod loaded by the gravity forces $\mathbf{g}_g$. The coordinate frame $\{\mathbf{e}_j\}$ is attached to the axial line of a stiff wire loaded by the gravity forces. The axial lines of the two wires, the absolutely flexible wire and the stiff one, differ slightly, consequently, the wire's stiffness has little effect on the shape of its axial line but it may be useful in the determination of the real stress–strain state of the wire and thus estimate the wire's strength and reliability essentially (especially for vibrating wires).

In scalar form, nonlinear equations of equilibrium of a wire (Fig. 8.12*b*) represented in the attached basis are as follows (Svetlitsky (2000a)):

$$\frac{\mathrm{d}Q_{10}}{\mathrm{d}\eta} - æ_{30}Q_{20} = g_g \sin\vartheta_{30} ,$$

$$\begin{aligned}
&\frac{\mathrm{d}Q_{20}}{\mathrm{d}\eta} + æ_{30}Q_{10} = g_g \cos\vartheta_{30}\,,\\
&\frac{\mathrm{d}M_{30}}{\mathrm{d}\eta} + Q_{20} = 0\,,\\
&\frac{\mathrm{d}\vartheta_{30}^{(1)}}{\mathrm{d}\eta} - \frac{M_{30}}{A_{33}} = 0\,,\\
&\frac{\mathrm{d}u_{10}}{\mathrm{d}\eta} - æ_{30}u_{20} = 0\,,\\
&\frac{\mathrm{d}u_{20}}{\mathrm{d}\eta} + æ_{30}u_{10} - \Delta\vartheta_{30} = 0\,,
\end{aligned} \tag{8.116}$$

where

$$æ_{30} = \frac{\mathrm{d}\vartheta_{30}^{(1)}}{\mathrm{d}\eta}\,, \qquad g_g = \frac{m_0\, g\, l^3}{A_{33}}\,.$$

The angle $\vartheta_{30}^{(1)}$ (see Fig. 8.12*b*) is

$$\vartheta_{30}^{(1)} = \vartheta_{300} + \Delta\vartheta_{30}\,.$$

The angle $\Delta\vartheta_{30}^{(1)}$ may be considered small. Consequently, the last equation in the system of equations (8.116) may be represented as follows:

$$\frac{\mathrm{d}u_{20}}{\mathrm{d}\eta} + æ_{30}u_{10} - \vartheta_{30} = -\vartheta_{300}\,. \tag{8.117}$$

Consider a wire with the following value of the parameters: $l = 20\,\mathrm{m}$, $x_{10} = 0$, $x_{20} = 0$, $x_{1k} = 15\,\mathrm{m}$, $x_{2k} = 5\,\mathrm{m}$, $d = 0.02\,\mathrm{m}$, $E = 0.675 \cdot 10^5\,\mathrm{MPa}$, and $m_0 = 0.847\,\mathrm{kg/m}$; here, d is the diameter of the wire and E is its elasticity modulus.

To solve the nonlinear system of equations (8.116) numerically, we may employ the method of iterations (or the Newton–Raphson procedure). As the initial approximation, we may put

$$æ_{30} = æ_{300} = \frac{\mathrm{d}\vartheta_{300}}{\mathrm{d}\eta}\,, \qquad \vartheta_{30} = \vartheta_{300}$$

in the right-hand side of the first two equations. The functions ϑ_{300} and $æ_{300}$ can be obtained by solving the system of equilibrium equations for an absolutely flexible rod (see Svetlitsky (2000b)). Thus, we have

$$\vartheta_{300}(\eta) = \arctan\left(\frac{\eta}{c_1} + \sinh\gamma\right)\,,$$

$$æ_{300} = \frac{\mathrm{d}\vartheta_{300}}{\mathrm{d}\eta} = \frac{1}{c_1\left[1 + \left(\frac{\eta}{c_1} + \sinh\gamma\right)^2\right]}\,, \qquad \gamma = \frac{x_{1k}}{c_1} + c_2\,.$$

For the particular value of the parameters of the wire under consideration, the values of the arbitrary constants are: $c_1 = 0.295$ and $c_2 = -1.019$ If the functions $æ_{300}$ and ϑ_{300} are known, the system of equations (8.116)–(8.117) is linear. Consequently, to solve the problem, we can use the method of initial parameters. At the first step, we get $\vartheta_{30}^{(1)}$ and

$$æ_{30}^{(1)} = \frac{d\vartheta_{30}^{(1)}}{d\eta} .$$

At the second step of the iteration procedure, the right-hand sides depend on $\vartheta_{30}^{(1)}$. The left-hand sides of the equations contain the known value of the curvature $æ_{30}^{(1)}$, that is, at the second step of the iteration procedure we have the linear equations again. Solving this system of equation, we find $\vartheta_{30}^{(11)}$ and

$$æ_{30}^{(11)} = \frac{d\vartheta_{30}^{(11)}}{d\eta} .$$

Thus, we can make as many iterations as required.

The number of iterations depends on the required accuracy of the solution. For the "stiff" wire, the solution of the system of equations (8.116)–(8.117) obtained at the second iteration is practically the same as the solution corresponding to the first iteration. The graphs of Q_{10}, Q_{20}, M_{30}, $\vartheta_{30}(æ_{30})$, u_{10}, and u_{20} corresponding to the second iteration are shown in Figs. 8.14 and 8.15.

8.5.2 Equations of Small Oscillations of a Wire

Consider a wire that was at rest in the plane x_1Ox_2 before loading. When the wire is suddenly loaded by aerodynamic forces (Fig. 8.13*a*,*b*), the vibrations of the wire relative to its state of equilibrium occur. The general equations of small oscillations relative to an equilibrium configuration of a rod were derived in Chapter 3 (equations (3.35)).

Recall that in the problem under consideration, the axial line of the rod in equilibrium is a plane curve. In this case, using (3.35), we can obtain two independent sets of equations: equations of oscillations in the plane x_1Ox_2 (in-plane vibrations, equations (3.36)) and equations of oscillations of the wire relative to the plane x_1Ox_2 (out-of-plane vibrations, equations (3.37)).

The systems of equations (3.36) and (3.37) are the equations of free oscillations; consequently, we should introduce aerodynamic forces and the projections of the gravity forces onto the axes of the attached frame into the right-hand sides of these equations. Besides, to simplify the notation, we put $\Delta Q_j = Q_j$ and $\Delta M_j = M_j$. Let us neglect the rotating inertia of a rod segment. For the wire of constant cross-section, we have $n_1 = 1$ and $A_{33} = 1$. As a result, we arrive at the following equations in the projections with respect to the attached basis of the moving wire $\{\mathbf{e}_j^{(1)}\}$:

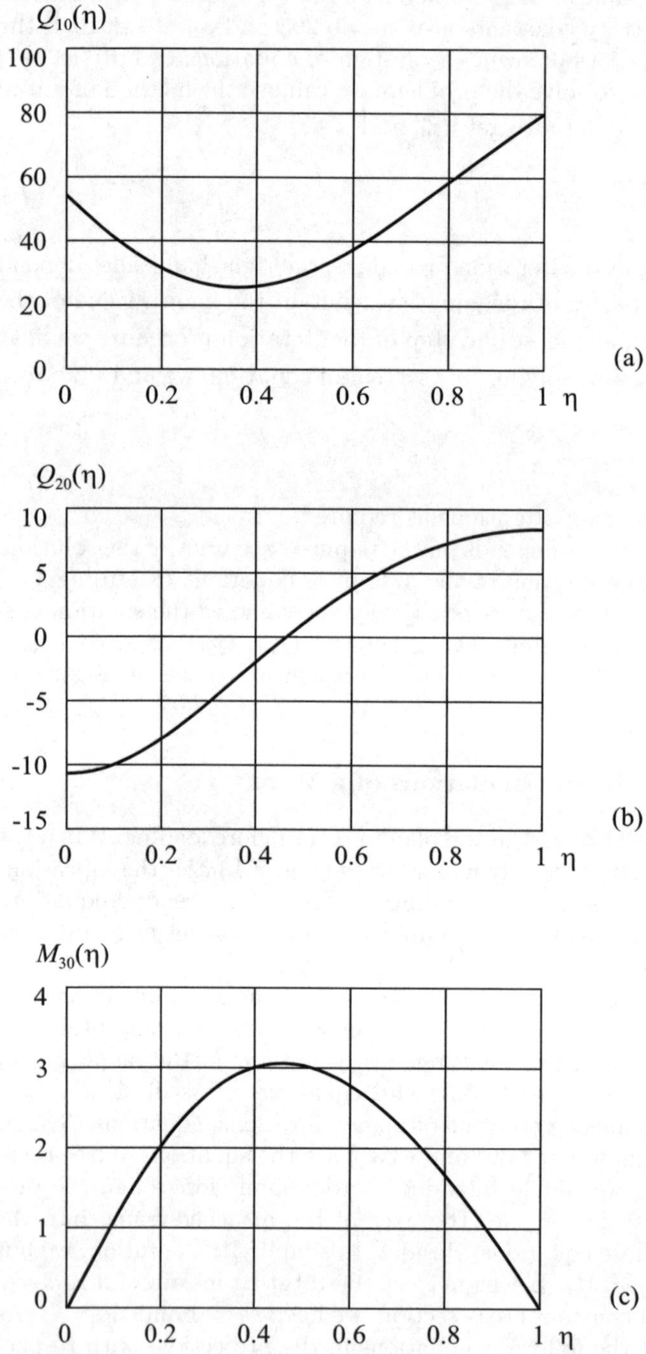
$Q_{10}(\eta)$
100
80
60
40
20
0
0
0.2
0.4
0.6
0.8
1 η
(a)
$Q_{20}(\eta)$
10
5
0
-5
-10
-15
0
0.2
0.4
0.6
0.8
1 η
(b)
$M_{30}(\eta)$
4
3
2
1
0
0
0.2
0.4
0.6
0.8
1 η
(c)

Fig. 8.14.

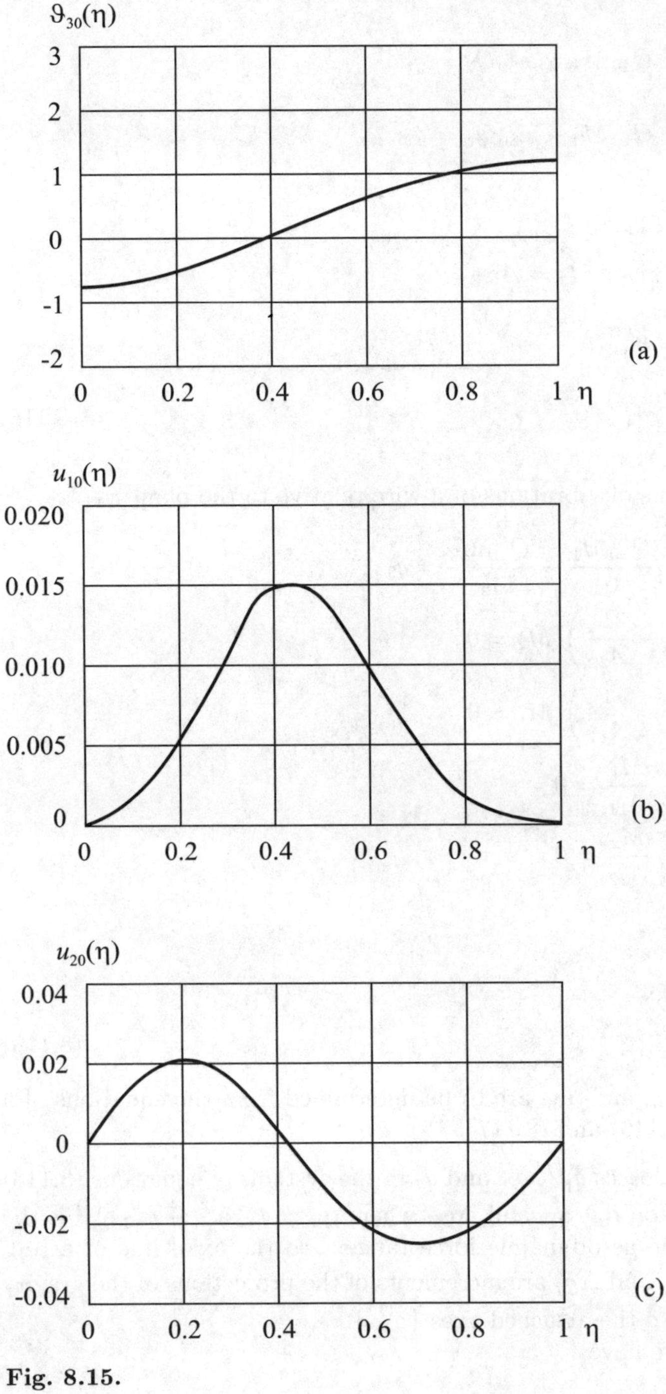
$\vartheta_{30}(\eta)$
3
2
1
0
-1
-2
0
0.2
0.4
0.6
0.8
1 η
(a)
$u_{10}(\eta)$
0.020
0.015
0.010
0.005
0
0
0.2
0.4
0.6
0.8
1 η
(b)
$u_{20}(\eta)$
0.04
0.02
0
-0.02
-0.04
0
0.2
0.4
0.6
0.8
1 η
(c)

Fig. 8.15.

(1) The equations of vibrations in the plane x_1Ox_2.

$$\frac{\partial^2 u_1}{\partial \tau^2} - \frac{\partial Q_1}{\partial \eta} + Q_{20}M_3 + æ_{30}Q_2 = q_1^{(1)} ,$$
$$\frac{\partial^2 u_2}{\partial \tau^2} - \frac{\partial Q_2}{\partial \eta} - Q_{10}M_3 - æ_{30}Q_1 = q_2 ,$$
$$\frac{\partial M_3}{\partial \eta} + Q_2 = 0 ,$$
$$\frac{\partial \vartheta_3}{\partial \eta} - M_3 = 0 , \qquad M_3 = \Delta æ_3 ,$$
$$\frac{\partial u_1}{\partial \eta} - æ_{30}u_2 = 0 ,$$
$$\frac{\partial u_2}{\partial \eta} + æ_{30}u_1 - \vartheta_3 = 0 . \tag{8.118}$$

(2) The equations of vibrations of a wire relative to the plane x_1Ox_2.

$$\frac{\partial^2 u_3}{\partial \tau^2} - \frac{\partial Q_3}{\partial \eta} - \frac{Q_{20}M_1}{A_{11}} - \frac{Q_{10}M_2}{A_{22}} = q_3 ,$$
$$\frac{\partial M_1}{\partial \eta} - æ_{30}\left(1 - \frac{1}{A_{22}}\right) M_2 = 0 ,$$
$$\frac{\partial M_2}{\partial \eta} + æ_{30}\left(1 - \frac{1}{A_{11}}\right) M_1 = 0 ,$$
$$\frac{\partial \vartheta_1}{\partial \eta} - æ_{30}\vartheta_2 - \frac{M_1}{A_{11}} = 0 ,$$
$$\frac{\partial \vartheta_2}{\partial \eta} + æ_{30}\vartheta_1 - \frac{M_2}{A_{22}} = 0 ,$$
$$\frac{\partial u_3}{\partial \eta} + \vartheta_2 = 0 ,$$
$$M_1 = A_{11}\Delta æ_1 ,$$
$$M_2 = A_{22}\Delta æ_2 . \tag{8.119}$$

Here, Q_{10}, Q_{20}, M_{30}, and $æ_{30}$ are to be determined from the equations of a wire equilibrium (8.116) and (8.117).

The distributed forces $q_1^{(1)}$, q_2, and q_3 in the systems of equations (8.118) and (8.119) depend on $\mathbf{q}_n$, $\mathbf{q}_1$, and Δg_j, where $\mathbf{q}_n = q_{n2}\,\mathbf{e}_2^{(2)} + q_{n3}\,\mathbf{e}_3^{(3)}$ is the drag force, $\mathbf{q}_1$ is the aerodynamic force tangent to the axial line of a rod, that is, $\mathbf{q}_1 = q_1\,\mathbf{e}_1^{(1)}$, and Δg_j are increments of the projections of the gravity force $\mathbf{q} = -g_\xi\,\mathbf{i}_2$ onto the attached axes $\{\mathbf{e}_j^{(1)}\}$.

Consequently, we have

$$q_1^{(1)} = q_1 + \Delta g_1 ,$$
$$q_2 = q_{n2} + \Delta g_2 ,$$
$$q_3 = q_{n3} + \Delta g_3 . \qquad (8.120)$$

To determine Δg_j, we should refer the vector $\mathbf{i}_2$ to the basis $\{\mathbf{e}_j^{(1)}\}$. Let us use the matrix of transformation of the basis $\{\mathbf{i}_j\}$ to the basis $\{\mathbf{e}_j^{(1)}\}$. From Svetlitsky (2000a) it follows that this matrix is

$$\mathrm{L}^{(1)} = \left[l_{ij}^{(1)} \right] = \Delta \mathrm{L}\, \mathrm{L}_0 ,$$

where ΔL is the matrix of transformation from the basis $\{\mathbf{e}_j\}$ to the basis $\{\mathbf{e}_j^{(1)}\}$ and L_0 is the matrix of transformation from the basis $\{\mathbf{i}_j\}$ to the basis $\{\mathbf{e}_j\}$. These matrixes are

$$\Delta \mathrm{L} = \begin{bmatrix} 1 & \vartheta_3 & -\vartheta_2 \\ -\vartheta_3 & 1 & \vartheta_1 \\ \vartheta_2 & -\vartheta_1 & 1 \end{bmatrix}, \qquad \mathrm{L}_0 = \begin{bmatrix} \cos\vartheta_{30} & \sin\vartheta_{30} & 0 \\ -\sin\vartheta_{30} & \cos\vartheta_{30} & 0 \\ 0 & 0 & 1 \end{bmatrix} .$$

On rearrangement, we arrive at

$$\mathbf{i}_2 = (\sin\vartheta_{30} + \cos\vartheta_{30}\vartheta_3)\, \mathbf{e}_1^{(1)}$$
$$+ (\cos\vartheta_{30} - \sin\vartheta_{30}\vartheta_3)\, \mathbf{e}_2^{(1)} + (-\cos\vartheta_{30}\vartheta_1 + \sin\vartheta_{30}\vartheta_2)\, \mathbf{e}_3^{(1)} .$$

Consequently, the projections Δg_j are as follows:

$$\Delta g_1 = -g_\xi \cos\vartheta_{30}\vartheta_3 ,$$
$$\Delta g_2 = g_\xi \sin\vartheta_{30}\vartheta_3 ,$$
$$\Delta g_3 = -g_\xi(-\cos\vartheta_{30}\vartheta_1 + \sin\vartheta_{30}\vartheta_2 . \qquad (8.121)$$

The second term of the system (8.119) includes the multiplier $(1-1/A_{22})$. For a wire of circular cross section, we have $A_{22} = A_{33} = 1$, consequently, in this case, this multiplier equals zero and the second equation becomes uncoupled from the other equations of the system (8.119). Hence, setting, for example, $A_{22} = 0.99$, we arrive at the system of six dependent equations, which can be solved easily. The solution obtained is just approximate and may be evaluated more accurately by setting $A_{22} = 0.999$, etc.

We can represent the systems of equations (8.118) and (8.119) in vector form.

(1) The equations of vibrations of a wire in the plane x_1Ox_2 are

$$\mathrm{A}^{(1)} \frac{\partial^2 \mathbf{Z}}{\partial \tau^2} - \frac{\partial \mathbf{Z}}{\partial \eta} + \mathrm{A}^{(2)} \mathbf{Z} = \mathbf{b}_1 , \qquad (8.122)$$

where

$$\mathbf{Z} = (Q_1,\, Q_2,\, M_3,\, \vartheta_3,\, u_1,\, u_2)^{\mathrm{T}}\,,$$

$$\mathbf{b}_1 = (q_1,\, q_{n2},\, 0,\, 0,\, 0,\, 0)^{\mathrm{T}}\,,$$

$$\mathrm{A}^{(1)} = \begin{bmatrix} 0 & 0 & 0 & 0 & 1 & 0 \\ 0 & 0 & 0 & 0 & 0 & 1 \\ 0 & 0 & 0 & 0 & 0 & 0 \\ 0 & 0 & 0 & 0 & 0 & 0 \\ 0 & 0 & 0 & 0 & 0 & 0 \\ 0 & 0 & 0 & 0 & 0 & 0 \end{bmatrix},$$

$$\mathrm{A}^{(2)} = \begin{bmatrix} 0 & æ_{30} & Q_{20} & g_\xi \cos\vartheta_{30} & 0 & 0 \\ -æ_{30} & 0 & -Q_{10} & -g_\xi \sin\vartheta_{30} & 0 & 0 \\ 0 & -1 & 0 & 0 & 0 & 1 \\ 0 & 0 & 1 & 0 & 0 & 0 \\ 0 & 0 & 0 & 0 & 0 & æ_{30} \\ 0 & 0 & 0 & 1 & -æ_{30} & 0 \end{bmatrix}.$$

(2) The equations of vibrations of a wire relative to the plane x_1Ox_2 are

$$\mathrm{B}^{(1)}\frac{\partial^2 \mathbf{Z}}{\partial \tau^2} - \frac{\partial \mathbf{Z}}{\partial \eta} + \mathrm{B}^{(2)}\mathbf{Z} = \mathbf{b}_2\,, \tag{8.123}$$

where

$$\mathbf{Z} = (Q_3,\, M_1,\, M_2,\, \vartheta_1,\, \vartheta_2,\, u_3)^{\mathrm{T}}\,,$$

$$\mathbf{b}_2 = (q_{n3},\, 0,\, 0,\, 0,\, 0,\, 0)^{\mathrm{T}}\,,$$

$$\mathrm{B}^{(1)} = \begin{bmatrix} 0 & 0 & 0 & 0 & 0 & 1 \\ 0 & 0 & 0 & 0 & 0 & 0 \\ 0 & 0 & 0 & 0 & 0 & 0 \\ 0 & 0 & 0 & 0 & 0 & 0 \\ 0 & 0 & 0 & 0 & 0 & 0 \\ 0 & 0 & 0 & 0 & 0 & 0 \end{bmatrix}, \qquad \mathrm{B}^{(2)} =$$

$$\begin{bmatrix} 0 & -\dfrac{Q_{20}}{A_{11}} & -Q_{10} & -g_\xi \cos\vartheta_{30} & g_\xi \sin\vartheta_{30} & 0 \\ 0 & 0 & æ_{30}\left(1 - \dfrac{1}{A_{22}}\right) & 0 & 0 & 0 \\ 1 & -æ_{30}\left(1 - \dfrac{1}{A_{11}}\right) & 0 & 0 & 0 & 0 \\ 0 & \dfrac{1}{A_{11}} & 0 & 0 & æ_{30} & 0 \\ 0 & 0 & \dfrac{1}{A_{22}} & -æ_{30} & 0 & 1 \\ 0 & 0 & 0 & 0 & -1 & 0 \end{bmatrix}.$$

The solutions of equations (8.122) and (8.123) may be obtained by using numerical approximate methods only, for example, the methods represented in Chap. 4 and in Chap. 5.

We seek an approximate solution of the vector equations (8.122) and (8.123) in the form

$$\tilde{\mathbf{Z}} = \sum_{j-1}^{n} f_j(\tau)\,\mathbf{Z}_0^{(j)}(\eta)\,, \tag{8.124}$$

where $\mathbf{Z}_0^{(j)}(\eta)$ are eigenvectors of the boundary value problems for the homogeneous equations (8.122) and (8.123) and $f_j(\tau)$ are unknown functions.

Let us consider in detail the procedure of determination of the eigenvectors when a wire is in free vibrations in the plane x_1Ox_2. (The procedure for determination of eigenvalues and eigenvectors for the general case of vibrations of space-curved rods is presented in Chap. 4.)

We seek a solution to the homogeneous equation (8.122) in the form

$$\mathbf{Z} = \mathbf{Z}_0\,\mathrm{e}^{\mathrm{i}\,\beta\tau}\,. \tag{8.125}$$

Substituting (8.125) into (8.122) for $\mathbf{b}_1 = 0$, we get

$$\frac{\mathrm{d}\mathbf{Z}_0}{\mathrm{d}\eta} + \left(\mathrm{A}^{(1)}\beta^2 - \mathrm{A}^{(2)}\right)\mathbf{Z}_0 = 0\,. \tag{8.126}$$

Taking an arbitrary value of $\beta^{(\rho)}$ and employing the method of initial parameters, we obtain the following solution to the vector equation (8.126):

$$\mathbf{Z}_0 = \mathrm{K}(\eta,\,\beta^{(\rho)})\,\mathbf{c}\,, \qquad \mathrm{K}(0,\,\beta^{(\rho)}) = \mathrm{E}\,. \tag{8.127}$$

The components of the vector $\mathbf{Z}_0$ must satisfy the following boundary conditions.

(1) At $\eta = 0$,

$$Z_{03} = M_{30} = 0\,, \qquad Z_{05} = u_{10} = 0\,, \qquad Z_{06} = u_{20} = 0\,.$$

(2) At $\eta = 1$,

$$Z_{03} = Z_{05} = Z_{06} = 0\,.$$

From the boundary conditions at $\eta = 0$ it follows that

$$c_3 = c_5 = c_6 = 0\,.$$

The system of equations for some values of $\beta^{(\rho)}$ ($\rho = 1,\,2,\,\ldots$) follows from the boundary conditions at $\eta = 1$:

$$k_{31}(1)\,c_1 + k_{32}(1)\,c_2 + k_{34}(1)\,c_4 = 0\,,$$
$$k_{51}c_1 + k_{52}c_2 + k_{54}c_4 = 0\,,$$
$$k_{61}c_1 + k_{62}c_2 + k_{64}c_4 = 0\,. \tag{8.128}$$

Equating the determinant of this system to zero, we obtain the equation to be solved for β_j and the corresponding eigenvectors $\mathbf{Z}_0^{(j)}$ (see Chap. 4). The eigenvectors may be determined from the vector equation (8.126):

$$\frac{\mathrm{d}\mathbf{Z}_0^{(j)}}{\mathrm{d}\eta} + \left(\mathrm{A}^{(1)}\beta_j^2 - \mathrm{A}^{(2)}\right)\mathbf{Z}_0^{(j)} = 0\,. \tag{8.129}$$

The solution of (8.129) is

$$\mathbf{Z}_0^{(j)} = \mathrm{K}(\eta,\,\beta_j)\,c^{(j)}\,, \qquad c_3^{(j)} = c_5^{(j)} = c_6^{(j)} = 0\,.$$

From the boundary conditions at $\eta = 1$ it follows that

$$k_{31}(1,\,\beta_j)\,c_1^{(j)} + k_{32}(1,\,\beta_j)\,c_2^{(j)} + k_{34}(1,\,\beta_j)\,c_4^{(j)} = 0\,,$$
$$k_{51}c_1^{(j)} + k_{52}c_2^{(j)} + k_{54}c_4^{(j)} = 0\,,$$
$$k_{61}c_1^{(j)} + k_{62}c_2^{(j)} + k_{64}c_4^{(j)} = 0\,. \tag{8.130}$$

The determinant of the system of equations (8.130) is equal to zero. Thus, it is sufficient to use any two equations of this system. For example, from the first two equations of the system we may obtain $c_1^{(j)}$ and $c_2^{(j)}$ as functions of $c_4^{(j)}$:

$$c_1^{(j)} = \gamma_1^{(j)}c_4^{(j)}\,, \qquad c_2^{(j)} = \gamma_2^{(j)}c_4^{(j)}\,. \tag{8.131}$$

Here, it is assumed that

$$\varDelta = \begin{bmatrix} k_{31} & k_{32} \\ k_{51} & k_{52} \end{bmatrix} \neq 0\,.$$

The arbitrary multiplier $c_4^{(j)}$ may be equal to unity. As a result, we arrive at the set of the eigenvectors

$$\mathbf{Z}_0^{(j)} = \mathrm{K}(\eta,\,\beta_j)\,\mathbf{C}^{(j)}\,, \tag{8.132}$$

where

$$\mathbf{C}^{(j)} = (\gamma_1^{(j)},\,\gamma_2^{(j)},\,0,\,1,\,0,\,0)^{\mathrm{T}}\,.$$

The graphs of the dimensionless components of the first two eigenvectors $\mathbf{Z}_0^{(1)}$ and $\mathbf{Z}_0^{(2)}$ are shown in Figs. 8.16–8.19. These eigenvectors correspond to the first two dimensionless eigenvalues (frequencies) $\beta_1 = 45.060$ and $\beta_2 = 99.340$, respectively. The dimensional frequencies $\tilde{\beta}_i$ are as follows:

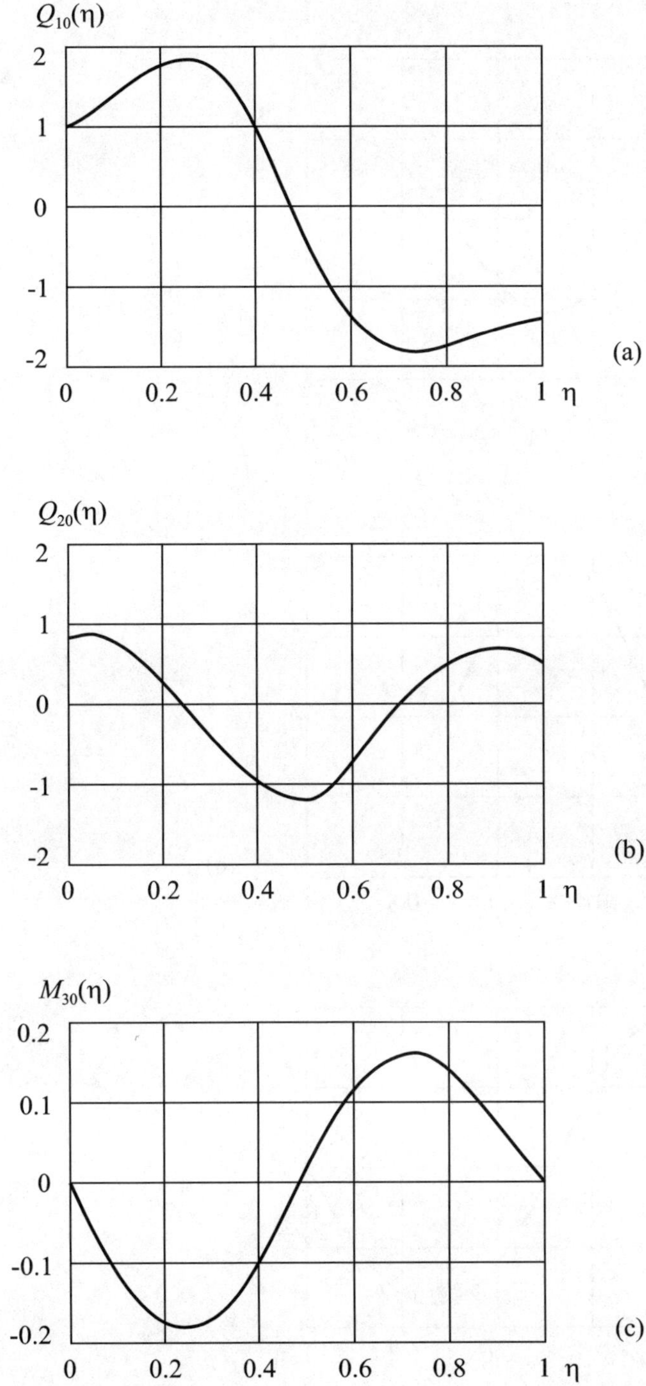

$Q_{10}(\eta)$
2
1
0
-1
-2
0
0.2
0.4
0.6
0.8
1 η
(a)
$Q_{20}(\eta)$
2
1
0
-1
-2
0
0.2
0.4
0.6
0.8
1 η
(b)
$M_{30}(\eta)$
0.2
0.1
0
-0.1
-0.2
0
0.2
0.4
0.6
0.8
1 η
(c)

Fig. 8.16.

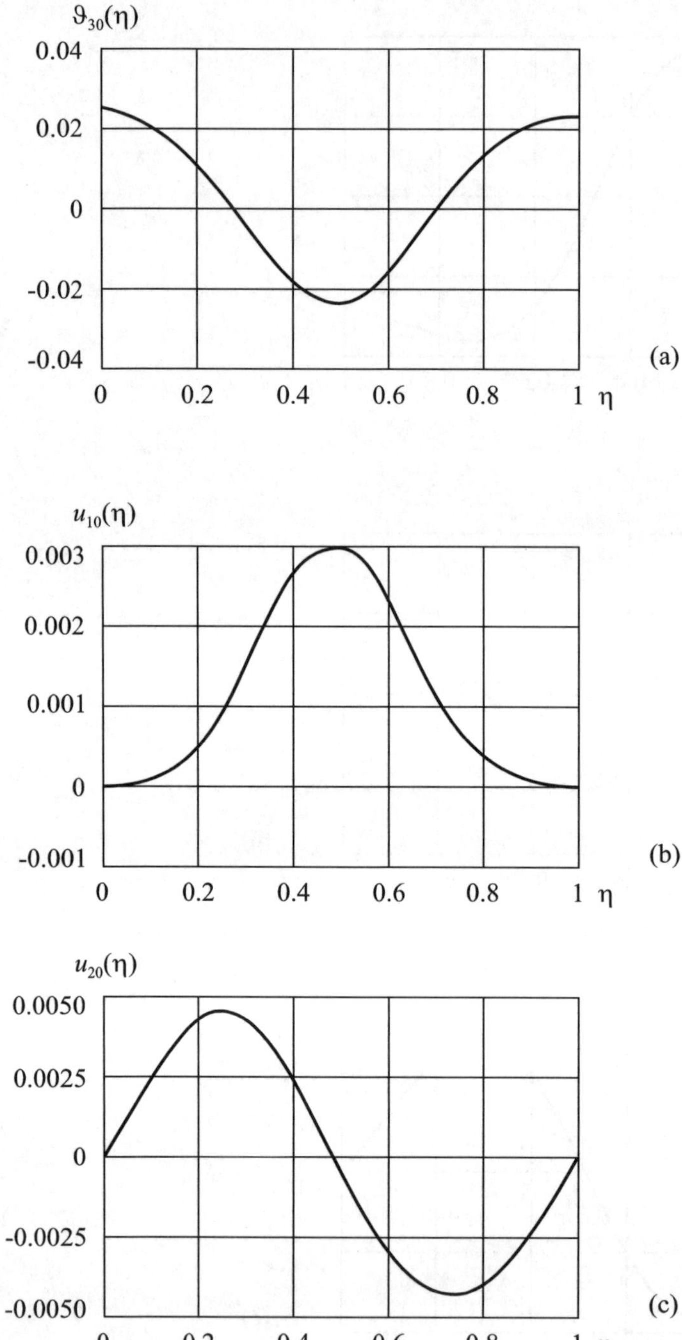
$\vartheta_{30}(\eta)$
0.04
0.02
0
-0.02
-0.04
0
0.2
0.4
0.6
0.8
1 η
(a)
$u_{10}(\eta)$
0.003
0.002
0.001
0
-0.001
0
0.2
0.4
0.6
0.8
1 η
(b)
$u_{20}(\eta)$
0.0050
0.0025
0
-0.0025
-0.0050
0
0.2
0.4
0.6
0.8
1 η
(c)

Fig. 8.17.

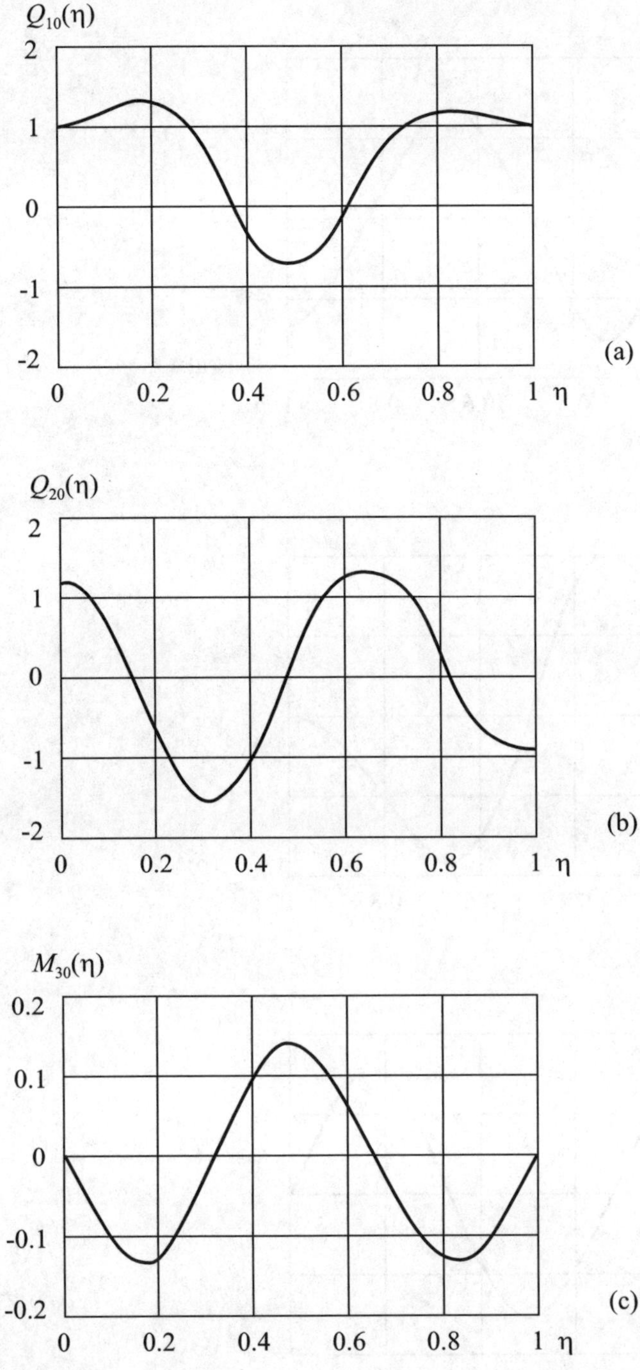
$Q_{10}(\eta)$
2
1
0
-1
-2
0
0.2
0.4
0.6
0.8
1 η
(a)
$Q_{20}(\eta)$
2
1
0
-1
-2
0
0.2
0.4
0.6
0.8
1 η
(b)
$M_{30}(\eta)$
0.2
0.1
0
-0.1
-0.2
0
0.2
0.4
0.6
0.8
1 η
(c)

Fig. 8.18.

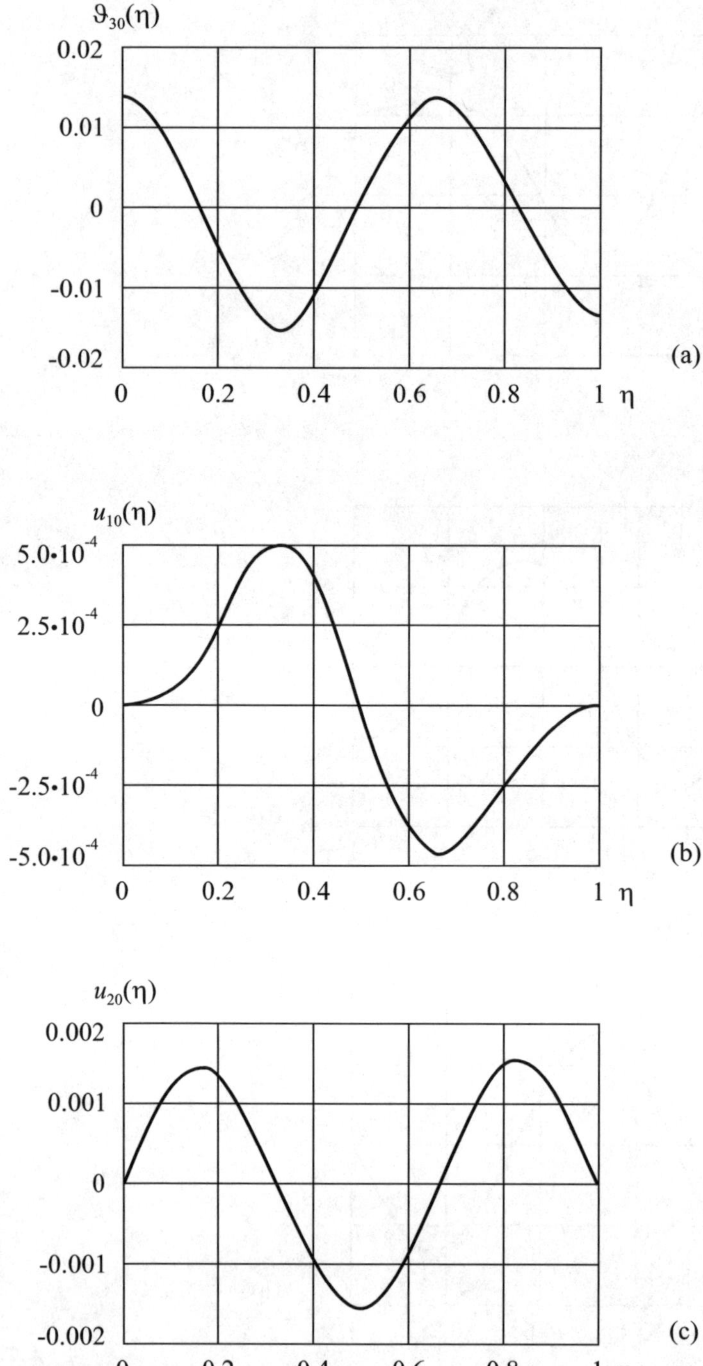
ϑ30(η)
0.02
0.01
0
-0.01
-0.02
0
0.2
0.4
0.6
0.8
1 η
(a)
u10(η)
5.0•10-4
2.5•10-4
0
-2.5•10-4
-5.0•10-4
0
0.2
0.4
0.6
0.8
1 η
(b)
u20(η)
0.002
0.001
0
-0.001
-0.002
0
0.2
0.4
0.6
0.8
1 η
(c)

Fig. 8.19.

$$\tilde{\beta}_i = \beta_i \sqrt{\frac{A_{33}}{m_0 l^4}}, \qquad \tilde{\beta}_1 = 2.838\,\mathrm{s}^{-1}, \qquad \beta_2 = 6.288\,\mathrm{s}^{-1}.$$

The eigenvalues and the eigenvectors corresponding to free vibrations relative to the plane x_1Ox_2 (the vector equation (8.123)) may be obtained similarly. The values of the first two dimensionless frequencies corresponding to free vibrations of a wire relative to the vertical plane are

$$\beta_1 = 23.511, \qquad \beta_2 = 50.202.$$

The dimensional frequencies are

$$\tilde{\beta}_1 = 1.482\,\mathrm{s}^{-1}, \qquad \tilde{\beta}_2 = 3.155\,\mathrm{s}^{-1}.$$

The graphs of the dimensionless components of the vectors $\mathbf{Z}_0^{(1)}$ and $\mathbf{Z}_0^{(2)}$ are shown in Figs. 8.20–8.21 and 8.22–8.23, respectively.

Let us now return to the main problem under consideration, the problem of small vibrations of a wire under impulsive loading generated by an air flow (see the vector equations (8.122) and (8.123)).

First, we consider vibrations of the wire out of plane (8.123). Assuming that the time of duration of the pulse, $\Delta\tau$, is small as compared to the period corresponding to the first frequency, we may admit that the points of the axial line of the rod will start to move with some velocities after the end of the pulse but their displacements will be still zero. (These velocities can be found using a balance of linear momentum.) Besides, we may assume that the aerodynamic forces due to the motion of the wire after the pulse terminated are small and these forces may be disregarded. Consequently, the motion of the wire after the pulse of aerodynamic forces terminated may be described by the following homogeneous equation with known initial conditions:

$$\mathbf{L}_2 = B^{(1)}\frac{\partial^2 \mathbf{Z}}{\partial \tau^2} - \frac{\partial \mathbf{Z}}{\partial \eta} + B^{(2)}\mathbf{Z} = 0. \tag{8.133}$$

Solving this equation, we confine ourselves to the two-term approximation

$$\tilde{\mathbf{Z}} = f_1(\tau)\,\mathbf{Z}_0^{(1)}(\eta) + f_2(\tau)\,\mathbf{Z}_0^{(2)}(\eta). \tag{8.134}$$

Substituting (8.134) into (8.133), we get

$$\mathbf{L}_2(\tilde{\mathbf{Z}}) = \boldsymbol{\delta}_2(\eta, \tau). \tag{8.135}$$

Let us now employ the generalized principle of virtual displacements (see Sect. 4.4). For the two-term approximation, the virtual displacements that satisfy all the boundary conditions of the problem are as follows:

$$\delta\mathbf{Z} = \delta \mathrm{A}_1 \mathrm{E}_{02}\mathbf{Z}_0^{(1)} + \delta \mathrm{A}_2 \mathrm{E}_{02}\mathbf{Z}_0^{(2)}, \tag{8.136}$$

where δA_j are small independent variables and

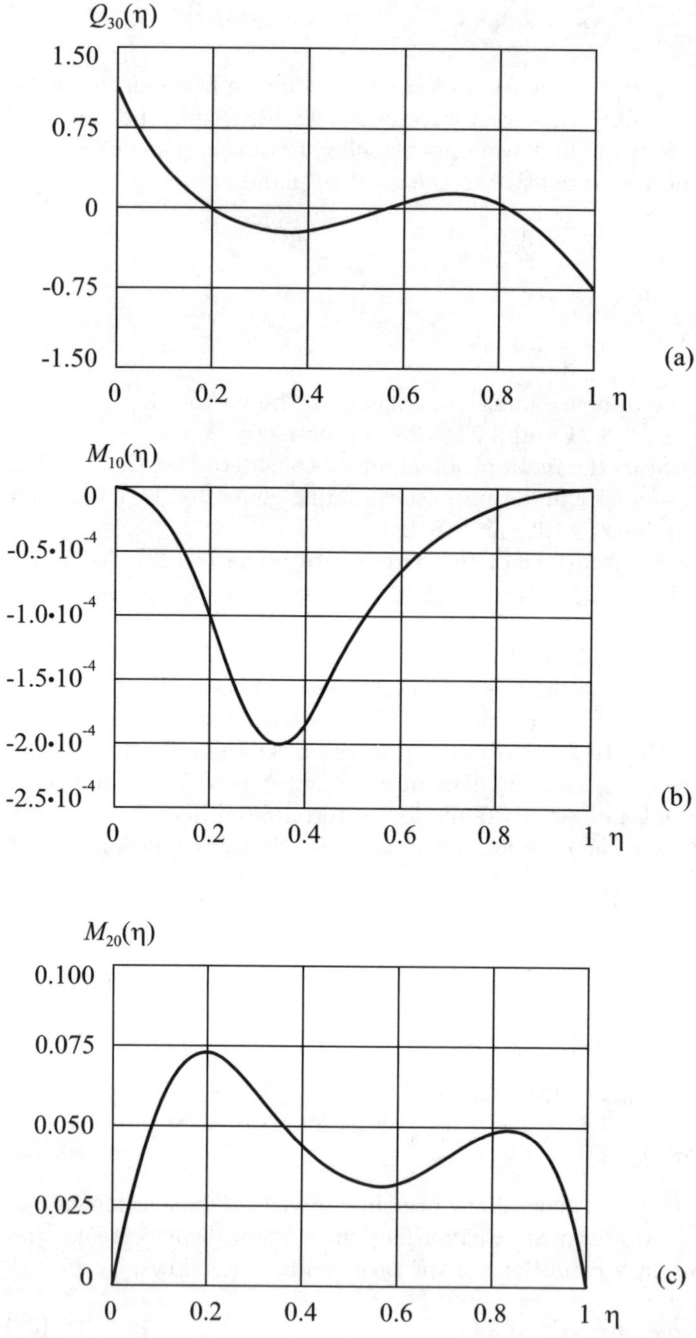
$Q_{30}(\eta)$
1.50
0.75
0
-0.75
-1.50
0 0.2 0.4 0.6 0.8 1 η
(a)
$M_{10}(\eta)$
0
-0.5•10^-4
-1.0•10^-4
-1.5•10^-4
-2.0•10^-4
-2.5•10^-4
0 0.2 0.4 0.6 0.8 1 η
(b)
$M_{20}(\eta)$
0.100
0.075
0.050
0.025
0
0 0.2 0.4 0.6 0.8 1 η
(c)

Fig. 8.20.

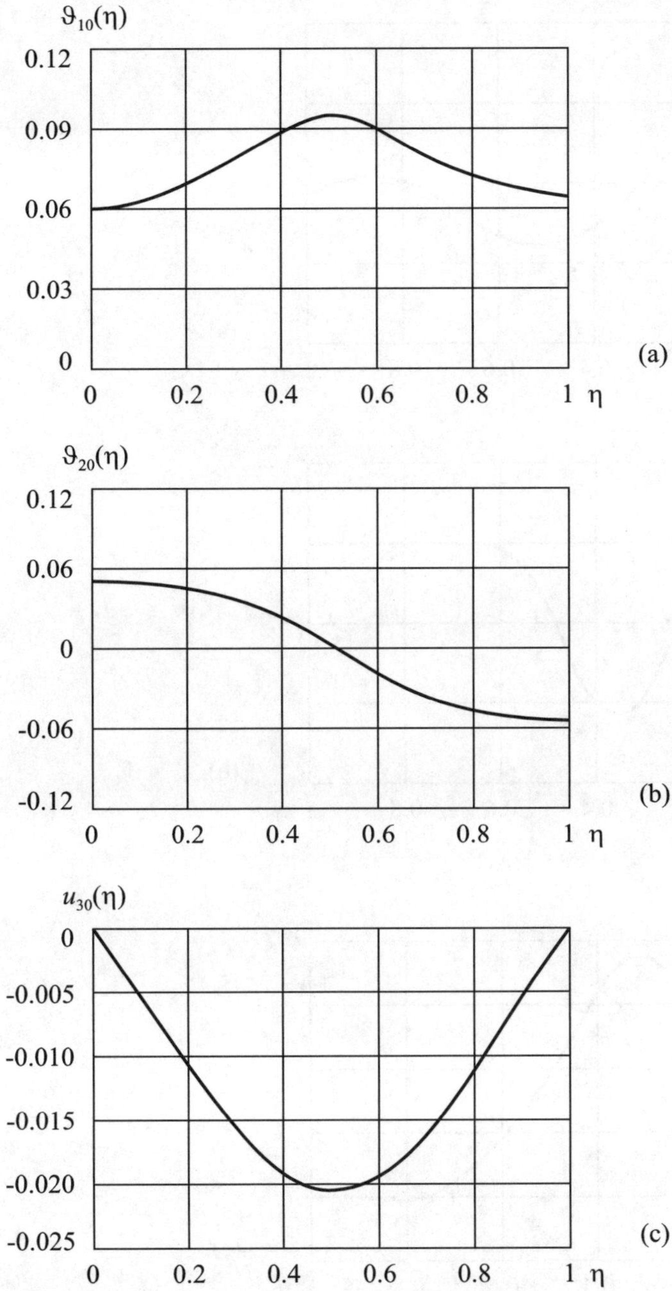
$\vartheta_{10}(\eta)$
0.12
0.09
0.06
0.03
0
(a)
0 0.2 0.4 0.6 0.8 1 η
$\vartheta_{20}(\eta)$
0.12
0.06
0
-0.06
-0.12
(b)
0 0.2 0.4 0.6 0.8 1 η
$u_{30}(\eta)$
0
-0.005
-0.010
-0.015
-0.020
-0.025
(c)
0 0.2 0.4 0.6 0.8 1 η

Fig. 8.21.

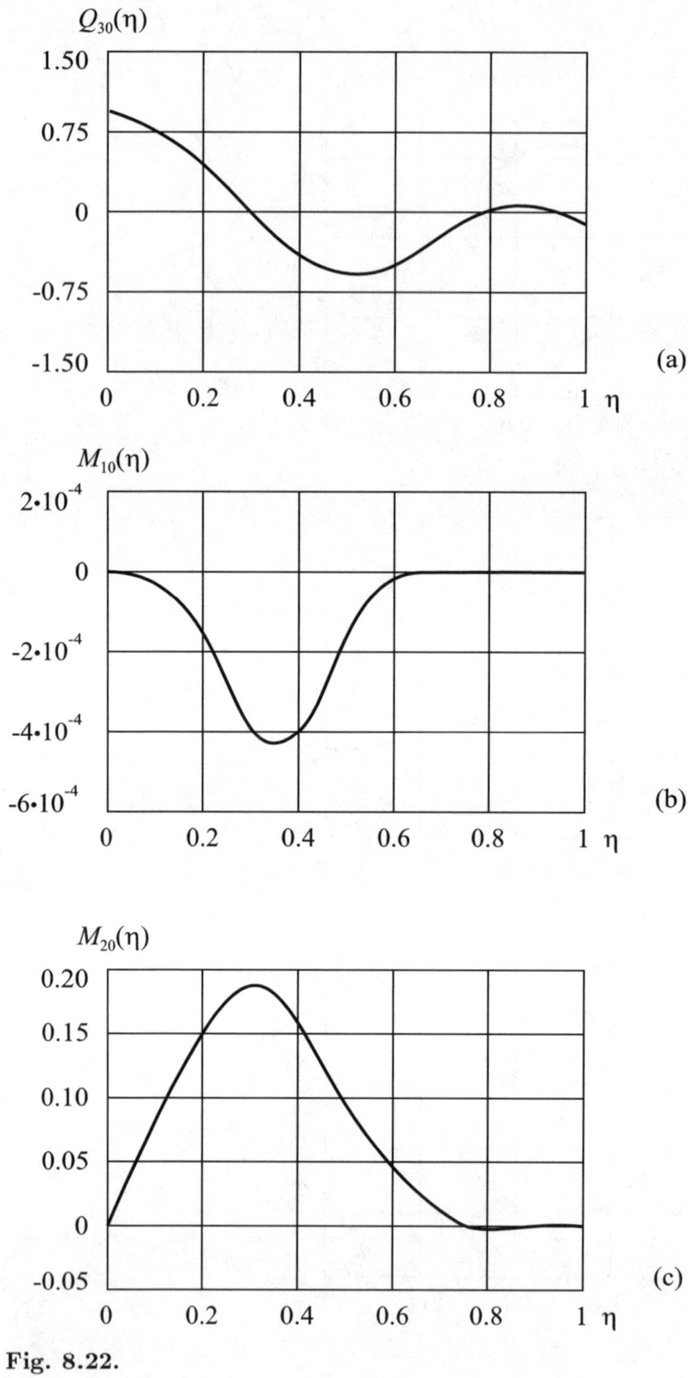
$Q_{30}(\eta)$
1.50
0.75
0
-0.75
-1.50
0
0.2
0.4
0.6
0.8
1 η
(a)
$M_{10}(\eta)$
2•10⁻⁴
0
-2•10⁻⁴
-4•10⁻⁴
-6•10⁻⁴
0
0.2
0.4
0.6
0.8
1 η
(b)
$M_{20}(\eta)$
0.20
0.15
0.10
0.05
0
-0.05
0
0.2
0.4
0.6
0.8
1 η
(c)

Fig. 8.22.

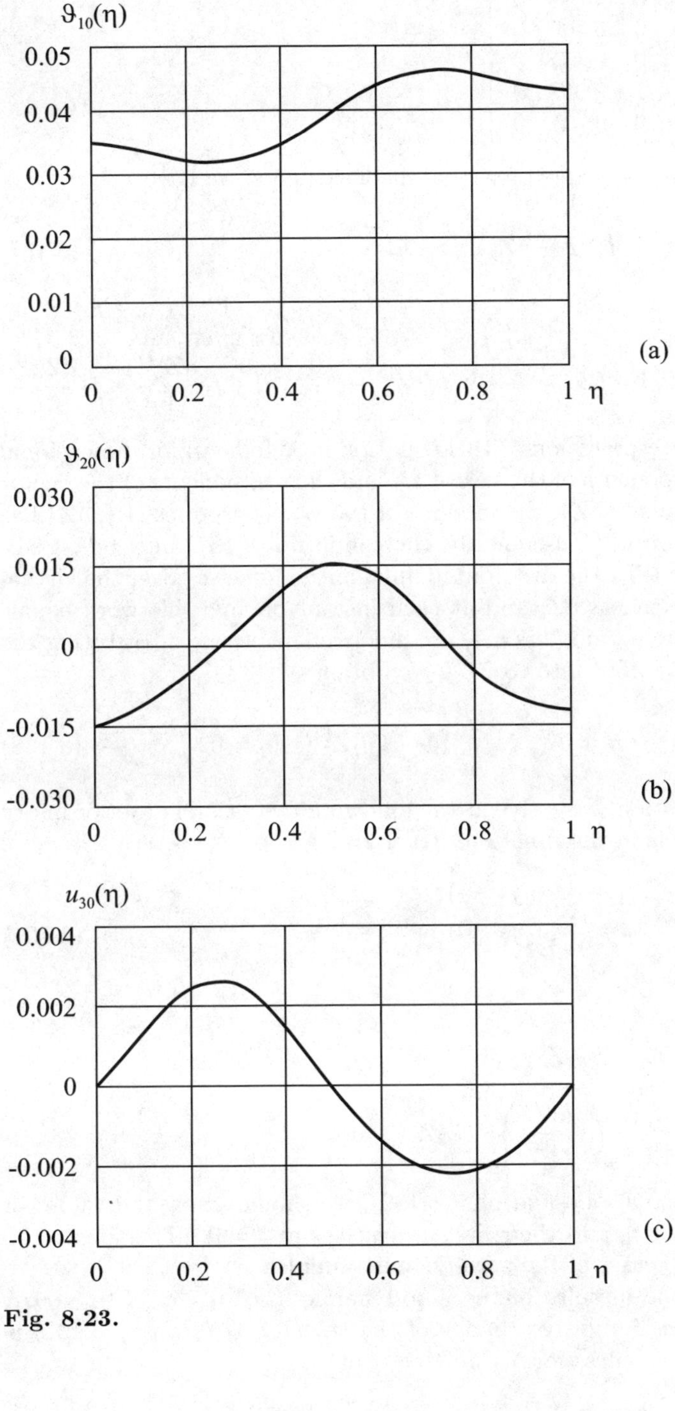
ϑ10(η)
0.05
0.04
0.03
0.02
0.01
0
0 0.2 0.4 0.6 0.8 1 η
(a)
ϑ20(η)
0.030
0.015
0
-0.015
-0.030
0 0.2 0.4 0.6 0.8 1 η
(b)
u30(η)
0.004
0.002
0
-0.002
-0.004
0 0.2 0.4 0.6 0.8 1 η
(c)

Fig. 8.23.

$$
\mathrm{E}_{02} = \begin{bmatrix} 0 & 0 & 0 & 0 & 0 & 1 \\ 0 & 0 & 0 & 1 & 0 & 0 \\ 0 & 0 & 0 & 0 & 1 & 0 \\ 0 & 1 & 0 & 0 & 0 & 0 \\ 0 & 0 & 1 & 0 & 0 & 0 \\ 1 & 0 & 0 & 0 & 0 & 0 \end{bmatrix} .
$$

According to the principle of virtual displacements, we have

$$
\int_0^1 (\mathbf{L}_2 \cdot \delta \mathbf{Z})\, \mathrm{d}\eta = \int_0^1 (\boldsymbol{\delta}_2 \cdot \delta \mathbf{Z})\, \mathrm{d}\eta = 0 \,, \tag{8.137}
$$

where

$$
\boldsymbol{\delta}_2 = B^{(1)}(\ddot{f}_1 \mathbf{Z}_0^{(1)} + \ddot{f}_2 \mathbf{Z}_0^{(2)}) + B^{(2)}(f_1 \mathbf{Z}_0^{(1)} + f_2 \mathbf{Z}_0^{(2)}) - f_1 \frac{\mathrm{d}\mathbf{Z}_0^{(1)}}{\mathrm{d}\eta} - f_2 \frac{\mathrm{d}\mathbf{Z}_0^{(2)}}{\mathrm{d}\eta} \,.
$$

The physical interpretation of this equation is as follows: for dimensional functions (the components of the vector $\mathbf{L}_2$ and the components of the vector of virtual displacements $\delta\mathbf{Z}$), the integral of the scalar product $(\mathbf{L}_2, \delta\mathbf{Z})$ has the dimensions of work. Consequently, the condition (8.137) may be considered as the work done by the distributed unbalanced forces $\boldsymbol{\delta}_2$ over the virtual generalized displacements $\delta\mathbf{Z}$, and at each instant of time this work equals zero. Recall that the parameters δA_j are independent; then, substituting the expressioin for $\delta\mathbf{Z}$ (8.136) into (8.137), we obtain

$$
\int_0^1 (\boldsymbol{\delta}_2 \cdot E_{02}\mathbf{Z}_0^{(1)})\, \mathrm{d}\eta = 0 \,, \qquad \int_0^1 (\boldsymbol{\delta}_2 \cdot E_{02}\mathbf{Z}_0^{(2)})\, \mathrm{d}\eta = 0 \,. \tag{8.138}
$$

Analyzing this equation, we arrive at the following two second order ordinary differential equations in the functions $f_1(\tau)$ and $f_2(\tau)$:

$$
\begin{aligned}
&a_{11}\ddot{f}_1 + a_{12}\ddot{f}_2 + c_{11} f_1 + c_{12} f_2 = 0 \,, \\
&a_{21}\ddot{f}_1 + a_{22}\ddot{f}_2 + c_{21} f_1 + c_{22} f_2 = 0 \,;
\end{aligned} \tag{8.139}
$$

here,

$$
\begin{aligned}
a_{ij} &= \int_0^1 (B^{(1)}\mathbf{Z}_0^{(i)} \cdot E_{02}\mathbf{Z}_0^{(j)})\, \mathrm{d}\eta \,, \\
c_{ij} &= \int_0^1 \left[\left(B^{(2)}\mathbf{Z}_0^{(i)} - \frac{\mathrm{d}\mathbf{Z}_0^{i}}{\mathrm{d}\eta} \right) \cdot E_{02}\mathbf{Z}_0^{(j)} \right] \mathrm{d}\eta \,.
\end{aligned}
$$

To solve the system of equations (8.139), we should know the values of the functions f_1 and f_2 and their first derivatives at $t = 0$. (The time $t = 0$ corresponds to the ceases of the aerodynamic impulse.)

The vector of the impulse of the aerodynamic force $\mathbf{q}_a$, $\mathbf{q}_a = \mathbf{q}_n + \mathbf{q}_1$, corresponding to the impulsive change of the velocity v_0 (see Fig. 8.13a) is as follows (in dimensionless form):

$$\mathbf{J} = \int_0^{\Delta\tau} \mathbf{q}_a \mathrm{d}\tau^{(1)} \cong \mathbf{q}_n \Delta\tau + \mathbf{q}_1 \Delta\tau \,.$$

Because

$$\mathbf{q}_n = q_{n2}\mathbf{e}_2 + q_{n3}\mathbf{e}_3 \,,$$

we finally arrive at

$$\mathbf{J} = \sum_{j=1}^{3} J_j \mathbf{e}_j \,, \tag{8.140}$$

where

$$\begin{aligned}
&J_1 = q_1 \Delta\tau \,, \qquad J_2 = q_{n2} \Delta\tau \,, \qquad J_3 = q_{n3} \Delta\tau \,, \\
&q_1 = q_{10} \cos^2 \varphi_{a0} = q_{10} \cos^2 \alpha \cos^2 \vartheta_{30}^{(1)} \,, \\
&q_{n2} = -q_{n0} \sin \varphi_{a0} \left(l_{210} \cos\alpha + l_{230} \sin\alpha\right) , \\
&q_{n3} = q_{n0} \sin \varphi_{a0} \left(l_{310} \cos\alpha + l_{330} \sin\alpha\right) , \\
&q_{10} = \frac{c_1 \rho d v_0^2 l^3}{2A_{33}} \,, \qquad q_{n0} = \frac{c_n \rho d v_0^2 l^3}{2A_{33}} \,.
\end{aligned}$$

In the case under consideration, the axial line of the wire before loading is a plane curve, consequently, we have

$$\begin{aligned}
&l_{210} = -\sin\vartheta_{30}^{(1)} \,, \qquad l_{230} = 0 \,, \qquad l_{310} = 0 \,, \qquad l_{330} = 1 \,, \\
&\sin\varphi_{a0} = \sqrt{1 - \cos^2\alpha \cos^2 \vartheta_{30}^{(1)}} \,.
\end{aligned}$$

The graphs of the aerodynamic forces q_1, q_{n2}, and q_{n3} versus the dimensionless axial coordinate η are shown in Fig. 8.24; these graphs correspond to the impulsive loading at $\alpha = 45°$ and $v_0 = 50\,\mathrm{m/s}$.

When the pulse terminates, the points of the axial line of the rod will be set in motion with velocities $\dot{\mathbf{u}}(\eta,\, 0)$ such that

$$\dot{\mathbf{u}}(\eta,\, 0) = \tilde{\mathbf{J}} \,, \tag{8.141}$$

where

$$\dot{\mathbf{u}} = \sum_{j=1}^{3} \dot{u}_j(\eta,\, 0)\, \mathbf{e}_j \,.$$

From (8.141) it follows that

$$\dot{u}_j(\eta,\, 0) = J_j \,. \tag{8.142}$$

The equation of vibrations of the wire relative to the vertical plane (8.119) includes the projection of the drag force q_{n3} only. Consequently, the initial conditions are as follows:

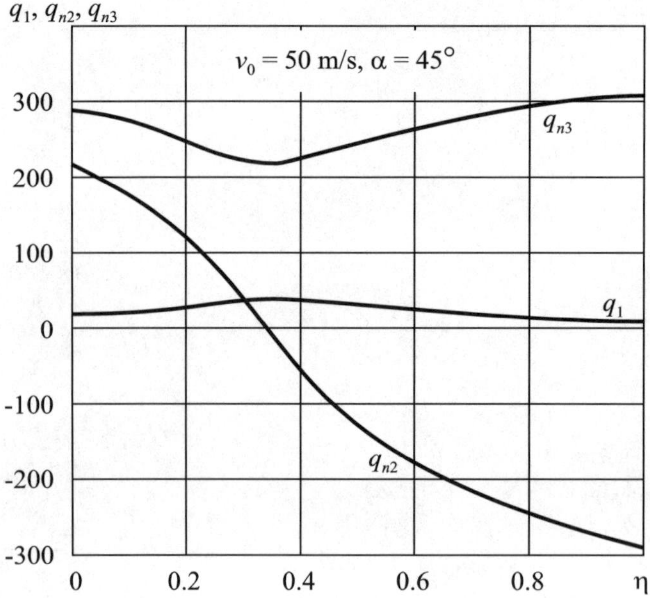

Fig. 8.24.

$$u_3(\eta,\, 0) = 0\,, \qquad \dot{u}_3(\eta,\, 0) = J_3 = q_{n0} \varDelta\tau \sin\varphi_{a0} \sin\alpha\,.$$

Using the approximate representation of the vector **Z** (8.134), we obtain

$$u_3(\eta,\, \tau) = f_1(\tau)\, Z_{06}^{(1)}(\eta) + f_2(\tau)\, Z_{06}^{(2)}(\eta)\,. \tag{8.143}$$

Because $u_3(\eta,\, \tau) = 0$ at $\tau = 0$, from (8.143) it follows that

$$f_1(0) = f_2(0) = 0\,.$$

Differentiating (8.143) with respect to τ and substituting $\tau = 0$ into the result, we obtain

$$\dot{u}_3(\eta,\, 0) = J_3 = \dot{f}_1(0)\, Z_{06}^{(1)}(\eta) + \dot{f}_2(0)\, Z_{06}^{(2)}(\eta)\,. \tag{8.144}$$

We can obtain $\dot{f}_1(0)$ and $\dot{f}_2(0)$ from (8.144) just approximately using the following condition of the minimum of the functional:

$$I = \int_0^1 \left(\dot{u}_3(\eta,\, 0) - \sum \dot{f}_j(0)\, Z_{06}^{(j)} \right)^2 \mathrm{d}\eta = \min\,. \tag{8.145}$$

To determine $\dot{f}_1(0)$ and $\dot{f}_2(0)$, we get two equations

$$\frac{\partial I}{\partial \dot{f}_1} = 0\,, \qquad \frac{\partial I}{\partial \dot{f}_2} = 0$$

or

$$a_{11}\dot{f}_1(0) + a_{12}\dot{f}_2(0) = b_1\,,$$
$$a_{21}\dot{f}_1(0) + a_{22}\dot{f}_2(0) = b_2\,, \tag{8.146}$$

where

$$a_{ij} = \int_0^1 (Z_{06}^{(i)} \cdot Z_{06}^{(j)})\, \mathrm{d}\eta\,,$$
$$b_i = \int_0^1 (\dot{u}_3 \cdot Z_{06}^{(i)})\, \mathrm{d}\eta = q_{n0}\Delta\tau \int_0^1 Z_{06}^{(i)} \sin\varphi_{a0} \cos\alpha\, \mathrm{d}\eta\,.$$

Having found the values of $f_j(0)$ and $\dot{f}_j(0)$, $j = 1, 2$, from the system of equations (8.139), we can determine $f_1(\tau)$, $f_2(\tau)$, $\dot{f}_1(\tau)$, and $\dot{f}_2(\tau)$. As a result, we arrive at the approximate solution of the vector equation (8.133), corresponding to the two-term approximation:

$$Z_j(\eta,\,\tau) = f_1(\tau)\, Z_{0j}^{(1)}(\eta) + f_2(\tau)\, Z_{0j}^{(2)}(\eta)\,. \tag{8.147}$$

Following the procedure similar to the described one, we can determine the approximate solution of (8.133) corresponding to the three-term approximation, etc.

The graphs of the components $Z_j(\eta,\,\tau)$ of the vector $\mathbf{Z}(\eta,\,\tau)$ versus η for the three values of dimensionless time τ ($\tau_1 = 0.02$, $\tau_2 = 0.04$, and $\tau_3 = 0.06$) are shown in Figs. 8.25 and 8.26. The value of dimensional parameters for numerical calculations were taken as follows: $v_0 = 50\,\mathrm{m/s}$, $\alpha = 45°$, $c_n = 1.2$, and $\Delta\tau = 0.028$ (the last value corresponds to the dimensional time $\Delta t = 0.466\,\mathrm{s}$; whereas, the period T_1 equals $4.34\,\mathrm{s}$).

Let us now consider vibrations of a wire in the vertical plane after the pulse terminated. Setting $\mathbf{b}_1 = 0$ in the vector equations (8.122)), we have

$$\mathbf{L}_1 = A^{(1)}\frac{\partial^2 \mathbf{Z}}{\partial \tau^2} - \frac{\partial \mathbf{Z}}{\partial \eta} + A^{(2)}\mathbf{Z} = 0\,. \tag{8.148}$$

The procedure of approximate solution of the vector equation (8.148) is practically the same as the procedure for determination of an approximate solution of the vector equation (8.133). The only difference is that we should use the matrix

$$\mathrm{E}_{01} = \begin{bmatrix} 0 & 0 & 0 & 0 & 1 & 0 \\ 0 & 0 & 0 & 0 & 0 & 1 \\ 0 & 0 & 0 & 1 & 0 & 0 \\ 0 & 0 & 1 & 0 & 0 & 0 \\ 1 & 0 & 0 & 0 & 0 & 0 \\ 0 & 1 & 0 & 0 & 0 & 0 \end{bmatrix}$$

instead of the matrix E_{02} in the vector of generalized virtual displacements $\delta\mathbf{Z}$ (8.136).

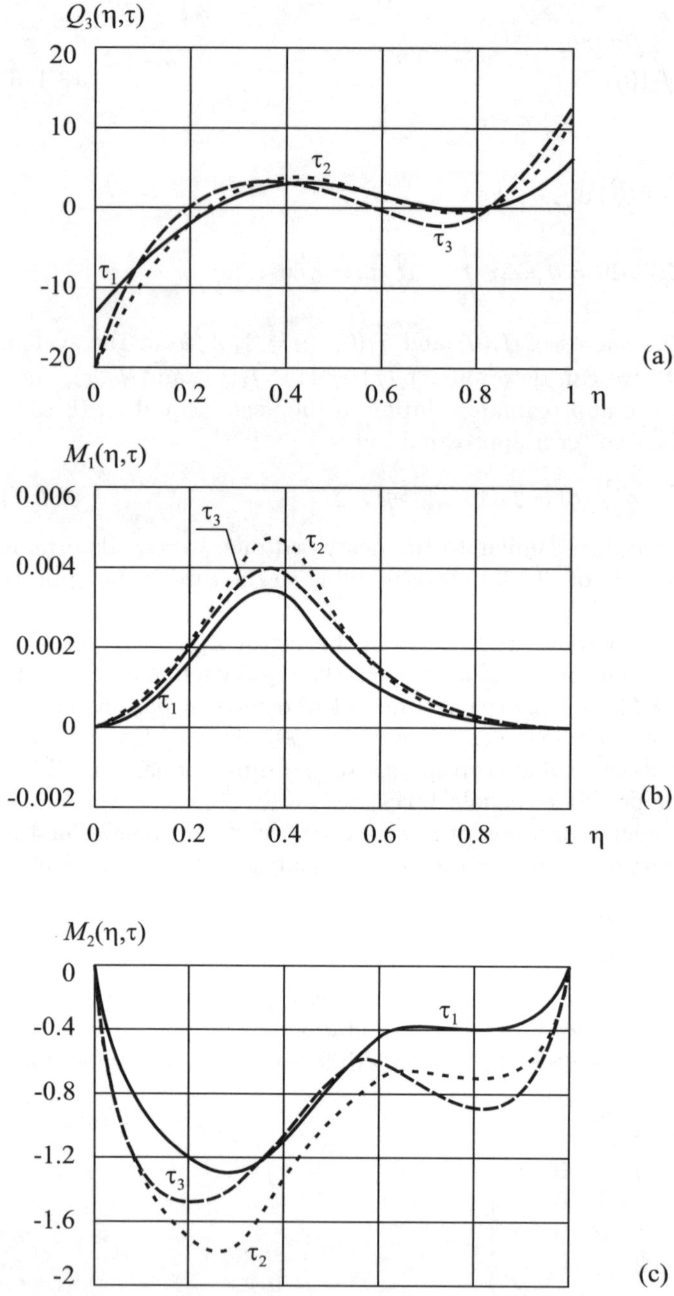

Q3(η,τ)
20
10
0
-10
-20
τ2
τ3
τ1
(a)
0
0.2
0.4
0.6
0.8
1 η
M1(η,τ)
0.006
0.004
0.002
0
-0.002
τ3
τ2
τ1
(b)
0
0.2
0.4
0.6
0.8
1 η
M2(η,τ)
0
-0.4
-0.8
-1.2
-1.6
-2
τ1
τ3
τ2
(c)
0
0.2
0.4
0.6
0.8
1 η

Fig. 8.25.

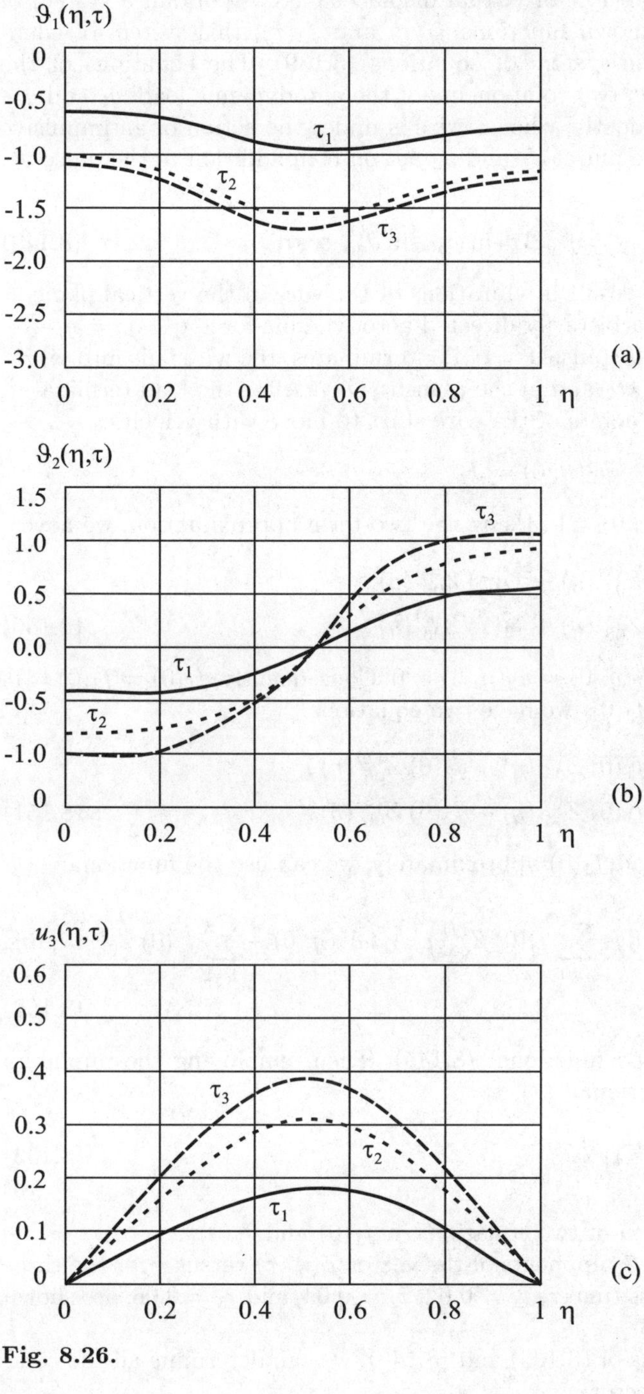

ϑ1(η,τ)
0
-0.5
-1.0
-1.5
-2.0
-2.5
-3.0
τ1
τ2
τ3
(a)
0
0.2
0.4
0.6
0.8
1 η
ϑ2(η,τ)
1.5
1.0
0.5
0.0
-0.5
-1.0
0
τ3
τ1
τ2
(b)
0
0.2
0.4
0.6
0.8
1 η
u3(η,τ)
0.6
0.5
0.4
0.3
0.2
0.1
0
τ3
τ2
τ1
(c)
0
0.2
0.4
0.6
0.8
1 η

Fig. 8.26.

Employing the principle of virtual displacements, we obtain a system of equations in the unknown functions $f_1(\tau)$ and $f_2(\tau)$; this system of equations is similar to the system of equations (8.139). The equations of the system (8.118) include two components of the aerodynamic load, q_1 and q_{n2} (see (8.120)). Consequently, when a wire is under the action of an impulsive aerodynamic load, the pulses J_1 and J_2 act on it in addition to the pulse J_3. We have

$$J_1 = q_1 \Delta\tau\,, \qquad J_2 = q_{n0} \Delta\tau \sin\varphi_{a0} \sin\vartheta_{30}^{(1)} \cos\alpha\,. \tag{8.149}$$

The pulses J_1 and J_2 result in vibrations of the wire in the vertical plane.

Consider now an arbitrarily directed aerodynamic force $\mathbf{q}_a$, $\mathbf{q}_a = q_1\mathbf{e}_1 + q_{n2}\mathbf{e}_2 + q_{n3}\mathbf{e}_3$. After the pulse $\mathbf{J} = \mathbf{q}_a \Delta\tau$ terminates, the wire falls into vibrations in the plane x_1Ox_2 and in the plane x_1Ox_3. After the load terminates, according to (8.142), points of the wire start to move with velocities

$$\dot{u}_1(\eta,\,0) = J_1\,, \qquad \dot{u}_2(\eta,\,0) = J_2\,;$$

at that, we have $\dot{u}_3(\eta,\,0) = 0$. Using the two-term approximation, we have

$$\begin{aligned} u_1(\eta,\,\tau) &= f_1(\tau)\, Z_{05}^{(1)}(\eta) + f_2(\tau)\, Z_{05}^{(2)}(\eta)\,,\\ u_2(\eta,\,\tau) &= f_1(\tau)\, Z_{06}^{(1)}(\eta) + f_2(\tau)\, Z_{06}^{(2)}(\eta)\,. \end{aligned} \tag{8.150}$$

At $\tau = 0$, we get $u_1(\eta,\,0) = u_2(\eta,\,0) = 0$. Consequently, $f_1(0) = f_2(0) = 0$. To obtain $\dot{f}_1(0)$ and $\dot{f}_2(0)$, we have two equations:

$$\begin{aligned} \dot{u}_1(\eta,\,0) = J_1 &= \dot{f}_1(0)\, Z_{05}^{(1)}(\eta) + \dot{f}_2(0)\, Z_{06}^{(2)}(\eta)\,,\\ \dot{u}_2(\eta,\,0) = J_2 &= \dot{f}_1(0)\, Z_{05}^{(1)}(\eta) + \dot{f}_2(0)\, Z_{06}^{(2)}(\eta)\,. \end{aligned} \tag{8.151}$$

To determine $\dot{f}_1(0)$ and $\dot{f}_2(0)$ approximately, we can use the functional

$$J = \int_0^1 \left[\left(\dot{u}_1(\eta,\,0) - \sum_{j=1}^{2} \dot{f}_j(0)\, Z_{05}^{(j)}\right)^2 + \left(\dot{u}_2(\eta,\,0) - \sum_{j=1}^{2} \dot{f}_j(0)\, Z_{06}^{(j)}\right)^2\right] \mathrm{d}\eta\,, \tag{8.152}$$

which is similar to the functional (8.145). Then, employing the minimum conditions of the functional J,

$$\frac{\partial J}{\partial \dot{f}_j} = 0\,, \qquad j = 1,\,2\,. \tag{8.153}$$

we arrive at the system of two equations in $\dot{f}_1(0)$ and $\dot{f}_2(0)$.

The graphs of the components of the vector $\mathbf{Z}(\eta,\,\tau)$ versus η for the three values of dimentionless time τ, $\tau_1 = 0.02$, $\tau_2 = 0.04$, and $\tau_3 = 0.06$, are shown in Figs. 8.27 and 8.28.

Using the solutions of (8.133) and (8.148), we can determine all the component of the state vector

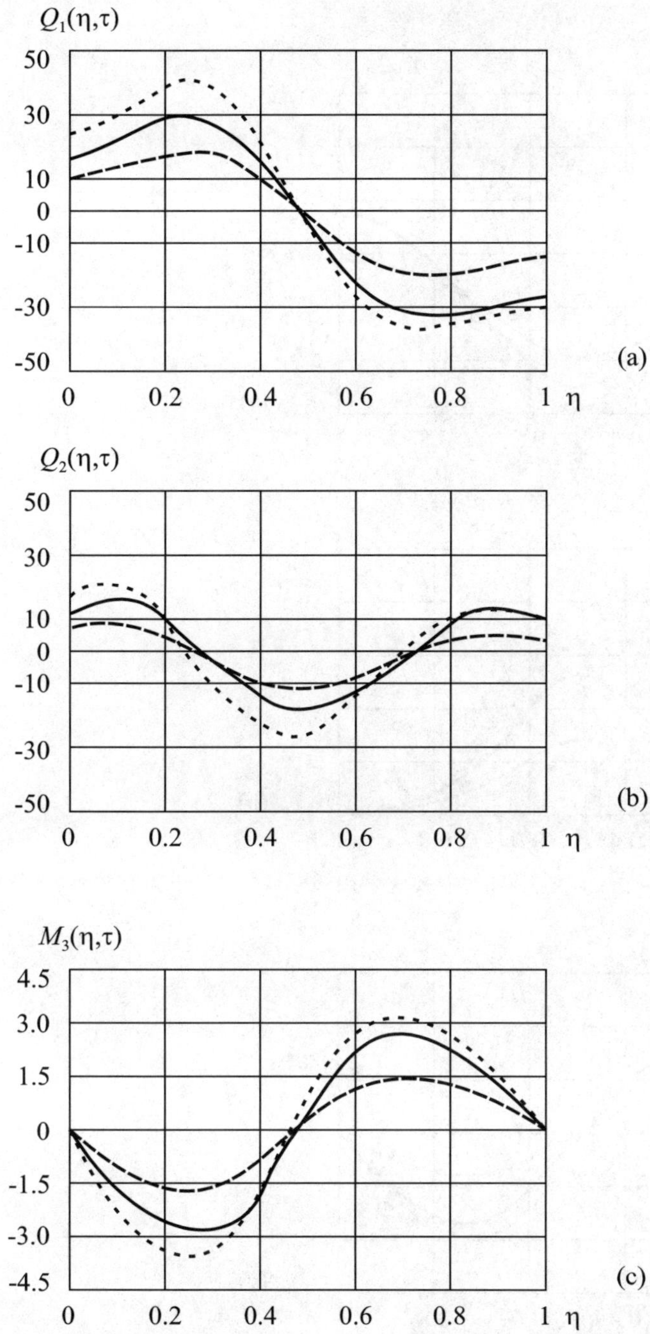

$Q_1(\eta,\tau)$
50
30
10
0
-10
-30
-50
(a)
0
0.2
0.4
0.6
0.8
1 η
$Q_2(\eta,\tau)$
50
30
10
0
-10
-30
-50
(b)
0
0.2
0.4
0.6
0.8
1 η
$M_3(\eta,\tau)$
4.5
3.0
1.5
0
-1.5
-3.0
-4.5
(c)
0
0.2
0.4
0.6
0.8
1 η

Fig. 8.27.

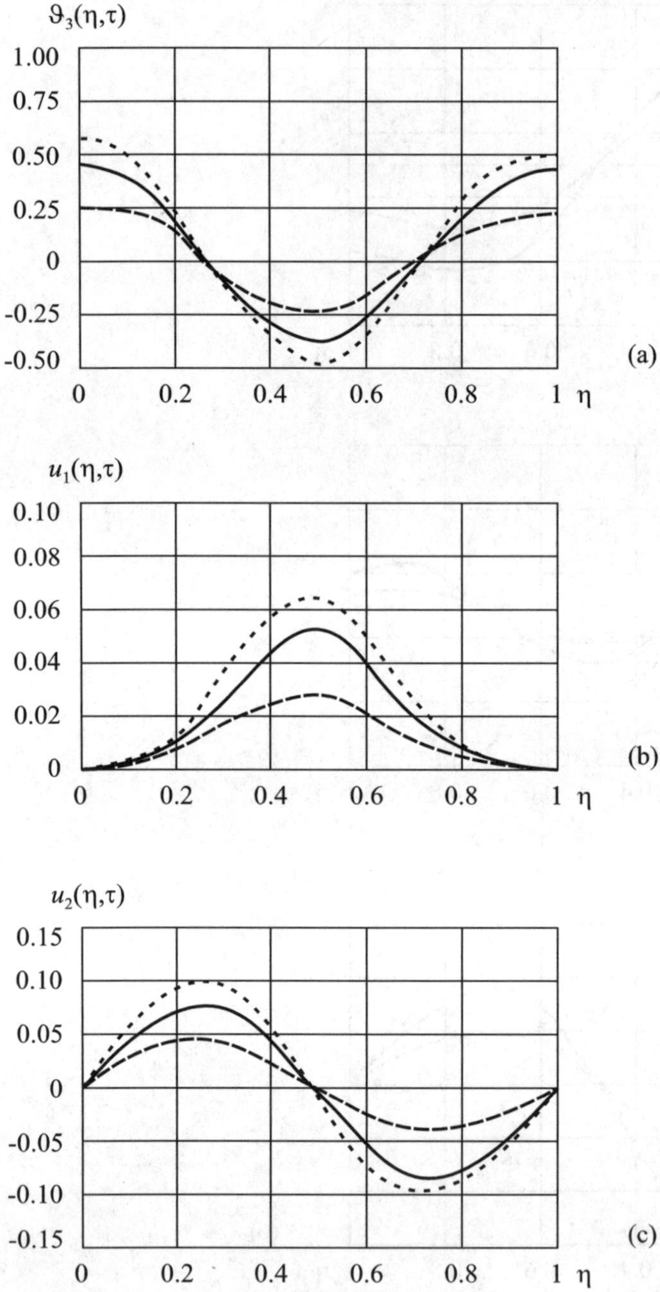

$\vartheta_3(\eta,\tau)$
1.00
0.75
0.50
0.25
0
-0.25
-0.50
(a)
0
0.2
0.4
0.6
0.8
1 η
$u_1(\eta,\tau)$
0.10
0.08
0.06
0.04
0.02
0
(b)
0
0.2
0.4
0.6
0.8
1 η
$u_2(\eta,\tau)$
0.15
0.10
0.05
0
-0.05
-0.10
-0.15
(c)
0
0.2
0.4
0.6
0.8
1 η

Fig. 8.28.

$$\mathbf{Z} = (Q_1,\, Q_2,\, Q_3,\, M_1,\, M_2,\, M_3,\, \vartheta_1,\, \vartheta_2,\, \vartheta_3,\, u_1,\, u_2,\, u_3)^{\mathrm{T}}$$

for nonstationary vibrations of the wire due to the impulsive loading caused by aerodynamic forces. If we are mostly interested in "strength" of a wire (for example, if we need to estimate the limiting conditions), the forces $Q_j(\eta,\, \tau)$ and the moments $M_j(\eta,\, \tau)$ play the main role. Knowing $Q_j(\eta,\, \tau)$ and $M_j(\eta,\, \tau)$, we can determine the equivalent stress, σ_{eq}, and its maximum amplitude values in all cross-sections of the wire as functions of time. For example, when considering a single short-time loading (an impulsive loading or a gust of wind), we are interesting to estimate the possibility of plastic deformations; this may take place when the amplitude values of the equivalent stresses in a "weak" cross sections exceed the yield stress σ_{y} of the wire's material:

$$\max_{0 \le \eta \le 1} \sigma_{...}(\eta,\, \tau) \ge \sigma_{\mathrm{T}}\,. \tag{8.154}$$

The Mohr criterion is the most popular one for determining the limiting state of plastic deformations appearence (see Feodosiev (1999)). In accordance with this criterion, the equivalent stress at a weak point of a cross-section of a rod is determined by the following formula:

$$\sigma_{\mathrm{y}} = \sqrt{\sigma^2 + 4\tau^2}\,; \tag{8.155}$$

here, σ is the maxmum normal stress at the weak point and τ is the tangential stress at this point. Using dimensional parameters, for a wire of circular cross-section we can write

$$\sigma = \frac{\sqrt{M_2^2 + M_3^2}}{W}\,\frac{A_{33}}{e} + \frac{Q_1}{F}\,\frac{A_{33}}{l^2}\,, \qquad \tau = \frac{M_1}{W_\rho}\,\frac{A_{33}}{e}\,, \tag{8.156}$$

where

$$W = \frac{\pi \mathrm{d}^3}{32}\,, \qquad W_\rho = \frac{\pi \mathrm{d}^3}{16}\,.$$

The graphs of the equivalent stress versus η for the three values of dimensionless time, $\tau_1 = 0.02$, $\tau_2 = 0.04$, and $\tau_3 = 0.06$, are shown in Fig. 8.29 (the dimensional time is $t = \tau\, p_0^{-1}$, where $p_0 = 0.063\,\mathrm{s}^{-1}$). From Fig. 8.29 it follows that in the case under consideration the equivalent stress takes it maximum value at $t = 0.635\,\mathrm{s}$ in the cross-section corresponding to the dimensional axial coordinate $s = 4.6\,\mathrm{m}$. The maximum value of equivalent stress is $\max \sigma_{\mathrm{eq}} = 1.346 \cdot 10^8\,\mathrm{Pa}$; this value is less than the yield stress of low-carbon steel ($\sigma_{\mathrm{y}} = 2.2 \ldots 3.0 \cdot 10^8\,\mathrm{Pa}$) and the yield stress of duralumin ($\sigma_{\mathrm{y}} = 2.1 \ldots 3.1 \cdot 10^8\,\mathrm{Pa}$). That is, in the example under consideration, there is no plastic deformation in the wire after the application of an impulsive loading at $v_0 = 50\,\mathrm{m/s}$.

Let us now consider a wire loaded abruptly by a stationary airflow (Fig. 8.13*b*). Then, the vectors $\mathbf{b}_1$ and $\mathbf{b}_2$ in the vector equations (8.122)

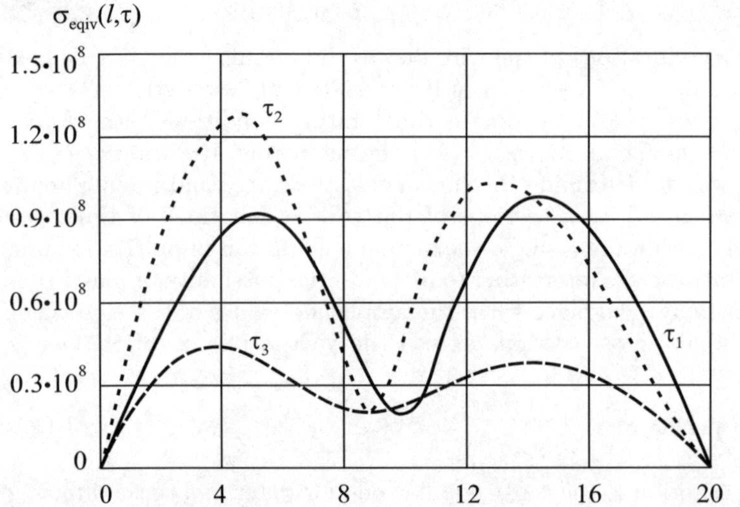

Fig. 8.29.

and (8.123) are not equal to zero. Disregarding the influence of vibration of the wire on the aerodynamic forces q_1, q_{n2}, and q_{n3}, we can assume that these vectors are of constant magnitude and direction.

As a result, we may represent the equations of vibrations relative to the vertical plane (8.123) as follows:

$$\mathbf{L}_2 = \mathrm{B}^{(1)} \frac{\partial^2 \mathbf{Z}}{\partial \tau^2} - \frac{\partial \mathbf{Z}}{\partial \eta} + \mathrm{B}^{(2)} \mathbf{Z} - \mathbf{b}_2 = 0 . \tag{8.157}$$

Restricting our consideration to the two-term approximation (8.134) and employing the principle of virtual displacements, we arrive at the following system of non-homogeneous second order ordinary differential equations in the unknown functions $f_1(\tau)$ and $f_2(\tau)$:

$$\sum_{j=1}^{2} (a_{ij} \ddot{f}_j + c_{ij} f_j) = b_{2i} , \qquad i = 1, 2 , \tag{8.158}$$

where the coefficients a_{ij} and c_{ij} are the same as the coefficients in the system of equations (8.139). The coefficients b_{2i} in the right-hand side of (8.158) can be determined from the following relationships:

$$b_{2i} = \int_0^1 (\mathbf{b}_2 \cdot E_{02} \mathbf{Z}_0^{(i)}) \, \mathrm{d}\eta , \qquad i = 1, 2 . \tag{8.159}$$

On rearrangements, we have

$$b_{21} = \int_0^1 q_{n3}\, u_{30}^{(1)}\, \mathrm{d}\eta = \int_0^1 q_{n0}\, u_{30}^{(1)} \sin\varphi_{a0} \sin\alpha\, \mathrm{d}\eta\,,$$

$$b_{22} = \int_0^1 q_{n3}\, u_{30}^{(2)}\, \mathrm{d}\eta = \int_0^1 q_{n0}\, u_{30}^{(2)} \sin\varphi_{a0} \sin\alpha\, \mathrm{d}\eta\,.$$

The system of equations (8.158) should be solved under zero initial conditions.

Knowing the functions $f_1(\tau)$ and $f_2(\tau)$, we obtain the vector $\mathbf{Z}$ corresponding to the two-term approximation:

$$\tilde{\mathbf{Z}} = f_1(\tau)\, \mathbf{Z}_0^{(1)} + f_2(\tau)\, \mathbf{Z}_0^{(2)}\,. \tag{8.160}$$

The graphs of amplitude values of the components of the vector $\mathbf{Z}$ for $v_0 = 15\,\mathrm{m/s}$ and $\alpha = 45°$ at the three values of time, $\tau_1 = 0.045$, $\tau_2 = 0.07$, and $\tau_3 = 0.085$, are shown in Figs. 8.30 and 8.31.

For the two-term approximation of vibrations of the wire in the vertical plane (equations (8.122)), we get a system of equations is similar to the system (8.158):

$$\sum_{j=1}^{2} (a_{ij}^{(1)}\, \ddot{f}_j + c_{ij}^{(1)}\, f_j) = b_{1i}\,, \qquad i = 1,\, 2\,, \tag{8.161}$$

here, the coefficients b_{1i} should be determined from the following relationships:

$$b_{1i} = \int_0^1 (\mathbf{b}_1 \cdot E_{01} \mathbf{Z}_0^{(i)})\, \mathrm{d}\eta\,, \qquad \mathbf{b}_1 = (q_1,\, q_{n2},\, 0,\, 0,\, 0,\, 0)^{\mathrm{T}}\,.$$

On rearrangement, we get

$$\begin{aligned} b_{1i} &= \int_0^1 (q_1\, u_1^{(i)} + q_{n2}\, u_2^{(i)})\, \mathrm{d}\eta \\ &= \int_0^1 (q_{10} \cos^2\alpha \cos^2\vartheta_{30}^{(1)} + q_{n0}\, u_2^{(i)} \sin\varphi_{a0} \cos\alpha \sin\vartheta_{30}^{(1)})\, \mathrm{d}\eta\,. \end{aligned} \tag{8.162}$$

Solving for the functions $f_1(\tau)$ and $f_2(\tau)$ from (8.161) under the zero initial data, we thereby obtain an approximate solution of the vector equation (8.122).

The graphs of amplitude values of the components of the vector $\mathbf{Z}$ for $v_0 = 15\,\mathrm{m/s}$ and $\alpha = 45°$ at the three values of time, $\tau_1 = 0.045$, $\tau_2 = 0.07$, and $\tau_3 = 0.085$, are shown in Figs. 8.32 and 8.33.

The graphs of amplitude values of equivalent stresses at the three values of time, $\tau_1 = 0.045$, $\tau_2 = 0.07$, and $\tau_3 = 0.085$, are shown in Fig. 8.34. These graphs are similar to the graphs in Fig. 8.29 for impulsive loadings. It follows from these graphs that the maximum value of the equivalent stress in the interval $0 \le \tau \le 0.1$ is attained at the cross-section $s = 5\,\mathrm{m}$ at $\tau \approx 0.07$ (the

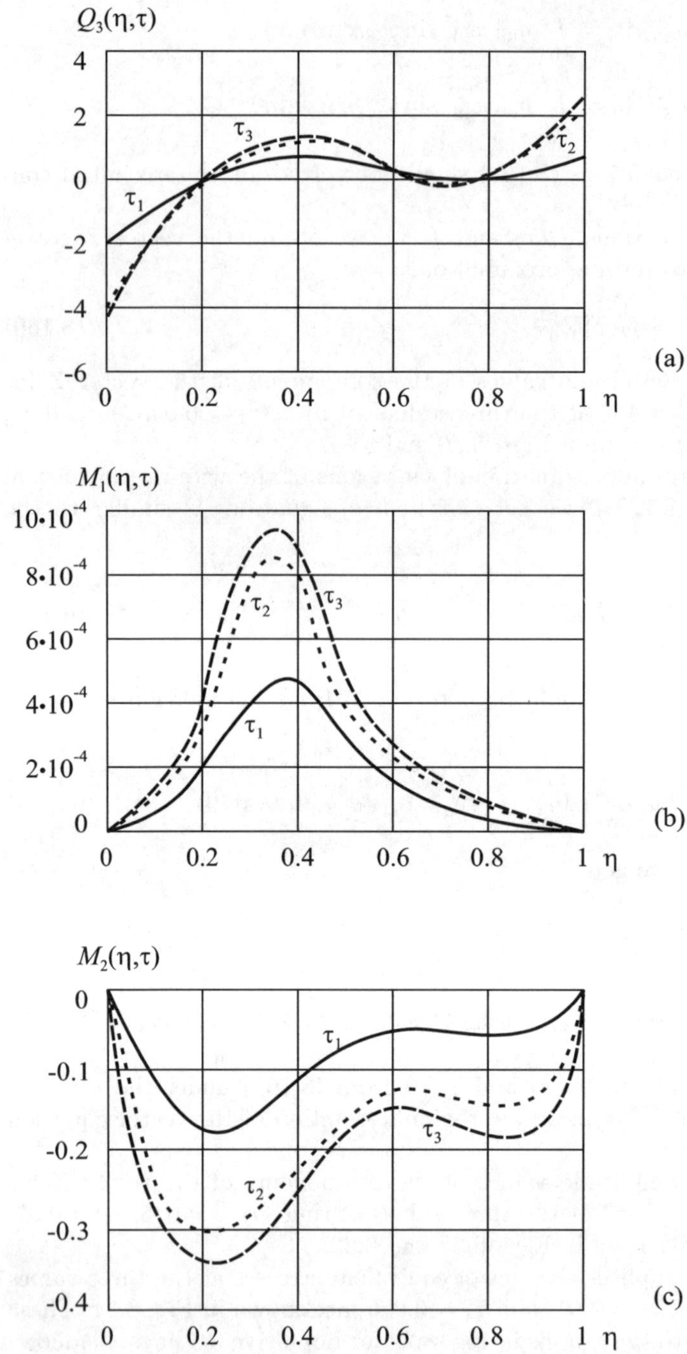
Q3(η,τ)
4
2
0
-2
-4
-6
τ3
τ2
τ1
0
0.2
0.4
0.6
0.8
1 η
(a)
M1(η,τ)
10•10-4
8•10-4
6•10-4
4•10-4
2•10-4
0
τ2
τ3
τ1
0
0.2
0.4
0.6
0.8
1 η
(b)
M2(η,τ)
0
-0.1
-0.2
-0.3
-0.4
τ1
τ3
τ2
0
0.2
0.4
0.6
0.8
1 η
(c)

Fig. 8.30.

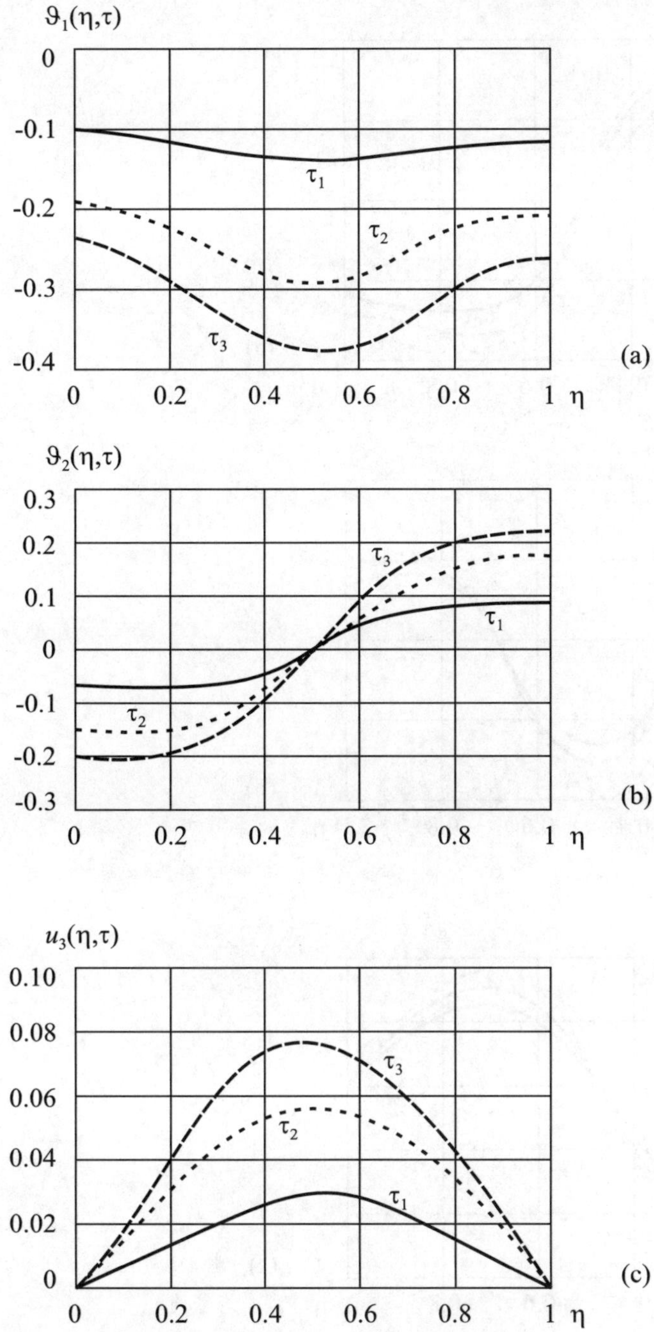

ϑ1(η,τ)
0
-0.1
-0.2
-0.3
-0.4
τ1
τ2
τ3
(a)
0 0.2 0.4 0.6 0.8 1 η
ϑ2(η,τ)
0.3
0.2
0.1
0
-0.1
-0.2
-0.3
τ3
τ1
τ2
(b)
0 0.2 0.4 0.6 0.8 1 η
u3(η,τ)
0.10
0.08
0.06
0.04
0.02
0
τ3
τ2
τ1
(c)
0 0.2 0.4 0.6 0.8 1 η

Fig. 8.31.

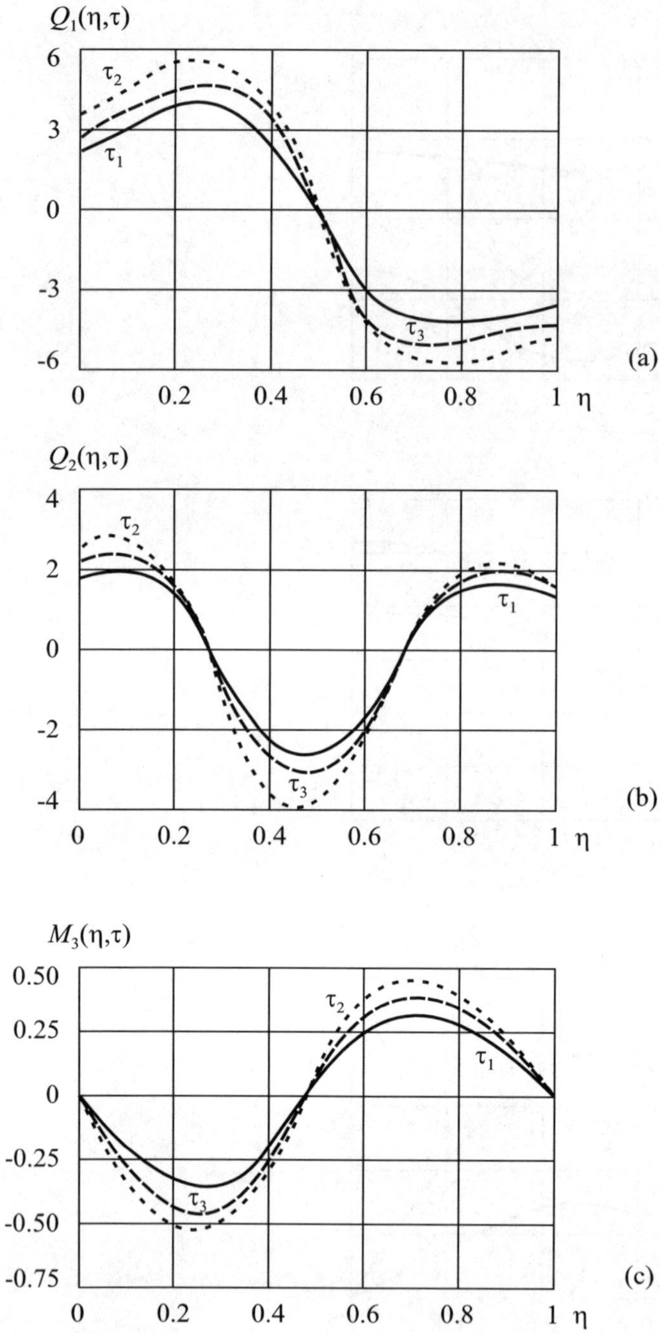
$Q_1(\eta,\tau)$
6
3
0
-3
-6
τ_2
τ_1
τ_3
(a)
0
0.2
0.4
0.6
0.8
1 η
$Q_2(\eta,\tau)$
4
2
0
-2
-4
τ_2
τ_1
τ_3
(b)
0
0.2
0.4
0.6
0.8
1 η
$M_3(\eta,\tau)$
0.50
0.25
0
-0.25
-0.50
-0.75
τ_2
τ_1
τ_3
(c)
0
0.2
0.4
0.6
0.8
1 η

Fig. 8.32.

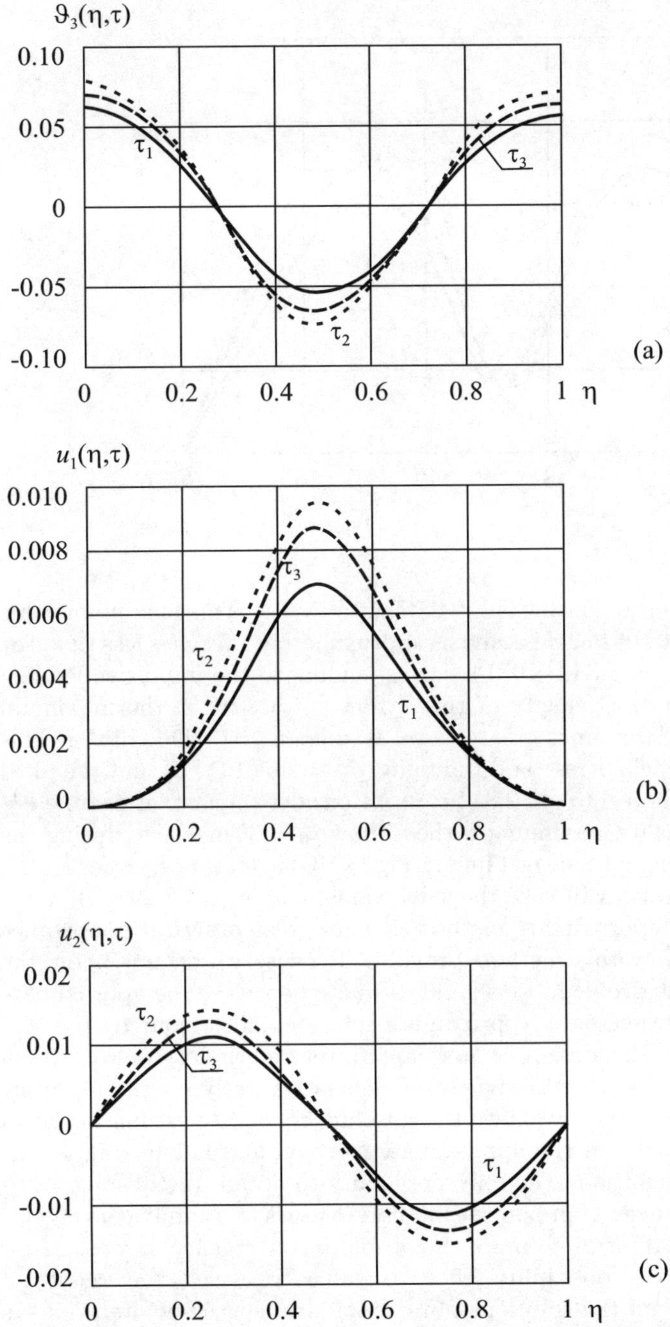
ϑ3(η,τ)
0.10
0.05
0
-0.05
-0.10
τ1
τ3
τ2
(a)
0
0.2
0.4
0.6
0.8
1 η
u1(η,τ)
0.010
0.008
0.006
0.004
0.002
0
τ3
τ2
τ1
(b)
0
0.2
0.4
0.6
0.8
1 η
u2(η,τ)
0.02
0.01
0
-0.01
-0.02
τ2
τ3
τ1
(c)
0
0.2
0.4
0.6
0.8
1 η

Fig. 8.33.

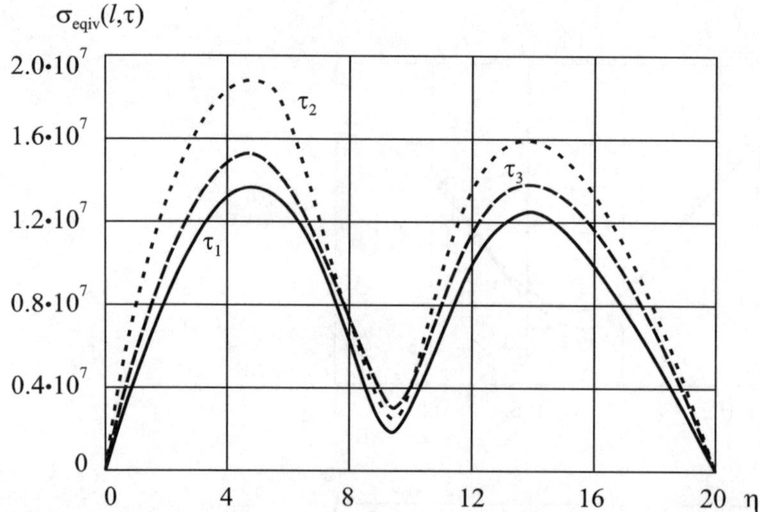

Fig. 8.34.

corresponding dimensional time is $t = 0.07\,p_0^{-1} = 1.1\,\mathrm{s}$). This maximum value σ_{eq} is equal to $0.824{\cdot}10^8$ Pa; this value is approximately 1.5 times less than the maximum value $\sigma_{\mathrm{eq}} = 1.346 \cdot 10^8$ Pa corresponding to the impulsive loading (see Fig. 8.29). For the velocity of the airflow $v_0 = 30\,\mathrm{m/s}$, the maximum equivalent stress is four times greater, that is, max $\sigma_{\mathrm{eq}} = 3.296 \cdot 10^8$ Pa. This stress exceeds the yield stress of duralumin ($\sigma_{\mathrm{y}} = 2.6 \cdot 10^8$ Pa) and the yield stress of low-carbon steel ($\sigma_{\mathrm{y}} = 2.8 \cdot 10^8$ Pa). Consider the instant of time $t^{(1)}$ at which σ_{eq} attains its maximum; if the airflow $v_0 = 30\,\mathrm{m/s}$ acts during the finite interval of time t_{f} (a dashed line in Fig. 8.13*b*) such that t_{f} exceeds $t^{(1)}$, than the parameter σ_{eq} will take the same value, max $\sigma_{\mathrm{eq}} = 3.296 \cdot 10^8$ Pa.

When using an approximate method, it is of great practical importance to know the accuracy this method provides. Because we do not know the exact solution of the problem, we should estimate how close the approximate solution is to the exact one. Approximate solutions are a source of actual dimensions used in the design of new apparatus, machinery, aeronautical vehicles, etc. The reliability of structures depends strictly on the accuracy of calculations. We cannot evaluate the absolute reliability estimation for a structure basing solely on the approximate solution and not having at our disposal the exact solution. The only possibility to obtain the absolute estimate is to proof that the approximate method in use is absolutely convergent. Unfortunately, it is difficult to proof the absolute convergence in most cases. Consequently, the only possibility left is to watch "relative" convergence: if the difference between the solutions obtained at two successive iterations is decreasing continuously, we may say that the approximate solution obtained is sufficiently accurate.

To clarify the concept of relative convergence, let us solve approximately the vector equation (8.157). We will obtain estimate solutions of (8.157) using one-term, two-term, and three-term approximations:

$$\tilde{\mathbf{Z}}(\eta,\,\tau) = f_1(\tau)\,\mathbf{Z}_0^{(1)}(\eta)\,,$$
$$\tilde{\mathbf{Z}}(\eta,\,\tau) = f_1(\tau)\,\mathbf{Z}_0^{(1)}(\eta) + f_2(\tau)\,\mathbf{Z}_0^{(2)}(\eta)\,,$$
$$\tilde{\mathbf{Z}}(\eta,\,\tau) = f_1(\tau)\,\mathbf{Z}_0^{(1)}(\eta) + f_2(\tau)\,\mathbf{Z}_0^{(2)}(\eta) + f_3(\tau)\,\mathbf{Z}_0^{(3)}(\eta)\,.$$

The plots of the components of $\mathbf{Z}(\eta,\,\tau)$ versus η at $\tau_1 = 0.3$ are shown in Figs. 8.35 and 8.36. This figure clearly illustrates that the internal force and moments Q_3, M_1, and M_2 obtained by using the one-term approximation differ significantly from those obtained via the two-term and three-term approximations. The graphs of ϑ_1, ϑ_2, and u_3 corresponding to the one-term approximation differ from those corresponding to the two-term and three-term approximations also; this difference, however, is not so significant as for the internal force components Q_3, M_1, and M_2. The graphs of ϑ_1, ϑ_2, and u_3 obtained by using the two-term and three-term approximations are practically the same. To determine Q_3, M_1, and M_2 more accurately, we can use the four-term approximation, etc.

The method presented here is based on the "generalized" principle of virtual displacements. This method allows us to analyze just about any problem of rod dynamics governing by linear partial differential equations.

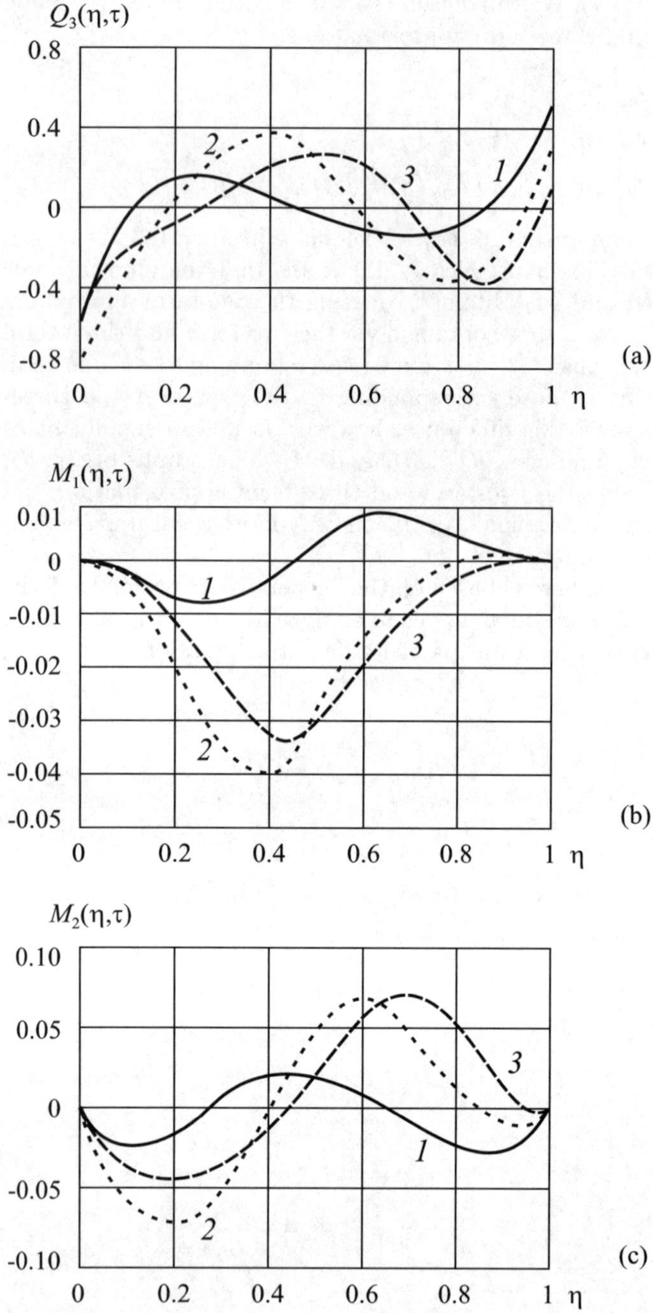

$Q_3(\eta,\tau)$
0.8
0.4
0
-0.4
-0.8
1
2
3
(a)
0
0.2
0.4
0.6
0.8
1 η
$M_1(\eta,\tau)$
0.01
0
-0.01
-0.02
-0.03
-0.04
-0.05
1
2
3
(b)
0
0.2
0.4
0.6
0.8
1 η
$M_2(\eta,\tau)$
0.10
0.05
0
-0.05
-0.10
1
2
3
(c)
0
0.2
0.4
0.6
0.8
1 η

Fig. 8.35.

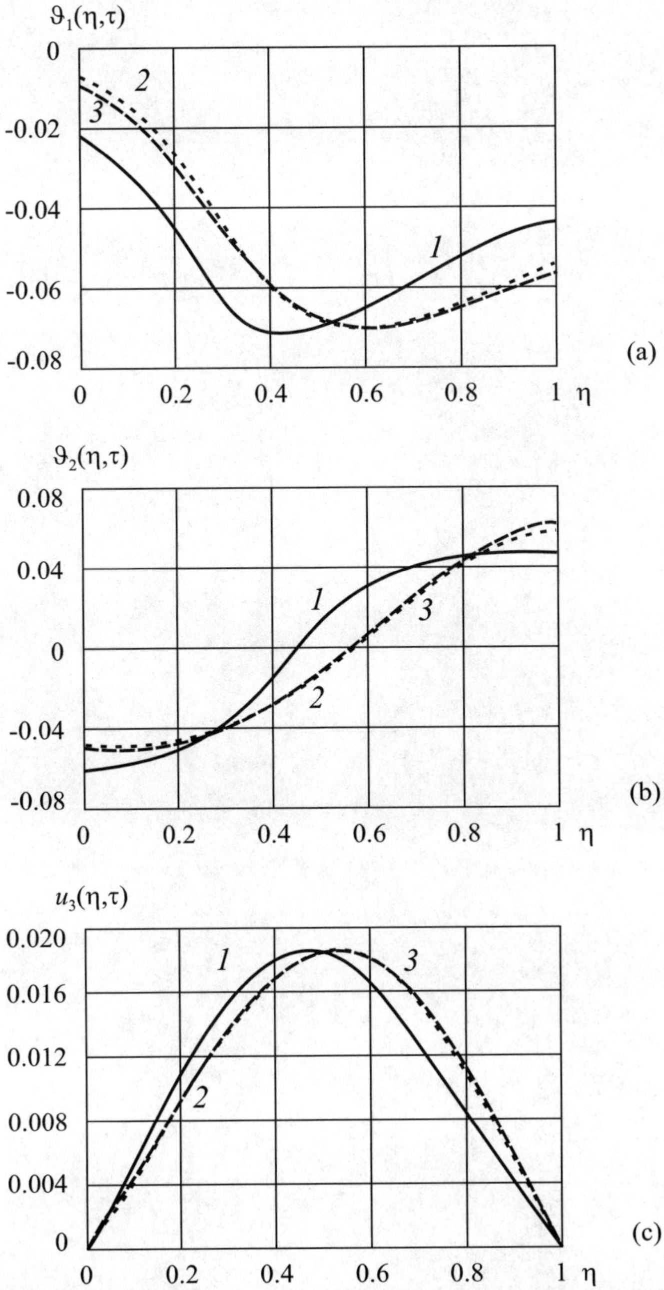
ϑ1(η,τ)
0
-0.02
-0.04
-0.06
-0.08
0
0.2
0.4
0.6
0.8
1 η
1
2
3
(a)
ϑ2(η,τ)
0.08
0.04
0
-0.04
-0.08
0
0.2
0.4
0.6
0.8
1 η
1
2
3
(b)
u3(η,τ)
0.020
0.016
0.012
0.008
0.004
0
0
0.2
0.4
0.6
0.8
1 η
1
2
3
(c)

Fig. 8.36.

9. Rods Exerted by an Internal Flow of Liquid

In Chap. 6 of Part 1 (see Svetlitsky (2000a)) we considered most significant problems of statics of hollow rods containing a flow of an ideal incompressible fluid. Applications of rods with an internal flow of liquid were illustrated by examples drawn from various branches of engineering (Figs. 0.13–0.15 in Part 1 (Svetlitsky (2000a))). Some problems of dynamics of rods are considered in Paidoussis (1998) and Paidoussis (2003).

To obtain equations of small vibrations for a rod, it is necessary to take into consideration the stress-strain state of the rod caused by the internal flow. The fact is that the rod's frequencies depend on its stress-strain state. Consider an example to make this point clear.

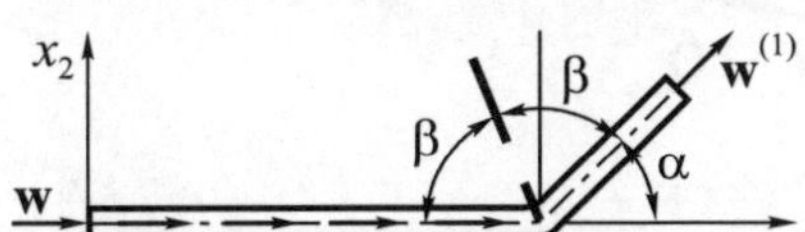

Fig. 9.1.

A straight-line pipeline with a bent segment near its right end is shown in Fig. 9.1. Because of this bend, the direction of the flow changes and, as a result, a concentrated force, **P**, appears. In the figure, this force is shown as a dashed line. The force **P** can be found using a balance of linear momentum (the change in momentum of the liquid is caused by an abrupt change of the vector **w**). In this example, the pressure of liquid is neglected. The change in momentum of the liquid in a unit of time equals the impulse of the force **P**, that is,

$$m_2\mathbf{w}^{(1)}|\mathbf{w}|\,\mathrm{d}t - m_2\mathbf{w}|\mathbf{w}|\,\mathrm{d}t = \mathbf{P}\,\mathrm{d}t$$

or

$$\mathbf{P} = m_2\mathbf{w}^{(1)}|\mathbf{w}| - m_2\mathbf{w}|\mathbf{w}|\,, \qquad |\mathbf{w}^{(1)}| = |\mathbf{w}|\,.$$

The direction of the vector **P** is indicated in Fig. 9.1. The magnitude of **P** is

$$|\mathbf{P}| = P = 2m_2w^2 \sin\frac{\alpha}{2}\,.$$

The horizontal component of the force **P**, which is the axial force in the rod, reads

$$Q_{10} = |\mathbf{P}| \cos\beta = m_2 w^2 (1 - \cos\alpha)\,.$$

Thus, we see that the rod is subject to the tensile force Q_{10}, which depends on the flow's velocity **w**. At the same time, the tensile force affects the frequencies of the rod. This example will be considered in Sect. 9.2 in greater detail.

9.1 Nonlinear Equations of Motion for a Space-Curved Pipeline

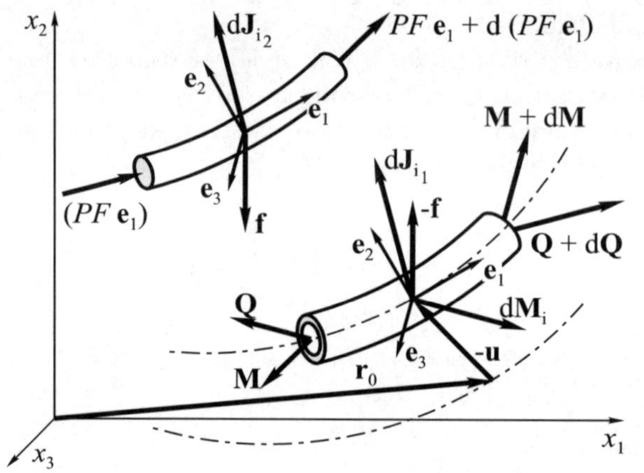

Fig. 9.2.

Consider separately an element of a pipeline of constant cross section and the volume of liquid it contains (Fig. 9.2). The following equations of motion can be obtained:

$$m_1 \frac{\partial \mathbf{v}}{\partial t} = \frac{\partial \mathbf{Q}}{\partial s} + \mathbf{P} + \mathbf{f} + m_1 \mathbf{g}\,, \tag{9.1}$$

$$m_2 \left(\frac{\partial \mathbf{v}}{\partial t} + \frac{\partial \mathbf{v}}{\partial s} w + \frac{\partial \mathbf{w}}{\partial t} + \frac{\partial \mathbf{w}}{\partial s} w \right) = -\frac{\partial (pF_2 \mathbf{e}_1)}{\partial s} - \mathbf{f}\,; \tag{9.2}$$

here, m_1 is the mass per unit length of the rod and m_2 is the mass of the liquid within a unit length of the rod; **v** is the vector of the absolute velocity of the center of mass of the element of the rod; $\mathbf{w} = w\,\mathbf{e}_1$ is the vector of the relative velocity of the liquid; **Q** is the vector of internal force in the rod; **P** denotes the vector of the external forces; p is the pressure of the liquid; and the vector **f** stands for the force of interaction of the rod with the liquid.

In (9.2), the rod's acceleration is written in terms of the Eulerian variables (see Sect. 1.1). The speed $w(t)$ is independent of the motion of the rod (this is true for incompressible liquids only). At the same time, the pressure of the liquid is not known because inertial forces that occur when the rod is in motion affect it.

In the case when the determination of $\mathbf{f}$ is not our goal, we can eliminate it from (9.1)–(9.2) and thereby obtain

$$(m_1+m_2)\,\frac{\partial \mathbf{v}}{\partial t}+m_2\left(\frac{\partial \mathbf{v}}{\partial s}\,w+\frac{\partial \mathbf{w}}{\partial t}\right)=\frac{\partial \mathbf{Q}^{(1)}}{\partial s}+\mathbf{P}+(m_1+m_2)\,\mathbf{g}\,, \tag{9.3}$$

$$\mathbf{Q}^{(1)}=\mathbf{Q}-(pF_2+m_2w^2)\,\mathbf{e}_1\,, \tag{9.4}$$

where F_2 is the channel's cross section area. In the general case where the rod's motion about an equilibrium configuration is considered, the vector $\mathbf{Q}$ can be written as a sum $\mathbf{Q}^{(1)}=\mathbf{Q}_0+\mathbf{Q}_{\mathrm{d}}$, where $\mathbf{Q}_0$ coincides with $\mathbf{Q}^{(1)}$ in the static configuration and $\mathbf{Q}_{\mathrm{d}}$ is the dynamical component of $\mathbf{Q}^{(1)}$.

To represent equation (9.3) in nondimensional form, we put

$$\tilde{w}=\frac{w}{l\,p_0}\,,\qquad \tilde{p}=\frac{p\,A_{33}}{F_2l^2}\,,\qquad p_0=\sqrt{\frac{A_{33}}{(m_1+m_2)\,l^4}}\,,$$

where w and p are the (dimensional) flow speed and pressure. From here on, when it will make no misunderstanding, we shall omit tildes in dimensionless values without any supplementing references. On rearrangement, we get

$$\frac{\partial \mathbf{v}}{\partial \tau}+n_{11}\left(\frac{\partial \mathbf{v}}{\partial \eta}\,w+\frac{\partial \mathbf{w}}{\partial \tau}\right)-\frac{\partial \mathbf{Q}^{(1)}}{\partial \eta}-\mathbf{P}-\boldsymbol{\gamma}=0\,, \tag{9.5}$$

where

$$\mathbf{Q}^{(1)}=\mathbf{Q}-(P^{(1)}+n_{11}w^2)\,\mathbf{e}_1\,,\qquad P^{(1)}=\frac{pF_2l^2}{A_{33}}\,,$$

$$n_{11}=\frac{m_2}{m_1+m_2}\,,\qquad \boldsymbol{\gamma}=\boldsymbol{\gamma}_1+\boldsymbol{\gamma}_2=\frac{(m_1+m_2)\,\mathbf{g}l^3}{A_{33}}\,.$$

By using in (9.5) the following formulas of local derivatives

$$\frac{\partial \mathbf{v}}{\partial \eta}=\frac{\tilde{\partial} \mathbf{v}}{\partial \eta}+\text{æ}\times\mathbf{v}\,,$$

$$\frac{\partial \mathbf{v}}{\partial \tau}=\frac{\tilde{\partial} \mathbf{v}}{\partial \tau}+\boldsymbol{\omega}\times\mathbf{v}\,,$$

$$\frac{\partial \mathbf{w}}{\partial \tau}=\frac{\tilde{\partial} \mathbf{w}}{\partial \tau}+\boldsymbol{\omega}\times\mathbf{w}\,,\qquad \mathbf{w}=w\,\mathbf{e}_1\,,$$

$$\frac{\partial \mathbf{Q}}{\partial \eta}=\frac{\tilde{\partial} \mathbf{Q}}{\partial \eta}+\text{æ}\times\mathbf{Q}\,,$$

we obtain the equation of translational motion for the rod element with respect to the coordinate frame of reference attached to the rod axis (tildes over the local derivative terms are omitted)

$$\frac{\partial \mathbf{v}}{\partial \tau} + \boldsymbol{\omega} \times \mathbf{v} + 2n_{11} w (\boldsymbol{\omega} \times \mathbf{e}_1) + n_{11} \frac{\partial w}{\partial \tau} \mathbf{e}_1$$
$$- \frac{\partial \mathbf{Q}^{(1)}}{\partial \eta} - æ \times \mathbf{Q}^{(1)} - \mathbf{P} - \boldsymbol{\gamma} = 0 . \tag{9.6}$$

Equation (9.6) is identical to equation (2.43), which was derived in Chap. 2 when a rod in a longitudal motion was considered. If we put $n_{11} = 1$ in (9.6) (this is the case for $m_1 = 0$), then equation (9.6) is exactly the same as (2.43). The other equations coincide with equations (2.21), (2.23)–(2.25) at $J = 0$:

$$\frac{\partial \mathbf{M}}{\partial \eta} + æ \times \mathbf{M} + \mathbf{e}_1 \times \mathbf{Q}^{(1)} + \mathbf{T} = 0 , \tag{9.7}$$

$$\mathrm{L}_1 \frac{\partial \boldsymbol{\vartheta}}{\partial \eta} + \mathrm{L} æ_0^{(1)} - æ = 0 , \tag{9.8}$$

$$\frac{\partial \mathbf{v}}{\partial \eta} + æ \times \mathbf{v} - \boldsymbol{\omega} \times \mathbf{e}_1 = 0 , \tag{9.9}$$

$$\mathrm{L}_1 \frac{\partial \boldsymbol{\vartheta}}{\partial \tau} - \boldsymbol{\omega} = 0 . \tag{9.10}$$

Equation (2.22) with the static stress-strain state caused by the internal flow taken into consideration becomes

$$\mathbf{M}(\eta, \tau) = \mathrm{A} \left[æ(\eta, \tau) - æ_0^{(1)}(\eta) \right] + \mathbf{M}_0(\eta) . \tag{9.11}$$

Here, $\mathbf{M}_0(\eta)$ is the static moment; the vector $æ_0^{(1)}(\eta)$ describes the geometry of the rod's axis in the static configuration, and the vector $æ(\eta, \tau)$ describes the geometry of the rod's axis when the rod is in motion. To account for the rotary inertia of the rod element, the term

$$-J \frac{\partial \boldsymbol{\omega}}{\partial \tau} - \boldsymbol{\omega} \times J \boldsymbol{\omega}$$

should be added to the left-hand side of equation (9.7).

9.1.1 Equations in the Attached Axes

It follows from (9.6) that

$$\frac{\partial v_1}{\partial \tau} + \omega_2 v_3 - \omega_3 v_2 + n_{11} \frac{\partial w}{\partial \tau} - \frac{\partial Q_1^{(1)}}{\partial \eta} - æ_2 Q_3 + æ_3 Q_2 - P_1 + \gamma_1 = 0 ,$$
$$\frac{\partial v_2}{\partial \tau} + \omega_3 v_1 - \omega_1 v_3 + 2n_{11} w \omega_3 - \frac{\partial Q_2}{\partial \eta} - æ_3 Q_1^{(1)} - æ_1 Q_3 - P_2 + \gamma_2 = 0 ,$$
$$\frac{\partial v_3}{\partial \tau} + \omega_1 v_2 - \omega_2 v_1 - 2n_{11} w \omega_2 - \frac{\partial Q_3}{\partial \eta} - æ_1 Q_2 + æ_2 Q_1^{(1)} - P_3 + \gamma_3 = 0 ,$$

where γ_j are the projections of the total weight and $Q_1^{(1)} = Q_1 - (P + n_{11}w^2)$.

The system of six equations (9.6)–(9.11) includes six unknowns $\mathbf{Q}$, $\mathbf{M}$, $\mathbf{v}$, $\boldsymbol{\vartheta}$, æ, and $\boldsymbol{\omega}$. By solving the system of equations (9.6)–(9.11), we obtain the unknown vectors and their components, including the component $Q_1^{(1)}$ of the vector $\mathbf{Q}^{(1)}$, which includes the axial force in the rod Q_1. To find this axial force Q_1, one should first calculate the pressure of the liquid p because the force P is controlled by p:

$$Q_1 = Q_1^{(1)} + P + n_{11}w^2\,, \qquad P = \frac{pl^2F_2}{A_{33}}\,. \tag{9.12}$$

Consider the projection of the nondimensional equation (9.2) onto the tangent to the rod's axis. The vector $\mathbf{f}$ for a channel of constant cross section filled with an ideal fluid is orthogonal to the vector $\mathbf{e}_1$, consequently, we have

$$n_{11}\left(\frac{\partial v_1}{\partial\tau} + \omega_2 v_3 - \omega_3 v_2\right) + n_{11}\frac{\partial w}{\partial\tau} + \frac{\partial P}{\partial\tau} - \gamma_{21}(\eta) = 0\,, \tag{9.13}$$

where γ_{21} is the projection of the nondimensional weight of the liquid; the quantities v_j and ω_i can be found from the system of equations (9.6)–(9.11).

Then, from (9.13) it follows that

$$P = \int_0^\eta \left\{-n_{11}\left[\frac{\partial v_1}{\partial\tau} + (v_3\omega_2 - v_2\omega_3)\right] - n_{11}\frac{\partial w}{\partial\tau} + \gamma_{21}\right\} \mathrm{d}\eta^{(1)} + P_{10}\,. \tag{9.14}$$

The nondimensional force P_{10} can be found using the boundary conditions for the liquid flow. If the pressure at the exit of the pipeline is maintained at a constant value P_f, then, using (9.14) and assuming that the speed w is known, the pressure at the entrance can be found as follows:

$$P_{10} = P_\mathrm{f} - \int_0^1 \left\{\gamma - n_{11}\left[\frac{\partial v_1}{\partial\tau} + (v_3\omega_2 - v_2\omega_3)\right] \mathrm{d}\eta \right.$$
$$\left. + \gamma_{21} - n_{11}\frac{\partial w}{\partial\tau}\right\} \mathrm{d}\eta^{(1)}\,. \tag{9.15}$$

The expression for the nondimensional force now looks like

$$P(\eta,\,\tau) = P_{10} - n_{11}\frac{\partial w}{\partial\tau}\eta$$
$$- \int_0^\eta \gamma_2(\eta^{(1)})\,\mathrm{d}\eta^{(1)} - n_{11}\int_0^\eta \left[\frac{\partial v_1}{\partial\tau} + \omega_2 v_3 - \omega_3 v_2\right] \mathrm{d}\eta^{(1)}\,. \tag{9.16}$$

After calculating $P(\eta,\,\tau)$, the axial force Q_1 can now be found from the relationship (9.12).

9.2 Small Vibrations of a Space-Curved Pipeline

9.2.1 Derivation of the Governing Equations

Assuming that the flow is not stationary, let us derive the equations of small vibrations about an equilibrium configuration. Let

$$\mathbf{Q}^{(1)} = \mathbf{Q}_0^{(1)} + \Delta\mathbf{Q}^{(1)}, \qquad \mathbf{M} = \mathbf{M}_0 + \Delta\mathbf{M},$$
$$\text{æ} = \text{æ}_0^{(1)} + \Delta\text{æ}, \qquad \mathbf{v} = \frac{\partial \mathbf{u}}{\partial \tau}.$$

The components of the vectors $\boldsymbol{\omega}$ and $\boldsymbol{\vartheta}$ are assumed to be small. Then, using (9.6)–(9.11), the following equations of small vibrations for the rod exerted by a flow can be obtained:

$$\begin{aligned}
&\frac{\partial^2 \mathbf{u}}{\partial \tau^2} + 2n_{11}w(\boldsymbol{\omega} + \mathbf{e}_1) + n_{11}\frac{\partial w}{\partial \tau}\,\mathbf{e}_1 \\
&\qquad -\frac{\partial \Delta\mathbf{Q}^{(1)}}{\partial \eta} - \Delta\text{æ} \times \mathbf{Q}_0^{(1)} - \text{æ}_0 \times \Delta\mathbf{Q}^{(1)} - \Delta\mathbf{P} = 0, \\
&\frac{\partial \Delta\mathbf{M}}{\partial \eta} + \text{æ}_0 \times \Delta\mathbf{M} + \Delta\text{æ} \times \mathbf{M}_0 + \mathbf{e}_1 \times \Delta\mathbf{Q}^{(1)} - \Delta\mathbf{T} = 0, \\
&\frac{\partial \boldsymbol{\vartheta}}{\partial \eta} + \text{æ}_0 \times \boldsymbol{\vartheta} - \Delta\text{æ} = 0, \\
&\frac{\partial \mathbf{u}}{\partial \eta} + \text{æ}_0 \times \mathbf{u} - \vartheta_2\mathbf{e}_2 + \vartheta_2\mathbf{e}_3 = 0, \\
&\frac{\partial \boldsymbol{\vartheta}}{\partial \tau} = \boldsymbol{\omega}, \qquad \Delta\mathbf{M} = \mathrm{A}\,\Delta\text{æ};
\end{aligned} \tag{9.17}$$

here,

$$\Delta\mathbf{Q}^{(1)} = [\Delta Q_1 - \Delta(P + n_{11}w^2)]\,\mathbf{e}_1 + \Delta Q_2\mathbf{e}_2 + \Delta Q_3\mathbf{e}_3,$$
$$\Delta\mathbf{P} = \Delta\mathbf{q} + \sum_{i=1}^{\eta} \Delta\mathbf{P}^{(i)}\delta(\eta - \eta_i),$$
$$\Delta\mathbf{T} = \Delta\boldsymbol{\mu} + \sum_{\nu=1}^{\rho} \Delta\mathbf{T}^{(\nu)}\delta(\eta - \eta_\nu).$$

For an incompressible fluid, we have Δw^2, hence $\Delta Q_1^{(1)} = \Delta Q_1 - \Delta P$. Eliminating Δæ from the equations and rearranging the vector products just as we did in Sect. 3.1 (the relations (3.9)), we get the following equations (the rotary inertia of the rod is neglected):

$$\frac{\partial^2 \mathbf{u}}{\partial \tau^2} + \mathrm{A}_w\frac{\partial \boldsymbol{\vartheta}}{\partial \tau} + n_{11}\frac{\partial w}{\partial \tau}\,\mathbf{e}_1 - \frac{\partial \Delta\mathbf{Q}^{(1)}}{\partial \eta} - \mathrm{A}_Q\mathrm{A}^{-1}\Delta\mathbf{M} - \mathrm{A}_{\text{æ}}\Delta\mathbf{Q}^{(1)} - \Delta\mathbf{P} = 0, \tag{9.18}$$

$$\frac{\partial \Delta \mathbf{M}}{\partial \eta} + \mathrm{A}_{æ} \Delta \mathbf{M} + \mathrm{A}_M \mathrm{A}^{-1} \Delta \mathbf{M} + \mathrm{A}_1 \Delta \mathbf{Q}^{(1)} - \Delta \mathbf{T} = 0\,, \tag{9.19}$$

$$\frac{\partial \boldsymbol{\vartheta}}{\partial \eta} + \mathrm{A}_{æ} \boldsymbol{\vartheta} - \mathrm{A}^{-1} \Delta \mathbf{M} = 0\,, \tag{9.20}$$

$$\frac{\partial \mathbf{u}}{\partial \eta} + \mathrm{A}_{æ} \mathbf{u} + \mathrm{A}_1 \boldsymbol{\vartheta} = 0\,, \tag{9.21}$$

$$\mathrm{A}_w = \begin{bmatrix} 0 & 0 & 0 \\ 0 & 0 & 2n_{11}w \\ 0 & -2n_{11}w & 0 \end{bmatrix}.$$

In the case of stationary flow, we have $w = w_0 =$ const. Consider now equations (9.18) for this case. With respect to the coordinate frame attached to the rod axis, the equations become

$$\frac{\partial^2 u_1}{\partial \tau^2} - \frac{\partial \Delta Q_1^{(1)}}{\partial \eta} - \frac{Q_{30}}{A_{22}} \Delta M_2 + Q_{20} \Delta M_3 + æ_{30} \Delta Q_3 - æ_{20} \Delta Q_3 - \Delta P_1 = 0\,, \tag{9.22}$$

$$\frac{\partial^2 u_2}{\partial \tau^2} + 2n_{11}w_0 \frac{\partial \vartheta_3}{\partial \tau} - \frac{\partial \Delta Q_2}{\partial \eta} + \frac{Q_{30}}{A_{11}} \Delta M_1 - Q_{10}^{(1)} \Delta M_3 + æ_{10} \Delta Q_3 - æ_{30} \Delta Q_1^{(1)} - \Delta P_2 = 0\,, \tag{9.23}$$

$$\frac{\partial^2 u_3}{\partial \tau^2} - 2n_{11}w_0 \frac{\partial \vartheta_2}{\partial \tau} - \frac{\partial \Delta Q_3}{\partial \eta} - \frac{Q_{20}}{A_{11}} \Delta M_1 + \frac{Q_{10}^{(1)}}{A_{22}} \Delta M_2 + æ_{20} \Delta Q_1^{(1)} - æ_{10} \Delta Q_2 - \Delta P_3 = 0\,, \tag{9.24}$$

where

$$\Delta Q_1^{(1)} = \Delta Q - \Delta P\,, \qquad Q_{10}^{(1)} = Q_{10} - (P_0 + n_{11} w_0^2)\,.$$

Using (9.18)–(9.21), we can obtain the equations of small vibrations for a straight rod of constant cross section ($A_{33} = 1$), which are identical to equations (7.17)–(7.20) and (7.21)–(7.24), obtained in Sect. 7.1. For example, for a rod subject only to an axial force (Fig. 9.1), we arrive at two independent systems of equations.

(1) Equations of small vibrations in the plane x_1Ox_2 :

$$\frac{\partial^2 u_2}{\partial \tau^2} + 2n_{11}w_0 \frac{\partial \vartheta_3}{\partial \tau} - Q_{10}^{(1)} \Delta M_3 - \frac{\partial \Delta Q_2}{\partial \eta} = 0\,,$$
$$\frac{\partial \Delta M_3}{\partial \eta} + \Delta Q_2 = 0\,, \qquad \frac{\partial \vartheta_3}{\partial \eta} - \Delta M_3 = 0\,, \qquad \frac{\partial u_2}{\partial \eta} - \vartheta_3 = 0\,, \tag{9.25}$$

(2) Equations of small vibrations relative to the plane x_1Ox_2

$$\frac{\partial^2 u_3}{\partial \tau^2} - 2n_{11}w_0\frac{\partial \vartheta_2}{\partial \tau} + \frac{Q_{10}^{(1)}}{A_{22}}\Delta M_2 - \frac{\partial \Delta Q_3}{\partial \eta} = 0\,,$$

$$\frac{\partial \Delta M_2}{\partial \eta} - \Delta Q_3 = 0\,, \qquad \frac{\partial \vartheta_2}{\partial \eta} - \frac{\Delta M_2}{A_{22}} = 0\,, \qquad \frac{\partial u_3}{\partial \eta} + \vartheta_2 = 0\,. \tag{9.26}$$

As an illustration, let us determine the first two frequencies for the oscillating rod shown in Fig. 9.1. Setting $P_0 = 0$ in (9.25), we obtain

$$\frac{\partial^4 u_2}{\partial \eta^4} + 2n_{11}w_0\frac{\partial^2 u_2}{\partial \eta \partial \tau} - Q_{10}^{(1)}\frac{\partial^2 u_2}{\partial \eta^2} + \frac{\partial^2 u_2}{\partial \tau^2} = 0\,, \tag{9.27}$$

where

$$Q_{10}^{(1)} = Q_{10} - n_{11}w_0^2\,.$$

At the beginning of the chapter, an expression of the axial force Q_{10} as a function of the flow rate was established. The nondimensional form of this force is $Q_{10} = n_{11}w_0^2(1 - \cos\alpha)$, therefore,

$$Q_{10}^{(1)} = -m_{11}w_0^2\cos\alpha\,.$$

We seek a solution to (9.27) in the form

$$u_2 = u_{20}(\eta)\,\mathrm{e}^{\mathrm{i}\lambda\tau}\,. \tag{9.28}$$

Substituting (9.28) into (9.27) yields

$$L = u_{20}'''' + n_{11}w_0^2 u_{20}''\cos\alpha + 2n_{11}w_0\lambda\,\mathrm{i}\,u_{20}' - \lambda^2 u_{20} = 0\,.$$

Here, primes denote differentiation with respect to η.

An approximate solution is assumed of the form

$$u_{20} = A_1\sin\pi\eta + A_2\sin 2\pi\eta\,.$$

Using the Galerkin method, we obtain two equations

$$\int_0^1 L\sin\pi\eta\,\mathrm{d}\eta = 0\,, \qquad \int_0^1 L\sin 2\pi\eta\,\mathrm{d}\eta = 0\,,$$

or

$$(\pi^4 - n_{11}w_0^2\pi^2\cos\alpha - \lambda^2)\,A_1 - \mathrm{i}\,\frac{8}{3}\,n_{11}w_0\lambda A_2 = 0\,,$$

$$\mathrm{i}\,\frac{8}{3}\,n_{11}w_0\lambda A_1 + (16\pi^4 - n_{11}w_0^2 4\pi^2\cos\alpha - \lambda^2)\,A_2 = 0\,. \tag{9.29}$$

Equating the determinant of the system of equations (9.29) to zero, we obtain the equation

$$\lambda^4 - \alpha_1\lambda^2 + a_2 = 0\,, \tag{9.30}$$

from which the rod frequencies λ_1 and λ_2 can be found. The coefficients a_1 and a_2 are given by

$$a_1 = 17\pi^4 - 5\pi^2 n_{11} w_0^2 \cos\alpha - \frac{64}{9} n_{11}^2 w_0^2 \,,$$
$$a_2 = (\pi^4 - n_{11} w_0^2 \pi^2 \cos\alpha)(16\pi^4 - n_{11} w_0^2 4\pi^2 \cos\alpha) \,.$$

From (9.30) it follows that

$$\lambda_{1,2}^2 = \frac{a_1 \mp \sqrt{a_1^2 - 4a_2}}{2} \,.$$

For a fixed value of the angle α, the critical magnitude of the flow velocity w_{0*} can be found using the expression of λ_1^2. (When the flow speed takes its critical value, the loss of static stability of the rod may happen.) The critical value of w_{0*} can be found from the condition $\lambda_1^2 = 0$. Thus, we have

$$a_2 = (\pi^4 - n_{11} w_{0*}^2 \pi^2 \cos\alpha)(16\pi^4 - n_{11} w_{0*}^2 4\pi^2 \cos\alpha) = 0 \,.$$

The nondimensional form of the least critical value is

$$w_{0*} = \frac{\pi}{\sqrt{n_{11}\cos\alpha}} \,.$$

Methods for determination of the eigenvalues in boundary value problems for the general equations of vibrations for a rod containing a stationary flow of liquid will be presented in Sect. 9.3.

For a nonstationary flow, we have

$$w(\tau) = w_0 + w_1(\tau) \,, \qquad P(\eta,\tau) = P_0(\eta) + P_1(\eta,\tau) + \Delta P(\eta,\tau) \,, \tag{9.31}$$

where $w_1(\tau)$ and $P_1(\eta,\tau)$ are the time dependent components of the speed and the pressure that do not depend on the motion of the rod; $\Delta P(\eta,\tau)$ is the additional pressure of the liquid that results from the action of the forces of inertia of the liquid due to the rod vibrations.

Similarly, the vector $\mathbf{Q}^{(1)}$ can be decomposed as

$$\mathbf{Q}^{(1)} = \mathbf{Q}_0^{(1)} - \mathbf{Q}_{11}^{(1)} + \Delta\mathbf{Q}^{(1)} \,,$$

where

$$\mathbf{Q}_0^{(1)} = \mathbf{Q}_0 - (P_0 + n_{11} w_0^2)\,\mathbf{e}_1 \,,$$
$$\mathbf{Q}_{11}^{(1)} = Q_{11}\mathbf{e}_1 = [P_1 + n_{11}(w_0 w_1 + w_1^2)]\,\mathbf{e}_1 \,,$$
$$\Delta\mathbf{Q}^{(1)} = \Delta\mathbf{Q} - \Delta P\,\mathbf{e}_1 \,. \tag{9.32}$$

Using (9.31) in (9.13), we get the following equations in w and P:

$$\frac{\partial P_0}{\partial \eta} = \gamma_{21,0}(\eta)\,, \qquad \gamma_{21,0}(\eta) = \frac{m_2 g}{A_{33}}\, l^3 x_{20}{}'(\eta)\,, \tag{9.33}$$

$$n_{11}\frac{\partial w_1}{\partial \tau} + \frac{\partial P_1}{\partial \eta} = 0\,, \tag{9.34}$$

$$n_{11}\frac{\partial^2 u_1}{\partial \tau^2} + \frac{\partial \Delta P}{\partial \eta} = \Delta\gamma_{21}(\eta,\,\tau)\,; \tag{9.35}$$

here,

$$\Delta\gamma_{21}(\eta,\,\tau) = \frac{m_2 g}{A_{33}}\, l^3 u_{x_2}{}'(\eta,\,\tau)\,.$$

Equation (9.35) refers to the coordinate frame attached to the rod axis, therefore, u'_{x_2} is to be written in terms of the projections of the vector of displacements $\mathbf{u}_e$ on the axes of this frame. Because $\mathbf{u}_x = \mathrm{L}^{(0)\,\mathrm{T}}\mathbf{u}_e$, we have

$$u_{x_2} = \sum_{j=1}^{3} l^0_{2j} u_j\,,$$

therefore,

$$u_{x_2}{}' = \sum_{j=1}^{3} (l^{(0)\prime}_{2j} u_j + l^{(0)}_{2j} u'_j)\,.$$

After solving the system of equations (9.18)–(9.21), the change in the force which depends on the pressure ΔP can be found from (9.35). Moreover, upon solution of the system (9.18)–(9.21), the components $u_j(\eta,\,\tau)$ of the vector of displacements, which appear in (9.35), can also be found. It follows from (9.34) and (9.35) that

$$P_1 = -n_{11}\frac{\partial w_1}{\partial \tau}\eta + P_{10}(\tau)\,,$$

$$\Delta P = \gamma_0 \sum_{j=1}^{3} \int_0^{\eta} (l^{(0)\prime}_{2j} u_j + l^{(0)}_{2j} u'_j)\,\mathrm{d}\eta^{(1)}$$

$$- n_{11}\int_0^{\eta} \frac{\partial^2 u_1(\tau,\,\eta^{(1)})}{\partial \tau^2}\,\mathrm{d}\eta^{(1)} + \Delta P_0(\tau)\,.$$

Arbitrary functions $P_{10}(\tau)$ and $\Delta P_0(\tau)$ can be found from boundary conditions. For example, if the pressure at the exit is constant, then $P_1(1,\,\tau)$ and $\Delta P(1,\,\tau)$ are zero. In this case, we have

$$P_{10} = n_{11}\frac{\partial w_1}{\partial \tau}\,,$$

$$\Delta P_0(\tau) = n_{11}\int_0^1 \frac{\partial^2 u_1}{\partial \tau^2}\,\mathrm{d}\eta - \gamma_0 \sum_{j=1}^{3} \int_0^1 (l^{(0)\prime}_{2j} u_j + l^{(0)}_{2j} u'_j)\,\mathrm{d}\eta\,.$$

Taking into consideration the non-stationary flow (9.2) and using (9.6), we can obtain an equation of small vibrations identical to (9.18):

$$\frac{\partial^2 \mathbf{u}}{\partial \tau^2} + \mathrm{A}_w \frac{\partial \boldsymbol{\vartheta}}{\partial \tau} + n_{11} \frac{\partial w_1}{\partial \tau} \mathbf{e}_1 - \frac{\partial \Delta \mathbf{Q}^{(1)}}{\partial \eta} - \mathrm{A}_Q \mathrm{A}^{-1} \Delta \mathbf{M} - \mathrm{A}_{æ} \Delta \mathbf{Q}^{(1)}$$
$$-Q_{11} \mathrm{A}_1 \mathrm{A}^{-1} \Delta \mathbf{M} - \Delta \mathbf{P} = \frac{\partial \mathbf{Q}_{11}^{(1)}}{\partial \eta} + \mathrm{A}_{æ} \mathbf{Q}_{11}^{(1)}, \tag{9.36}$$

where

$$\mathrm{A}_w = \begin{bmatrix} 0 & 0 & 0 \\ 0 & 0 & 2n_{11}(w_0 + w_1) \\ 0 & -2n_{11}(w_0 + w_1) & 0 \end{bmatrix}.$$

The other equations, (9.19)–(9.21), describing small vibrations of the rod, remain the same. If the non-stationary components w_1 and P_1 are periodic functions of time, then the vibrations of the rod are forced parametric vibrations.

9.3 Determination of Eigenvalues

9.3.1 Exact Numerical Method

To find the complex eigenvalues, we will make use of the system of equations for the case of a stationary flow ($w = w_0 = \text{const}$ and $P_0 = \text{const}$) (9.18)–(9.21), in which we put $\Delta \mathbf{P} = \Delta \mathbf{T} = 0$. As a result, we have

$$\Delta \mathbf{Q} = (\Delta \mathbf{Q}_0^{(1)} + \mathrm{i}\, \mathbf{Q}_0^{(2)})\, \mathrm{e}^{(\alpha + \mathrm{i}\,\beta)\,\tau},$$
$$\Delta \mathbf{M} = (\Delta \mathbf{M}_0^{(1)} + \mathrm{i}\, \Delta \mathbf{M}_0^{(2)})\, \mathrm{e}^{(\alpha + \mathrm{i}\,\beta)\,\tau},$$
$$\boldsymbol{\vartheta} = (\boldsymbol{\vartheta}_0^{(1)} + \mathrm{i}\, \boldsymbol{\vartheta}_0^{(2)})\, \mathrm{e}^{(\alpha + \mathrm{i}\,\beta)\,\tau},$$
$$\mathbf{u} = (\mathbf{u}_0^{(1)} + \mathrm{i}\, \mathbf{u}_0^{(2)})\, \mathrm{e}^{(\alpha + \mathrm{i}\,\beta)\,\tau}.$$

On rearrangement, separating real and imaginary parts, we arrive at the following system of equations:

$$\frac{\mathrm{d}\Delta \mathbf{Q}_0^{(1)}}{\mathrm{d}\eta} + \mathrm{A}_Q \mathrm{A}^{-1} \Delta \mathbf{M}_0^{(1)} + \mathrm{A}_{æ} \Delta \mathbf{Q}_0^{(1)} - \mathrm{A}_w \alpha \boldsymbol{\vartheta}_0^{(1)} + \mathrm{A}_w \beta \boldsymbol{\vartheta}_0^{(2)}$$
$$-n_{11}(\alpha^2 - \beta^2)\, \mathbf{u}_0^{(1)} + n_{11} 2\alpha\beta \mathbf{u}_0^{(2)} = 0,$$
$$\frac{\mathrm{d}\Delta \mathbf{Q}_0^{(2)}}{\mathrm{d}\eta} + \mathrm{A}_Q \mathrm{A}^{-1} \Delta \mathbf{M}_0^{(2)} + \mathrm{A}_{æ} \Delta \mathbf{Q}_0^{(2)} - \mathrm{A}_M \alpha \boldsymbol{\vartheta}_0^{(2)} + \mathrm{A}_w \beta \boldsymbol{\vartheta}_0^{(1)}$$
$$-n_{11}(\alpha^2 - \beta^2)\, \mathbf{u}_0^{(2)} - 2n_{11} \alpha\beta \mathbf{u}_0^{(1)} = 0,$$
$$\frac{\mathrm{d}\Delta \mathbf{M}_0^{(1)}}{\mathrm{d}\eta} + \mathrm{A}_{æ} \Delta \mathbf{M}_0^{(1)} + \mathrm{A}_M \mathrm{A}^{-1} \Delta \mathbf{M}_0^{(1)} + \mathrm{A}_Q \Delta \mathbf{Q}_0^{(1)} = 0,$$

$$\frac{\mathrm{d}\Delta\mathbf{M}_0^{(2)}}{\mathrm{d}\eta} + \mathrm{A}_{æ}\Delta\mathbf{M}_0^{(2)} + \mathrm{A}_M\mathrm{A}^{-1}\Delta\mathbf{M}_0^{(2)} + \mathrm{A}_Q\Delta\mathbf{Q}_0^{(2)} = 0\,,$$

$$\frac{\mathrm{d}\boldsymbol{\vartheta}_0^{(1)}}{\mathrm{d}\eta} + \mathrm{A}_{æ}\boldsymbol{\vartheta}_0^{(1)} - \mathrm{A}^{-1}\Delta\mathbf{M}_0^{(1)} = 0\,,$$

$$\frac{\mathrm{d}\boldsymbol{\vartheta}_0^{(2)}}{\mathrm{d}\eta} + \mathrm{A}_{æ}\boldsymbol{\vartheta}_0^{(2)} - \mathrm{A}^{-1}\Delta\mathbf{M}_0^{(2)} = 0\,,$$

$$\frac{\mathrm{d}\mathbf{u}_0^{(1)}}{\mathrm{d}\eta} + \mathrm{A}_{æ}\mathbf{u}_0^{(1)} + \mathrm{A}_1\boldsymbol{\vartheta}_0^{(1)} = 0\,,$$

$$\frac{\mathrm{d}\mathbf{u}_0^{(2)}}{\mathrm{d}\eta} + \mathrm{A}_{æ}\mathbf{u}_0^{(2)} + \mathrm{A}_1\boldsymbol{\vartheta}_0^{(2)} = 0\,. \tag{9.37}$$

In vector-matrix notation this reads

$$\mathbf{Z}' + \mathrm{B}(\eta,\, \alpha,\, \beta)\,\mathbf{Z} = 0\,, \tag{9.38}$$

where

$$\mathbf{Z} = (\Delta\mathbf{Q}_0^{(1)},\, \Delta\mathbf{Q}_0^{(2)},\, \Delta\mathbf{M}_0^{(1)},\, \Delta\mathbf{M}_0^{(2)},\, \boldsymbol{\vartheta}_0^{(1)},\, \boldsymbol{\vartheta}_0^{(2)},\, \mathbf{u}_0^{(1)},\, \mathbf{u}_0^{(2)})^{\mathrm{T}}\,,$$

$$\mathrm{B}(\alpha,\, \beta) = \left[\begin{array}{cccccccccccc} \mathrm{A}_{æ} & 0 & \mathrm{A}_Q\mathrm{A}^{-1} & 0 & -\mathrm{A}_w\alpha & \mathrm{A}_w\beta & -\mathrm{E}n_{11}(\alpha^2-\beta^2) & \mathrm{E}2n_{11}\alpha\beta \\ 0 & \mathrm{A}_{æ} & 0 & \mathrm{A}_Q\mathrm{A}^{-1} & -\mathrm{A}_w\beta & -\mathrm{A}_w\alpha & -\mathrm{E}2n_{11}\alpha\beta & -\mathrm{E}n_{11}(\alpha^2-\beta^2) \\ \mathrm{A}_1 & 0 & \mathrm{A}_{æ}+\mathrm{A}_M\mathrm{A}^{-1} & 0 & 0 & 0 & 0 & 0 \\ 0 & \mathrm{A}_1 & 0 & \mathrm{A}_{æ}+\mathrm{A}_M\mathrm{A}^{-1} & 0 & 0 & 0 & 0 \\ 0 & 0 & \mathrm{A}^{-1} & 0 & \mathrm{A}_{æ} & 0 & 0 & 0 \\ 0 & 0 & 0 & \mathrm{A}^{-1} & 0 & \mathrm{A}_{æ} & 0 & 0 \\ 0 & 0 & 0 & 0 & \mathrm{A}_1 & 0 & \mathrm{A}_{æ} & 0 \\ 0 & 0 & 0 & 0 & 0 & \mathrm{A}_1 & 0 & \mathrm{A}_{æ} \end{array}\right]. \tag{9.39}$$

The solution of the system looks like

$$\mathbf{Z} = \mathrm{K}(\eta,\, \alpha,\, \beta)\,\mathbf{C}\,, \qquad \mathrm{K}(0,\, \alpha,\, \beta) = \mathrm{E}\,.$$

Because the unknowns are complex numbers, we thereby get 24 boundary condition equations (12 at each end). From the boundary conditions at $\eta = 1$ the homogeneous equations in 12 unknowns c_ρ follow:

$$\sum_{\rho=1}^{12} k_{j\rho}(1,\, \alpha,\, \beta)\, c_\rho = 0\,, \qquad j = 1,\, \ldots\,,\, 12\,.$$

The determinant of this system of equations must vanish, that is, we have $D(1, w_0, P_0, \alpha, \beta) = 0$. The values α_j and β_j for which the determinant vanishs define a complex eigenvalue $\lambda_j = \alpha_j + \mathrm{i}\,\beta_j$. A more detailed calculation of eigenvalues was presented in Sect. 4.1.

Thus we obtain α_j and β_j as functions of the speed w_0 and pressure P_0. The critical value of the flow parameters w_0^* and P_0^* correspond to the case $\alpha_j = 0$. In applications, the critical values of the flow speed are of most practical interest. To find them, we put $\alpha = 0$. Then, setting some characteristics of the stationary flow, w_0 and P_0, (related by the Bernoulli equation (6.20), see Svetlitsky (2000a)), we proceed numerical determination by the selection of β_j for which $D(1, w_0^*, P_0^*, 0, \beta) = 0$.

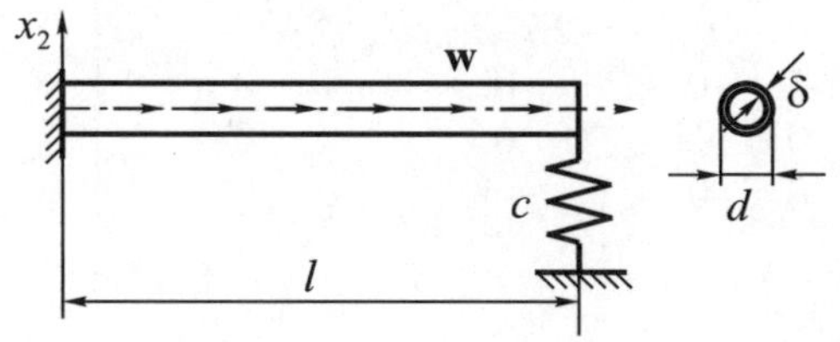

Fig. 9.3.

To illustrate the determination of eigenvalues, consider a straight-line section of a pipeline with an elastic support (Fig. 9.3). By setting $Q_{10} = 0$ in the equation of small vibrations for a straight-line pipeline (9.27), we get the equation of vibrations for the rod shown in Fig. 9.3. For numerical analysis, it would be more convenient to use the first order equations (9.25). We put

$$\Delta Q_2 = \Delta Q_{20}\,\mathrm{e}^{(\alpha+\mathrm{i}\,\beta)\,\tau}, \qquad \Delta M_3 = \Delta M_{30}\,\mathrm{e}^{(\alpha+\mathrm{i}\,\beta)\,\tau},$$
$$\vartheta_3 = \vartheta_{30}\,\mathrm{e}^{(\alpha+\mathrm{i}\,\beta)\,\tau}, \qquad u_2 = u_{20}\,\mathrm{e}^{(\alpha+\mathrm{i}\,\beta)\,\tau},$$

where

$$\Delta Q_{20} = \Delta Q_{20}^{(1)} + \mathrm{i}\,\Delta Q_{20}^{(2)}, \qquad \Delta M_{30} = \Delta M_{30}^{(1)} + \mathrm{i}\,\Delta M_{30}^{(2)},$$
$$\vartheta_{30} = \vartheta_{30}^{(1)} + \mathrm{i}\,\vartheta_{30}^{(2)}, \qquad u_{20} = u_{20}^{(1)} + \mathrm{i}\,u_{20}^{(2)},$$

On rearrangement, using (9.25) and separating real and imaginary parts, we arrive at

$$\mathbf{Z}' + \mathrm{B}(\eta, \alpha, \beta)\,\mathbf{Z} = 0; \tag{9.40}$$

here,

$$\mathbf{Z} = (\Delta Q_{20}^{(1)}, \Delta M_{30}^{(1)}, \vartheta_{30}^{(1)}, u_{20}^{(1)}, \Delta Q_{20}^{(2)}, \Delta M_{30}^{(2)}, \vartheta_{30}^{(2)}, u_{20}^{(2)})^{\mathrm{T}},$$

$$
\mathrm{B}(\eta,\,\alpha,\,\beta) = \begin{bmatrix}
0 & Q_{10}^{(1)} & -2n_{11}w_0\alpha & -(\alpha^2-\beta^2) & 0 & 0 & 2n_{11}w_0\beta & 2\alpha\beta \\
1 & 0 & 0 & 0 & 0 & 0 & 0 & 0 \\
0 & -1 & 0 & 0 & 0 & 0 & 0 & 0 \\
0 & 0 & -1 & 0 & 0 & 0 & 0 & 0 \\
0 & 0 & -2n_{11}w_0\beta & -2\alpha\beta & 0 & Q_{10}^{(1)} & -2n_{11}w_0\alpha & -(\alpha^2-\beta^2) \\
0 & 0 & 0 & 0 & 1 & 0 & 0 & 0 \\
0 & 0 & 0 & 0 & 0 & -1 & 0 & 0 \\
0 & 0 & 0 & 0 & 0 & 0 & -1 & 0
\end{bmatrix},
$$

where

$$Q_{10}^{(1)} = -n_{11}w_0^2\,.$$

The solution to (9.40) looks like

$$\mathbf{Z} = \mathrm{K}(\eta)\,\mathbf{C}\,. \tag{9.41}$$

This solution must satisfy the boundary conditions

$$
\begin{aligned}
&(1)\quad \eta = 0\,, \qquad \vartheta_{30}^{(1)} = \vartheta_{30}^{(2)} = u_{20}^{(1)} = u_{20}^{(2)} = 0\,,\\
&(2)\quad \eta = 1\,, \qquad \Delta Q_{20}^{(1)} + \tilde c_1 u_{20}^{(1)} = 0\,, \qquad \Delta Q_{20}^{(2)} + \tilde c_1 u_{20}^{(2)} = 0\,,\\
&\qquad\qquad\qquad\qquad \Delta M_{30}^{(1)} = \Delta M_{30}^{(2)} = 0\,,
\end{aligned}
$$

where

$$\tilde c_1 = \frac{c\,l^3}{A_{33}}$$

is a nondimensional stiffness.

From the boundary conditions at $\eta = 0$ it follows that

$$c_3 = c_4 = c_7 = c_8 = 0\,.$$

The boundary conditions at $\eta = 1$ yield the following system of equations:

$$
\begin{aligned}
&(k_{11} + c_1 k_{41})\,c_1 + (k_{12} + \tilde c_1 k_{42})\,c_2 + (k_{15} + \tilde c_1 k_{45})\,c_5 + (k_{16} + \tilde c_1 k_{46})\,c_6 = 0\,,\\
&(k_{51} + \tilde c_1 k_{81})\,c_1 + (k_{52} + \tilde c_1 k_{82})\,c_2 + (k_{55} + \tilde c_1 k_{85})\,c_5\\
&\qquad\qquad\qquad\qquad + k_{31}c_1 + k_{32}c_2 + k_{35}c_5 + k_{36}c_6 = 0\,,\\
&k_{61}c_1 + k_{62}c_2 + k_{65}c_5 + k_{66}c_6 = 0\,.
\end{aligned}
$$

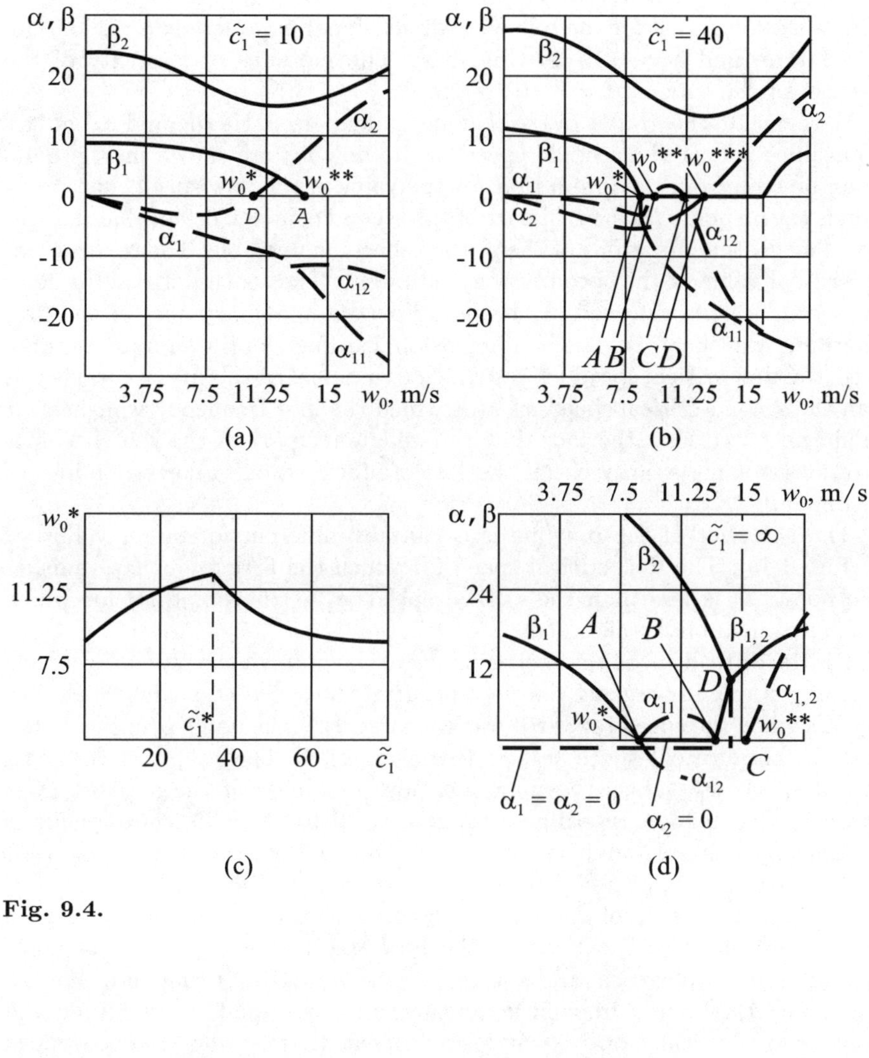
α, β
β₂
c̃₁ = 10
20
10
β₁
α₂
w₀*
w₀**
0
D
A
α₁
-10
α₁₂
-20
α₁₁
3.75 7.5 11.25 15 w₀, m/s
(a)
α, β
c̃₁ = 40
β₂
20
10
β₁
w₀** w₀***
α₂
α₁
w₀*
0
α₂
α₁₂
-10
-20
α₁₁
AB CD
3.75 7.5 11.25 15 w₀, m/s
(b)
w₀*
11.25
7.5
c̃₁*
20 40 60 c̃₁
(c)
3.75 7.5 11.25 15 w₀, m/s
α, β
c̃₁ = ∞
β₂
24
β₁
A
B
β₁,₂
12
D
α₁,₂
α₁₁
w₀*
w₀**
C
α₁₂
α₁ = α₂ = 0
α₂ = 0
(d)

Fig. 9.4.

The real and imaginary parts of the eigenvalues can be found from the equation $D(\alpha, \beta, w_0) = 0$, D is the determinant of the system of equations.

The quantities α_j and β_j can be calculated using the steepest descent method or any other method of locating extremums of multivariable functions. Numerically obtained graphs of α_j and β_j versus the dimensional speed of the flow are shown in Fig. 9.4(*a–d*).

In the example under consideration, the nondimensional parameter in (9.40) was taken to be $n_{11} = m_2/(m_2+m_1) = 0.64$; the nondimensional quantities were calculated for the following dimensional parameters: $l = 200.0\,\text{cm}$, $d = 4.0\,\text{cm}$, and $\delta = 0.15\,\text{cm}$ (Fig. 9.3). The modulus of elasticity of the material of the tube is $E = 3 \cdot 10^5\,\text{N/cm}^2$.

For $\tilde{c}_1 = 10$, the graphs of the real and imaginary parts (α_j and β_j) of two eigenvalues versus dimensional speed of the flow w_0 are shown in Fig. 9.4*a*. It can be seen from the graph that for the value of the flow speed that corresponds to the point D, the real part of the second complex eigenvalue changes sign, that is, the vibrations of the pipeline become unstable. The corresponding critical value of the speed is denoted by w_0^*. The second critical value of the speed, w_0^{**}, is attained at the point A (the imaginary part of the first eigenvalue vanishes). For the nondimensional stiffness of the support $\tilde{c}_1$ equal to 10, the first critical speed w_0^* (at which dynamical instability occurs) is less than the second critical speed w_0^{**} (at which the first frequency vanishes). It should be noted that the fact that the imaginary part of the root becomes zero does not necessarily entail the loss of static stability corresponding to this mode.

The behavior of the pipeline is essentially different for larger values of $\tilde{c}_1$ (Fig. 9.4*b*). The first critical speed for which the first frequency vanishes (the point A) is less than the critical speed w_0^{***} (the point D) for which dynamical instability takes place.

For the critical speed that corresponds to the point A the pipeline remains statically stable (divergence does not occur at the speed w_0^*). The divergence appears when the speed takes the critical value w_0^{**} (the point B in Fig. 9.4*b*). At that, the critical speed w_0^{**} is less then w_0^{***}. The graph of the least critical speed w_0^* versus the nondimensional stiffness of the support $\tilde{c}_1$ is shown in Fig. 9.4*c*. In a stiffness range $\tilde{c}_1 = 0$ to $\tilde{c}_1^* = 35$, the pipeline is dynamically unstable, and for $\tilde{c}_1 > \tilde{c}_1^*$, the pipeline is statically stable, that is, the divergence appears before the loss of dynamical stability.

In Fig. 9.4*d*, graphs of the real and imaginary parts of the roots are shown for the limiting case $\tilde{c}_1 = \infty$. As the flow speed increases, the imaginary parts β_1 and β_2 decrease, and the real parts α_1 and α_2 equal zero. For w_0^* (the point A), the first frequency vanishes, and two equal in magnitude and opposite in sign real roots, α_{11} and α_{12}, appear, that is, when the flow speed equals w_0^*, the divergence takes place. At the point B, the real roots α_{11} and α_{12} become zero with β_1 emerging again. The real part of the root α_1 remains zero until the speed is below w_0^{**} (the point C). At the point C, the imaginary parts of the two complex roots merge (the point D) and the real part $\alpha_{1,2}$ becomes positive. The point C corresponds to the critical value w_0^{**} for which the tube becomes dynamically unstable.

9.4 Parametric Vibrations of a Pipeline

9.4.1 Basic Problems

In Sect. 9.2, the general equations of small vibrations (9.19)–(9.21) for a space-curved pipeline exerted by an internal non-stationary flow of ideal incompressible liquid were obtained. Restrict our consideration to the case of pipelines of constant cross section (F_2 = const). When the non-stationary characteristics of the flow (see (9.32)) vary with time periodically, we have $w_1(t+T) = w_1(t)$ and $P_1(t+T) = P_1(t)$. It was mentioned in Sect. 9.2 that the most important thing about small vibrations of pipelines when the internal flow is non-stationary is that these vibrations (for curved pipelines only) are forced vibrations because of the presence of the terms $A_{æ}Q_{11}^{(1)}$ and $\partial Q_{11}/\partial\eta$ in (9.36). If w_1 and P_1 change periodically, then the component of axial force Q_{11} (9.32) is also periodic in time, therefore, the vector equation (9.36) governs forced parametric vibrations of the pipeline. The analysis of parametric vibrations of pipelines involves two independent problems: (1) analysis of stability of parametric vibrations and location of the regions of instability; (2) determination of amplitude values of the components of the state vector for steady-state forced parametric vibrations.

9.4.2 Dynamical Stability Analysis

To investigate the stability of parametric vibrations of a rod of constant cross section, we consider the system of linear equations (9.36), (9.19)–(9.21), which with regard for viscous damping can be written as a vector equation

$$\mathbf{L}(\mathbf{Z}) = \mathrm{A}^{(1)}\frac{\partial^2\mathbf{Z}}{\partial\tau^2} + \mathrm{A}^{(3)}\frac{\partial\mathbf{Z}}{\partial\tau} + \frac{\partial\mathbf{Z}}{\partial\eta} + \mathrm{A}^{(2)}\mathbf{Z} = 0\,, \tag{9.42}$$

where

$$\mathbf{Z} = (\Delta\mathbf{Q},\, \Delta\mathbf{M},\, \boldsymbol{\vartheta},\, \mathbf{u})^{\mathrm{T}}\,,$$

$$\mathrm{A}^{(1)} = \begin{bmatrix} 0 & 0 & 0 & -\mathrm{E} \\ 0 & 0 & 0 & 0 \\ 0 & 0 & 0 & 0 \\ 0 & 0 & 0 & 0 \end{bmatrix},$$

$$\mathrm{A}^{(2)} = \begin{bmatrix} \mathrm{A}_{æ} & \mathrm{A}_Q^{(1)}\mathrm{A}^{-1} & 0 & 0 \\ \mathrm{A}_1 & \mathrm{A}_M\mathrm{A}^{-1} + \mathrm{A}_{æ} & 0 & 0 \\ 0 & -\mathrm{A}^{-1} & \mathrm{A}_{æ} & 0 \\ 0 & 0 & \mathrm{A}_1 & \mathrm{A}_{æ} \end{bmatrix},$$

$$\mathrm{A}^{(3)} = \begin{bmatrix} 0 & 0 & 0 & -\alpha_0 \mathrm{E} \\ 0 & 0 & 0 & 0 \\ 0 & 0 & -\mathrm{A}_w & 0 \\ 0 & 0 & 0 & 0 \end{bmatrix},$$

$$\mathrm{A}_Q^{(1)} = (\mathrm{A}_Q + Q_{11}\mathrm{A}_1)\,\mathrm{A}^{-1},$$

$$Q_{11} = P_1 + n_{11}(w_0 w_1 + w_1^2).$$

To solve the vector equation (9.42) approximately, we will use the principle of virtual displacements presented in Sect. 4.4. We seek an approximate solution in the form

$$\tilde{\mathbf{Z}} = \sum_{j=1}^{n} f^{(j)}(\tau)\mathbf{Z}_0^{(j)}(\eta). \tag{9.43}$$

As generalized displacements we take (see Sect. 4.4)

$$\delta\mathbf{Z}^{(j)} = \delta b_j \mathrm{E}_0 \mathbf{Z}_0^{(j)}. \tag{9.44}$$

Applying the principle of virtual displacements yields a system of equations

$$\int_0^1 (\mathbf{L}(\tilde{\mathbf{Z}}) \cdot \mathrm{E}_0 \mathbf{Z}_0^{(j)})\, \mathrm{d}\eta = 0, \qquad j = 1, 2, \ldots, n. \tag{9.45}$$

On rearrangement, from (9.45) we get the system of equations to determine the functions $f^{(j)}(\tau)$:

$$\sum_{j=1}^{n} \left(h_{ij} \ddot{f}^{(j)} + b_{ij} \dot{f}^{(j)} + c_{ij} f^{(j)} \right) = 0, \qquad i = 1, 2, \ldots, n.$$

In vector-matrix form this reads

$$\mathrm{H}\ddot{\mathbf{f}} + \mathrm{B}(\tau)\,\dot{\mathbf{f}} + \mathrm{C}(\tau)\,\mathbf{f} = 0. \tag{9.46}$$

The components of the matrices $\mathrm{B}(\tau)$ and $\mathrm{C}(\tau)$ are periodic functions of τ. The Rayleigh method (see Sect. 7.7) can be used for an approximate determination of the regions of instability.

9.4.3 Parametric Vibrations of Straight-Line Pipelines

The equations of small parametric vibrations for a straight-line pipeline subject to a non-stationary flow of liquid (in the presence of viscous damping) can be obtained by direct specialization of the system of equations (9.18)–(9.21) and adding a term to the first equation to account for viscous damping. Thus, we have

$$\frac{\partial^2 u}{\partial \tau^2} + 2n_{11}(w_0 + w_1)\frac{\partial^2 u}{\partial \tau \partial \eta} - \left[Q_{10} - P_1 - n_{11}(2w_0 w_1 + w_1^2)\right]\frac{\partial^2 u}{\partial \eta^2} + \alpha_0 \frac{\partial u}{\partial \tau} + \frac{\partial^4 u}{\partial \eta^4} = 0\,, \quad (9.47)$$

where (for harmonically changing w_1 and P_1)

$$P_1 = P_{100}(1-\eta)\cos\omega_0\tau\,, \qquad w_1 = w_{10}\sin\omega_0\tau\,, \qquad P_{100} = n_{11}\omega_0 w_{10}\,.$$

Let

$$u = \sum_{j=1}^{n} \varphi^{(j)}(\eta)\, f_j(\tau)\,.$$

Then, using the principle of virtual displacements, we get

$$\sum_{j=1}^{n} \left[h_{ij}\ddot{f}_j + b_{ij}(\tau)\,\dot{f}_j + c_{ij}(\tau)\,f_j\right] = 0\,, \qquad i = 1, \ldots, n\,, \quad (9.48)$$

where

$$h_{ij} = \int_0^1 \varphi^{(i)}\varphi^{(j)}\,\mathrm{d}\eta\,,$$

$$b_{ij} = \alpha_0 \int_0^1 \varphi^{(j)}\varphi^{(i)}\,\mathrm{d}\eta + 2n_{11}(w_0 + w_1)\int_0^1 \varphi^{(j)\prime}\varphi^{(i)}\,\mathrm{d}\eta\,,$$

$$c_{ij} = -\left[Q_{10}^{(1)} - P_{100}\cos\omega_0\tau - n_{11}(2w_0 w_1 + w_1^2)\right]\int_0^1 \varphi^{(j)\prime\prime}\varphi^{(i)}\,\mathrm{d}\eta + \int_0^1 \varphi^{(j)}\eta\varphi^{(i)}\,\mathrm{d}\eta - \cos\omega_0\tau \int_0^1 \eta\varphi^{(j)\prime\prime}\varphi^{(i)}\,\mathrm{d}\eta\,.$$

The matrices B and C can be written as

$$\mathrm{B} = \mathrm{B}_0 + \mathrm{B}^{(1)}\sin\omega_0\tau\,,$$
$$\mathrm{C} = \mathrm{C}_0 + \mathrm{C}^{(1)}\sin\omega_0\tau + \mathrm{C}^{(2)}\cos\omega_0\tau + \mathrm{C}^{(3)}\cos 2\omega_0\tau\,. \quad (9.49)$$

Using the Rayleigh method (see Sect. 7.7), the system of equations (9.49) can be solved approximately. This solution takes place at the boundary of the stability region. We put

$$\mathbf{f} = \sum_{k=1,3,5}\left(\mathbf{a}_j \cos\frac{\omega_0 k}{2}\tau + \mathbf{b}_j \sin\frac{\omega_0 k}{2}\tau\right)\,. \quad (9.50)$$

Substituting (9.50) into the system of equations (9.49) and collecting the terms with the similar trigonometric functions $\cos(\omega_0 k\tau/2)$ and $\sin(\omega_0 k\tau/2)$, we arrive at

$$\sum_{k=1,3,5}\left(\mathbf{d}_k \cos\frac{\omega_0 k}{2}\tau + \mathbf{c}_k \sin\frac{\omega_0 k}{2}\tau\right) = 0\,. \tag{9.51}$$

From (9.51) we obtain a system of equations $\mathbf{d}_k = 0$, $\mathbf{c}_k = 0$ ($k = 1, 3, 5, \ldots$), which in expanded notation yields the following equations in the vectors $\mathbf{a}_j$ and $\mathbf{b}_j$:

$$\begin{aligned}&\alpha_{11}\mathbf{a}_1 + \alpha_{12}\mathbf{b}_1 + \alpha_{13}\mathbf{a}_2 + \alpha_{14}\mathbf{b}_2 + \ldots = 0\,,\\&\alpha_{\rho 1}\mathbf{a}_1 + \alpha_{\rho 2}\mathbf{b}_1 + \alpha_{\rho 3}\mathbf{a}_2 + \alpha_{\rho 4}\mathbf{b}_2 + \ldots = 0\,;\end{aligned} \tag{9.52}$$

here, the matrices α_{ij} are linear combinations of the matrices H, B_0, B, C_0, $\mathrm{C}^{(1)}$, $\mathrm{C}^{(2)}$, and $\mathrm{C}^{(3)}$. Equating the determinant of the system of equations (9.52) to zero, we obtain an equation establishing the relationship between characteristics of the flow (that is, between w_{10} and ω_0); this relationship defines the boundaries of the regions of instability.

As an example, consider vibrations of a simply supported section of a pipeline (Fig. 9.1). Let $\alpha_0 = 0$. We seek a two-term approximate solution. For this kind of support, the functions $\varphi^{(1)}(\eta)$ and $\varphi^{(2)}(\eta)$ that are members of the approximate solution are as follows: $\varphi^{(1)}(\eta) = \sin \pi\eta$ and $\varphi^{(2)}(\eta) = \sin 2\pi\eta$.

We assume the solution of the form

$$f_1 = a_1 \cos\frac{\omega_0}{2}\tau + b_1 \sin\frac{\omega_0}{2}\tau\,, \qquad f_2 = a_2 \cos\frac{\omega_0}{2}\tau + b_2 \sin\frac{\omega_0}{2}\tau\,.$$

The system of equations (9.52) now looks like

$$\begin{aligned}&\alpha_{11}a_1 + \alpha_{12}b_1 + \alpha_{13}a_2 + \alpha_{14}b_2 = 0\,,\\&\ldots\\&\alpha_{41}a_1 + \alpha_{42}b_1 + \alpha_{43}a_2 + \alpha_{44}b_2 = 0\,.\end{aligned}$$

The boundaries of the regions of instability were determined for the following value of the parameters: $l = 400.0\,\mathrm{cm}$, $Q_{10} = 0$, $w_{0\,\mathrm{d}} = 600.0\,\mathrm{cm/s}$, $A_{33} = 2\cdot 10^8\,\mathrm{kg/cm^2}$, $m_1 = 50\cdot 10^{-6}\,\mathrm{kg}\cdot\mathrm{s}^2\cdot\mathrm{cm}^{-2}$, $m_2 = 75\cdot 10^{-6}\,\mathrm{kg}\cdot\mathrm{s}^2\cdot\mathrm{cm}^{-2}$, and $p_0 = 10.0\,\mathrm{N/cm^2}$; the corresponding nondimensional parameters from (9.48) are as follows: $n_{11} = 0.85$, $w_0 = 0.18$, $Q^{(1)} = -P_e - n_{11}w_0^2 = -6\cdot 10^{-2}$. The regions of instability for two cases, $\alpha_0 = 0$ and $\alpha_0 = 10^{-1}$, are shown in Fig. 9.5.

From the numerical results obtained (see Fig. 9.5) it follows that for real pipelines with large bending stiffness unstable (with regard to viscous damping) parametric vibrations may occur for comparatively large amplitudes of the periodically changing characteristic of the flow w_{10}. In the example under consideration such vibrations occur for dimensional amplitudes larger than 150 cm/s. Note that this value is very close to the stationary component of the flow speed, $w_{0\,\mathrm{d}}$. In practice, forced parametric vibrations causing an accumulation of fatigue damage are the most undesirable as they reduce the pipeline's lifetime.

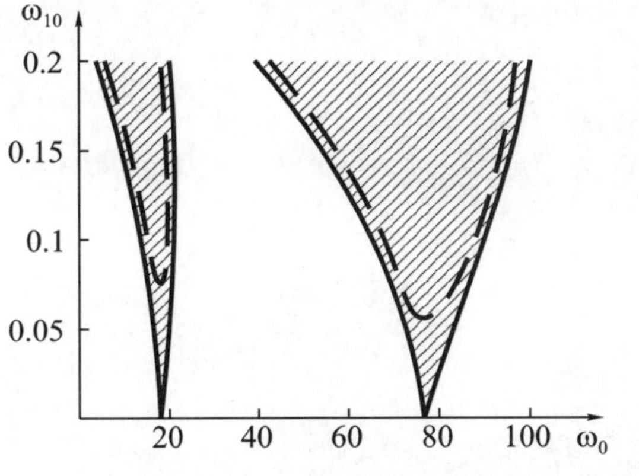

Fig. 9.5.

9.4.4 Forced Parametric Vibrations of Pipelines

In Sect. 9.2, equations of small, forced parametric vibrations of pipelines were derived (see the system of equations (9.19)–(9.21) and (9.36)). Stability of small parametric vibrations was discussed in Sect. 9.4. Dynamical stability was studied using the homogeneous-equation system (9.19)–(9.21) and (9.36). When considering forced parametric vibrations, non-homogeneous equations (9.19)–(9.21) and (9.36) with $\Delta\mathbf{P} = \Delta\mathbf{T} = 0$ are to be used. By analogy with (5.50), the system of equations (9.19)–(9.21) and (9.36) may be written as

$$\frac{\partial \mathbf{Z}}{\partial \eta} + \mathrm{A}^{(1)} \frac{\partial^2 \mathbf{Z}}{\partial \tau^2} + \mathrm{A}_w^{(1)}(\tau) \frac{\partial \mathbf{Z}}{\partial \tau} + \mathrm{A}^{(2)}(\tau)\, \mathbf{Z} = \Delta\boldsymbol{\Phi}(\tau)\,. \tag{9.53}$$

The components of the matrices $\mathrm{A}_w^{(1)}$ and $\mathrm{A}^{(2)}$ and the vector $\Delta\boldsymbol{\Phi}$ are time periodic functions. They depend on w_1 and P_1 (see (9.32)), which are assumed to be periodic functions (for example, periodic harmonic functions $w_1 = w_{10}(\eta) \sin \omega\tau$ and $P_1 = P_{10}(\eta) \cos \omega\tau$). We seek an approximate one-term solution of (9.53) in the form $\mathbf{Z} = \mathbf{Z}_0^{(1)}(\eta)\, f^{(1)}(\tau)$. Using the principle of virtual displacements and putting $\delta\mathbf{Z}_0^{(1)} = \delta b_1\, \mathrm{E}_0\, \mathbf{Z}_0^{(1)}$, on rearrangement, it follows from (9.53) that

$$\ddot{f}^{(1)} + a_1(\tau)\, \dot{f}^{(1)} + a_2(\tau)\, f^{(1)} = b(\tau)\,, \tag{9.54}$$

where a_1, a_2, and b are periodic functions that can be expanded in a Fourier series. Suppose now that steady state parametric vibrations of the pipeline take place; then, an approximate solution of (9.54) can be assumed of the form

$$f^{(1)} = b_0 + b_1 \cos \omega\tau + b_2 \sin \omega\tau\,. \tag{9.55}$$

Substituting (9.55) into (9.54) and retaining only the coefficients of $\sin \omega\tau$ and $\cos \omega\tau$, we arrive at a system of non-homogeneous equations to determine b_0, b_1, and b_2.

A. Elements of Some Matrixes

A.1 Elements of the Matrixes $\mathbf{N}^{(1)}$ and $\mathbf{N}^{(2)}$

$$
\begin{aligned}
n_{11}^{(1)} &= d_1^{(1)(1)} q_{n00} \,, \\
n_{12}^{(1)} &= (d_2^{(1)(1)} - \Delta_{30}^{(1)} + g_{12})\, q_{n00} \,, \\
n_{13}^{(1)} &= (d_3^{(1)(1)} + \Delta_{20}^{(1)} + g_{11})\, q_{n00} \,,
\end{aligned}
$$

$$
\begin{aligned}
n_{21}^{(1)} &= d_1^{(1)(2)} q_{n00} \,, \\
n_{22}^{(1)} &= (d_2^{(1)(2)} - \Delta_{30}^{(2)} + g_{22})\, q_{n00} \,, \\
n_{23}^{(1)} &= (d_3^{(1)(2)} + \Delta_{20}^{(2)} + g_{21})\, q_{n00} \,,
\end{aligned}
$$

$$
\begin{aligned}
n_{31}^{(1)} &= d_1^{(1)(3)} q_{n00} \,, \\
n_{32}^{(1)} &= (d_2^{(1)(3)} - \Delta_{30}^{(3)} + g_{32})\, q_{n00} \,, \\
n_{33}^{(1)} &= (d_3^{(1)(3)} + \Delta_{20}^{(3)} + g_{31})\, q_{n00} \,,
\end{aligned}
$$

$$
\begin{aligned}
n_{1j}^{(2)} &= (-\Delta_{1j}^{(1)} + c_{1j} + d_j^{(2)(1)} - a_{1j}^{(1)})\, q_{n00} \,, \\
n_{2j}^{(2)} &= (-\Delta_{1j}^{(2)} + c_{2j} + d_j^{(2)(2)} - a_{1j}^{(2)})\, q_{n00} \,, \\
n_{3j}^{(2)} &= (-\Delta_{1j}^{(3)} + c_{3j} + d_j^{(2)(3)} - a_{1j}^{(3)})\, q_{n00} \,, \qquad j = 1,\, 2,\, 3 \,,
\end{aligned}
$$

where

$$
\begin{aligned}
q_{n00} &= \frac{\rho\, V_0^2 d\, l^3}{2\, A_{33}(0)} \,, \\
\Delta_{20}^{(i)} &= (\mathbf{e}_\nu \cdot \mathbf{e}_{20})\, \cot\varphi_{a0} \cdot c_{n_{10}}\, (e_{\nu x_i} - l_{1i}^{(0)} \cos\varphi_{a0}) \,, \\
\Delta_{30}^{(i)} &= (\mathbf{e}_\nu \cdot \mathbf{e}_{30})\, \cot\varphi_{a0} \cdot c_{n_{10}}\, (e_{\nu x_i} - l_{1i}^{(0)} \cos\varphi_{a0}) \,, \\
\Delta_{1j}^{(i)} &= c_{n_{10}}\, (l_{1j}^{(0)} \cos\varphi_{a0} - e_{\nu x_j})\, \cot\varphi_{a0} \cdot (e_{\nu x_i} - l_{1i}^{(0)} \cos\varphi_{a0}) \,, \\
& \qquad\qquad i,\, j = 1,\, 2,\, 3 \,,
\end{aligned}
$$

$$c_{1j} = \left[-l_{11}^{(0)}\left(l_{1j}^{(0)} \cos\varphi_{a0} - e_{\nu x_j}\right) + b_{1j}\right] c_{n_{10}} \sin\varphi_{a0}\,,$$
$$c_{2j} = \left[-l_{12}^{(0)}\left(l_{1j}^{(0)} \cos\varphi_{a0} - e_{\nu x_j}\right) + b_{2j}\right] c_{n_{10}} \sin\varphi_{a0}\,,$$
$$c_{3j} = \left[-l_{13}^{(0)}\left(l_{1j}^{(0)} cos\varphi_{a0} - e_{\nu x_j}\right) + b_{3j}\right] c_{n_{10}} \sin\varphi_{a0}\,, \qquad j = 1, 2, 3\,,$$

$$b_{11} = -[1-(x'_{10})^2]\,, \qquad b_{12} = x'_{10}x'_{20}\,, \qquad b_{13} = x'_{10}x'_{30}\,,$$
$$b_{21} = x'_{10}x'_{20}\,, \qquad b_{22} = -[1-(x'_{20})^2]\,, \qquad b_{23} = x'_{20}x'_{30}\,,$$
$$b_{31} = x'_{10}x'_{30}\,, \qquad b_{32} = x'_{20}x'_{30}\,, \qquad b_{33} = -[1-(x'_{30})^2]\,.$$

$$d_j^{(i)(1)} = d_j^{(i)} \frac{\partial c_{n_1}}{\partial \alpha_a} (e_{\nu x_1} - l_{11}^{(0)} \cos\varphi_{a0}) \sin\varphi_{a0}\,,$$
$$d_j^{(i)(2)} = d_j^{(i)} \frac{\partial c_{n_1}}{\partial \alpha_a} (e_{\nu x_2} - l_{12}^{(0)} \cos\varphi_{a0}) \sin\varphi_{a0}\,, \qquad i = 1, 2\,,$$
$$d_j^{(i)(3)} = d_j^{(i)} \frac{\partial c_{n_1}}{\partial \alpha_a} (e_{\nu x_3} - l_{13}^{(0)} \cos\varphi_{a0}) \sin\varphi_{a0}\,, \qquad j = 1, 2, 3\,,$$

$$a_{1i}^{(1)} = \beta_0 e_{\nu x_i} c_{n_{10}} (e_{\nu x_1} - l_{11}^{(0)} \cos\varphi_{a0}) \sin\varphi_{a0}\,,$$
$$a_{1i}^{(2)} = \beta_0 e_{\nu x_i} c_{n_{10}} (e_{\nu x_2} - l_{12}^{(0)} \cos\varphi_{a0}) \sin\varphi_{a0}\,,$$
$$a_{1i}^{(3)} = \beta_0 e_{\nu x_i} c_{n_{10}} (e_{\nu x_3} - l_{13}^{(0)} \cos\varphi_{a0}) \sin\varphi_{a0}\,, \qquad i = 1, 2, 3\,,$$

$$g_{11} = (-l_{11}^{(0)}(\mathbf{e}_\nu \cdot \mathbf{e}_{20}) - l_{21}^{(0)} \cos\varphi_{a0})\, c_{n_{10}} \sin\varphi_{a0}\,,$$
$$g_{12} = (l_{11}^{(0)}(\mathbf{e}_\nu \cdot \mathbf{e}_{30}) - l_{31}^{(0)} \cos\varphi_{a0})\, c_{n_{10}} \sin\varphi_{a0}\,,$$
$$g_{21} = (-l_{12}^{(0)}(\mathbf{e}_\nu \cdot \mathbf{e}_{20}) - l_{22}^{(0)} \cos\varphi_{a0})\, c_{n_{10}} \sin\varphi_{a0}\,,$$
$$g_{22} = (l_{12}^{(0)}(\mathbf{e}_\nu \cdot \mathbf{e}_{30}) - l_{32}^{(0)} \cos\varphi_{a0})\, c_{n_{10}} \sin\varphi_{a0}\,,$$
$$g_{31} = (-l_{13}^{(0)}(\mathbf{e}_\nu \cdot \mathbf{e}_{20}) - l_{23}^{(0)} \cos\varphi_{a0})\, c_{n_{10}} \sin\varphi_{a0}\,,$$
$$g_{32} = (l_{13}^{(0)}(\mathbf{e}_\nu \cdot \mathbf{e}_{30}) - l_{33}^{(0)} \cos\varphi_{a0})\, c_{n_{10}} \sin\varphi_{a0}\,,$$

$$\cos\varphi_{a0} = l_{11}^{(0)} \cos\alpha\, \cos\beta + l_{12}^{(0)} \sin\beta + l_{13}^{(0)} \sin\alpha\, \cos\beta\,,$$
$$\sin\varphi_{a0} = \sqrt{1-\cos^2\varphi_{a0}}\,,$$
$$l_{1j}^{(0)} = x'_{j0}\,, \qquad j = 1, 2, 3\,,$$
$$\mathbf{e}_\nu = \cos\alpha\, \cos\beta\, \mathbf{i}_1 + \sin\beta\, \mathbf{i}_2 + \sin\alpha\, \cos\beta\, \mathbf{i}_3\,,$$
$$\mathbf{e}_{20} = \sum_{j=1}^{3} l_{2j}^{(0)}\, \mathbf{i}_j\,, \qquad \mathbf{e}_{30} = \sum_{j=1}^{3} l_{3j}^{(0)}\, \mathbf{i}_j\,,$$

A.2 Elements of the Matrixes $\mathbf{P}^{(1)}$ and $\mathbf{P}^{(2)}$

$$
\begin{aligned}
p_{11}^{(1)} &= \left(\frac{\partial c_L}{\partial \alpha_a} A_1 d_1^{(1)} \sin\varphi_{a0}\right) q_{L00}\,,\\
p_{12}^{(1)} &= \left(c_{L0} g_{12} \sin\varphi_{a0} + c_{L0} A_1 \Delta_{30}' + \frac{\partial c_L}{\partial \alpha_a} A_1 d_2^{(1)} \sin\varphi_{a0}\right) q_{L00}\,,\\
p_{13}^{(1)} &= \left(c_{L0} g_{13} \sin\varphi_{a0} - c_{L0} A_1 \Delta_{20}' + \frac{\partial c_L}{\partial \alpha_a} A_1 d_3^{(1)} \sin\varphi_{a0}\right) q_{L00}\,,
\end{aligned}
$$

$$
\begin{aligned}
p_{21}^{(1)} &= \left(\frac{\partial c_L}{\partial \alpha_a} A_2 d_1^{(1)} \sin\varphi_{a0}\right) q_{L00}\,,\\
p_{22}^{(1)} &= \left(c_{L0} g_{22} \sin\varphi_{a0} + c_{L0} A_2 \Delta_{30}' + \frac{\partial c_L}{\partial \alpha_a} A_2 d_2^{(1)} \sin\varphi_{a0}\right) q_{L00}\,,\\
p_{23}^{(1)} &= \left(c_{L0} g_{23} \sin\varphi_{a0} - c_{L0} A_2 \Delta_{20}' + \frac{\partial c_L}{\partial \alpha_a} A_2 d_3^{(1)} \sin\varphi_{a0}\right) q_{L00}\,,
\end{aligned}
$$

$$
\begin{aligned}
p_{31}^{(1)} &= \left(\frac{\partial c_L}{\partial \alpha_a} A_3 d_1^{(1)} \sin\varphi_{a0}\right) q_{L00}\,,\\
p_{32}^{(1)} &= \left(c_{L0} g_{32} \sin\varphi_{a0} + c_{L0} A_3 \Delta_{30}' + \frac{\partial c_L}{\partial \alpha_a} A_3 d_2^{(1)} \sin\varphi_{a0}\right) q_{L00}\,,\\
p_{33}^{(1)} &= \left(c_{L0} g_{33} \sin\varphi_{a0} - c_{L0} A_3 \Delta_{20}' + \frac{\partial c_L}{\partial \alpha_a} A_3 d_3^{(1)} \sin\varphi_{a0}\right) q_{L00}\,,
\end{aligned}
$$

$$
\begin{aligned}
p_{11}^{(2)} &= \left(-c_{L0} A_1 \Delta_{11}' + \frac{\partial c_L}{\partial \alpha_a} A_1 d_1^{(2)} \sin\varphi_{a0} - c_{L0} a_{11} A_1 \sin\varphi_{a0}\right) q_{L00}\,,\\
p_{12}^{(2)} &= \Big(c_{L0} f_{12} \sin\varphi_{a0} - c_{L0} A_1 \Delta_{12}'\\
&\qquad + \frac{\partial c_L}{\partial \alpha_a} A_1 d_2^{(2)} \sin\varphi_{a0} - c_{L0} a_{12} A_1 \sin\varphi_{a0}\Big) q_{L00}\,,\\
p_{13}^{(2)} &= \Big(c_{L0} f_{13} \sin\varphi_{a0} - c_{L0} A_1 \Delta_{13}'\\
&\qquad + \frac{\partial c_L}{\partial \alpha_a} A_1 d_3^{(2)} \sin\varphi_{a0} - c_{L0} a_{13} A_1 \sin\varphi_{a0}\Big) q_{L00}\,,
\end{aligned}
$$

$$
\begin{aligned}
p_{21}^{(2)} &= \Big(c_{L0} f_{21} \sin\varphi_{a0} - c_{L0} A_2 \Delta_{11}'\\
&\qquad + \frac{\partial c_L}{\partial \alpha_a} A_2 d_1^{(2)} \sin\varphi_{a0} - c_{L0} a_{11} A_2 \sin\varphi_{a0}\Big) q_{L00}\,,\\
p_{22}^{(2)} &= \left(-c_{L0} A_2 \Delta_{12}' + \frac{\partial c_L}{\partial \alpha_a} A_2 d_2^{(2)} \sin\varphi_{a0} - c_{L0} a_{12} A_2 \sin\varphi_{a0}\right) q_{L00}\,,
\end{aligned}
$$

$$p_{23}^{(2)} = \Big(c_{L0} f_{23} \sin\varphi_{a0} - c_{L0} A_2 \Delta'_{13} + \frac{\partial c_L}{\partial \alpha_a} A_2 d_3^{(2)} \sin\varphi_{a0} - c_{L0} a_{13} A_2 \sin\varphi_{a0}\Big) q_{L00} ,$$

$$p_{31}^{(2)} = \Big(c_{L0} f_{31} \sin\varphi_{a0} - c_{L0} A_3 \Delta'_{11} + \frac{\partial c_L}{\partial \alpha_a} A_3 d_1^{(2)} \sin\varphi_{a0} - c_{L0} a_{11} A_3 \sin\varphi_{a0}\Big) q_{L00} ,$$

$$p_{32}^{(2)} = \Big(c_{L0} f_{32} \sin\varphi_{a0} - c_{L0} A_3 \Delta'_{12} + \frac{\partial c_L}{\partial \alpha_a} A_3 d_2^{(2)} \sin\varphi_{a0} - c_{L0} a_{12} A_3 \sin\varphi_{a0}\Big) q_{L00} ,$$

$$p_{33}^{(2)} = \Big(-c_{L0} A_1 \Delta'_{13} + \frac{\partial c_L}{\partial \alpha_a} A_3 d_3^{(2)} \sin\varphi_{a0} - c_{L0} a_{13} A_3 \sin\varphi_{a0}\Big) q_{L00} ,$$

where

$$q_{L00} = \frac{\rho V_0^2 d\, l^3}{2\, A_{33}(0)} , \qquad \Delta'_{11} = \Big(l_{11}^{(0)} \cos\varphi_{a0} - e_{\nu x_1}\Big) \cot\varphi_{a0} ,$$

$$\Delta'_{12} = \Big(l_{12}^{(0)} \cos\varphi_{a0} - e_{\nu x_2}\Big) \cot\varphi_{a0} ,$$

$$\Delta'_{13} = \Big(l_{13}^{(0)} \cos\varphi_{a0} - e_{\nu x_3}\Big) \cot\varphi_{a0} ,$$

$$\Delta'_{20} = (\mathbf{e}_\nu \cdot \mathbf{e}_{20}) \cot\varphi_{a0} ,$$

$$\Delta'_{30} = (\mathbf{e}_\nu \cdot \mathbf{e}_{30}) \cot\varphi_{a0} ,$$

$$a_{11} = \beta_0\, e_{\nu x_1} , \qquad a_{12} = \beta_0\, e_{\nu x_2} , \qquad a_{13} = \beta_0\, e_{\nu x_3} ,$$

$$A_1 = e_{\nu x_3} l_{12}^{(0)} - e_{\nu x_2} l_{13}^{(0)} ,$$

$$A_2 = e_{\nu x_3} l_{11}^{(0)} - e_{\nu x_1} l_{13}^{(0)} ,$$

$$A_3 = e_{\nu x_1} l_{12}^{(0)} - e_{\nu x_2} l_{11}^{(0)} ,$$

$$f_{12} = \beta_0 l_{13}^{(0)} , \qquad f_{13} = -\beta_0 l_{12}^{(0)} ,$$

$$f_{21} = \beta_0 l_{13}^{(0)} , \qquad f_{23} = -\beta_0 l_{11}^{(0)} ,$$

$$f_{31} = -\beta_0 l_{12}^{(0)} , \qquad f_{32} = \beta_0 l_{11}^{(0)} ,$$

$$g_{13} = e_{\nu x_3} l_{22}^{(0)} - e_{\nu x_2} l_{23}^{(0)}\,, \qquad g_{12} = -e_{\nu x_3} l_{32}^{(0)} + e_{\nu x_2} l_{33}^{(0)}\,,$$
$$g_{23} = e_{\nu x_3} l_{21}^{(0)} - e_{\nu x_1} l_{23}^{(0)}\,, \qquad g_{22} = -e_{\nu x_3} l_{31}^{(0)} + e_{\nu x_1} l_{33}^{(0)}\,,$$
$$g_{33} = e_{\nu x_1} l_{22}^{(0)} - e_{\nu x_2} l_{21}^{(0)}\,, \qquad g_{32} = -e_{\nu x_1} l_{32}^{(0)} + e_{\nu x_2} l_{31}^{(0)}\,,$$

A.3 Elements of the Matrixes M$^{(1)}$ and M$^{(2)}$

$$m_{j1}^{(1)} = \mu_{10} \gamma_3\, l_{1j}^{(0)} d_1^{(1)} \;\mathrm{sgn}\,\alpha_a,\,,$$
$$m_{j2}^{(1)} = \mu_{10} \left[\gamma_3\, l_{1j}^{(0)} d_2^{(1)} + \gamma_1 \left(\sum_{\rho=1}^{3} e_{\nu x_\rho} l_{3\rho}^{(0)}\right) l_{1j}^{(0)} - \gamma_4\, l_{3j}^{(0)}\right] \mathrm{sgn}\,\alpha_a\,,$$
$$m_{j3}^{(1)} = \mu_{10} \left[\gamma_3\, l_{1j}^{(0)} d_3^{(1)} - \gamma_1 \left(\sum_{\rho=1}^{3} e_{\nu x_\rho} l_{2\rho}^{(0)}\right) l_{1j}^{(0)} + \gamma_4\, l_{2j}^{(0)}\right] \mathrm{sgn}\,\alpha_a\,,$$

$$\begin{aligned}
m_{j1}^{(2)} &= \mu_{10}\Big[-\Big(\gamma_1\, \beta_0\, \cos\varphi_{a0} + \gamma_2\Big)\, l_{1j}^{(0)} e_{\nu x_1} \\
&\qquad + \gamma_1\, l_{1j}^{(0)} \beta_0 l_{11}^{(0)} + \gamma_3\, l_{3j}^{(0)} d_1^{(2)}\Big] \;\mathrm{sgn}\,\alpha_a\,, \\
m_{j2}^{(2)} &= \mu_{10}\Big[-\Big(\gamma_1\, \beta_0\, \cos\varphi_{a0} + \gamma_2\Big)\, l_{1j}^{(0)} e_{\nu x_2} \\
&\qquad + \gamma_1\, l_{1j}^{(0)} \beta_0 l_{12}^{(0)} + \gamma_3\, l_{3j}^{(0)} d_2^{(2)}\Big] \;\mathrm{sgn}\,\alpha_a\,, \\
m_{j3}^{(2)} &= \mu_{10}\Big[-\Big(\gamma_1\, \beta_0\, \cos\varphi_{a0} + \gamma_2\Big)\, l_{1j}^{(0)} e_{\nu x_3} \\
&\qquad + \gamma_1\, l_{1j}^{(0)} \beta_0 l_{13}^{(0)} + \gamma_3\, l_{3j}^{(0)} d_3^{(2)}\Big] \;\mathrm{sgn}\,\alpha_a\,, \qquad j = 1,\, 2,\, 3\,,
\end{aligned}$$

where

$$\mu_{10} = c_{m0} \mu_{100} \sin^2 \varphi_{a0}\,, \qquad \mu_{100} = \frac{\rho\, V_0^2\, b\, l^2}{2\, A_{33}(0)}\,,$$
$$\gamma_1 = 2\, c_{m0}\, \cos\varphi_{a0}\,, \qquad \gamma_2 = 2\, \beta_0 c_{m0}\, \sin^2 \varphi_{a0}\,,$$
$$\gamma_3 = \frac{\partial c_m}{\partial \alpha_a}\, \sin^2 \varphi_{a0}\,, \qquad \gamma_4 = c_{m0}\, \sin^2 \varphi_{a0}\,.$$

B. Solution of the Problems

B.1 To Chapter 3

3.1. The rod oscillates about its equilibrium position, therefore, its stress strain state in static should be determined first. Because the rod is inextensible and located symmetrically with respect to the axis of rotation, we have

$$\Delta æ \equiv 0\,, \qquad \mathbf{M}_0 = 0\,, \qquad q_1 = q_3 = 0\,, \qquad q_2 = -\omega_0^2 R_0\,.$$

From the equilibrium equations, we get

$$Q_{20} = 0\,, \qquad Q_{10} = -\frac{q_2}{æ_{30}}\,, \qquad æ_{30} = \frac{1}{R_0}\,.$$

Let us evaluate analytical expressions for the dynamical loads caused by oscillations of the rod. For this purpose, it is useful to refer to Fig. B.1 and to recall the derivation of an expression for the acceleration of a material point in a rotating frame of reference.

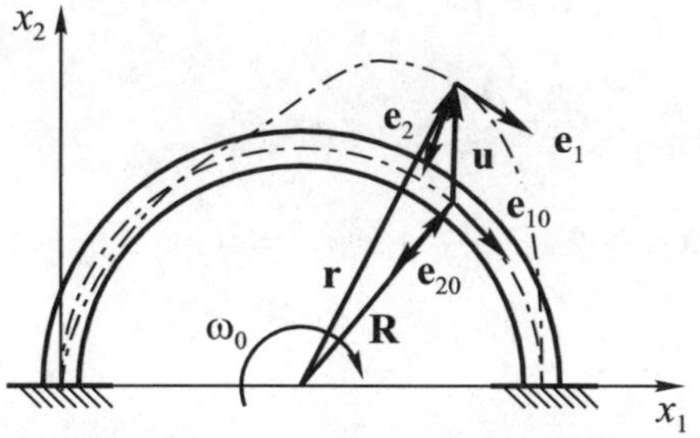

Fig. B.1.

The dash-dotted line in Fig. B.1 indicates the rod's axis during oscillations. From Fig. B.1 it can be seen that the absolute velocity of the point $O^{(1)}$ is

$$\mathbf{V}_{O^{(1)}} = \frac{\mathrm{d}}{\mathrm{d}\tau}\left(\mathbf{R}_0 + \mathbf{u}\right).$$

The change of the absolute velocity of the point $O^{(1)}$ reads

$$\mathbf{v} = \mathbf{V}_{O^{(1)}} - \mathbf{V}_0 = \frac{\mathrm{d}\mathbf{u}}{\mathrm{d}\tau}\,, \qquad \mathbf{V}_0 = \frac{\mathrm{d}\mathbf{R}_0}{\mathrm{d}\tau}\,.$$

Using local derivative notation, we get

$$\mathbf{v} = \frac{\tilde{\partial}\mathbf{u}}{\partial\tau} + \boldsymbol{\omega}_0 \times \mathbf{u}\,, \qquad |\boldsymbol{\omega}_0| = \text{const}\,.$$

The change of the absolute acceleration of the point $O^{(1)}$ (the tildes indicating the local derivatives are dropped) is

$$\frac{\mathrm{d}\mathbf{v}}{\mathrm{d}\tau} = \frac{\partial}{\partial\tau}\left(\frac{\partial\mathbf{u}}{\partial\tau} + \boldsymbol{\omega}_0 \times \mathbf{u}\right) + \boldsymbol{\omega}_0 \times \left(\frac{\partial\mathbf{u}}{\partial\tau} + \boldsymbol{\omega}_0 \times \mathbf{u}\right)\,,$$

$$\frac{\mathrm{d}\mathbf{v}}{\mathrm{d}\tau} = \frac{\partial^2\mathbf{u}}{\partial\tau^2} + 2\,\boldsymbol{\omega}_0 \times \frac{\partial\mathbf{u}}{\partial\tau} - \omega_0^2 u_1\mathbf{e}_1 - \omega_0^2 u_2\mathbf{e}_2\,. \tag{1}$$

From (1) it follows that the additional dynamic load $\Delta\mathbf{q}$ on the rod fixed on a rotating disk can be written as

$$\Delta\mathbf{q} = -2\left(\boldsymbol{\omega}_0 \times \frac{\partial\mathbf{u}}{\partial\tau}\right) + \omega_0^2 u_1\mathbf{e}_1 - \omega_0^2 u_2\mathbf{e}_2\,,$$

or (since $\boldsymbol{\omega}_0 = \omega_0\mathbf{e}_3$)

$$\Delta q_1 = 2\,\omega_0\,\frac{\partial u_2}{\partial\tau} + \omega_0^2 u_1\,,$$
$$\Delta q_2 = -2\,\omega_0\,\frac{\partial u_1}{\partial\tau} + \omega_0^2 u_2\,. \tag{2}$$

Using, for example, equations (3.68) corresponding to a rod of constant cross section together with (2), we finally arrive at

$$\frac{\partial^2 u_1}{\partial\tau^2} - \frac{\partial\Delta Q_1}{\partial\eta} + Q_{20}\Delta æ_3 + æ_{30}\Delta Q_2 - 2\,\omega_0\frac{\partial u_2}{\partial\tau} - \omega_0^2 u_1 = 0\,,$$
$$\frac{\partial^2 u_2}{\partial\tau^2} - \frac{\partial\Delta Q_2}{\partial\eta} - Q_{10}\Delta æ_3 - æ_{30}\Delta Q_1 + 2\,\omega_0\frac{\partial u_1}{\partial\tau} - \omega_0^2 u_2 = 0\,,$$
$$J_{33}\,\frac{\partial^2\vartheta_3}{\partial\tau^2} - \frac{\partial\Delta M_3}{\partial\eta} - \Delta Q_2 = 0\,,$$
$$\frac{\partial\vartheta_3}{\partial\eta} - \Delta æ_3 = 0\,,$$
$$\frac{\partial u_1}{\partial\eta} - æ_{30}u_2 = 0\,,$$
$$\frac{\partial u_2}{\partial\eta} + æ_{30}u_1 - \vartheta_3 = 0\,, \qquad \Delta M_3 = \Delta æ_{33}\,. \tag{3}$$

3.2. The ring is exerted by a static distributed load, hence, we have

$$Q_{10} = -q_0 R_0 , \qquad Q_{20} = Q_{30} = 0 , \qquad M_{10} = M_{20} = M_{30} = 0 .$$

The equations of free oscillations for a rod whose static axial line is a plane curve split out into two independent sets of equations, (3.68) and (3.69). In the case under consideration (the case of oscillations of the rod in the plane of drawing), the system of equations (3.69) should be used. Because the load is follower and the equations of small oscillations (3.69) are associated with the rod-fixed frame, we have $\Delta q_3 = 0$ (and $\Delta q_1 = \Delta q_2 = 0$). For a rod of constant cross section, the system of equations (3.69) take the form (here, the initial stress state is taken into consideration)

$$\begin{aligned}
&\frac{\partial^2 u_3}{\partial \tau^2} - \frac{\partial \Delta Q_3}{\partial \eta} - q_0 R_0 \Delta M_2 = 0 , \\
&J_{11} \frac{\partial^2 \vartheta_1}{\partial \tau^2} - \frac{\partial \Delta M_1}{\partial \eta} + \frac{1}{R_0} \Delta M_2 = 0 , \\
&J_{22} \frac{\partial^2 \vartheta_2}{\partial \tau^2} - \frac{\partial \Delta M_2}{\partial \eta} - \frac{1}{R_0} \Delta M_1 + \Delta Q_3 = 0 , \\
&\frac{\partial \vartheta_1}{\partial \eta} - \frac{1}{R_0} \vartheta_2 - \frac{1}{A_{11}} \Delta M_1 = 0 , \\
&\frac{\partial \vartheta_2}{\partial \eta} + \frac{1}{R_0} \vartheta_1 - \Delta M_2 = 0 , \\
&\frac{\partial u_3}{\partial \eta} + \vartheta_2 = 0 .
\end{aligned}$$

For a rod of circular cross section, the nondimensional moments of inertia and the stiffness coefficients are as follows:

$$J_{11} = \frac{d^2}{8\,l^2} , \qquad J_{22} = \frac{d}{16\,l} , \qquad A_{11} = \frac{2\,G}{E} , \qquad A_{22} = 1 .$$

3.4. Because the static stress strain state can be neglected, in the equations of small oscillations (say in (3.38) and (3.39)) we can put $Q_{j0} = M_{j0} = 0$, $j = 1, 2$. Let us consider in greater detail the right-hand sides of the systems of equations (3.38) and (3.39), which contain the concentrated forces of inertia. In this problem, there are no other forces acting on the rod but the concentrated forces of inertia produced by the masses m_1 and m_2, thus we obtain

$$\begin{aligned}
\Delta P_1 &= -\tilde{m}_1 \frac{\partial^2 u_1}{\partial \tau^2} \,\delta(\eta - \eta_1) - \tilde{m}_2 \frac{\partial^2 u_1}{\partial \tau^2} \,\delta(\eta - \eta_2) , \\
\Delta P_2 &= -\tilde{m}_1 \frac{\partial^2 u_2}{\partial \tau^2} \,\delta(\eta - \eta_1) - \tilde{m}_2 \frac{\partial^2 u_2}{\partial \tau^2} \,\delta(\eta - \eta_2) , \\
\Delta P_3 &= -\tilde{m}_1 \frac{\partial^2 u_3}{\partial \tau^2} \,\delta(\eta - \eta_1) - \tilde{m}_2 \frac{\partial^2 u_3}{\partial \tau^2} \,\delta(\eta - \eta_2) , \qquad (1)
\end{aligned}$$

$$
\begin{aligned}
\Delta T_1 &= -J_{01}\frac{\partial^2\vartheta_1}{\partial\tau^2}\,\delta(\eta-\eta_2)\,,\\
\Delta T_2 &= -J_{02}\frac{\partial^2\vartheta_2}{\partial\tau^2}\,\delta(\eta-\eta_2)\,,\\
\Delta T_3 &= -J_{03}\frac{\partial^2\vartheta_3}{\partial\tau^2}\,\delta(\eta-\eta_2)\,.
\end{aligned}
\tag{2}
$$

The nondimensional quantities $\tilde{m}_i$ and J_{0i} are as follows

$$
\tilde{m}_j = \frac{m_j}{m_0\,l}\,, \qquad J_{0j} = \frac{J_j^{(0)}}{m_0\,l^3}\,, \qquad j = 1,\,2\,,
$$

where $J_j^{(0)}$ are the principal moments of inertia of the mass m_2. Equations (3.40)–(3.43) remain the same.

3.5. We examine oscillations in the plane x_1Ox_2 (Fig. 3.13) relative to the rod's natural configuration, therefore, we have $Q_{10} = Q_{20} = M_{30} = 0$. During oscillations, the rod is exerted by two concentrated forces, $\mathbf{R}_1$ and $\mathbf{R}_2$, applied at the cross sections with coordinates η_1 and η_2, respectively. These forces are

$$
\mathbf{R}_1 = R_2\,\mathbf{e}_2\,, \qquad \mathbf{R}_2 = R_{x_2}\,\mathbf{i}_2\,,
$$

where $R_{x_2} = -c\,(\mathbf{u}\cdot\mathbf{i}_2)$. Therefore, we have

$$
\begin{aligned}
\Delta P_1 &= -\tilde{c}\,(\mathbf{u}\cdot\mathbf{i}_2)\,(\mathbf{i}_2\cdot\mathbf{e}_1)\,\delta(\eta-\eta_2)\,,\\
\Delta P_2 &= R_2\,\delta(\eta-\eta_1) - \tilde{c}\,(\mathbf{u}\cdot\mathbf{i}_2)\,(\mathbf{i}_2\cdot\mathbf{e}_2)\,\delta(\eta-\eta_2)\,,\\
\mathbf{i}_2 &= \sin(\vartheta_{30}+\Delta\vartheta_3)\,\mathbf{e}_1 + \cos(\vartheta_{30}+\Delta\vartheta_3)\,\mathbf{e}_2\,, \qquad \tilde{c} = \frac{c\,l^3}{A_{33}}\,,
\end{aligned}
\tag{1}
$$

where ϑ_{30} is the angle between the vectors $\mathbf{e}_1$ and $\mathbf{i}_1$. Because of smallness of the oscillations, on rearrangement, we get

$$
\begin{aligned}
\Delta P_1 &= -\tilde{c}\,(u_1\,\sin^2\vartheta_{30} + u_2\,\cos\vartheta_{30}\sin\vartheta_{30})\,\delta(\eta-\eta_2)\,,\\
\Delta P_2 &= R_2\,\delta(\eta-\eta_1) - \tilde{c}\,(u_1\,\cos\vartheta_{30}\sin\vartheta_{30} + u_2\,\cos^2\vartheta_{30})\,\delta(\eta-\eta_2)\,.
\end{aligned}
\tag{2}
$$

Using the system of equations (3.68) and taking into account the right-hand sides of the first two equations, we arrive at the following system of equations:

$$
\begin{aligned}
&n_1\frac{\partial^2 u_1}{\partial\tau^2} - \frac{\partial\Delta Q_1}{\partial\eta} + æ_{30}\Delta Q_2\\
&= -\tilde{c}\,\sin^2\vartheta_{30}\cdot u_1\,\delta(\eta-\eta_2) - \tilde{c}\,\cos^2\vartheta_{30}\sin\vartheta_{30}\cdot u_2\,\delta(\eta-\eta_2)\,,\\
&n_1\frac{\partial^2 u_2}{\partial\tau^2} - \frac{\partial\Delta Q_2}{\partial\eta} - æ_{30}\Delta Q_1\\
&= R_2\,\delta(\eta-\eta_1) - \tilde{c}\,\cos\vartheta_{30}\,\sin\vartheta_{30}\cdot\delta(\eta-\eta_2) - \tilde{c}\,\cos^2\vartheta_{30}\cdot u_2\,\delta(\eta-\eta_2)\,.
\end{aligned}
\tag{3}
$$

The other equations of the system (3.68) remain the same.

3.6. Small free oscillations of a spiral in the plane of drawing are governed by the system of equations (3.70). However, to be able to use these equations in numerical simulation, we should establish how the curvature $æ_{30}(\eta)$ depends on the axial coordinate η. Without this information, the equations cannot be solved (for example, the frequencies cannot be found). Thus, let us find the curvature of the spiral as a function of η. In polar coordinates (Fig. B.2), the spiral of Archimedes is given by

$$r = a\varphi\,, \qquad x_1 = r\cos\varphi\,, \qquad x_2 = r\sin\varphi\,. \tag{1}$$

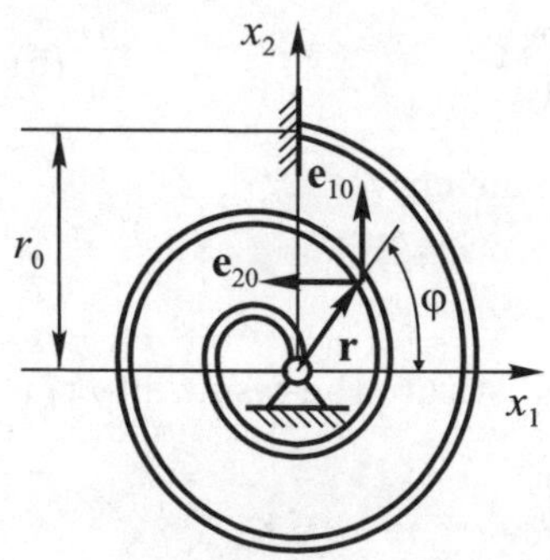

Fig. B.2.

The parameter a will be defined later. The spiral length is l. From the end fixity conditions ($\varphi = 0$ at $r = 0$ and $\varphi = a\pi/2$ at $r = r_0$) it follows that the angle φ is varied through the range $0 \le \varphi \le a\pi/2$. The value of r_0 is unknown because the length of the spiral but not its shape is known. The differential of the arc length in polar coordinates has the representation

$$\mathrm{d}\eta^2 = \tilde{a}\left(\frac{\mathrm{d}\varphi}{\mathrm{d}\eta}\right)^2 + \tilde{a}^2\varphi^2\left(\frac{\mathrm{d}\varphi}{\mathrm{d}\eta}\right)^2, \tag{2}$$

(here, η is nondimensional axial coordinate), therefore, we have

$$\mathrm{d}\eta = \tilde{a}\sqrt{1+\varphi^2}\,\mathrm{d}\varphi\,, \qquad \tilde{a} = \frac{a}{l}\,. \tag{3}$$

Integrating (3) from 0 to 1 yields

$$1 = \tilde{a}\int_0^{a\pi/2} \sqrt{1+\varphi^2}\,\mathrm{d}\varphi\,. \tag{4}$$

Then, from this equation we solve for $\tilde{a}$. Dividing (1) by l, we obtain the following nondimensional relationships:

$$r = \tilde{a}\varphi\,, \qquad \tilde{x}_1 = \tilde{r}\cos\varphi\,, \qquad \tilde{x}_r = \tilde{r}\sin\varphi\,.$$

Having found $\tilde{a}$, we then calculate the nondimensional value $\tilde{r}_0 = 9\tilde{a}\pi/2$. Hereinafter we drop the tildes ornamenting the nondimensional variables.

It follows from (2) that

$$\frac{\mathrm{d}\varphi}{\mathrm{d}\eta} = \frac{1}{a\sqrt{1+\varphi^2}}\,. \tag{5}$$

Upon numerical integration of (5), we obtain the relationship $\varphi(\eta)$ ($\varphi = 0$ when $\eta = 0$). Using (5), we find the derivatives $x_1'(\eta)$ and $x_2'(\eta)$ to be

$$\begin{aligned} x_1'(\eta) &= a\varphi' \cos\varphi - a\varphi\varphi' \sin\varphi = \frac{\cos\varphi - \varphi\sin\varphi}{\sqrt{1+\varphi^2}}\,, \\ x_2'(\eta) &= a\varphi' \sin\varphi + a\varphi\varphi' \cos\varphi = \frac{\sin\varphi + \varphi\cos\varphi}{\sqrt{1+\varphi^2}}\,. \end{aligned} \tag{6}$$

Recall the expression for the curvature of a plane curve

$$æ_{30}(\eta) = \sqrt{(x_1'')^2 + (x_2'')^2}\,. \tag{7}$$

Then, we differentiate (6) with respect to η and substitute the result into (7). On rearrangement and elimination of φ' and

$$\varphi'' = -\frac{\varphi\varphi'}{a(1+\varphi^2)^{3/2}}\,,$$

we get

$$æ_{30}(\eta) = \frac{1}{a}\,\frac{(2+\varphi^2)}{(1+\varphi^2)^{3/2}}\,. \tag{8}$$

From (5) we can find φ as a function of η and then, using (8), find $æ_{30}$ as a function of η.

B.2 To Chapter 5

5.1. To solve this problem, first we should determine the static stress strain state. That is, we should find the displacements of the points of the rod's axial line when the rod is subjected to the dead force **P** (recall that we consider small displacements). The solution of the equilibrium equations for a rod was considered in Part 1 of Svetlitsky (2000a) (for example, see Chapter 2). After the force **P** ceases, we have $\mathbf{Q}_0 = \mathbf{M}_0 = 0$. By solving the equilibrium equations, we obtain, among other things, the components u_{01} and u_{02} of the displacement vector.

Let us use the system of equations (5.57), which governs small oscillation of the rod in the plane of drawing. In these equations we put $Q_{10} = Q_{20} = M_{30} = 0$. Since the rod is of a circular constant cross section, we have

$$æ_{30} = \frac{\pi}{2}, \qquad n_1 = 1, \qquad A_{33} = 1, \qquad J_{33} = \frac{d}{16\,l},$$

where d is the diameter of the rod. As a result, we obtain the following equations:

$$\begin{aligned}
&\frac{\partial^2 u_1}{\partial \tau^2} - \frac{\partial \Delta Q_1}{\partial \eta} + \frac{\pi}{2} \Delta Q_2 = 0, \\
&\frac{\partial^2 u_2}{\partial \tau^2} - \frac{\partial \Delta Q_2}{\partial \eta} - \frac{\pi}{2} \Delta Q_1 = 0, \\
&J_{33} \frac{\partial^2 \vartheta_3}{\partial \tau^2} - \frac{\partial \Delta M_3}{\partial \eta} - \Delta Q_2 = 0, \\
&\frac{\partial \vartheta_3}{\partial \eta} - \Delta M_3 = 0, \\
&\frac{\partial u_1}{\partial \eta} - \frac{\pi}{2} u_2 = 0, \\
&\frac{\partial u_2}{\partial \eta} + \frac{\pi}{2} u_1 - \vartheta_3 = 0.
\end{aligned} \tag{1}$$

Setting $u_1 = u_{10}(\eta)\,\mathrm{e}^{\mathrm{i}\,\lambda\tau}$, $u_2 = u_{20}(\eta)\,\mathrm{e}^{\mathrm{i}\,\lambda\tau}$, and so forth, we arrive at the vector equation (similar to (5.60))

$$\mathbf{Z}_0' + \mathbf{A}\mathbf{Z}_0 = 0, \tag{2}$$

where

$$\mathbf{A} = \begin{bmatrix} 0 & -\frac{\pi}{2} & 0 & 0 & \lambda^2 & 0 \\ \frac{\pi}{2} & 0 & 0 & 0 & 0 & \lambda^2 \\ 0 & 1 & 0 & J_{33}\lambda^2 & 0 & 0 \\ 0 & 0 & -1 & 0 & 0 & 0 \\ 0 & 0 & 0 & 0 & 0 & -\frac{\pi}{2} \\ 0 & 0 & 0 & -1 & \frac{\pi}{2} & 0 \end{bmatrix}, \qquad \mathbf{Z}_0 = \begin{vmatrix} \Delta Q_{10} \\ \Delta Q_{20} \\ \Delta M_{30} \\ \vartheta_{30} \\ u_{10} \\ u_{20} \end{vmatrix}.$$

Using the boundary conditions (1) at $\eta = 0$, $u_{10} = u_{20} = \vartheta_{30} = 0$, and (2) at $\eta = 1$, $\Delta Q_1 = \Delta Q_2 = \Delta M_3 = 0$, and restricting our consideration to the first three terms, $j = 1, 2, 3$, we find λ_j. For each λ_j we determine a corresponding eigenvector (a method for determination of eigenvectors was discussed in Sect. 4.4). To solve this problem, we should know only $u_{k0}^{(j)}(\eta)$, where $k = 1, 2$ and $j = 1, 2, 3$.

The initial conditions for this problem are as follows: (1) at $\tau = 0$, $\mathbf{u} = \mathbf{u}_0$, and (2) at $\tau = 0$, $\dot{\mathbf{u}} = 0$.

For the three λ_j, the components of the vector $\mathbf{u}$ read

$$u_1 = c^{(1)} u_{10}^{(1)} \cos\lambda_1\tau + c^{(2)} u_{10}^{(2)} \cos\lambda_2\tau + c^{(3)} u_{10}^{(3)} \cos\lambda_3\tau\,,$$
$$u_2 = c^{(1)} u_{20}^{(1)} \cos\lambda_1\tau + c^{(2)} u_{20}^{(2)} \cos\lambda_2\tau + c^{(3)} u_{20}^{(3)} \cos\lambda_3\tau\,. \qquad (3)$$

In vector form, we have

$$\mathbf{u} = \sum c^{(j)} \mathbf{u}^{(j)} \cos\lambda_j\tau\,.$$

Denoting $\mathbf{u}_j = \mathbf{Z}_u^{(j)}$, for $\tau = 0$, we have

$$\mathbf{u}_0 = \sum_{j=1}^{3} c^{(j)} \mathbf{Z}_u^{(j)}\,,$$

where

$$\mathbf{Z}_u^{(j)} = \begin{vmatrix} u_{10}^{(j)} \\ u_{20}^{(j)} \end{vmatrix}\,.$$

We have three equations (similar to (5.16)) to be solved for $c^{(j)}$:

$$\int_0^1 (\mathbf{u}_0 \cdot \mathbf{Z}_u^{(1)})\,\mathrm{d}\eta$$
$$= c^{(1)} \int_0^1 (\mathbf{Z}_u^{(1)})^2\,\mathrm{d}\eta + c^{(2)} \int_0^1 (\mathbf{Z}_u^{(2)} \cdot \mathbf{Z}_u^{(1)})\,\mathrm{d}\eta + c^{(3)} \int_0^1 (\mathbf{Z}_u^{(3)} \cdot \mathbf{Z}_u^{(1)})\,\mathrm{d}\eta\,,$$
$$\int_0^1 (\mathbf{u}_0 \cdot \mathbf{Z}_u^{(2)})\,\mathrm{d}\eta$$
$$= c^{(1)} \int_0^1 (\mathbf{Z}_u^{(1)} \cdot \mathbf{Z}_u^{(2)})\,\mathrm{d}\eta + c^{(2)} \int_0^1 (\mathbf{Z}_u^{(2)})^2\,\mathrm{d}\eta + c^{(3)} \int_0^1 (\mathbf{Z}_u^{(3)} \cdot \mathbf{Z}_u^{(2)})\,\mathrm{d}\eta\,,$$
$$\int_0^1 (\mathbf{u}_0 \cdot \mathbf{Z}_u^{(3)})\,\mathrm{d}\eta$$
$$= c^{(1)} \int_0^1 (\mathbf{Z}_u^{(1)} \cdot \mathbf{Z}_u^{(3)})\,\mathrm{d}\eta + c^{(2)} \int_0^1 (\mathbf{Z}_u^{(2)} \cdot \mathbf{Z}_u^{(3)})\,\mathrm{d}\eta + c^{(3)} \int_0^1 (\mathbf{Z}_u^{(3)})^2\,\mathrm{d}\eta\,.$$

Since our goal is to establish the law of motion of the point K (at $\eta = 1$) along the x_1-axis, let us find the projection u_{x_1} of the total displacement of the point K on the x_1-axis. Assuming the angles of rotation of the end cross section to be small, we finally arrive at

$$u_{x1} = (u_K \cdot \mathbf{i}_1) = (u_{1K}\mathbf{e}_1 + u_{2K}\mathbf{e}_2) \cdot \mathbf{i}_1 = -u_2$$

or

$$u_{x1} = -(c^{(1)} u_{20}^{(1)}(1) \cos\lambda_1\tau + c^{(2)} u_{20}^{(2)}(1) \cos\lambda_2\tau + c^{(3)} u_{20}^{(3)}(1) \cos\lambda_3\tau)\,.$$

5.2. We put $Q_{10} = Q_{20} = 0$, $n_1 = 1$, and $A_{33} = 1$ in (5.57). In vector-matrix form, the system of equations (5.57) reads

$$\mathbf{L}_1 = \mathrm{A}^{(1)} \frac{\partial^2 \mathbf{Z}}{\partial \tau^2} + \frac{\partial \mathbf{Z}}{\partial \eta} + \mathrm{A}^{(2)} \mathbf{Z} - \Delta \boldsymbol{\Phi} = 0\,, \tag{1}$$

where

$$\mathrm{A}^{(1)} = \begin{bmatrix} 0 & 0 & 0 & 0 & 1 & 0 \\ 0 & 0 & 0 & 0 & 0 & 1 \\ 0 & 0 & 0 & J_{33} & 0 & 0 \\ 0 & 0 & 0 & 0 & 0 & 0 \\ 0 & 0 & 0 & 0 & 0 & 0 \\ 0 & 0 & 0 & 0 & 0 & 0 \end{bmatrix},$$

$$\mathrm{A}^{(2)} = \begin{bmatrix} 0 & æ_{30} & 0 & 0 & 0 & 0 \\ -æ_{30} & 0 & 0 & 0 & 0 & 0 \\ 0 & -1 & 0 & 0 & 0 & 0 \\ 0 & 0 & -1 & 0 & 0 & 0 \\ 0 & 0 & 0 & 0 & 0 & -æ_{30} \\ 0 & 0 & 0 & -1 & æ_{30} & 0 \end{bmatrix},$$

$$\mathbf{Z} = (\Delta Q_1,\, \Delta Q_2,\, \Delta M_3,\, \vartheta_3,\, u_1,\, u_2)^{\mathrm{T}}\,,$$
$$\Delta \boldsymbol{\Phi} = (0,\, 0,\, T_3,\, 0,\, 0,\, 0)^{\mathrm{T}}\, \delta(\eta - \eta_1)\,.$$

To obtain a two-term approximate solution to (1), we should determine preliminary the eigenfunctions $\mathbf{Z}_0^{(1)}$ and $\mathbf{Z}_0^{(2)}$. The equations used for determining the eigenvalues and eigenfunctions for a rod of circular cross section, executing oscillations in the plane of drawing, are given in the solution of Problem 5.1. Once the eigenfunctions (eigenvectors) have been found, we seek a solution to the vector equation (1) of the form

$$\mathbf{Z} = f^{(1)}(\tau)\, \mathbf{Z}_0^{(1)}(\eta) + f^{(2)}(\tau)\, \mathbf{Z}_0^{(2)}(\eta)\,. \tag{2}$$

Next, we use the generalized principal of virtual displacements (see Chap. 4) and put

$$\delta \mathbf{Z}_0^{(1)} = \delta b_1\, \mathrm{E}_0 \mathbf{Z}_0^{(1)}\,, \qquad \delta \mathbf{Z}_0^{(2)} = \delta b_2\, \mathrm{E}_0 \mathbf{Z}_0^{(2)}\,,$$

where

$$\mathrm{E}_0 = \begin{bmatrix} 0 & 0 & 0 & 0 & 1 & 0 \\ 0 & 0 & 0 & 0 & 0 & 1 \\ 0 & 0 & 0 & 1 & 0 & 0 \\ 0 & 0 & 1 & 0 & 0 & 0 \\ 1 & 0 & 0 & 0 & 0 & 0 \\ 0 & 1 & 0 & 0 & 0 & 0 \end{bmatrix}, \qquad \mathrm{E}_0 \mathbf{Z}_0^{(j)} = \begin{bmatrix} u_{10}^{(j)} \\ u_{20}^{(j)} \\ \vartheta_{30}^{(j)} \\ \Delta M_{30}^{(j)} \\ \Delta Q_{10}^{(j)} \\ \Delta Q_{20}^{(j)} \end{bmatrix}.$$

The matrix E_0 is a particular case of the matrix (4.136). According to the algorithm of deriving the equations in f_1 and f_2, we obtain two equations

$$\int_0^1 (\mathbf{L}_1 \cdot \mathrm{E}_0 \mathbf{Z}_0^{(1)})\, \mathrm{d}\eta = 0\,, \qquad \int_0^1 (\mathbf{L}_1 \cdot \mathrm{E}_0 \mathbf{Z}_0^{(2)})\, \mathrm{d}\eta = 0\,,$$

or

$$\begin{aligned} h_{11}\ddot{f}^{(1)} + h_{12}\ddot{f}^{(2)} + b_{11}f^{(1)} + b_{12}f^{(2)} &= b_1\,, \\ h_{21}\ddot{f}^{(1)} + h_{22}\ddot{f}^{(2)} + b_{21}f^{(1)} + b_{22}f^{(2)} &= b_2\,, \end{aligned} \tag{3}$$

where

$$h_{11} = \int_0^1 (\mathrm{A}^{(1)}\mathbf{Z}_0^{(1)} \cdot \mathrm{E}_0\mathbf{Z}_0^{(1)})\, \mathrm{d}\eta\,, \qquad h_{12} = \int_0^1 (\mathrm{A}^{(1)}\mathbf{Z}_0^{(2)} \cdot \mathrm{E}_0\mathbf{Z}_0^{(1)})\, \mathrm{d}\eta\,,$$

$$h_{21} = \int_0^1 (\mathrm{A}^{(1)}\mathbf{Z}_0^{(1)} \cdot \mathrm{E}_0\mathbf{Z}_0^{(2)})\, \mathrm{d}\eta\,, \qquad h_{22} = \int_0^1 (\mathrm{A}^{(1)}\mathbf{Z}_0^{(2)} \cdot \mathrm{E}_0\mathbf{Z}_0^{(2)})\, \mathrm{d}\eta\,,$$

$$b_{11} = \int_0^1 (\mathbf{Z}_0^{(1)\prime} + \mathrm{A}^{(2)}\mathbf{Z}_0^{(1)})\, \mathrm{E}_0\mathbf{Z}_0^{(1)}\, \mathrm{d}\eta\,,$$

$$b_{12} = \int_0^1 (\mathbf{Z}_0^{(2)\prime} + \mathrm{A}^{(2)}\mathbf{Z}_0^{(2)})\, \mathrm{E}_0\mathbf{Z}_0^{(1)}\, \mathrm{d}\eta\,,$$

$$b_{21} = \int_0^1 (\mathbf{Z}_0^{(1)\prime} + \mathrm{A}^{(2)}\mathbf{Z}_0^{(1)})\, \mathrm{E}_0\mathbf{Z}_0^{(2)}\, \mathrm{d}\eta\,,$$

$$b_{22} = \int_0^1 (\mathbf{Z}_0^{(2)\prime} + \mathrm{A}^{(2)}\mathbf{Z}_0^{(2)})\, \mathrm{E}_0\mathbf{Z}_0^{(2)}\, \mathrm{d}\eta\,.$$

As an example, consider the expressions for two coefficients h_{11} and b_{11} in greater detail:

$$h_{11} = \int_0^1 \left[\left(u_{10}^{(1)}\right)^2 + \left(u_{20}^{(1)}\right)^2 + J_{33}\left(\vartheta_{30}^{(1)}\right)^2\right] \mathrm{d}\eta\,,$$

$$\begin{aligned} b_{11} = \int_0^1 \Big[& \Delta Q_{10}^{(1)\prime} u_{10}^{(1)} + \Delta Q_{20}^{(1)\prime} u_{20}^{(1)} + \Delta M_{30}^{(1)\prime} \vartheta_{30}^{(1)} + \vartheta_{30}^{(1)\prime} \Delta M_{30}^{(1)} \\ & + u_{10}^{(1)\prime} \Delta Q_{10}^{(1)} + u_{20}^{(1)\prime} \Delta Q_{20}^{(1)} + æ_{30}\Delta Q_{20}^{(1)} u_{10}^{(1)} - æ_{30}\Delta Q_{10}^{(1)} u_{20}^{(1)} \\ & - \Delta Q_{20}^{(1)} \vartheta_{30}^{(1)} - \left(\Delta M_{30}^{(1)}\right)^2 - æ_{30} u_{20}^{(1)} \Delta Q_{10}^{(1)} \\ & + \left(æ_{30} u_{10}^{(1)} - \vartheta_{30}^{(1)}\right) \Delta Q_{20}^{(1)} \Big]\, \mathrm{d}\eta\,. \end{aligned}$$

The right-hand sides of (3) are as follows:

$$b_1 = \int_0^1 \left(\Delta\boldsymbol{\Phi} \cdot \mathrm{E}_0\mathbf{Z}_0^{(1)}\right) \mathrm{d}\eta = T_3\vartheta_{30}^{(1)}(\eta_1)\,,$$

$$b_2 = \int_0^1 \left(\Delta\boldsymbol{\Phi} \cdot \mathrm{E}_0\mathbf{Z}_0^{(2)}\right) \mathrm{d}\eta = T_3\vartheta_{30}^{(2)}(\eta_1)\,.$$

In vector-matrix form, the system of equations (3) reads

$$\ddot{\mathbf{f}} + \mathrm{H}^{-1}\mathrm{B}\mathbf{f} = \mathrm{H}^{-1}\mathbf{b}. \tag{4}$$

By setting $\dot{\mathbf{f}} = \mathbf{F}_1$ and $\mathbf{f} = \mathbf{F}_2$, the vector equation (4) can be written as a system of the first order equations

$$\dot{\mathbf{F}}_1 + \mathrm{H}^{-1}\mathrm{B}\mathbf{F}_2 = \mathrm{H}^{-1}\mathbf{b},$$
$$\dot{\mathbf{F}}_2 - \mathbf{F}_1 = 0,$$

or

$$\dot{\mathbf{F}} + \mathrm{A}\mathbf{F} = \mathbf{b}^{(1)}, \tag{5}$$

where

$$\mathbf{F} = \begin{vmatrix} \dot{\mathbf{f}} \\ \mathbf{f} \end{vmatrix}, \qquad \mathrm{A} = \begin{bmatrix} 0 & \mathrm{H}^{-1}\mathrm{B} \\ -\mathrm{E} & 0 \end{bmatrix},$$
$$\mathbf{b}^{(1)} = \begin{vmatrix} \mathrm{H}^{-1}\mathbf{b} \\ 0 \end{vmatrix}, \qquad \mathbf{b} = \begin{vmatrix} b_1 \\ b_2 \end{vmatrix}.$$

The solution of (5) takes the form

$$\mathbf{F} = \mathrm{K}(\tau)\,\mathbf{C} + \int_0^\tau \mathrm{G}(\tau, \tau_1)\,\mathbf{b}^{(1)}\,\mathrm{d}\tau, \tag{6}$$

where $\mathrm{K}(\tau)$ is the fundamental matrix for the homogeneous equation (5) ($\mathrm{K}(0) = \mathrm{E}$).

In our case, we have the zero initial conditions $\mathbf{f}(0) = 0$ and $\dot{\mathbf{f}}(0) = 0$, hence,

$$\mathbf{F}(\tau) = \int_0^\tau \mathrm{G}(\tau, \tau_1)\,\mathbf{b}^{(1)}\,\mathrm{d}\tau_1, \qquad \mathrm{G}(\tau, \tau_1) = \mathrm{K}(\tau)\,\mathrm{K}^{-1}(\tau_1), \tag{7}$$

or (because the vector $\mathbf{b}^{(1)}$ does not depend on τ)

$$\mathbf{F}(\tau) = \int_0^\tau \mathrm{G}(\tau, \tau_1)\,\mathrm{d}\tau_1 \cdot \mathbf{b}^{(1)}. \tag{8}$$

If the right-hand side of (5) is independent of τ, the solution of (5) has the form

$$\mathbf{F} = \mathrm{K}(\tau)\,\mathbf{C} + \mathbf{F}_0, \tag{9}$$

where $\mathbf{F}_0$ is a partial solution to the inhomogeneous equation (5). We seek a partial solution of the form

$$\mathbf{F}_0 = \mathbf{C}_1, \tag{10}$$

where $\mathbf{C}_1$ is a vector with constant components. Substituting (10) into (5) yields $\mathrm{A}\,\mathbf{C}_1 = \mathbf{b}^{(1)}$, therefore, $\mathbf{C}_1 = \mathrm{A}^{-1}\,\mathbf{b}^{(1)}$ and

$$\mathbf{F} = \mathrm{K}(\tau)\,\mathbf{C} + \mathrm{A}^{-1}\mathbf{b}^{(1)} .$$

Because $\mathbf{F} = 0$ and $\mathrm{K}(0) = \mathrm{E}$ for $\tau = 0$, we have $\mathbf{C} = -\mathrm{A}^{-1}\,\mathbf{b}^{(1)}$ and therefore

$$\mathbf{F} = (\mathrm{E} - \mathrm{K}(\tau))\,\mathbf{A}^{-1}\mathbf{b}^{(1)} = \mathbf{P}\mathbf{b}^{(1)}$$

or

$$\mathbf{F} = \left| \begin{array}{c} \dot{\mathbf{f}} \\ \mathbf{f} \end{array} \right| = \left[\begin{array}{cc} P_{11} & P_{12} \\ P_{21} & P_{22} \end{array} \right] \cdot \left| \begin{array}{c} \mathrm{H}^{-1}\mathbf{b} \\ 0 \end{array} \right| . \tag{11}$$

From (11) we can solve for the vector $\mathbf{f}$ whose components, $f^{(1)}$ and $f^{(2)}$, enter into the approximate solution (2):

$$\mathbf{f} = \left| \begin{array}{c} f^{(1)} \\ f^{(2)} \end{array} \right| = P_{11}(\tau)\,\mathrm{H}^{-1}\mathbf{b} . \tag{12}$$

From (12) it follows that

$$f^{(1)} = d_{11}(\tau)\,b_1 + d_{12}(\tau)\,b_2 \,, \qquad f^{(2)} = d_{21}(\tau)\,b_1 + d_{22}(\tau)\,b_2 \,,$$

where d_{ij} are the components of the matrix $\mathrm{P}_{11}(\tau)\,\mathrm{H}^{-1}$. As a result, we obtain the solution

$$Z = (d_{11}b_1 + d_{12}b_2)\,\mathbf{Z}_0^{(1)} + (d_{21}b_1 + d_{22}b_2)\,\mathbf{Z}_0^{(2)} . \tag{13}$$

In this problem, we have to find the moment ΔM_3 at the clamped end ($\eta = 0$). From (13), we finally arrive at

$$Z_3(0,\,\tau) = \Delta M_3 = (d_{11}b_1 + d_{12}b_2)\,\Delta M_{30}^{(1)}(0) + (d_{21}b_1 + d_{22}b_2)\,\Delta M_{30}^{(2)}(0) .$$

B.3 To Chapter 7

7.1. To solve the equations of free oscillations, we should know the static deflection $u_{x_2}^{(0)}$ for the rod subjected to the force P. In other words, we must solve the following nondimensional equation:

$$\frac{\mathrm{d}^4 u_{x_2}^{(0)}}{\mathrm{d}\eta^4} = P\,\delta(\eta - 0.5) . \tag{1}$$

From (1) it follows that

$$u_{x_2}^{(0)} = \frac{(\eta - 0.5)^3}{6}\,\mathrm{PH}(\eta - 0.5) - \frac{1.375}{12}\,P\eta^3 + \frac{0.375}{4}\,P\eta^2 .$$

There are two ways to solve this problem. The first way associated with the use of the Krylov functions, the second one is numerical.

The method of the Krylov functions is relevant only for rods of constant cross section. However, going with this way allows us to obtain solutions in analytical form. Some examples of solutions of equations for oscillations of straight rods are considered in Sect. 7.5.

Considering the inertial force produced by the point mass as a concentrated force applied to the rod, we thus get

$$\frac{\partial^4 u_{x_2}}{\partial \eta^4} + \frac{\partial^2 u_{x_2}}{\partial \tau^2} = \Delta P_1 = -n_2 \frac{\partial^2 u_{x_2}}{\partial \tau^2} \delta(\eta - 0.5) , \qquad n_2 = \frac{m}{m_0 l} . \tag{2}$$

In terms of the Krylov functions, the solution to (2) can be written as

$$u_{x20} = c_1 \mathrm{K}_1 + c_2 \mathrm{K}_2 + c_3 \mathrm{K}_3 + c_4 \mathrm{K}_4 - \frac{\mathrm{K}_4(\eta - 0.5)}{\lambda_0^3} n_2 \lambda_0 u_{x_2}(0.5) \, \mathrm{H}(\eta - 0.5) . \tag{3}$$

Because $u_{x20} = u'_{x20} = 0$ at $\eta = 0$, we have $c_1 = c_2 = 0$ and thus

$$u_{x20}(0.5) = c_3 \mathrm{K}_3(0.5) + c_4 \mathrm{K}_4(0.5) . \tag{4}$$

Eliminating $u_{x_2}(0.5)$ from (3), we obtain

$$\begin{aligned} u_{x20} = {} & [\mathrm{K}_3(\eta) - \lambda_0 n_2 \mathrm{K}_4(\eta - 0.5) \, \mathrm{K}_3(0.5) \, \mathrm{H}(\eta - 0.5)] \, c_3 \\ & + [\mathrm{K}_4(\eta) - \lambda_0 n_2 \mathrm{K}_4(\eta - 0.5) \, \mathrm{K}_4(0.5) \, \mathrm{H}(\eta - 0.5)] \, c_4 . \end{aligned} \tag{5}$$

The boundary conditions at $\eta = 1$ are $u_{x20}(1) = (u_{x20})''(1) = 0$. Thus, we arrive at the following system of two homogeneous equations:

$$\begin{aligned} & [\mathrm{K}_3(\lambda_0) - \lambda_0 n_2 \mathrm{K}_4(\lambda_0 \cdot 0.5) \, \mathrm{K}_3(\lambda_0 \cdot 0.5)] \, c_3 \\ & \qquad + [\mathrm{K}_4(\lambda_0) - \lambda_0 n_2 \mathrm{K}_4^2(\lambda_0 \cdot 0.5)] \, c_4 = 0 , \\ & [\mathrm{K}_1(\lambda_0) - \lambda_0 n_2 \mathrm{K}_2(\lambda_0 \cdot 0.5) \, \mathrm{K}_3(\lambda_0 \cdot 0.5)] \, c_3 \\ & \qquad + [\mathrm{K}_2(\lambda_0) - \lambda_0 n_2 \mathrm{K}_2(\lambda_0 \cdot 0.5) \, \mathrm{K}_4(\lambda_0 \cdot 0.5)] \, c_4 = 0 . \end{aligned} \tag{6}$$

From the condition $D = 0$ (D is the determinant of the system of equations (6)) we find $\lambda_j = \lambda_{0j}^2$. By putting $c_4^{(j)} = 1$, we obtain $c_3^{(j)}$ for each λ_j, that is,

$$c_3^{(j)} = - \frac{\mathrm{K}_4(\lambda_{0j} \cdot 1) - \lambda_{0j} n_2 \mathrm{K}_2^2(\lambda_{0j} \cdot 0.5)}{\mathrm{K}_3(\lambda_{0j} \cdot 1) - \lambda_{0j} n_2 \mathrm{K}_4(\lambda_{0j} \cdot 0.5) \, \mathrm{K}_3(\lambda_{0j} \cdot 0.5)} .$$

As a result, we obtain the eigenfunctions

$$\begin{aligned} \varphi^{(j)}(\eta) & = u_{x_2}^{(j)} \\ & = \{\mathrm{K}_3(\lambda_{0j}, \eta) - \lambda_{0j} n_2 \mathrm{K}_4[(\lambda_{0j}(\eta - 0.5)] \, \mathrm{K}_3(\lambda_{0j} \cdot 0.5) \, \mathrm{H}(\eta - 0.5)\} \, c_3^{(j)} \\ & \quad + \{\mathrm{K}_4(\lambda_{0j}, \eta) - \lambda_{0j} n_2 \mathrm{K}_4[(\lambda_{0j}(\eta - 0.5)] \, \mathrm{K}_4(\lambda_{0j} \cdot 0.5) \, \mathrm{H}(\eta - 0.5)\} . \end{aligned}$$

Recall that we put $u_{x_2} = u_{x20}\,\mathrm{e}^{\mathrm{i}\,\lambda\tau}$ (see Sect.7.2) when solving (1), the general solution of (1) looks like

$$u_{x_2} = \sum_{j=1}^{n} (C^{(j)} \cos\lambda_j\tau + B^{(j)} \sin\lambda_j\tau)\,\varphi^{(j)}(\eta)\,.$$

The initial conditions at $\tau = 0$ are as follows:

$$(1)\quad u_{x_2}(0,\,\eta) = u_{x_2}^{(0)}(\eta) = \sum_{j=0}^{n} C^{(j)}\varphi^{(j)}(\eta)\,,$$

$$(2)\quad u_{x_2}(0,\,\eta) = 0\,.$$

From the second condition it follows that $B^{(j)} = 0$; the arbitrary coefficients $C^{(j)}$ may be determined from the equations

$$\int_0^1 u_{x_2}^{(0)}\varphi^{(k)}\,\mathrm{d}\eta = \sum_{j=2}^{n} \mathrm{C}^{(j)} \int_0^1 \varphi^{(l)}\varphi^{(k)}\,\mathrm{d}\eta\,, \qquad k = 1,\,2,\,\ldots,\,n\,.$$

Having found $C^{(j)}$ and $B^{(j)}$, we obtain the solution to (1), satisfying the initial and boundary conditions:

$$u_{x_2}(\tau,\,\eta) = \sum_{j=1}^{n} C^{(j)}\varphi^{(j)}(\eta)\,\cos\lambda_j\tau\,. \tag{7}$$

The reaction in the hinge support is

$$R = \Delta Q_{x_2}(\tau,\,1) = -u_{x_2}'''(\tau,\,1) = -\sum_{j=1}^{n} C^{(j)}\varphi^{(j)'''}(1)\,\cos\lambda_j\tau\,, \tag{8}$$

where

$$\begin{aligned}\varphi^{(j)'''}(1) = {} & [\lambda_{0j}^3\mathrm{K}_4(\lambda_{0j},\,1) - \lambda_{0j}^4 n_2 \mathrm{K}_1(\lambda_{0j}0.5)\,\mathrm{K}_3(\lambda_{0j}0.5)]\,c_3^{(j)} \\ & + \lambda_{0j}^3\mathrm{K}_1(\lambda_{0j},\,1) - \lambda_{0j}^4 n_2 \mathrm{K}_1(\lambda_{0j}0.5)\,\mathrm{K}_4(\lambda_{0j}0.5)\,.\end{aligned}$$

Thus, we obtain an analytical solution to the problem (exept that λ_j have been found numerically). Therein lies the advantage of the method. Unfortunately, this method cannot be used for rods of variable cross section.

The second way is purely numerical and allows us to solve almost any problem. When solving the problem numerically, we use (say, for a straight rod of variable cross section and the rotary inertia taken into consideration) the following equations (the system (7.31) in which $Q_{x_10} = \Delta T_{x_3} = 0$):

$$n_1(\eta)\,\frac{\partial^2 u_{x2}}{\partial\tau^2}-\frac{\partial\Delta Q_{x2}}{\partial\eta}=-n_2\,\frac{\partial^2 u_{x2}}{\partial\tau^2}\,\delta(\eta-0.5)\,,$$
$$J_{33}(\eta)\,\frac{\partial^2\vartheta_3}{\partial\tau^2}-\frac{\partial\Delta M_{x3}}{\partial\eta}-\Delta Q_{x_2}=0\,,$$
$$\frac{\partial\vartheta_3}{\partial\eta}-\frac{1}{A_{33}(\eta)}\,\Delta M_{x_3}=0\,,$$
$$\frac{\partial u_{x_2}}{\partial\eta}-\vartheta_3=0\,, \tag{9}$$

In vector-matrix form, this reads

$$\mathrm{A}^{(1)}\,\frac{\partial^2\mathbf{Z}}{\partial\tau^2}+\frac{\partial\mathbf{Z}}{\partial\eta}+\mathrm{A}^{(2)}\mathbf{Z}=\Delta\boldsymbol{\Phi}\,, \tag{10}$$

where

$$\mathrm{A}^{(1)}=\begin{bmatrix}0&0&0&-n_1\\0&0&-J_{33}&0\\0&0&0&0\\0&0&0&0\end{bmatrix},\qquad \mathrm{A}^{(2)}=\begin{bmatrix}0&0&0&0\\1&0&0&0\\0&-\dfrac{1}{A_{33}}&0&0\\0&0&-1&0\end{bmatrix},$$

$$\mathbf{Z}=\begin{vmatrix}\Delta Q_{x_2}\\\Delta M_{x_3}\\\vartheta_3\\u_{x_2}\end{vmatrix},\qquad \Delta\boldsymbol{\Phi}=\begin{vmatrix}-n_2\,\dfrac{\partial^2 u_{x_2}}{\partial\tau^2}\\0\\0\\0\end{vmatrix}\delta(\eta-0.5)\,.$$

Writing $\mathbf{Z}=\mathbf{Z}_0\,\mathrm{e}^{\mathrm{i}\,\lambda\tau}$ gives an equation similar to (4.14)

$$\mathbf{Z}_0'+\mathrm{B}(\lambda,\,\eta)\,\mathbf{Z}_0=\Delta\boldsymbol{\Phi}_0\,,\qquad \mathrm{B}(\lambda,\,\eta)=\mathrm{A}^{(2)}-\lambda^2\mathrm{A}^{(1)}\,, \tag{11}$$

where

$$\Delta\boldsymbol{\Phi}_0=\left(n_2\,\lambda^2 u_{x_2}\,\delta(\eta-0.5),\,0,\,0,\,0\right)^{\mathrm{T}}.$$

The algorithm for determining the eigenvalues λ_j and the eigenfunctions $Z_{0k}^{(j)}$ (the components of the eigenvectors $\mathbf{Z}_0^{(j)}$) was discussed in Chap. 4 (see Sects 4.2 and 4.3).

Having found λ_j and $Z_{0k}^{(j)}$, we then obtain a solution to (10) similar to (5.8) from Chap. 5:

$$\mathbf{Z}(\tau,\,\eta)=\sum_{j=1}^{n}\left(C^{(j)}\mathbf{Z}_0^{(j)}\cos\lambda_j\tau+B^{(j)}\mathbf{Z}_0^{(j)}\sin\lambda_j\tau\right). \tag{12}$$

From (12) the expression for u_{x_2} follows:

$$u_{x2} = \sum_{j=1}^{n} \left(C^{(j)} Z_{04}^{(j)} \cos \lambda_j \tau + B^{(j)} Z_{04}^{(j)} \sin \lambda_j \tau \right) .$$

The initial conditions yield:

(1) $B^{(j)} = 0$,

(2) $u_{x2}(0, \eta) = u_{x2}^{(0)} = \sum_{j=1}^{n} C^{(j)} Z_{04}^{(j)}$.

The arbitrary constants $C^{(j)}$ can be found numerically from the system of equations

$$\int_0^1 u_{x2} Z_{04}^{(k)} \, d\eta = \sum_{j=1}^{n} C^{(j)} Z_{04}^{(j)} Z_{04}^{(k)} \, d\eta , \qquad k = 1, 2, \ldots, n .$$

Because

$$\Delta Q_{x2} = \sum_{j=1}^{n} C^{(j)} Z_{01}^{(j)} \cos \lambda_j \tau ,$$

the reaction in the hinge is

$$R = \Delta Q_{x2}|_{\eta=1} = \sum_{j=1}^{n} C^{(j)} \mathbf{Z}_{01}^{(j)}(1) \cos \lambda_j \tau .$$

7.2. The rod is subjected to the impulsive load P_{x_2} and the moment M_{x_3} about the center of mass O given by

$$P_{x_2} = -J \cos \alpha \cdot \mathbf{i}_2 , \qquad M_{x_3} = -Jh \sin \alpha \cdot \mathbf{i}_3 .$$

The projection of $\mathbf{J}$ on the x_1-axis is not taken into consideration because the rod is assumed to be inextensible. After cessation of the impulse, the linear and angular velocities of the mass are

$$\dot{u}_{x_2}(0, 1) = -I \cos \alpha , \qquad \dot{\vartheta}_3(0, 1) = -\frac{I \sin \alpha}{J_0^{(1)}} , \tag{1}$$

where I and $J_0^{(1)}$ are nondimensional parameters given by

$$I = \frac{J}{mlP_0} , \qquad J_0^{(1)} = \frac{J_0}{mlh} .$$

For free oscillations that emerge after the impulse I ceases, we obtain a solution similar to (5.58) (in greater detail, the solution of equations for free oscillations is discussed in the solution of Problem 7.1). From (5.11) and (5.12), we get

$$\vartheta_3(\tau,\, \eta) = \sum_{j=1}^{n} (C^{(j)} \cos \lambda_j \tau + B^{(j)} \sin \lambda_j \tau)\, Z_{03}^{(j)}\,,$$

$$u_{x2}(\tau,\, \eta) = \sum_{j=1}^{n} (C^{(j)} \cos \lambda_j \tau + B^{(j)} \sin \lambda_j \tau)\, Z_{04}^{(j)}\,. \tag{2}$$

In the problem under consideration, the initial conditions at $\tau = 0$ are

$$\begin{aligned}
&(1) \quad u_{x2}(0,\, \eta) = 0\,,\\
&(2) \quad \dot{u}_{x2}(0,\, \eta) = -I \cos\alpha\, \delta(\eta - 1)\,, \qquad \dot{\vartheta}_3(0,\, \eta) = -\frac{I \sin\alpha}{J_0^{(1)}}\, \delta(\eta - 1)\,.
\end{aligned}$$

where $\delta(\eta - 1)$ is the nondimensional Dirac function.

The first boundary condition equation yields $C^{(j)} = 0$. The second equation leads to two relationships

$$\sum_{j=1}^{n} B^{(j)} \lambda_j Z_{03}^{(j)} = -\frac{I \sin\alpha}{J_0^{(1)}}\, \delta(\eta - 1)\,, \tag{3}$$

$$\sum_{j=1}^{n} B^{(j)} \lambda_j Z_{04}^{(j)} = -I \cos\alpha\, \delta(\eta - 1)\,. \tag{4}$$

It is required to find $B^{(j)}$ for which (3) and (4) are satisfied most closely. Thus, consider the integral of the sum of squares of errors

$$\gamma = \int_0^1 \left[\left(\frac{I \sin\alpha}{J_0^{(1)}}\, \delta(\eta - 1) + \sum_j B^{(j)} \lambda_j Z_{03}^{(j)} \right)^2 + \left(I \cos\alpha\, \delta(\eta - 1) + \sum_j B^{(j)} \lambda_j Z_{04}^{(j)} \right)^2 \right] \mathrm{d}\eta\,.$$

Let us find $B^{(j)}$ from the condition of minimum of γ. We obtain the system of equations

$$\frac{\partial \gamma}{\partial B^{(k)}} = 0\,, \qquad k = 1,\, 2,\, \ldots,\, n\,,$$

or

$$\sum_{j=1}^{n} B^{(j)} \lambda_j \int_0^1 \left(Z_{03}^{(j)} Z_{03}^{(k)} + Z_{04}^{(j)} Z_{04}^{(k)} \right) \mathrm{d}\eta + \frac{I \sin\alpha}{J_0^{(1)}}\, Z_{0k}^{(k)}(1) + I \cos\alpha Z_{04}^{(k)}(1) = 0\,.$$

Once the arbitrary coefficients $B^{(j)}$ have been found, we obtain from (2) the angle of rotation of the mass m:

$$\vartheta_3(\tau,\,1) = \sum_{j=1}^{n} B^{(j)} Z_{03}^{(j)}(1)\,\sin\gamma_j\tau\,.$$

7.3. In the equations for transversal oscillations (7.105), in addition to the term

$$\frac{\partial}{\partial\eta}\left(Q_1\frac{\partial u}{\partial\eta}\right)$$

we consider the term

$$-q_1\,\frac{\partial u}{\partial\eta}$$

which appears due to the follower axial distributed load q_1, that is,

$$-\frac{\partial}{\partial\eta}\left(Q_1\frac{\partial u}{\partial\eta}\right) + q_1\frac{\partial u}{\partial\eta}\,, \tag{1}$$

Therefore, equation (7.105) takes the form

$$\frac{\partial^4 u}{\partial\eta^4} - \frac{\partial}{\partial\eta}\left(Q_1(\eta)\,\frac{\partial u}{\partial\eta}\right) + q_1\frac{\partial u}{\partial\eta} + 2w_0\frac{\partial^2 u}{\partial\tau\partial\eta} + w_0^2\frac{\partial^2 u}{\partial\eta^2} + \frac{\partial^2 u}{\partial\tau^2} = 0\,. \tag{2}$$

Because we are asked to find the first two frequencies, let us consider a two-term approximate solution of (7.105) of the form

$$u = f^{(1)}(\tau)\,\varphi^{(1)}(\eta) + f^{(2)}(\tau)\,\varphi^{(2)}(\eta)\,, \tag{3}$$

where $\varphi^{(1)}$ and $\varphi^{(2)}$ are any independent functions satisfying the boundary conditions of the problem. In the problem under consideration, the boundary conditions are as follows: (1) at $\eta = 0$, $u = u' = 0$ and (2) at $\eta = 1$, $u = u' = 0$. As such functions, we can take eigenfunctions satisfying the boundary conditions and equation (7.67). In terms of the Krylov functions, the eigenfunctions $\varphi^{(j)}(\eta)$ can be written as

$$\varphi^{(j)}(\eta) = K_4(\lambda_{0j},\,\eta) - \frac{K_4(\lambda_{0j},\,1)}{K_3(\lambda_{0j},\,1)}\,K_3(\lambda_{0j},\,\eta)\,.$$

Substituting (3) into (2), we get

$$\begin{aligned} L = {} & \ddot f^{(1)}\varphi^{(1)} + \ddot f^{(2)}\varphi^{(2)} + 2w_0\left(\dot f^{(1)}\varphi^{(1)\prime} + \dot f^{(2)}\varphi^{(2)\prime}\right) \\ & + w_0^2\left(f^{(1)}\varphi^{(1)\prime\prime} + f^{(2)}\varphi^{(2)\prime\prime}\right) - f^{(1)}\left(Q_1\varphi^{(1)\prime}\right)' - f^{(2)}\left(Q_1\varphi^{(2)\prime}\right)' \\ & + f^{(1)}q_1\varphi^{(1)\prime} + f^{(2)}q_1\varphi^{(2)\prime} + f^{(1)}(\varphi^{(1)})^{\mathrm{IV}} + f^{(2)}(\varphi^{(2)})^{\mathrm{IV}} = 0\,. \end{aligned} \tag{4}$$

Using the principal of virtual displacements, we obtain the two equations

$$\int_0^1 L\varphi^{(1)}\,\mathrm{d}\eta = 0\,, \qquad \int_0^1 L\varphi^{(2)}\,\mathrm{d}\eta = 0\,,$$

or

$$\begin{aligned} h_{11}\ddot{f}^{(1)} + h_{12}\ddot{f}^{(2)} + b_{11}\dot{f}^{(1)} + b_{12}\dot{f}^{(2)} + c_{11}f^{(1)} + c_{12}f^{(2)} = 0\,,\\ h_{21}\ddot{f}^{(1)} + h_{22}\ddot{f}^{(2)} + b_{21}\dot{f}^{(1)} + b_{22}\dot{f}^{(2)} + c_{21}f^{(1)} + c_{22}f^{(2)} = 0\,, \end{aligned}$$

where

$$h_{ij} = \int_0^1 \varphi^{(i)}\varphi^{(j)}\,\mathrm{d}\eta\,, \qquad b_{ij} = 2w_0\int_0^1 (\varphi^{(j)\prime}\varphi^{(i)})\,\mathrm{d}\eta\,,$$

$$c_{ij} = \int_0^1 \left[w_0^2\varphi^{(j)\prime\prime} + (\varphi^{(j)})^{\mathrm{IV}} - \left(Q_1\varphi^{(j)\prime}\right)' + q_1\varphi^{(j)\prime}\right]\varphi^{(i)}\,\mathrm{d}\eta\,.$$

The coefficients b_{ij} satisfy the relations $b_{ij} = -b_{ji}$, hence $b_{11} = b_{22} = 0$. Consider the coefficients c_{ij} in greater detail. Integrating each of the terms by parts and using the homogeneous boundary conditions imposed on the functions $\varphi^{(j)}(\eta)$ and their first derivatives, we get

$$c_{ij} = c_{ij}^{(0)} + c_{ij}^{(1)}\,,$$

where

$$c_{ij}^{(0)} = \int_0^1 \left(\varphi^{(j)\prime\prime}\varphi^{(i)\prime\prime} - w_0^2\varphi^{(j)\prime}\varphi^{(i)\prime} + Q_1\varphi^{(j)\prime}\varphi^{(i)\prime}\right)\mathrm{d}\eta\,,$$

$$c_{ij}^{(1)} = -c_{ji}^{(1)} = q_1\int_0^1 \varphi^{(j)\prime}\varphi^{(i)}\,\mathrm{d}\eta\,, \qquad c_{11}^{(1)} = c_{22}^{(1)} = 0\,.$$

We seek the solution to (4) of the form

$$f^{(1)} = f_{10}\,\mathrm{e}^{\lambda\tau}\,, \qquad f^{(2)} = f_{20}\,\mathrm{e}^{\lambda\tau}\,. \tag{5}$$

Upon substitution of (5) into (4) and some algebra, we arrive at the characteristic equation

$$\det\begin{bmatrix} h_{11}\lambda^2 + c_{11}^{(0)} & h_{12}\lambda^2 + c_{12}^{(0)} + c_{12}^{(1)} + b_{12}\lambda \\ h_{12}\lambda^2 + c_{12}^{(0)} - c_{12}^{(1)} - b_{12}\lambda & h_{22}\lambda^2 + c_{22}^{(0)} \end{bmatrix} = 0\,. \tag{6}$$

Here, we used some properties of the coefficients, namely, the relations $b_{12} = h_{21}$, $c_{12}^{(0)} = c_{21}^{(0)}$, $b_{12} = -b_{21}$, and $c_{12}^{(1)} = -c_{21}^{(1)}$. Expanding the determinant (6), we obtain

$$a_0\lambda^4 + a_2\lambda^2 + a_3\lambda + a_4 = 0\,, \tag{7}$$

where

$$a_0 = h_{11} \cdot h_{22} - h_{12}^2\,, \qquad a_1 = 0\,,$$
$$a_2 = h_{11} c_{22}^0 + h_{22} c_{11}^0 - 2h_{12} c_{12}^0 + b_{12}^2\,,$$
$$a_3 = 2c_{12}^{(1)} b_{12}\,, \qquad a_4 = c_{11}^0 c_{22}^0 - (c_{12}^0)^2 + (c_{12}^{(1)})^2\,.$$

From (7) it follow that the eigenvalues are complex,

$$\lambda_{1,2} = \alpha_1 \pm i\beta_1\,, \qquad \lambda_{3,4} = \alpha_2 \pm i\beta_2\,,$$

that is, the problem is non-conservative.

7.5. In nondimensional form, the equation for forced oscillations looks like

$$\frac{\partial^2 u}{\partial \tau^2} + \frac{\partial^4 u}{\partial \eta^4} = q_0 \cos \omega_0 \tau \, H(\eta - 0.5)\,, \qquad \omega_0 = \frac{\omega}{p_0}\,. \tag{1}$$

Putting $u = u_0 \cos \omega_0 \tau$, we get

$$u_0^{\mathrm{IV}} - \omega_0^2 u_0 = q_0 \, H(\eta - 0.5)\,. \tag{2}$$

Let us consider a more general case assuming that the right-hand side of (2) is an arbitrary function. In this case, we should use the original system of four first-order equations. For a rod of constant cross section, we have

$$u_0' = \vartheta_{30}\,, \qquad u_0'' = \Delta M\,, \qquad u_0''' = \Delta Q\,,$$

and therefore (2) can be written as a system of equations

$$\begin{aligned} &\Delta Q'_{x_2 0} + \omega_0^2 u_0 = b_1(\eta)\,,\\ &\Delta M'_{x_3 0} + \Delta Q_{x_2 0} = 0\,,\\ &\vartheta'_{30} - \Delta M_{x_3 0} = 0\,,\\ &u_0' - \vartheta_{30} = 0\,, \end{aligned} \tag{3}$$

or

$$\mathbf{Z}_0' + \mathrm{A}\mathbf{Z}_0 = \mathbf{b}(\eta)\,. \tag{4}$$

Equation (4) is an equation with constant coefficients, therefore, we can write its general solution as

$$\mathbf{Z}_0 = \mathrm{K}(\eta)\,\mathbf{C} + \int_0^\eta \mathrm{K}(\eta - \tilde{\eta})\,\mathbf{b}(\tilde{\eta})\,\mathrm{d}\tilde{\eta}\,. \tag{5}$$

In Sect. 7.2 we obtained the fundamental matrix $\mathrm{K}^{(1)}(\eta)$ (7.70) in which now we should replace λ_0 by ω_0. From (5) we obtain the displacement of the rod's axis points

$$u_0 = \sum_{j=1}^{4} k_{4j} c_j + \int_0^\eta k_{41}(\eta - \tilde{\eta})\, b_1(\tilde{\eta})\,\mathrm{d}\tilde{\eta}\,. \tag{6}$$

In the problem being examined, we have

$$b_1(\tilde{\eta}) = q_0\, H(\tilde{\eta} - 0.5)\,, \qquad k_{41}(\eta - \tilde{\eta}) = \mathrm{e}^{\omega_0(\eta - \tilde{\eta})}\,,$$

therefore, upon integration, we get

$$\int_0^{\eta} k_{41}(\eta - \tilde{\eta}) b_1(\tilde{\eta})\, \mathrm{d}\tilde{\eta} = -\frac{q_0}{\omega_0}\,(1 - \mathrm{e}^{\omega_0(\eta - 0.5)}) \cdot H(\eta - 0.5)\,. \tag{7}$$

From the boundary conditions at $\eta = 0$ it follows that $c_1 = c_2 = 0$. The arbitrary constants c_3 and c_4 can be found from the boundary conditions at $\eta = 1$:

$$\begin{aligned} \Delta M_{x_2 0}(1) &= u_0''(1) = k_{23}(1)\, c_3 + k_{24}(1)\, c_4 - \omega_0 q_0 (1 - \mathrm{e}^{0.5\omega_0}) = 0\,, \\ \Delta Q_{x_2 0}(1) &= u_0'''(1) = k_{13}(1)\, c_3 + k_4(1)\, c_4 - \omega_0^2 q_0 (1 - \mathrm{e}^{0.5\omega_0}) = 0\,. \end{aligned} \tag{8}$$

Solving for c_3 and c_4 from the system of equations (8), we find the amplitude of steady oscillations of the point K to be

$$u_{0K}(1) = k_{43}(1)\, c_3 + k_{44}(1)\, c_4 - \frac{q_0}{\omega_0}\,(1 - \mathrm{e}^{0.5\omega_0})\,.$$

7.6. Let us suppose that an unknown torque $T_0 \cos\omega\tau$ is applied at the cross section K. Therefore, the nondimensional equation of forced oscillations, which is similar to (7.156), reads

$$\frac{\partial^2 u_{x_2}}{\partial \tau^2} + \frac{\partial^4 u_{x_2}}{\partial \eta^4} = T_0 \cos\omega\tau\, \delta'(\eta - 1)\,. \tag{1}$$

For steady oscillations, we put

$$u_{x_2} = u_0(\eta)\, \cos\omega\tau\,.$$

It follows from (1) that

$$u_0(\eta) = \sum_{j=1}^{4} c_j K_j(\sqrt{\omega}\eta) + \frac{T_0}{\omega}\, K_3[\sqrt{\omega}(\eta - 1)]\, H(\eta - 1)\,. \tag{2}$$

From the boundary conditions at $\eta = 0$ it follows that $c_1 = c_2 = 0$. At $\eta = 1$, the solution (2) must meet the three conditions: (1) $u_0 = 0$, (2) $u_0' = \vartheta_{30}$, and (3) $u_0'' = T_0$. Consequently, we have

$$\begin{aligned} & c_3 K_3(1) + c_4 K_4(1) = 0\,, \\ & c_3 \sqrt{\omega} K_2(1) + c_4 \sqrt{\omega} K_3(1) = \vartheta_{30}\,, \\ & c_3 \omega K_1(1) + c_4 \omega K_2(1) - T_0 = 0\,. \end{aligned} \tag{3}$$

This system of equations can be solved for c_3, c_4, and T_0 as functions of ϑ_{30} as follows:

$$c_3 = \alpha_1 \vartheta_{30}\,, \qquad c_4 = \alpha_2 \vartheta_{30}\,, \qquad T_0 = \alpha_3 \vartheta_{30}\,; \tag{4}$$

therefore, we get

$$u_0(\eta) = \left[\alpha_1 K_3(\sqrt{\omega}\eta) + \alpha_2 K_4(\sqrt{\omega}\eta)\right] \vartheta_{30} + \frac{\alpha_3 \vartheta_{30}}{\omega} K_3[\sqrt{\omega}(\eta - 1)] . \quad (5)$$

The amplitude value of the moment at the clamped end is

$$M_0 = u_0''(0) = \omega \alpha_1 \vartheta_{30} .$$

7.7. The equation for forced oscillations of a rod in nondimensional form is

$$L = \frac{\partial^2 u}{\partial \tau^2} + \frac{\partial^4 u}{\partial \eta^4} - P(\tau)\,\delta(\eta - 0.5) = 0 . \quad (1)$$

We seek a two-term approximate solution, that is,

$$u = \varphi^{(1)}(\eta)\, f^{(1)}(\tau) + \varphi^{(2)}(\eta)\, f^{(2)}(\tau) . \quad (2)$$

As functions $\varphi^{(j)}(\eta)$, we can take the first two free oscillation modes. For the boundary conditions of the problem, we have

$$\varphi^{(j)} = K_4(\eta,\, \lambda_{0j}) - \frac{K_4(1,\, \lambda_{0j})}{K_3(1,\, \lambda_{0j})} K_3(\eta,\, \lambda_{0j}) , \quad (3)$$

where K_j are the Krylov functions.

We assume the virtual displacements of the form

$$\delta u^{(j)} = \delta b_j\, \varphi^{(j)} .$$

Upon substitution of (2) into (1) and rearrangement based on the principal of virtual displacements, we get the two equations

$$\ddot{f}^{(1)} + \lambda_{01}^4 f^{(1)} = \frac{P(\tau)}{h_{11}} \varphi^{(1)}(0,\, 5) , \quad (4)$$

$$\ddot{f}^{(2)} + \lambda_{02}^4 f^{(2)} = \frac{P(\tau)}{h_{22}} \varphi^{(2)}(0,\, 5) , \quad (5)$$

where

$$h_{11} = \int_0^1 (\varphi^{(1)})^2 \, \mathrm{d}\eta , \qquad h_{22} = \int_0^1 (\varphi^{(2)})^2 \, \mathrm{d}\eta .$$

In the problem under consideration, the equations in $f^{(i)}$ are uncoupled because of the orthogonality property of $\varphi^{(j)}$. Therefore, we consider only one of these equations, say (4). The solution of (4) for an arbitrary right-hand side looks like

$$f^{(1)} = c_1 \cos \lambda_1 \tau + c_2 \sin \lambda_1 \tau + \frac{\varphi^{(1)}(0,\, 5)}{h_{11} \lambda_1} \int_0^\tau \sin \lambda_1 (\tau - \tau_1)\, P \, \mathrm{d}\tau_1 ,$$

$$\dot{f}^{(1)} = -\lambda_1 c_1 \sin \lambda_1 \tau + \lambda_1 c_2 \cos \lambda_1 \tau + \frac{\varphi^{(1)}}{h_{11}} \int_0^\tau \cos \lambda_1 (\tau - \tau_1)\, P \, \mathrm{d}\tau_1 . \quad (6)$$

By applying the Douffing method, we find the functions $\dot{f}^{(1)}$ and $f^{(1)}$ at $\tau = 0$ and $\tau = T$:

$$
\begin{aligned}
&f^{(1)}(0) = c_1\,, \qquad \dot{f}^{(1)}(0) = c_2\lambda_1\,, \qquad \lambda_1 = \lambda_{01}^2\,,\\
&f^{(1)}(T) = f^{(1)}(0)\cos\lambda_1 T + \frac{\dot{f}^{(1)}(0)}{\lambda_1}\sin\lambda_1 T + \frac{a_1}{\lambda_1}\int_0^T \sin\lambda_1(T-\tau)\,P\,\mathrm{d}\tau\,,\\
&\dot{f}^{(1)}(T) = -\lambda_1 f^{(1)}(0)\sin\lambda_1 T + \dot{f}^{(1)}(0)\cos\lambda_1 T + a_1\int_0^T \cos\lambda_1(T-\tau)\,P\,\mathrm{d}\tau\,,
\end{aligned} \tag{7}
$$

where $a_1 = \varphi^{(1)}(0.5)/h_{11}$.

For stationary oscillations, the periodicity conditions must hold, consequently, we have $f^{(1)}(0) = f^{(1)}(T)$ and $\dot{f}^{(1)}(0) = \dot{f}^{(1)}(T)$. From (7) we obtain two inhomogeneous equations to be solved for $f^{(1)}(0)$ and $\dot{f}^{(1)}(0)$:

$$
\begin{aligned}
&(1-\cos\lambda_1 T)\,f^{(1)}(0) - \frac{\sin\lambda_1 T}{\lambda_1}\,\dot{f}^{(1)}(0) = b_1\,,\\
&\sin\lambda_1 T\, f^{(1)}(0) + (1-\cos\lambda_1 T)\,\dot{f}^{(1)}(0) = b_2\,;
\end{aligned} \tag{8}
$$

here,

$$
\begin{aligned}
b_1 &= \frac{a_1 P_0}{\lambda_1}\left(\int_0^{T/2} \sin\lambda_1(T-\tau)\,\mathrm{d}\tau - \int_{T/2}^{T} \sin\lambda_1(T-\tau)\,\mathrm{d}\tau\right)\,,\\
b_2 &= a_1 P_0\left(\int_0^{T/2} \cos\lambda_1(T-\tau)\,\mathrm{d}\tau - \int_{T/2}^{T} \cos\lambda_1(T-\tau)\,\mathrm{d}\tau\right)\,.
\end{aligned}
$$

After determining $f^{(1)}(0)$ and $\dot{f}^{(1)}(0)$, we obtain the following solution of (4) on the interval $0 \le \tau \le T$:

$$
\begin{aligned}
f^{(1)}(\tau) &= f^{(1)}(0)\cos\lambda_1\tau + \frac{\dot{f}^{(1)}(0)}{\lambda_1}\sin\lambda_1\tau\\
&+\begin{cases}
\dfrac{a_1 P_0}{\lambda_1^2}\,(1-\cos\lambda_1\tau)\,, & 0 \le \tau \le \dfrac{T}{2}\,,\\[2ex]
\dfrac{a_1 P_0}{\lambda_1^2}\,[(1-\cos\dfrac{\lambda_1 T}{2}) - (1-\cos\lambda_1(\tau-\dfrac{T}{2}))]\,, & \dfrac{T}{2} \le \tau \le T\,.
\end{cases}
\end{aligned}
$$

By analogy, we obtain the expression for the function $f^{(2)}(\tau)$:

$$
\begin{aligned}
f^{(2)}(\tau) &= f^{(2)}(0)\cos\lambda_2\tau + \frac{\dot{f}^{(2)}(0)}{\lambda_2}\sin\lambda_2\tau\\
&+\begin{cases}
\dfrac{a_2 P_0}{\lambda_2^2}\,(1-\cos\lambda_2\tau)\,, & 0 \le \tau \le \dfrac{T}{2}\,,\\[2ex]
\dfrac{a_2 P_0}{\lambda_2^2}\,[(1-\cos\dfrac{\lambda_2 T}{2}) - (1-\cos\lambda_2(\tau-\dfrac{T}{2}))]\,, & \dfrac{T}{2} \le \tau \le T\,.
\end{cases}
\end{aligned}
$$

The solutions obtained are valid provided the determinants of the system of equations (8) and its counterpart for $f^{(2)}(0)$ and $\dot{f}^{(2)}(0)$, namely,

$$D = \begin{vmatrix} (1-\cos\lambda_j T) & -\sin\lambda_j T \\ \sin\lambda_j T & (1-\cos\lambda_j T) \end{vmatrix}, \qquad j = 1, 2,$$

are not equal to zero.

7.9. In nondimensional form, the equation for parametric oscillations of the rod is

$$L = \frac{\partial^2 u}{\partial \tau^2} + \frac{\partial^4 u}{\partial \eta^4} + P_0 \cos\omega_0\tau \, [H(\eta) - H(\eta - 0.5)] \, \frac{\partial^2 u}{\partial \eta^2} = 0 . \qquad (1)$$

We seek an approximate solution to (1) of the form

$$u = f^{(1)}(\tau) \sin \pi\eta . \qquad (2)$$

Using the principal of virtual displacements, we get

$$\int_0^1 L \sin \pi\eta \, \mathrm{d}\eta = 0$$

or

$$\ddot{f}^{(1)} + (\pi^4 - \pi^2 0.5 P_0 \cos\omega_0\tau) \, f^{(1)} = 0 . \qquad (3)$$

According to the Rayleigh method, we put

$$f^{(1)} = a_1 \cos \frac{\omega_0 \tau}{2} + b_1 \sin \frac{\omega_0 \tau}{2} . \qquad (4)$$

Substituting (4) into (3), we arrive at two relationships

$$\left(\pi^4 + \frac{\pi^2}{4} P_0 - \frac{\omega_0^2}{4}\right) b_1 = 0 , \qquad (5)$$

$$\left(\pi^4 - \frac{\pi^2}{4} P_0 - \frac{\omega_0^2}{4}\right) a_1 = 0 . \qquad (6)$$

Equating the expressions in brackets to zero, we find the boundaries of the main region of instability.

7.11. Consider the most general case when the gravity force acting on the mass m is taken into consideration. The rod is thus exerted by the two forces: the force due to gravity and the inertial force given by

$$mg\,\delta(x_1 - vt)\,, \qquad -m\left(\frac{\partial^2 u_{x_2}}{\partial t^2} + 2v\,\frac{\partial^2 u_{x_2}}{\partial t \partial x_1} + v^2 \frac{\partial^2 u_{x_2}}{\partial x_1^2}\right)\delta(x_1 - vt)\,.$$

Restricting our consideration to the first term of the inertial force,

$$-m\,\frac{\partial^2 u_{x_2}}{\partial t^2}\,,$$

we obtain the dimensional equation for small oscillations of the rod:

$$m_0 \frac{\partial^2 u_{x_2}}{\partial t^2} + A_{33} \frac{\partial^4 u_{x_2}}{\partial x_1^4} + k_1 u_{x_2} - \left(mg - m\,\frac{\partial^2 u_{x_2}}{\partial t^2}\right)\delta(x_1 - vt) = 0\,; \quad (1)$$

here, k_1 is the stiffness coefficient of the foundation. By putting

$$u_{x_2} = ul\,, \qquad x_1 = \eta l\,, \qquad \tau = p_0 t\,, \qquad v = v_0 l p_0\,, \qquad p_0 = \sqrt{\frac{A_{33}}{m_0 l^4}}\,,$$

we obtain the equation in nondimensional form

$$L = \frac{\partial^2 u}{\partial \tau^2} + \frac{\partial^2 u}{\partial \eta^4} + n_2 \frac{\partial^2 u}{\partial \tau^2}\,\delta(\eta - v_0\tau) + ku - n_3\,\delta(\eta - v_0\tau) = 0\,, \quad (2)$$

where

$$n_2 = \frac{m}{m_0 l}\,, \qquad n_3 = \frac{mgl^2}{A_{33}}\,, \qquad k = \frac{k_1 l^4}{A_{33}}\,.$$

We assume a two-term approximate solution of (2) of the form

$$u = f^{(1)}(\tau)\sin \pi\eta + f^{(2)}(\tau)\sin 2\,\pi\eta\,. \quad (3)$$

Applying the principal of virtual displacements, we arrive at two equations

$$\int_0^1 L \sin \pi\eta \,\mathrm{d}\eta = 0\,, \qquad \int_0^1 L \sin 2\,\pi\eta \,\mathrm{d}\eta = 0\,,$$

from which, upon manipulation, we get

$$\begin{aligned} h_{11}\ddot{f}^{(1)} + h_{12}\ddot{f}^{(2)} + (\pi^4 + k)\,f^{(1)} &= 2\,n_2 \sin \pi v_0\tau\,, \\ h_{21}\ddot{f}^{(1)} + h_{22}\ddot{f}^{(2)} + (16\pi^4 + k)\,f^{(2)} &= 2\,n_2 \sin 2\,\pi v_0\tau\,, \end{aligned} \quad (4)$$

where

$$\begin{aligned} h_{11} &= 1 + 2\,n_1 \sin^2\,\pi v_0\tau\,, \\ h_{22} &= 1 + 2\,n_1 \sin^2 2\,\pi v_0\tau\,, \\ h_{12} &= h_{21} = 2\,n_1 \sin \pi v_0\tau \cdot \sin 2\,\pi v_0\tau\,. \end{aligned}$$

We have thus obtained a system of equations with periodic coefficients. The main peculiarity of the problem is that the rod is being acted upon by the point mass only over a finite time interval (the time the mass takes to move along the rod). Hence, the oscillation of the rod is not stationary, and thus we have no opportunity to apply the methods presented in Sect. 7.7. The point mass m moves along the rod during time $t_\mathrm{f} = l/v$. The nondimensional time is $\tau_\mathrm{f} = 1/v_0$. The system of equations (4) can be written as

$$\ddot{\mathbf{f}} + \mathrm{H}^{-1}\mathrm{B}\mathbf{f} = \mathrm{H}^{-1}\mathbf{b}\,, \tag{5}$$

where

$$\mathrm{H} = \begin{bmatrix} 1 + 2\,n_1 \sin^2\,\pi v_0 \tau & 2\,n_1 \sin \pi v_0 \tau \sin 2\,\pi v_0 \tau \\ 2\,n_1 \sin \pi v_0 \tau \sin 2\,\pi v_0 \tau & 1 + 2\,n_1 \sin^2\,\pi v_0 \tau \end{bmatrix},$$

$$\mathrm{B} = \begin{bmatrix} \pi^4 + k & 0 \\ 0 & 16\pi^4 + k \end{bmatrix}, \qquad \mathbf{b} = \begin{vmatrix} 2\,n_2 \sin \pi v_0 \tau \\ 2\,n_2 \sin 2\,\pi v_0 \tau \end{vmatrix}.$$

The determinant of the matrix H,

$$D = 1 + 2\,n_1 \sin^2\,\pi v_0 \tau + 2\,n_1 \sin^2\,2\,\pi v_0 \tau\,,$$

is always greater than zero meaning that the matrix H is non-degenerate. Then, equation (5) should be solved numerically under the zero initial conditions. As a result, at $\tau = \tau_\mathrm{f}$, we have $f^{(1)}(\tau_\mathrm{f})$ and $f^{(2)}(\tau_\mathrm{f})$. Considering two-term approximation, we can write the angle of rotation of a cross section as follows:

$$\vartheta_3 = \frac{\partial u}{\partial \eta} = \pi \cos \pi\eta f^{(1)}(\tau) + 2\,\pi \cos 2\,\pi\eta f^{(2)}(\tau)\,,$$

Finally, the angle of rotation of the cross section K at the moment the mass m leaves the rod is (at $\eta = 1$)

$$\vartheta_{3\mathrm{K}} = -\pi f^{(1)}(\tau_\mathrm{f}) + 2\,\pi f^{(2)}(\tau_\mathrm{f})\,.$$

7.12. The equation for random oscillations of the rod is

$$\frac{\partial^2 u}{\partial \tau^2} + \frac{\partial^4 u}{\partial \eta^4} - P(\tau)\,\delta(\eta - 0.5) = 0\,. \tag{1}$$

Putting $u = \varphi_1(\eta) f_1(\tau)$, we obtain, after some rearrangement, the equation

$$\ddot{f}_1 + a_{11} f_1 = \frac{P(\tau)}{c_1}\varphi_1(0.5)\,, \qquad c_1 = \int_0^1 \varphi_1^2\,\mathrm{d}\eta\,. \tag{2}$$

The spectral density of input S_P and the spectral density of the output S_f satisfy the equation

$$S_f(\omega) = \frac{\varphi_1^2(0.5)\,S_p(\omega)}{c_1^2\,\left|(\mathrm{i}\,\omega)^2 + a_{11}\right|^2}\,. \tag{3}$$

Since the mathematical expectation of the force P is zero, the mathematical expectation of f_1 is also zero. The reaction in the hinge is

$$R(\tau) = u'''(1,\, \tau) = f_1(\tau)\, \varphi_1'''(1)\,. \tag{4}$$

The correlation function K_R is

$$K_R(\tau_1) = M\left[\varphi_1''' f_1(\tau) \cdot \varphi_1''' f^*(\tau')\right] = (\varphi_1''')^2 K_f(\tau_1)\,, \qquad \tau_1 = \tau - \tau'\,, \tag{5}$$

Because

$$K_f = \int_{-\infty}^{\infty} \mathrm{e}^{\mathrm{i}\,\omega\tau_1} S_f\, \mathrm{d}\omega\,,$$

substituting K_f into (5), we obtain

$$K_R(\tau_1) = (\varphi_1''')^2 \int_{-\infty}^{\infty} \frac{\varphi_1^2(0.5)\, \mathrm{e}^{\mathrm{i}\,\omega\tau_1} S_P(\omega)\, \mathrm{d}\omega}{c_1^2 \left|(\mathrm{i}\,\omega)^2 + a_{11}\right|^2}\,.$$

The variance of the reaction R reads

$$D_R = K_R(0) = \left(\frac{\varphi_1'''\varphi_1(0.5)}{c_1}\right)^2 \int_{-\infty}^{\infty} \frac{S_P(\omega)\, \mathrm{d}\omega}{\left|(\mathrm{i}\,\omega)^2 + a_{11}\right|^2}\,.$$

Assuming that the reaction R is normally distributed and applying the three sigma rule, we find the maximum value to be

$$\max R = 3\sigma_R\,, \qquad \sigma_R = \sqrt{D_R}\,.$$

References

Biderman, V.L. (1980): Theory of mechanical oscillations. Vysshaya Shkola, Moscow (Russian)

Bisplinghoff, R.L., Ashley, H., Halftman, R.L. (1955): Aeroelasticity. Addison–Wesley Publishing Company Inc, Cambridge

Bolotin, V.V. (1956): Dynamic stability of elastic systems. Holden Day, San Francisco

Bolotin, V.V. (1963): Nonconservative problems in the theory of elastic stability. Pergamon Press, London

Bolotin, V.V. (1979): Random vibrations of elastic systems. Nauka, Moscow (Russian)

Devnin, S.I. (1975): Hydroelasticity of structures at separation streamlining. Sudostroenie, Leningrad (Russian)

Grafskii, I.Yu., Kazakevich, M.I. (1983): Aerodynamics of bodies. Naukova Dumka, Dnepropetrovsk (Russian)

Gusev, A.S., Svetlitsky, V.A. (1984): Design of structures under random influences. Nauka, Moscow (Russian)

Feodosiev, V.I. (1999): Strength of materials. BSTU Publishers, Moscow (Russian)

Försching, H.W. (1974): Grundlagen der Aeroelastizität. Springer-Verlag, Berlin, Heidelberg

Fung, Y.C. (1959): An introduction to the theory of aeroelasticity, John Wiley & Sons, New York; Chapman & Hall, London

Kazakevich, M.I. (1977): Aerodynamic stability of overhead and suspended pipelines. Nedra, Moscow (Russian)

Kazakevich, M.I. (1987): Aerodynamics of bridges. Transport, Moscow (Russian)

Lankaster, P. (1969): Theory of matrices. Academy Press, New York, London

Leipholz, H (1980): Stability of elastic systems. Sijthoff and Noordhoff, Alphen van den Rijn

Mises, R. (1945): Theory of flight. McGraw–Hill, New York

Paidoussis, M.P. (1998): Fluid-structure interactions: Slender structures and axial flow, Volume 1. Academic Press, London

Paidoussis, M.P. (2003): Fluid-structure interactions: Slender structures and axial flow, Volume 2. Elsevier, Academic Press, London

Panovko, Ya.G. (1985): Mechanics of solids. Nauka, Moscow (Russian)

Prandtl, L., Tietjens, O. (1931): Hydro-Aeromechanik **2**. Springer-Verlag, Berlin

Sveshnikov, A.A. (1968): Applied methods of theory of random functions. Nauka, Moscow (Russian)

Svetlitsky, V.A. (1982): Mechanics of pipelines and hoses. Mashinostroenie, Moscow (Russian)

Svetlitsky, V.A. (2000a): Statics of rods. Springer-Verlag, Berlin, Heidelberg

Svetlitsky, V.A. (2000b): Mechanics of absolutely flexible rods. MAI Publishers, Moscow (Russian)

Svetlitsky, V.A. (2003): Statistical dynamics and reliability theory for mechanical structures. Springer-Verlag, Berlin, Heidelberg
Ventsel, E.S. (1964): Theory of probability. Nauka, Moscow (Russian)

Index

Printing: Krips bv, Meppel
Binding: Litges & Dopf, Heppenheim